Convex Functions: Constructions, Characterizations and Counterexamples

Like differentiability, convexity is a natural and powerful property of functions that plays a significant role in many areas of mathematics, both pure and applied. It ties together notions from topology, algebra, geometry and analysis, and is an important tool in optimization, mathematical programming and game theory. This book, which is the product of a collaboration of over 15 years, is unique in that it focuses on convex functions themselves, rather than on convex analysis. The authors explore the various classes and their characteristics, treating convex functions in both Euclidean and Banach spaces.

They begin by demonstrating, largely by way of examples, the ubiquity of convexity. Chapter 2 then provides an extensive foundation for the study of convex functions in Euclidean (finite-dimensional) space, and Chapter 3 reprises important special structures such as polyhedrality, selection theorems, eigenvalue optimization and semidefinite programming. Chapters 4 and 5 play the same role in (infinite-dimensional) Banach space. Chapter 6 discusses a number of other basic topics, such as selection theorems, set convergence, integral and trace class functionals, and convex functions on Banach lattices.

Chapters 7 and 8 examine Legendre functions and their relation to the geometry of Banach spaces. The final chapter investigates the application of convex functions to (maximal) monotone operators through the use of a recently discovered class of convex representative functions of which the Fitzpatrick function is the progenitor.

The book can either be read sequentially as a graduate text, or dipped into by researchers and practitioners. Each chapter contains a variety of concrete examples and over 600 exercises are included, ranging in difficulty from early graduate to research level.

Encyclopedia of Mathematics and Its Applications

All the titles listed below can be obtained from good booksellers or from Cambridge University Press. For a complete series listing visit http://www.cambridge.org/uk/series/sSeries.asp?code=EOM

73 M. Stern *Semimodular Lattices*
74 I. Lasiecka and R. Triggiani *Control Theory for Partial Differential Equations I*
75 I. Lasiecka and R. Triggiani *Control Theory for Partial Differential Equations II*
76 A. A. Ivanov *Geometry of Sporadic Groups I*
77 A. Schinzel *Polynomials with Special Regard to Reducibility*
78 T. Beth, D. Jungnickel and H. Lenz *Design Theory II, 2nd edn*
79 T. W. Palmer *Banach Algebras and the General Theory of *-Algebras II*
80 O. Stormark *Lie's Structural Approach to PDE Systems*
81 C. F. Dunkl and Y. Xu *Orthogonal Polynomials of Several Variables*
82 J. P. Mayberry *The Foundations of Mathematics in the Theory of Sets*
83 C. Foias, O. Manley, R. Rosa and R. Temam *Navier–Stokes Equations and Turbulence*
84 B. Polster and G. Steinke *Geometries on Surfaces*
85 R. B. Paris and D. Kaminski *Asymptotics and Mellin–Barnes Integrals*
86 R. McEliece *The Theory of Information and Coding, 2nd edn*
87 B. A. Magurn *An Algebraic Introduction to K-Theory*
88 T. Mora *Solving Polynomial Equation Systems I*
89 K. Bichteler *Stochastic Integration with Jumps*
90 M. Lothaire *Algebraic Combinatorics on Words*
91 A. A. Ivanov and S. V. Shpectorov *Geometry of Sporadic Groups II*
92 P. McMullen and E. Schulte *Abstract Regular Polytopes*
93 G. Gierz et al. *Continuous Lattices and Domains*
94 S. R. Finch *Mathematical Constants*
95 Y. Jabri *The Mountain Pass Theorem*
96 G. Gasper and M. Rahman *Basic Hypergeometric Series, 2nd edn*
97 M. C. Pedicchio and W. Tholen (eds.) *Categorical Foundations*
98 M. E. H. Ismail *Classical and Quantum Orthogonal Polynomials in One Variable*
99 T. Mora *Solving Polynomial Equation Systems II*
100 E. Olivieri and M. Eulália Vares *Large Deviations and Metastability*
101 A. Kushner, V. Lychagin and V. Rubtsov *Contact Geometry and Nonlinear Differential Equations*
102 L. W. Beineke and R. J. Wilson (eds.) with P. J. Cameron *Topics in Algebraic Graph Theory*
103 O. Staffans *Well-Posed Linear Systems*
104 J. M. Lewis, S. Lakshmivarahan and S. K. Dhall *Dynamic Data Assimilation*
105 M. Lothaire *Applied Combinatorics on Words*
106 A. Markoe *Analytic Tomography*
107 P. A. Martin *Multiple Scattering*
108 R. A. Brualdi *Combinatorial Matrix Classes*
109 J. M. Borwein and J. D. Vanderwerff *Convex Functions*
110 M.-J. Lai and L. L. Schumaker *Spline Functions on Triangulations*
111 R. T. Curtis *Symmetric Generation of Groups*
112 H. Salzmann, T. Grundhöfer, H. Hähl and R. Löwen *The Classical Fields*
113 S. Peszat and J. Zabczyk *Stochastic Partial Differential Equations with Lévy Noise*
114 J. Beck *Combinatorial Games*
115 L. Barreira and Y. Pesin *Nonuniform Hyperbolicity*
116 D. Z. Arov and H. Dym *J-Contractive Matrix Valued Functions and Related Topics*
117 R. Glowinski, J.-L. Lions and J. He *Exact and Approximate Controllability for Distributed Parameter Systems*
118 A. A. Borovkov and K. A. Borovkov *Asymptotic Analysis of Random Walks*
119 M. Deza and M. Dutour Sikirić *Geometry of Chemical Graphs*
120 T. Nishiura *Absolute Measurable Spaces*
121 M. Prest *Purity, Spectra and Localisation*
122 S. Khrushchev *Orthogonal Polynomials and Continued Fractions: From Euler's Point of View*
123 H. Nagamochi and T. Ibaraki *Algorithmic Aspects of Graph Connectivity*
124 F. W. King *Hilbert Transforms I*
125 F. W. King *Hilbert Transforms II*
126 O. Calin and D.-C. Chang *Sub-Riemannian Geometry*
127 M. Grabisch, J.-L. Marichal, R. Mesiar and E. Pap *Aggregation Functions*
128 L. W. Beineke and R. J. Wilson (eds) with J. L. Gross and T. W. Tucker *Topics in Topological Graph Theory*
129 J. Berstel, D. Perrin and C. Reutenauer *Codes and Automata*
130 T. G. Faticoni *Modules over Endomorphism Rings*

Convex Functions: Constructions, Characterizations and Counterexamples

Jonathan M. Borwein
*University of Newcastle,
New South Wales*

Jon D. Vanderwerff
*La Sierra University,
California*

CAMBRIDGE UNIVERSITY PRESS
Cambridge, New York, Melbourne, Madrid, Cape Town, Singapore,
São Paulo, Delhi, Dubai, Tokyo

Cambridge University Press
The Edinburgh Building, Cambridge CB2 8RU, UK

Published in the United States of America by Cambridge University Press, New York

www.cambridge.org
Information on this title: www.cambridge.org/9780521850056

© J. M. Borwein and J. D. Vanderwerff 2010

This publication is in copyright. Subject to statutory exception
and to the provisions of relevant collective licensing agreements,
no reproduction of any part may take place without
the written permission of Cambridge University Press.

First published 2010

A catalogue record for this publication is available from the British Library

ISBN 978-0-521-85005-6 Hardback

Additional resources for this publication at http://projects.cs.dal.ca/ddrive/ConvexFunctions/

Cambridge University Press has no responsibility for the persistence or
accuracy of URLs for external or third-party internet websites referred to
in this publication, and does not guarantee that any content on such
websites is, or will remain, accurate or appropriate.

*To our wives
Judith and Judith*

Contents

Preface		*page* ix
1	**Why convex?**	**1**
	1.1 Why 'convex'?	1
	1.2 Basic principles	2
	1.3 Some mathematical illustrations	8
	1.4 Some more applied examples	10
2	**Convex functions on Euclidean spaces**	**18**
	2.1 Continuity and subdifferentials	18
	2.2 Differentiability	34
	2.3 Conjugate functions and Fenchel duality	44
	2.4 Further applications of conjugacy	64
	2.5 Differentiability in measure and category	77
	2.6 Second-order differentiability	83
	2.7 Support and extremal structure	91
3	**Finer structure of Euclidean spaces**	**94**
	3.1 Polyhedral convex sets and functions	94
	3.2 Functions of eigenvalues	99
	3.3 Linear and semidefinite programming duality	107
	3.4 Selections and fixed points	111
	3.5 Into the infinite	117
4	**Convex functions on Banach spaces**	**126**
	4.1 Continuity and subdifferentials	126
	4.2 Differentiability of convex functions	149
	4.3 Variational principles	161
	4.4 Conjugate functions and Fenchel duality	171
	4.5 Čebyšev sets and proximality	186
	4.6 Small sets and differentiability	194
5	**Duality between smoothness and strict convexity**	**209**
	5.1 Renorming: an overview	209
	5.2 Exposed points of convex functions	232
	5.3 Strictly convex functions	238
	5.4 Moduli of smoothness and rotundity	252
	5.5 Lipschitz smoothness	267

viii   Contents

6	**Further analytic topics**	**276**
	6.1 Multifunctions and monotone operators	276
	6.2 Epigraphical convergence: an introduction	285
	6.3 Convex integral functionals	301
	6.4 Strongly rotund functions	306
	6.5 Trace class convex spectral functions	312
	6.6 Deeper support structure	317
	6.7 Convex functions on normed lattices	329
7	**Barriers and Legendre functions**	**338**
	7.1 Essential smoothness and essential strict convexity	338
	7.2 Preliminary local boundedness results	339
	7.3 Legendre functions	343
	7.4 Constructions of Legendre functions in Euclidean space	348
	7.5 Further examples of Legendre functions	353
	7.6 Zone consistency of Legendre functions	358
	7.7 Banach space constructions	368
8	**Convex functions and classifications of Banach spaces**	**377**
	8.1 Canonical examples of convex functions	377
	8.2 Characterizations of various classes of spaces	382
	8.3 Extensions of convex functions	392
	8.4 Some other generalizations and equivalences	400
9	**Monotone operators and the Fitzpatrick function**	**403**
	9.1 Monotone operators and convex functions	403
	9.2 Cyclic and acyclic monotone operators	413
	9.3 Maximality in reflexive Banach space	433
	9.4 Further applications	439
	9.5 Limiting examples and constructions	445
	9.6 The sum theorem in general Banach space	449
	9.7 More about operators of type (NI)	450
10	**Further remarks and notes**	**460**
	10.1 Back to the finite	460
	10.2 Notes on earlier chapters	470

List of symbols 483
References 485
Index 508

Preface

This book on *convex functions* emerges out of 15 years of collaboration between the authors. It is far from being the first on the subject nor will it be the last. It is neither a book on *convex analysis* such as Rockafellar's foundational 1970 book [369] nor a book on *convex programming* such as Boyd and Vandenberghe's excellent recent text [128]. There are a number of fine books – both recent and less so – on both those subjects or on *convexity* and relatedly on *variational analysis*. Books such as [371, 255, 378, 256, 121, 96, 323, 332] complement or overlap in various ways with our own focus which is to explore the interplay between the structure of a normed space and the properties of convex functions which can exist thereon. In some ways, among the most similar books to ours are those of Phelps [349] and of Giles [229] in that both also straddle the fields of geometric functional analysis and convex analysis – but without the convex function itself being the central character.

We have structured this book so as to accommodate a variety of readers. This leads to some intentional repetition. Chapter 1 makes the case for the ubiquity of convexity, largely by way of examples, many but not all of which are followed up in later chapters. Chapter 2 then provides a foundation for the study of convex functions in Euclidean (finite-dimensional) space, and Chapter 3 reprises important special structures such as polyhedrality, eigenvalue optimization and semidefinite programming.

Chapters 4 and 5 play the same role in (infinite-dimensional) Banach space. Chapter 6 comprises a number of other basic topics such as Banach space selection theorems, set convergence, integral functionals, trace-class spectral functions and functions on normed lattices.

The remaining three chapters can be read independently of each other. Chapter 7 examines the structure of *Legendre functions* which comprises those barrier functions which are essentially smooth and essentially strictly convex and considers how the existence of such barrier functions is related to the geometry of the underlying Banach space; as always the nicer the space (e.g. is it reflexive, Hilbert or Euclidean?) the more that can be achieved. This coupling between the space and the convex functions which may survive on it is attacked more methodically in Chapter 8.

Chapter 9 investigates (maximal) monotone operators through the use of a specialized class of convex *representative functions* of which the *Fitzpatrick function* is the progenitor. We have written this chapter so as to make it more usable as a stand-alone source on convexity and its applications to monotone operators.

In each chapter we have included a variety of concrete examples and exercises – often guided, some with further notes given in Chapter 10. We both believe strongly that general understanding and intuition rely on having fully digested a good cross-section of particular cases. Exercises that build required theory are often marked with *, those that include broader applications are marked with † and those that take excursions into topics related – but not central to – this book are marked with **.

We think this book can be used as a text, either primary or secondary, for a variety of introductory graduate courses. One possible half-course would comprise Chapters 1, 2, 3 and the finite-dimensional parts of Chapters 4 through 10. These parts are listed at the end of Chapter 3. Another course could encompass Chapters 1 through 6 along with Chapter 8, and so on. We hope also that this book will prove valuable to a larger group of practitioners in mathematical science; and in that spirit we have tried to keep notation so that the infinite-dimensional and finite-dimensional discussion are well comported and so that the book can be dipped into as well as read sequentially. This also requires occasional intentional redundancy. In addition, we finish with a 'bonus chapter' revisiting the boundary between Euclidean and Banach space and making comments on the earlier chapters.

We should like to thank various of our colleagues and students who have provided valuable input and advice and particularly Miroslav Bacak who has assisted us greatly. We should also like to thank Cambridge University Press and especially David Tranah who has played an active and much appreciated role in helping shape this work. Finally, we have a companion web-site at `http://projects.cs.dal.ca/ddrive/ConvexFunctions/` on which various related links and addenda (including any subsequent errata) may be found.

1
Why convex?

The first modern formalization of the concept of convex function appears in J. L. W. V. Jensen, "Om konvexe funktioner og uligheder mellem midelvaerdier." Nyt Tidsskr. Math. B 16 (1905), pp. 49–69. Since then, at first referring to "Jensen's convex functions," then more openly, without needing any explicit reference, the definition of convex function becomes a standard element in calculus handbooks. (A. Guerraggio and E. Molho)[1]

Convexity theory ... reaches out in all directions with useful vigor. Why is this so? Surely any answer must take account of the tremendous impetus the subject has received from outside of mathematics, from such diverse fields as economics, agriculture, military planning, and flows in networks. With the invention of high-speed computers, large-scale problems from these fields became at least potentially solvable. Whole new areas of mathematics (game theory, linear and nonlinear programming, control theory) aimed at solving these problems appeared almost overnight. And in each of them, convexity theory turned out to be at the core. The result has been a tremendous spurt in interest in convexity theory and a host of new results. (A. Wayne Roberts and Dale E. Varberg)[2]

1.1 Why 'convex'?

This introductory polemic makes the case for a study focusing on convex functions and their structural properties. We highlight the centrality of convexity and give a selection of salient examples and applications; many will be revisited in more detail later in the text – and many other examples are salted among later chapters. Two excellent companion pieces are respectively by Asplund [15] and by Fenchel [212]. A more recent survey article by Berger has considerable discussion of convex geometry [53].

It has been said that most of number theory devolves to the Cauchy–Schwarz inequality and the only problem is deciding 'what to Cauchy with'. In like fashion, much mathematics is tamed once one has found the right convex 'Green's function'. Why convex? Well, because ...

- For convex sets topological, algebraic, and geometric notions often coincide; one sees this in the study of the simplex method and of continuity of convex functions. This allows one to draw upon and exploit many different sources of insight.

---

[1] A. Guerraggio and E. Molho, "The origins of quasi-concavity: a development between mathematics and economics," *Historia Mathematica*, **31**, 62–75, (2004).
[2] Quoted by Victor Klee in his review of [366], *SIAM Review*, **18**, 133–134, (1976).

- In a computational setting, since the *interior-point revolution* [331] in linear optimization it is now more or less agreed that 'convex' = 'easy' and 'nonconvex' = 'hard' – both theoretically and computationally. A striking illustration in combinatorial optimization is discussed in Exercise 3.3.9. In part this easiness is for the prosaic reason that local and global minima coincide.
- 'Differentiability' is understood and has been exploited throughout the sciences for centuries; 'convexity' less so, as the opening quotations attest. It is not emphasized in many instances in undergraduate courses – convex principles appear in topics such as the second derivative test for a local extremum, in linear programming (extreme points, duality, and so on) or via Jensen's inequality, etc. but often they are not presented as part of any general corpus.
- Three-dimensional convex pictures are surprisingly often realistic, while two-dimensional ones are frequently not as their geometry is too special. (Actually in a convex setting even two-dimensional pictures are much more helpful compared to those for nonconvex functions, still three-dimensional pictures are better. A good illustration is Figure 2.16. For example, working two-dimensionally, one may check convexity along lines, while seeing equal right-hand and left-hand derivatives in all directions implies differentiability.)

1.2 Basic principles

First we define some of the fundamental concepts. This is done more methodically in Chapter 2. Throughout this book, we will typically use E to denote the finite-dimensional real vector space $\mathbb{R}^n$ for some $n \in \mathbb{N}$ endowed with its usual norm, and typically X will denote a real infinite-dimensional Banach space – and sometimes merely a normed space. In this introduction we will tend to state results and introduce terminology in the setting of the Euclidean space E because this more familiar and concrete setting already illustrates their power and utility.

A set $C \subset E$ is said to be *convex* if it contains all line segments between its members: $\lambda x + (1-\lambda)y \in C$ whenever $x, y \in C$ and $0 \le \lambda \le 1$. Even in two dimensions this deserves thought: every set S with $\{(x,y) : x^2+y^2 < 1\} \subset S \subset \{(x,y) : x^2+y^2 \le 1\}$ is convex.

The *lower level sets* of a function $f : E \to [-\infty, +\infty]$ are the sets $\{x \in E : f(x) \le \alpha\}$ where $\alpha \in \mathbb{R}$. The *epigraph* of a function $f : E \to [-\infty, +\infty]$ is defined by

$$\mathrm{epi}\, f := \{(x,t) \in E \times \mathbb{R} : f(x) \le t\}.$$

We will see a function as *convex* if its epigraph is a convex set; and we will use ∞ and $+\infty$ interchangeably, but we prefer to use $+\infty$ when $-\infty$ is nearby.

Consider a function $f : E \to [-\infty, +\infty]$; we will say f is *closed* if its epigraph is closed; whereas f is *lower-semicontinuous* (lsc) if $\liminf_{x \to x_0} f(x) \ge f(x_0)$ for all $x_0 \in E$. These two concepts are intimately related for convex functions. Our primary focus will be on *proper functions*, those functions $f : E \to [-\infty, +\infty]$ that do not take the value $-\infty$ and whose *domain* of f, denoted by $\mathrm{dom}\, f$, is defined by $\mathrm{dom}\, f := \{x \in E : f(x) < \infty\}$. The *indicator function* of a nonempty set D

is the function δ_D defined by $\delta_D(x) := 0$ if $x \in D$ and $\delta_D(x) := +\infty$ otherwise. These notions allow one to study convex functions and convex sets interchangeably, however, our primary focus will center on convex functions.

A sketch of a real-valued differentiable convex function very strongly suggests that the derivative of such a function is monotone increasing, in fact this is true more generally – but in a nonobvious way. If we denote the derivative (or *gradient*) of a real function g by ∇g, then using the inner product the monotone increasing property of ∇g can be written as

$$\langle \nabla g(y) - \nabla g(x), y - x \rangle \geq 0 \text{ for all } x \text{ and } y.$$

The preceding inequality leads to the definition of the *monotonicity* of the gradient mapping on general spaces. Before stating our first basic result, let us recall that a set $K \subset E$ is a *cone* if $tK \subset K$ for every $t \geq 0$; and an *affine mapping* is a translate of a linear mapping.

We begin with a recapitulation of the useful preservation and characterization properties convex functions possess:

Lemma 1.2.1 (Basic properties). *The convex functions form a convex cone closed under pointwise suprema: if f_γ is convex for each $\gamma \in \Gamma$ then so is $x \mapsto \sup_{\gamma \in \Gamma} f_\gamma(x)$.*

(a) *A function g is convex if and only if* epi g *is convex if and only if $\delta_{\text{epi} g}$ is convex.*
(b) *A differentiable function g is convex on an open convex set D if and only if ∇g is a monotone operator on D, while a twice differentiable function g is convex if and only if the Hessian $\nabla^2 g$ is a positive semidefinite matrix for each value in D.*
(c) $g \circ \alpha$ *and* $m \circ g$ *are convex when g is convex, α is affine and m is monotone increasing and convex.*
(d) *For $t > 0$, the function $(x,t) \mapsto tg(x/t)$ is convex if and only if the function g is convex.*

Proof. See Lemma 2.1.8 for (a), (c) and (d). Part (b) is developed in Theorem 2.2.6 and Theorem 2.2.8, where we are more precise about the form of differentiability used. In (d) one may be precise also about the lsc hulls, see [95] and Exercise 2.3.9. □

Before introducing the next result which summarizes many of the important continuity and differentiability properties of convex functions, we first introduce some crucial definitions. For a proper function $f : E \to (-\infty, +\infty]$, the *subdifferential* of f at $\bar{x} \in E$ where $f(\bar{x})$ is finite is defined by

$$\partial f(\bar{x}) := \{\phi \in E : \langle \phi, y - \bar{x} \rangle \leq f(y) - f(\bar{x}), \text{ for all } y \in E\}.$$

If $f(\bar{x}) = +\infty$, then $\partial f(\bar{x})$ is defined to be empty. Moreover, if $\phi \in \partial f(\bar{x})$, then ϕ is said to be a *subgradient* of f at $\bar{x}$. Note that, trivially but importantly, $0 \in \partial f(x)$ – and we call x a *critical point* – if and only if x is a minimizer of f.

While it is possible for the subdifferential to be empty, we will see below that very often it is not. An important consideration for this is whether $\bar{x}$ is in the boundary of the domain of f or in its interior, and in fact, in finite dimensions, the *relative interior*

AN ESSENTIALLY STRICTLY CONVEX FUNCTION WITH NONCONVEX SUBGRADIENT DOMAIN AND WHICH IS NOT STRICTLY CONVEX

max{(x−2)^2+y^2−1, −(x*y)^(1/4)}

Figure 1.1 A subtle two-dimensional function from Chapter 6.

(i.e. the interior relative to the affine hull of the set) plays an important role. The function f is *Fréchet differentiable* at $\bar{x} \in \text{dom} f$ with Fréchet derivative $f'(\bar{x})$ if

$$\lim_{t \to 0} \frac{f(\bar{x} + th) - f(\bar{x})}{t} = \langle f'(\bar{x}), h \rangle$$

exists uniformly for all h in the unit sphere. If the limit exists only pointwise, f is *Gâteaux differentiable* at $\bar{x}$. With these terms in mind we are now ready for the next theorem.

Theorem 1.2.2. *In Banach space, the following are central properties of convexity:*

(a) *Global minima and local minima coincide for convex functions.*
(b) *Weak and strong closures coincide for convex functions and convex sets.*
(c) *A convex function is locally Lipschitz if and only if it is continuous if and only if it is locally bounded above. A finite lsc convex function is continuous; in finite dimensions lower-semicontinuity is not automatic.*
(d) *In finite dimensions, say n=dim E, the following hold.*
  (i) *The relative interior of a convex set always exists and is nonempty.*
  (ii) *A convex function is differentiable if and only if it has a unique subgradient.*
  (iii) *Fréchet and Gâteaux differentiability coincide.*
  (iv) *'Finite' if and only if 'n + 1' or 'n' (e.g. the theorems of Radon, Helly, Carathéodory, and Shapley–Folkman stated below in Theorems 1.2.3, 1.2.4, 1.2.5, and 1.2.6). These all say that a property holds for all finite sets as soon as it holds for all sets of cardinality of order the dimension of the space.*

1.2 Basic principles

Proof. For (a) see Proposition 2.1.14; for (c) see Theorem 2.1.10 and Proposition 4.1.4. For the purely finite-dimensional results in (d), see Theorem 2.4.6 for (i); Theorem 2.2.1 for (ii) and (iii); and Exercises 2.4.13, 2.4.12, 2.4.11, and 2.4.15, for Helly's, Radon's, Carathéodory's and Shapley–Folkman theorems respectively. □

Theorem 1.2.3 (Radon's theorem). *Let $\{x_1, x_2, \ldots, x_{n+2}\} \subset \mathbb{R}^n$. Then there is a partition $I_1 \cup I_2 = \{1, 2, \ldots, n+2\}$ such that $C_1 \cap C_2 \neq \emptyset$ where $C_1 = \mathrm{conv}\{x_i : i \in I_1\}$ and $C_2 = \mathrm{conv}\{x_i : i \in I_2\}$.*

Theorem 1.2.4 (Helly's theorem). *Suppose $\{C_i\}_{i \in I}$ is a collection of nonempty closed bounded convex sets in $\mathbb{R}^n$, where I is an arbitrary index set. If every subcollection consisting of $n+1$ or fewer sets has a nonempty intersection, then the entire collection has a nonempty intersection.*

In the next two results we observe that when positive as opposed to convex combinations are involved, '$n+1$' is replaced by 'n'.

Theorem 1.2.5 (Carathéodory's theorem). *Suppose $\{a_i : i \in I\}$ is a finite set of points in E. For any subset J of I, define the cone*

$$C_J = \left\{ \sum_{i \in J} \mu_i a_i : \mu_i \in [0, +\infty), \; i \in J \right\}.$$

(a) *The cone C_I is the union of those cones C_J for which the set $\{a_j : j \in J\}$ is linearly independent. Furthermore, any such cone C_J is closed. Consequently, any finitely generated cone is closed.*
(b) *If the point x lies in $\mathrm{conv}\{a_i : i \in I\}$ then there is a subset $J \subset I$ of size at most $1 + \dim E$ such that $x \in \mathrm{conv}\{a_i : i \in J\}$. It follows that if a subset of E is compact, then so is its convex hull.*

Theorem 1.2.6 (Shapley–Folkman theorem). *Suppose $\{S_i\}_{i \in I}$ is a finite collection of nonempty sets in $\mathbb{R}^n$, and let $S := \sum_{i \in I} S_i$. Then every element $x \in \mathrm{conv}\, S$ can be written as $x = \sum_{i \in I} x_i$ where $x_i \in \mathrm{conv}\, S_i$ for each $i \in I$ and moreover $x_i \in S_i$ for all except at most n indices.*

Given a nonempty set $F \subset E$, the *core* of F is defined by $x \in \mathrm{core}\, F$ if for each $h \in E$ with $\|h\| = 1$, there exists $\delta > 0$ so that $x + th \in F$ for all $0 \leq t \leq \delta$. It is clear from the definition that the *interior* of a set F is contained in its core, that is, $\mathrm{int}\, F \subset \mathrm{core}\, F$. Let $f : E \to (-\infty, +\infty]$. We denote the set of points of continuity of f by $\mathrm{cont}\, f$. The *directional derivative* of f at $\bar{x} \in \mathrm{dom}\, f$ in the direction h is defined by

$$f'(\bar{x}; h) := \lim_{t \to 0^+} \frac{f(\bar{x} + th) - f(\bar{x})}{t}$$

if the limit exists – and it always does for a convex function. In consequence one has the following simple but crucial result.

Theorem 1.2.7 (First-order conditions). *Suppose $f : E \to (-\infty, +\infty]$ is convex. Then for any $x \in \operatorname{dom} f$ and $d \in E$,*

$$f'(x; d) \leq f(x + d) - f(x). \tag{1.2.1}$$

In consequence, f is minimized (locally or globally) at x_0 if and only if $f'(x_0; d) \geq 0$ for all $d \in E$ if and only if $0 \in \partial f(x_0)$.

The following fundamental result is also a natural starting point for the so-called Fenchel duality/Hahn–Banach theorem circle. Let us note, also, that it directly relates differentiability to the uniqueness of subgradients.

Theorem 1.2.8 (Max formula). *Suppose $f : E \to (-\infty, +\infty]$ is convex (and lsc in the infinite-dimensional setting) and that $\bar{x} \in \operatorname{core}(\operatorname{dom} f)$. Then for any $d \in E$,*

$$f'(\bar{x}; d) = \max\{\langle \phi, d \rangle : \phi \in \partial f(\bar{x})\}. \tag{1.2.2}$$

In particular, the subdifferential $\partial f(\bar{x})$ is nonempty at all core points of $\operatorname{dom} f$.

Proof. See Theorem 2.1.19 for the finite-dimensional version and Theorem 4.1.10 for infinite-dimensional version. □

Building upon the Max formula, one can derive a quite satisfactory calculus for convex functions and linear operators. Let us note also, that for $f : E \to [-\infty, +\infty]$, the *Fenchel conjugate* of f is denoted by f^* and defined by $f^*(x^*) := \sup\{\langle x^*, x \rangle - f(x) : x \in E\}$. The conjugate is always convex (as a supremum of affine functions) while $f = f^{**}$ exactly if f is convex, proper and lsc. A very important case leads to the formula $\delta_C^*(x^*) = \sup_{x \in C} \langle x^*, x \rangle$, the *support function* of C which is clearly continuous when C is bounded, and usually denoted by σ_C. This simple conjugate formula will play a crucial role in many places, including Section 6.6 where some duality relationships between Asplund spaces and those with the Radon–Nikodým property are developed.

Theorem 1.2.9 (Fenchel duality and convex calculus). *Let E and Y be Euclidean spaces, and let $f : E \to (-\infty, +\infty]$ and $g : Y \to (-\infty, +\infty]$ and a linear map $A : E \to Y$, and let $p, d \in [-\infty, +\infty]$ be the primal and dual values defined respectively by the Fenchel problems*

$$p := \inf_{x \in E}\{f(x) + g(Ax)\} \tag{1.2.3}$$

$$d := \sup_{\phi \in Y}\{-f^*(A^*\phi) - g^*(-\phi)\}. \tag{1.2.4}$$

1.2 Basic principles

Then these values satisfy the weak duality inequality $p \geq d$. If, moreover, f and g are convex and satisfy the condition

$$0 \in \text{core}(\text{dom}\, g - A \,\text{dom}\, f) \tag{1.2.5}$$

or the stronger condition

$$A \,\text{dom}\, f \cap \text{cont}\, g \neq \emptyset \tag{1.2.6}$$

then $p = d$ and the supremum in the dual problem (1.2.4) is attained if finite.

At any point $x \in E$, the subdifferential sum rule,

$$\partial(f + g \circ A)(x) \supset \partial f(x) + A^* \partial g(Ax) \tag{1.2.7}$$

holds, with equality if f and g are convex and either condition (1.2.5) or (1.2.6) holds.

Proof. The proof for Euclidean spaces is given in Theorem 2.3.4; a version in Banach spaces is given in Theorem 4.4.18. □

A nice application of Fenchel duality is the ability to obtain primal solutions from dual ones; this is described in Exercise 2.4.19.

Corollary 1.2.10 (Sandwich theorem). *Let $f : E \to (-\infty, +\infty]$ and $g : Y \to (-\infty, +\infty]$ be convex, and let $A : E \to Y$ be linear. Suppose $f \geq -g \circ A$ and $0 \in \text{core}(\text{dom}\, g - A \,\text{dom}\, f)$ (or $A \,\text{dom}\, f \cap \text{cont}\, g \neq \emptyset$). Then there is an affine function $\alpha : E \to \mathbb{R}$ satisfying $f \geq \alpha \geq -g \circ A$.*

It is sometimes more desirable to symmetrize this result by using a *concave function* g, that is a function for which $-g$ is convex, and its *hypograph*, hyp g, as in Figure 1.2.

Using the sandwich theorem, one can easily deduce Hahn–Banach extension theorem (2.1.18) and the max formula to complete the so-called Fenchel duality/Hahn–Banach circle.

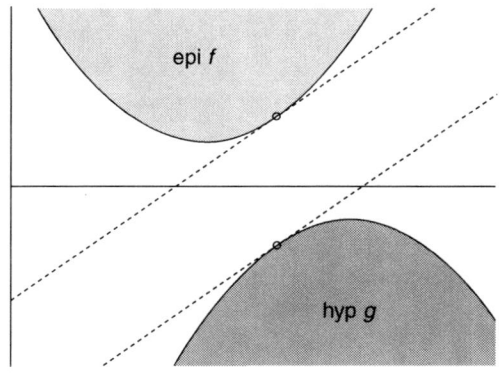

Figure 1.2 A sketch of the sandwich theorem.

A final key result is the capability to reconstruct a convex set from a well defined set of boundary points, just as one can reconstruct a convex polytope from its corners (extreme points). The basic result in this area is:

Theorem 1.2.11 (Minkowski). *Let E be a Euclidean space. Any compact convex set $C \subset E$ is the convex hull of its extreme points. In Banach space it is typically necessary to take the closure of the convex hull of the extreme points.*

Proof. This theorem is proved in Euclidean spaces in Theorem 2.7.2. □

With these building blocks in place, we use the following sections to illustrate some diverse examples where convex functions and convexity play a crucial role.

1.3 Some mathematical illustrations

Perhaps the most forcible illustration of the power of convexity is the degree to which the theory of *best approximation,* i.e. existence of *nearest points* and the study of nonexpansive mappings, can be subsumed as a convex optimization problem. For a closed set S in a Hilbert space X we write $d_S(x) := \inf_{x \in S} \|x - s\|$ and call d_S the (metric) *distance function* associated with the set S. A set C in X such that each $x \in X$ has a unique nearest point in C is called a *Čebyšev set*.

Theorem 1.3.1. *Let X be a Euclidean (resp. Hilbert) space and suppose C is a nonempty (weakly) closed subset of X. Then the following are equivalent.*

(a) *C is convex.*
(b) *C is a Čebyšev set.*
(c) *d_C^2 is Fréchet differentiable.*
(d) *d_C^2 is Gâteaux differentiable.*

Proof. See Theorem 4.5.9 for the proof. □

We shall use the necessary condition for $\inf_C f$ to deduce that the projection on a convex set is nonexpansive; this and some other properties are described in Exercise 2.3.17.

Example 1.3.2 (Algebra). *Birkhoff's theorem* [57] says the doubly stochastic matrices (those with nonnegative entries whose row and column sum equal one) are convex combinations of permutation matrices (their extreme points).

A proof using convexity is requested in Exercise 2.7.5 and sketched in detail in [95, Exercise 22, p. 74].

Example 1.3.3 (Real analysis). The following very general construction links convex functions to nowhere differentiable continuous functions.

Theorem 1.3.4 (Nowhere differentiable functions [145]). *Let $a_n > 0$ be such that $\sum_{n=1}^{\infty} a_n < \infty$. Let $b_n < b_{n+1}$ be integers such that $b_n | b_{n+1}$ for each n, and the* sequence $a_n b_n$ does not converge to 0. For each index $j \geq 1$, let f_j be a continuous function mapping the real line onto the interval [0, 1] such that $f_j = 0$ at each even integer and $f_j = 1$ at each odd integer. For each integer k and each index j, let f_j be convex on the interval $(2k, 2k + 2)$.

Then the continuous function $\sum_{j=1}^{\infty} a_j f_j(b_j x)$ has neither a finite left-derivative nor a finite right-derivative at any point.

In particular, for a convex nondecreasing function f mapping [0, 1] to [0, 1], define $f(x) = f(2 - x)$ for $1 < x < 2$ and extend f periodically. Then $F_f(x) := \sum_{j=1}^{\infty} 2^{-j} f(2^j x)$ defines a continuous nowhere differentiable function.

Example 1.3.5 (Operator theory). The Riesz–Thorin convexity theorem informally says that if T induces a bounded linear operator between Lebesgue spaces L^{p_1} and L^{p_2} and also between L^{q_1} and L^{q_2} for $1 < p_1, p_2 < \infty$ and $1 < q_1, q_2 < \infty$ then it also maps L^{r_1} to L^{r_2} whenever $(1/r_1, 1/r_2)$ is a convex combination of $(1/p_1, 1/p_2)$ and $(1/q_1, 1/q_2)$ (all three pairs lying in the unit square).

A precise formulation is given by Zygmund in [451, p. 95].

Example 1.3.6 (Real analysis). The Bohr–Mollerup theorem characterizes the gamma-function $x \mapsto \int_0^\infty t^{x-1} \exp(-t)\, dt$ as the unique function f mapping the positive half line to itself such that (a) $f(1) = 1$, (b) $xf(x) = f(x+1)$ and (c) $\log f$ is convex function

A proof of this is outlined in Exercise 2.1.24; Exercise 2.1.25 follows this by outlining how this allows for computer implementable proofs of results such as $\beta(x, y) = \Gamma(x)\Gamma(y)/\Gamma(x, y)$ where β is the classical beta-function. A more extensive discussion of this topic can be found in [73, Section 4.5].

Example 1.3.7 (Complex analysis). *Gauss's theorem* shows that the roots of the derivative of a polynomial lie inside the convex hull of the zeros.

More precisely one has the *Gauss–Lucas theorem*: For an arbitrary not identically constant polynomial, the zeros of the derivative lie in the smallest convex polygon containing the zeros of the original polynomial. While Gauss originally observed: *Gauss's theorem*: The zeros of the derivative of a polynomial P that are not multiple zeros of P are the positions of equilibrium in the field of force due to unit particles situated at the zeros of P, where each particle repels with a force equal to the inverse distance. *Jensen's sharpening* states that if P is a real polynomial not identically constant, then all nonreal zeros of P' lie inside the Jensen disks determined by all pairs of conjugate nonreal zeros of P. See Pólya–Szegő [273].

Example 1.3.8 (Levy–Steinitz theorem (combinatorics)). The rearrangements of a series with values in Euclidean space always is an affine subspace (also called a flat).

Riemann's rearrangement theorem is the one-dimensional version of this lovely result. See [382], and also *Pólya-Szegő* [272] for the complex (planar) case.

We finish this section with an interesting example of a convex function whose convexity, established in [74, §1.9], seems hard to prove directly (a proof is outlined in Exercise 4.4.10):

Example 1.3.9 (Concave reciprocals). Let $g(x) > 0$ for $x > 0$. Suppose $1/g$ is concave (which implies $\log g$ and hence g are convex) then

$$(x,y) \mapsto \frac{1}{g(x)} + \frac{1}{g(y)} - \frac{1}{g(x+y)},$$

$$(x,y,z) \mapsto \frac{1}{g(x)} + \frac{1}{g(y)} + \frac{1}{g(z)} - \frac{1}{g(x+y)} - \frac{1}{g(y+z)} - \frac{1}{g(x+z)} + \frac{1}{g(x+y+z)}$$

and all similar n-fold alternating combinations are reciprocally concave on the strictly positive orthant. The foundational case is $g(x) := x$. Even computing the Hessian in a computer algebra system in say six dimensions is a Herculean task.

1.4 Some more applied examples

Another lovely advertisement for the power of convexity is the following reduction of the classical *Brachistochrone problem* to a tractable convex equivalent problem. As Balder [29] recalls

> 'Johann Bernoulli's famous 1696 brachistochrone problem asks for the optimal shape of a metal wire that connects two fixed points A and B in space. A bead of unit mass falls along this wire, without friction, under the sole influence of gravity. The shape of the wire is defined to be optimal if the bead falls from A to B in as short a time as possible.'

Example 1.4.1 (Calculus of variations). *Hidden convexity* in the Brachistochrone problem. The standard formulation, requires one to minimize

$$T(f) := \int_0^{x_1} \frac{\sqrt{1+f'^2(x)}}{\sqrt{gf(x)}} \, dx \tag{1.4.1}$$

over all positive smooth arcs f on $(0, x_1)$ which extend continuously to have $f(0) = 0$ and $f(x_1) = y_1$, and where we let $A = (0,0)$ and $B := (x_1, y_1)$, with $x_1 > 0, y_1 \geq 0$. Here g is the gravitational constant.

A priori, it is not clear that the minimum even exists – and many books slough over all of the hard details. Yet, it is an easy exercise to check that the substitution $\phi := \sqrt{f}$ makes the integrand *jointly convex*. We obtain

$$S(\phi) := \sqrt{2gT}(\phi^2) = \int_0^{x_1} \sqrt{1/\phi^2(x) + 4\phi'^2(x)} \, dx. \tag{1.4.2}$$

One may check elementarily that the solution ψ on $(0, x_1)$ of the differential equation

$$\left(\psi'(x)\right)^2 \psi^2(x) = C/\psi(x)^2 - 1, \qquad \psi(0) = 0,$$

where C is chosen to force $\psi(x_1) = \sqrt{y_1}$, exists and satisfies $S(\phi) > S(\psi)$ for all other feasible ϕ. Finally, one unwinds the transformations to determine that the original problem is solved by a cardioid.

1.4 Some more applied examples

It is not well understood when one can make such convex transformations in variational problems; but, when one can, it always simplifies things since we have immediate access to Theorem 1.2.7, and need only verify that the first-order necessary condition holds. Especially for hidden convexity in quadratic programming there is substantial recent work, see e.g. [50, 440].

Example 1.4.2 (Spectral analysis). There is a beautiful *Davis–Lewis theorem* characterizing convex functions of eigenvalues of symmetric matrices. We let $\lambda(S)$ denote the (real) eigenvalues of an n by n symmetric matrix S in nonincreasing order. The theorem shows that if $f : E \to (-\infty, +\infty]$ is a symmetric function, then the 'spectral function' $f \circ \lambda$ is (closed) and convex if and only if f is (closed) and convex. Likewise, differentiability is inherited.

Indeed, what Lewis (see Section 3.2 and [95, §5.2]) established is that the convex conjugate which we shall study in great detail satisfies

$$(f \circ \lambda)^* = f^* \circ \lambda,$$

from which much more actually follows. Three highly illustrative applications follow.

I. (Log determinant) Let $\mathrm{lb}(x) := -\log(x_1 x_2 \cdots x_n)$ which is clearly symmetric and convex. The corresponding spectral function is $S \mapsto -\log \det(S)$.

II. (Sum of eigenvalues) Ranging over permutations π, let

$$f_k(x) := \max_{\pi}\{x_{\pi(1)} + x_{\pi(2)} + \cdots + x_{\pi(k)}\} \text{ for } k \le n.$$

This is clearly symmetric, continuous and convex. The corresponding spectral function is $\sigma_k(S) := \lambda_1(S) + \lambda_2(S) + \cdots + \lambda_k(S)$. In particular the largest eigenvalue, σ_1, is a continuous convex function of S and is differentiable if and only if the eigenvalue is simple.

III. (k-th largest eigenvalue) The k-th largest eigenvalue may be written as

$$\mu_k(S) = \sigma_k(S) - \sigma_{k-1}(S).$$

In particular, this represents μ_k as the difference of two convex continuous, hence locally Lipschitz, functions of S and so we discover the very difficult result that for each k, $\mu_k(S)$ is a locally Lipschitz function of S. Such *difference convex functions* appear at various points in this book (e.g. Exercises 3.2.11 and 4.1.46). Sometimes, as here, they inherit useful properties from their convex parts.

Harder analogs of the Davis–Lewis theorem exists for *singular values, hyperbolic polynomials, Lie algebras*, and the like.

Lest one think most results on the real line are easy, we challenge the reader to prove the empirical observation that

$$p \mapsto \sqrt{p} \int_0^\infty \left|\frac{\sin x}{x}\right|^p dx$$

is difference convex on $(1, \infty)$.

Another lovely application of modern convex analysis is to the theory of two-person zero-sum games.

Example 1.4.3 (Game theory). The seminal result due to von Neumann shows that

$$\mu := \min_C \max_D \langle Ax, y \rangle = \max_D \min_C \langle Ax, y \rangle, \tag{1.4.3}$$

where $C \subset E$ and $D \subset F$ are compact convex sets (originally sets of finite probabilities) and $A : E \mapsto F$ is an arbitrary *payoff matrix*. The common value μ is called the *value* of the game.

Originally, Equation (1.4.3) was proved using fixed point theory (see [95, p. 201]) but it is now a lovely illustration of the power of Fenchel duality since we may write $\mu := \inf_E \{\delta_D^*(Ax) + \delta_C(x)\}$; see Exercise 2.4.21.

One of the most attractive extensions is due to Sion. It asserts that

$$\min_C \max_D f(x, y) = \max_D \min_C f(x, y)$$

when C, D are compact and convex in Banach space while $f(\cdot, y), -f(x, \cdot)$ are required only to be lsc and *quasi-convex* (i.e. have convex lower level sets). In the convex-concave proof one may use compactness and the max formula to achieve a very neat proof. We shall see substantial applications of reciprocal concavity and log convexity to the construction of barrier functions in Section 7.4.

Next we turn to entropy:

> 'Despite the narrative force that the concept of entropy appears to evoke in everyday writing, in scientific writing entropy remains a thermodynamic quantity and a mathematical formula that numerically quantifies disorder. When the American scientist Claude Shannon found that the mathematical formula of Boltzmann defined a useful quantity in information theory, he hesitated to name this newly discovered quantity entropy because of its philosophical baggage. The mathematician John von Neumann encouraged Shannon to go ahead with the name entropy, however, since "no one knows what entropy is, so in a debate you will always have the advantage."[3]

Example 1.4.4 (Statistics and information theory). The function of finite probabilities

$$\vec{p} \mapsto \sum_{i=1}^n p_i \log(p_i)$$

defines the (negative of) *Boltzmann–Shannon entropy*, where $\sum_{i=1}^n p_i = 1$ and $p_i \geq 0$, and where we set $0 \log 0 = 0$. (One maximizes entropy and minimizes convex functions.)

I. (Extended entropy.) We may extend this function (minus 1) to the nonnegative orthant by

$$\vec{x} \mapsto \sum_{i=1}^n (x_i \log(x_i) - x_i). \tag{1.4.4}$$

---

[3] *The American Heritage Book of English Usage*, p. 158.

(See Exercise 2.3.25 for some further properties of this function.) It is easy to check that this function has Fenchel conjugate

$$\vec{y} \mapsto \sum \exp(y_i),$$

whose conjugate is given by (1.4.4) which must therefore be convex – of course in this case it is also easy to check that $x \log x - x$ has second derivative $1/x > 0$ for $x > 0$.

II. (Divergence estimates.) The function of two finite probabilities

$$(\vec{p}, \vec{q}) \mapsto \sum_{i=1}^{n} \left\{ p_i \log \left(\frac{p_i}{q_i} \right) - (p_i - q_i) \right\},$$

is called the Kullback–Leibler *divergence* and measures how far $\vec{q}$ deviates from $\vec{p}$ (care being taken with $0 \div 0$). Somewhat surprisingly, this function is jointly convex as may be easily seen from Lemma 1.2.1 (d), or more painfully by taking the second derivative. One of the many attractive features of the divergence is the beautiful inequality

$$\sum_{i=1}^{n} p_i \log \left(\frac{p_i}{q_i} \right) \geq \frac{1}{2} \left(\sum_{i=1}^{n} |p_i - q_i| \right)^2, \qquad (1.4.5)$$

valid for any two finite probability measures. Note that we have provided a lower bound in the 1-norm for the divergence (see Exercise 2.3.26 for a proof and Exercise 7.6.3 for generalizations). Inequalities bounding the divergence (or generalizations as in Exercise 7.6.3) below in terms of the 1-norm are referred to as *Pinsker-type* inequalities [228, 227].

III. (Surprise maximization.) There are many variations on the current theme. We conclude this example by describing a recent one. We begin by recalling the *Paradox of the Surprise Exam*:

> 'A teacher announces in class that an examination will be held on some day during the following week, and moreover that the examination will be a surprise. The students argue that a surprise exam cannot occur. For suppose the exam were on the last day of the week. Then on the previous night, the students would be able to predict that the exam would occur on the following day, and the exam would not be a surprise. So it is impossible for a surprise exam to occur on the last day. But then a surprise exam cannot occur on the penultimate day, either, for in that case the students, knowing that the last day is an impossible day for a surprise exam, would be able to predict on the night before the exam that the exam would occur on the following day. Similarly, the students argue that a surprise exam cannot occur on any other day of the week either. Confident in this conclusion, they are of course totally surprised when the exam occurs (on Wednesday, say). The announcement is vindicated after all. Where did the students' reasoning go wrong?' ([151])

This paradox has a grimmer version involving a hanging, and has a large literature [151]. As suggested in [151], one can leave the paradox to philosophers and ask, more pragmatically, the information-theoretic question *what distribution of events*

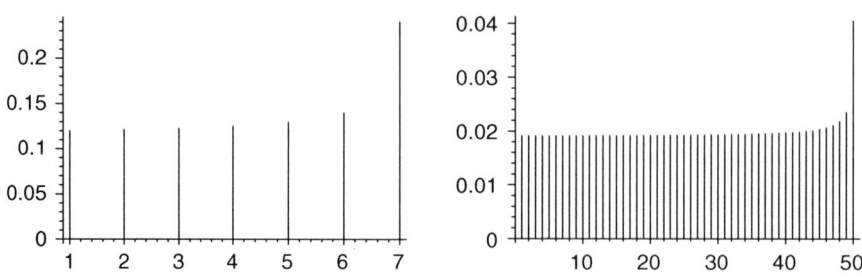

Figure 1.3 Optimal distributions: $m = 7$ (L) and $m = 50$ (R).

will maximize group surprise? This question has a most satisfactory resolution. It leads naturally (see [95, Exercise 28, p. 87]) to the following optimization problem involving S_m, the *surprise function*, given by

$$S_m(\vec{p}) := \sum_{j=1}^{m} p_j \log\left(\frac{p_j}{\frac{1}{m}\sum_{i \geq j} p_i}\right),$$

with the explicit constraint that $\sum_{j=1}^{m} p_j = 1$ and the implicit constraint that each $p_i \geq 0$.

From the results quoted above the reader should find it easy to show S_m is convex. Remarkably, the optimality conditions for maximizing surprise can be solved beautifully recursively as outlined in [95, Exercise 28, p. 87]. Figure 1.3 shows examples of optimal probability distributions, for $m = 7$ and $m = 50$.

1.4.1 Further examples of hidden convexity

We finish this section with two wonderful 'hidden convexity' results.

I. (Aumann integral) The integral of a multifunction $\Omega : T \mapsto E$ over a finite measure space T, denoted $\int_T \Omega$, is defined as the set of all points of the form $\int_T \phi(t) d\mu$, where μ is a finite positive measure and $\phi(\cdot)$ is an integrable measurable *selection* $\phi(t) \in \Omega(t)$ a.e. We denote by conv Ω the multifunction whose value at t is the convex hull of $\Omega(t)$. Recall that Ω is *measurable* if $\{t : \Omega(t) \cap W \neq \emptyset\}$ is measurable for all open sets W and is *integrably bounded* if $\sup_{\sigma \in \Omega} \|\sigma(t)\|$ is integrable; here σ ranges over all integrable selections.

Theorem 1.4.5 (Aumann convexity theorem). *Suppose (a) E is finite-dimensional and μ is a nonatomic probability measure. Suppose additionally that (b) Ω is measurable, has closed nonempty images and is integrably bounded. Then*

$$\int_T \Omega = \int_T \text{conv}\,\Omega,$$

and is compact.

In the fine survey by Zvi Artstein [9] compactness follows from the Dunford–Pettis criterion (see §6.3); and the exchange of convexity and integral from an extreme point argument plus some measurability issues based on Filippov's lemma. We refer the reader to [9, 155, 156, 157] for details and variants.[4]

In particular, since the right-hand side of Theorem 1.4.5 is clearly convex we have the following weaker form which is easier to prove – directly from the Shapley–Folkman theorem (1.2.6) – as outlined in Exercise 2.4.16 and [415]. Indeed, we need not assume (b).

Theorem 1.4.6 (Aumann convexity theorem (weak form)). *If E is finite-dimensional and μ is a nonatomic probability measure then*

$$\int_T \Omega = \operatorname{conv} \int_T \Omega.$$

The simplicity of statement and the potency of this result (which predates Aumann) means that it has attracted a large number of alternative proofs and extensions, [155]. An attractive special case – originating with Lyapunov – takes

$$\Omega(t) := \{-f(t), f(t)\}$$

where f is any continuous function. This is the genesis of so-called 'bang-bang' control since it shows that in many settings control mechanisms which only take extreme values will recapture all behavior. More generally we have:

Corollary 1.4.7 (Lyapunov convexity theorem). *Suppose E is finite-dimensional and μ is a nonatomic finite vector measure $\mu := (\mu_1, \mu_2, \ldots, \mu_n)$ defined on a sigma-algebra, Σ, of subsets of T, and taking range in E. Then $R(\mu) := \{\mu(A) : A \in \Sigma\}$ is convex and compact.*

We *sketch* the proof of convexity (the most significant part). Let $\nu := \sum |\mu_k|$. By the Radon–Nikodým theorem, as outlined in Exercise 6.3.6, each μ_i is absolutely continuous with respect to ν and so has a Radon–Nikodým derivative f_k. Let $f := (f_1, f_2, \ldots, f_n)$. It follows, with $\Omega(t) := \{0, f(t)\}$ that we may write

$$R(\mu) = \int_T \Omega \, d\nu.$$

Then Theorem 1.4.5 shows the convexity of the range of the vector measure. (See [260] for another proof.)

II. (Numerical range) As a last taste of the ubiquity of convexity we offer the beautiful hidden convexity result called the *Toeplitz–Hausdorff theorem* which establishes the convexity of the *numerical range*, $W(A)$, of a complex square matrix A (or indeed of a bounded linear operator on complex Hilbert space). Precisely,

$$W(A) := \{\langle Ax, x \rangle : \langle x, x \rangle = 1\},$$

---

[4] [156] discusses the general statement.

so that it is not at all obvious that $W(A)$ should be convex, though it is clear that it must contain the spectrum of A.

Indeed much more is true. For example, for a normal matrix the numerical range is the convex hull of the eigenvalues. Again, although it is not obvious there is a tight relationship between the Toeplitz–Hausdorff theorem and Birkhoff's result (of Example 1.3.2) on doubly stochastic matrices.

Conclusion Another suite of applications of convexity has not been especially highlighted in this chapter but will be at many places later in the book. Wherever possible, we have illustrated a convexity approach to a piece of pure mathematics. Here is one of our favorite examples.

Example 1.4.8 (Principle of uniform boundedness). The principle asserts that *pointwise bounded families of bounded linear operators between Banach spaces are uniformly bounded*. That is, we are given bounded linear operators $T_\alpha \colon X \to Y$ for $\alpha \in \mathcal{A}$ and we know that $\sup_{\alpha \in \mathcal{A}} \|T_\alpha(x)\| < \infty$ for each x in X. We wish to show that $\sup_{\alpha \in \mathcal{A}} \|T_\alpha\| < \infty$. Here is the convex analyst's proof:

Proof. Define a function f_A by

$$f_A(x) := \sup_{\alpha \in \mathcal{A}} \|T_\alpha(x)\|$$

for each x in X. Then, as observed in Lemma 1.2.1, f_A is convex. It is also closed since each mapping $x \mapsto \|T_\alpha(x)\|$ is (see also Exercise 4.1.5). Hence f_A is a finite, closed convex (actually sublinear) function. Now Theorem 1.2.2 (c) (Proposition 4.1.5) ensures f_A is continuous at the origin. Select $\varepsilon > 0$ with $\sup\{f_A(x) : \|x\| \leq \varepsilon\} \leq 1$. It follows that

$$\sup_{\alpha \in \mathcal{A}} \|T_\alpha\| = \sup_{\alpha \in \mathcal{A}} \sup_{\|x\| \leq 1} \|T_\alpha(x)\| = \sup_{\|x\| \leq 1} \sup_{\alpha \in \mathcal{A}} \|T_\alpha(x)\| \leq 1/\varepsilon.$$

$\square$

We give a few other examples:

- The Lebesgue–Radon–Nikodým decomposition theorem viewed as a convex optimization problem (Exercise 6.3.6).
- The Krein–Šmulian or Banach–Dieudonné theorem derived from the von Neumann minimax theorem (Exercise 4.4.26).
- The existence of Banach limits for bounded sequences illustrating the Hahn–Banach extension theorem (Exercise 5.4.12).
- Illustration that the full axiom of choice is embedded in various highly desirable convexity results (Exercise 6.7.11).
- A variational proof of Pitt's theorem on compactness of operators in ℓ_p spaces (Exercise 6.6.3).

1.4 Some more applied examples 17

- The whole of Chapter 9 in which convex Fitzpatrick functions are used to attack the theory of maximal monotone operators – not to mention Chapter 7.

Finally we would be remiss not mentioned the many lovely applications of convexity in the study of partial differential equations (especially elliptic) see [195] and in the study of control systems [157]. In this spirit, Exercises 3.5.17, 3.5.18 and Exercise 3.5.19 make a brief excursion into differential inclusions and convex Lyapunov functions.

2
Convex functions on Euclidean spaces

The early study of Euclid made me a hater of geometry. (J.J. Sylvester)[1]

2.1 Continuity and subdifferentials

In this chapter we will let E denote the *Euclidean* vector space $\mathbb{R}^n$ endowed with its usual norm, unless we specify otherwise. One of the reasons for doing so is that the coordinate free vector notation makes the transition to infinite-dimensional Banach spaces more transparent, another is that it lends itself to studying other vector spaces – such as the symmetric $m \times m$ matrices – that isomorphically identify with some $\mathbb{R}^n$.

2.1.1 Basic properties of convex functions

A set $C \subset E$ is said to be *convex* if $\lambda x + (1-\lambda)y \in C$ whenever $x, y \in C$ and $0 \le \lambda \le 1$. A subset of S of a vector space is said to be *balanced* if $\alpha S \subset S$ whenever $|\alpha| \le 1$. The set S is *symmetric* if $-x \in S$ whenever $x \in S$. Consequently a convex subset C of E is balanced provided $-x \in C$ whenever $x \in C$. Therefore, in E – or any real vector space – we will typically use the term *symmetric* in the context of convex sets.

Suppose $C \subset E$ is convex. A function $f : C \to \mathbb{R}$ is said to be *convex* if

$$f(\lambda x + (1-\lambda)y) \le \lambda f(x) + (1-\lambda)f(y) \tag{2.1.1}$$

for all $0 \le \lambda \le 1$ and all $x, y \in C$.

Given an extended real-valued function $f : E \to (-\infty, +\infty]$, we shall say the *domain* of f is $\{x \in E : f(x) < \infty\}$, and we denote this by $\text{dom} f$. Moreover, we say such a function f is *convex* if (2.1.1) is satisfied for all $x, y \in \text{dom} f$. If $\text{dom} f$ is not empty and the inequality in (2.1.1) is strict for all distinct $x, y \in \text{dom} f$ and all $0 < \lambda < 1$, then f is said to be *strictly convex*. For example, the single variable functions $f(t) := t^2$ and $g(t) := |t|$ are both convex, while f is additionally strictly convex but g is not.

The following basic geometric lemma is useful in studying properties of convex functions both on the real line, and on higher-dimensional spaces.

---

[1] James Joseph Sylvester, 1814–1897, Second President of the London Mathematical Society, quoted in D. MacHale, *Comic Sections*, Dublin, 1993.

2.1 Continuity and subdifferentials

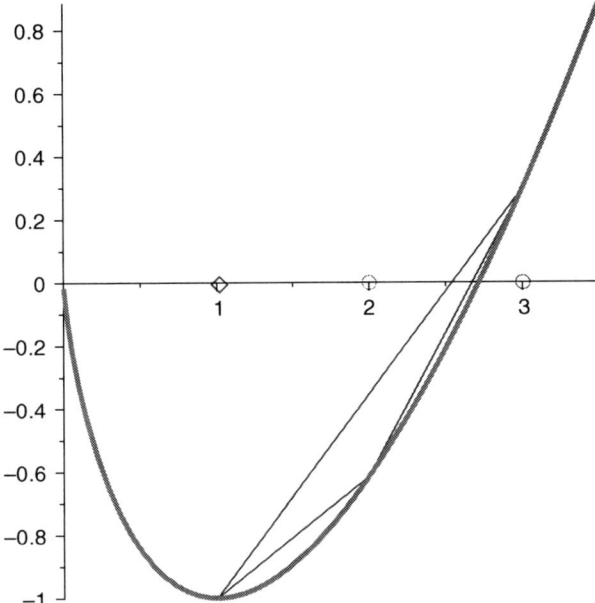

Figure 2.1 Three-slope inequality (2.1.1) for $x \log(x) - x$.

Fact 2.1.1 (Three-slope inequality). *Suppose $f : \mathbb{R} \to (-\infty, +\infty]$ is convex and $x < y < z$. Then*

$$\frac{f(y) - f(x)}{y - x} \leq \frac{f(z) - f(x)}{z - x} \leq \frac{f(z) - f(y)}{z - y},$$

whenever $x, y, z \in \operatorname{dom} f$.

Proof. Observe that $y = \frac{z - y}{z - x} x + \frac{y - x}{z - x} z$. Then the convexity of f implies

$$f(y) \leq \frac{z - y}{z - x} f(x) + \frac{y - x}{z - x} f(z).$$

Both inequalities can now be deduced easily. $\square$

We will say a function $f : E \to \mathbb{R}$ is *Lipschitz* on a subset D of E if there is a constant $M \geq 0$ so that $|f(x) - f(y)| \leq M \|x - y\|$ for all $x, y \in D$ and M is a *Lipschitz constant* for f on D. If for each $x_0 \in D$, there is an open set $U \subset D$ with $x_0 \in U$ and a constant M so that $|f(x) - f(y)| \leq M \|x - y\|$ for all $x, y \in U$, we will say f is *locally Lipschitz on D*. If D is the entire space, we simply say f is Lipschitz or locally Lipschitz respectively.

The following properties of convex functions on the real line foreshadow many of the important properties that convex functions possess in much more general settings; as is standard, f'_+ and f'_- represent the right-hand and left-hand derivatives of the real function f.

Theorem 2.1.2 (Properties of convex functions on $\mathbb{R}$). *Let $I \subset \mathbb{R}$ be an open interval and suppose $f : I \to \mathbb{R}$ is convex. Then*

(a) *$f'_+(x)$ and $f'_-(x)$ exist and are finite at each $x \in I$;*
(b) *f'_+ and f'_- are nondecreasing functions on I;*
(c) *$f'_+(x) \leq f'_-(y) \leq f'_+(y)$ for $x < y$, $x, y \in I$;*
(d) *f is differentiable except at possibly countably many points of I;*
(e) *if $[a, b] \subset I$ and $M = \max\{|f'_+(a)|, |f'_-(b)|\}$, then*

$$|f(x) - f(y)| \leq M|x - y| \quad \text{for all } x, y \in [a, b];$$

(f) *f is locally Lipschitz on I.*

Proof. First (a), (b) and (c) follow from straightforward applications of the three-slope inequality (2.1.1). To prove (d), suppose $x_0 \in I$ is a point of continuity of the monotone function f'_+. Then we have by (c) and the continuity of f'_+

$$f'_+(x_0) = \lim_{x \to x_0^-} f'_+(x) \leq f'_-(x_0) \leq f'_+(x_0).$$

The result now follows because as a monotone function f'_+ has at most countably many discontinuities on I. Now (e), is again, an application of the three-slope inequality (2.1.1), and (f) is a direct consequence of (e). The full details are left as Exercise 2.1.1. □

The function f defined by $f(x, y) := |x|$ fails to be differentiable at a continuum of points in $\mathbb{R}^2$, thus Theorem 2.1.2(d) fails for convex functions on $\mathbb{R}^2$. Still, as we will see later, the points where a continuous convex function on E fails to be differentiable is both measure zero (Theorem 2.5.1) and first category (Corollary 2.5.2). The following observation is a one-dimensional version of the Max formula.

Corollary 2.1.3 (Max formula on the real line). *Let $I \subset \mathbb{R}$ be an open interval, $f : I \to \mathbb{R}$ be convex and $x_0 \in I$. If $f'_-(x_0) \leq \lambda \leq f'_+(x_0)$, then*

$$f(x) \geq f(x_0) + \lambda(x - x_0) \text{ for all } x \in I.$$

Moreover, $f'_+(x_0) = \max\{\lambda : \lambda(x - x_0) \leq f(x) - f(x_0), \text{ for all } x \in I\}$.

Proof. See Exercise 2.1.2. □

We now turn our attention to properties of convex functions on Euclidean spaces. The *lower level sets* of a function $f : E \to [-\infty, +\infty]$ are the sets $\{x \in E : f(x) \leq \alpha\}$ where $\alpha \in \mathbb{R}$. The *epigraph* of a function $f : E \to [-\infty, +\infty]$ is defined by

$$\text{epi} f := \{(x, t) \in E \times \mathbb{R} : f(x) \leq t\}.$$

We will say a function $f : E \to [-\infty, +\infty]$ is *closed* if its epigraph is closed in $X \times \mathbb{R}$. The function f is said to be *lower-semicontinuous (lsc)* at x_0 if $\liminf_{x \to x_0} f(x) \geq f(x_0)$, and f is said to be *lsc* if f is lsc at all $x \in E$. It is easy to check that a function f is closed if and only if it is lsc. Moreover, for $f : E \to [-\infty, +\infty]$ it

2.1 Continuity and subdifferentials

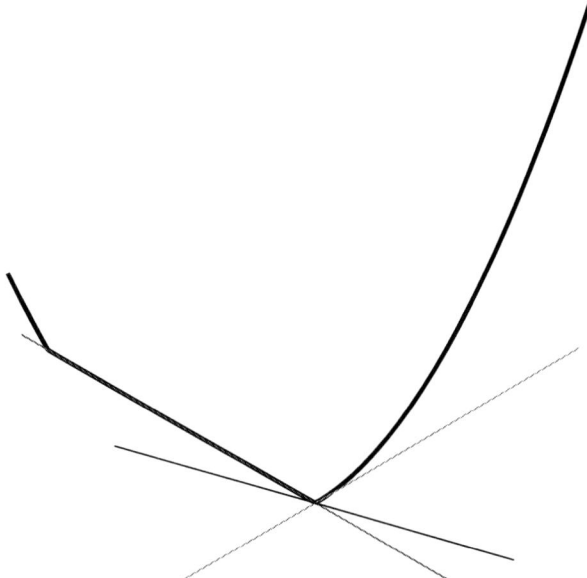

Figure 2.2 The Max formula (2.1.3) for $\max\{x^2, -x\}$.

is easy to check that the closure of $\operatorname{epi} f$ is also the epigraph of a function. We thus define *closure of f* as the function $\operatorname{cl} f$ whose epigraph is the closure of f; that is $\operatorname{epi}(\operatorname{cl} f) = \operatorname{cl}(\operatorname{epi} f)$. Similarly, we may then say a function is *convex* if its epigraph is convex. This geometric approach has the advantage that convexity is then defined naturally for functions $f : E \to [-\infty, +\infty]$. Furthermore, for a set $S \subset E$, the *convex hull* of S, denoted by $\operatorname{conv} S$ is the intersection of all convex sets that contain S. Analogously, but with a little more care, for a (nonconvex) function f on E, the *convex hull of the function f*, denoted by $\operatorname{conv} f$, is defined by

$$\operatorname{conv} f(x) := \inf\{\mu : (x, \mu) \in \operatorname{conv} \operatorname{epi} f\}$$

and it is not hard to check that $\operatorname{conv} f$ is the largest convex function minorizing f; see Exercise 2.1.15 for further related information.

Our primary focus will be on *proper functions*, i.e. those functions $f : E \to (-\infty, +\infty]$ such that $\operatorname{dom} f \neq \emptyset$. However, one can see that for $f := -|t|$, $\operatorname{conv} f \equiv -\infty$ is not proper (i.e. *improper*). Moreover, some of the natural operations we will study later, such as conjugation or infimal convolutions when applied to proper convex functions may result in improper functions.

A set K is a *cone* if $tK \subset K$ for every $t \geq 0$. In other words, $\mathbb{R}_+ K \subset K$ where $\mathbb{R}_+ := [0, \infty)$; we will also use the notation $\mathbb{R}_{++} = (0, \infty)$. The *indicator function* of a nonempty set D is the function δ_D which is defined by $\delta_D(x) := 0$ if $x \in D$ and $\delta_D(x) := \infty$ otherwise. Let E and F be Euclidean spaces; a mapping $\alpha : E \to F$ is said to be *affine* if $\alpha(\lambda x + (1 - \lambda)y) = \lambda \alpha(x) + (1 - \lambda)\alpha(y)$ for all $\lambda \in \mathbb{R}, x, y \in E$. If the range space $F = \mathbb{R}$, we will use the terminology *affine function* instead of affine mapping. The next fact shows that affine mappings differ from linear mappings by constants.

Lemma 2.1.4. *Let E and F be Euclidean spaces. Then a mapping $\alpha : E \to F$ is affine if and only if $\alpha = y_0 + T$ where $y_0 \in F$ and $T : E \to F$ is linear.*

Proof. See Exercise 2.1.3. □

Suppose $f : E \to (-\infty, +\infty]$. If $f(\lambda x) = \lambda f(x)$ for all $x \in E$ and $\lambda \geq 0$, then f is said to be *positively homogeneous*. A *subadditive* function f satisfies the property that $f(x+y) \leq f(x) + f(y)$ for all $x, y \in E$. The function is said to be *sublinear* if

$$f(\alpha x + \beta y) \leq \alpha f(x) + \beta f(y) \text{ for all } x, y \in E, \text{ and } \alpha, \beta \geq 0.$$

For this, and unless stated otherwise elsewhere, we use the convention $0 \cdot (+\infty) = 0$.

Fact 2.1.5. *A function $f : E \to (-\infty, +\infty]$ is sublinear if and only if it is positively homogeneous and subadditive.*

Proof. See Exercise 2.1.4. □

Some of the most important examples of sublinear functions on vector spaces are *norms*, where we recall a nonnegative function $\|\cdot\|$ on a vector space X is called a *norm* if

(a) $\|x\| \geq 0$ for each $x \in X$,
(b) $\|x\| = 0$ if and only if $x = 0$,
(c) $\|\lambda x\| = |\lambda| \|x\|$ for every $x \in X$ and scalar λ,
(d) $\|x + y\| \leq \|x\| + \|y\|$ for every $x, y \in X$.

The condition in (d) is often referred to as the *triangle inequality*. A vector space X endowed with a norm is said to be a *normed linear space*. A *Banach space* is a complete normed linear space. Consequently, Euclidean spaces are finite-dimensional Banach spaces. Unless we specify otherwise, $\|\cdot\|$ will denote the Euclidean norm on E and $\langle \cdot, \cdot \rangle$ denotes the inner product. With this notation, the *Cauchy–Schwarz inequality* can be written $|\langle x, y \rangle| \leq \|x\| \|y\|$ for all $x, y \in E$.

Bounded sets and neighborhoods play important role in convex functions, and two of the most important such sets are the *closed unit ball* $B_E := \{x \in E : \|x\| \leq 1\}$ and the *unit sphere* $S_E := \{x \in E : \|x\| = 1\}$. Figure 2.3 shows three spheres for the p-norm in the plane and the balls $\{(x,y,z) : |x| + |y| + |z| \leq 1\}$ and $\{(x,y,z) : \max |x|, |y|, |z| \leq 1\}$ in three-dimensional space.

For a convex set $C \subset E$, we define the *gauge function* of C, denoted by γ_C, by $\gamma_C(x) := \inf\{\lambda \geq 0 : x \in \lambda C\}$. When $C = B_E$, one can easily see that γ_C is just the norm on E. Some fundamental properties of this function, which is also known as the *Minkowski functional* of C, are given in Exercise 2.1.13.

Given a nonempty set $S \subset E$, the *support function* of S is denoted by σ_S and defined by $\sigma_S(x) := \sup\{\langle s, x \rangle : s \in S\}$; notice that the support function is convex, proper and $0 \in \operatorname{dom} \sigma_S$. There is also a naturally associated (metric) *distance function*, that is

$$d_S(x) := \inf\{\|x - y\| : y \in S\}.$$

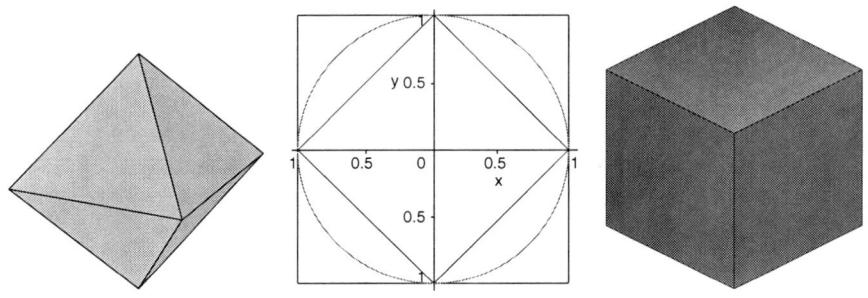

Figure 2.3 The 1-ball in $\mathbb{R}^3$, spheres in $\mathbb{R}^2$ $(1, 2, \infty)$, and ∞-ball in $\mathbb{R}^3$.

Distance functions play a central role in convex analysis, both in theory and algorithmically. The following important fact concerns the convexity of metric distance functions.

Fact 2.1.6. *Suppose $C \subset E$ is a nonempty closed convex set. Then $d_C(\cdot)$ is a convex function with Lipschitz constant 1.*

Proof. Let $x, y \in E$, and $0 < \lambda < 1$. Let $\varepsilon > 0$, and choose $x_0, y_0 \in C$ so that $\|x - x_0\| < d_C(x) + \varepsilon$ and $\|y - y_0\| < d_C(y) + \varepsilon$. Then

$$d_C(\lambda x + (1 - \lambda)y) \leq \|\lambda x + (1 - \lambda)y - (\lambda x_0 + (1 - \lambda)y_0)\|$$
$$\leq \lambda \|x - x_0\| + (1 - \lambda)\|y - y_0\|$$
$$< \lambda d_C(x) + (1 - \lambda)d(y) + \varepsilon.$$

Because $\varepsilon > 0$ was arbitrary, this establishes the convexity. We leave the Lipschitz assertion as an exercise. □

Fact 2.1.7. *Suppose $f : E \to (-\infty, +\infty]$ is a convex function, then f has convex lower level sets (i.e. is quasi-convex) and the domain of f is convex.*

Proof. See Exercise 2.1.5. □

Some useful facts concerning convex functions are listed as follows.

Lemma 2.1.8 (Basic properties). *The convex functions on E form a convex cone closed under taking pointwise suprema: if f_γ is convex for each $\gamma \in \Gamma$ then so is $x \mapsto \sup_{\gamma \in \Gamma} f_\gamma(x)$.*

(a) *Suppose $g : E \to [-\infty, +\infty]$, then g is convex if and only if $\operatorname{epi} g$ is a convex set if and only if $\delta_{\operatorname{epi} g}$ is convex.*
(b) *Suppose $\alpha : E \to F$ is affine, and $g : F \to (-\infty, +\infty]$, then $g \circ \alpha$ is convex when g is convex.*
(c) *Suppose $g : E \to (-\infty, +\infty]$ is convex, and $m : (-\infty, +\infty] \to (-\infty, +\infty]$ is monotone increasing and convex, then $m \circ g$ is convex (see also Exercise 2.4.31).*
(d) *For $t \geq 0$, the function $(x, t) \mapsto tg(x/t)$ is convex if and only if g is a convex function.*

Proof. The proof of (a) is straightforward. For (b), observe that

$$(g \circ \alpha)(\lambda x + (1-\lambda)y) = g(\lambda \alpha(x) + (1-\lambda)\alpha(y)) \leq \lambda g(\alpha(x)) + (1-\lambda)g(\alpha(y)).$$

(d) Consider the function $h : E \times \mathbb{R} \to (-\infty, +\infty]$ defined by $h(x,t) := tg(x/t)$. Then for $\lambda > 0$, $h(\lambda(x,t)) = \lambda t g(\lambda x/(\lambda t)) = \lambda t g(x/t)$. Also, the convexity of g implies

$$\frac{t}{s+t}g(x/t) + \frac{s}{s+t}g(y/s) \geq g\left(\frac{t}{s+t} \cdot \frac{x}{t} + \frac{s}{s+t} \cdot \frac{y}{s}\right)$$

$$= g\left(\frac{x+y}{s+t}\right).$$

Therefore, $h(x+y, t+s) = (s+t)g((x+y)/(t+s)) \leq tg(x/t) + sg(y/s) = h(x,t) + h(y,s)$. This shows h is positively homogeneous and subadditive.

The remainder of the proof is left as Exercise 2.1.6. $\square$

Proposition 2.1.9. *Suppose $f : C \to (-\infty, +\infty]$ is a proper convex function. Then f has bounded lower level sets if and only if*

$$\liminf_{\|x\| \to \infty} \frac{f(x)}{\|x\|} > 0. \tag{2.1.2}$$

Proof. $\Rightarrow$: By shifting f and C appropriately, we may assume for simplicity that $0 \in C$ and $f(0) = 0$. Suppose f has bounded lower level sets, but that (2.1.2) fails. Then there is a sequence $(x_n) \subset C$ such that $\|x_n\| \geq n$ and $f(x_n)/\|x_n\| \leq 1/n$. Then

$$f\left(\frac{nx_n}{\|x_n\|}\right) = f\left(\frac{\|x_n\| - n}{\|x_n\|} 0 + \frac{n}{\|x_n\|} x_n\right) \leq \frac{n}{\|x_n\|} f(x_n) \leq 1.$$

Hence we have the contradiction that $\{x : f(x) \leq 1\}$ is unbounded. The converse is left for Exercise 2.1.19. $\square$

A function $f : E \to (-\infty, +\infty]$ is said to be *coercive* if $\lim_{\|x\| \to \infty} f(x) = \infty$. For proper convex functions, coercivity is equivalent to (2.1.2), but different for nonconvex functions as seen in functions such as $f := \|\cdot\|^{1/2}$.

2.1.2 Continuity and subdifferentials

We now show that local boundedness properties of convex functions imply local Lipschitz conditions.

Theorem 2.1.10. *Suppose $f : E \to (-\infty, +\infty]$ is a convex function. Then f is locally Lipschitz around a point x in its domain if and only if it is bounded above on a neighborhood of x.*

Proof. Sufficiency is clear. For necessity, by scaling and translating, we can without loss of generality take $x = 0, f(0) = 0$ and suppose $f \leq 1$ on $2B_E$, and then we will show f is Lipschitz on B_E.

First, for any $u \in 2B_E, 0 = f(0) \leq \frac{1}{2}f(-u) + \frac{1}{2}f(u)$ and so $f(u) \geq -1$. Now for any two distinct points u and v in B_E, we let $\lambda = \|u-v\|$ and consider $w = v + \lambda^{-1}(v-u)$. Then $w \in 2B_E$, and the convexity of f implies

$$f(v) - f(u) \leq \frac{1}{1+\lambda}f(u) + \frac{\lambda}{1+\lambda}f(w) - f(u) \leq \frac{2\lambda}{1+\lambda} \leq 2\|v-u\|.$$

The result follows by interchanging u and v. □

Lemma 2.1.11. *Let Δ be the simplex $\{x \in \mathbb{R}^n_+ : \sum x_i \leq 1\}$. If the function $f : \Delta \to \mathbb{R}$ is convex, then it is continuous on int Δ.*

Proof. According to Theorem 2.1.10, it suffices to show that f is bounded above on Δ. For this, let $x \in \Delta$. Then

$$f(x) = f\left(\sum_{i=1}^n x_i e_i + \left(1 - \sum x_i\right)0\right) \leq \sum_{i=1}^n x_i f(e_i) + \left(1 - \sum x_i\right)f(0)$$
$$\leq \max\{f(e_1), f(e_2), \ldots, f(e_n), f(0)\},$$

where $\{e_1, e_2, \ldots, e_n\}$ represents the standard basis of $\mathbb{R}^n$. □

Theorem 2.1.12. *Let $f : E \to (-\infty, +\infty]$ be a convex function. Then f is continuous (in fact locally Lipschitz) on the interior of its domain.*

Proof. For any point $x \in \text{int dom} f$ we can choose a neighborhood of $x \in \text{dom} f$ that is a scaled and translated copy of the simplex. The result now follows from Theorem 2.1.10 and the proof of Lemma 2.1.11. □

Given a nonempty set $M \subset E$, the *core* of M is defined by $x \in \text{core } M$ if for each $h \in S_E$, there exists $\delta > 0$ so that $x + th \in M$ for all $0 \leq t \leq \delta$. It is clear from the definition that, int $M \subset$ core M.

Proposition 2.1.13. *Suppose $C \subset E$ is convex. Then $x_0 \in \text{core } C$ if and only if $x_0 \in \text{int } C$. However, this need not be true if C is not convex (see the nonconvex apple in Figure 2.4).*

Proof. See Exercise 2.1.7. □

The utility of the core arises in the convex context because it is often easier to check than interior, and it is naturally suited for studying directionally defined concepts, such as the directional derivative which now introduce. The *directional derivative* of f at $\bar{x} \in \text{dom} f$ in the direction h is defined by

$$f'(\bar{x}; h) := \lim_{t \to 0^+} \frac{f(\bar{x} + th) - f(\bar{x})}{t}$$

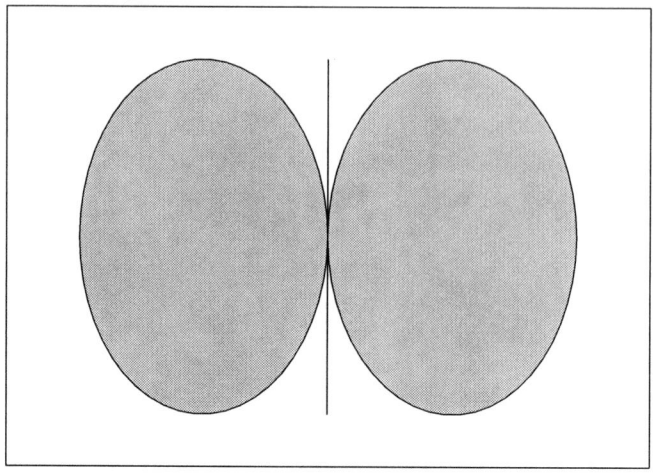

Figure 2.4 A nonconvex set with a boundary core point.

if the limit exists. We use the term directional derivative with the understanding that it is actually a *one-sided* directional derivative. Moreover, when f is a function on the real line, the directional derivatives are related to the usual one-sided derivatives by $f'_-(x) = -f(x; -1)$ and $f'_+(x) = f'(x; 1)$.

The *subdifferential* of f at $\bar{x} \in \text{dom} f$ is defined by

$$\partial f(\bar{x}) := \{\phi \in E : \langle \phi, y - \bar{x} \rangle \leq f(y) - f(\bar{x}), \text{ for all } y \in E\}. \quad (2.1.3)$$

When $\bar{x} \notin \text{dom} f$, we define $\partial f(\bar{x}) = \emptyset$. Even when $\bar{x} \in \text{dom} f$, it is possible that $\partial f(\bar{x})$ may be empty. However, if $\phi \in \partial f(\bar{x})$, then ϕ is said to be a *subgradient* of f at $\bar{x}$. An important example of a subdifferential is the *normal cone* to a convex set $C \subset E$ at a point $x \in C$ which is defined by $N_C(x) := \partial \delta_C(x)$.

Proposition 2.1.14 (Critical points). *Let $f : E \to (-\infty, +\infty]$ be a convex function. Then the following are equivalent.*

(a) f has a local minimum at $\bar{x}$.
(b) f has a global minimum at $\bar{x}$.
(c) $0 \in \partial f(\bar{x})$.

Proof. (a) $\Rightarrow$ (b): Let $y \in X$, if $y \notin \text{dom} f$, there is nothing to do. Otherwise, let $g(t) := f(\bar{x} + t(y - \bar{x}))$ and observe that $g'_+(0) \geq 0$ and then apply the three-slope inequality (2.1.1) to conclude $g(1) \geq g(0)$; or in other words, $f(y) \geq f(\bar{x})$.

(b) $\Rightarrow$ (c) and (c) $\Rightarrow$ (a) are straightforward exercises. □

The following provides a method for recognizing convex functions via the subdifferential.

Proposition 2.1.15. *Let $U \subset E$ be an open convex set, and let $f : U \to \mathbb{R}$. If $\partial f(\bar{x}) \neq \emptyset$ for each $\bar{x} \in U$, then f is a convex function.*

2.1 Continuity and subdifferentials

Proof. Let $x, y \in U$, $0 \leq \lambda \leq 1$ and let $\phi \in \partial f(\lambda x + (1-\lambda)y)$. Now let $a := f(\lambda x + (1-\lambda)y) - \phi(\lambda x + (1-\lambda)y)$. Then the subdifferential inequality implies $a + \phi(u) \leq f(u)$ for all $u \in U$, and so

$$\begin{aligned} f(\lambda x + (1-\lambda)y) &= \phi(\lambda x + (1-\lambda)y) + a \\ &= \lambda(\phi(x) + a) + (1-\lambda)(\phi(y) + a) \\ &\leq \lambda f(x) + (1-\lambda)f(y) \end{aligned}$$

as desired. $\square$

An elementary relationship between subgradients and directional derivatives is recorded as follows.

Fact 2.1.16. *Suppose $\phi \in \partial f(\bar{x})$. Then $\langle \phi, d \rangle \leq f'(\bar{x}; d)$ whenever the right-hand side is defined.*

Proof. This follows by taking the limit as $t \to 0^+$ in the following inequality.

$$\phi(d) = \frac{\phi(td)}{t} \leq \frac{f(t\bar{x} + d) - f(\bar{x})}{t}.$$

$\square$

Proposition 2.1.17. *Suppose the function $f : E \to (-\infty, +\infty]$ is convex. Then for any point $\bar{x} \in \operatorname{core} \operatorname{dom} f$, the directional derivative $f'(\bar{x}; \cdot)$ is everywhere finite and sublinear.*

Proof. Let $d \in E$ and $t \in \mathbb{R} \setminus \{0\}$, define

$$g(d; t) := \frac{f(\bar{x} + td) - f(\bar{x})}{t}.$$

The three-slope inequality (2.1.1) implies

$$g(d; -s) \leq g(d; -t) \leq g(d; t) \leq g(d; s) \text{ for } 0 < t < s.$$

Since $\bar{x}$ lies in $\operatorname{core} \operatorname{dom} f$, for small $s > 0$ both $g(d; -s)$ and $g(d; s)$ are finite, consequently as $t \downarrow 0$ we have

$$+\infty > g(d; s) \geq g(d; t) \downarrow f'(\bar{x}; d) \geq g(d; -s) > -\infty. \tag{2.1.4}$$

The convexity of f implies that for any directions $d, e \in E$ and for any $t > 0$, one has

$$g(d + e; t) \leq g(d; 2t) + g(e; 2t).$$

Letting $t \downarrow 0$ establishes the subadditivity of $f'(\bar{x}; \cdot)$. It is left for the reader to check the positive homogeneity. $\square$

One of the several natural ways to prove that subdifferentials are nonempty for convex functions at points of continuity uses the Hahn–Banach extension theorem which we now present.

Proposition 2.1.18 (Hahn–Banach extension). *Suppose $p : E \to \mathbb{R}$ is a sublinear function and $f : S \to \mathbb{R}$ is linear where S is a subspace of E. If $f(x) \leq p(x)$ for all $x \in S$, then there is a linear function $\phi : E \to \mathbb{R}$ such that $\phi(x) = f(x)$ for all $x \in S$ and $\phi(v) \leq p(v)$ for all $v \in E$.*

Proof. If $S \neq E$ choose $x_1 \in E \setminus S$ and let S_1 be the linear span of $\{x_1\} \cup S$. Observe that for all $x, y \in S$

$$f(x) + f(y) = f(x+y) \leq p(x+y) \leq p(x - x_1) + p(x_1 + y)$$

and consequently

$$f(x) - p(x - x_1) \leq p(y + x_1) - f(y) \quad \text{for all } x, y \in S. \tag{2.1.5}$$

Let α be the supremum for $x \in S$ of the left-hand side of (2.1.5). Then

$$f(x) - \alpha \leq p(x - x_1) \quad \text{for all } x \in S, \tag{2.1.6}$$

and

$$f(y) + \alpha \leq p(y + x_1) \quad \text{for all } y \in S. \tag{2.1.7}$$

Define f_1 on S_1 by

$$f_1(x + tx_1) := f(x) + t\alpha \quad \text{for all } x \in S, t \in \mathbb{R}. \tag{2.1.8}$$

Then $f_1 = f$ on S and f_1 is linear on S_1. Now let $t > 0$, and replace x with $t^{-1}x$ in (2.1.6), and replace y with $t^{-1}y$ in (2.1.7). Combining this with (2.1.8) will show that $f_1 \leq p$ on S_1. Since E is finite-dimensional, repeating the above process finitely many times yields ϕ as desired. □

We are now ready to relate subgradients more precisely to directional derivatives.

Theorem 2.1.19 (Max formula). *Suppose $f : E \to (-\infty, +\infty]$ is convex and $\bar{x} \in \text{core dom} f$. Then for any $d \in E$,*

$$f'(\bar{x}; d) = \max\{\langle \phi, d \rangle : \phi \in \partial f(\bar{x})\}. \tag{2.1.9}$$

In particular, the subdifferential $\partial f(\bar{x})$ is nonempty.

Proof. Fix $d \in S_E$ and let $\alpha = f'(\bar{x}; d)$, then α is finite because $\bar{x} \in \text{core dom} f$ (Proposition 2.1.17). Let $S = \{td : t \in \mathbb{R}\}$ and define the linear function $\Lambda : S \to \mathbb{R}$

by $\Lambda(t\alpha) := t\alpha$ for $t \in \mathbb{R}$. Then $\Lambda(\cdot) \leq f'(\bar{x}; \cdot)$ on S; according to the Hahn–Banach extension theorem (2.1.18) there exists $\phi \in E$ such that

$$\phi = \Lambda \text{ on } S, \quad \phi(\cdot) \leq f'(\bar{x}; \cdot) \text{ on } E.$$

Then $\phi \in \partial f(\bar{x})$ and $\phi(sd) = f'(\bar{x}; sd)$ for all $s \geq 0$. □

Corollary 2.1.20. *Suppose $f : E \to (-\infty, +\infty]$ is a proper convex function that is continuous at $\bar{x}$. Then $\partial f(\bar{x})$ is a nonempty, closed, bounded and convex subset of E.*

Proof. See Exercise 2.1.8. □

Corollary 2.1.21 (Basic separation). *Suppose $C \subset E$ is a closed nonempty convex set, and suppose $x_0 \notin C$. Then there exists $\phi \in E$ such that*

$$\sup_C \phi < \langle \phi, x_0 \rangle.$$

Proof. Let $f : E \to \mathbb{R}$ be defined by $f(\cdot) = d_C(\cdot)$. Then f is Lipschitz and convex (Fact 2.1.6) and so by the max formula (2.1.19) there exists $\phi \in \partial f(x_0)$. Then $\langle \phi, x - x_0 \rangle \leq f(x) - f(x_0)$ for all $x \in E$. In particular, if $x \in C$, $f(x) = 0$, and so the previous inequality implies $\phi(x) + d_C(x_0) \leq \phi(x_0)$ for all $x \in C$. □

Exercises and further results

2.1.1.* Fill in the necessary details for the proof of Theorem 2.1.2.
2.1.2.* Prove Corollary 2.1.3.
2.1.3.* Prove Lemma 2.1.4.
2.1.4.* Prove Fact 2.1.5.
2.1.5.* Prove Fact 2.1.7.
2.1.6.* Prove the remaining parts of Lemma 2.1.8.
2.1.7.* Prove Proposition 2.1.13.

Hint. Suppose $x_0 \in \text{core } C$; then there exists $\delta > 0$ so that $x_0 + te_i \in C$ for all $|t| \leq \delta$ and $i = 1, 2, \ldots, n$ where $\{e_i\}_{i=1}^n$ is the usual basis of $\mathbb{R}^n$. Use the convexity of C to conclude $\text{int } C \neq \emptyset$. A conventional example of a nonconvex set F with $0 \in \text{core } F \setminus \text{int } F$ is $F = \{(x, y) \in \mathbb{R}^2 : |y| \geq x^2 \text{ or } y = 0\}$; see also Figure 2.4. □

2.1.8.* Prove Corollary 2.1.20.
2.1.9.* Prove Theorem 1.2.7.
2.1.10 (Jensen's Inequality). Let $\phi : I \to \mathbb{R}$ be a convex function where $I \subset \mathbb{R}$ is an open interval. Suppose $f \in L_1(\Omega, \mu)$ where μ is a probability measure and $f(x) \in I$ for all $x \in \Omega$. Show that

$$\int \phi(f(t)) d\mu \geq \phi\left[\int f d\mu\right]. \tag{2.1.10}$$

Hint. Verify that $\phi \circ f$ is measurable. Let $a := \int f d\mu$ and note $a \in I$. Apply Corollary 2.1.3 to obtain $\lambda \in \mathbb{R}$ such that $\phi(t) \geq \phi(a) + \lambda(t - a)$ for all $t \in I$.

Then integrate both sides of $\phi(f(t)) \geq \lambda(f(t) - a) + \phi(a)$. See [384, p. 62] for further details. □

2.1.11. (a) [Converse to Jensen] Suppose g is a real function on $\mathbb{R}$ such that

$$g\left(\int_0^1 f(x)dx\right) \leq \int_0^1 g(f)dx$$

for every bounded measurable function f on $[0, 1]$. Show that g is convex.
(b) Use Jensen's inequality to show the *arithmetic-geometric mean* inequality:

$$(x_1 x_2 \cdots x_n)^{1/n} \leq \frac{1}{n}(x_1 + x_2 + \ldots + x_n) \text{ where } x_i > 0, \ i = 1, 2, \ldots, n.$$

Hint. (a) For arbitrary $a, b \in \mathbb{R}$ and $0 < \lambda < 1$, let f be defined by $f(x) := a$ for $0 \leq x \leq \lambda$ and $f(x) := b$ for $\lambda < x \leq 1$. For (b), apply Jensen's inequality when $\phi := \exp(\cdot)$ and $f(p_i) := y_i$ where $\mu(p_i) = 1/n$, $i = 1, 2, \ldots, n$ and y_i is chosen so that $x_i = \exp(y_i)$. See [384, p. 63] for more details. There are several other proofs, and the reader should note equality occurs if and only if $x_1 = x_2 = \ldots = x_n$. □

2.1.12. Suppose f is a convex function on E (or any normed linear space) that is globally bounded above. Show that f is constant.

Hint. If f is not identically $-\infty$, fix x_0 where f is finite-valued and use the definition of convexity to show $f(x) = f(x_0)$ for all x. □

2.1.13.* Let C be a convex subset of E and let γ_C denote its gauge function.

(a) Show that γ_C is sublinear.
(b) Show that if $x \in \text{core } C$ then $\text{dom } \gamma_{C-x} = E$. Deduce that γ_C is continuous.
(c) Suppose $0 \in \text{core } C$. Prove that $\text{cl } C \subset \{x \in E : \gamma_C(x) \leq 1\}$.

2.1.14. Suppose $f : \mathbb{R} \to (-\infty, +\infty]$ is convex, and strictly convex on the interior of its domain. Show that f is strictly convex. Provide an example showing this can fail if $\mathbb{R}$ is replaced with $\mathbb{R}^2$.

2.1.15 (Convex hull of a family). Show that for a family of proper convex functions $(f_i)_{i \in I}$, the largest convex minorant $f := \text{conv}_{i \in I} f_i$ is given by

$$f(x) := \inf\left\{\sum_{i \in I} \lambda_i f(x_i) : \sum_{i \in I} \lambda_i x_i = x, \sum_{i \in I} \lambda_i = 1, \lambda_i \geq 0 \text{ finitely nonzero}\right\}.$$

(See [369, Theorem 5.6].) Find an example where $\text{epi conv} f$ is not $\text{conv epi} f$.

Hint. For the example, let $f : \mathbb{R} \to \mathbb{R}$ be defined by $f(t) = t^2$ if $t \neq 0$ and $f(0) = 1$. □

2.1.16 (Open mapping theorem). A mapping $T : X \to Y$ between normed linear spaces is *open* if it maps open sets onto open sets. Show that linear mapping between Euclidean spaces X and Y is onto if and only if it is open.

Hint. Suppose $T : X \to Y$ is onto. For any open convex set $U \subset X$, and $x \in T(U)$, show $x \in \text{core } T(U)$. □

2.1.17 (Midpoint convexity). For $0 < \alpha < 1$, an extended real-valued function f on a Euclidean space E is α-*convex* if $f(\alpha x + (1-\alpha)y) \leq \alpha f(x) + (1-\alpha)f(y)$ for all $x, y \in E$. When $\alpha = 1/2$ the function is said to be *midpoint convex*.

(a) Show that any finite-valued measurable α-convex function is convex. In particular, a nonconvex midpoint convex function cannot be bounded above on any nonempty open set.
(b) In consequence, every finite lsc midpoint convex function is convex; this remains true in any normed space.
(c) Use the existence for a *Hamel basis* for $\mathbb{R}$ over $\mathbb{Q}$ to construct a nonmeasurable function such both f and $-f$ are nonconvex and midpoint convex.

Hint. Note also that finite is required in the 'measurable' assertion. For the example, let $f(t) := 0$ if t is rational and $f(t) := \infty$ otherwise. □

2.1.18 (Lower-semicontinuity and closure). Suppose $f : E \to [-\infty, +\infty]$ is convex. Show it is lsc at points x where it is finite if and only if $f(x) = (\mathrm{cl}\, f)(x)$. In this case f is proper.

2.1.19.★

(a) Show that any function $f : E \to [-\infty, +\infty]$ which satisfies (2.1.2) has bounded lower level sets.
(b) Consider $f : \mathbb{R} \to \mathbb{R}$ defined by $f(t) := \sqrt{|t|}$. Show that f is not convex and f has convex bounded lower level sets, but f does not satisfy the growth condition (2.1.2).

2.1.20. Suppose $f : \mathbb{R} \to (-\infty, +\infty]$ is convex and proper. Show that

$$\lim_{t \to +\infty} f(t)/t \text{ exists in } (-\infty, +\infty].$$

Hint. Use the three-slope inequality (2.1.1). □

2.1.21 (Almost sublinear functions). A real-valued function q on a Euclidean space E is *almost sublinear* (also called almost homogeneous) if there is a sublinear function $p : E \to \mathbb{R}$ with $\sup_{x \in E} |p(x) - q(x)| < \infty$. Show that q is almost sublinear if and only if it has an affine minorant representation

$$q(x) = \sup_{n \in \mathbb{N}} \langle a_n, x \rangle + b_n$$

for a bounded set $\{(a_n, b_n) \in E \times \mathbb{R} : n \in \mathbb{N}\}$.

2.1.22 (Uniform convergence).★ Let S be a closed subset of E, and let $f, f_1, f_2, f_3 \ldots$ be continuous real functions on S. For (a) and (c), suppose $f_n \to f$ pointwise.

(a) Use the Baire category theorem to show that there is a relatively open set $U \subset S$ such that $(f_n)_{n=1}^\infty$ is uniformly bounded on the set U.
For the next two parts, suppose additionally $f, f_1, f_2, f_3, \ldots$ are convex.

(b) Suppose $C \subset \mathbb{R}$ is a bounded convex set, and $|f|, |f_1|, |f_2|, |f_3|, \ldots$ are uniformly bounded on $C + rB_E$ for some $r > 0$. Show there exists $K > 0$ so that $f, f_1, f_2, f_3, \ldots$ have Lipschitz constant K on C.

(c) Suppose further S is a convex set with nonempty interior. Show that for each $x \in \mathrm{int}\, S$, there is an open set U so that $x \in U$, and $(f_n)_{n=1}^{\infty}$ is uniformly bounded on U. Deduce further that the convergence is uniform on compact subsets in int S.

2.1.23 (Boundedness properties of subdifferentials).* Let U be an open convex subset of E, and suppose $f : U \to \mathbb{R}$ is convex.

(a) Show that $\partial f(U) \subset KB_E$ if and only if f has Lipschitz constant K on U. In particular, for each $x \in U$, there is a neighborhood V of x such that $\partial f(V)$ is a bounded subset of E; in other words ∂f is *locally bounded* on U.

(b) Suppose $\mathrm{dom}\, f = E$. Show that $\partial f(S)$ is a bounded subset of E if S is a bounded subset of E.

For the next two exercises, recall that the *Gamma function* (or Γ-function) is usually defined for $\mathrm{Re}(x) > 0$ as

$$\Gamma(x) := \int_0^{\infty} e^{-t} t^{x-1}\, dt. \tag{2.1.11}$$

The following exercise provides a fine convexity characterization of the Γ-function.

2.1.24 (Bohr–Mollerup theorem).† Show the gamma-function is the unique function f mapping the positive half-line to itself such that (a) $f(1) = 1$, (b) $xf(x) = f(x+1)$ and (c) $\log f$ is a convex function.

Hint. First use Hölder's inequality (see Exercise 2.3.3) to show Γ is as desired. Conversely, let $g := \log f$. Then $g(n+1) = \log(n!)$ and convexity of g implies that $x \log(n) \leq (n+1+x) - g(n+1) \leq x \log(n+1)$. Thus,

$$0 \leq g(x) - \log \frac{n!\, n^x}{x(x+1)\cdots(x+n)} \leq \log\left(1 + \frac{1}{n}\right).$$

Take limits to show

$$f(x) = \lim_{n \to \infty} \frac{n!\, n^x}{x(x+1)\cdots(x+n)} = \Gamma(x).$$

Note we have discovered an important product representation to boot. □

2.1.25 (Blaschke–Santalo theorem).† Application of the Bohr–Mollerup theorem is often *automatable* in a computer algebra system, as we now illustrate. The *β-function* is defined by

$$\beta(x, y) := \int_0^1 t^{x-1}(1-t)^{y-1}\, dt \tag{2.1.12}$$

for $\mathrm{Re}(x), \mathrm{Re}(y) > 0$. As is often established using polar coordinates and double integrals

$$\beta(x, y) = \frac{\Gamma(x)\Gamma(y)}{\Gamma(x+y)}. \tag{2.1.13}$$

(a) Use the Bohr–Mollerup theorem (Exercise 2.1.24) with

$$f := x \to \beta(x,y)\Gamma(x+y)/\Gamma(y)$$

to prove (2.1.13) for real x, y. Now (a) and (b) from Exercise 2.1.24 are easy verify. For (c) show f is log-convex via Hölder's inequality (Exercise 2.3.3). Thus $f = \Gamma$ as required.

(b) Show that the volume of the ball in the $\|\cdot\|_p$-norm, $V_n(p)$ is

$$V_n(p) = 2^n \frac{\Gamma(1+\frac{1}{p})^n}{\Gamma(1+\frac{n}{p})}. \tag{2.1.14}$$

as was first determined by Dirichlet. When $p = 2$, this gives

$$V_n = 2^n \frac{\Gamma(\frac{3}{2})^n}{\Gamma(1+\frac{n}{2})} = \frac{\Gamma(\frac{1}{2})^n}{\Gamma(1+\frac{n}{2})},$$

which is more concise than that usually recorded in texts. *Maple code derives this formula as an iterated integral for arbitrary p and fixed n.*

(c) Let C be *convex body* in $\mathbb{R}^n$, that is, C is a closed bounded convex set with nonempty interior. Denoting n-dimensional Euclidean volume of $S \subseteq \mathbb{R}^n$ by $V_n(S)$, the *Blaschke–Santalo* inequality says

$$V_n(C)\,V_n(C^\circ) \leq V_n(E)\,V_n(E^\circ) = V_n^2(B_n(2)) \tag{2.1.15}$$

where maximality holds (only) for *any* ellipsoid E and $B_n(2)$ is the Euclidean unit ball. It is conjectured the minimum is attained by the 1-ball and the ∞-ball. Here as always the *polar set* is defined by $C^\circ := \{y \in \mathbb{R}^n : \langle y, x \rangle \leq 1 \text{ for all } x \in C\}$.

(d) Deduce the p-ball case of (2.1.15) by proving the following convexity result:

Theorem 2.1.22 (Harmonic-arithmetic log-concavity). *The function $V_\alpha(p) := 2^\alpha \Gamma(1+\frac{1}{p})^\alpha / \Gamma(1+\frac{\alpha}{p})$ satisfies*

$$V_\alpha(p)^\lambda\, V_\alpha(q)^{1-\lambda} < V_\alpha\left(\frac{1}{\frac{\lambda}{p} + \frac{1-\lambda}{q}}\right), \tag{2.1.16}$$

for all $\alpha > 1$, if $p, q > 1$, $p \neq q$, and $\lambda \in (0,1)$.

Hint. Set $\alpha := n$, $\frac{1}{p} + \frac{1}{q} = 1$ with $\lambda = 1 - \lambda = 1/2$ to recover the p-norm case of the Blaschke–Santalo inequality. □

(e) Deduce the corresponding lower bound.

This technique extends to various substitution norms (see Exercise 6.7.6). Further details may be found in [73, §5.5].

2.2 Differentiability

A function $f : E \to [-\infty, +\infty]$ is said to be *Fréchet differentiable* at $x_0 \in \text{dom} f$ if there exists $\phi \in E$ such that for each $\varepsilon > 0$ there exists $\delta > 0$ so that

$$|f(x_0 + h) - f(x_0) - \phi(h)| \leq \varepsilon \|h\| \quad \text{whenever } \|h\| < \delta.$$

In this case, we write $f'(x_0) = \phi$ or $\nabla f(x_0) = \phi$ and we say that ϕ is the *Fréchet derivative* of f at x_0. Notice that f is Fréchet differentiable at x_0 if and only if for some $f'(x_0) \in E$ the following limit exists uniformly for $h \in B_E$

$$\lim_{t \to 0} \frac{f(x_0 + th) - f(x_0)}{t} = \langle f'(x_0), h \rangle.$$

If the above limit exists for each $h \in S_E$, but not necessarily uniformly, then f is said to be *Gâteaux differentiable* at x_0. In other words, f is Gâteaux differentiable at $x_0 \in E$ if for each $\varepsilon > 0$ and each $h \in S_E$, there exists $\delta > 0$ so that

$$|f(x_0 + th) - f(x_0) - \phi(th)| \leq \varepsilon t \quad \text{if } 0 \leq t < \delta.$$

In either of the above cases, we may write $\phi = f'(x_0)$, or $\phi = \nabla f(x_0)$.

We should note that Gâteaux differentiability is precisely the form of differentiability that ensures the validity of using the dot product of a unit vector with the gradient for computing the standard (two-sided) directional derivative of a function f of several variables from elementary calculus. That is, for a unit vector $u \in S_E$, the definition of Gâteaux differentiability ensures that

$$\langle \nabla f(\bar{x}), u \rangle = D_u f(\bar{x}) \quad \text{where } D_u f(\bar{x}) = \lim_{t \to 0} \frac{f(\bar{x} + tu) - f(\bar{x})}{t}.$$

One should also note that Gâteaux differentiability at $\bar{x}$ is stronger than demanding the existence of $D_u f(\bar{x})$ for every $u \in S_E$, because Gâteaux differentiability requires that $f'(\bar{x}) : E \to \mathbb{R}$ be a linear mapping; see Exercise 2.2.10.

As a consequence of the max formula (2.1.19), we can express differentiability of convex functions in several different but equivalent ways.

Theorem 2.2.1 (Characterizations of differentiability). *Suppose $f: \mathbb{R}^n \to (-\infty, +\infty]$ is convex and $x_0 \in \text{dom} f$. Then the following are equivalent.*

(a) f is Fréchet differentiable at x_0.
(b) f is Gâteaux differentiable at x_0.
(c) $\dfrac{\partial f}{\partial x_i}(x_0)$ exist for $i = 1, 2, \ldots, n$.
(d) $\partial f(x_0)$ is a singleton.

Moreover, when any of the above conditions is true, $x_0 \in \text{core dom} f$ and

$$\nabla f(x_0) = \left(\frac{\partial f}{\partial x_1}(x_0), \ldots, \frac{\partial f}{\partial x_n}(x_0) \right) \quad \text{and} \quad \partial f(x_0) = \{\nabla f(x_0)\}.$$

2.2 Differentiability

Proof. We first suppose that $x_0 \in \operatorname{core} \operatorname{dom} f$; later we will outline how each of (a)–(d) implies $x_0 \in \operatorname{core} \operatorname{dom} f$. Clearly (a) $\Rightarrow$ (b) $\Rightarrow$ (c). To see that (c) $\Rightarrow$ (d), we know from the Max formula (2.1.19) that $\partial f(x_0) \neq \emptyset$. Suppose there are distinct $\phi_1, \phi_2 \in \partial f(x_0)$, then there is a basis element e_i with, say, $\phi_1(e_i) > \phi_2(e_i)$. Then the Max formula (2.1.19) implies $f'(x_0; e_i) \geq \phi_1(e_i)$ and $f'(x_0; -e_i) \geq \phi_2(-e_i) > -f'(x_0; e_i)$. This implies that $\dfrac{\partial f}{\partial x_i}(x_0)$ does not exist.

(d) $\Rightarrow$ (b): Suppose $\partial f(x_0) = \{\phi\}$, then the max formula (2.1.19) shows

$$\phi(d) = \lim_{t \to 0^+} \frac{f(x_0 + td) - f(x_0)}{t}$$

and

$$\phi(-d) = \lim_{t \to 0^+} \frac{f(x_0 - td) - f(x_0)}{t} = -\lim_{t \to 0^-} \frac{f(x_0 + td) - f(x_0)}{t}.$$

Consequently, $\phi(d) = \lim_{t \to 0} \dfrac{f(x_0 + td) - f(x_0)}{t}$ for all $d \in E$ as desired.

(b) $\Rightarrow$ (a): Follows because on $\mathbb{R}^n$ a locally Lipschitz function is Fréchet differentiable whenever it is Gâteaux differentiable (Exercise 2.2.9).

For the moreover part, notice that the proof of (d) $\Rightarrow$ (b) shows $\partial f(x_0) = \{f'(x_0)\}$; and the definition of Gâteaux differentiability ensures $\langle f'(x_0), e_i \rangle = \dfrac{\partial f}{\partial x_i}(x_0)$ where $\{e_i\}_{i=1}^n$ is the standard basis for $\mathbb{R}^n$.

Finally, we need to verify that each of the above conditions implies $x_0 \in \operatorname{core} \operatorname{dom} f$; indeed, this is easy to verify for (a), (b) and (c). So we now do so for (d); indeed, suppose x_0 is in the boundary of the domain of f. In the event $\partial f(x_0) = \emptyset$, there is nothing further to do. Thus we may assume $\phi \in \partial f(x_0)$. Now let C be the closure of the domain of f, and define $g : \mathbb{R}^n \to (-\infty, +\infty]$, by $g(x) := f(x_0) + \phi(x - x_0)$ for $x \in \mathbb{R}^n$. Define the function $h : \mathbb{R}^n \to \mathbb{R}$ by $h(x) := g(x) + d_C(x)$. Then h is Lipschitz and convex, $h \leq f$, and $h(x_0) = f(x_0)$. Therefore, $\partial h(x_0) \subset \partial f(x_0)$, and one can check that $\partial h(x_0)$ is not a singleton using the max formula (2.1.19) since we already know $\phi \in \partial h(x_0)$. See Exercise 2.2.23 for a stronger result. $\square$

An important property about differentiable convex functions is that they are automatically continuously differentiable.

Theorem 2.2.2. *Suppose $f : E \to (-\infty, +\infty]$ is a proper convex function that is differentiable on an open set U. Then $x \mapsto \nabla f(x)$ is continuous on U.*

Proof. This and more is outlined in Exercise 2.2.22. $\square$

However, in the absence of convexity, Gâteaux differentiability is far from Fréchet differentiability, and, in fact, does not even ensure continuity of the function.

Example 2.2.3. Let $f(x, y) := \dfrac{xy^4}{x^2 + y^8}$ if $(x, y) \neq (0, 0)$ and let $f(0, 0) := 0$; let $g(x, y) := \dfrac{xy^3}{x^2 + y^4}$ if $(x, y) \neq (0, 0)$ and $g(0, 0) := 0$. Then, both functions are C^∞-smooth on $\mathbb{R}^2 \setminus \{(0, 0)\}$. Moreover, f is Gâteaux differentiable at $(0, 0)$, but not

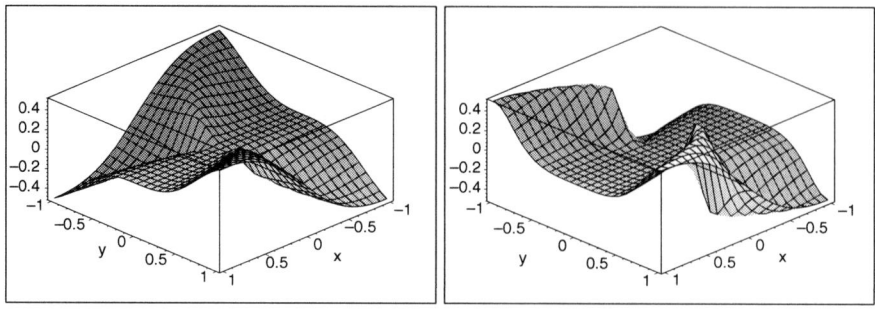

Figure 2.5 Functions $\frac{xy^3}{x^2+y^4}$ and $\frac{xy^4}{x^2+y^8}$ with a Gâteaux derivative at the origin that is not Fréchet.

continuous there, while g is Gâteaux differentiable and continuous at $(0,0)$, but not Fréchet differentiable there.

Proof. The reader can check that the Gâteaux derivatives of both functions at $(0,0)$ are the zero functional. One can check that f is not continuous at $(0,0)$ by examining the limit of f along the path $x = y^4$. To see that the derivative of g is not a Fréchet derivative, observe $\lim_{t\to 0^+}[g((t^2,t)) - g(0)]/\|(t^2,t)\| = 1/2$. See Figure 2.5 for graphs of these functions. □

In contrast, we note that if a function has a continuous Gâteaux derivative, as is guaranteed by the existence of continuous partial derivatives, it is necessarily a Fréchet derivative – as the mean value theorem shows. A more general result is given in Exercise 2.2.9.

2.2.1 Recognizing convex functions via differentiability

This section examines briefly how differentiable convex functions can be recognized through properties of first or second derivatives.

Proposition 2.2.4. *Let I be an interval in $\mathbb{R}$.*

(a) *A differentiable function $f : I \to \mathbb{R}$ is convex (resp. strictly convex) if and only if f' is nondecreasing (resp. strictly increasing) on I.*
(b) *A twice differentiable function $f : I \to \mathbb{R}$ is convex if and only if $f''(t) \geq 0$ for all $t \in I$.*

Proof. We need only prove (a), because (b) is a consequence of (a). In (a) we prove only the convex case because the strictly convex case is similar. Suppose f is convex and differentiable, then Theorem 2.1.2(b) implies that f' is nondecreasing on I.

Conversely, suppose f fails to be convex on I. Then it is easy to verify there exist $x < y < z$ in the interval I such that

$$f(y) > \frac{z-y}{z-x}f(x) + \frac{y-x}{z-x}f(z).$$

But this implies
$$\frac{f(y)-f(x)}{y-x} > \frac{f(z)-f(y)}{z-y},$$
which, together with the mean value theorem, shows that f' is not nondecreasing on I. □

Example 2.2.5. The functions $f(t) := e^t$, $g(t) := t\log t$ for $t > 0$, $h(t) := |t|^p$ for $p > 1$ are strictly convex. Note that $t \mapsto t^4$ is strictly convex on the line but the second derivative vanishes at zero. Thus strict positivity of the second derivative is sufficient but not necessary for strict convexity.

Let $f : I \to \mathbb{R}$ where I is an interval. Then its derivative f' is nondecreasing on I if and only if
$$(f'(t) - f'(s))(t - s) \geq 0$$
for all $s, t \in I$. Proposition 2.2.4(a) then has the following natural generalization.

Theorem 2.2.6 (Monotone gradients and convexity). *Suppose $f : U \to \mathbb{R}$ is Gâteaux differentiable and $U \subset E$ is an open convex set. Then f is convex if and only if $\langle \nabla f(x) - \nabla f(y), x - y \rangle \geq 0$ for all $x, y \in U$. Moreover, f is strictly convex if and only if the preceding inequality is strict for distinct x and y.*

Proof. ⇒: This is a consequence of Exercise 2.2.21 which establishes the *monotonicity* of the subdifferential map.

⇐: Let $\bar{x} \in U$, $h \in S_E$ and define g by $g(t) := f(\bar{x} + th)$. It suffices to show that any such g is convex. Let s and t be chosen so that $g(s)$ and $g(t)$ are finite. Then for $u = \bar{x} + th$ and $v = \bar{x} + sh$ we have
$$\begin{aligned}(g'(t) - g'(s))(t - s) &= (t - s)(\langle \nabla f(\bar{x} + th), h \rangle - \langle \nabla f(\bar{x} + sh), h \rangle) \\ &= \langle \nabla f(\bar{x} + th) - \nabla f(\bar{x} + sh), th - sh \rangle \\ &= \langle \nabla f(u) - \nabla f(v), u - v \rangle \geq 0.\end{aligned}$$

Hence g is convex because g' is nondecreasing (Proposition 2.2.4). Therefore, f is convex. The moreover part is left to the reader. □

We now define differentiability for mappings between Euclidean spaces, this will allow us to view second-order derivatives in a natural fashion. Let U be an open subset of E, let $\bar{x} \in U$ and let $f : U \to Y$ where Y is a Euclidean space. We will say f is *Gâteaux differentiable* at $\bar{x}$ if there is a linear mapping $A : E \to Y$ so that
$$\lim_{t \to 0} \frac{f(\bar{x} + th) - f(\bar{x})}{t} = Ah$$
for each $h \in E$. If the limit exists uniformly for $\|h\| = 1$, then f is said to be *Fréchet differentiable* at $\bar{x}$. In either case, A is the derivative of f at x, and we denote it by $f'(x)$ or $\nabla f(x)$. Consequently, a function $f : E \to (-\infty, +\infty]$ is *twice Gâteaux differentiable* at $\bar{x} \in E$ if there is an open set U such that $\bar{x} \in U$ and the mapping $f' : U \to E$ is Gâteaux differentiable at $\bar{x}$. The second-order derivative of f at $\bar{x}$

is often denoted by $f''(\bar{x})$ (or by $\nabla^2 f(\bar{x})$), and it is a linear mapping from E to E. Second-order Fréchet differentiability is defined analogously.

A standard result from multivariable calculus states that a function $f : E \to \mathbb{R}$ is continuously twice Fréchet differentiable (i.e. $x \mapsto f''(x)$ is continuous) if it has continuous second partials. Let us note that many important second-order results, such as Alexandrov's theorem (see Theorem 2.6.4 below) provide second-order results for convex functions even though they need not be everywhere first-order differentiable. Section 2.6 will study this and some more general notions of second-order differentiability on Euclidean spaces.

When the second partials of a function $f : \mathbb{R}^n \to \mathbb{R}$ are defined at $\bar{x}$, we define the *Hessian* matrix of f at $\bar{x}$ by

$$H = \begin{bmatrix} f_{x_1 x_1}(\bar{x}) & f_{x_1 x_2}(\bar{x}) & \cdots & f_{x_1 x_n}(\bar{x}) \\ f_{x_2 x_1}(\bar{x}) & f_{x_2 x_2}(\bar{x}) & \cdots & f_{x_2 x_n}(\bar{x}) \\ \vdots & \vdots & \vdots & \vdots \\ f_{x_n x_1}(\bar{x}) & f_{x_n x_2}(\bar{x}) & \cdots & f_{x_n x_n}(\bar{x}) \end{bmatrix} \quad \text{where } f_{x_i x_j}(\bar{x}) = \frac{\partial^2 f(\bar{x})}{\partial x_j \partial x_i}.$$

Now suppose f is twice Gâteaux differentiable at $\bar{x}$. Then $f''(\bar{x})$ identifies naturally with the Hessian. Indeed, viewing $k \in \mathbb{R}^n$ and $f''(\bar{x})(k)$ as column vectors, we have

$$f''(\bar{x})(k) = \lim_{t \to 0} \frac{\nabla f(\bar{x} + tk) - \nabla f(\bar{x})}{t} \tag{2.2.1}$$

$$= \left(\lim_{t \to 0} \frac{f_{x_1}(\bar{x} + tk) - f_{x_1}(\bar{x})}{t}, \ldots, \lim_{t \to 0} \frac{f_{x_n}(\bar{x} + tk) - f_{x_n}(\bar{x})}{t} \right)^T$$

$$= (\langle \nabla f_{x_1}(\bar{x}), k \rangle, \langle \nabla f_{x_2}(\bar{x}), k \rangle, \ldots, \langle \nabla f_{x_n}(\bar{x}), k \rangle)^T = Hk.$$

Consequently, the linear operator $f''(\bar{x}) = H$ as expected. Furthermore, when f is twice Gâteaux differentiable at $\bar{x}$, we define $f''(\bar{x})(\cdot, \cdot)$ by $f''(\bar{x})(h, k) := \langle h, Hk \rangle$. This leads naturally to identifying second-order derivatives with bilinear mappings; see Exercise 2.2.24. Using the continuity of the inner product on $\mathbb{R}^n$ we compute the following useful limit representation for $f''(\cdot, \cdot)$:

$$f''(\bar{x})(h, k) = \langle Hk, h \rangle = \left\langle \lim_{t \to 0} \frac{\nabla f(\bar{x} + tk) - \nabla f(\bar{x})}{t}, h \right\rangle$$

$$= \lim_{t \to 0} \frac{\langle \nabla f(\bar{x} + tk), h \rangle - \langle \nabla f(\bar{x}), h \rangle}{t}. \tag{2.2.2}$$

We now state a classical result on the symmetry of second derivatives; a proof of this in a more general setting can be found in [185, Section VII.12].

Theorem 2.2.7 (Symmetry of second-order derivatives). *Let $U \subset E$ be an open set and $x \in U$. If $f : U \to \mathbb{R}$ is twice Fréchet differentiable at x, then $\nabla^2 f(x)$ is a symmetric matrix. In particular, if f has continuous second partial derivatives at x, then $\nabla^2 f(x)$ is symmetric.*

2.2 Differentiability

In fact, Theorem 2.6.1 below shows that a twice Gâteaux differentiable convex function has a symmetric second derivative. That, together with the next theorem combine to show that a twice Gâteaux differentiable function is convex if and only if its Hessian is a symmetric positive semidefinite matrix.

Theorem 2.2.8 (Hessians and convexity). *Let U be an open convex subset of E, and suppose that $f : U \to \mathbb{R}$ is twice Gâteaux differentiable on U. Then f is convex if and only if $\nabla^2 f(\bar{x})(h, h) \geq 0$ for all $\bar{x} \in U, h \in E$. Moreover, if $\nabla^2 f(\bar{x})(h, h) > 0$ for all $\bar{x} \in U$, and $h \in S_E$ then f is strictly convex.*

Proof. For $\bar{x} \in U$ and $h \in E \setminus \{0\}$, define $g(t) = f(\bar{x} + th)$. The Gâteaux differentiability of f implies

$$g'(t) = \lim_{\Delta t \to 0} \frac{f(\bar{x} + (t + \Delta t)h) - f(\bar{x} + th)}{\Delta t} = \langle \nabla f(\bar{x} + th), h \rangle,$$

and then with the assistance of (2.2.2) one has

$$g''(t) = \lim_{\Delta t \to 0} \frac{\langle \nabla f(\bar{x} + (t + \Delta t)h, h \rangle - \langle \nabla f(\bar{x} + th), h \rangle}{\Delta t}$$
$$= \nabla^2 f(\bar{x} + th)(h, h).$$

The result now follows from Proposition 2.2.4 because f is convex (resp. strictly convex) if and only if every such g is convex (resp. strictly convex). □

The proof of the previous theorem illustrates the usefulness of functions of one variable in the the study of convex functions on vector spaces. Clearly, it was crucial that $\bar{x}$ was freely chosen in U, so that the convexity of f on any segment $[\bar{x}, y]$ where $y = \bar{x} + h$ could be checked. For indeed, otherwise one could consider the function on $\mathbb{R}^2$ defined by $f(x, y) := xy$ if $x > 0$ and $y > 0$, and $f(x, y) := 0$ otherwise. Then f is convex along every line through $(0, 0)$, but clearly f is not convex.

Example 2.2.9. Let $f : \mathbb{R}^n \to (-\infty, +\infty]$ be defined by

$$f(x_1, x_2, \ldots, x_n) := \begin{cases} -\sqrt[n]{x_1 x_2 \cdots x_n} & \text{if } x_1, \ldots, x_n \geq 0; \\ +\infty & \text{otherwise.} \end{cases}$$

Then f is convex. Moreover,

$$\sqrt[n]{(x_1 + y_1) \cdots (x_n + y_n)} \geq \sqrt[n]{x_1 \cdots x_n} + \sqrt[n]{y_1 \cdots y_n}$$

for all $x_1, \ldots x_n, y_1, \ldots, y_n \geq 0$.

Proof. This is left as Exercise 2.2.1. □

Exercises and further results

2.2.1.* Verify Example 2.2.9.

2.2.2. Suppose $f : \mathbb{R} \to (0, \infty)$, and consider the following three properties.

(a) $1/f$ is concave.
(b) f is log-convex, that is, $\log \circ f$ is convex.
(c) f is convex.

Show that (a) $\Rightarrow$ (b) $\Rightarrow$ (c), but that none of these implications reverse.

Hint. Use basic properties of convex functions; see [34, Corollary 2.2]; to see the implications don't reverse, consider $g(t) = e^t$ and $h(t) = t$ respectively. $\square$

2.2.3. Prove that the *Riemann zeta* function, $\zeta(s) := \sum_{n=1}^{\infty} \dfrac{1}{n^s}$ is log-convex on $(1, \infty)$.

2.2.4. Suppose $h : I \to (0, \infty)$ is a differentiable function. Prove the following assertions.

(a) $1/h$ is concave $\Leftrightarrow h(y) + h'(y)(y - x) \geq (h(y))^2/h(x)$, for all x, y in I.
(b) $1/h$ is affine $\Leftrightarrow h(y) + h'(y)(y - x) = (h(y))^2/h(x)$, for all x, y in I.
(c) $1/h$ is concave $\Rightarrow h$ is log-convex $\Rightarrow h$ is convex.
(d) If h is twice differentiable, then: $1/h$ is concave $\Leftrightarrow hh'' \geq 2(h')^2$.

2.2.5. Let $U \subset E$ be an open convex set, and let $f : U \to \mathbb{R}$ be a convex function. Suppose $x, x + h \in U$, $g(t) = f(x + th)$ for $0 \leq t \leq 1$ and $\phi_t \in \partial f(x + th)$. Show that $g'(t) = \langle \phi_t, h \rangle$ except for possibly countably many $t \in [0, 1]$.

Hint. First, g is differentiable except at possibly countably many $t \in [0, 1]$, and at points of differentiability $\nabla g(t) = \{\partial g(t)\}$. For each $t \in [0, 1]$, observe that

$$\langle \phi_t, sh \rangle \leq f(x + (s + t)h) - f(x + th) = g(t + s) - g(t).$$

Hence $\langle \phi_t, h \rangle \in \partial g(t)$. $\square$

2.2.6 (A compact maximum formula [95]). Let T be a compact Hausdorff space and let $f : E \times T \to \mathbb{R}$ be closed and convex for in $x \in E$ and continuous in $t \in T$. Consider the convex continuous function

$$f_T(x) := \max_{t \in T} f_t(x) \quad \text{where we write } f_t(x) := f(x, t),$$

and let $T(x) := \{t \in T : f_T(x) = f_t(x)\}$.

(a) Show
$$\partial f_T(x) = \overline{\text{conv}}\{\partial f_t(x) : t \in T(x)\}.$$

(b) Show that f_T is differentiable at x if and only if $T(x)$ is singleton.

Hint. First show that with no compactness assumption $\partial f_t(x) \subset \partial f_T(x)$ whenever $t \in T(x)$. Then show that $f'_T(x; h) \leq \sup_{t \in T(x)} f'_t(x; h)$, for all $h \in X$. Now apply a separation theorem. Compare Exercise 4.1.44. $\square$

2.2 Differentiability

2.2.7 (A bad convex function [376]). Consider the function $f : \mathbb{R}^2 \to \mathbb{R}$ defined by

$$f(x,y) := \max_{(u,v) \in W} \{ux + vy - u^2/2\}$$

where

$$W := \{(u,v) \in \mathbb{R}^2 : v = \sin(1/(2u)), 0 < u \le 1\} \cup \{(u,v) \in \mathbb{R}^2 : |v| \le 1, u = 0\}.$$

Show that f is convex and continuous and that for $0 < |x| < 1$, $\nabla f(x, 0) = (x, \sin(1/(2x))$ but $\partial f(0,0) = \{0\} \times [-1,1]$. Hence, $\lim_{x \downarrow 0} \nabla f(x, 0)$ does not exist even though the derivative does for $x \ne 0$.

Hint. Apply the compact maximum formula of Exercise 2.2.6. Note that no such convex function can exist on $\mathbb{R}$. □

2.2.8.* Suppose $f : E \to [-\infty, +\infty]$ is Fréchet differentiable at x_0, show that f is continuous at x_0.

2.2.9.* Suppose $f : E \to [-\infty, +\infty]$ is Lipschitz in a neighborhood of x_0. If f is Gâteaux differentiable at x_0, then it is Fréchet differentiable at x_0.

2.2.10. Provide an example of a continuous function $f : \mathbb{R}^2 \to \mathbb{R}$ such that $D_u f(0,0)$ exists for all $u \in S_{\mathbb{R}^2}$, but that f is not Gâteaux differentiable at $(0,0)$.

Hint. Let $f(x,y) = \sqrt{xy}$ if $x > 0, y > 0$ and $f(x,y) = -\sqrt{xy}$ if $x < 0, y < 0$; while $f(x,y) = 0$ otherwise. □

The next entry illustrates how convexity can resolve seemingly difficult inequalities.

2.2.11 (Knuth's problem, MAA #11369 (2008), 567). Prove that for all real t, and all $\alpha \ge 2$,

$$e^{\alpha t} + e^{-\alpha t} - 2 \le (e^t + e^{-t})^\alpha - 2^\alpha.$$

Hint. Let $f(t) := e^{\alpha t} + e^{-\alpha t} - 2 - (e^t + e^{-t})^\alpha + 2^\alpha$. Since $f(0) = 0$ it suffices to show $f'(t) \ge 0$ for $t > 0$, or the equivalent pretty inequality:

$$(1 + e^{-2t})^{\alpha - 1} \ge \frac{1 - e^{-2\alpha t}}{1 - e^{-2t}} \quad \text{for } \alpha \ge 2, t > 0. \tag{2.2.3}$$

Write $g := (y, a) \mapsto \log(1 - y) + (a - 1) \log(1 + y) - \log(1 - y^a)$. Then (2.2.3) is equivalent to $g(y, a) \ge 0$ for $0 < y < 1$ and $a > 2$. Fix y in $(0, 1)$. Now $g_{aa}(y, a) = \frac{y^a \log^2(y)}{(-1 + y^a)^2} > 0$. Thus, $a \mapsto g(y, a)$ is convex for $a > 0$. As $g(y, 1) = g(y, 2) = 0$, the minimum $a(y)$ occurs at a point in $(1, 2)$ and for $0 < y < 1, a > 2$ we have $g(y, a) > 0$ as required. □

2.2.12. Show that for $a \ge 1$ both

$$x \mapsto \log\left(\frac{\sinh(ax)}{\sinh(x)}\right) \quad \text{and} \quad x \mapsto \log\left(\frac{e^{ax} - 1}{e^x - 1}\right)$$

are convex on $\mathbb{R}$.

The following seven exercises are culled from [332].

2.2.13. Show that a real-valued function f defined on an interval I is convex if and only if, for every compact subinterval $J \subset I$ and for every affine function m, $\sup_J (f + m)$ is attained at an endpoint of J. [332, p. 9].

2.2.14. Consider $x_1 > 0, x_2 > 0, \ldots, x_n > 0$ with $n > 1$. For $1 \leq k \leq n$ set

$$A_k := \frac{x_1 + x_2 + \cdots + x_k}{k} \quad \text{and} \quad G_k := (x_1 \cdot x_2 \cdots x_k)^{1/k}.$$

Show for $2 \leq k \leq n$ that $(A_k/G_k)^k \geq (A_{k-1}/G_{k-1})^{k-1}$ (Popoviciu) and that $k(A_k - G_k) \geq (k-1)(A_{k-1}/G_{k-1})$ (Rado).

Hint. Use the convexity of $-\log$ and $\exp$ respectively.

2.2.15. Let $D := \{r_1, r_2, \ldots, r_n, \ldots\}$ be an arbitrary countable set of real numbers. Show that

$$f(x) := \int_0^x \sum_{r_k \leq t} 2^{-k} \, dt$$

is continuous and convex and is not differentiable exactly at members of D.

2.2.16. Show that if f is convex on a closed bounded interval $[a, b]$, then

$$f\left(\frac{a+b}{2}\right) \leq \frac{1}{b-a} \int_a^b f(t) \, dt \leq \frac{f(a) + f(b)}{2}.$$

2.2.17. Let $I \subset \mathbb{R}$ be an interval and suppose $f : I \to (0, +\infty)$. Show the following are equivalent: (a) f is log-convex; (b) $x \mapsto e^{\alpha x} f(x)$ is convex on I for all $\alpha \in \mathbb{R}$; (c) $x \mapsto f^\alpha(x)$ is convex on I for all $\alpha > 0$.

Hint. $\log(x) = \lim_{\alpha \to 0^+} [f^\alpha(x) - 1]/\alpha$. □

2.2.18. Show that

$$f(x, y, z) := \frac{1}{xy - z^2}$$

is convex on $\{(x, y, z) : x > 0, y > 0, xy > z^2\}$.

2.2.19. Show that the sum of two log-convex functions is log-convex.

Hint. $a^\alpha b^\beta + c^\alpha d^\beta \leq (a+c)^\alpha (b+d)^\beta$, for positive $\alpha, \beta, a, b, c, d$. □

2.2.20. Let

$$f(x, y) := \begin{cases} \frac{x^3}{y^2}, & \text{if } x^2 \leq y \leq x \text{ and } 0 < x \leq 1; \\ 0 & \text{if } (x, y) = (0, 0); \\ \infty, & \text{otherwise.} \end{cases}$$

Show that f is a closed convex function that fails to be continuous at some points in its domain (where continuity is for f restricted to its domain).

2.2.21 (Monotonicity of subdifferentials).* Suppose $U \subset E$ is a convex set, and $f : U \to \mathbb{R}$ is a convex function. Show that $\partial f : U \to 2^E$ is a *monotone mapping*, that is, $\langle y^* - x^*, y - x \rangle \geq 0$ whenever $x, y \in U$ and $x^* \in \partial f(x), y^* \in \partial f(y)$. In particular, if U is open and f is Gâteaux differentiable on U, then

$$\langle \nabla f(y) - \nabla f(x), x - y \rangle \geq 0 \text{ for all } x, y \in U.$$

2.2 Differentiability

Hint. Use the subdifferential inequality to show

$$\langle y^* - x^*, y - x \rangle = y^*(y - x) + x^*(x - y)$$
$$\geq f(y) - f(x) + f(x) - f(y) = 0.$$

$\square$

2.2.22 (Continuity properties of subdifferentials).* Suppose $f : E \to (-\infty, +\infty]$ is proper a convex function.

(a) Suppose f is continuous at x_0, and suppose $\varepsilon > 0$. Then there exists $\delta > 0$ so that $\partial f(u) \subset \partial f(x_0) + \varepsilon B_E$ whenever $\|u - x_0\| < \delta$.
(b) Suppose f is differentiable at x_0, and suppose $\phi_n \in \partial f(x_n)$ where $x_n \to x_0$. Show that $\phi_n \to f'(x_0)$. In particular, if a convex function is differentiable on an open set U in E, then it is automatically continuously differentiable on U.
(c) Suppose f is differentiable and (f_n) is a sequence of convex functions that converges uniformly to f on bounded sets. Show that for any bounded set $W \subset E$ and $\varepsilon > 0$ there exists $N \in \mathbb{N}$ so that $\partial f_n(w) \subset \nabla f(w) + \varepsilon B_E$ whenever $w \in W$ and $n > N$.
(d) Show that part (c) can fail when f is merely continuous, but show that for any bounded set W and any $\varepsilon > 0$ and $\delta > 0$, there exists $N > 0$ such that $\partial f_n(w) \subset \bigcup_{\|x-w\|<\delta} \partial f(x) + \varepsilon B_E$ whenever $n > N$.

Hint. (a) Suppose not, then there exists $x_n \to x_0$ and $\varepsilon > 0$ so that $\phi_n \in \partial f(x_n)$, but $\phi_n \notin \partial f(x_0) + \varepsilon B_E$. Use the local Lipschitz property of f (Theorem 2.1.12) to deduce that $(\|\phi_n\|)_n$ is bounded. Then use compactness to find convergent subsequence, say $\phi_{n_k} \to \phi$. Show that $\phi \in \partial f(x_0)$ to arrive at a contradiction.
 (b) This follows from (a) and the fact $\partial f(x_0) = \{f'(x_0)\}$ (Theorem 2.2.1).
 (c) Suppose not, then there is a subsequence (n_k) and $\varepsilon > 0$ such that $\phi_{n_k} \in \partial f_{n_k}(w_{n_k})$, $w_{n_k} \in W$ but $\phi_{n_k} \notin \nabla f(w_{n_k}) + \varepsilon B_E$. Because $f_n \to f$ uniformly on bounded sets, it follows that (f_n) is eventually uniformly Lipschitz on bounded sets. Hence by passing to a further subsequence, if necessary, we may assume $w_{n_k} \to w_0$, and $\phi_{n_k} \to \phi$ for some w_0 and ϕ in E. Now show that $\phi \in \partial f(w_0)$, however, by (b), $\nabla f(w_k) \to \nabla f(w_0) = \phi$ which yields a contradiction.
 (d) For example, let $f_n := \max\{|\cdot| - 1/n, 0\}$ and $f := |\cdot|$ on $\mathbb{R}$. Then $\partial f_n(1/n) \not\subset \partial f(1/n) + \frac{1}{2} B_\mathbb{R}$ for any $n \in \mathbb{N}$. Suppose no such N exists. As in (c), choose $\phi_{n_k} \in \partial f_{n_k}(w_{n_k})$ but $\phi_{n_k} \notin \partial f(w) + \varepsilon B_E$ for $\|w - w_{n_k}\| < \delta$, where as in (c), $w_{n_k} \to w_0$ and $\phi_{n_k} \to \phi$ and $\phi \in \partial f(w_0)$. This produces a contradiction. $\square$

2.2.23 (Subdifferentials at boundary points).* Suppose $f : E \to (-\infty, +\infty]$ is a proper convex function.

(a) If $x_0 \in \text{bdry dom} f$, then $\partial f(x_0)$ is either empty or unbounded.
(b) Given examples showing that either case in (a) may occur.

Hint. (a) Suppose $\partial f(x_0) \neq \emptyset$. With notation as in the last part of the proof of Theorem 2.2.1, let $h_n(x) := g(x) + nd_C(x)$ and show $\partial h_n(x_0) \subset \partial f(x_0)$, find a direction d so that $h'_n(x_0; d) \to \infty$, and use the max formula (2.1.19) to conclude there are $\phi_n \in \partial f(x_0)$ so that $\|\phi_n\| \to \infty$.

(b) Consider $f(x) := -\sqrt{x}$ and $g(x) := \delta_{[0,+\infty)}$.   □

Combining Exercises 2.2.23 with Exercise 2.4.7 given below, one sees that the subdifferential map is not locally bounded at any x in the boundary of the domain of f.

2.2.24 (Reformulation of second-order derivatives). Show that f is twice Gâteaux differentiable at $x \in E$ if and only if f is Gâteaux differentiable in a neighborhood U of x and there is a *bilinear* mapping $b : E \times E \to \mathbb{R}$ (i.e. $b(h, k)$ is linear in h and k separately) such that

$$b(h,k) = \lim_{t \to 0} \left\langle \frac{\nabla f(x + tk) - \nabla f(x)}{t}, h \right\rangle \quad \text{for all } h, k \in E. \tag{2.2.4}$$

Moreover, in this case $b(h, k) = \langle Hk, h \rangle$ where H is the Hessian of f at x.

Hint. Suppose f is twice Gâteaux differentiable at x with Hessian H. Verify that $(h, k) \mapsto \langle h, Hk \rangle$ equals the right-hand side in (2.2.4). Conversely

$$k \mapsto \lim_{t \to 0} t^{-1}[\nabla f(x + tk) - \nabla f(x)]$$

maps $E \to E$; the bilinearity of b implies this map is linear, and hence f is twice Gâteaux differentiable at x.   □

2.3 Conjugate functions and Fenchel duality

The *Fenchel conjugate* (also called the *Legendre-Fenchel conjugate* or *transform*) of a function $f : E \to [-\infty, +\infty]$ is the function $f^* : E \to [-\infty, +\infty]$ defined by

$$f^*(\phi) := \sup_{x \in E} \{\langle \phi, x \rangle - f(x)\}.$$

A useful and instructive example is $\sigma_D = \delta_D^*$. As one expects, f^{**} denotes the conjugate of f^*, and it is called the *biconjugate of f*. The conjugate function f^* is always convex and if the domain of f is nonempty, then f^* never takes the value $-\infty$. A direct consequence of the definition is that for $f, g : E \to [-\infty, +\infty]$, the inequality $f \geq g$ implies $f^* \leq g^*$. As we shall see throughout this book, the conjugate plays a role in convex analysis in many ways analogous to the role played by the Fourier transform in harmonic analysis.

Proposition 2.3.1 (Fenchel–Young). *Any points $\phi \in E$ and $x \in \text{dom} f$ where $f : E \to (-\infty, +\infty]$ satisfy the inequality*

$$f(x) + f^*(\phi) \geq \langle \phi, x \rangle. \tag{2.3.1}$$

Equality holds if and only if $\phi \in \partial f(x)$.

2.3 Conjugate functions and Fenchel duality

Proof. The inequality follows directly from the definition of conjugate functions, while

$$\phi \in \partial f(x) \Leftrightarrow \phi(y) - f(y) \leq \phi(x) - f(x) \text{ for all } y \in E$$
$$\Leftrightarrow f^*(\phi) = \phi(x) - f(x).$$

establishes the equality. □

Some basic examples and facts concerning one-dimensional conjugate functions are collected in Tables 2.1 and 2.2. Several further properties of conjugate functions will be given in Section 4.4.

Table 2.1 *Conjugate pairs of convex functions on $\mathbb{R}$.*

$f(x) = g^*(x)$	dom f	$g(y) = f^*(y)$	dom g
0	$\mathbb{R}$	0	$\{0\}$
0	$[0, +\infty)$	0	$(-\infty, 0]$
0	$[-1, 1]$	$|y|$	$\mathbb{R}$
0	$[0, 1]$	y^+	$\mathbb{R}$
$|x|^p/p,\ p>1$	$\mathbb{R}$	$|y|^q/q\ \left(\frac{1}{p} + \frac{1}{q} = 1\right)$	$\mathbb{R}$
$|x|^p/p,\ p>1$	$[0, +\infty)$	$|y^+|^q/q\ \left(\frac{1}{p} + \frac{1}{q} = 1\right)$	$\mathbb{R}$
$-x^p/p,\ 0<p<1$	$[0, +\infty)$	$-(-y)^q/q\ \left(\frac{1}{p} + \frac{1}{q} = 1\right)$	$(-\infty, 0)$
$\sqrt{1+x^2}$	$\mathbb{R}$	$-\sqrt{1-y^2}$	$[-1, 1]$
$-\log x$	$(0, +\infty)$	$-1 - \log(-y)$	$(-\infty, 0)$
$\cosh x$	$\mathbb{R}$	$y \sinh^{-1}(y) - \sqrt{1+y^2}$	$\mathbb{R}$
$-\log(\cos x)$	$\left(-\frac{\pi}{2}, \frac{\pi}{2}\right)$	$y \tan^{-1}(y) - \frac{1}{2}\log(1+y^2)$	$\mathbb{R}$
$\log(\cosh x)$	$\mathbb{R}$	$y \tanh^{-1}(y) + \frac{1}{2}\log(1-y^2)$	$(-1, 1)$
e^x	$\mathbb{R}$	$\begin{cases} y \log y - y & (y > 0) \\ 0 & (y = 0) \end{cases}$	$[0, +\infty)$
$\log(1 + e^x)$	$\mathbb{R}$	$\begin{cases} y \log y + (1-y)\log(1-y) & (y \in (0,1)) \\ 0 & (y = 0, 1) \end{cases}$	$[0, 1]$
$-\log(1 - e^x)$	$(-\infty, 0]$	$\begin{cases} y \log y - (1+y)\log(1+y) & (y > 0) \\ 0 & (y = 0) \end{cases}$	$[0, +\infty)$

Table 2.2 *Transformed conjugates.*

$f = g^*$	$g = f^*$
$f(x)$	$g(y)$
$h(ax)\ (a \neq 0)$	$h^*(y/a)$
$h(x+b)$	$h^*(y) - by$
$ah(x)\ (a > 0)$	$ah^*(y/a)$

Aspects of the growth rate of a convex function are often reflected in properties of its conjugate as seen in the following.

Theorem 2.3.2 (Moreau–Rockafellar). *A closed convex proper function on E has bounded lower level sets if and only if its conjugate is continuous at 0.*

Proof. According to Proposition 2.1.9, a convex function $f : E \to (-\infty, +\infty]$ has bounded lower level sets if and only if it satisfies the growth condition

$$\liminf_{\|x\| \to \infty} \frac{f(x)}{\|x\|} > 0.$$

Because f is closed, it is bounded below on compact sets and thus the previous inequality is equivalent to the existence of a minorant of the form $\varepsilon \|\cdot\| + k \leq f(\cdot)$ for some constants $\varepsilon > 0$ and k. Taking conjugates, this is equivalent to f^* being bounded above on a neighborhood of 0, which in turn is equivalent to f^* being continuous at 0 (Theorem 2.1.10). □

In order to prepare for the important Fenchel duality theorem, we present the following lemma.

Lemma 2.3.3. *Suppose $f : E \to [-\infty, +\infty]$ is convex and that some point $x_0 \in$ core dom f satisfies $f(x_0) > -\infty$. Then f never takes the value $-\infty$.*

Proof. Suppose $f(x_0 - h) = -\infty$ for some $h \in X$. The convexity of the epigraph of f then implies $f(x_0 + th) = \infty$ for all $t > 0$. This contradicts that $x_0 \in$ core dom f. □

Let E and Y be Euclidean spaces and $A : E \to Y$ a linear mapping. The *adjoint* of A, denoted by A^*, is the linear mapping from Y to E defined by $\langle A^* y, x \rangle = \langle y, Ax \rangle$ for all $x \in E$.

Theorem 2.3.4 (Fenchel duality theorem). *Let E and Y be Euclidean spaces, and let $f : E \to (-\infty, +\infty]$ and $g : Y \to (-\infty, +\infty]$ and a linear map $A : E \to Y$, and let $p, d \in [-\infty, +\infty]$ be the primal and dual values defined respectively by the Fenchel problems*

$$p := \inf_{x \in E} \{f(x) + g(Ax)\} \tag{2.3.2}$$

$$d := \sup_{\phi \in Y} \{-f^*(A^*\phi) - g^*(-\phi)\}. \tag{2.3.3}$$

2.3 Conjugate functions and Fenchel duality

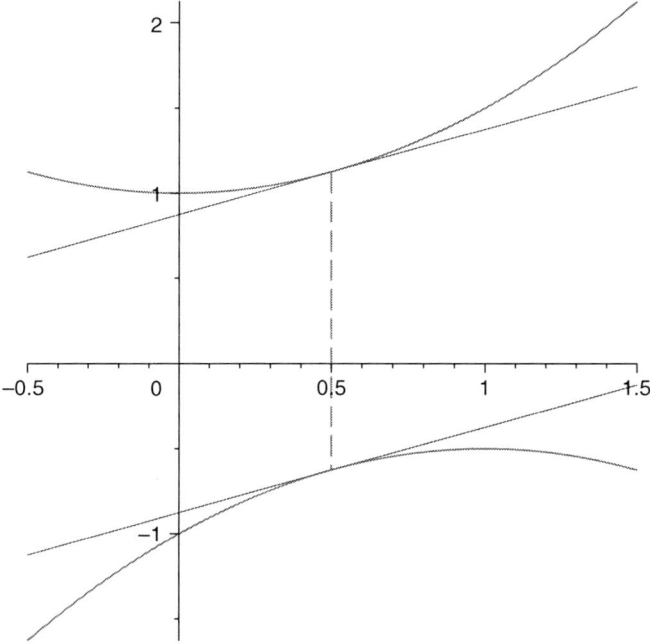

Figure 2.6 Fenchel duality (Theorem 2.3.4) illustrated for $x^2/2 + 1$ and $-(x-1)^2/2 - 1/2$. The minimum gap occurs at $1/2$ with value $7/4$.

Then these values satisfy the weak duality inequality $p \geq d$. If, moreover, f and g are convex and satisfy the condition

$$0 \in \text{core}(\text{dom}\, g - A\, \text{dom}\, f) \qquad (2.3.4)$$

or the stronger condition

$$A\, \text{dom}\, f \cap \text{cont}\, g \neq \emptyset \qquad (2.3.5)$$

then $p = d$ and the supremum in the dual problem (2.3.3) is attained whenever it is finite.

Proof. The weak duality inequality $p \leq d$ follows immediately from the Fenchel–Young inequality (2.3.1).

To prove equality, we first define an optimal value function $h : Y \to [-\infty, +\infty]$ by

$$h(u) = \inf_{x \in E}\{f(x) + g(Ax + u)\}.$$

It is easy to check that h is convex and $\text{dom}\, h = \text{dom}\, g - A\, \text{dom}\, f$. If $p = -\infty$, there is nothing to prove. If condition (2.3.4) holds and p is finite, then Lemma 2.3.3 and the max formula (2.1.19) show there is a subgradient $-\phi \in \partial h(0)$. Consequently, for all $u \in Y$ and all $x \in E$, one has

$$h(0) \leq h(u) + \langle \phi, u \rangle$$
$$\leq f(x) + g(Ax + u) + \langle \phi, u \rangle$$
$$= \{f(x) - \langle A^*\phi, x \rangle\} + \{g(Ax + u) - \langle -\phi, Ax + u \rangle\}.$$

Taking the infimum over all points u, and then over all points x yield the inequalities

$$h(0) \leq -f^*(A^*\phi) - g^*(-\phi) \leq d \leq p = h(0).$$

Thus ϕ attains the supremum in problem (2.3.3), and $p = d$. We leave it as exercises to show that condition (2.3.5) implies condition (2.3.4). □

The following is an important subdifferential sum rule, which is sometimes called the *convex calculus theorem*.

Theorem 2.3.5 (Subdifferential sum rule). *Let E and Y be Euclidean spaces, and let $f : E \to (-\infty, +\infty]$ and $g : Y \to (-\infty, +\infty]$ and a linear map $A : E \to Y$. At any point $x \in E$, the calculus rule*

$$\partial(f + g \circ A)(x) \supset \partial f(x) + A^*\partial g(Ax) \tag{2.3.6}$$

holds, with equality if f and g are convex and either condition (2.3.4) or (2.3.5) holds.

Proof. This is a consequence of the Fenchel duality theorem (2.3.4). See Exercise 2.3.12 and also Exercise 2.4.1. □

When g is simply the indicator function of a point, one has the following useful corollary to the Fenchel duality theorem (2.3.4).

Corollary 2.3.6 (Fenchel duality for linear constraints). *Given any function $f : E \to (-\infty, +\infty]$, any linear map $A : E \to Y$, and any element b of Y, the weak duality inequality*

$$\inf_{x \in E}\{f(x) : Ax = b\} \geq \sup_{\phi \in Y}\{\langle b, \phi \rangle - f^*(A^*\phi)\} \tag{2.3.7}$$

holds. If f is convex and b belongs to $\operatorname{core}(A \operatorname{dom} f)$ then equality holds, and the supremum is attained when finite.

In general an *ordinary convex program* is the problem of minimizing a convex objective function f subject to a finite number of convex inequalities, $g_j(x) \leq b_j, j \in J$, and linear equalities, $\langle a_k, x \rangle = b_k, k \in K$, called *constraints*. Additionally a set constraint $x \in C$ may be given. We write

$$\inf_{x \in C}\{f(x) : g_j(x) \leq c_j, j \in J, \quad \langle a_k, x \rangle = b_k, k \in K\}.$$

Since a set constraint $x \in C$ can be incorporated by adding an indicator function δ_C to the objective function, a great many problems can be subsumed in (2.3.7)

2.3 Conjugate functions and Fenchel duality

and extensions to be met later on. Indeed the fact that f often takes infinite values blurs the distinction between objective and constraint. Thus, the *feasible set* is $C \cap \text{dom} f \cap_{j \in J} \{x : g_j(x) \leq c_j\} \cap_{k \in K} \{x : \langle a_k, x \rangle = b_k\}$. We analyze the Lagrangian dual of an ordinary convex program in Exercise 2.3.10.

Exercises and further results

2.3.1. Table 2.1 shows some elegant examples of conjugates on $\mathbb{R}$, and Table 2.2 describes some simple transformations of these examples. Check the calculation of f^* and check $f = f^{**}$ for functions in Table 2.1. Verify the formulas in Table 2.2.

2.3.2 (Young inequality). Use calculus to derive the special case of the Fenchel–Young inequality (2.3.1). For $p, q \geq 1$ with $1/p + 1/q = 1$, show that $|x|^p/p + |y|^q/q \geq xy$ for all real x and y.

2.3.3 (Hölder inequality). Let f and g be measurable on a measure space (X, μ). Show for $p, q \geq 1$ with $1/p + 1/q = 1$ that

$$\int_X fg \, d\mu \leq \|f\|_p \|g\|_q.$$

When $p = q = 2$ this is the *Cauchy–Schwarz inequality*.

Hint. Assume $\|f\|_p = \|g\|_q = 1$ and apply the Young inequality of Exercise 2.3.2. □

2.3.4 (Convexity of p-norm). Show, for $1 \leq p \leq \infty$, that the p-norm, $\|\bar{x}\|_p := \left(\sum_{k=1}^N |x_k|^p\right)^{1/p}$, on $\mathbb{R}^N$ is indeed a norm.

Hint. (a) First show that the unit ball $B_p = \{\bar{x} : \sum_{k=1}^N |x_k|^p \leq 1\}$ in N-space is convex and then use the gauge construction of Exercise 2.1.13.

(b) Alternatively, one may apply Hölder's inequality of Exercise 2.3.3 to the sequences

$$\left(|x_k| \, |(x_k + y_k)|^{p-1}\right)_{k=1}^N, \quad \left(|y_k| \, |(x_k + y_k)|^{p-1}\right)_{k=1}^N$$

to estimate $\|\bar{x} + \bar{y}\|_p^p \leq (\|\bar{x}\|_p + \|\bar{y}\|_p) \|\bar{x} + \bar{y}\|_p^{p/q}$. Here as always $1/p + 1/q = 1$ and $q \geq 1$. □

2.3.5 (Favard inequality). Let f be concave and nonnegative on $[0, 1]$. Show that for $p \geq 1$

$$\int_0^1 f(t) \, dt \geq \frac{(p+1)^{1/p}}{2} \|f\|_p.$$

2.3.6 (Grüss–Barnes inequality). Let f and g be concave and nonnegative on $[0, 1]$. Show for $p, q \geq 1$ that

$$\int_0^1 f(t)g(t) \, dt \geq \frac{(p+1)^{1/p}(q+1)^{1/q}}{6} \|f\|_p \|g\|_q.$$

When $1/p + 1/q = 1$ this provides a pretty if restricted converse to Hölder's inequality.

Hint. First prove this for $p = q = 1$ (due to Grüss) and then apply Favard's inequality of Exercise 2.3.5. □

2.3.7. Let f be convex on a product space $X \times Y$. Let $h(y) := \inf_{x \in X} f(x, y)$. Show that $h^*(y^*) = f^*(0, y^*)$. Hence $h(0) = \inf_{x \in X} f(x, 0)$, while $h^{**}(0) = -\inf_{y^* \in Y^*} f^*(0, y^*)$.

2.3.8 (Recession functions). For a closed convex function f, the *recession function* is defined by
$$0^+ f(x) := \lim_{t \to +\infty} \frac{f(z + tx) - f(z)}{t}$$
for $z \in \text{dom} f$. Show that $0^+ f$ is well defined (independent of z) and is sublinear and closed. Show that for any $\alpha > \inf f$, one has $0^+ f(d) \leq 0$ if and only if $f(y) \leq \alpha$ implies $f(y + td) \leq \alpha$ for all $t > 0$. In other words, d is in the *recession cone* of each such lower level set: $\{d : C + \mathbb{R}_+ d \subset C\}$ [95, p. 83]. In particular, $0^+ f(d) \leq 0$ for some $d \neq 0$ if and only if f has unbounded lower level sets.

2.3.9 (Homogenization). Recall from the proof of Lemma 2.1.8 that a function f is convex if and only if the function defined by $g(x, t) := tf(x/t)$ is sublinear for $t > 0$. Hence deduce that y^2/x is convex on $\mathbb{R} \times \mathbb{R}_{++}$. Determine the closure of $(x, y) \mapsto y^2/x$. Show more generally when f is closed and convex on $\mathbb{R}$, that $g^{**}(x, t) = tf(x/t)$ for $t > 0$, that in the language of Exercise 2.3.8 $g^{**}(x, 0) = 0^+ f(x)$, and that $g^{**}(x, t) = \infty$ otherwise [95, p. 84].

2.3.10 (Lagrangian duality). Consider the ordinary convex program
$$h(c) := \inf\{f(x) : g_j(x) \leq c_j, j = 1, \ldots, n\} \text{ where } c := (c_1, c_2, \ldots, c_n) \in \mathbb{R}^n$$
with Lagrangian $L(x, \lambda) := f(x) + \langle \lambda, g(x) \rangle$. Use Exercise 2.3.7 to show that h is convex when f and each g_j is. Then

(a) Confirm that
$$h^{**}(0) = \sup_{\lambda \in \mathbb{R}_+^n} \inf_{x \in E} f(x) + \langle \lambda, g(x) \rangle.$$

Hence,
$$\inf\{f(x) : g_j(x) \leq c_j, j = 1, \ldots, n\} = \sup_{\lambda \in \mathbb{R}_+^n} \inf_{x \in E} f(x) + \langle \lambda, g(x) \rangle.$$
if and only if h is lsc at zero.

(b) In particular, h is continuous at zero when *Slater's condition* holds: there is $\widehat{x} \in \text{dom} f$ with $g_j(\widehat{x}) < 0$ for $j = 1, \ldots, n$.

(c) Moreover, the set of optimal Lagrangian dual solutions is precisely $-\partial h(0)$.

(d) Extend the analysis to the case where there are also affine constraints.

Hint. First apply the previous analysis to $f + \delta_P$ where P is the affine part of the feasible set. Then use the ideas of the mixed Fenchel duality corollary (3.1.9). □

2.3.11 (Duffin's duality gap). Consider the following problem (for real b):
$$v(b) := \inf \left\{ e^{x_2} : \sqrt{x_1^2 + x_2^2} - x_1 \leq b, \, x \in \mathbb{R}^2 \right\}. \tag{2.3.8}$$

(a) Draw the feasible region for $b > 0$ and $b = 0$. Then plot the value function v and compute v^{**}. Compare the results to those promised by Exercise 2.3.10.
(b) Repeat the above steps with the objective function e^{x_2} replaced by x_2.

2.3.12.* Fill in the details for the proof of the Fenchel duality (2.3.4) and subdifferential sum (2.3.5) theorems as follows.

(a) Fenchel duality theorem (2.3.4):
  (i) Use the Fenchel–Young inequality (2.3.1) to prove the weak duality inequality.
  (ii) Prove the inclusion in (2.3.6).
     For the remaining parts, assume f and g are convex.
  (iii) Prove the function h defined in the proof is convex with $\operatorname{dom} h = \operatorname{dom} g - A \operatorname{dom} f$.
  (iv) Prove the condition (2.3.5) implies condition (2.3.4).
(b) Subdifferential sum theorem (2.3.5): assume additionally that (2.3.4) holds.
  (i) Suppose $\phi \in \partial(f + g \circ A)(\bar{x})$. Use the Fenchel duality theorem (2.3.4) and the fact $\bar{x}$ is an optimal solution to the problem
  $$\inf_{x \in E}\{f(x) - \langle \phi, x \rangle + g(Ax)\}$$
  to deduce equality in (2.3.6).
  (ii) Prove points $\bar{x} \in E$ and $\bar{\phi} \in Y$ are optimal for problems (2.3.2) and (2.3.3), respectively, if and only if they satisfy the conditions $A^*\bar{\phi} \in \partial f(\bar{x})$ and $-\bar{\phi} \in \partial g(A\bar{x})$.

2.3.13.* Suppose $f : E \to (-\infty, +\infty]$ is a proper convex function. Show that f is Lipschitz with constant k if and only if $\operatorname{dom} f^* \subset kB_E$. Give an example of a continuous nonconvex function on the real line that is not Lipschitz whose conjugate has bounded domain.

Hint. See Proposition 4.4.6 and [369, Corollary 13.3.3] and try an example like $f := \sqrt{|\cdot|}$. □

2.3.14 (Infimal convolutions).* Let f and g be proper extended real-valued functions E. The *infimal convolution* of f and g is defined by
$$(f \square g)(x) := \inf_{y \in E} f(y) + g(x - y)$$

See Figure 2.7 for an illustration of infimal convolutions.

(a) Show that, geometrically, the infimal convolution of f and g is the largest extended real-valued function whose epigraph contains the sum of epigraphs of f and g.
(b) Provide an example showing that the sum of two epigraphs of continuous convex functions may not be closed.

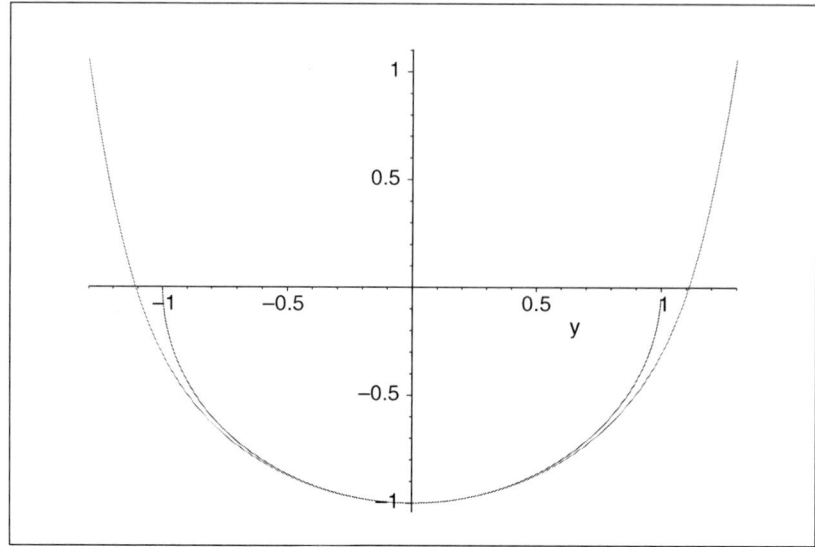

Figure 2.7 Infimal convolution of $-\sqrt{1-x^2}$ and $\frac{25}{2}|x|^2$.

(c) Provide an example showing that the infimal convolution of two continuous convex functions on $\mathbb{R}$ need not be proper.
(d) Provide an example showing that even if the infimal convolution of two lsc convex functions is proper, it need not be lsc.
(e) Show that the infimal convolution of two convex functions is convex.
(f) Suppose f, g are convex functions on E that are bounded below. If $f : E \to \mathbb{R}$ is continuous (resp. bounded on bounded, Lipschitz), then $f \square g$ is a convex function that is continuous (resp. bounded on bounded sets, Lipschitz).
(g) Suppose C is a closed convex set. Show that $d_C = \|\cdot\| \square \delta_C$. Conclude that d_C is a Lipschitz convex function (hence this provides an alternate proof to Fact 2.1.6).

Hint. For (b) consider $f(x) := e^x$ and $g(x) := 0$; for (c) let $f(x) := x$ and $g(x) := 0$; for (d) use the indicator functions of the epigraphs of $f(x) := e^x$ and $g(x) := 0$. The infimal convolution is the indicator function of the sum of the epigraphs, i.e. the indicator function of open half plane $\{(x, y) : y > 0\}$ which is not lsc; for (e) use (a); for the first two parts of (f) notice that for

$$\inf f + \inf g \leq (f \square g)(x) \leq f(x - x_0) + g(x_0) \text{ for each } x \in E$$

where $x_0 \in \operatorname{dom} g$ is fixed. Continuity then follows from Theorem 2.1.12, and boundedness on bounded sets follows from bounds on f. Given that $f \square g$ is everywhere finite, an easy estimate shows the Lipschitz constant for f is valid for the convolution. □

2.3.15 (Infimal convolutions and conjugation).*

(a) Let f and g be proper functions on E, then $(f \square g)^* = f^* + g^*$.

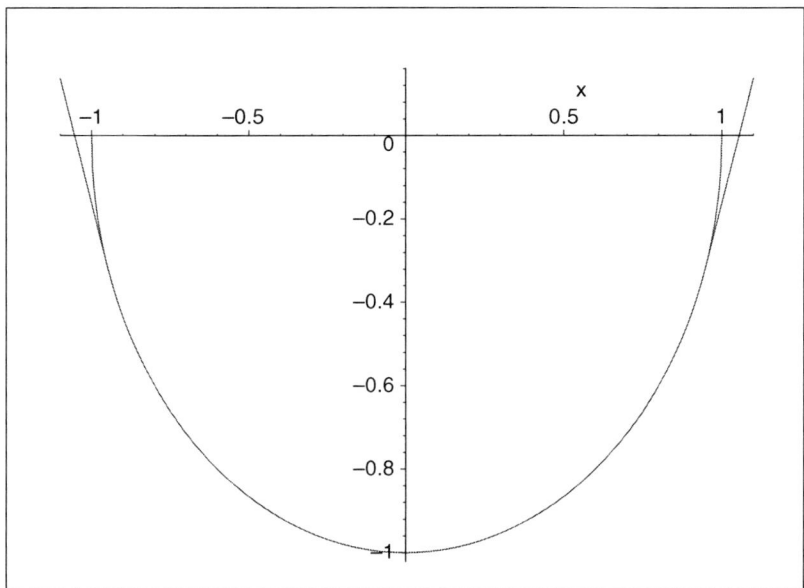

Figure 2.8 Infimal convolution of $3|x|$ and $-\sqrt{1-x^2}$. Note the curves diverge when the slope reaches ± 3.

(b) Suppose $f : E \to \mathbb{R}$ is a convex function bounded on bounded sets. If $f_n \leq f$ for each n, and $f_n^* \to f^*$ uniformly on bounded subsets of the domain of f^*, then $f_n \to f$ uniformly on bounded subsets of E.
(c) Let $f : X \to \mathbb{R}$ be convex and bounded on bounded sets. Then both $f \square n \|\cdot\|^2$ and $f \square n \|\cdot\|$ converge uniformly to f on bounded sets.
(d) If f and g are convex and $\text{dom} f \cap \text{cont} g \neq \emptyset$, show that $(f+g)^* = f^* \square g^*$. Is convexity needed?

Hint. For (a) see the proof of Lemma 4.4.15; for (b) see the proof of Lemma 4.4.16; and for (c), use (a) and (b), cf. Corollary 4.4.17. □

2.3.16 (Lipschitz extension).* Suppose the real function f has Lipschitz constant k on the set $C \subset E$. By considering the infimal convolution of the functions $f + \delta_C$ and $k\|\cdot\|$, prove there is a function $\tilde{f} : E \to \mathbb{R}$ with Lipschitz constant k that agrees with f on C. Prove furthermore that if f and C are convex then $\tilde{f}$ can be assumed convex. See Figure 2.8 for an example illustrating this.

2.3.17 (Nearest points). (a) Suppose that $f : X \to (-\infty, +\infty]$ is strictly convex. Show that it has at most one global minimizer.
(b) Prove that the function $f(x) := \|x - y\|^2/2$ is strictly convex on E for any point $y \in E$.
(c) Suppose C is a nonempty closed convex subset of E.
   (i) Let $y \in E$; show that there is a unique nearest point $P_C(y)$ in C to y, characterized by

$$\langle y - P_C(y), x - P_C(y) \rangle \leq 0 \quad \text{for all } x \in C.$$

The mapping $x \to P_C(x)$ is called the *nearest point mapping*.

(ii) For any point $\bar{x} \in C$, deduce that $d \in N_C(\bar{x})$ holds if and only if $\bar{x} = P_C(\bar{x} + d)$.

(iii) Deduce, furthermore, that any points y and z in E satisfy

$$\|P_C(y) - P_C(z)\| \le \|y - z\|;$$

that is, the projection $P_C : E \to C$ is a Lipschitz mapping.

(d) Give an example of a nonconvex set $S \subset \mathbb{R}$ where P_S is discontinuous.

2.3.18 (Nearest point to an ellipse). Consider the ellipse

$$E := \left\{ (x,y) : \frac{x^2}{a^2} + \frac{y^2}{b^2} = 1 \right\}$$

in standard form. Show that the best approximation $P_E(u,v) = \left(\frac{a^2 u}{a^2 - t}, \frac{b^2 v}{b^2 - t} \right)$ where t solves $\frac{a^2 u^2}{(a^2-t)^2} + \frac{b^2 v^2}{(b^2-t)^2} = 1$. Generalize this to a hyperbola and to an arbitrary ellipsoid.

2.3.19 (Nearest point to the p-sphere). For $0 < p < \infty$, consider the p-sphere in two dimensions

$$S_p := \{(x,y) : |x|^p + |y|^p = 1\}.$$

Let $z^* := (1 - z^p)^{1/p}$. Show that, for $uv \neq 0$, the best approximation $P_{S_p}(u,v) = (\operatorname{sign}(u)z, \operatorname{sign}(v)z^*)$ where either $z = 0, 1$ or $0 < z < 1$ solves

$$z^{*p-1}(z - |u|) - z^{p-1}(z^* - |v|) = 0.$$

(Then compute the two or three distances and select the point yielding the least value. It is instructive to make a plot, say for $p = 1/2$.) Extend this to the case where $uv = 0$. Note that this also yields the nearest point formula for the p-ball.

Hint. One approach is to apply the method of multipliers [96, Theorem 7.7.9] and then eliminate the multiplier. □

2.3.20 (Distance functions). Suppose $C, S \subset E$ are nonempty and C is convex.

(a) Prove that d_S^2 is a difference of convex functions by showing

$$d_S^2(x) = \frac{\|x\|^2}{2} - \left(\frac{\|\cdot\|^2}{2} + \delta_S \right)^*(x).$$

(b) Recall that δ_C is convex and show that $d_C^* = \delta_{B_E} + \delta_C^*$, that is $d_C^* = \delta_{B_E} + \sigma_C$.

(c) For $x \in C$, show that $\partial d_C(x) = B_E \cap N_C(x)$.

(d) If C is closed and $x \notin C$, show that

$$\nabla d_C(x) = \frac{1}{d_C(x)}(x - P_C(x)),$$

where $P_C(x)$ is the nearest point to x in C.

(e) If C is closed, show that

$$\frac{1}{2} \nabla d_C^2(x) = x - P_C(x) \quad \text{for all } x \in E.$$

2.3.21 (First-order conditions for linear constraints). Let C be a convex subset of E, $f : C \to \mathbb{R}$ be a function and $A : E \to Y$ be a linear map where Y is a Euclidean space. For $b \in Y$, consider the optimization problem

$$\inf\{f(x) : x \in C, Ax = b\}. \tag{2.3.9}$$

Suppose $\bar{x} \in \text{int } C$ satisfies $A\bar{x} = b$.

(a) Suppose that $\bar{x}$ is a local minimizer for (2.3.9) and f is differentiable at $\bar{x}$. Show that $\nabla f(\bar{x}) \in A^*Y$.
(b) Conversely, suppose that $\nabla f(\bar{x}) \in A^*Y$ and f is convex. Show that $\bar{x}$ is a global minimizer for (2.3.9).

Hint. Show that the normal cone to the set $\{x \in E : Ax = b\}$ at any point in it is A^*Y, and deduce the result from this. □

2.3.22 (Hidden convexity, I).[†] We have used many special cases of the simple fact that $x \mapsto \inf_Y f(x, y)$ is convex whenever f is jointly convex on a vector space $X \times Y$. It is, however, possible for the infimum to be convex without f being so. A useful example is given by the Fan minimax theorem. Recall that for arbitrary sets, $f : X \times Y \to \mathbb{R}$ is said to be *convex-concave like* on $X \times Y$ if, for $0 < t < 1$, (a) for x_1, x_2 in X there exists x_3 in X with $f(x_3, y) \leq tf(x_1, y) + (1-t)f(x_2, y)$ for all y in Y; and (b) for y_1, y_2 in Y there exists y_3 in Y with $f(x, y_3) \geq tf(x, y_1) + (1-t)f(x, y_2)$ for all x in X. The general form of Fan's minimax theorem is:

Theorem 2.3.7 (Fan minimax theorem). *Suppose that X, Y are nonempty sets with f convex-concave like on $X \times Y$. Suppose that X is compact and $f(\cdot, y)$ is lsc on X for each y in Y. Then*

$$p := \min_X \sup_Y f(x, y) = \sup_Y \min_X f(x, y).$$

If Y is actually compact and $f(x, \cdot)$ is usc on Y for each x in X, then 'sup' may be replaced by 'max' above.

Hint. We follow [123]. Consider, for any finite set $y_1, y_2 \cdots, y_n$ in Y

$$h(z) := \inf_X \{r : f(x, y_i) \leq r, i = 1, 2, \ldots, n\}$$

and observe that, while f is not convex, h is since $f(\cdot, y_i)$ is convex-like. Also h is continuous at zero and so possesses a subgradient which justifies the existence of a Lagrange multiplier [123]. Hence, one has $\lambda_1, \ldots, \lambda_n \geq 0$ with

$$r + \sum_{i=1}^{n} \lambda_i (f(x, y_i) - r) \geq h(0),$$

for all real r and all x in X. Deduce that $\sum_{i=1}^{n} \lambda_i = 1$ and hence, by concave-likeness, there is an element $\bar{y}$ of Y with

$$\min_X \sup_Y f(x,y) \geq \min_X f(x,\bar{y}) \geq h(0) \geq \max_{1 \leq i \leq n} \min_X f(x,y).$$

Now appeal to compactness to show that

$$\min_X \sup_Y f(x,y) \geq \sup_Y \min_X f(x,y).$$

Weak duality provides the other inequality. □

2.3.23 (Hidden convexity, II).[†] Use Lyapunov's theorem (1.4.7) to show that h given by

$$(HC) \quad h(\bar{b}) := \inf \int_T \phi_0(x(t),t)\,\mu(dt) \quad \text{subject to}$$

$$\int_T \phi_0(x(t),t)\,\mu(dt) = b_1, \cdots, \int_T \phi_n(x(t),t)\,\mu(dt) = b_n$$

for $1 \leq k \leq n$ is convex whenever each ϕ_k is a normal (convex or nonconvex) integrand, in the language of the discussion before Theorem 6.3.6, and μ is a purely nonatomic, complete finite measure. In particular, this holds when the constraints are linear functionals and the objective function is of the form $\int_T \phi(x(t))\,\mu(dt)$ for an arbitrary lsc ϕ.

Hint. Let $\bar{\phi} := (\phi_0, \phi_1, \ldots, \phi_n)$. Fix two vectors $\bar{b}$ and $\bar{c}$ and feasible functions x and y with $\int_T \phi_i(x(t),t)\,d\mu = b_i$, $\int_T \phi_i(y(t),t)\,d\mu = c_i$, and $h(b) < \int_T \phi_0(x(t),t)\,d\mu + \varepsilon$, $h(\bar{c}) < \int_T \phi_0(y(t),t)\,d\mu + \varepsilon$. By translation we may assume $\bar{c} = 0, y = 0$, and $\phi(y) = 0$. Use the vector measure given by $m(E) := \int_E \phi(x(t)\,d\mu$ for $E \subset T$ to show $h(s\bar{b} + (1-s)\bar{c}) \leq sh(\bar{b}) + (1-s)h(\bar{c}) + \varepsilon$ for $0 < s < 1$. □

2.3.24 (Direct and indirect utility).[†] In mathematical economics the relationship between production and consumption is often modeled by the following dual pair. Begin with a real-valued, nondecreasing concave or quasi-concave function on the nonnegative orthant in Euclidean space, E, that is supposed to represent a consumer's preference for consumption of some amount $y \in E$ of consumables. Fix a nonnegative vector p of prices and consider the primal or *utility maximization problem*

$$u(p) := \sup_{0 \leq y \in E} \{v(y) : \langle p, y \rangle \leq r\} \qquad (2.3.10)$$

of working out how one maximizes ones individual utility given a fixed budget $r > 0$. The function u is called the *indirect utility function* and expresses the consumer utility as a function of prices. Under conditions discussed in [163, 186] the *direct utility function* can be reconstructed from

$$v(y) = \inf_{0 \leq p \in E} \{u(p/r) : \langle p, y \rangle \leq r\}. \qquad (2.3.11)$$

(a) Show u is nonincreasing on the positive orthant.
(b) It can be shown u is quasi-convex and continuous on the positive orthant when v is quasi-concave and continuous on the nonnegative orthant. Moreover, u satisfies (2.3.11).
(c) Show that the *Cobb–Douglas functions* defined for $y \geq 0$ by

$$v(y) := \prod_{i=1}^{n} y_i^{\alpha_i},$$

with $\alpha_i > 0$, are quasi-concave and continuous on the nonnegative orthant and that they have log-convex indirect utility functions

$$u(p) = \prod_{i=1}^{n} \left(\frac{\alpha_i}{\alpha x_i}\right)^{\alpha_i},$$

infinite on the boundary of the orthant, where $\alpha := \sum_{i=1}^{n} \alpha_i$. (Compare [95, Example 13].)

Hint. e^g is quasi-concave when g is.

2.3.25 (Maximum entropy [100]).[†] Define a convex function $p : \mathbb{R} \to (-\infty, +\infty]$ by

$$p(t) := \begin{cases} t \log t - t, & \text{if } t > 0; \\ 0 & \text{if } t = 0; \\ +\infty, & \text{otherwise}; \end{cases}$$

and a convex function $f : \mathbb{R}^n \to (-\infty, +\infty]$ by

$$f(x) := \sum_{i=1}^{n} p(x_i).$$

Suppose $\hat{x}$ is in the interior of $\mathbb{R}^n_+$.

(a) Prove f is strictly convex on $\mathbb{R}^n_+$ with compact lower level sets.
(b) Prove $f'(x; \hat{x} - x) = -\infty$ for any x on the boundary of $\mathbb{R}^n_+$.
(c) Suppose the map $G : \mathbb{R}^n \to \mathbb{R}^m$ is linear with $G\hat{x} = b$. Prove for any vector $c \in \mathbb{R}^n$ that the problem

$$\inf\{f(x) + \langle c, x \rangle : Gx = b, x \in \mathbb{R}^n\}$$

has a unique optimal solution $\bar{x}$ lying in the interior of $\mathbb{R}^n_+$.
(d) Use Exercise 2.3.21 to prove that some vector $\lambda \in \mathbb{R}^m$ satisfies $\nabla f(\bar{x}) = G^* \lambda - c$ and deduce $\bar{x}_i = \exp(G^* \lambda - c)_i$.

Maximum entropy leads natural to the notion of divergence estimates or *relative entropy*.

2.3.26 (Divergence estimates [304]).[†]

(a) Prove the real function
$$t \mapsto 2(2+t)(\exp^* t + 1) - 3(t-1)^2$$
is convex and minimized when $t = 1$.

(b) For $v \in \text{int } \mathbb{R}_+$ and $u \in \mathbb{R}_+$, deduce the inequality
$$3(u-v)^2 \leq 2(u+2v)\left(u \log\left(\frac{u}{v}\right) - u + v\right).$$

Now suppose the vector $p \in \text{int } \mathbb{R}_+^n$ satisfies $\sum p_i = 1$.

(c) If the vector $q \in \text{int } \mathbb{R}_+^n$ satisfies $\sum_{i=1}^n q_i$, use the Cauchy–Schwarz inequality to prove the inequality
$$\left(\sum_{i=1}^n |p_i - q_i|\right)^2 \leq 3 \sum_{i=1}^n \frac{(p_i - q_i)^2}{p_i + 2q_i},$$
and deduce the inequality
$$\sum_{i=1}^n p_i \log\left(\frac{p_i}{q_i}\right) \geq \frac{1}{2}\left(\sum_{i=1}^n |p_i - q_i|\right)^2.$$

2.3.27. Let C be a convex subset of E, and let $k_C := \gamma_C$ be its associated gauge function. The *polar function* is defined by
$$(k_C)^o(x) := \inf\{v \geq 0 : \langle x^*, x \rangle \leq v k_C(x)\}.$$
Note that if $k_C(x) > 0$ for $x \neq 0$, then
$$k_C^o(x^*) = \sup_{x \neq 0} \frac{\langle x^*, x \rangle}{k_C(x)}.$$

(a) Show that $k_C^o = k_{C^o}$ and that $k_C^{oo} = \text{cl } k_C$. Deduce that polar functions are support functions.

(b) Deduce that the polar of a norm is the dual norm.

(c) Observe the *polar inequality*
$$\langle x^*, x \rangle \leq k_C(x) k_C^o(x^*),$$
for all x and x^*. Hence rederive Hölder's inequality.

(d) (Experimental design duality).[†] An elegant illustration of the use of polar functions is to be found in [359, 358]. The setup, from linear model theory in statistics, is that we let E_k be the Euclidean space of $k \times k$ symmetric matrices, S_k the cone of positive semidefinite matrices and let $\mathcal{M} \subset E_k$ be a compact, convex subset of the nonnegative semidefinite matrices containing some positive definite element

2.3 Conjugate functions and Fenchel duality

(this is Slater's condition in this context). Fix $0 \leq s \leq k$, let K be a full-rank $k \times s$ matrix and define a mapping – not the duality map – by $J : S_k \to S_s$ by $J(A) = (K'A^-K)^{-1}$ if $A \in \mathcal{A}(K) := \{B \in N_k : \text{range } B \subset \text{range } K\}$, and $J(A) = 0$ otherwise. Here A^- denotes the Moore–Penrose *pseudo-inverse* matrix.

An *information functional* is a concave, positively homogeneous function $j : S_s \to \mathbb{R}$ which is isotone with respect to the semidefinite or Loewner ordering (see (2.3.30)). Then the *optimal (experimental) design problem* in statistics asks for the solutions to

$$\mu := \sup_{M \in \mathcal{M}} (j \circ J)(M),$$

and the *dual design problem* is

$$\nu := \min_{N \in \mathcal{M}^\circ} 1/j^\circ(K'NK).$$

(i) Show that if j is strictly concave then it is strictly increasing – but not necessarily conversely as $j = \text{tr}$, the trace, shows.
(ii) Characterize subgradients of $-\log \circ j \circ J$.
(iii) Show that $j \circ J$ is usc if and only if $\text{dom } j \subset \text{int } S_s$, in which case μ is attained. Note that this gives a 'natural' class of nonclosed convex functions, such as $\text{tr} \circ J$ when $0 < s < k$.
(iv) Show that strong duality obtains: $\mu = \nu$.

2.3.28 (The location problem [280, 130]).† The classical *location problem*, also known as the *Fermat–Weber* problem or the *Steiner* problem asks for the point least distance from m given points $a_1, a_2, \ldots, a_m$ – the locations – in a Euclidean space E. That is we seek the point $\bar{x}$ minimizing the coercive convex function

$$F(x) := \sum_{i=1}^{m} w_i \|x - a_i\|,$$

for given positive weights w_i. Needless to say there are many generalizations. For example, the Euclidean norm could be replaced by a different norm or gauge function for each index.

(a) Fermat posed the question, answered by Torricelli, for the triangle with equal weights: the solution is the *Fermat point*. For triangles with no angle exceeding $120°$ this is achieved by constructing an equilateral triangle on each side of the given triangle. For each new vertex one draws a line to the opposite triangle's vertex. See Figure 2.9. The three lines intersect at the Fermat point. In the remaining case, a vertex is the solution.
(b) Show the minimum exists and is unique (in the Euclidean case).
(c) Use the proposition on critical points (2.1.14) and the subdifferential sum rule (2.3.5) to show that a nonlocation $\bar{x}$ is minimal if and only if

$$\sum_{i=1}^{m} w_i \frac{\bar{x} - a_i}{\|\bar{x} - a_i\|} = 0,$$

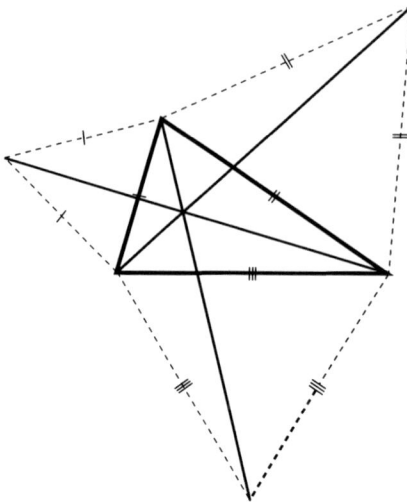

Figure 2.9 The Fermat point of a triangle.

and write down the optimality condition in the general case. Deduce that the optimum is a fixed point of the mapping defined, for x not a location, by

$$T(x) := \frac{\sum_{i=1}^m w_i a_i / \|x - a_i\|}{\sum_{i=1}^m w_i / \|x - a_i\|}$$

with continuity assured by setting $T(a_i) = a_i$ for each location.

(d) Analyze Wieszfeld's proposed 1937 algorithm which is just to iterate T. Convince yourself that, assuming the locations span the space, it usually converges to a solution – typically at linear rate [280, 130]

2.3.29 (Joint and separate convexity on $\mathbb{R}$ [34]).** Suppose that I is a nonempty open interval in $\mathbb{R}$, and that $f \in C^3(I)$ with $f'' > 0$ on I. Let D_f denote the *Bregman distance* defined as follows

$$D_f : I \times I \to [0, \infty) : (x, y) \mapsto f(x) - f(y) - f'(x)(x - y).$$

The convexity of f implies that $x \mapsto D_f(x, y)$ is convex for every $y \in I$. We will say D_f is *jointly convex* if $(x, y) \mapsto D_f(x, y)$ is convex on $I \times I$, and we will say D_f is *separately convex* if $y \mapsto D_f(x, y)$ is convex for every $x \in I$.

(a) Let $h := f''$. Show that D_f is jointly convex $\Leftrightarrow 1/h$ is concave $\Leftrightarrow$

$$h(y) + h'(y)(y - x) \geq \bigl(h(y)\bigr)^2 / h(x), \quad \text{for all } x, y \text{ in } I. \tag{j}$$

In particular, if f'''' exists, then: D_f is jointly convex $\Leftrightarrow hh'' \geq 2(h')^2$.

(b) Let $h := f''$. Show that D_f is separately convex $\Leftrightarrow$

$$h(y) + h'(y)(y - x) \geq 0, \quad \text{for all } x, y \text{ in } I. \tag{s}$$

2.3 Conjugate functions and Fenchel duality

(c) Let $f(t) := e^{-t}$ on $I = (0, 1)$. Use part (a) to show D_f is not jointly convex on I, and use part (b) to show that D_f is separately convex on I. Hence joint convexity is genuinely stronger than separate convexity.

(d) (Fermi–Dirac entropy) Let $f(t) := t \log t + (1-t) \log(1-t)$ on $I = (0, 1)$. Use part (a) to show that D_f is jointly convex.

(e) Let $I = (-\infty, +\infty)$. Show that D_f is separately convex if and only if f is a quadratic function.

(f) Let $I = (0, \infty)$. Show that D_f is separately convex if and only if $f''(x) + x f'''(x) \geq 0 \geq f'''(x)$ for every $x > 0$.

(g) Suppose $I = (0, \infty)$ and f'''' exists. Let $\psi := \log \circ (f'')$. Show that:
   (i) D_f is jointly convex $\Leftrightarrow \psi''(x) \geq (\psi'(x))^2$, for all $x > 0$.
   (ii) D_f is separately convex $\Leftrightarrow 0 \geq \psi'(x) \geq -1/x$, for all $x > 0$.
   (iii) f'' is log-convex $\Leftrightarrow \psi''(x) \geq 0$, for all $x > 0$.
   (iv) f'' is convex $\Leftrightarrow \psi''(x) \geq -(\psi'(x))^2$, for all $x > 0$.

(h) (Boltzmann–Shannon entropy) Let $f(t) := t \log t - t$ on $(0, \infty)$. Show D_f is jointly convex, but D_{f^*} is not separately convex.

(i) (Burg entropy) Let $f(t) := -\log t$ on $(0, \infty)$. Show that D_f is not separately convex, but f'' is log convex.

(k) (Energy) Let $f(t) := t^2/2$ on $\mathbb{R}$. Show that $D_f(x, y) = |x - y|^2/2$ recovers the usual distance (squared).

Hint. For (a), use the Hessian characterization of twice differentiable convex functions, along with Exercise 2.2.4. For (b), observe the second derivative of $y \mapsto D_f(x, y)$ is $h(y) + h'(y)(y - x)$. For (e), use (b) and let $x \to \pm\infty$ to conclude $f'''(x) = 0$ for all $x \in \mathbb{R}$. For (f), use (b). For (g), use (a) for (i), use (f) for (ii), and use basic properties of convex and log-convex functions for (iii) and (iv). Use appropriate parts of (g) to show (h) and (i). For more details on the proof and for further results see [34]. □

2.3.30 (Joint and separate convexity on E [34]).** Let U be a convex nonempty open subset of E, and let $f \in C^3(U)$ with f'' positive definite on U. Then f is strictly convex, and we let $H = \nabla^2 f$ be the Hessian of f. Recall that the real symmetric matrices can be partially ordered by the *Loewner ordering*

$$H_1 \succeq H_2 \text{ if and only if } H_1 - H_2 \text{ is positive semidefinite}$$

and form a Euclidean space with the inner product $\langle H_1, H_2 \rangle := \text{tr}(H_1 H_2)$. For more information see [261, Section 7.7] and [303, Section 16.E]. Observe that the Bregman distance associated with the Euclidean norm $1/2\|\cdot\|_2^2$ is $(x, y) \mapsto 1/2\|x - y\|_2^2$. Extend the definitions of D_f, separately convex and jointly convex given Exercise 2.3.29 to E and show that:

(a) D_f is jointly convex if and only if

$$H(y) + (\nabla H(y))(y - x) \succeq H(y) H^{-1}(x) H(y), \quad \text{for all } x, y \in U. \quad (J)$$

(b) D_f is separately convex if and only if

$$H(y) + (\nabla H(y))(y - x) \succeq 0, \quad \text{for all } x, y \in U. \tag{S}$$

(c) Show that the following are equivalent:
  (i) D_f is jointly convex.
  (ii) $\nabla H^{-1}(y)(x - y) \succeq H^{-1}(x) - H^{-1}(y)$, for all $x, y \in U$.
  (iii) H^{-1} is *(matrix) concave*, i.e.

$$H^{-1}(\lambda x + \mu y) \succeq \lambda H^{-1}(x) + \mu H^{-1}(y),$$

  for all $x, y \in U$ and $\lambda, \mu \in [0, 1]$ with $\lambda + \mu = 1$.
  (iv) $x \mapsto \langle P, H^{-1}(x)\rangle$ is concave, for every $P \succeq 0$.

Hint. (a) The Hessian of the function $U \times U \to [0, +\infty) : (x, y) \mapsto D_f(x, y)$ is the block matrix

$$\nabla^2 D_f(x, y) = \begin{pmatrix} H(x) & -H(y) \\ -H(y) & H(y) + (\nabla H(y))(y - x) \end{pmatrix}.$$

Using standard criteria for positive semidefiniteness of block matrices (see [261, Section 7.7]) and remembering that H is positive definite, we obtain that $\nabla^2 D_f(x, y)$ is positive semidefinite for all x, y if and only if (J) holds.

(b) For fixed $x \in U$, similarly discuss positive semidefiniteness of the Hessian of the map $y \mapsto D_f(x, y)$.

(c) Consider the mapping $U \to \mathbb{R}^{N \times N} : y \mapsto H(y)H^{-1}(y)$. It is constant, namely the identity matrix. Take the derivative with respect to y. Using an appropriate product rule (see, for instance, [185]) yields $0 = H(y)\big((\nabla H^{-1}(y))(z)\big) + \big((\nabla H(y))(z)\big)H^{-1}(y)$, for every $z \in \mathbb{R}^N$. In particular, after setting $z = x - y$, multiplying by $H^{-1}(y)$ from the left, and rearranging, we obtain

$$H^{-1}(y)\big((\nabla H(y))(y - x)\big)H^{-1}(y) = (\nabla H^{-1}(y))(x - y).$$

(i)$\Leftrightarrow$(ii): The equivalence follows readily from the last displayed equation and (a).

(ii)$\Leftrightarrow$(iii): The proof of [366, Theorem 42.A] works without change in the present positive semidefinite setting.

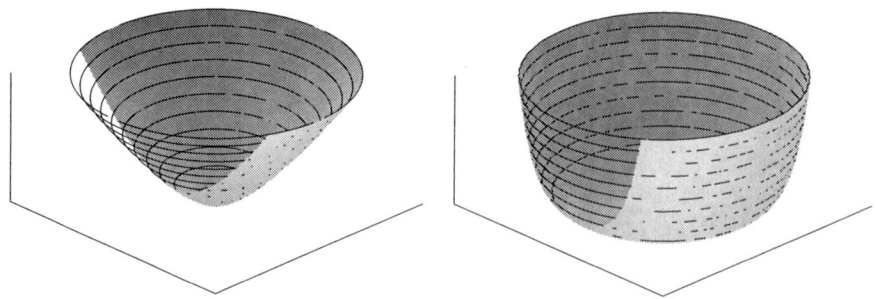

Figure 2.10 The NMR entropy and its conjugate.

2.3 Conjugate functions and Fenchel duality

(iii)⇔(iv): is clear, since the cone of positive semidefinite matrices is self-dual. □

2.3.31 (NMR entropy).[†] Let $z = (x,y)$ and let $|z|$ denote the Euclidean norm. Show that $z \mapsto \cosh(|z|)$ and $z \mapsto |z|\log\left(|z| + \sqrt{1+|z|^2}\right) - \sqrt{1+|z|^2}$ are mutually conjugate convex function on $\mathbb{R}^2$ (or $\mathbb{C}$). They are shown in Figure 2.10. The latter is the building block of the *Hoch–Stern information function* introduced for entropy-based magnetic resonancing [102].

2.3.32 (Symbolic convex analysis).[†] It is possible to perform a significant amount of convex analysis in a computer algebra system – both in one and several dimensions. Some of the underlying ideas are discussed in [87]. The computation of $f_5 := x \mapsto |x| - 2\sqrt{1-x}$ in Chris Hamilton's *Maple* package SCAT (http://flame.cs.dal.ca/~chamilto/files/scat.mpl) is shown in Figure 2.11. Figure 2.12 and Figure 2.13 illustrate computing $sdf5 = \partial f_5$ and $g_5 = f_5^*$ respectively. Note the need to deal carefully with piecewise smooth functions. Figure 2.13 also confirms the convexity of f_5. The plots of f_5 and ∂f_5 are shown in Figure 2.14.

```
piecewise(-3<=x and x<=1,abs(x)-2*sqrt(1-x),infinity):
f5 := convert(%,PWF);
```

$$f5 := \begin{cases} \infty, & x < -3 \\ 1, & x = -3 \\ -2\sqrt{1-x} - x, & (-3 < x) \text{ and } (x < 0) \\ -2, & x = 0 \\ -2\sqrt{1-x} + x, & (0 < x) \text{ and } (x < 1) \\ 1, & x = 1 \\ \infty, & 1 < x \end{cases}$$

Figure 2.11 Symbolic convex analysis of f_5.

```
Plot(sdf5,-3..1,view=[-3..1,-3..5],axes=none), yielding
```

$$sdf5 := \begin{cases} \{\}, & x < -3 \\ [-\infty, -\frac{1}{2}], & x = -3 \\ \{\frac{(-1+\sqrt{1-x})\sqrt{1-x}}{x-1}\}, & (-3 < x) \text{ and } (x < 0) \\ [0,2], & x = 0 \\ \{-\frac{(1+\sqrt{1-x})\sqrt{1-x}}{x-1}\}, & (0 < x) \text{ and } (x < 1) \\ \{\}, & x = 1 \\ \{\}, & 1 < x \end{cases}$$

Figure 2.12 Symbolic convex analysis of ∂f_5.

64 Convex functions on Euclidean spaces

$$g5 := \begin{cases} -3y+1, & y < -\frac{1}{2} \\ \frac{5}{2}, & y = \frac{-1}{2} \\ \frac{y^2+2y+2}{1+y}, & (\frac{-1}{2} < y) \text{ and } (y < 0) \\ 2, & y = 0 \\ 2, & (0 < y) \text{ and } (y < 2) \\ 2, & y = 2 \\ \frac{y^2-2y+2}{-1+y}, & 2 < y \end{cases}$$

```
> F5 := Conj(g5,x):
  Equal(f5,F5);
```

$$true$$

Figure 2.13 Symbolic convex analysis of $g_5 = f_5^*$.

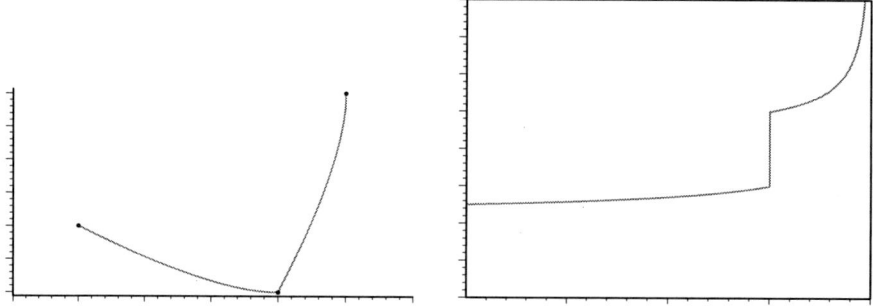

Figure 2.14 Plots f_5 and ∂f_5.

2.4 Further applications of conjugacy

We start with some accessible but important further geometric and functional analytic corollaries to the foundational results of the previous section.

Corollary 2.4.1 (Sandwich theorem). *Let $f : E \to (-\infty, +\infty]$ and $g : Y \to (-\infty, +\infty]$ be convex, and let $A : E \to Y$ be linear. Suppose $f \geq -g \circ A$ and $0 \in \text{core}(\text{dom} \, g - A \, \text{dom} f)$ (or $A \, \text{dom} f \cap \text{cont} \, g \neq \emptyset$). Then there is an affine function $\alpha : E \to \mathbb{R}$ of the form $\alpha(x) = \langle A^*\phi, x \rangle + r$ satisfying $f \geq \alpha \geq -g \circ A$. Moreover, for any $\bar{x}$ satisfying $f(\bar{x}) = -g \circ A(\bar{x})$, we have $-\phi \in \partial g(A\bar{x})$.*

Proof. See Exercise 2.4.1 for two suggested proofs. □

The sandwich theorem makes a very satisfactory alternative starting point rather than a corollary. Figure 2.15 illustrates the failure of the sandwich theorem in the absence of the constraint qualification.

Corollary 2.4.2 (Separation theorem). *Suppose $C_1, C_2 \subset E$ are convex sets such that C_1 has nonempty interior and $\text{int} \, C_1 \cap C_2 = \emptyset$. Then there exists $\phi \in E \setminus \{0\}$ such that $\langle \phi, x \rangle \leq \langle \phi, y \rangle$ for all $x \in C_1$ and all $y \in C_2$.*

2.4 Further applications of conjugacy

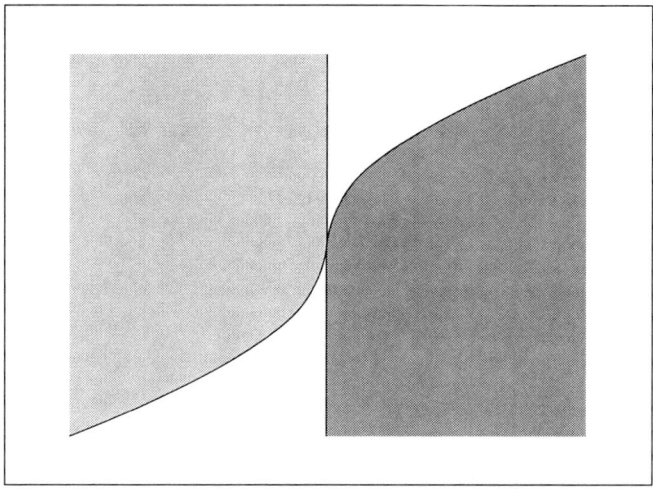

Figure 2.15 Failure of Fenchel duality for $\sqrt{x}$ and $-\sqrt{-x}$.

Proof. By translation we may without loss of generality assume $0 \in \operatorname{int} C_1$. Let $f := \gamma_{C_1}$ be the gauge of C_1 and let $g := \gamma_{C_2} - 1$. Then f is continuous because $0 \in \operatorname{int} C_1$ (Exercise 2.1.13). According to the sandwich theorem (2.4.1) there exist an affine function α, (say $\alpha = \phi + k$ where $\phi \in E$ and $k \in \mathbb{R}$) such that $f \geq \alpha \geq -g$. Thus $\phi(x) + k \leq 1$ for all $x \in C_1$, and $\phi(x) + k \geq 1$ for all $x \in C_2$, moreover $k = \phi(0) + k \leq \gamma_{C_1}(0) = 0$, and so ϕ is not the 0 functional. □

Corollary 2.4.3 (Supporting hyperplane). *Suppose that $C \subset E$ is a convex set with nonempty interior and that $\bar{x} \in \operatorname{bdry} C$. Then there exists $\phi \in E$ such that $\langle \phi, x \rangle \leq \langle \phi, \bar{x} \rangle$ for all $x \in C$ (the set $\{x \in E : \langle \phi, x \rangle = \langle \phi, \bar{x} \rangle\}$ is called a supporting hyperplane).*

Proof. This follows directly by applying the separation theorem (2.4.2) to $C_1 := C$ and $C_2 := \{\bar{x}\}$. □

Theorem 2.4.4 (Fenchel biconjugation). *For any function $f : E \to (-\infty, +\infty]$, the following are equivalent.*

(a) *f is closed and convex.*
(b) *$f = f^{**}$.*
(c) *For all points x in E,*

$$f(x) = \sup\{\alpha(x) : \alpha \text{ is an affine minorant of } f\}.$$

Hence the conjugation operation induces an order-inverting bijection between proper closed convex functions.

Proof. (a) ⇒ (c): Let $\bar{x} \in \operatorname{dom} f$ and fix $\bar{t} < f(\bar{x})$. Because f is lsc we choose $r > 0$ so that $f(x) > \bar{t}$ whenever $\|x - \bar{x}\| \leq r$. Then apply the sandwich theorem (2.4.1) to f and $g := -\delta_{B_r(\bar{x})} + \bar{t}$ to conclude f has an affine minorant, say α_0. It suffices to show the result for $h := f - \alpha_0$. Suppose $\bar{x} \in X$ and $h(\bar{x}) > \bar{t}$. Choose $\delta > 0$ so that $h(x) > \bar{t}$ for $\|x - \bar{x}\| \leq \delta$. Then choose $M > 0$ so large that $\bar{t} - M\delta < 0$ and let

$g := \bar{t} - M\|\cdot - \bar{x}\|$. The definition of g along with the fact $h \geq 0$ implies $g \leq h$. Thus one can apply the sandwich theorem (2.4.1) to h and g to obtain an affine minorant α of h so that $\alpha(\bar{x}) > \bar{t}$ as desired.

(c) $\Rightarrow$(b): Because $f^{**} \leq f$, it suffices to show $f^{**} \geq \alpha$ for any affine minorant α of f. Indeed, write $\alpha := y + k$ for some $y \in E$ and $k \in \mathbb{R}$. Then $\alpha^* = \delta_{\{y\}} - k$ and so $\alpha^{**} = \alpha$. Then $\alpha \leq f$ implies $\alpha^* \geq f^*$ implies $\alpha^{**} \leq f^{**}$, and so $\alpha \leq f^{**}$ as desired.

(b) $\Rightarrow$ (a): This follows because f^{**} is closed. □

Conditions on finite-dimensional separation theorems can be relaxed using the notion of relative interior introduced below. First, a set L in E is said to be *affine* if the entire line through any distinct points x and y in L lies in L; algebraically this means $\lambda x + (1 - \lambda) y \in L$ for any real λ. The *affine hull* of a set D, denoted by aff D is the smallest affine set containing D. An *affine combination* of points $x_1, x_2, \ldots, x_m$ is a point of the form $\sum_{i=1}^{m} \lambda_i x_i$ for real numbers λ_i summing to one.

Lemma 2.4.5. *Let D be a nonempty subset of E. For each point $x \in D$, aff $D = x + \mathrm{span}(D - x)$, and consequently the linear subspace $\mathrm{span}(D - x)$ is independent of the choice of $x \in D$.*

Proof. Let $x \in D$; then $v \in x + \mathrm{span}(D - x)$ implies

$$v = x + \sum_{i=1}^{k} a_i(d_i - x) = \left(1 - \sum_{i=1}^{k} a_i\right) x + \sum_{i=1}^{k} a_i d_i$$

$$= \sum_{i=0}^{k} \lambda_i d_i \text{ where } \lambda_0 = 1 - \sum_{i=1}^{k} a_i, \lambda_i = a_i, d_0 = x \text{ and so } \sum_{i=0}^{k} \lambda_i = 1,$$

this shows $x + \mathrm{span}(D - x) \subset$ aff D. Reversing the steps above shows the reverse inclusion.

Because $y + \mathrm{span}(D - y) =$ aff $D = x + \mathrm{span}(D - x)$ for any $x, y \in D$, it is another elementary exercise to show that $\mathrm{span}(D - x)$ is independent of the choice of $x \in D$. □

The *relative interior* of a convex set C in E denoted ri C is its interior relative to its affine hull. In other words, $x \in$ ri C if there exists $\delta > 0$ so that $(x + \delta B_E) \cap$ aff $C \subset C$. The utility of the relative interior is in part that it depends only on the convex set and not on the space in which it lives: a disc drawn on the blackboard has the same relative interior when viewed in three dimensions. Although there are analogs in infinite dimensions, the relative interior is primarily of use in Euclidean space.

Theorem 2.4.6. *Suppose $\dim E > \{0\} \subsetneq C$, $0 \in C$ and aff $C = Y$ and C is a convex set. Then Y is a subspace of E and C contains a basis $\{y_1, y_2, \ldots, y_m\}$ of Y and has nonempty interior relative to Y. In particular, any nonempty convex set in E has nonempty relative interior.*

Proof. Because $0 \in C$, it follows from Lemma 2.4.5 that Y is a subspace of E. Because C contains a minimal spanning set for Y, C contains a basis, say $y_1, y_2, \ldots, y_m$ of Y.

Then $U = \{y \in Y : y = \sum_{i=1}^{m} \lambda_i y_i, 0 < \lambda_i < 1/m\}$ is open in Y. Moreover, $U \subset C$ and is nonempty because $(1 - \sum_{i=1}^{m} \lambda_i)0 + \sum_{i=1}^{m} \lambda_i y_i \in C$ whenever $0 < \lambda_i < 1/m$. The 'in particular' statement follows from a standard translation argument. □

Further facts related to relative interior or affine sets can be found in Exercises 2.4.4 and 2.4.5. One of the useful consequences of relative interior is that it allows us to relax interiority conditions in certain Euclidean space results, two examples of this are as follows.

Theorem 2.4.7 (Separation theorem). *Suppose $C_1, C_2 \subset E$ are convex and ri $C_1 \cap C_2 = \emptyset$. Then there exists $\phi \in E \setminus \{0\}$ such that $\langle \phi, x \rangle \leq \langle \phi, y \rangle$ for all $x \in C_1$ and all $y \in C_2$.*

Proof. Let $C = C_1 - C_2$ and let $Y = \text{aff } C$. It suffices to show there is a nontrivial $\phi \in E$ that separates C from 0. Suppose $0 \notin Y$. Then by elementary linear algebra, or by the separation theorem (2.4.2), there is a $\phi \in E \setminus \{0\}$ that separates 0 from Y, and hence separates C from 0. In the case $0 \in Y$, we may apply the separation theorem (2.4.2) to 0 and C in Y, and then extend the separating functional to all of E. □

Theorem 2.4.8 (Nonempty subdifferential). *Suppose $f : E \to (-\infty, +\infty]$ is a proper convex function. Then $\partial f(x) \neq \emptyset$ whenever $x \in \text{ri}(\text{dom} f)$.*

Proof. See Exercise 2.4.6. □

A pretty application of the Fenchel duality circle of ideas is the calculation of polar cones. The *(negative) polar cone* of the set $K \subset E$ is the convex cone

$$K^- := \{\phi \in E : \langle \phi, x \rangle \leq 0 \text{ for all } x \in K\}.$$

Analogously, $K^+ := \{\phi \in E : \langle \phi, x \rangle \geq 0, \text{ for all } x \in K\}$ is the *(positive) polar cone*. The cone $K^{--} = (K^-)^-$ is called the *bipolar*. A particularly important example of the polar cone is the *normal cone* to a convex set $C \subset E$ at a point x in C which we recall is defined by $N_C(x) := \partial \delta_C(x)$; indeed one can check $N_C(x) = (C - x)^-$.

We use the following two examples extensively.

Proposition 2.4.9 (Self-dual cones). *Let S_+^n denote the set of positive semidefinite n by n matrices viewed as a subset of the $n \times n$ symmetric matrices, and let $\mathbb{R}_+^n$ denote the positive orthant in $\mathbb{R}^n$. Then*

$$(\mathbb{R}_+^n)^- = -\mathbb{R}_+^n \quad \text{and} \quad (S_+^n)^- = -S_+^n.$$

Proof. See Exercise 2.4.22. □

The next result shows how the calculus rules above can be used to derive geometric consequences.

Corollary 2.4.10 (Krein–Rutman polar cone calculus). *Let E and Y be Euclidean spaces. Any cones $H \subset Y$ and $K \subset E$ and linear map $A : E \to Y$ satisfy*

$$(K \cap A^{-1}H)^- \supset A^*H^- + K^-.$$

Equality holds if H and K are convex and satisfy $H - AK = Y$ (or in particular $AK \cap \mathrm{int}\, H \neq \emptyset$).

Proof. Rephrasing the definition of the polar cone shows that for any cone $K \subset E$, the polar cone K^- is just $\partial \delta_K(0)$. The result now follows from the subdifferential sum rule (2.3.5). □

The polarity operation arises naturally from Fenchel conjugation, since for any cone $K \subset E$ we have $\delta_{K^-} = \delta_K^*$, whence $\delta_{K^{--}} = \delta_K^{**}$. The next result, which is an elementary application of the basic separation theorem (2.1.21), will allow one to identify K^{--} as the closed convex cone generated by K.

Theorem 2.4.11 (Bipolar cone). *The bipolar cone of any nonempty set $K \subset E$ is given by $K^{--} = \mathrm{cl}(\mathrm{conv}(\mathbb{R}_+ K))$.*

Proof. See Exercise 2.4.23. □

From this, one can deduce that the normal cone $N_C(x)$ to a convex set C at a point x in C, and the *tangent cone* to C at x defined by $T_C(x) = \mathrm{cl}\, \mathbb{R}_+(C - x)$, are polars of each other. The next result characterizes *pointed* cones (i.e. those closed convex cones K satisfying $K \cap -K = \{0\}$).

Theorem 2.4.12 (Pointed cones). *If $K \subset E$ is a closed convex cone, then K is pointed if and only if there is an element y of E for which the set*

$$C := \{x \in K : \langle x, y \rangle = 1\}$$

is compact and generates K (that is, $K = \mathbb{R}_+ C$).

Proof. See Exercise 2.4.27. □

Exercises and further results

2.4.1. This exercise looks at the interrelationships between various results presented so far.

(a) Proofs for the Hahn–Banach/Fenchel duality circle of ideas.
   (i) Prove the sandwich theorem (2.4.1) using the max formula (2.1.19).
   (ii) Use the Fenchel duality theorem (2.3.4) to prove the sandwich theorem (2.4.1).
   (iii) Use the sandwich theorem (2.4.1) to prove the subdifferential sum rule (2.3.6).
   (iv) Use the sandwich theorem (2.4.1) to prove Hahn–Banach extension theorem (2.1.18).
(b) Find a sequence of proofs to show that each of (i) through (iv) is equivalent (in the strong sense that they are easily inter-derivable) to the nonemptyness of the subgradient at a point of continuity as ensured by Corollary 2.1.20.

2.4 Further applications of conjugacy

Hint. (a) For (i), define h as in the proof of the Fenchel duality theorem (2.3.4), and proceed as in the last paragraph of the proof of Theorem 4.1.18.

For (ii), with notation as in the Fenchel duality theorem (2.3.4) observe $p \geq 0$ because $f(x) \geq -g(Ax)$, and then the Fenchel duality theorem (2.3.4) says $d = p$ and because the supremum in d is attained, deduce that there exist $\phi \in Y$ such that

$$0 \leq p \leq [f(x) - \langle \phi, Ax \rangle] + [g(y) + \langle \phi, y \rangle].$$

Then for any $z \in E$, setting $y = Az$, in the previous inequality, we obtain

$$a := \inf_{x \in E}[f(x) - \langle A^*\phi, x \rangle] \geq b := \sup_{z \in E}[-g(Az) - \langle A^*\phi, z \rangle].$$

Now choose $r \in [a, b]$ and let $\alpha(x) = \langle A^*y^*, x \rangle + r$.

For (iii), see the proof Theorem 4.1.19, while (iv) is straightforward. □

2.4.2 (A limiting sandwich example from [297]). Let f and g be defined by

$$f(u, v) := \begin{cases} 1 - \sqrt{uv}, & \text{if } u, v \geq 0; \\ \infty, & \text{otherwise} \end{cases}$$

and

$$g(u, v) := \begin{cases} 1 - \sqrt{-uv}, & \text{if } -u, v \geq 0; \\ \infty, & \text{otherwise.} \end{cases}$$

Show that although $\inf_{\mathbb{R}^2} (f + g)$ is strictly positive there is no affine separator m with $f \geq m \geq -g$.

2.4.3 (Accessibility lemma).* Suppose C is a convex set in E.

(a) Prove $\text{cl}\, C \subset C + \varepsilon B_E$ for any $\varepsilon > 0$.
(b) For sets D and F in E with D open, prove $D + F$ is open.
(c) For $x \in \text{int}\, C$ and $0 < \lambda \leq 1$, prove $\lambda x + (1 - \lambda)\text{cl}\, C \subset C$. Deduce $\lambda \text{int}\, C + (1 - \lambda)\text{cl}\, C \subset \text{int}\, C$.
(d) Deduce $\text{int}\, C$ is convex.
(e) Deduce further that if $\text{int}\, C$ is nonempty, then $\text{cl}(\text{int}\, C) = \text{cl}\, C$. Is convexity necessary?

2.4.4 (Affine sets).* Establish the following facts for affine sets.

(a) Prove the intersection of an arbitrary collection of affine sets is affine.
(b) Prove that a set is affine if and only if it is a translate of a linear subspace.
(c) Prove $\text{aff}\, D$ is the set of all affine combinations of elements in D.
(d) Prove $\text{cl}\, D \subset \text{aff}\, D$ and deduce $\text{aff}\, D = \text{aff}(\text{cl}\, D)$.

2.4.5 (The relative interior).* Let C be a convex subset of E.

(a) Find convex sets $C_1 \subset C_2$ with $\text{ri}\, C_1 \not\subset \text{ri}\, C_2$.
(b) Prove that for $0 < \lambda \leq 1$, one has $\lambda \text{ri}\, C + (1 - \lambda)\text{cl}\, C \subset \text{ri}\, C$, and hence $\text{ri}\, C$ is convex with $\text{cl}(\text{ri}\, C) = \text{cl}\, C$.
(c) Prove that for a point $x \in C$, the following are equivalent.

(i) $x \in \mathrm{ri}\, C$.

(ii) For any $y \in C$, there exists $\varepsilon > 0$ with $x + \varepsilon(x - y) \in C$.

(iii) The set $\{\lambda(c - x) : \lambda \geq 0, c \in C\}$ is a linear subspace.

(d) If F is another Euclidean space and the map $T : E \to F$ is linear, then $\mathrm{ri}\, TC \supset T(\mathrm{ri}\, C)$.

For more extensive information on relative interior and its applications, including to separation theorems, see [369].

2.4.6.* Prove Theorem 2.4.8.

Hint. Assume $0 \in \mathrm{dom}\, f$, and let $Y = \mathrm{span}(\mathrm{dom}\, f)$. For $x \in \mathrm{ri}(\mathrm{dom}\, f)$, show that $\partial f|_Y(x) \neq \emptyset$, and use the fact Y is complemented in E to deduce the conclusion. □

2.4.7 (Subdifferentials near boundary points).* Suppose $f : E \to (-\infty, +\infty]$ is a closed proper convex function, and $x_0 \in \mathrm{dom}\, f$. Show that if $\partial f(x_0) = \emptyset$, then there exist $x_n \to x_0$ and $\phi_n \in \partial f(x_n)$ such that $\|\phi_n\| \to \infty$. Conclude that ∂f is not bounded on any neighborhood of a boundary point of $\mathrm{dom}\, f$. This phenomenon is studied in detail in Chapter 7.

Hint. First, there exists a sequence $(x_n) \subset \mathrm{ri}(\mathrm{dom}\, f)$ converging to x_0, and hence by Theorem 2.4.8 there exist $\phi_n \in \partial f(x_n)$. If a subsequence of (ϕ_n) were bounded, then it would have a convergent subsequence with limit ϕ. Show that $\phi \in \partial f(x_0)$ which is a contradiction. Combining this with the result of Exercise 2.2.23 provides the conclusion. □

2.4.8.* Show that if $f : E \to (-\infty, +\infty]$ is a proper convex function, then $\mathrm{cl}\, f$ is a proper convex function.

Hint. For properness use Theorem 2.4.8. □

2.4.9. Let C be the convex hull of the circle $\{(x, y, z) : x^2 + y^2 = 1, z = 0\}$ and the line segment $\{(1, 0, z) : -1 \leq z \leq 1\}$ in $\mathbb{R}^3$. Deduce that C is closed but its set of extreme points is not closed. See Figure 2.16 for a sketch of this set.

2.4.10 (Theorems of the alternative). (a) (Gordan [237]) Let $x_0, x_1, \ldots, x_m \in E$. Show that exactly one of the following systems has a solution.

$$\sum_{i=1}^{m} \lambda_i x_i = 0, \quad \sum_{i=0}^{m} \lambda_i = 1, \ \lambda_0, \ldots, \lambda_m \in [0, \infty) \qquad (2.4.12)$$

$$\langle x_i, x \rangle < 0 \quad \text{for } 0, 1, \ldots, m, \text{ and } x \in E. \qquad (2.4.13)$$

(b) (Farkas [209]) Let $x_1, x_2, \ldots, x_m$ and c be in E. Show that exactly one of the following has a solution.

$$\sum_{i=1}^{m} \mu_i x_i = c, \quad \mu_1, \mu_2, \ldots, \mu_m \in [0, \infty) \qquad (2.4.14)$$

$$\langle x_i, x \rangle \leq 0 \quad \text{for } i = 1, 2, \ldots, m, \ \langle c, x \rangle > 0, x \in E. \qquad (2.4.15)$$

Hint. For (a), if (2.4.12) has a solution, then clearly (2.4.13) does not. Let $C = \{x \in E : \sum_{i=0}^{m} \lambda_i x_i = x, \lambda_i \geq 0, \sum \lambda_i = 1\}$. If $0 \in C$, then (2.4.12) has a solution; if not,

2.4 Further applications of conjugacy

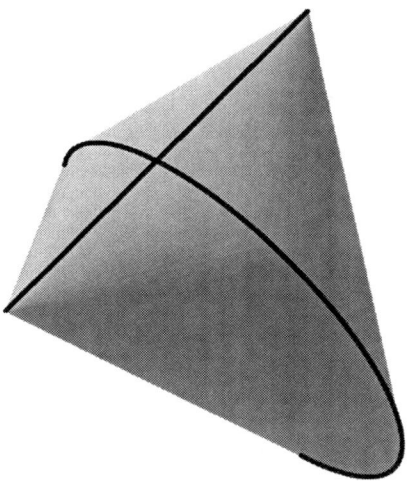

Figure 2.16 A convex set whose extreme (or exposed) points are not closed.

use the basic separation theorem (2.1.21) to find $x \in E$ so that $\sup_C x < \langle x, 0 \rangle = 0$, and conclude (2.4.13) has a solution.

(b) This can also be shown using the basic separation theorem (2.1.21) and the fact that finitely generated cones are closed by Exercise 2.4.11. An alternative analytic approach to both (a) and (b) was given by Hiriart-Urruty [254]; see also [95, Section 2.2]. □

We turn to the proofs of several foundational results discussed in the first chapter.

2.4.11 (Carathéodory's theorem).* Suppose $\{a_i : i \in I\}$ is a finite set of points in E. For any subset J of I, define the cone

$$C_J := \left\{ \sum_{i \in J} \mu_i a_i : \mu_i \in [0, \infty), \ i \in J \right\}.$$

(a) Prove the cone C_I is the union of those cones C_J for which the set $\{a_j : j \in J\}$ is linearly independent. Furthermore, prove directly that any such cone C_J is closed.
(b) Deduce that any finitely generated cone is closed.
(c) If the point x lies in $\mathrm{conv}\{a_i : i \in I\}$ prove that in fact there is a subset $J \subset I$ of size at most $1 + \dim E$ such that $x \in \mathrm{conv}\{a_i : i \in J\}$.
(d) Use part (c) to prove that the convex hull of a compact subset of E is compact.

2.4.12.* Prove Radon's theorem (1.2.3).

Hint. First, find real numbers $\{a_i\}_{i=1}^{n+2}$ not all 0 so that

$$\sum_{i=1}^{n+2} a_i x_i = 0 \quad \text{and} \quad \sum_{i=1}^{n+2} a_i = 0. \tag{2.4.16}$$

Indeed, the collection $\{x_i - x_1\}_{i=2}^{n+2}$ is linearly dependent in $\mathbb{R}^n$, hence find $\{a_i\}_{i=2}^{n+2}$ not all 0 so that $\sum_{i=2}^{n+2} a_i(x_i - x_1) = 0$, and set $a_1 = -\sum_{i=2}^{n+2} a_i$. Then let $I_1 = \{i : a_i > 0\}$ and $I_2 = \{i : a_i \leq 0\}$, and let C_1 and C_2 be as in the statement of the theorem. Let $a = \sum_{i \in I_1} a_i$ and let $\bar{x} = \sum_{i \in I_1} \frac{a_i}{a} x_i$. Then $\bar{x} \in C_1$, and it follows from (2.4.16) that $\bar{x} = \sum_{i \in I_2} -\frac{a_i}{a} x_i$ and $\sum_{i \in I_2} -\frac{a_i}{a} = 1$, and so $\bar{x} \in C_2$. □

2.4.13.* Prove Helly's theorem (1.2.4).

Hint. First show the case when I is finite (in this case we need not assume the sets C_i are closed and bounded). If $|I| = n + 1$, the result is trivial, so suppose $|I| = n + 2$ and we have sets $C_1, C_2, \ldots, C_{n+2}$ so that every subcollection of $n + 1$ sets has nonempty intersection. Choose $\bar{x}_i \in \bigcap_{j \in I, j \neq i} C_j$. If $\bar{x}_{j_1} = \bar{x}_{j_2}$ for some $j_1 \neq j_2$, then $\bar{x}_{j_1} \in \bigcap_{i \in I} C_i$. If the $\bar{x}_i's$ are all distinct use Radon's theorem (1.2.3) to get a point in the intersection.

Now suppose $|I| = k > n + 2$, and the assertion is true whenever $|I| \leq k - 1$ the above argument shows every subcollection of $n + 2$ sets will have nonempty intersection. Replace C_1 and C_2 with $C_1 \cap C_2$. Then every collection of $n + 1$ sets will have nonempty intersection, and use the induction hypothesis.

For an infinite collection of sets, use the finite intersection property for compact sets. □

Note. Helly's theorem is valid for much weaker assumptions on the closed convex sets C_i; in fact the conclusion remains valid if the C_i have no common direction of recession; see [369, p. 191ff] for this and more.

2.4.14 (Kirchberger's theorem). Let A and B be finite sets in a Euclidean space of dimension n. Use Helly's theorem (1.2.4) to show that *if for every finite subset $F \subseteq A \cup B$ with cardinality not exceeding $n + 2$ the sets $F \cap A$ and $F \cap B$ can be strictly separated, then A and B can be strictly separated.*

2.4.15 (Shapley–Folkman theorem).* Suppose $\{S_i\}_{i \in I}$ is a finite collection of nonempty sets in $\mathbb{R}^n$, and let $S := \sum_{i \in I} S_i$. Then every element $x \in \text{conv } S$ can be written as $x = \sum_{i \in I} x_i$ where $x_i \in \text{conv } S_i$ for each $i \in I$ and, moreover, $x_i \in S_i$ for all except at most n indices.

Hint. As in [448] consider writing the convexity conditions for x as positivity requirements in $\mathbb{R}^{m+n}$ where m is the cardinality of I. □

We now return to the Aumann convexity theorem also highlighted in Chapter 1.

2.4.16 (Aumann convexity theorem (weak form)).* If E is finite-dimensional and μ is a nonatomic probability measure then

$$\int_T \Omega = \text{conv} \int_T \Omega.$$

Hint. (a) Recall that a measure is *nonatomic* if every set of strictly positive measurable measure contains a subset of strictly smaller positive measure. First establish that every nonatomic probability measure μ satisfies the following *Darboux property*: for each measurable set of A with $\mu(A) > 0$ and each $0 < \alpha < 1$ there is a measurable subset $B \subset A$ with $\mu(B) = \alpha \mu(A)$.

(b) Now suppose $x_1 = \int_T f_1(t) \, d\mu$ and $x_2 = \int_T f_2(t) \, d\mu$. Then fix x on the line segment between these two points. Consider $\Omega := \{f_1, f_2\}$ and observe that $x \in \text{conv} \int_T \Omega(t) \, d\mu$. Use part (a) to partition T into $2n$ subsets $\{A_i\}$ each of measure $\mu(T)/(2n)$. Then write $x \in \text{conv} \sum_{i=1}^{2n} \int_{A_i} \Omega$, and apply the Shapley–Folkman theorem to write $x \in \sum_{i \notin I} \int_{A_i} \Omega + \text{conv} \sum_{i \in I} \int_{A_i} \Omega$, for some subset I with *exactly* n members. Hence

$$x \in \int_{S_1} \Omega + \text{conv} \sum_{i \in I} \int_{T \setminus S_1} \Omega,$$

with $\mu(S_1) = \mu(T)/2$.

(c) Repeat this argument inductively for each $m = 1, 2, \ldots$ to obtain $x_i \in \int_{S_i} \Omega$, and $r_m \in \text{conv} \int_{T \setminus \bigcup_{i=1}^m S_i} \Omega$ with $\mu(S_i) = \mu(T)/2^i$ for $1 \le i \le m$ and

$$x = \sum_{i=1}^m x_i + r_m.$$

Now take limits, observing that $\mu\{T \setminus \bigcup_{i=1}^m S_i\} \to_m 0$, to conclude that $x \in \int_T \Omega$. □

2.4.17. Suppose (T, Σ, μ) is a probability space while E is finite-dimensional and $\Omega: T \to E$ is measurable with closed nonempty images. Show that $\int_T \Omega(t) \, d\mu$ is nonempty if and only if $\sigma(t) := \inf \|\Omega(t)\|$ is integrable.

2.4.18. Suppose (T, Σ, μ) is a nonatomic probability space while E is finite-dimensional and $\Omega: T \to E$ is measurable with closed nonempty images. Show that

$$\text{conv} \int_T \Omega(t) \, d\mu \neq \int_T \text{conv} \, \Omega(t) \, d\mu$$

is possible.

2.4.19 (Primal solutions from dual solutions). Suppose the conditions for the Fenchel duality theorem (2.3.4) hold, and additionally that the functions f and g are lsc.

(a) Prove that if a point $\bar{\phi} \in Y$ is an optimal dual solution then the point $\bar{x} \in E$ is optimal for the primal problem if and only if it satisfies the two conditions $\bar{x} \in \partial f^*(A^*\bar{\phi})$ and $A\bar{x} \in \partial g^*(-\bar{\phi})$.

(b) Deduce that if f^* is differentiable at the point $A^*\bar{\phi}$ then the only possible primal optimal solution is $\bar{x} = \nabla f^*(A^*\bar{\phi})$.

(c) Interpret this in the context of linear programming as say in (3.3.1).

2.4.20 (Pshenichnii–Rockafellar conditions [357]).* Suppose the convex set C in E satisfies the condition $C \cap \text{cont} f \neq \emptyset$ (or the condition $\text{int} \, C \cap \text{dom} f \neq \emptyset$), where f is convex and bounded below on C. Use the sandwich theorem (2.4.1) to prove there is an affine function $\alpha \le f$ with $\inf_C f = \inf_C \alpha$. Deduce that a point $\bar{x}$ minimizes f on C if and only if it satisfies $0 \in \partial f(\bar{x}) + N_C(\bar{x})$.

Apply this to the following two cases:

(a) C a single point $\{x^0\} \subset E$;
(b) C a polyhedron $\{x : Ax \le b\}$, where $b \in \mathbb{R}^n$ and $A : E \to \mathbb{R}^n$ is linear.

2.4.21 (von Neumann's minimax theorem). Suppose Y is a Euclidean space. Suppose that the sets $C \subset E$ and $D \subset Y$ are nonempty and convex with D closed and that the map $A : E \to Y$ is linear.

(a) By considering the Fenchel problem

$$\inf_{x \in E} \{\delta_C(x) + \delta_D^*(Ax)\}$$

prove $\inf_{x \in C} \sup_{y \in D} \langle y, Ax \rangle = \max_{y \in D} \inf_{x \in C} \langle y, Ax \rangle$, where the max is attained if finite under the assumption

$$0 \in \text{core}(\text{dom } \delta_D^* - AC). \tag{2.4.17}$$

(b) Prove property (2.4.17) holds in either of the two cases
   (i) D is bounded, or
   (ii) A is surjective and $0 \in \text{int } C$ (you may use the open mapping theorem (Exercise 2.1.16)).

(c) Suppose both C and D are compact. Prove

$$\min_{x \in C} \max_{y \in D} \langle y, Ax \rangle = \max_{y \in D} \min_{x \in C} \langle y, Ax \rangle.$$

2.4.22 (Self-dual cones).* Prove Proposition 2.4.9.

Hint. Recall that $\langle X, Y \rangle := \text{tr}(XY)$ defines an inner product on the space of n by n symmetric matrices. This space will be used extensively in Section 3.2. See also [95] for further information. □

2.4.23.* Use the basic separation theorem (2.1.21) to prove Theorem 2.4.11.

2.4.24.* Suppose $C \subset E$ is closed and convex, and that $D \subset E$ is compact and convex. Show that the sets $D - C$ and $D + C$ are closed and convex.

2.4.25 (Sums of closed cones).*

(a) Prove that any cones $H, K \subset E$ satisfy $(H + K)^- = H^- \cap K^-$.
(b) Deduce that if H and K are closed convex cones then they satisfy $(H \cap K)^- = \text{cl}(H^- + K^-)$, and prove that the closure can be omitted under the condition $K \cap \text{int } H \neq \emptyset$.

In $\mathbb{R}^3$ define the sets

$$H := \{x : x_1^2 + x_2^2 \leq x_3^2, \, x_3 \leq 0\}, \text{ and}$$

$$K := \{x : x_2 = -x_3\}.$$

(c) Prove that H and K are closed convex cones.
(d) Calculate the polar cones H^-, K^-, and $(H \cap K)^-$.
(e) Prove $(1, 1, 1) \in (H \cap K)^- \setminus (H^- + K^-)$, and deduce that the sum of two closed convex cones is not necessarily closed.

2.4.26 (Pointed cones and bases).* Consider a closed convex cone K in E. A *base* for K is a convex set C with $0 \notin \operatorname{cl} C$ and $K = \mathbb{R}_+ C$. Using Exercise 2.4.25, prove the following are equivalent.

(a) K is pointed.
(b) $\operatorname{cl}(K^- - K^-) = E$.
(c) $K^- - K^- = E$.
(d) K^- has nonempty interior.
(e) There exists a vector $y \in E$ a real $\varepsilon > 0$ with $\langle y, x \rangle \geq \varepsilon \|x\|$ for all $x \in K$.
(f) K has a bounded base.

Hint. Use Exercise 2.4.25 to show the cycle of implications (a) $\Rightarrow$ (b) $\Rightarrow$ (c) $\Rightarrow$ (d) $\Rightarrow$ (e) $\Rightarrow$ (f) $\Rightarrow$ (a). □

2.4.27.* With the help of Exercise 2.4.26, prove the Theorem 2.4.12.

2.4.28 (Cone convex functions). A function $F : A \subset E \to Y$ is said to be convex with respect to a convex cone S (*S-convex*) if

$$tF(x) + (1-t)F(y) - F(tx + (1-t)y) \in S$$

for each x, y in A and $0 < t < 1$. We write $x \geq y$ or $x \geq_S y$ when $x - y \in S$. Show the following.

(a) Every function is Y-convex and every affine function is 0-convex.
(b) When S is closed then F is S-convex if and only if $s^+ F$ is convex on A for each s^+ in S^+.
(c) Suppose S has a closed bounded base and A is open. Then F is Fréchet (resp. Gâteaux) differentiable at $a \in A$ if and only if $s^+ F$ is Fréchet (resp. Gâteaux) differentiable at $a \in A$ for some s^+ in the norm-interior of S^+.
(d) Suppose $f : E \times \Omega \to \mathbb{R}$ is convex in $x \in E$ and continuous for $t \in \Omega$ a compact topological space. Then the formula $F_f(x)(t) := f(x, t)$ induces a convex operator $F_f : E \to \mathbb{C}(\Omega)$ in the pointwise-order. This construction has many variations and applications, as it allows one to study parameterized convex functions as a single operator.

2.4.29 (Composition of convex functions). The composition result of Lemma 2.1.8 has many extensions. Suppose that the set $C \subset E$ is convex and that the functions $f_1, f_2, \ldots, f_n : C \to \mathbb{R}$ are convex, and define a function $f : C \to \mathbb{R}^n$ with components f_i. Suppose further that $D \supset f(C)$ is convex and that the function $g : D \to (-\infty, +\infty]$ is convex and *isotone*: any points $y \leq z$ in D satisfy $g(y) \leq g(z)$. Prove that the composition $g \circ f$ is convex. Generalize this result to the composition of a convex and S-isotone function with a S-convex function.

2.4.30. Let W be the *Lambert W function*; that is, the inverse of $x \mapsto x \exp(x)$, which is concave with real domain of $(-1/e, \infty)$. A remarkable application of the Lambert function to divergence estimates is given in [228].

(a) Show that the conjugate of

$$x \mapsto \frac{x^2}{2} + x \log x$$

is
$$y \mapsto W(e^y) + \frac{W^2(e^y)}{2}.$$

(b) Show that the conjugate of $(r,s) \mapsto r^2 e^{s^2}$ with $|s| \leq 1/\sqrt{2}, r \in \mathbb{R}$ is
$$(x,y) \mapsto \frac{|y|}{\sqrt{2}} \frac{1 - W(-8y^2/x^4)}{\sqrt{-W(-8y^2/x^4)}},$$
where implicitly $y^2 \leq ex^4$.

2.4.31 (Conjugates of compositions). Consider the composition $g \circ f$ of a nondecreasing convex function $g : \mathbb{R} \to (-\infty, +\infty]$ with a convex function $f : E \to (-\infty, +\infty]$. Assume there is a point $\hat{x}$ in E satisfying $f(\hat{x}) \in \operatorname{int}(\operatorname{dom} g)$. Use the Lagrangian duality (Exercise 2.3.10) to prove the formula, for ϕ in E,
$$(g \circ f)^*(\phi) = \inf_{t \in \mathbb{R}_+} \left\{ g^*(t) + tf^*\left(\frac{\phi}{t}\right) \right\},$$
where we interpret $g(+\infty) = +\infty$ and
$$0f^*\left(\frac{\phi}{0}\right) = \delta^*_{\operatorname{dom} f}(\phi).$$

Specialize this result to $\exp \circ f$ and then to $\exp \circ \exp$.

Hint. It helps to express the final conjugate in terms of the Lambert W function defined in Exercise 2.4.30. □

2.4.32 (Hahn–Banach theorem in ordered spaces).** The Hahn–Banach theorem in extension form has an extension to the case when the mappings take range in an ordered vector space and convexity is in the sense of Exercise 2.4.28. Show that the following are equivalent.

(a) Let a linear space X and a subspace $Z \subset X$ be given, along with a cone sublinear operator $p : X \to Y$ (i.e. convex and positively homogeneous) and a linear operator $T : Z \to Y$ such that $Tz \leq_S p(z)$ for all $z \in Z$. Then there exists a linear operator $\overline{T}$ such that $\overline{T}Z = T$ and $\overline{T}x \leq p(x)$ for all $x \in X$.
(b) Suppose f, g are S-convex functions which take values in $Y \cup \{\infty\}$ and that $f \geq_S -g$. while one of f and g is continuous at a point at which the other is finite. Then there is a linear operator $T : X \to Y$ and a vector $b \in y$ such that $f(x) \geq_S Tx + b \geq -g(x)$ for all $x \in X$.
(c) (Y, S) is an order complete vector space.

Hint. For (c) implies (b) implies (a) imitate the real case as in [62]. To see that (a) implies (c) we refer the reader to Exercise 4.1.52 and [64]. □

2.4.33 (Subgradients in ordered spaces).**

(a) Show that a cone-convex function has a nonempty subgradient at any point of continuity as soon as the order cone has the *monotone net property* (MNP): every order-increasing net with an upper bound has a supremum.
(b) Show in finite dimensions that the MNP holds if it holds for sequences and hence for every closed pointed convex cone; but that any order complete generating cone induces a lattice order. Hence, subgradients exist for orders like the semidefinite matrices (the Loewner ordering) which are far from order complete.

Hint. To show existence of subgradients, mimic the proof we gave in the scalar case. □

2.5 Differentiability in measure and category

We begin with a classical result which is a special easy case of Rademacher's theorem given later in this section as Theorem 2.5.4.

Theorem 2.5.1. *Suppose $f : U \to \mathbb{R}$ is a convex function where U is an open convex subset of $\mathbb{R}^n$. Then f is differentiable almost everywhere on U with respect to Lebesgue measure on $\mathbb{R}^n$.*

Proof. Let $\{e_k\}_{k=1}^n$ represent the standard basis of $\mathbb{R}^n$. Then since f is convex and continuous on U, the one-sided derivative exists, see Proposition 2.1.17. Since

$$f'(x; e_k) = \lim_{m \to \infty} \frac{f(x + e_k/m) - f(x)}{1/m} \text{ for each } k,$$

it follows that $f'(x; e_k)$ is a Borel measurable function as a pointwise limit of continuous functions. Similarly, $f'(x; -e_k)$ is Borel measurable. Hence $E_k = \{x : f'(x; e_k) + f'(x; -e_k) > 0\}$ is measurable, and moreover, $x \in E_k$ if and only if $\frac{\partial}{\partial x_k} f(x)$ does not exist. Now consider $E_{k,m} = E_k \cap mB_E$ where $m \in \mathbb{N}$. Then denoting the Lebesgue measure by μ and using Fubini's theorem

$$\mu(E_{k,m}) = \int_{\mathbb{R}^n} \chi_{E_{k,m}} \, d\mu$$

$$= \int_{\mathbb{R}} \cdots \left(\int_{\mathbb{R}} \chi_{E_{k,m}} \, dx_k \right) dx_1 \cdots dx_{k-1} \, dx_{k+1} \cdots dx_n = 0$$

because the inner integral is 0 since f is convex as a function of x_k (and thus differentiable except possibly a countable set of points by Theorem 2.1.2). Then $E_k = \bigcup_{m=1}^\infty E_{k,m}$ is a null set because it is a countable union of null sets. Thus the partial derivatives of f exist, except possibly on a null subset of U, and so by Theorem 2.2.1, f is differentiable almost everywhere on U. □

Corollary 2.5.2. *Suppose $f : U \to \mathbb{R}$ is convex where U is an open convex subset of E. Then f is differentiable precisely on the points of a dense G_δ-subset of U.*

Proof. Theorem 2.5.1 shows that f is differentiable on a dense set of U, and then Exercise 2.5.2 shows the set of differentiability points is a countable intersection of open sets. □

Subgradients of continuous convex functions may be reconstructed from their nearby derivatives as we now show. There is a more subtle formula (4.3.13) which applies at boundary points of the domain of a continuous convex function.

Corollary 2.5.3. *Suppose $f : U \to \mathbb{R}$ is a convex function where U is an open convex subset of $\mathbb{R}^n$ and $\bar{x} \in U$. Then*

$$\partial f(\bar{x}) = \mathrm{conv}\{y : y = \lim_n \nabla f(x_n), x_n \to \bar{x}, \nabla f(x_n) \text{ exists}\}.$$

Proof. The right-hand side is a compact convex set (call it C) as the convex hull of a closed bounded set; see Carathéodory's theorem (1.2.5). One can check the left-hand side contains the right-hand side. Suppose by way of contradiction that there exists $\phi \in \partial f(\bar{x}) \setminus C$. By the basic separation theorem (2.1.21), we choose $h_0 \in S_E$ so that $\phi(h_0) > \sup_{y \in C} \langle y, h_0 \rangle$. Now there is an open neighborhood, say V of h_0, and $M \in \mathbb{R}$ so that

$$\phi(h) > M > \sup_{y \in C} \langle y, h_0 \rangle \text{ for all } h \in V.$$

According to Theorem 2.5.1 we may choose $h_n \in V$ with $h_n \to h_0$ and $t_n \to 0^+$ so that $y_n = \nabla f(\bar{x} + t_n h_n)$ exists. Because f is locally Lipschitz on U (Theorem 2.1.12), the sequence (y_n) is bounded. Passing to a subsequence if necessary, we may assume $y_n \to \bar{y}$, and hence $\bar{y} \in C$. Now

$$\frac{f(\bar{x} + t_n h_n) - f(\bar{x})}{t_n} \geq \langle \phi, h_n \rangle > M.$$

Then

$$\frac{\langle y_n, \bar{x} \rangle - \langle y_n, \bar{x} + t_n h_n \rangle}{t_n} \leq \frac{f(\bar{x}) - f(\bar{x} + t_n h_n)}{t_n} < -M.$$

Thus $\langle y_n, -h_n \rangle < -M$. Consequently,

$$\langle \bar{y}, h_0 \rangle = \lim_n \langle y_n, h_n \rangle \geq M$$

which is a contradiction. □

The next theorem is an extension of Theorem 2.5.1 to the setting of locally Lipschitz functions. While it is not specifically about convex functions, we will present it in full because of its broader interest and its application to second-order derivatives of convex functions.

Theorem 2.5.4 (Rademacher). *Any locally Lipschitz map between Euclidean spaces is Fréchet differentiable almost everywhere.*

2.5 Differentiability in measure and category

Proof. Without loss of generality (Exercise 2.5.3), we can consider a locally Lipschitz function $f : \mathbb{R}^n \to \mathbb{R}$. In fact, we may as well further suppose that f has Lipschitz constant L throughout $\mathbb{R}^n$, by Exercise 2.3.16.

Fix a direction h in $\mathbb{R}^n$. For any $t \neq 0$, the function g_t defined on $\mathbb{R}^n$ by

$$g_t(x) := \frac{f(x+th) - f(x)}{t}$$

is continuous, and takes values in the interval $I = L\|h\|[-1, 1]$, by the Lipschitz property. Hence, for $k = 1, 2, \ldots$, the function $p_k : \mathbb{R}^n \to I$ defined by

$$p_k(x) := \sup_{0 < |t| < 1/k} g_t(x)$$

is lsc and therefore Borel measurable. Consequently, the *upper Dini derivative* $D_h^+ f : \mathbb{R}^n \to I$ defined by

$$D_h^+ f(x) := \limsup_{t \to 0} g_t(x) = \inf_{k \in \mathbb{N}} p_k(x)$$

is measurable, being the infimum of a sequence of measurable functions. Similarly, the *lower Dini derivative* $D_h^- f : \mathbb{R}^n \to I$ defined by

$$D_h^- f(x) := \liminf_{t \to 0} g_t(x)$$

is also measurable.

The subset of $\mathbb{R}^n$ where f is not differentiable along the direction h, namely

$$A_h = \{x \in \mathbb{R}^n : D_h^- f(x) < D_h^+ f(x)\},$$

is therefore also measurable. Given any point $x \in \mathbb{R}^n$, the function $t \mapsto f(x+th)$ is absolutely continuous (being Lipschitz), so the fundamental theorem of calculus implies this function is differentiable (or equivalently, $x + th \notin A_h$) almost everywhere on $\mathbb{R}$.

Consider the nonnegative measurable function $\phi : \mathbb{R}^n \times \mathbb{R} \to \mathbb{R}$ defined by $\phi(x, t) := \delta_{A_h}(x + th)$. By our observation above, for any fixed $x \in \mathbb{R}^n$ we know $\int_{\mathbb{R}} \phi(x, t)\, dt = 0$. Denoting Lebesgue measure on $\mathbb{R}^n$ by μ, Fubini's theorem shows

$$0 = \int_{\mathbb{R}^n} \left(\int_{\mathbb{R}} \phi(x, t)\, dt \right) d\mu$$

$$= \int_{\mathbb{R}} \left(\int_{\mathbb{R}^n} \phi(x, t)\, d\mu \right) dt = \int_{\mathbb{R}} \mu(A_h)\, dt$$

so the set A_h has measure zero. Consequently, we can define a measurable function $D_h f : \mathbb{R}^n \to \mathbb{R}$ having the property that $D_h f = D_h^+ f = D_h^- f$ almost everywhere.

Denote the standard basis vectors in $\mathbb{R}^n$ by $e_1, e_2, \ldots, e_n$. The function $G : \mathbb{R}^n \to \mathbb{R}^n$ with components defined almost everywhere by

$$G_i := D_{e_i} f = \frac{\partial f}{\partial x_i}$$

for each $i = 1, 2, \ldots, n$ is the only possible candidate for the derivative of f. As we shall show, a standard integration by parts argument establishes that it is the derivative.

Consider any continuously differentiable function $\psi : \mathbb{R}^n \to \mathbb{R}$ that is zero except on a bounded set. For our fixed direction h, if $t \neq 0$ we have

$$\int_{\mathbb{R}^n} g_t(x) \psi(x) \, d\mu = -\int_{\mathbb{R}^n} f(x) \frac{\psi(x+th) - \psi(x)}{t} \, d\mu.$$

As $t \to 0$, the Bounded convergence theorem applies, since both f and ψ are Lipschitz, so

$$\int_{\mathbb{R}^n} D_h f(x) \psi(x) \, d\mu = -\int_{\mathbb{R}^n} f(x) \langle \nabla \psi(x), h \rangle \, d\mu.$$

Setting $h = e_i$ in the above equation, multiplying by h_i, and adding over $i = 1, 2, \ldots, n$, where we now represent an arbitrary $h \in E$ as $h = \sum_{i=1}^n h_i e_i$, yields

$$\int_{\mathbb{R}^n} \langle h, G(x) \rangle \psi(x) \, d\mu = -\int_{\mathbb{R}^n} f(x) \langle \nabla \psi(x), h \rangle \, d\mu = \int_{\mathbb{R}^n} D_h f(x) \psi(x) \, d\mu.$$

Since ψ was arbitrary, we deduce $D_h f = \langle h, G \rangle$ almost everywhere.

Now extend the basis $e_1, e_2, \ldots, e_n$ to a dense sequence of unit vectors $\{h_k\}$ in the unit sphere $S_{\mathbb{R}^n}$. Define the set $A \subset \mathbb{R}^n$ to consist of those points where each function $D_{h_k} f$ is defined and equals $\langle h_k, G \rangle$. Our argument above shows A^c has measure zero. We aim to show, at each point $x \in A$, that f has Fréchet derivative $G(x)$.

Fix any $\varepsilon > 0$. For any $t \neq 0$, define a function $r_t : \mathbb{R}^n \to \mathbb{R}$ by

$$r_t(h) := \frac{f(x+th) - f(x)}{t} - \langle G(x), h \rangle.$$

It is easy to check that r_t has Lipschitz constant $2L$. Furthermore, for each $k = 1, 2, \ldots$, there exists $\delta_k > 0$ such that

$$|r_t(h_k)| < \frac{\varepsilon}{2} \quad \text{whenever } 0 < |t| < \delta_k.$$

Since the sphere $S_{\mathbb{R}^n}$ is compact, there is an integer M such that

$$S_{\mathbb{R}^n} \subset \bigcup_{k=1}^M \left(h_k + \frac{\varepsilon}{4L} B_{\mathbb{R}^n} \right).$$

2.5 Differentiability in measure and category

If we define $\delta = \min\{\delta_1, \delta_2, \ldots, \delta_M\} > 0$, we then have

$$|r_t(h_k)| < \frac{\varepsilon}{2} \text{ whenever } 0 < |t| < \delta, \ k = 1, 2 \ldots, M.$$

Finally, consider any unit vector h. For some positive integer $k \leq M$ we know $\|h - h_k\| \leq \varepsilon/4L$, so whenever $0 < |t| < \delta$ we have

$$|r_t(h)| \leq |r_t(h) - r_t(h_k)| + |r_t(h_k)| \leq 2L\frac{\varepsilon}{4L} + \frac{\varepsilon}{2} = \varepsilon.$$

Hence $G(x)$ is the Fréchet derivative of f at x, as we claimed. □

Remark 2.5.5. (a) In contrast to the convex case, the points of differentiability of a Lipschitz function on the real line need not contain a dense G_δ-set. Indeed, following [373, p. 97], we let $S \subset \mathbb{R}$ be such that S is measurable with $\mu(I \cap S) > 0$ and $\mu(I \cap (\mathbb{R} \setminus S)) > 0$ for each nonempty open interval I in $\mathbb{R}$. Let

$$f(x) := \int_0^x (\chi_S - \chi_{\mathbb{R} \setminus S})(t) \, dt.$$

Then f has Lipschitz constant 1, but the points of differentiability of S is a set of first category; see Exercise 2.5.8.
(b) The paper [329] presents a simple approach to Rademacher's theorem.
(c) Priess's paper [354] shows the disturbing result that for $n > 1$ there is a null dense G_δ-set in $\mathbb{R}^n$ on which every Lipschitz function from $\mathbb{R}^n$ to $\mathbb{R}$ is densely differentiable.

Some infinite-dimensional versions of results in this section and the next will be discussed in Section 4.6.

Exercises and further results

2.5.1.* Suppose $f : E \to \mathbb{R}$ is convex. Show that f is differentiable at $x \in E$ if and only if the symmetric error

$$\lim_{\|h\| \to 0} \frac{f(x+h) + f(x-h) - 2f(x)}{\|h\|} = 0.$$

Hint. For the 'if' assertion assume $\phi, \Lambda \in \partial f(x)$, and that $\phi(h_0) > \Lambda(h_0)$ for some $h_0 \in S_E$. Then for $t > 0$, $f(x + th_0) + f(x - th_0) - 2f(x) \geq t(\phi(h_0) - \Lambda(h_0))$. □

2.5.2.* Suppose $f : U \to \mathbb{R}$ is convex where U is a nonempty open convex subset of E. Show that the points at which f is differentiable is a G_δ-subset of E, that is a countable intersection of open sets.
Hint. Consider, for example, the sets

$$G_{n,m} := \left\{ x \in U : \sup_{\|h\|} \leq \frac{1}{m} f(x+h) + f(x-h) - 2f(x) < \frac{1}{mn} \right\}.$$

Use Exercise 2.5.1 to show that f is differentiable at x if and only if $x \in \bigcap_n O_n$ where $O_n := \bigcup_{m \geq n} G_{nm}$. □

2.5.3.* Assuming Rademacher's theorem with range $\mathbb{R}$, prove the general vectorial version.

2.5.4. Suppose $f : \mathbb{R}^n \to \mathbb{R}$ is locally Lipschitz and $\bar{x} \in \mathbb{R}^n$. Suppose there exists $y \in \mathbb{R}^n$ and dense set of directions $\{h_k\}_{k=1}^\infty \subset S_{\mathbb{R}^n}$ such that

$$\lim_{t \to 0} \frac{f(\bar{x} + th_k) - f(\bar{x})}{t} = \langle y, h_k \rangle \quad \text{for each } k \in \mathbb{N}.$$

Show that f is Fréchet differentiable at $\bar{x}$ and $\nabla f(\bar{x}) = y$.

Hint. This is another proof of the last part of the proof of Rademacher's theorem. □

2.5.5. Let $C \subset E$ be a closed convex set with nonempty interior. Show that the boundary of C is a Lebesgue-null set.

Hint. Since every boundary point of C is a support point, $d_C(\cdot)$ is not differentiable at boundary points; now apply Theorem 2.5.1. □

2.5.6 (The Clarke derivative).** The *Clarke directional derivative* is defined by

$$f^\circ(x; h) := \limsup_{y \to x, t \downarrow 0} \frac{f(y + th) - f(y)}{t};$$

and the *Clarke subdifferential* is defined by

$$\partial_\circ f(x) := \{\phi \in E : \langle \phi, h \rangle \leq f^\circ(x; h) \text{ for all } h \in E\}.$$

Let $f : E \to \mathbb{R}$ have a Lipschitz constant K on a open neighborhood of $x \in E$.

(a) Show that $f^\circ(x; \cdot)$ is sublinear and $f^\circ(x; \cdot) \leq K \|\cdot\|$.
(b) Show that $\partial_\circ f(x)$ is a nonempty compact and convex subset of $K B_E$ and

$$f^\circ(x; h) = \max\{\langle \phi, h \rangle : \phi \in \partial_\circ f(x)\}, \text{ for any direction } h \in E.$$

(c) Show that $0 \in \partial_\circ f(x)$ if f has a local minimum at x.
(d) Use Rademacher's theorem (2.5.4) to show that

$$\partial_\circ f(x) = \text{conv}\{\lim_r \nabla f(x^r) : x^r \to x, \nabla f(x^r) \text{ exists}\}.$$

(e) Compute $\partial_\circ f(x)$ for f as given in Remark 2.5.5(a); notice this shows the converse of (c) fails very badly, and it also shows the failure of integration to recover f from its Clarke subdifferential.
(f) If, additionally, f is convex, show that $\partial f(x) = \partial_\circ f(x)$.

Hint. See [95, Chapter 6] a brief introduction to generalized derivatives which includes the above and other information. See [156] for a more comprehensive exposition on the Clarke derivative and its many applications. □

2.5.7 ([34, Theorem 2.6]). Suppose U is a convex nonempty open set in $\mathbb{R}^N$ and $g : U \to \mathbb{R}$ is continuous. Let $A := \{x \in U : \nabla g(x) \text{ exists}\}$. Show that the following are equivalent.

(a) g is convex.
(b) $U \setminus A$ is a set of measure zero, and $\nabla g(x)(y - x) \leq g(y) - g(x)$, for all $x \in A$ and $y \in U$.
(c) A is dense in U, and $\nabla g(x)(y - x) \leq g(y) - g(x)$, for all $x \in A$ and $y \in U$.

Hint. (a)$\Rightarrow$(b): g is almost everywhere differentiable by Theorem 2.5.1. On the other hand, the subgradient inequality holds on U. Together, these facts imply (b).

(b)$\Rightarrow$(c): trivial.

(c)$\Rightarrow$(a): Fix u, v in U, $u \neq v$, and $t \in (0, 1)$. It is not hard to see that there exist sequences (u_n) in U, (v_n) in U, (t_n) in $(0,1)$ with $u_n \to u$, $v_n \to v$, $t_n \to t$, and $x_n := t_n u_n + (1 - t_n)v_n \in A$, for every n. By assumption, $\nabla g(x_n)(u_n - x_n) \leq g(u_n) - g(x_n)$ and $\nabla g(x_n)(v_n - x_n) \leq g(v_n) - g(x_n)$. Equivalently, $(1 - t_n)\nabla g(x_n)(v_n - u_n) \leq g(u_n) - g(x_n)$ and $t_n \nabla g(x_n)(u_n - v_n) \leq g(v_n) - g(x_n)$. Multiply the former inequality by t_n, the latter by $1 - t_n$, and add. We obtain that $g(x_n) \leq t_n g(u_n) + (1 - t_n)g(v_n)$. Let n tend to $+\infty$ to deduce that $g(tu + (1 - t)v) \leq tg(u) + (1 - t)g(v)$. The convexity of g follows and the proof is complete. $\square$

2.5.8.* Show that the points where the function f in Remark 2.5.5 is differentiable is a set of first category.

Hint. Let $f_n(x) := n[f(x+1/n) - f(x)]$. Let $G := \{x : f'(x) \text{ exists}\}$; then $f_n(x) \to f'(x)$ for $x \in G$. Let $F_n := \{x : |f_j(x) - f_k(x)| \leq 1/2$ for all $j, k \geq n\}$. Then $\bigcup F_n \supset G$. Suppose F_{n_0} contains an open interval I for some $n_0 \in \mathbb{N}$. Then $|f_{n_0}(x) - f'(x)| \leq 1/2$ almost everywhere on I. Conclude that $f_{n_0} \geq 1/2$ on a dense subset of I, and $f_{n_0} \leq -1/2$ on a dense subset of I to contradict the continuity of f_{n_0}. $\square$

2.6 Second-order differentiability

The notions of twice Gâteaux and twice Fréchet differentiability were introduced earlier in this chapter. In this section we will connect those notions with some other and more general approaches to second-order differentiability. First we introduce second-order Taylor expansions.

Let $U \subset E$ be an open convex set and let $f : U \to \mathbb{R}$ be a convex function. Then f is said to have a *weak second-order Taylor expansion* at $x \in U$ if there exists $y^* \in \partial f(x)$ and a matrix $A : E \to E$ such that f has a representation of the form

$$f(x + th) = f(x) + t\langle y^*, h\rangle + \frac{t^2}{2}\langle Ah, h\rangle + o(t^2) \quad (t \to 0), \tag{2.6.1}$$

for each $h \in E$ (with the order term possibly depending on h); if more arduously,

$$f(x + h) = f(x) + \langle y^*, h\rangle + \frac{1}{2}\langle Ah, h\rangle + o(\|h\|^2) \quad (\|h\| \to 0). \tag{2.6.2}$$

for $h \in E$, we say that f has a *strong second-order Taylor expansion* at x.

On replacing A with $(A + A^T)/2$, we may assume without loss of generality the matrix A is symmetric in the preceding Taylor expansions. Also, once we know A is symmetric, then it is uniquely determined by the values $\langle Ah, h\rangle$ (Exercise 2.6.1).

Furthermore, we will say ∇f has a *generalized Fréchet derivative at $x \in U$* and f has a *generalized second-order Fréchet derivative at x* if there exist $\phi \in E$ and a matrix $A : E \to E$ such that

$$\partial f(x+h) \subset \phi + Ah + o(\|h\|)B_E.$$

Analogously, we will say ∇f has a *generalized Gâteaux derivative at $x \in U$* and f has a *generalized second-order Gâteaux derivative at x* if

$$\partial f(x+th) \subset \phi + A(th) + o(|t|)B_E \text{ for each } h \in S_E.$$

In either case above, $\phi = \nabla f(x)$, and the matrix A must be unique; see Exercise 2.6.4 for these facts and more. Therefore, with the understanding we are speaking of a generalized derivative, we will write $\nabla^2 f(x) = A$.

The generalized notions of second-order Fréchet and Gâteaux differentiability are formally different, but they are in fact equivalent for convex functions; this and more is recorded in the following theorem.

Theorem 2.6.1 (Equivalence of second-order properties). *Let $U \subset E$ be an open convex set, and let $f : U \to \mathbb{R}$ be convex, and let $x \in U$. Then the following are equivalent:*

(a) f has a generalized second-order Fréchet derivative at x.
(b) f has a generalized second-order Gâteaux derivative at x.
(c) f has a strong second-order Taylor expansion at x.
(d) f has a weak second-order Taylor expansion at x.

Moreover, in (a) and (b) the matrix $\nabla^2 f(x)$ representing the generalized second-derivative is symmetric. Additionally, if ∂f is single-valued on a neighborhood of x, then each of the above conditions is equivalent to f being twice Fréchet differentiable at x in the usual sense.

Before proving this theorem, we introduce various notions of difference quotients. The *difference quotient* of f is defined by

$$\Delta_t f(x) : h \mapsto \frac{f(x+th) - f(x)}{t}; \qquad (2.6.3)$$

and the *second-order difference quotient* of f by

$$\Delta_t^2 f(x) : h \mapsto \frac{f(x+th) - f(x) - t\langle \nabla f(x), h \rangle}{\frac{1}{2}t^2}. \qquad (2.6.4)$$

Notice that if f is Fréchet differentiable at x, then $\lim_{t \to 0} \Delta_t f(x) \to \nabla f(x)$ uniformly on bounded sets; if f has a strong second-order Taylor expansion at x, then $\Delta_t^2 f(x) \to \langle Ah, h \rangle$ uniformly on bounded sets. In order to study extended second-derivatives, it is

2.6 Second-order differentiability

helpful to consider the following set-valued difference quotients of the subdifferential

$$\Delta_t[\partial f](x) : h \mapsto \frac{\partial f(x+th) - \nabla f(x)}{t}. \tag{2.6.5}$$

Two propositions will ease our path. First:

Proposition 2.6.2. *The convex function f has a generalized second-order Fréchet derivative at x if and only if for every bounded set $W \subset E$ and every $\varepsilon > 0$, there exists $\delta > 0$ such that*

$$\Delta_t[\partial f](x)(h) - Ah \subset \varepsilon B_E \text{ for all } h \in W, t \in (0, \delta).$$

Proof. This is left as Exercise 2.6.5. □

Second, and quite beautifully:

Proposition 2.6.3. *For any $t > 0$, the function $\Delta_t f(x) : \mathbb{R}^n \to (-\infty, +\infty]$ is closed, proper, convex and nonnegative. Moreover,*

$$\partial \left[\frac{1}{2} \Delta_t^2 f(x) \right] = \Delta_t[\partial f](x).$$

Proof. This is left as Exercise 2.6.6. □

Proof. (Theorem 2.6.1) First we prove (c) $\Rightarrow$ (a): Suppose that f has a second-order Taylor expansion with A symmetric. Let $q(h) := \frac{1}{2}\langle Ah, h\rangle$. Then q is differentiable with $\nabla q(h) = Ah$. According to Proposition 2.6.3 the functions $q_t := \frac{1}{2}\Delta_t^2 f(x)$ are convex and converge pointwise to q as $t \to 0^+$, and thus the convergence is uniform on bounded sets (Exercise 2.1.22). Consequently q is convex, and hence A is positive semidefinite.

Because q_t converges uniformly to q on bounded sets, according to Exercise 2.2.22(c) one has for every bounded set $W \subset E$ and every $\varepsilon > 0$, the existence of $\delta > 0$ so that

$$\partial q_t(h) \subset \partial q(h) + \varepsilon B_E \text{ whenever } t \in (0, \delta), \, w \in W,$$

where additionally δ can be chosen small enough that $\partial q_t(h) \neq \emptyset$ in these circumstances. According to Proposition 2.6.3, $\partial q_t(h) = \Delta_t[\partial f](h)$. Noting that $\partial q(h) = \{Ah\}$, it now follows from Proposition 2.6.2 that f has a generalized second-order Fréchet derivative at x with $\nabla^2 f(x) = A$.

(a) $\Rightarrow$ (c): Suppose that f has a generalized second-order Fréchet derivative at x. Given $\varepsilon > 0$, we choose $\delta > 0$ so that f is Lipschitz on $B_\delta(x)$ and

$$\partial f(x + th) \subset \nabla f(x) + A(th) + \varepsilon |t| B_E \text{ for } |t| \leq \delta, h \in S_E. \tag{2.6.6}$$

For $h \in S_E$ fixed, let $g_1(t) := f(x + th)$ and

$$g_2(t) := f(x) + t\langle \nabla f(x), h\rangle + \frac{1}{2}t^2 \langle Ah, h\rangle.$$

Then it is easy to check that $g_1'(t) = \langle \phi_t, h \rangle$ almost everywhere for $0 \leq t \leq \delta$, where $\phi_t \in \partial f(x+th)$ and $g_2'(t) = \langle \nabla f(x), h \rangle + t\langle Ah, h \rangle$. Thus using (2.6.6) we have

$$|g_1'(t) - g_2'(t)| \leq \varepsilon t, \quad a.e. \text{ for } 0 \leq t \leq \delta.$$

Therefore, by the fundamental theorem of calculus (for absolutely continuous functions), for $0 < t \leq \delta$, one has

$$|g_1(t) - g_2(t)| = \left| \int_0^t (g_1'(u) - g_2'(u)) \, du \right| \leq \int_0^t \varepsilon u \, du = \frac{1}{2} \varepsilon t^2.$$

Since $\varepsilon > 0$ was arbitrary and $\delta > 0$ was independent of $h \in S_E$, it follows that f has a strong Taylor expansion at x with matrix A as desired.

It is clear that (a) $\Rightarrow$ (b), and the proof of (a) $\Rightarrow$ (c) also works for (b) $\Rightarrow$ (d), noting now that the choice of δ depends on the direction h. The equivalence of (a) through (d) is then completed in Exercise 2.6.3 which outlines a proof of (d) $\Rightarrow$ (c).

The symmetry of the matrix $\nabla^2 f(x)$ is deduced as follows. First, (a) $\Rightarrow$ (c) shows that the Taylor expansion holds with the matrix $A = \nabla^2 f(x)$. Then the Taylor expansion holds with the symmetrization of $A_s = (A + A^T)/2$. Then the proof of (c) $\Rightarrow$ (a) shows that $\nabla^2 f(x) = A_s$. By the uniqueness of the second derivative matrix, we conclude that $\nabla^2 f(x)$ is symmetric. Finally, let us note that when ∂f is single-valued on a neighborhood of x, then f is differentiable on a neighborhood of x, and so the generalized notions of second-order differentiability reduce to the usual notions. □

Now, Alexandrov's famous theorem may be presented.

Theorem 2.6.4 (Alexandrov). *Every convex function $f : \mathbb{R}^n \to \mathbb{R}$ has a second-order Taylor expansion at almost every $x \in \mathbb{R}^n$, or equivealently, f has a generalied second-order Fréchet derivative almost everywhere.*

Proof. According to Theorem 2.1.12, f is continuous; consequently $\text{dom}(\partial f) = \mathbb{R}^n$ by the max formula (2.1.19). Letting

$$E_1 := \{x \in \mathbb{R}^n : \nabla f(x) \text{ exists}\}$$

we know that $\mathbb{R}^n \setminus E_1$ is a set of measure 0 (Theorem 2.5.1). Let $J := (I + \partial f)^{-1}$. Then J is a nonexpansive map of $\mathbb{R}^n$ into itself (this will be shown later in Exercise 3.5.8). Next consider the set

$$E_2 := \{J(x) : J \text{ is differentiable at } x \text{ and } \nabla J(x) \text{ is nonsingular}\}.$$

Rademacher's theorem (2.5.4) ensures that J is differentiable almost everywhere. Moreover, because J is Lipschitz, we may apply the area formula (see [198, Theorem 3.3.2]) to obtain

$$\int_B |\det(\nabla J(x))| \, dx = \int_{\mathbb{R}^n} \#(B \cap J^{-1}(y)) \, dy \quad \text{for all Borel sets } B \text{ in } \mathbb{R}^n,$$

2.6 Second-order differentiability

where # is the counting measure. By this formula along with the fact J is onto (since ∂f is nonempty for all $x \in \mathbb{R}^n$), we see that the complement of

$$\{x : J \text{ is differentiable at } x \text{ and } \nabla J(x) \text{ is nonsingular}\}$$

has measure 0. On the other hand, a Lipschitz function maps negligible sets into negligible sets (see [384, Lemma 7.25]). Consequently the complement of E_2 has measure 0. Thus E_3 has a negligible complement where $E_3 = E_2 \cap E_2$.

Fix x so that $J(x) \in E_3$. Then by the definition of J,

$$\nabla f(J(x)) = x - J(x). \qquad (2.6.7)$$

According to Exercise 2.6.14, there are $\delta > 0$ and $C > 0$ so that whenever $\|h\| \leq \delta$ there exists y such that

$$J(x + y) = J(x) + h \text{ and } \|h\| \leq \|y\| \leq C\|h\|. \qquad (2.6.8)$$

Moreover, the differentiability of J at x implies

$$J(x + h) = J(x) + \nabla J(x)h + o(\|h\|). \qquad (2.6.9)$$

Suppose now h is such that $\|h\| \leq \delta$ and $J(x) + h \in E_1$. Then using (2.6.8), (2.6.7) and (2.6.9) we have

$$\nabla f(J(x) + h) = \nabla f(J(x + y)) = x + y - J(x + y)$$
$$= \nabla f(J(x)) + (I - \nabla J(x))y + o(\|y\|). \qquad (2.6.10)$$

Now (2.6.8) and (2.6.9) imply

$$J(x) + h = J(x) + \nabla J(x)y + o(\|y\|). \qquad (2.6.11)$$

Because $\|y\|$ and $\|h\|$ are comparable, (2.6.11) implies

$$y = (\nabla J(x))^{-1}h + o(\|h\|). \qquad (2.6.12)$$

Finally, we show that f has a second-order Taylor expansion at $J(x)$ with matrix $A = (\nabla J(x))^{-1} - I$. For this, let $\psi(h) := f(J(x) + h)$ and let

$$\tilde{\psi}(h) := f(J(x)) + \langle \nabla f(J(x)), h \rangle + \frac{1}{2}\langle Ah, h \rangle.$$

We have $\psi(0) = \tilde{\psi}(0)$ and for almost all small h,

$$\nabla \psi(h) = \nabla f(J(x) + h)$$
$$= \nabla f(J(x)) + Ah + o(\|h\|) \text{ (by (2.6.10) and (2.6.12))}$$
$$= \nabla \tilde{\psi}(h) + o(\|h\|);$$

consequently, $\Psi(h) := \psi(h) - \tilde{\psi}(h)$ is locally Lipschitz and satisfies $\Psi(0) = 0$ and $\nabla\Psi(h) = o(\|h\|)$ for almost all small h. Now one can easily show that $\Psi(h) = o(\|h\|^2)$ as desired. $\square$

Exercises and further results

2.6.1.* Let ϕ be a symmetric bilinear map. Show that $\phi(x, y)$ is uniquely determined if $\phi(x, x)$ is known for all x. Note that any skew part $(T^* = -T)$ of a nonsymmetric form cannot be so detected.

Hint. Solve $\phi(x + y, x + y) = \phi(x, x) + 2\phi(x, y) + \phi(y, y)$ for $\phi(x, y)$. $\square$

2.6.2. Suppose in Euclidean space that

$$q(x) := \frac{1}{2}\langle Qx, x\rangle + \langle b, x\rangle + c.$$

Suppose Q is symmetric and positive definite. Show that

$$q^*(x) := \frac{1}{2}\langle Q^{-1}x, x\rangle + \langle d, x\rangle + e,$$

where $d := -Q^{-1}b$ and $e := \frac{1}{2}\langle Q^{-1}b, b\rangle - c$. What happens if Q is only positive semidefinite?

2.6.3.* Suppose $U \subset E$ is open and convex, let $f : U \to \mathbb{R}$ be convex and let $x \in U$.

(a) Show that f has a strong (resp. weak) second-order Taylor expansion at x if and only if there is a matrix A such that $\Delta_t^2 f(x)$ converges uniformly on bounded sets (resp. pointwise) to the function $h \mapsto \langle Ah, h\rangle$ as $t \to 0$.

(b) Use (a) and Exercise 2.1.22 to deduce f has a strong second-order Taylor expansion at x if it has a weak second-order Taylor expansion at x.

Hint. To apply Exercise 2.1.22, notice that the difference quotients $\Delta_t^2 f(x)$ are convex, and hence the function $h \mapsto \langle Ah, h\rangle$ is convex. $\square$

2.6.4.* Suppose $U \subset E$ is open and convex where dim E is finite, let $f : U \to \mathbb{R}$ be convex and let $x \in U$.

(a) Suppose f has a generalized second-order Fréchet derivative at x. Show that f is Fréchet differentiable at x with $\nabla f(x) = \phi$, where ϕ is as in the definition.

(b) Suppose that f has a generalized second-order Gâteaux derivative at x. Show that the matrix A in the definition must be unique.

(c) Show that f has a generalized second-order Fréchet derivative at x if and only if there exists an n by n matrix A such that

$$\lim_{t \to 0} \frac{\langle \phi_t - \nabla f(x), th\rangle}{t} = Ah \quad \text{where } \phi_t \in \partial f(x + th)$$

exists uniformly for $h \in B_E$. Consequently, when $\partial f(u)$ is single-valued in a neighborhood of U, such a function is twice Fréchet differentiable at x in the usual sense.

2.6 Second-order differentiability

2.6.5.* Prove Proposition 2.6.2.

Hint. This is a restatement of the definition; see [377, Proposition 2.6] for more details. □

2.6.6.* Prove Proposition 2.6.3.

Hint. Use the properties of convex functions and the definitions involved; see [377, Proposition 2.7]. □

2.6.7. Construct continuous real nonconvex functions on $\mathbb{R}^2$ that are continuously twice differentiable everywhere except $(0,0)$, and:

(a) twice Gâteaux differentiable at $(0,0)$, but not Fréchet differentiable at $(0,0)$;
(b) Fréchet differentiable and twice Gâteaux differentiable at $(0,0)$, but not twice Fréchet differentiable at $(0,0)$.

2.6.8. Let $f : \mathbb{R} \to \mathbb{R}$ be defined by $f(t) = t^3 \cos(1/t)$ if $t \neq 0$, and $f(0) = 0$. Show that f is continuously differentiable, but $f''(0)$ does not exist. Moreover, show that f has a second-order Taylor expansion at 0, i.e.

$$f(t) = f(0) + f'(0)t + \frac{1}{2}At^2 + o(t^2)$$

where $A = 0$. Hence the existence of a second-order Taylor expansion is not equivalent to second-order differentiability for nonconvex functions.

2.6.9. Construct a convex function $f : \mathbb{R} \to \mathbb{R}$ that is C^2-smooth and has a third-order Taylor expansion at 0, but whose third derivative does not exist at 0.

Hint. Construct a continuous function $h : [0, \infty) \to [0, \infty)$ such that

$$\limsup_{t \to 0^+} h(t)/t = 1, \quad \liminf_{t \to 0^+} h(t)/t = 0 \text{ and } \int_0^x h(t)\, dt \leq x^3 \text{ for } x \geq 0.$$

Visually h could be a function with sparse narrow peaks touching the line $y = x$ as $x \to 0^+$. Now let f be the function such that $f(t) = 0$ for $t \leq 0$, $f''(t) = h(t)$ for $t \geq 0$ and $f'(0) = 0$ and $f(0) = 0$ as already stipulated. The estimate $\int_0^x h(t)\, dt \leq x^3$ allows one to show $\lim_{x \to 0^+} f(x)/x^3 = 0$ which provides the desired third-order Taylor expansion. □

2.6.10. Consider $f : \mathbb{R}^2 \to \mathbb{R}$ by

$$f(x,y) := \begin{cases} xy\frac{x^2-y^2}{x^2+y^2} + 13(x^2 + y^2), & \text{if } x,y \neq 0; \\ 0, & \text{otherwise.} \end{cases}$$

Show that f is convex and its second mixed partials exist but are different at $(0,0)$. Verify directly that this function cannot have a weak second-order Taylor expansion at $(0,0)$. Additionally, the real function $g(t) := f(at,bt)$ has a second-order Taylor expansion at $t = 0$ for any $a, b \in \mathbb{R}$. Hence this example illustrates that requiring the same matrix A to work for every direction in the weak second-order Taylor expansion is stronger for convex functions than requiring the restrictions of f to lines through the origin to have second-order Taylor expansions.

90 Convex functions on Euclidean spaces

Hint. Note this exercise was motivated by [409, p. 152]. Checking convexity and computing the requisite partial derivatives takes some patience if done by hand: for this, note that

$$f_x(x,y) = \frac{y(x^4 + 4x^2y^2 - y^4)}{(x^2+y^2)^2} + 26x \quad f_y(x,y) = \frac{x(x^4 - 4x^2y^2 - y^4)}{(x^2+y^2)^2} + 26y.$$

Then $f_x(0,0) = 0$, $f_y(0,0) = 0$, $f_{xy}(0,0) = -1$, $f_{yx}(0,0) = 1$. On the Taylor expansion, note that since $f(0,0) = 0$, and since $f'(0,0)$ is the zero vector, the second-order Taylor expansion reduces to the quadratic $Q(x,y) = \frac{1}{2}[ax^2 + (b+c)xy + dy^2]$. If the expansion is valid along the lines $y = 0$, $x = 0$, and $y = x$, then $a = 26$, $b + c = 0 = b - c$ and $d = 26$. However, then the expansion is not valid along some other line, e.g. $y = 2x$. □

2.6.11. Let U be an open subset of E, and suppose that $f : U \to \mathbb{R}$ is Gâteaux differentiable. Then [332, p. 142] defines f to be twice Gâteaux differentiable at $\bar{x} \in U$ if the limit

$$f''(\bar{x};h,k) = \lim_{t \to 0} \frac{\langle \nabla f(\bar{x}+tk), h \rangle - \langle \nabla f(\bar{x}), h \rangle}{t}$$

exists for all $h, k \in E$. Show that the function f in Exercise 2.6.10 is twice Gâteaux differentiable at $(0,0) \in \mathbb{R}^2$ in the sense of [332]. Conclude that this definition is weaker than the definition of twice Gâteaux differentiability given earlier in this chapter. Nevertheless, it is important to note that it is still strong enough to deduce directional version of Taylor's formula and in turn to deduce that f is convex when $f''(x;h,h) \geq 0$ for $x \in U$ and $h \in E$ where U is an open convex subset of E; see [332, p. 143].

Hint. For this function, $\nabla f = (f_x, f_y)$ and can check $\nabla f(tx, ty) = t\nabla f(x,y)$. Thus, in vector notation, for $0, h, k \in \mathbb{R}^2$,

$$f''(0;h,k) = \lim_{t \to 0} \frac{\langle \nabla f(th) - \nabla f(0), k \rangle}{t} = \lim_{t \to 0} \frac{\langle t\nabla f(h), k \rangle}{t} = \langle \nabla f(h), k \rangle,$$

from which the conclusion follows easily. □

2.6.12. Suppose U is an open convex subset of $\mathbb{R}^n$ and $x \in U$. Show that f has a generalized second-order Fréchet derivative at x if and only if there is an n by n matrix A such that

$$\nabla f(x+h) \in \nabla f(x) + Ah + o(\|h\|)B_{\mathbb{R}^n} \quad \text{when } \nabla f(x+h) \text{ exists.}$$

Hint. The 'only if' portion is clear. For the 'if' portion, $f : U \to \mathbb{R}$ is differentiable almost everywhere. Choose $\delta > 0$ so that $\delta S_{\mathbb{R}^n} \subset U$. According to Fubini's theorem, for almost all $h \in \delta S_{\mathbb{R}^n}$ with respect to the surface measure $\nabla f(x+th)$ exists for almost all $0 \leq t \leq \delta$. Use an argument similar to the proof of (a) implies (c) in Theorem 2.6.1 to show f has a strong second-order Taylor expansion at x (some almost everywhere arguments will be needed). Then Theorem 2.6.1 shows that f has a generalized second-order derivative at x. □

2.6.13. Show that one cannot replace 'almost everywhere' with a residual set in the conclusion of Alexandrov's theorem (2.6.4).

Hint. One example: consider the function f given in Remark 2.5.5, let $g(x) := \int_0^x (f(t) + 2t)\,dt$ for $x \geq 0$ and observe $f(t) + 2t$ is nondecreasing. Another example: let $G \subset \mathbb{R}$ be a dense G_δ-set of measure 0. Then [133, Theorem 7.6, p. 288] ensures there is a continuous strictly increasing function f such that $g(t) = \infty$ for all $t \in G$. Let $f(x) = \int_0^x g(t)\,dt$. See [200, p. 117]. □

2.6.14.* Verify there exist $C > 0$ and $\delta > 0$ so that (2.6.8) holds.

Hint. First, if $F : \delta B_{\mathbb{R}^n} \to \mathbb{R}^n$ is continuous, $0 < \varepsilon < \delta$ and $\|F(x) - x\| < \varepsilon$ for all $x \in \delta S_{\mathbb{R}^n}$, then $\{x : \|x\| < \delta - \varepsilon\} \subset F(\{x : \|x\| < \delta\})$; see e.g. [384, Lemma 7.23]. Use this fact in conjunction with (2.6.9) and the fact $\nabla J(x)$ is nonsingular to deduce the existence of $C > 0$ and $\delta > 0$: for each $r > 0$ there exists $s > 0$ so that $\nabla J(x)(rB_{\mathbb{R}^n}) \supset \nabla J(x)(sB_{\mathbb{R}^n})$ then apply the preceding fact to $\nabla J(x)h \mapsto \nabla J(x)h + o(\|h\|)$ for $\|h\| \leq \delta$ for appropriate δ. Deduce $\|h\| \leq \|y\|$ because J is nonexpansive. □

2.6.15. Suppose U is a convex nonempty open set in $\mathbb{R}^N$ and $g : U \to \mathbb{R}$ is continuously differentiable on U. Let $A := \{x \in U : g''(x) \text{ exists}\}$. Show that g is convex $\Leftrightarrow U \setminus A$ is a set of measure zero, and g'' is positive semidefinite on A.

Hint. $\Rightarrow$: According to Alexandrov's theorem (2.6.4), the set $U \setminus A$ is of measure zero. Fix $x \in U$ and $y \in \mathbb{R}^N$ arbitrarily. The function $t \mapsto g(x + ty)$ is convex in a neighborhood of 0; consequently, its second derivative $\langle y, g''(x + ty)(y) \rangle$ is nonnegative whenever it exists. Since $g''(x)$ does exist, it must be that $g''(x)$ is positive semidefinite, for every $x \in A$.

$\Leftarrow$: Fix x and y in U. By assumption and a Fubini argument, we obtain two sequences (x_n) and (y_n) in U with $x_n \to x$, $y_n \to y$, and $t \mapsto g''(x_n + t(y_n - x_n))$ exists almost everywhere on $[0, 1]$. Integrating

$$t \mapsto \langle y_n - x_n, (g''(x_n + t(y_n - x_n)))(y_n - x_n) \rangle \geq 0 \text{ from 0 to 1,}$$

we deduce that $(g'(y_n) - g'(x_n))(y_n - x_n) \geq 0$. Taking limits and recalling that g' is continuous, we see that g' is monotone and so g is convex (Theorem 2.2.6). □

2.6.16.** Suppose the convex function $f : \mathbb{R}^n \to (-\infty, +\infty]$ is twice continuously differentiable in a neighborhood of x_0 and $\nabla^2 f(x_0)$ is nonsingular. Then f^* is twice continuously differentiable at $x_0^* = \nabla f(x_0)$ and

$$\nabla^2 f^*(x_0^*) = [\nabla^2 f(x_0)]^{-1}.$$

Hint. See [258, Corollary 4.8]. □

2.7 Support and extremal structure

We briefly introduce some fundamental notions to which we return in more detail in Section 5.2 and in Section 6.6.

An *extreme point* of a convex set $C \subset E$ is a point $x \in C$ whose complement $C \setminus \{x\}$ is convex; equivalently, this means x is not a convex combination of any $y, z \in C \setminus \{x\}$. The set of extreme points of C is denoted by ext C.

Lemma 2.7.1. *Given a supporting hyperplane H of a convex set $C \subset E$, any extreme point of $C \cap H$ is also an extreme point of C.*

Proof. Suppose x is an extreme point of $C \cap H$. If x is a nontrivial convex combination of $y, z \in C$, then y and z must both be in H. However, this cannot be because x is an extreme point of $C \cap H$. □

Theorem 2.7.2 (Minkowski). *Any compact convex set $C \subset E$ is the convex hull of its extreme points.*

Proof. The proof proceeds by induction on dim C; clearly the result holds when dim $C = 0$. Now assume the result holds for all sets of dimension less than dim C. By translating C we may assume $0 \in C$, and let $F = \text{span } C$. Then C has nonempty interior in F.

Given any point $x \in \text{bdry } C$, the supporting hyperplane theorem (2.4.3) shows that C has a supporting hyperplane H at x. By the induction hypotheses applied to the set $C \cap H$ we deduce using Lemma 2.7.1

$$x \in \text{conv}(\text{ext}(C \cap H)) \subset (\text{ext } C).$$

Thus we have proved bdry $C \subset \text{conv}(\text{ext } C))$ so $\text{conv}(\text{bdry } C) \subset \text{conv}(\text{ext } C)$. Because C is compact, it follows that $\text{conv}(\text{bdry } C) = C$ and the result now follows. □

Exercises and further results

2.7.1. Suppose C is a compact convex subset of E and $f : C \to \mathbb{R}$ is a continuous convex function. Show that f attains its maximum at an extreme point of C.

2.7.2. (Exposed points). A point x in a closed convex set C is an *exposed point* of C if there exists a vector y such that $\langle y, x \rangle > \langle y, c \rangle$ for every $c \in C, c \neq x$, and y is said to *expose* x in C. In other words $\sigma_C(y)$ is uniquely attained in C. Show that

(a) Every exposed point of a convex C is an extreme point but not conversely. (See also Example 2.4.9.)
(b) Show that every exposed point of a compact convex set in a Euclidean space is *strongly exposed* as defined on p. 211.
(c) Minkowski's theorem implies Straszewicz's' theorem [369, Theorem 18.6] that *a compact convex set in $\mathbb{R}^n$ is the closed convex hull of its exposed points*.

Hint. It suffices to show the exposed points are dense in the extreme points. □

2.7.3 (Milman converse theorem). Let C be a compact convex subset of a Euclidean space E and suppose that $C = \overline{\text{conv}}A$ for some closed $A \subseteq C$. Show that ext $C \subseteq A$.

2.7 Support and extremal structure

Hint. Fix $\varepsilon > 0$. Since A is precompact, there is a finite set $F \subset A$ with $A \subset F + \varepsilon B_E$. Then $C \subset \operatorname{conv} \bigcup_{x \in F} \operatorname{conv} \{A \cap (x + \varepsilon B_E)\}$. Then if $e \in \operatorname{ext} C$ it must be that $e \in x + \varepsilon B_E$ for some $x \in F$. $\square$

2.7.4 (Nested unions). For $n \in \mathbb{N}$, let $C_n \subseteq C_{n+1}$ be closed convex subsets of a ball in Euclidean space. Show that

$$\operatorname{ext} \overline{\bigcup_n C_n} \subseteq \overline{\bigcup_n \operatorname{ext} C_n}. \tag{2.7.1}$$

Compute both sides of (2.7.1) when (C_n) is a nested sequence of regular 2^n-gons inscribed in the unit ball in $\mathbb{R}^2$.

2.7.5 (Birkhoff's theorem [95]).[†] Recall that an entry-wise nonnegative $n \times n$ matrix is *doubly stochastic* if all row and column sums equal one, denoted $M \in SS_n$. Show that

(a) Every $n \times n$ permutation matrix is an extreme point of SS_n.
(b) Every extreme point of SS_n is a permutation matrix.
(c) Deduce from Theorem 2.7.2 that every doubly stochastic matrix is a convex combination of permutation matrices.

3
Finer structure of Euclidean spaces

The infinite we shall do right away. The finite may take a little longer. (Stanislaw Ulam)[1]

3.1 Polyhedral convex sets and functions

In the book [96, Section 2.3], theorems of the alternative (see Exercise 2.4.10) are used to observe that finitely generated cones are closed. Remarkably, a finite linear-algebraic assumption leads to a topological conclusion. In the section which follows [96, Section 5.1] we pursue the consequences of this type of assumption in convex analysis.

A cone C is *finitely generated* if there exist $x_1, x_2, \ldots, x_m \in E$ so that

$$C = \{x : x = \sum_{i=1}^{m} a_i x_i, \ a_i \geq 0\}. \tag{3.1.1}$$

There are two natural ways to impose a finite linear structure on the sets and functions we consider. The first we have already seen: a 'polyhedron' (or *polyhedral set*) is a finite intersection of closed half-spaces in E, and we say a function $f : E \to [-\infty, +\infty]$ is *polyhedral* if its epigraph is polyhedral.

On the other hand, a *polytope* is the convex hull of a finite subset of E, and we call a subset of E *finitely generated* if it is the sum of a polytope and a finitely generated cone (in the sense of formula (3.1.1)). Notice we do not yet know if a cone that is a finitely generated set in this sense is finitely generated in the sense of (3.1.1); we shall return to this point later in the section. The function f is *finitely generated* if its epigraph is finitely generated. A central result of this section is that polyhedra and finitely generated sets in fact coincide – this is sometimes called the *key theorem*.

We begin by collecting together some relatively easy observations in the following two results.

Proposition 3.1.1 (Polyhedral functions). *Suppose that the function $f : E \to [-\infty, +\infty]$ is polyhedral. Then f is closed and convex and can be decomposed*

---

[1] Stanislaw Marcin Ulam, 1909–1984, quoted from D. MacHale, *Comic Sections*, Dublin, 1993.

in the form
$$f = \max_{i \in I} g_i + \delta_P, \qquad (3.1.2)$$
where the index set I is finite (and possibly empty), the functions g_i are affine, and the set $P \subset E$ is polyhedral (and possibly empty).

Thus $\mathrm{dom}\, f$ is polyhedral and coincides with $\mathrm{dom}\, \partial f$ iff f is proper.

Proof. Since any polyhedron is closed and convex, so is f, and the decomposition (3.1.2) follows directly from the definition. If f is proper then both the sets I and P are nonempty in this decomposition. At any point x in $P\,(=\mathrm{dom}\,f)$ we know $0 \in \partial \delta_P(x)$, and the convex function $\max_i g_i$ certainly has a subgradient at x since it is everywhere finite. Hence we deduce the condition $\partial f(x) \neq \emptyset$. $\square$

Proposition 3.1.2 (Finitely generated functions). *Suppose the function $f : E \to [-\infty, +\infty]$ is finitely generated. Then f is closed and convex and $\mathrm{dom}\, f$ is finitely generated. Furthermore, f^* is polyhedral.*

Proof. Polytopes are compact and convex by Carathéodory's theorem (1.2.5), and finitely generated cones are closed and convex, so finitely generated sets (and therefore finitely generated functions) are closed and convex (by Exercise 2.4.24). We leave the remainder of the proof as an exercise. $\square$

An easy exercise shows that a set $P \subset E$ is polyhedral (respectively, finitely generated) if and only if δ_P is polyhedral (respectively, finitely generated).

To prove that polyhedra and finitely generated sets in fact coincide, we consider the two 'extreme' special cases: first, compact convex sets, and second, convex cones. Observe first that by definition compact, finitely generated sets are just polytopes.

Lemma 3.1.3. *A polyhedron has at most finitely many extreme points.*

Proof. Fix a finite set of affine functions $\{g_i : i \in I\}$ on E, determining a polyhedron
$$P := \{x \in E : g_i(x) \leq 0 \text{ for } i \in I\}.$$

For any point x in P, the 'active set' is the indices $\{i \in I : g_i(x) = 0\}$. Suppose two distinct extreme points x and y of P have the same active set. Then for any sufficiently small real ε the points $x \pm \varepsilon(y-x)$ both lie in P. But this contradicts the assumption that x is extreme. Hence different extreme points have different active sets, and the result follows. $\square$

This lemma together with Minkowski's theorem (2.7.2) reveals the nature of compact polyhedra.

Theorem 3.1.4. *Every compact polyhedron is a convex polytope.*

We next turn to cones.

Lemma 3.1.5. *A polyhedral cone is finitely generated in the sense of (3.1.1).*

Proof. Given a polyhedral cone $P \subset E$, consider the subspace $L := P \cap -P$ and the pointed polyhedral cone $K = P \cap L^\perp$. Observe that one has the decomposition $P = K \oplus L$. By the pointed cone theorem (2.4.12), there is an element y of E for which the set

$$C := \{x \in K : \langle x, y \rangle = 1\}$$

is compact and satisfies $K = \mathbb{R}_+ C$. Since C is polyhedral, the previous result shows it is a polytope. Thus K is finitely generated, whence so is P. □

Theorem 3.1.6 (Key theorem of polyhedrality). *A convex set or function is polyhedral if and only if it is finitely generated.*

Proof. For finite sets $\{a_i : i \in I\} \subset E$ and $\{b_i : i \in I\} \subset \mathbb{R}$, consider the polyhedron in E defined by

$$P = \{x \in E : \langle a_i, x \rangle \leq b_i \text{ for } i \in I\}.$$

The polyhedral cone in $E \times \mathbb{R}$ defined by

$$Q := \{(x, r) \in E \times \mathbb{R} : \langle a_i, x \rangle - b_i r \leq 0 \text{ for } i \in I\}$$

is finitely generated by the previous lemma, so there are finite subsets $\{x_j : j \in J\}$ and $\{y_t : t \in T\}$ of E with

$$Q = \left\{ \sum_{j \in J} \lambda_j(x_j, 1) + \sum_{t \in T} \mu_t(y_t, 0) : \lambda_j \in \mathbb{R}_+ \text{ for } j \in J, \mu_t \in \mathbb{R}_+ \text{ for } t \in T \right\}.$$

We deduce

$$P = \{x : (x, 1) \in Q\}$$
$$= \mathrm{conv}\{x_j : j \in J\} + \left\{ \sum_{t \in T} \mu_t y_y : \mu_t \in \mathbb{R}_+ \text{ for } t \in T \right\},$$

so P is finitely generated. We have thus shown that any polyhedral set (and hence function) is finitely generated.

Conversely, suppose the function $f : E \to [-\infty, +\infty]$ is finitely generated. Consider first the case when f is proper. By Proposition 3.1.2, f^* is polyhedral, and hence (by the above argument) finitely generated. But f is closed and convex, also by Proposition 3.1.2, so the Fenchel biconjugation theorem (2.4.4) implies $f = f^{**}$. By applying Proposition 3.1.2 once again we see f^{**} (and hence f) is polyhedral. We leave the improper case as an exercise. □

Notice these two results show our two notions of a finitely generated cone do indeed coincide.

The following list of properties shows that many linear-algebraic operations preserve polyhedrality. In each case one of finite generation or half-space representation is not obvious, but the other often is.

Proposition 3.1.7 (Polyhedral algebra). *Consider a Euclidean space Y and a linear map $A : E \to Y$.*

(a) *If the set $P \subset E$ is polyhedral then so is its image AP.*
(b) *If the set $K \subset Y$ is polyhedral then so is its inverse image $A^{-1}K$.*
(c) *The sum and pointwise maximum of finitely many polyhedral functions are polyhedral.*
(d) *If the function $g : Y \to [-\infty, +\infty]$ is polyhedral then so is the composite function $g \circ A$.*
(e) *If the function $q : E \times Y \to [-\infty, +\infty]$ is polyhedral then so is the perturbation function $h : Y \to [-\infty, +\infty]$ defined by $h(u) := \inf_{x \in E} q(x, u)$.*

We shall see the power of this proposition in the final section of this chapter.

Corollary 3.1.8 (Polyhedral Fenchel duality). *The conclusions of the Fenchel duality theorem (2.3.4) and of the subdifferential sum rule (2.3.5) remain valid if the regularity condition (2.3.4) is replaced by the assumption that the functions f and g are polyhedral with*

$$\mathrm{dom}\, g \cap A\, \mathrm{dom}\, f \neq \emptyset.$$

Proof. We follow the proof of the Fenchel duality theorem (2.3.4), simply observing that the value function h defined therein is polyhedral by the polyhedral algebra proposition (3.1.7). Thus, when the optimal value is finite, h has a subgradient at 0. □

We conclude this section with a result emphasizing the power of Fenchel duality for convex problems with linear constraints.

Corollary 3.1.9 (Mixed Fenchel duality). *The conclusions of the Fenchel duality theorem (2.3.4) and of the subdifferential sum rule (2.3.5) remain valid if the regularity condition (2.3.4) is replaced by the assumption that*

$$\mathrm{dom}\, g \cap A\, \mathrm{cont}\, f \neq \emptyset$$

while the function g is polyhedral.

Proof. Assume without loss that the primal optimal value

$$p := \inf_{x \in E} \{f(x) + g(Ax)\} = \inf_{x \in E,\, r \in \mathbb{R}} \{f(x) + r \;:\; g(Ax) \leq r\}$$

is finite. By assumption there is a feasible point for the problem on the right at which the objective function is continuous, so there is an affine function $\alpha : E \times \mathbb{R} \to \mathbb{R}$ minorizing the function $(x, r) \mapsto f(x) + r$ such that

$$p = \inf_{x \in E,\, r \in \mathbb{R}} \{\alpha(x, r) \;:\; g(Ax) \leq r\}$$

(see Exercise 2.4.20). Clearly α has the form $\alpha(x,r) = \beta(x) + r$ for some affine minorant β of f, so

$$p = \inf_{x \in E} \{\beta(x) + g(Ax)\}.$$

Now we apply polyhedral Fenchel duality (3.1.8) to deduce the existence of an element ϕ of $\mathbf{Y}$ such that

$$p = -\beta^*(A^*\phi) - g^*(-\phi) \le -f^*(A^*\phi) - g^*(-\phi) \le p$$

(using the weak duality inequality), and the duality result follows. The calculus rules follow as before. □

Exercises and further results

3.1.1.* Prove directly from the definition that any polyhedral function has a decomposition of the form (3.1.2).

3.1.2.* Fill in the details for the proof of the finitely generated functions proposition (3.1.2).

3.1.3.* Use Exercise 2.1.18 (lower-semicontinuity and closure) to show that if a finitely generated function f is not proper then it has the form

$$f(x) = \begin{cases} +\infty & \text{if } x \notin K \\ -\infty & \text{if } x \in K \end{cases}$$

for some finitely generated set K.

3.1.4. Prove a set $K \subset E$ is polyhedral (respectively, finitely generated) if and only if δ_K is polyhedral (respectively, finitely generated), without using the polyhedrality theorem (3.1.6).

3.1.5.* Complete the proof of the polyhedrality theorem (3.1.6) for improper functions using Exercise 3.1.3.

3.1.6 (Tangents to polyhedra). Prove the tangent cone to a polyhedron P at a point x in P is given by $T_P(x) = \mathbb{R}_+(P - x)$.

3.1.7 (Polyhedral algebra).* Prove Proposition 3.1.7 using the following steps.

(a) Prove parts (a)–(d).
(b) In the notation of part (e), consider the natural projection

$$P_{Y \times \mathbb{R}} : E \times Y \times \mathbb{R} \to Y \times \mathbb{R}.$$

Prove the inclusions

$$P_{Y \times \mathbb{R}}(\operatorname{epi} q) \subset \operatorname{epi} h \subset \operatorname{cl}(P_{Y \times \mathbb{R}}(\operatorname{epi} q)).$$

(c) Deduce part (e).

3.1.8. Suppose the function $f : E \to (-\infty, +\infty]$ is polyhedral. Show that the subdifferential of f at a point x in $\text{dom}\, f$ is a nonempty polyhedron and is bounded if and only if x lies in $\text{int}(\text{dom}\, f)$.

3.1.9 (Polyhedral cones). For polyhedral cones $H \subset Y$ and $K \subset E$ and a linear map $A : E \to Y$, prove that

$$(K \cap A^{-1} H)^- = A^* H^- + K^-$$

by using convex calculus.

3.1.10. Apply the mixed Fenchel duality corollary (3.1.9) to the problem

$$\inf\{f(x) : Ax \leq b\},$$

for a linear map $A : E \to \mathbb{R}^m$ and a point b in $\mathbb{R}^m$ with the orthant ordering.

3.1.11 (Generalized Fenchel duality). Consider convex functions

$$h_1, h_2, \ldots, h_m : E \to (-\infty, +\infty]$$

with $\bigcap_i \text{cont}\, h_i$ nonempty. By applying the mixed Fenchel duality corollary (3.1.9) to the problem

$$\inf_{x, x^1, x^2, \ldots, x^m \in E} \left\{ \sum_{i=1}^m h_i(x^i) : x^i = x \text{ for } i = 1, 2, \ldots, m \right\},$$

prove

$$\inf_{x \in E} \sum_i h_i(x) = -\inf \left\{ \sum_i h_i^*(\phi^i) : \phi^1, \phi^2, \ldots, \phi^m \in E, \sum_i \phi^i = 0 \right\}.$$

3.1.12 (Relativizing mixed Fenchel duality). In Corollary 3.1.9, prove that the condition $\text{dom}\, g \cap A \,\text{cont}\, f \neq \emptyset$ can be relaxed to $\text{dom}\, g \cap A \,\text{ri}(\text{dom}\, f) \neq \emptyset$.

3.2 Functions of eigenvalues

In this section we will frequently encounter S^n the symmetric n by n matrices as well its subclasses S^n_+ of positive semidefinite matrices and S^n_{++} of positive definite matrices. We also let O^n denote the orthogonal matrices. The mapping $\text{Diag}\, \mathbb{R}^n \to S^n$ is defined by letting $\text{Diag}(x)$ be the diagonal matrix with diagonal entries x_i. Also, $\lambda : S^n \to \mathbb{R}^n$ where $\lambda(A)$ maps to the vector whose components are the eigenvalues of A written in nonincreasing order. As usual, $\text{tr}(X)$ is the trace of X which is the sum of the entries on the diagonal. For matrices $X, Y \in S^n$ we write $X \preceq Y$ if $Y - X \in S^n_+$. The vector space S^n becomes a Euclidean space with the inner product $\langle X, Y \rangle = \text{tr}(XY)$. Further information is given in [95, Section 1.2].

The key to studying functions of eigenvalues is the following inequality.

Theorem 3.2.1 (Fan). *Any matrices X and Y in S^n satisfy the inequality*

$$\text{tr}(XY) \leq \lambda(X)^T \lambda(Y). \tag{3.2.1}$$

Equality holds if and only if X and Y have a simultaneous ordered spectral decomposition: there is a matrix U in O^n with

$$X = U^T(\text{Diag}\,\lambda(X))U \text{ and } Y = U^T(\text{Diag}\,\lambda(Y))U. \qquad (3.2.2)$$

A proof of Fan's inequality via Birkhoff's theorem is sketched in [95, p. 12]. The special case of Fan's inequality where both matrices are diagonal gives the following classical inequality, in which for a vector x in $\mathbb{R}^n$, we denote by $[x]$ the vector with the same components permuted into nonincreasing order. A direct proof of this result is left as an exercise.

Proposition 3.2.2 (Hardy–Littlewood–Pólya). *Any vectors x and y in $\mathbb{R}^n$ satisfy the inequality*

$$x^T y \leq [x]^T[y].$$

Fenchel conjugacy gives a concise and beautiful avenue to many eigenvalue inequalities in classical matrix analysis. In this section we outline this approach.

The two cones $\mathbb{R}^n_+$ and S^n_+ appear repeatedly in applications, as do their corresponding logarithmic barriers lb and ld, which are introduced in Proposition 3.2.3.

Proposition 3.2.3 (Log barriers). *The functions* $\text{lb} : \mathbb{R}^n \to (-\infty, +\infty]$ *and* $\text{ld} : S^n \to (-\infty, +\infty]$ *defined by*

$$\text{lb}(x) := \begin{cases} -\sum_{i=1}^n \log x_i & \text{if } x \in \mathbb{R}^n_{++} \\ +\infty & \text{otherwise} \end{cases}$$

and

$$\text{ld}(X) := \begin{cases} -\log \det X & \text{if } X \in S^n_{++} \\ +\infty & \text{otherwise} \end{cases}$$

are essentially smooth, and strictly convex on their domains as defined in Section 7.1. They satisfy the conjugacy relations

$$\text{lb}^*(x) = \text{lb}(-x) - n \text{ for all } x \in \mathbb{R}^n, \text{ and}$$
$$\text{ld}^*(X) = \text{ld}(-X) - n \text{ for all } X \in S^n.$$

The perturbed functions $\text{lb} + \langle c, \cdot \rangle$ *and* $\text{ld} + \langle C, \cdot \rangle$ *have compact lower level sets for any vector $c \in \mathbb{R}^n_{++}$ and matrix $C \in S^n_{++}$, respectively.*

We can relate the vector and matrix examples through the identities

$$\delta_{S^n_+} = \delta_{\mathbb{R}^n_+} \circ \lambda \text{ and } \text{ld} = \text{lb} \circ \lambda. \qquad (3.2.3)$$

We see in this section that these identities fall into a broader pattern.

Recall the function $[\cdot] : \mathbb{R}^n \to \mathbb{R}^n$ rearranges components into nonincreasing order. We say a function f on $\mathbb{R}^n$ is *symmetric* if $f(x) = f([x])$ for all vectors x in $\mathbb{R}^n$; in

3.2 Functions of eigenvalues

other words, permuting components does not change the function value. We call a symmetric function of the eigenvalues of a symmetric matrix a *spectral function*. The following formula is crucial.

Theorem 3.2.4 (Spectral conjugacy). *If* $f : \mathbb{R}^n \to [-\infty, +\infty]$ *is a symmetric function, it satisfies the formula*

$$(f \circ \lambda)^* = f^* \circ \lambda.$$

Proof. By Fan's inequality (3.2.1) any matrix Y in S^n satisfies the inequalities

$$(f \circ \lambda)^*(Y) = \sup_{X \in S^n} \{\text{tr}(XY) - f(\lambda(X))\}$$

$$\leq \sup_X \{\lambda(X)^T \lambda(Y) - f(\lambda(X))\}$$

$$\leq \sup_{x \in \mathbb{R}^n} \{x^T \lambda(Y) - f(x)\}$$

$$= f^*(\lambda(Y)).$$

On the other hand, fixing a spectral decomposition $Y = U^T(\text{Diag}\,\lambda(Y))U$ for some matrix U in O^n leads to the reverse inequality

$$f^*(\lambda(Y)) = \sup_{x \in \mathbb{R}^n} \{x^T \lambda(Y) - f(x)\}$$

$$= \sup_x \{\text{tr}((\text{Diag}\,x)UYU^T) - f(x)\}$$

$$= \sup_x \{\text{tr}(U^T(\text{Diag}\,x)UY) - f(\lambda(U^T \text{Diag}\,xU))\}$$

$$\leq \sup_{X \in S^n} \{\text{tr}(XY) - f(\lambda(X))\}$$

$$= (f \circ \lambda)^*(Y),$$

which completes the proof. □

This formula, for example, makes it very easy to calculate ld^* (see the log barriers proposition (3.2.3)) and to check the self-duality of the cone S_+^n.

Once we can compute conjugates easily, we can also recognize closed convex functions easily using the Fenchel biconjugation theorem (2.4.4).

Corollary 3.2.5 (Davis). *Suppose the function* $f : \mathbb{R}^n \to (-\infty, +\infty]$ *is symmetric. Then the 'spectral function'* $f \circ \lambda$ *is closed and convex if and only if* f *is closed and convex.*

We deduce immediately that the logarithmic barrier ld is closed and convex, as well as the function $X \mapsto \text{tr}(X^{-1})$ on S_{++}^n, for example.

Identifying subgradients is also easy using the conjugacy formula and the Fenchel–Young inequality (2.3.1).

Corollary 3.2.6 (Spectral subgradients, Lewis). *If $f : \mathbb{R}^n \to (-\infty, +\infty]$ is a symmetric function, then for any two matrices X and Y in S^n, the following properties are equivalent:*

(a) $Y \in \partial(f \circ \lambda)(X)$.
(b) *X and Y have a simultaneous ordered spectral decomposition and satisfy $\lambda(Y) \in \partial f(\lambda(X))$.*
(c) *$X = U^T(\text{Diag } x)U$ and $Y = U^T(\text{Diag } y)U$ for some matrix U in O^n and vectors x and y in $\mathbb{R}^n$ satisfying $y \in \partial f(x)$.*

Proof. Notice the inequalities

$$(f \circ \lambda)(X) + (f \circ \lambda)^*(Y) = f(\lambda(X)) + f^*(\lambda(Y)) \geq \lambda(X)^T \lambda(Y) \geq \text{tr}(XY).$$

The condition $Y \in \partial(f \circ \lambda)(X)$ is equivalent to equality between the left- and right-hand sides (and hence throughout), and the equivalence of properties (a) and (b) follows using Fan's inequality (3.2.1). For the remainder of the proof, see Exercise 3.2.9. □

Corollary 3.2.7 (Spectral differentiability, Lewis). *Suppose that the function $f : \mathbb{R}^n \to (-\infty, +\infty]$ is symmetric, closed, and convex. Then $f \circ \lambda$ is differentiable at a matrix X in S^n if and only if f is differentiable at $\lambda(X)$.*

Proof. If $\partial(f \circ \lambda)(X)$ is a singleton, so is $\partial f(\lambda(X))$, by the spectral subgradients corollary above. Conversely, suppose $\partial f(\lambda(X))$ consists only of the vector $y \in \mathbb{R}^n$. Using Exercise 3.2.9(b), we see the components of y are nonincreasing, so by the same corollary, $\partial(f \circ \lambda)(X)$ is the nonempty convex set

$$\{U^T(\text{Diag } y)U \; : \; U \in O^n, \; U^T \text{Diag}(\lambda(X))U = X\}.$$

But every element of this set has the same norm (namely $\|y\|$), so the set must be a singleton. □

Notice that the proof in fact shows that when f is differentiable at $\lambda(X)$ we have the formula

$$\nabla(f \circ \lambda)(X) = U^T(\text{Diag } \nabla f(\lambda(X)))U \tag{3.2.4}$$

for any matrix U in O^n satisfying $U^T(\text{Diag } \lambda(X))U = X$.

The pattern of these results is clear: many analytic and geometric properties of the matrix function $f \circ \lambda$ parallel the corresponding properties of the underlying function f. The following exercise provides another excellent example.

Corollary 3.2.8. *Suppose the function $f : \mathbb{R}^n \to (-\infty, +\infty]$ is symmetric, closed, and convex. Then, in the language of Section 7.1, $f \circ \lambda$ is essentially strictly convex (respectively, essentially smooth) if and only if f is essentially strictly convex (respectively, essentially smooth).*

3.2 Functions of eigenvalues

For example, the logarithmic barrier ld is both essentially smooth and essentially strictly convex.

Exercises and further results

3.2.1.* Prove the identities (3.2.3).

3.2.2. Use the spectral conjugacy theorem (3.2.4) to calculate ld^* and $\delta^*_{S^n_+}$.

3.2.3.* Prove the Davis characterization (Corollary 3.2.5) using the Fenchel biconjugation theorem (2.4.4).

3.2.4 (Square-root iteration). Suppose a symmetric positive semidefinite matrix A satisfies $0 \preceq A \preceq I$. Prove that the iteration

$$Y_0 = 0, \quad Y_{n+1} = \frac{1}{2}(A + Y_n^2) \quad (n = 0, 1, 2, \ldots)$$

is nondecreasing (i.e. $Y_n \preceq Y_{n+1}$ for all n) and converges to the matrix $I - (I - A)^{1/2}$. Deduce that all positive definite matrices have unique positive definite square roots and likewise in the semidefinite case.

Hint. Consider diagonal matrices A. □

3.2.5 (Examples of convex spectral functions). Use the Davis characterization (Corollary 3.2.5) to prove the following functions of a matrix $X \in S^n$ are closed and convex:

(a) $\mathrm{ld}(X)$.

(b) $\mathrm{tr}(X^p)$, for any nonnegative even integer p.

(c) $\begin{cases} -\mathrm{tr}(X^{1/2}) & \text{if } X \in S^n_+ \\ \infty & \text{otherwise.} \end{cases}$

(d) $\begin{cases} \mathrm{tr}(X^{-p}) & \text{if } X \in S^n_{++} \\ \infty & \text{otherwise} \end{cases}$

for any nonnegative integer p.

(e) $\begin{cases} \mathrm{tr}(X^{1/2})^{-1} & \text{if } X \in S^n_{++} \\ \infty & \text{otherwise.} \end{cases}$

(f) $\begin{cases} -(\det X)^{1/n} & \text{if } X \in S^n_+ \\ \infty & \text{otherwise.} \end{cases}$

Deduce from the sublinearity of the function in part (f) the property

$$0 \preceq X \preceq Y \Rightarrow 0 \leq \det X \leq \det Y$$

for matrices X and Y in S^n.

3.2.6. Calculate the conjugate of each of the functions in Exercise 3.2.5.

3.2.7. Use formula (3.2.4) to calculate the gradients of the functions in Exercise 3.2.5.

3.2.8 (Orthogonal invariance). A function $h : S^n \to (-\infty, +\infty]$ is *orthogonally invariant* if all matrices X in S^n and U in O^n satisfy the relation $h(U^T X U) = h(X)$; in other words, orthogonal similarity transformations do not change the value of h.

(a) Prove h is orthogonally invariant if and only if there is a symmetric function $f : \mathbb{R}^n \to (-\infty, +\infty]$ with $h = f \circ \lambda$.
(b) Prove that an orthogonally invariant function h is closed and convex if and only if $h \circ \text{Diag}$ is closed and convex.

3.2.9.* Suppose the function $f : \mathbb{R}^n \to (-\infty, +\infty]$ is symmetric.

(a) Prove f^* is symmetric.
(b) If vectors x and y in $\mathbb{R}^n$ satisfy $y \in \partial f(x)$, prove $[y] \in \partial f([x])$ using Proposition 3.2.2.
(c) Finish the proof of the spectral subgradients corollary (3.2.6).
(d) Deduce $\partial (f \circ \lambda)(X) = \emptyset \Leftrightarrow \partial f(\lambda(X)) = \emptyset$.
(e) Prove Corollary 3.2.8.

3.2.10 (Fillmore–Williams [214]). Suppose the set $C \subset \mathbb{R}^n$ is *symmetric*: that is, $PC = C$ holds for all permutation matrices P. Prove the set

$$\lambda^{-1}(C) = \{X \in S^n : \lambda(X) \in C\}$$

is closed and convex if and only if C is closed and convex.

3.2.11 (DC functions). We call a real function f on a convex set $C \subset E$ a *DC function* if it can be written as the difference of two real convex functions on C.

(a) Prove the set of DC functions is a vector space.
(b) If f is a DC function, prove it is locally Lipschitz on int C.
(c) Let σ_k denote the sum of the k largest eigenvalues of a symmetric matrix. Show σ_k is convex and continuous for $k = 1, 2, \ldots, n$. Hence prove that the k-th largest eigenvalue λ_k is a DC function on S^n for $k = 1, 2, \ldots, n$, and deduce it is locally Lipschitz.
(d) Let us say a vector-valued function *is DC* if each of its coefficient functions is. In [422] a short proof in Banach space is given of Hartman's [250] result:

Theorem 3.2.9. *Let X, Y, Z be Euclidean spaces with $A \subset X, B \subset Y$ open and convex. Let $F : A \to Y$ be DC on A with $F(A) \subset B$. Let $G : B \to Z$ be DC on B. Then $G \circ F$ is locally DC on A.*

Hint. For (d), let $F_i = f_i - h_i$, $i = 1, \ldots, n$ and $G_j = g_j - k_j$, $j = 1, \ldots, m$ where all four of families f_i, g_i, h_j, k_j are convex. Fix $a \in A$ and let M be a common Lipschitz constant for g_j, G_j on some neighborhood V of $F(a)$. Show that around a both $(g \circ F + 2Mf)$ and $G \circ F + (g \circ F + 2Mf)$ have locally convex coordinates. The same method shows that if F and G are globally Lipschitz then $G \circ F$ is globally DC on A. See [422] for more details. □

3.2.12 (Log-concavity). Show that the following are logarithmically concave:

(a) $x \mapsto \frac{e^x}{1+e^x}$ for all x in $\mathbb{R}$.
(b) $(x_1, x_2, \cdots, x_N) \mapsto \frac{\prod_{k=1}^{N} x_k}{\sum_{k=1}^{N} x_k}$ for all $x_1 > 0, x_2 > 0, \cdots, x_N > 0$.
(c) The corresponding spectral function $A \mapsto \frac{\det A}{\operatorname{tr} A}$ for A positive definite.

3.2 Functions of eigenvalues

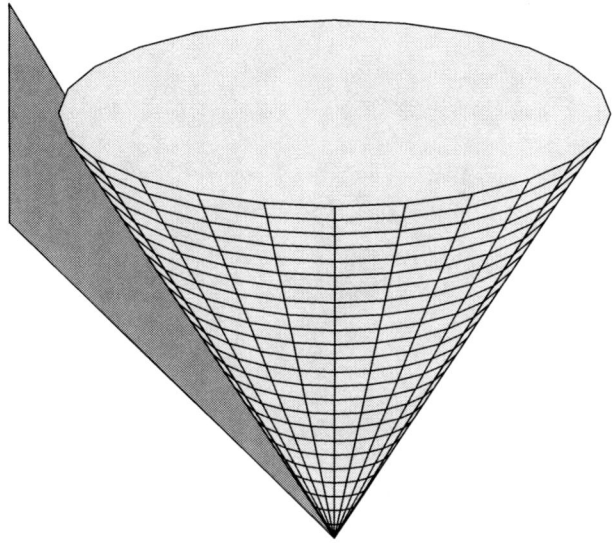

Figure 3.1 The semidefinite cone in 2-space with a boundary plane.

Other examples are to be found in [128, 96].

3.2.13 (Concavity and log-concavity). Show that the following are concave and so log-concave:

(a) $(x_1, x_2, \cdots, x_N) \mapsto \left(\prod_{k=1}^{M} \vec{x}_k \right)^{1/M}$ for all $x_1 > 0, x_2 > 0, \cdots, x_N > 0$. Here $1 \leq M \leq N$ and $\vec{x}$ denotes the sequence placed in nondecreasing order.

(b) The corresponding spectral function for A positive definite. This is the M-th root of the M smallest eigenvalues.

3.2.14 (2D semidefinite cone). Show that the cone of two-by-two semidefinite matrices is represented by $S_2 := \{(x, y, z) : xy \geq z^2, x + y \geq 0\}$. Prove, as illustrated in Figure 3.1, that the sum of S_2 and any hyperplane T tangent at a nonzero point of S_2 is never closed.

3.2.15 (Convex matrix functions). Let $C \in S_+^n$ be fixed. For matrices $A \in S_{++}^n$ and $D \in S^n$ show that

$$\frac{d^2}{dt^2} \operatorname{tr}\left(C(A + tD)^{-1} \right) \bigg|_{t=0} \geq 0.$$

(a) Deduce that (i) $A \mapsto \operatorname{tr}(CA^{-1})$, (ii) $A \mapsto \operatorname{tr}(CA^2)$, and (iii) $A \mapsto -\operatorname{tr}(CA^{1/2})$ are convex with domains S_{++}^n, S^n, and S_+^n respectively.

(b) Deduce from Exercise 2.4.28(b) that (i) $A \mapsto A^{-1}$, (ii) $A \mapsto A^2$, and (iii) $A \mapsto -A^{1/2}$ define Loewner-convex functions with domains S_{++}^n, S^n, and S_+^n respectively. Confirm that $A \mapsto A^4$ is not order-convex.

Hint. Consider $A := \begin{pmatrix} 4 & 2 \\ 2 & 1 \end{pmatrix}$ and $B := \begin{pmatrix} 4 & 0 \\ 0 & 8 \end{pmatrix}$. □

Hence $A \mapsto A^2$ cannot be order-monotone.

3.2.16 (Quadratic forms). Define $f_Q \colon S^n_+ \to \mathbb{R}$ via $f_Q(x) := \tfrac{1}{2}\langle Qx, x\rangle$.

(a) Show, that f_Q is convex and that $f_Q^* = f_{Q^{-1}}$ when $Q \in S^n_{++}$. (See also Exercise 2.6.2.) Deduce that $Q \mapsto Q^{-1}$ is order-reversing in the Loewner ordering.

(b) Show that for $A, B \in S^n_+$ one has

$$f_{\frac{A+B}{2}} \geq 2 f_A \square f_B,$$

and deduce that for $Q \in S^n_{++}$ the mapping $Q \mapsto Q^{-1}$ is order-convex – as shown by different methods in Exercise 3.2.15.

(c) Explore the above conjugacy formula and conclusions for $Q \in S^n_+$.

Hint. This entails using a generalized inverse Q^- [175, p. 179ff]. □.

3.2.17 (Matrix norms [261, 236]).** We restrict this discussion to real spaces. A *matrix norm* is a norm on a Euclidean space $\mathcal{M}$ of square $n \times n$ matrices that is also *submultiplicative*

$$\|AB\| \leq \|A\|\|B\| \text{ for all } A, B \in \mathcal{M}.$$

If $\|A^*\| = \|A\|$ for all $A \in \mathcal{M}$ we have a *-norm. All matrix norms on $\mathcal{M}$ are equivalent since the space is finite-dimensional.

An *induced norm* on a space of matrices is the one defined by $\|A\| := \sup\{\|Ax\| : \|x\| \leq 1\}$. (One can vary the norm on the range and domain spaces.) The *Frobenius* norm of a matrix A is the Euclidean norm we have already met, induced by the trace inner-product

$$\|A\|_F := \sqrt{\operatorname{tr} A^* A} = \sqrt{\sum_{i=1}^{n} \sum_{j=1}^{n} |a_{i,j}|^2}.$$

The eigenvalues of $\sqrt{A^*A}$ (or $\sqrt{AA^*}$) define the *singular values* of A, denoted $\sigma(A)$. Let $\rho(A) := \sup\{|\sigma| : \sigma \in \sigma(A)\}$ be the *spectral radius*. Show that:

(a) Every induced norm is a matrix norm.

 For all A in $\mathcal{M}$ show the following:

(b) $\|A\|_F = \sqrt{\sum_{i=1}^{n} \sigma_i^2}$, where $\sigma_i \in \sigma(A)$, is a matrix norm;

(c) $\|A^*\|_F = \|A\|_F = \|OA\|_F = \|AO\|_F$ for all orthogonal $O \in \mathcal{M}$;

(d) $\rho(A) \leq \|A\|$ for all matrix norms;

(e) $\rho(A) = \lim_{k \to \infty} \|A^k\|^{1/k}$ for all induced norms.

3.2.18 (Matrix *p*-norms [261, 236, 54]).** When the underlying norm in Exercise 3.2.17 is the *p*-norm, we write $\|A\|_p$ for the induced norm. Let n be the dimension of the underlying space. For all A in $\mathcal{M}$ show that

(a$_1$) $\|A\|_2 \leq \|A\|_F \leq \sqrt{n}\|A\|_2$;

(a$_2$) $\frac{1}{\sqrt{n}}\|A\|_1 \leq \|A\|_2 \leq \sqrt{n}\|A\|_1$;

(a$_3$) $\frac{1}{\sqrt{n}}\|A\|_\infty \leq \|A\|_2 \leq \sqrt{n}\|A\|_\infty$;

(b) $\|A\|_2^2 \leq \|A\|_1 \|A\|_\infty$;

(c$_1$) $\|A\|_1 = \max_{1 \leq j \leq n} \sum_{i=1}^{n} |a_{ij}|$;
(c$_2$) $\|A\|_\infty = \max_{1 \leq i \leq n} \sum_{j=1}^{n} |a_{ij}|$;
(d) $\|A\|_2 = \|OA\|_2 = \|AO\|_2$ for all orthogonal O in $\mathcal{M}$.

It is also common to see $\|A\|_p$ used to denote the *entrywise p-norm* of the matrix viewed as a rolled-out sequence. This leads to quite different results. Thence, $\|A\|_2 = \|A\|_F$, $\|A\|_1 = \text{tr} \sqrt{A^*A}$, is the *trace norm*, and $\|A\|_\infty = \max_{1 \leq i,j \leq n} |a_{ij}|$.

In greater generality the matrix norms become operator or Banach algebra norms and the entrywise norms become the Schatten norms of section 6.5. All matrix and entrywise norms on $\mathcal{M}$ are equivalent since the space is finite-dimensional. The ease of computation of the entrywise norms is one of the key reasons for so renorming within numerical linear algebra.

3.2.19 (Vector and matrix norms [261, 236, 54]).** Let $\|\cdot\|$ be any norm on $\mathcal{M}$. Show that there is a unique number $c > 0$ such that $c\|\cdot\|$ is a matrix norm. Need it be an induced norm?

A particularly nice discussion of matrix norms and many related issues can be found in [54, Chapter 9].

3.3 Linear and semidefinite programming duality

Linear programming (LP) is the study of convex optimization problems involving a linear objective function subject to linear constraints. This simple optimization model has proven extraordinarily powerful in both theory and practice, so we devote the first part of this section to deriving linear programming duality theory from our convex perspective.

We then contrast this theory with the corresponding results for *semidefinite programming* (SDP), a class of matrix optimization problems analogous to linear programs but involving the positive semidefinite cone.

Linear programs are inherently polyhedral, so our main development follows directly from the polyhedrality section (Section 3.1). Given vectors $a^1, a^2, \ldots, a^m$, and c in $\mathbb{R}^n$ and a vector b in $\mathbb{R}^m$, consider the *primal linear program*

$$\left.\begin{array}{rl} \inf & \langle c, x \rangle \\ \text{subject to} & \langle a^i, x \rangle - b_i \leq 0 \text{ for } i = 1, 2, \ldots, m \\ & x \in \mathbb{R}^n. \end{array}\right\} \quad (3.3.1)$$

Denote the primal optimal value by $p \in [-\infty, +\infty]$. In the Lagrangian duality framework, the dual problem is

$$\sup \left\{ -b^T \mu : \sum_{i=1}^{m} \mu_i a^i = -c, \ \mu \in \mathbb{R}_+^m \right\} \quad (3.3.2)$$

with dual optimal value $d \in [-\infty, +\infty]$.

This can be systematically analyzed using polyhedral theory. Suppose that Y is a Euclidean space, that the map $A : E \to Y$ is linear, and consider cones $H \subset Y$ and $K \subset E$. For given elements c of E and b of Y, consider the primal *abstract linear program*

$$\inf\{\langle c,x\rangle \,:\, Ax - b \in H,\ x \in K\}. \tag{3.3.3}$$

As usual, denote the optimal value by p. We can write this problem in Fenchel form (2.3.2) if we define functions f on E and g on Y by $f(x) := \langle c,x\rangle + \delta_K(x)$ and $g(y) := \delta_H(y - b)$. Then the Fenchel dual problem (4.4.1) is

$$\sup\{\langle b,\phi\rangle \,:\, A^*\phi - c \in K^-,\ \phi \in -H^-\} \tag{3.3.4}$$

with dual optimal value d. If we now apply the Fenchel duality theorem (2.3.4) first to problem (3.3.3), and then to problem (3.3.4) (using the bipolar cone theorem (2.4.11)), we obtain the following general result. Note that (ii) of either (a) or (b) below is usually referred to as Slater's condition since it asserts that an inequality holds strictly.

Corollary 3.3.1 (Cone programming duality). *Suppose the cones H and K in problem (3.3.3) are convex.*

(a) *If any of the conditions*

  (i) $b \in \text{int}(AK - H)$,
  (ii) $b \in AK - \text{int}\, H$, or
  (iii) $b \in A(\text{int}\, K) - H$, and either H is polyhedral or A is surjective

  hold then there is no duality gap ($p = d$) and the dual optimal value d is attained when finite.

(b) *Suppose H and K are also closed. If any of the conditions*

  (i) $-c \in \text{int}(A^*H^- + K^-)$,
  (ii) $-c \in A^*H^- + \text{int}\, K^-$, or
  (iii) $-c \in A^*(\text{int}\, H^-) + K^-$, and K is polyhedral or A^* is surjective

  hold then there is no duality gap and the primal optimal value p is attained when finite.

In both parts (a) and (b), the sufficiency of condition (iii) follows by applying the mixed Fenchel duality corollary (3.1.9). In the fully polyhedral case we obtain the following result.

Corollary 3.3.2 (Linear programming duality). *Suppose the cones H and K in the dual pair of problems (3.3.3) and (3.3.4) are polyhedral. If either problem has finite optimal value then there is no duality gap and both problems have optimal solutions.*

Proof. We may apply the polyhedral Fenchel duality corollary (3.1.8) to each problem in turn. □

Our earlier result for the linear program (3.3.1) is clearly just a special case of this corollary. The formulation allows many patterns to be unified. Suppose for instance, that the cone $K = \mathbb{R}^n_+ \times 0_m$ then $K^+ = \mathbb{R}^n_+ \times \mathbb{R}^m$ and equality variables are dual to unconstrained ones.

Linear programming has an interesting matrix analog. Given matrices $A_1, A_2, \ldots, A_m$, and C in S^n_+ and a vector b in $\mathbb{R}^m$, consider the primal *semidefinite*

3.3 Linear and semidefinite programming duality

program

$$\begin{aligned}\inf\quad & \operatorname{tr}(CX)\\ \text{subject to}\quad & \operatorname{tr}(A_iX) = b_i \text{ for } i=1,2,\ldots,m\\ & X \in S_+^n.\end{aligned} \quad (3.3.5)$$

This is a special case of the abstract linear program (3.3.3), so the dual problem is

$$\sup\left\{b^T\phi \,:\, C - \sum_{i=1}^m \phi_i A_i \in S_+^n,\ \phi \in \mathbb{R}^m\right\}, \quad (3.3.6)$$

since $(S_+^n)^- = -S_+^n$, by the self-dual cones proposition (2.4.9), and we obtain the following duality theorem from the general result above.

Corollary 3.3.3 (Semidefinite programming duality). *If the primal problem (3.3.5) has a positive definite feasible solution, there is no duality gap and the dual optimal value is attained when finite. On the other hand, if there is a vector ϕ in $\mathbb{R}^m$ with*

$$C - \sum_i \phi_i A_i \succ 0$$

then once again there is no duality gap and now the primal optimal value is attained when finite.

Unlike linear programming, we really do need a condition stronger than mere consistency to guarantee the absence of a duality gap. For example, if we consider the primal semidefinite program (3.3.5) with

$$n=2,\ m=1,\ C = \begin{bmatrix} 0 & 1 \\ 1 & 0 \end{bmatrix},\ A_1 = \begin{bmatrix} 1 & 0 \\ 0 & 0 \end{bmatrix},\ \text{and } b=0,$$

the primal optimal value is 0 (and is attained), whereas the dual problem (3.3.6) is inconsistent. Another related and flamboyant example is given in Exercise 10.1.6.

The semidefinite cone has many special properties that we have not exploited. One such property is mentioned in Exercise 3.3.8.

Exercises and further results

The fact that duality gaps can and do occur in nonpolyhedral linear programming was a discovery of the 1950s. It is still a topic of active interest especially since the emergence of semidefinite programming. This is well illustrated in [417] where the phenomenon is carefully analyzed and various strategies for repairing the gap are described, as are important concepts about the facial structure of nonpolyhedral cones.

3.3.1.* Verify the precise form of the dual problem for the linear program (3.3.1).

3.3.2 (Linear programming duality gap). Give an example of a linear program of the form (3.3.1) which is inconsistent ($p = +\infty$) with the dual problem (3.3.2) *also* inconsistent ($d = -\infty$).

3.3.3.* Verify the precise form of the dual problem for the abstract linear program (3.3.3).

3.3.4.* Fill in the details of the proof of the cone programming duality corollary (3.3.1). In particular, when the cones H and K are closed, show how to interpret problem (3.3.3) as the dual of problem (3.3.4).

3.3.5.* Fill in the details of the proof of the linear programming duality corollary 3.3.2.

3.3.6 (Complementary slackness). Suppose we know the optimal values of problems (3.3.3) and (3.3.4) are equal and the dual value is attained. Prove a feasible solution x for problem (3.3.3) is optimal if and only if there is a feasible solution ϕ for the dual problem (3.3.4) satisfying the conditions

$$\langle Ax - b, \phi \rangle = 0 = \langle x, A^*\phi - c \rangle.$$

3.3.7 (Semidefinite programming duality).* Prove Corollary 3.3.3.

3.3.8 (Exposedness in the semidefinite cone). Prove that every extreme point of the cone S_+^n is exposed, but that for $n > 2$ not every boundary point is extreme.

Hint. Cones in which all extreme faces are exposed are called *facially exposed*. See [340]. □

The now well-exploited links between semidefinite programming and combinatorial optimization [128] were made apparent by work by Lovász and others over the past two decades (and as is often the case in various earlier work), and especially by a spectacular 1995 result of Goemans and Williamson which applied the sort of ideas we now outline to illustrate how semidefinite relaxation works.

3.3.9 (Semidefinite relaxation).** Semidefinite programs provide an excellent way of relaxing hard combinatorial problems to more tractable convex problems [438, Ch. 13]. This is nicely highlighted by looking at a quadratic Boolean programming model

$$\mu := \max_{x \in \mathcal{F}} \left(\langle Qx, x \rangle - 2\langle c, x \rangle \right)$$

where Q is a $n \times n$ symmetric matrix and $\mathcal{F}$ denotes the Boolean constraint $x = \{\pm 1, \pm 1, \ldots, \pm 1\}$. Classically, such programs were solved by a coarse linear relaxation from $x \in \mathcal{F}$ to $-1 \leq x_k \leq 1, k = 1, \ldots, n$ and then reimposing the Boolean constraints by requiring various variables to be fixed and using branch-and-bound methods to determine which assignment is optimal. A superior semi-definite relaxation is produced as follows.

(a) By homogenization we may assume $c = 0$: replace Q by

$$Q_c := \begin{pmatrix} 0 & -c^T \\ -c & Q \end{pmatrix}$$

and replace x by (t, x). The original problem then has $x_1 = 1$.

(b) Now observe that, as the trace commutes,
$$\langle Qx, x \rangle = \operatorname{tr} x^T Qx = \operatorname{tr} Qxx^T = \operatorname{tr} QY$$
where $Y = xx^T$ is a rank one, positive semidefinite matrix with entries $y_{i,j} = x_i x_j$.

(c) Thus a semidefinite relaxation of the homogenized program is
$$\widehat{\mu} := \max_{Y \succeq 0} \{\operatorname{tr} QY \,:\, \operatorname{diag}(Y) = e\}$$
where $e := (1, 1, \ldots, 1)$, in which we have only relaxed the rank-one constraint.

Goemans and Williamson showed for the NP-hard 'min-cut/max-flow' problem that a randomized version of such a relaxation after dualization – which being a semidefinite program can approximately be solved in polynomial time – has expected performance exceeding 0.87856% of the exact solution [235].

3.4 Selections and fixed points

A *multifunction* is a set-valued mapping. The *range* of a multifunction $\Omega : E \to 2^Y$ is $\bigcup_{x \in E} \Omega(x)$. A *selection* for the multifunction Ω, is a single-valued function $f : E \to Y$ such that $f(x) \in \Omega(x)$ for all $x \in E$. We first present some standard terminology for multifunctions before presenting our main results.

A multifunction Ω between Euclidean spaces E and Y is is said to be *closed* if its *graph*, defined by $\operatorname{graph}(\Omega) := \{(x, y) \,:\, y \in \Omega(x)\}$, is closed in $E \times Y$. It is *upper-semicontinuous* (USC) or alternatively *inner-semicontinuous* (isc) at a point x if $\Omega^{-1}(V) := \{y \in E \,:\, \Omega(y) \subset V\}$ is open whenever V is an open set with $\Omega(x) \subset V$. Suppose U is an open subset in E. Then Ω is said to be an *usco* on U if Ω is USC and has nonempty compact images throughout U; and a *cusco* if the images are also convex.

Likewise Ω is is said to be *lower-semicontinuous* (LSC) or *outer-semicontinuous* (osc) at a point x if $\{y \in E \,:\, \Omega(y) \cap V \neq \emptyset\}$ is open whenever V is an open set with $\Omega(x) \cap V \neq \emptyset$. While isc and osc, as used in [378, Chapter 5], are arguably less ambiguous and more intuitive terms, USC and LSC are still more standard. A mapping that is both LSC and USC is sometimes called *continuous*.

We shall see in this and the next sections that cuscos arise in several important contexts. For the first such example, we shall say that a *fixed point* of a multifunction $\Omega \colon C \to C$ is an element $x \in C$ with $x \in \Omega(x)$. Although this section is presented in the context of Euclidean spaces, the key results extend easily as we shall indicate in Section 6.1.

Theorem 3.4.1 (Kakutani–Fan [274]). *If $C \subset E$ is nonempty, compact and convex, then any cusco $\Omega \colon C \to C$ has a fixed point.*

Before proving this result, we outline a little more topology. Given a finite open cover $\{O_1, O_2, \ldots, O_m\}$ of a set $K \subset E$, a *partition of unity subordinate to* this cover is a set of continuous functions $p_1, p_2, \ldots, p_m : K \to \mathbb{R}_+$ whose sum is identically equal to one and satisfying $p_i(x) = 0$ for all points x outside O_i (for each index i). We outline the proof of the next result, a central topological tool, in Exercise 3.4.6.

Theorem 3.4.2 (Partition of unity). *There is a partition of unity subordinate to any finite open cover of a compact subset of E.*

Topological spaces for which this result remains true are called *paracompact* spaces and include all compact spaces and all metric spaces.

The other theme of this section is the notion of a *continuous selection* of a multifunction Ω on a set $K \subset E$, by which we mean a continuous map f on K satisfying $f(x) \in \Omega(x)$ for all x in K. The central step in our proof of the Kakutani–Fan theorem is the following 'approximate selection' theorem.

Theorem 3.4.3 (Cellina approximate selection theorem [21]). *Given any compact set $K \subset E$, suppose the multifunction $\Omega : K \to Y$ is an usco. Then for any real $\varepsilon > 0$ there is a continuous map $f : K \to Y$ which is an 'approximate selection' of Ω:*

$$d_{G(\Omega)}(x, f(x)) < \varepsilon \text{ for all points } x \text{ in } K, \tag{3.4.1}$$

with range f contained in the closed convex hull of range Ω.

Proof. We use the norm on $E \times Y$ given by

$$\|(x,y)\|_{E \times Y} := \|x\|_E + \|y\|_Y \text{ for all } x \in E \text{ and } y \in Y.$$

As Ω is an usco, for each x in K there is a δ_x in $(0, \varepsilon/2)$ satisfying

$$\Omega(x + \delta_x B_E) \subset \Omega(x) + \frac{\varepsilon}{2} B_Y.$$

Since the sets $x + (\delta_x/2) \text{int } B_E$ (as the point x ranges over K) comprise an open cover of the compact set K, there is a finite subset $\{x_1, x_2, \ldots, x_m\}$ of K with the sets $x_i + (\delta_i/2) \text{int } B_E$ comprising a finite subcover (where δ_i is shorthand for δ_{x_i} for each index i).

Theorem 3.4.2 provides a partition of unity $p_1, p_2, \ldots, p_m : K \to \mathbb{R}_+$ subordinate to this subcover. We construct our approximate selection f by choosing a point y_i from $\Omega(x_i)$ for each i and defining

$$f(x) := \sum_{i=1}^{m} p_i(x) y_i \text{ for all } x \text{ in } K. \tag{3.4.2}$$

Fix any point x in K and set $I := \{i : p_i(x) \neq 0\}$. By definition, x satisfies $\|x - x_i\| < \delta_i/2$ for each i in I. If we choose an index j in I maximizing δ_j, the triangle inequality shows $\|x_j - x_i\| < \delta_j$, whence

$$y_i \in \Omega(x_i) \subset \Omega(x_j + \delta_j B_E) \subset \Omega(x_j) + \frac{\varepsilon}{2} B_Y$$

for all i in I. In other words, for each i in I we know $d_{\Omega(x_j)}(y_i) \leq \varepsilon/2$. Since this distance function is convex, equation (3.4.2) shows $d_{\Omega(x_j)}(f(x)) \leq \varepsilon/2$. Since

$\|x - x_j\| < \varepsilon/2$, this proves inequality (3.4.1). The final claim follows immediately from equation (3.4.2). □

Proof of the Kakutani–Fan theorem. For each positive integer r Cellina's result shows there is a continuous self map f_r of C satisfying

$$d_{G(\Omega)}(x, f_r(x)) < \frac{1}{r} \quad \text{for all points } x \text{ in } C.$$

By Brouwer's fixed point theorem [95, Theorem 8.1.3] each f_r has a fixed point x^r in C, with

$$d_{G(\Omega)}(x^r, x^r) < \frac{1}{r} \quad \text{for each } r.$$

Since C is compact, the sequence (x^r) has a convergent subsequence whose limit must be a fixed point of Ω since Ω is closed by Exercise 3.4.3(b). □

Cellina's result remains true with essentially the same proof in Banach space – as does the Kakutani–Fan theorem [208].

We end this section with an *exact* selection theorem parallel to Cellina's result but assuming the multifunction is *LSC* ($\{x : \Omega(x) \cap V \neq \emptyset\}$ is open whenever V is.) rather than an usco. The proof is outlined in Exercise 3.4.8.

Theorem 3.4.4 (Michael selection theorem [315]). *Given any closed set $K \subset E$, suppose the multifunction $\Omega : K \to Y$ is LSC with nonempty closed convex images. Then given any point $(\bar{x}, \bar{y})$ in $G(\Omega)$, there is a continuous selection f of Ω satisfying $f(\bar{x}) = \bar{y}$.*

Figure 3.2 illustrates the distinction between the Michael's and Cellina's results. The right-hand image clearly shows a better approximate selection, but the mapping is not LSC and any continuous mapping must stay bounded away from the maximal monotone mapping in supremum norm.

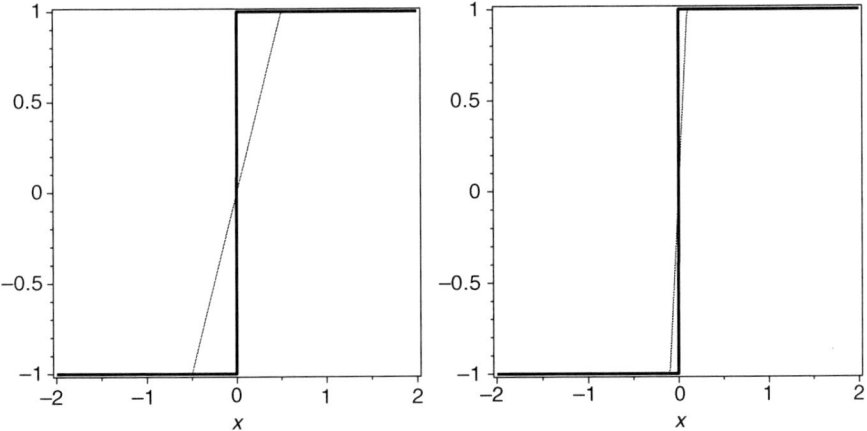

Figure 3.2 Two approximate selections to $\partial | \cdot |$.

Exercises and further results

3.4.1 (USC and continuity). Fix $\Omega : K \to 2^Y$ for a closed subset $K \subset E$.

(a) Prove the multifunction

$$x \in E \mapsto \begin{cases} \Omega(x) & \text{for } x \in K \\ \emptyset & \text{for } x \notin K \end{cases}$$

is USC if and only if Ω is USC.

(b) Prove a function $f : K \to Y$ is continuous if and only if the multifunction $x \in K \mapsto \{f(x)\}$ is USC.

(c) Prove a function $f : E \to [-\infty, +\infty]$ is lsc at a point x in E if and only if the multifunction whose graph is the epigraph of f is USC at x.

3.4.2 (Minimum norm selection). If $U \subset E$ is open and $\Omega : U \to Y$ is USC, prove the function $g : U \to Y$ defined by $g(x) = \inf\{\|y\| : y \in \Omega(x)\}$ is lsc.

3.4.3 (Closed versus USC).* If the multifunction $\Phi : E \to Y$ is closed and the multifunction $\Omega : E \to Y$ is USC at the point x in E with $\Omega(x)$ compact, prove the multifunction $z \in E \mapsto \Omega(z) \cap \Phi(z)$ is USC at x.

(a) Hence prove that any closed multifunction with compact range is USC.
(b) Prove any USC multifunction with closed images is closed.
(c) If a USC multifunction has compact images, prove it is locally bounded.

3.4.4 (Composition). If the multifunctions Φ and Ω are USC prove their composition $x \mapsto \Phi(\Omega(x))$ is also.

3.4.5 (USC images of compact sets). Fix a multifunction $\Omega : K \to Y$. Suppose K is compact and Ω is USC with compact images. Prove the range $\Omega(K)$ is compact by following the steps below.

(a) Consider an open cover $\{U_\gamma : \gamma \in \Gamma\}$ of $\Omega(K)$. For each point x in K, prove there is a finite subset Γ_x of Γ with

$$\Omega(x) \subset \bigcup_{\gamma \in \Gamma_x} U_\gamma.$$

(b) Construct an open cover of K by considering the sets

$$\left\{ z \in K : \Omega(z) \subset \bigcup_{\gamma \in \Gamma_x} U_\gamma \right\}$$

as the point x ranges over K.

(c) Hence construct a finite subcover of the original cover of $\Omega(K)$.

3.4.6 (Partitions of unity). Suppose the set $K \subset E$ is compact with a finite open cover $\{O_1, O_2, \ldots, O_m\}$.

(a) Show how to construct another open cover $\{V_1, V_2, \ldots, V_m\}$ of K satisfying cl $V_i \subset O_i$ for each index i.

3.4 Selections and fixed points

Hint. Each point x in K lies in some set O_i, so there is a real $\delta_x > 0$ with $x + \delta_x B \subset O_i$; now take a finite subcover of $\{x + \delta_x \text{ int } B : x \in K\}$ and build the sets V_i from it. □

(b) For each index i, prove the function $q_i : K \to [0, 1]$ given by

$$q_i := \frac{d_{K \setminus O_i}}{d_{K \setminus O_i} + d_{V_i}}$$

is well-defined and continuous, with q_i identically zero outside of O_i.

(c) Deduce that the set of functions $p_i : K \to \mathbb{R}_+$ defined by

$$p_i := \frac{q_i}{\sum_j q_j}$$

is a partition of unity subordinate to the cover $\{O_1, O_2, \ldots, O_m\}$.

3.4.7.* Prove the Kakutani–Fan theorem (3.4.1) is valid under the weaker assumption that the images of the cusco $\Omega : C \to E$ always intersect the set C using Exercise 3.4.3.

3.4.8 (Michael's theorem).* Suppose the assumptions of Michael's theorem (3.4.4) hold. Consider first the case with K compact.

(a) Fix a real $\varepsilon > 0$. By constructing a partition of unity subordinate to a finite subcover of the open cover of K consisting of the sets

$$O_y := \{x \in E : d_{\Omega(x)}(y) < \varepsilon\} \text{ for } y \text{ in } Y,$$

construct a continuous function $f : K \to Y$ satisfying

$$d_{\Omega(x)}(f(x)) < \varepsilon \text{ for all points } x \text{ in } K.$$

(b) Construct continuous functions $f_1, f_2, \ldots : K \to Y$ satisfying

$$d_{\Omega(x)}(f_i(x)) < 2^{-i} \text{ for } i = 1, 2, \ldots$$
$$\|f_{i+1}(x) - f_i(x)\| < 2^{1-i} \text{ for } i = 1, 2, \ldots$$

for all points x in K.

Hint. Construct f_1 by applying part (a) with $\varepsilon = 1/2$; then construct f_{i+1} inductively by applying part (a) to the multifunction

$$x \in K \mapsto \Omega(x) \cap (f_i(x) + 2^{-i} B_Y)$$

with $\varepsilon = 2^{-i-1}$. □

(c) The functions f_i of part (b) must converge uniformly to a continuous function f. Prove f is a continuous selection of Ω.

(d) Prove Michael's theorem by applying part (c) to the multifunction

$$\hat{\Omega}(x) := \begin{cases} \Omega(x) & \text{if } x \neq \bar{x} \\ \{\bar{y}\} & \text{if } x = \bar{x}. \end{cases}$$

(e) Now extend to the general case where K is possibly unbounded in the following steps. Define sets $K_n := K \cap nB_E$ for each $n = 1, 2, \ldots$ and apply the compact case to the multifunction $\Omega_1 = \Omega|_{K_1}$ to obtain a continuous selection $g_1 : K_1 \to Y$. Then inductively find a continuous selection $g_{n+1} : K_{n+1} \to Y$ from the multifunction built from

$$\Omega_{n+1}(x) := \begin{cases} \{g_n(x)\} & \text{for } x \in K_n \\ \Omega(x) & \text{for } x \in K_{n+1} \setminus K_n \end{cases}$$

and prove the function defined by

$$f(x) := g_n(x) \text{ for } x \in K_n, \ n = 1, 2, \ldots$$

is the required selection.

The result remains true with much the same proof if K is paracompact and Y is a Banach space. A lovely application is:

3.4.9 (Hahn–Katětov–Dowker sandwich theorem). Suppose $K \subset E$ is closed.

(a) For any two lsc functions $f, g : K \to \mathbb{R}$ satisfying $f \geq -g$, prove there is a continuous function $h : K \to \mathbb{R}$ satisfying $f \geq h \geq -g$ by considering the multifunction $x \mapsto [-g(x), f(x)]$. Observe the result also holds for extended-real-valued f and g.

(b) (Urysohn lemma). Suppose the closed set V and the open set U satisfy $V \subset U \subset K$. Use part (a) to prove there is a continuous function $f : K \to [0, 1]$ that is identically equal to one on V and to zero on U^c.

3.4.10 (Continuous extension theorem). Consider a closed subset K of E and a continuous function $f : K \to Y$. By considering the multifunction

$$\Omega(x) := \begin{cases} \{f(x)\} & \text{for } x \in K \\ \operatorname{cl}(\operatorname{conv} f(K)) & \text{for } x \notin K, \end{cases}$$

prove there is a continuous function $g : E \to Y$ satisfying $g|_K = f$ and $g(E) \subset \operatorname{cl}(\operatorname{conv} f(K))$.

3.4.11 (Generated cuscos). Suppose the multifunction $\Omega : K \to Y$ is locally bounded with nonempty images.

(a) Among those cuscos containing Ω, prove there is a unique one with minimal graph, given by

$$\Phi(x) := \bigcap_{\varepsilon > 0} \operatorname{cl} \operatorname{conv}(\Omega(x + \varepsilon B)) \text{ for } x \in K.$$

(b) If K is nonempty, compact, and convex, $Y = E$, and Ω satisfies the conditions $\Omega(K) \subset K$ and $x \in \Phi(x) \Rightarrow x \in \Omega(x)$ for $x \in K$, prove Ω has a fixed point.

3.4.12 (Multifunctions containing cuscos). Suppose $\Omega : K \to Y$ is closed with nonempty convex images, and the function $f : K \to Y$ is such that $f(x)$ is a point of minimum norm in $\Omega(x)$ for all points x in K. Prove Ω contains a cusco if and only if f is locally bounded.

Hint. Use Exercise 3.4.11 to consider the cusco generated by f. □

3.5 Into the infinite

Many of the results we shall meet in future chapters are of significance in Euclidean space and some admit much simpler proofs. To aid with browsing in chapters to follow we preview some of these results in their Euclidean avatars now. We begin with the simple but powerful Euclidean case of Ekeland's variational principle (4.3.1) which is given in a complete metric space.

Theorem 3.5.1 (Ekeland's variational principle). *Let E be Euclidean and let $f : E \to (-\infty, +\infty]$ be a lsc function bounded from below. Suppose that $\varepsilon > 0$ and $z \in E$ satisfy*

$$f(z) < \inf_E f + \varepsilon.$$

Suppose $\lambda > 0$ is given, then there exists $y \in E$ such that

*(a) $\|z - y\| \leq \lambda$, (b) $f(y) + (\varepsilon/\lambda)\|z - y\| \leq f(z)$, and
(c) $f(x) + (\varepsilon/\lambda)\|x - y\| > f(y)$, for all $x \in E \setminus \{y\}$.*

Proof. Let g be defined by $g(x) := f(x) + (\varepsilon/\lambda)\|x - z\|$. Then g is lsc and coercive and so achieves its minimum at a point y. Hence

$$f(x) + (\varepsilon/\lambda)\|x - z\| \geq f(y) + (\varepsilon/\lambda)\|z - y\| \tag{3.5.1}$$

for all $x \in E$. In particular $\inf_E f + \varepsilon > f(z) \geq f(y) + (\varepsilon/\lambda)\|z - y\|$, whence (a) and (b) follow. The triangle inequality applied to (3.5.1) gives (c). □

A potent application of Ekeland's principle is to the Brøndsted–Rockafellar theorem of Exercise 3.5.7. Another fine illustration of the variational principle at work is to rapidly establish that every Čebyšev set in Euclidean space is convex.

Recall that a set C in Euclidean space is said to be a *Čebyšev set* if every point in the space has a unique best approximation in C (see Theorem 1.3.1 and Exercise 2.3.20.) We will discuss this concept further and study another conjugate function approach in Section 4.5. We call $C \subset X$ *approximately convex* if, for any closed norm ball $D \subset X$ disjoint from C, there exists a closed ball $D' \supset D$ disjoint from C with arbitrarily large radius.

Theorem 3.5.2 (Čebyšev sets are convex [96, 229]). *Every Čebyšev set C in a Euclidean space is convex (and closed).*

Proof. By Exercise 3.5.2 it suffices to show C is approximately convex. Suppose not. We claim that for each $x \notin C$

$$\limsup_{y \to x} \frac{d_C(y) - d_C(x)}{\|y - x\|} = 1. \tag{3.5.2}$$

As a consequence of the mean value theorem (Exercise 3.5.5) for Lipschitz functions (since in the language of Section 9.2 all Fréchet (super-)gradients have norm-one off C). Consider any real $\alpha > d_C(x)$. Fix reals $\sigma \in (0, 1)$ and ρ satisfying

$$\frac{\alpha - d_C(x)}{\sigma} < \rho < \alpha - \beta.$$

Apply Theorem 3.5.1 to the function $-d_C + \delta_{x+\rho B}$, where $B := B_E$, to show there exists a point $v \in E$ satisfying the conditions

$$d_C(x) + \sigma \|x - v\| \leq d_C(v)$$
$$d_C(z) - \sigma \|z - v\| \leq d_C(v) \text{ for all } z \in x + \rho B.$$

Deduce $\|x - v\| = \rho$, and hence $x + \beta B \subset v + \alpha B$. □

We say a multifunction $T \colon E \to 2^{E^*}$ is *monotone* on a set U if $\langle y^* - x^*, y - x \rangle \geq 0$ whenever $x, y \in U$ and $x^* \in T(x)$, $y^* \in T(y)$. If no set is mentioned it is assumed to be the *domain* of T, that is: $\{x \colon T(x) \neq \emptyset\}$. A monotone multifunction – also called a *monotone operator*, especially if single-valued or in infinite dimensions – is *maximal monotone* if $\langle y^* - x^*, y - x \rangle \geq 0$ for all $x^* \in T(x)$ implies that $y^* \in T(y)$. We say such a pair (y, y^*) is *monotonically related* to the graph of T.

In Exercise 2.2.21 we verified monotonicity of the subgradient. It is worth noting that any positive definite matrix induces a (single-valued) monotone operator that is most often not a subgradient (Exercise 3.5.4). More subtly and significantly we have:

Theorem 3.5.3 (Rockafellar). *Suppose $f \colon E \to (-\infty, +\infty]$ is a lsc proper convex function. Then ∂f is a maximal monotone multifunction.*

Proof. The proof in Theorem 6.1.5 is applicable if X is replaced by E and Exercise 3.5.7 is used instead of the full Brøndsted–Rockafellar theorem (4.3.2). Another proof is given in Exercise 3.5.11. □

A function $f \colon E \to (-\infty, +\infty]$ is said to be *supercoercive* if $\lim_{\|x\| \to \infty} \frac{f(x)}{\|x\|} = +\infty$ whereas f is called *cofinite* if its conjugate is finite everywhere on E. In many sources, supercoercive functions are referred to as *strongly coercive*. A fundamental duality result is:

Proposition 3.5.4 (Moreau–Rockafellar). *A proper closed convex function f on Euclidean space is supercoercive if and only if it is cofinite.*

Proof. Let $\alpha > 0$ and $\beta \in \mathbb{R}$. Then Fenchel conjugation shows that $f \geq \alpha \| \cdot \| + \beta$ if and only if $f^* \leq -\beta$ on αB_E; see Theorem 2.3.2 for an alternate approach to this. After translation, this shows that f^* is continuous at x_0 if and only if $f - \langle x_0, \cdot \rangle$ grows

3.5 Into the infinite

as fast as some multiple of the norm. A simple argument shows that if this holds at all points then f is supercoercive. The converse follows more easily in the same way. □

For a discussion of this sort of result in Banach space see Fact 4.4.8 and Exercise 4.4.23.

Finally, we make a brief introduction to a remarkable convex function whose study is central to Chapter 9. For a monotone multifunction T, we associate the *Fitzpatrick function* $\mathcal{F}_T$ introduced in [215]. It is given by

$$\mathcal{F}_T(x, x^*) := \sup\{\langle x, y^*\rangle + \langle x^*, y\rangle - \langle y, y^*\rangle : y^* \in T(y)\}, \tag{3.5.3}$$

and it is clearly lsc and convex on $E \times E$ as it is an affine supremum. Moreover, see Exercise 3.5.10, one has:

Proposition 3.5.5. [215, 121] *The Fitzpatrick function associated with a maximal monotone multifunction $T: E \to E$ satisfies*

$$\mathcal{F}_T(x, x^*) \geq \langle x, x^*\rangle \tag{3.5.4}$$

for all (x, x^*) in $E \times E$ with equality if and only if $x^* \in T(x)$.

Proposition 3.5.6 (Minty condition). *A monotone operator T on a Euclidean space E is maximal if and only if* range$(T + I) = E$.

Proof. We prove the 'if'. Assume (w, w^*) is monotonically related to the graph of T. By hypothesis, we may solve $w^* + w \in (T + I)(x + w)$. Thus $w^* = t^* + x$ where $t^* \in T(x + w)$,. Hence

$$0 \leq \langle w - (w + x), w^* - t^*\rangle = -\langle x, w^* - t^*\rangle = -\langle x, x\rangle = -\|x\|^2 \leq 0.$$

Thus, $x = 0$. So $w^* \in T(w)$ and we are done.

The 'only if' is a special case of Theorem 3.5.7. □

Theorem 3.5.7 (Maximality of sums). *Let T be maximal monotone on a Euclidean space E and let f be closed and convex on E. Suppose the following constraint qualification (CQ) holds:*

$$0 \in \text{core}\{\text{conv dom}(T) - \text{conv dom}(\partial f)\}.$$

Then

(a) $(\partial f + T) + I$ *is surjective.*
(b) $\partial f + T$ *is maximal monotone.*

Proof. (a) We use the Fitzpatrick function $\mathcal{F}_T(x, x^*)$ and further introduce $f_1(x) := f(x) + 1/2\|x\|^2$. Let $G(x, x^*) := -f_1(x) - f_1^*(-x^*)$. Observe that for all $x, x^* \in E$, by the Fenchel–Young inequality (2.3.1)

$$\mathcal{F}_T(x, x^*) \geq \langle x, x^*\rangle \geq G(x, x^*)$$

along with Proposition 3.5.5. Now, $0 \in \operatorname{core}\{\operatorname{conv} \operatorname{dom}(T) - \operatorname{conv} \operatorname{dom}(\partial f)\}$ guarantees that the sandwich theorem (2.4.1) applies to $\mathcal{F}_T \geq G$ since f_I^* is everywhere finite (by Exercise 3.5.11). So there are $w, w^* \in E$ with

$$\mathcal{F}_T(x, x^*) - G(z, z^*) \geq w(x^* - z^*) + w^*(x - z) \qquad (3.5.5)$$

for all x, x^* and all z, z^*. In particular, for $x^* \in T(x)$ and for all $z^*, z \in E$ we have

$$\langle x - w, x^* - w^* \rangle + \left[f_I(z) + f_I^*(-z^*) + \langle z, z^* \rangle \right] \geq \langle w - z, w^* - z^* \rangle.$$

Now we use Exercise 3.5.11 again to solve $-w^* \in \operatorname{dom}(\partial f_I^*)$, and so to deduce that $-w^* \in \partial f_I(v)$ for some z. Thus

$$\langle v - w, x^* - w^* \rangle + [f_I(v) + f_I^*(-w^*) + \langle v, w^* \rangle] \geq \langle w - v, w^* - w^* \rangle = 0.$$

The second term on the left is zero and so $w^* \in T(w)$ by maximality. Substitution of $x = w$ and $x^* = w^*$ in (3.5.5), and rearranging yields

$$\langle w, w^* \rangle + \{ \langle -z^*, w \rangle - f_I^*(-z^*) \} + \{ \langle z, -w^* \rangle - f_I(z) \} \leq 0,$$

for all z, z^*. Taking the supremum over z and z^* produces $\langle w, w^* \rangle + f_I(w) + f_I^*(-w^*) \leq 0$. This shows $-w^* \in \partial f_I(w) = \partial f(w) + w$ on using the sum formula for subgradients.

Thus, $0 \in (T + \partial f_I)(w)$. Since all translations of $T + \partial f$ may be used, as the (CQ) is undisturbed by translation, $(\partial f + T) + I$ is surjective which completes (a); and (b) follows from Proposition 3.5.6. □

Exercises and further results

3.5.1 (Approximately critical points [193, 121]). Let g be continuous, Gâteaux differentiable and bounded below on a Euclidean space.

(a) Use Ekeland's variational principle (3.5.1) to show that, for each $\varepsilon > 0$, there is a point $x_\varepsilon \in E$ with $\|\nabla g(x_\varepsilon)\| \leq \varepsilon$.
(b) Suppose more stringently that g is supercoercive and continuously differentiable. Show that range $\nabla g = E$.

Hint. First, fix $y \in E$ and apply part (a) to $x \mapsto g(x) - \langle y, x \rangle$ to deduce that range g is dense. Then appeal again to coercivity to show the range is closed. □

3.5.2 (Approximate convexity [96, 229]).** Show every convex set in a finite-dimensional Banach space is approximately convex. If the space has a smooth norm (equivalently a rotund dual norm) then every approximately convex set is convex.

Hint. The first assertion follows from the Hahn–Banach theorem. Conversely, suppose C is approximately convex but not convex. Then there exist points $a, b \in C$ and a closed ball D centered at the point $c := (a + b)/2$ and disjoint from C. Hence, there exists a sequence of points $x_1, x_2, \ldots$ such that the balls $B_r = x_r + rB$ are disjoint from C and satisfy $D \subset B_r \subset B_{r+1}$ for all $r = 1, 2, \ldots$. The set $H := \operatorname{cl} \bigcup_r B_r$ is closed and convex, and its interior is disjoint from C but contains c. It remains to

3.5 Into the infinite

confirm that H is a half-space. Suppose the unit vector u lies in the polar set H°. By considering the quantity $\langle u, \|x_r - x\|^{-1}(x_r - x)\rangle$ as $r \uparrow \infty$, we discover H° must be a ray. This means H is a half-space. □

3.5.3. Show that a single-valued continuous monotone operator T on a Euclidean space is maximal.

Hint. $\langle T(x + th), h\rangle \geq 0$ for all $t > 0$ and all $h \in E$. □

3.5.4. Let A be a $n \times n$ semidefinite matrix, i.e. $\langle Ax, x\rangle \geq 0$ for all $x \in \mathbb{R}^n$. Show that $x \mapsto Ax$ defines a monotone operator, maximal (by Exercise 3.5.3), which is a subgradient if and only if A is symmetric.

Hint. $x \mapsto \langle Ax, x\rangle \geq 0$ is convex. □

3.5.5 (Convex mean value theorem). Suppose f is convex and everywhere continuous on a Euclidean space. Show that for $u, v \in E$ there exists z interior to the line segment $[u, v]$ with

$$f(u) - f(v) \in \langle \partial f(z), u - v\rangle. \qquad (3.5.6)$$

Hint. Define a convex function by $g(t) := f(u + t(v - u)) - t(f(v) - f(u))$ for $0 \leq t \leq 1$. Write $z(t) := u + t(v - u)$. As $g(0) = f(u) = g(1)$ there is an interior minimum at some s in $(0, 1)$. Set $z := z(s)$. Thus $0 \leq g'(s; \pm 1) = f'(z; \pm(v - u)) - (\pm)(f(v) - f(u))$. Now use the max formula (2.1.19). □

If f is merely locally Lipschitz, Exercise 2.5.6 allows one to show that (3.5.6) still holds if $\partial_o f(z)$ is used.

3.5.6. Use Exercise 3.5.5 to provide a simple proof that ∂f is maximal monotone when f is convex and everywhere continuous on a Euclidean space.

3.5.7 (Brøndsted–Rockafellar). Given $f : E \to (-\infty, +\infty]$ and $x_0 \in \text{dom} f$ and $\varepsilon > 0$, the ε-subdifferential of f at x_0 is defined by

$$\partial_\varepsilon f(x_0) = \{\phi \in X^* : \phi(x) - \phi(x_0) \leq f(x) - f(x_0) + \varepsilon, x \in E\}. \qquad (3.5.7)$$

Suppose f is a proper lsc convex function on the Euclidean space E. Then given any $x_0 \in \text{dom} f$, $\varepsilon > 0$, $\lambda > 0$ and $x_0^* \in \partial_\varepsilon f(x_0)$, there exist $x \in \text{dom} f$ and $x^* \in E$ such that

$$x^* \in \partial f(x), \|x - x_0\| \leq \varepsilon/\lambda \text{ and } \|x^* - x_0^*\| \leq \lambda.$$

Hint. First, $\langle x_0^*, x - x_0\rangle \leq f(x) - f(x_0)$ for all $x \in E$. Now define $g(x) := f(x) - \langle x_0^*, x\rangle$ for $x \in E$. Note that $g(x_0) \leq \inf_X g + \varepsilon$, and apply Theorem 3.5.1. A full proof is given in Theorem 4.3.2. □

3.5.8.* Suppose E is a Euclidean space and $T : E \to E$ is a maximal monotone mapping. Show that $(I + T)^{-1}$ has domain E and is nonexpansive. In particular, Rockafellar's theorem (3.5.3) ensures this applies when $T = \partial f$ where $f : E \to \mathbb{R}$ is convex.

Hint. Proposition 3.5.6 implies that $\mathrm{dom}(I+T)^{-1} = H$. Suppose $y_1 \in (I+T)(x_1)$ and $y_2 \in (I+T)(x_2)$. Then $y_1 - x_1 \in Tx_1$ and $y_2 - x_2 \in Tx_2$. By the monotonicity of T and inner product norm

$$0 \leq \langle (y_1 - x_1) - (y_2 - x_2), x_1 - x_2 \rangle$$
$$= \langle y_1 - y_2, x_1 - x_2 \rangle - \|x_1 - x_2\|^2$$

and so $\|x_1 - x_2\| \leq \|y_1 - y_2\|$ as desired. □

3.5.9 (Local boundedness). In Theorem 2.1.10 and Exercise 2.1.23 we examined the boundedness behavior of convex functions and subdifferentials. This behavior extends to monotone operators.

(a) Show that every (maximal) monotone multifunction is *locally bounded* at each point of the interior of its domain: for each such point x_0 there is a neighbourhood $U(x_0)$ of x_0 such that $\{y : y \in T(x), x \in U(x_0)\}$ is a bounded set.

Hint. See Exercise 6.1.6. □

(b) Show that Exercise 2.2.23 also extends to maximal monotone operators.

3.5.10.* Prove Proposition 3.5.5.

3.5.11. Use Exercise 2.3.15 to show that for a closed convex function f on E and $f_I := f + \frac{1}{2}\|\cdot\|^2$ we have $(f + \frac{1}{2}\|\cdot\|^2)^* = f^* \square \frac{1}{2}\|\cdot\|_*^2$ is everywhere continuous. Also $v^* \in \partial f(v) + v \Leftrightarrow f_I^*(v^*) + f_I(v) - \langle v, v^* \rangle \leq 0$.

3.5.12 (Maximality of subgradients). Apply Theorem 3.5.7 with $T := 0$ to deduce that the subgradient of a closed convex function on a Euclidean space is maximal monotone.

3.5.13 (Maximality of sums [369]). Apply Theorem 3.5.7 deduce that the sum of two maximal monotone multifunctions T_1 and T_2 on a Euclidean space is maximal monotone if $0 \in \mathrm{core}[\mathrm{conv}\,\mathrm{dom}(T_1) - \mathrm{conv}\,\mathrm{dom}(T_2)]$.

Hint. Apply Theorem 3.5.7 to the maximal monotone product mapping $T(x,y) := (T_1(x), T_2(y))$ and the indicator function $f(x,y) := \delta_{\{x=y\}}$ of the diagonal in $E \times E$. Check that the given condition implies the (CQ) needed to deduce that $T + I_{E \times E} + \partial \delta_{\{x=y\}}$ is surjective. Thus, so is $T_1 + T_2 + 2I$ and we are done, the constant 2 being no obstruction. □

3.5.14 (Cuscos [378]). Prove the following assertions for multifunctions between Euclidean spaces.

(a) A locally bounded multifunction with a closed graph is an usco on the interior of its domain.

(b) A multifunction Ω with nonempty compact images is USC at x if and only if for each $\varepsilon > 0$ there is $\delta > 0$ with $\Omega(y) \subset \Omega(x) + \varepsilon B_E$ for $\|y - x\| < \delta$.

(c) Any monotone operator T, and so the subgradient of a convex function, is single-valued at any point where it is LSC.

Hint. Suppose $y_1 \neq y_2 \in T(x)$. Fix $h \in E$ with $\langle y_2, h \rangle - \langle y_1, h \rangle > \varepsilon > 0$. For all small $t > 0$ one has $T(x + th) \cap \{y : \langle y, h \rangle < \langle y_1, h \rangle + \varepsilon\} = \emptyset$. □

3.5 Into the infinite

We shall see later in the book – starting with Exercises 3.5.15 and 3.5.16 that – in our context, in large part because of (c) – cuscos are usually the most appropriate extension of the notion of continuity to multifunctions.

3.5.15 (Clarke subdifferential cusco [156]).** Let $U \subseteq E$ be an open subset of a Euclidean space. In the language of Exercise 2.5.6 show that:

(a) The Clarke derivative $f°(x; v)$ is usc as a function of (x, v).
(b) The Clarke subdifferential $\partial_\circ f : U \to 2^E$ is a cusco on U.
(c) Thus, the subdifferential of a continuous convex function is a cusco.

3.5.16 (Maximal monotone cusco). Let $U \subseteq E$ be an open subset of a Euclidean space. Show that if T is maximal monotone on E and $U \subseteq \text{dom } T$ then T is a cusco on U.

3.5.17 (Differential inclusions [157]).† A *differential inclusion* requests a solution to

$$x'(t) \in F(t, x(t)) \text{ a.e.}, \quad a \leq t \leq b, \quad (3.5.8)$$

where a solution is taken to be an absolutely continuous mapping from $[a, b]$ to $\mathbb{R}^n$. This provides a very flexible way of unifying apparently different problems such as ordinary differential equations, differential inequalities and various control systems. Under mild conditions (see Filippov's lemma [157, p. 174]) one can replace $x'(t) = f(t, x(t), u(t))$ where the control is $u(t) \in U$ by the inclusion $x'(t) \in F(t, x) := f(t, x, U)$.

From now on we shall consider the following autonomous problem

$$x'(t) \in F(x(t)) \text{ a.e.}, \quad 0 \leq t < \infty, \quad (3.5.9)$$

where $F : \mathbb{R}^n \to \mathbb{R}^n$ is a cusco with constants γ, δ such that $\sup \|F(x)\| \leq \gamma \|x\| + \delta$ for $x \in \mathbb{R}^n$. A zero x^* of F is – in an obvious way – an *equilibrium point* of the system. If for every $\alpha \in \mathbb{R}^n$ there is a solution with $x(0) = \alpha$ on the entire positive half-line such that $\|x(t) - x^*\| \to 0$ as $t \to \infty$ then the equilibrium is said to be *globally asymptotically stable*. This is clearly a desirable property and is conveniently verified by the existence of a *(smooth) Lyapunov pair* (V, W) of smooth functions on $\mathbb{R}^n$ such that for all x in $\mathbb{R}^n$ one has (i) $V(x) \geq 0$, $W(x) \geq 0$ with $W(x) = 0$ if and only if $x = x^*$; (ii) V has compact lower level sets; and (iii)

$$\min_{v \in F(x)} \langle \nabla V(x), v \rangle \leq -W(x).$$

(a) When F is singleton, show that $t \mapsto V(x(t)) + \int_0^t W(x(s)) \, ds$ is decreasing and so bounded. Hence, so is $t \mapsto V(x(t))$. Thus, $x(t)$ remains bounded. In particular, the ODE has a solution on the whole positive half-line and $t \mapsto W(x(t))$ is globally Lipschitz on $[0, \infty)$.
(b) Deduce that $W(x(t))$ converges to zero and so $x(t) \to x^*$.
(c) A more technical version of the same argument works for a cusco with linear growth [157, p. 209].

Lyapunov functions are used very broadly and provide ways of validating the behavior of trajectories without having to compute them explicitly. The function V measures 'energy' of the system and $-W$ gives a rate of dissipation.

3.5.18 (Duality of Lyapunov functions [234]).[†] The generality of Exercise 3.5.17 perhaps obscures some remarkable convex duality structure. Let us consider the case of a *linear differential inclusion* or (LDI) where we set

$$F(x) := \text{conv}\{A_1, A_2, \ldots, A_m\}(x)$$

for $n \times n$ matrices $A_1, A_2, \ldots, A_m$. There is a natural *adjoint inclusion* with $G(x) := \text{conv}\{A_1^*, A_2^*, \ldots, A_m^*\}(x)$.

A pretty piece of convex analysis is the following whose proof is left as an exercise in convex analysis.

Theorem 3.5.8 (Lyapunov duality inequalities [234]).[†] *Let $1/p + 1/q = 1$. Suppose $V: \mathbb{R}^n \to \mathbb{R}$ is convex and everywhere finite, is positively homogeneous of degree $p > 1$ and has $V(x) > 0$ for $x \neq 0$. Then V^* is convex and everywhere finite, is positively homogeneous of degree $q > 1$ and has $V^*(x) > 0$ for $x \neq 0$. Moreover, the inequality*

$$\langle \partial V(x), Ax \rangle \leq -\gamma p V(x) \qquad \text{for all } x \in \mathbb{R}^n \qquad (3.5.10)$$

holds if and only if the adjoint inequality holds

$$\langle \partial V^*(y), A^* y \rangle \leq -\gamma q V^*(y) \qquad \text{for all } y \in \mathbb{R}^n. \qquad (3.5.11)$$

3.5.19 (More on Lyapunov duality [234]).[†] When $p = q = 2$, Theorem 3.5.8 leads to interesting results such as a neat proof that asymptotic stability at the origin of F and of G coincide [234]. This relies on understanding what happens for a single differential equation and building an appropriate Lyapunov pair or inequality.

(a) For a linear equation, global asymptotic stability implies *globally exponential stability*: for some $c \geq 1, \beta > 0$ every trajectory satisfies $\|x(t)\| \leq c\|x(0)\|e^{-\beta t}$. Indeed, in this case $V(z) := \sup\{\|x(t)\|^2 e^{2\beta t} : t \geq 0, x(0) = z, x' = Ax\}$ is convex and satisfies (3.5.10) with $p = 2$. This extends to the general (LDI).

(b) Conversely, the existence of V solving (3.5.10) with $p = 2$ assures global exponential stability.

(c) Concretely, one can solve the *Lyapunov equation* $A^T P + PA = -Q$ where P, Q are positive definite matrices and so construct $V(z) := \langle z, Pz \rangle$ for which $\frac{d}{dt} V(x(t))|_{t=0} = -\langle z, Qz \rangle$ where $x(t)$ is a solution of the ODE with $x(0) = z$. In this case the role of $W(z)$ in Exercise 3.5.17 is taken by $\langle z, Qz \rangle$.

3.5.1 How to browse in the rest of the book

In many cases replacing X by E and ignoring more technical conditions will cause little problem. More precisely we advertise the following.

3.5 Into the infinite 125

Material appropriate in Euclidean space. Exercises 4.1.22, 4.4.19, 4.4.23, 4.6.4, 5.1.34, Remark 5.3.9, Exercises 6.2.10, 6.2.11, Theorem 6.2.13, Exercises 6.2.15, 7.2.3, 7.3.2, Exercises 7.3.6, 7.3.7, Theorem 7.3.8, Sections 7.4, 7.5, Lemma 7.6.3, Theorem 8.2.2, Exercise 8.3.1, Theorem 8.4.2, Exercise 8.4.3, Exercises 9.1.1, 9.1.2, 9.1.3, Theorem 9.2.21, Section 10.1.

Material appropriate in Hilbert space. Exercises 4.1.27, 4.1.46, 4.1.15, 4.1.17, 4.4.5, Fact 4.5.6, Theorems 4.5.7, 4.5.9, Exercises 4.5.6, 4.5.8, 4.5.10, 4.5.11, 5.1.34, Theorem 5.1.36, Exercise 5.3.5, 5.5.6, Lemma 6.1.13, Theorem 6.1.14, Section 6.5, Subsection 6.7.1, Example 7.6.2, Exercises 8.1.2, 9.3.1, Example 9.5.3, Exercise 9.7.11.

Material appropriate in reflexive space. Exercises 4.1.25, 4.1.26, 4.1.46, Theorem 4.1.27, Exercises 4.3.3, 4.4.1, 4.4.7, 4.4.15, 4.4.20, Section 4.5, Exercise 4.6.16, Theorem 5.1.20, Remark 5.1.31, Exercises 5.1.13, 5.1.17, 5.1.28, 5.3.14, 5.3.15, 5.3.17, Theorem 6.2.9, Exercises 6.2.6, 6.2.10, 6.2.14, 6.2.25, 6.4.4, 6.7.6, 7.2.2, Theorems 7.3.2, 7.3.6, 7.3.7, Examples 7.5.3, 7.5.5, Section 7.6, Exercise 7.7.5, Theorem 8.4.3, Exercises 8.4.4–8.4.6, Proposition 9.1.14, Section 9.3, Theorems 9.4.6, 9.4.9, 9.5.1, 9.5.2, Exercise 9.6.3.

4
Convex functions on Banach spaces

Banach contributed to the theory of orthogonal series and made innovations in the theory of measure and integration, but his most important contribution was in functional analysis. Of his published works, his Théorie des opérations linéaires *(1932; "Theory of Linear Operations") is the most important. Banach and his coworkers summarized the previously developed concepts and theorems of functional analysis and integrated them into a comprehensive system. Banach himself introduced the concept of [complete] normed linear spaces, which are now known as Banach spaces. He also proved several fundamental theorems in the field, and his applications of theory inspired much of the work in functional analysis for the next few decades.* (Encyclopedia Britannica)[1]

4.1 Continuity and subdifferentials

Throughout the rest of the book X denotes a normed or a Banach space. We assume familiarity with basic results and language from functional analysis, and we will often draw upon the classical spaces ℓ_p, c_0 and others for examples. Notwithstanding, we will sketch some of the classic convexity theorems in our presentation.

We write X^* for the real dual space of continuous linear functionals, and denote the unit ball and sphere by $B_X := \{x \in X : \|x\| \leq 1\}$ and $S_X := \{x \in X : \|x\| = 1\}$, respectively. We say that a subset C of X is *convex* if, for any $x, y \in C$ and any $\lambda \in [0, 1]$, $\lambda x + (1 - \lambda)y \in C$. The *convex hull* of a set $S \subset X$ is the intersection of all convex sets containing S.

We say an extended-valued function $f : X \to (-\infty, +\infty]$ is *convex* if for any $x, y \in \operatorname{dom} f$ and any $\lambda \in [0, 1]$, one has

$$f(\lambda x + (1 - \lambda)y) \leq \lambda f(x) + (1 - \lambda)f(y).$$

A proper convex function $f : X \to (-\infty, +\infty]$ is said to be *strictly convex* if the preceding inequality is strict for all $0 < \lambda < 1$ and all distinct x and y in the domain of f. We call a function $f : X \to [-\infty, +\infty)$ *concave* if $-f$ is convex. Moreover, given a convex set $C \subset X$ and we may refer to $f : C \to \mathbb{R}$ as *convex* when the extended real-valued function $\tilde{f} : X \to (-\infty, +\infty]$ is convex where $\tilde{f}(x) := f(x)$ for $x \in C$ and $\tilde{f}(x) := +\infty$ otherwise.

---

[1] From www.britannica.com/eb/article-9012092/Stefan-Banach

In many regards, convex functions are the simplest natural class of functions next to affine functions and sublinear functions. Convex functions and convex sets are intrinsically related. For example, if C is a convex set then the *indicator function* δ_C and the *distance function* d_C are convex functions, with the latter function being *Lipschitz* (the definitions given in Chapter 2 for Euclidean spaces of the concepts mentioned in this paragraph extend naturally to Banach spaces). On the other hand if f is a convex function, then epi f and the *lower level sets* $f^{-1}(-\infty, a]$, $a \in \mathbb{R}$ are convex sets (Exercises 4.1.3 and 4.1.4). Two other important functions related to a convex set C are the *gauge function* γ_C defined by

$$\gamma_C(x) := \inf\{r \geq 0 : x \in rC\},$$

and the *support function* σ_C defined on the dual space X^* by

$$\sigma_C(x^*) = \sigma(C; x^*) := \sup\{\langle x, x^* \rangle : x \in C\}.$$

Several properties of the gauge function and the support function are discussed in Exercises 4.1.30 and 4.1.31.

Some extremely useful properties of functions convex and otherwise are as follows. A function $f : X \to [-\infty, +\infty]$ is said to be *lower-semicontinuous (lsc)* at $x_0 \in X$ if $\liminf_{x \to x_0} f(x) \geq f(x_0)$, and if this is the case for all $x_0 \in X$, then f is said to be *lower-semicontinuous*. Likewise, f is *upper-semicontinuous (usc)* at x_0 if $\limsup_{x \to x_0} f(x) \leq f(x_0)$, and if this is the case for all $x_0 \in X$, then f is *upper-semicontinuous*. We will say f is *closed* if its epigraph is a closed subset of $X \times \mathbb{R}$, where we recall the *epigraph* of f is defined by epi $f := \{(x,t) \in X \times \mathbb{R} : f(x) \leq t\}$. Lower-semicontinuity and closure are naturally related.

Proposition 4.1.1 (Closed functions). *Let $f : X \to [-\infty, +\infty]$ be a function. Then the following are equivalent.*

(a) f is lsc.
(b) f has closed lower level sets, i.e. $\{x : f(x) \leq \alpha\}$ is closed for each $\alpha \in \mathbb{R}$.
(c) f is closed.

Proof. This is left as Exercise 4.1.1. □

Consider a function $f : X \to [-\infty, +\infty]$. Then one can readily check that the closure of its epigraph is also the epigraph of a function. We thus define *closure of f* as the function cl f whose epigraph is the closure of f; that is epi(cl f) = cl(epi f).

Proposition 4.1.2. *Let $f : X \to [-\infty, +\infty]$ be a function. Then:*

(a) cl f is the largest lsc function minorizing f;
(b) cl $f(x_0) = \lim_{\delta \downarrow 0} \inf_{\|x - x_0\| < \delta} f(x)$.

Proof. See Exercise 4.1.2(a). □

As in finite dimensions, one can define a function $f : X \to [-\infty, +\infty]$ to be convex if epi f is a convex subset of $X \times \mathbb{R}$; see Exercise 4.1.4. Likewise, the *convex hull* of a function f, $\operatorname{conv} f$ is the largest convex function minorizing f (see p. 21). One of the remarkable – yet elementary properties – of convex functions is the equivalence of local boundedness and local Lipschitz properties without any additional assumptions on the function.

Lemma 4.1.3. *Let $f : X \to (-\infty, +\infty]$ be a convex function, and suppose C is a bounded convex set. If f is bounded above on $C + \delta B_X$ for some $\delta > 0$, then f is Lipschitz on C.*

Proof. By translation, we may without loss of generality assume $0 \in C$. Let $r > 0$ be such that $C \subset rB_X$, and let M be such that $f(x) \leq M$ for all $x \in C + \delta B_X$. Now let $x \in C \setminus \{0\}$, and let $\lambda = \|x\|$. Then by the convexity of f,

$$f(0) \leq \frac{\lambda}{\lambda + \delta} f\left(-\frac{\delta}{\lambda} x\right) + \frac{\delta}{\lambda + \delta} f(x).$$

Therefore, $f(x) \geq \frac{\lambda + \delta}{\delta}(f(0) - M) \geq \frac{r + \delta}{\delta}(f(0) - M)$. Consequently, f is bounded below on C.

Now let $K \in \mathbb{R}$ be such that $f(x) \geq K$ for all $x \in C$, and let $x \neq y \in C$ be arbitrary, and let $\lambda = \|x - y\|$. Then $y = \frac{\delta}{\delta + \lambda} x + \frac{\lambda}{\delta + \lambda}\left(y + \frac{\delta}{\lambda}(y - x)\right)$. By the bounds and convexity of f,

$$f(y) \leq \frac{\delta}{\delta + \lambda} f(x) + \frac{\lambda}{\delta + \lambda} M = f(x) - \frac{\lambda}{\delta + \lambda} f(x) + \frac{\lambda}{\delta + \lambda} M.$$

Therefore, $f(y) - f(x) \leq \frac{M - K}{\delta} \|x - y\|$. The same inequality works with x and y reversed. Thus, f has Lipschitz constant $\frac{M - K}{\delta}$ on C. □

Proposition 4.1.4. *Suppose $f : X \to (-\infty, +\infty]$ is proper and convex. Let $x \in \operatorname{dom} f$; then the following are equivalent:*

(a) f is Lipschitz on some neighborhood of x;
(b) f is continuous at x;
(c) f is bounded on some neighborhood of x;
(d) f is bounded above on some neighborhood of x.

Proof. First, (a) $\Rightarrow$ (b) $\Rightarrow$ (c) $\Rightarrow$ (d) are obvious. Let us now suppose that (d) is true; say f is bounded above on $B_{2\delta}(x)$ for some $\delta > 0$; by Lemma 4.1.3, f is Lipschitz on $B_\delta(x)$. This proves (a). □

Unlike the finite-dimensional case, a finite-valued convex function need not be continuous. Indeed, discontinuous linear functionals can be defined on any infinite-dimensional Banach space using a *Hamel basis*, i.e. an algebraic basis for the underlying vector space; see Exercise 4.1.22. However, when the convex function is lsc, things are quite nice as we now observe. Given a nonempty set $F \subset X$, the *core* of F is defined by $x \in \operatorname{core} F$ if for each $h \in S_X$, there exists $\delta > 0$ so that

4.1 Continuity and subdifferentials

$x + th \in F$ for all $0 \le t \le \delta$. It is clear from the definition that, $\operatorname{int} F \subset \operatorname{core} F$. See Exercise 4.1.20 for further relations. The following is a 'core' result in this subject.

Proposition 4.1.5. *Suppose X is a Banach space and $f : X \to (-\infty, +\infty]$ is lsc, proper and convex. Then the following are equivalent:*

(a) f is continuous at x;
(b) $x \in \operatorname{int} \operatorname{dom} f$;
(c) $x \in \operatorname{core} \operatorname{dom} f$.

In particular, f is continuous at points in the core of its domain.

Proof. The implications (a) $\Rightarrow$ (b) $\Rightarrow$ (c) are clear. Now suppose that (c) holds. By shifting f we may assume $x = 0$. Then consider $F_n := \{y \in X : \max\{f(y), f(-y)\} \le n\}$. Then $\bigcup n F_n = X$, and so the Baire category theorem implies that for some n_0, $n_0 F_{n_0}$ has nonempty interior. The convexity and symmetry of $n_0 F_{n_0}$ imply that 0 is in the interior of $n_0 F_{n_0}$, and hence 0 is in the interior of F_{n_0}. Consequently, f is continuous at 0 as desired (Proposition 4.1.4). □

The usefulness of the preceding result, is that it is often easier to check that something is in the core of the domain of a convex function than in its interior. Also, a lsc convex function whose domain lies in a finite-dimensional space need not be continuous as a function restricted to its domain (see Exercise 2.2.20). The reader can check this is the case for the function $f : \mathbb{R}^2 \to (-\infty, +\infty]$ defined by

$$f(x, y) := \begin{cases} \frac{x^3}{y^2}, & \text{if } x^2 \le y \le x \text{ and } 0 < x \le 1 \\ 0 & \text{if } (x, y) = (0, 0) \\ \infty, & \text{otherwise.} \end{cases}$$

The weak and weak* topologies and their basic properties are extremely useful tools in the study of convex functions on Banach spaces. Let X be a normed space, the *weak topology* on X is generated by a basis consisting of sets of the form

$$\{x \in X : f_i(x - x_0) < \varepsilon \text{ for } i = 1, \ldots, n\}$$

for all choices of $x_0 \in X, f_1, \ldots, f_n \in X^*$ and $\varepsilon > 0$.

Given a Banach space X with norm $\|\cdot\|$, the *dual space* of continuous linear functionals on X is denoted by X^*. The canonical *dual norm* on X^* is defined by $\|\phi\| := \sup\{\phi(x) : \|x\| \le 1\}$ for $\phi \in X^*$; additionally any norm $\|\cdot\|$ on X^* that is the dual norm to some equivalent norm on X is called a *dual norm* on X^*. The *weak*-topology* on the dual X^* of X is generated by a basis consisting of sets of the form

$$\{\phi \in X^* : (\phi - \phi_0)(x_i) < \varepsilon \text{ for } i = 1, \ldots, n\}$$

for all choices of $\phi_0 \in X^*, x_1, \ldots, x_n \in X$ and $\varepsilon > 0$. The following is a central but basic fact concerning the weak*-topology. Its proof, which we omit, is an application of Tikhonov's compactness theorem.

Theorem 4.1.6 (Alaoglu). *Let X be a Banach space, with dual space X^*. Then the dual unit ball B_{X^*} is weak*-compact.*

Thus, a net $(x_\alpha) \subset X$ converges weakly to $x \in X$, if $\phi(x_\alpha) \to \phi(x)$ for each $\phi \in X^*$. Similarly, a net $(\phi_\alpha) \subset X^*$ converges weak* to $\phi \in X^*$, if $\phi_\alpha(x) \to \phi(x)$ for each $x \in X$. It is also possible to define the concept of lower-semicontinuity for different topologies. Indeed, given a Hausdorff topology τ on X, we say f is τ-lsc at x_0 if $\liminf_\alpha f(x_\alpha) \geq f(x_0)$ whenever $x_\alpha \to x$ in the topology τ; in particular, a function will be *weakly lower-semicontinuous* (resp. *weak*-lower-semicontinuous*) when τ is the weak (resp. weak*) topology. Using separation theorems, we will see shortly that a lsc convex function is automatically weakly lsc.

4.1.1 Separation theorems and subdifferentials

We will begin with the Hahn–Banach theorem, and then derive the max formula and various separation theorems from it.

Theorem 4.1.7 (Hahn–Banach extension). *Let X be a Banach space and let $f: X \to \mathbb{R}$ be a continuous sublinear function with $\mathrm{dom}\, f = X$. Suppose that L is a linear subspace of X and the function $h: L \to \mathbb{R}$ is linear and dominated by f, that is, $f \geq h$ on L. Then there exists $x^* \in X^*$, dominated by f, such that*

$$h(x) = \langle x^*, x \rangle, \quad \text{for all } x \in L.$$

Proof. We leave it to the reader to check that this can be proved with a maximality argument using Zorn's lemma, for example, in conjunction with the proof given for finite-dimensional spaces (Proposition 2.1.18) where the reader should also check the domination property ensures the continuity of the linear functionals. □

Given a function $f: X \to (-\infty, +\infty]$ and $\bar{x} \in \mathrm{dom}\, f$, the *subdifferential* of f at $\bar{x}$ is defined by

$$\partial f(\bar{x}) := \{\phi \in X^* : \phi(x) - \phi(\bar{x}) \leq f(x) - f(\bar{x}), \text{ for all } x \in X\};$$

when $\bar{x} \notin \mathrm{dom}\, f$, we define $\partial f(\bar{x}) := \emptyset$. An element of $\partial f(x)$ is called a *subgradient* of f at x. It is easy to see that the subdifferential of a lsc convex function may be empty at some points in its domain. For example, let $f(x) := -\sqrt{1-x^2}$ if $-1 \leq x \leq 1$. Then $\partial f(x) = \emptyset$ for $x = \pm 1$. Although the domain of a convex function is always convex, this is not necessarily the case for $\mathrm{dom}\, \partial f$ as seen in Exercise 4.1.18. The following easy observation suggests the fundamental significance of subdifferential in optimization.

Proposition 4.1.8 (Subdifferential at optimality). *Let X be a Banach space and let $f: X \to (-\infty, +\infty]$ be a proper convex function. Then the point $\bar{x} \in X$ is a (global) minimizer of f if and only if the condition $0 \in \partial f(\bar{x})$ holds.*

Proof. Exercise 4.1.38. □

4.1 Continuity and subdifferentials

Alternatively put, minimizers of f correspond exactly to 'zeros' of ∂f. Later we will see that derivatives of a convex function must be subgradients – as was the case on finite-dimensional spaces.

Let $f: X \to (-\infty, +\infty]$, $x \in \text{dom} f$ and $d \in X$. The *directional derivative* of f at x in the direction of d is defined by

$$f'(x;d) := \lim_{t \to 0+} \frac{f(x+td) - f(x)}{t}$$

when this limit exists. We use this terminology with the understanding that this is a *one-sided* directional derivative.

Proposition 4.1.9. *Suppose the function $f: X \to (-\infty, +\infty]$ is convex. Then for any point $\bar{x} \in \text{core}(\text{dom} f)$, the directional derivative $f'(\bar{x}; \cdot)$ is everywhere finite and sublinear. Moreover, if $x \in \text{cont} f$, then the directional derivative is Lipschitz.*

Proof. The proof for Euclidean spaces (Proposition 2.1.17) establishes everything except for the moreover part. Suppose $\bar{x} \in \text{cont} f$. Then Proposition 4.1.4 implies there exists $\delta > 0$ so that f has Lipschitz constant K on $B_\delta(\bar{x})$. Then for any $h \in X \setminus \{0\}$, and $0 < t < \delta / \|h\|$,

$$f'(\bar{x}; h) = \lim_{t \to 0+} \frac{f(\bar{x} + th) - f(\bar{x})}{t} \leq \frac{Kt\|h\|}{t} = K\|h\|.$$

Because $f'(\bar{x}; \cdot)$ is convex, this implies $f'(\bar{x}; \cdot)$ satisfies a Lipschitz condition with constant K (Exercise 4.1.28). □

Theorem 4.1.10 (Max formula). *Let X be a Banach space, $d \in X$ and let $f: X \to (-\infty, +\infty]$ be a convex function. Suppose that $\bar{x} \in \text{cont} f$. Then, $\partial f(\bar{x}) \neq \emptyset$ and*

$$f'(\bar{x}; d) = \max\{\langle x^*, d \rangle : x^* \in \partial f(\bar{x})\}. \tag{4.1.1}$$

Proof. Because $\bar{x} \in \text{cont} f$, $f'(\bar{x}; h)$ is a sublinear and Lipschitz functional by Proposition 4.1.9. Now use the Hahn–Banach extension theorem (4.1.7) as in the proof of the max formula (2.1.19) for Euclidean spaces. □

The condition $\partial f(\bar{x}) \neq \emptyset$ can be stated alternatively as there exists a linear functional x^* such that $f - x^*$ attains its minimum at $\bar{x}$. When we study exposed points in Section 5.2, we will look at when $f - x^*$ attains its minimum in some stronger senses.

The condition $\bar{x} \in \text{cont} f$ (or equivalently, $\bar{x} \in \text{core dom} f$ when f is lsc) is crucial in ensuring $\partial f(\bar{x}) \neq \emptyset$. Without this condition the subdifferential may be empty. A simple example of this is $\partial f(0) = \emptyset$ for function $f: \mathbb{R} \to (-\infty, +\infty]$ defined by $f(x) := -\sqrt{x}, x \geq 0$ and $+\infty$ otherwise. The following is a systematical scheme for generating such functions in infinite-dimensional spaces.

Example 4.1.11. Let X be an infinite-dimensional separable Banach space and let C be a symmetric compact convex set whose core is empty but whose span is dense. (The Hilbert cube in ℓ_2 is a typical example of such a set, see Exercise 4.1.15.) Let

$\bar{x} \notin \text{span}(C)$. Define $f : X \to (-\infty, +\infty]$ by $f(x) := \min\{\lambda \in \mathbb{R} : x + \lambda \bar{x} \in C\}$, where we use the convention that $\min(\emptyset) = +\infty$. It is easy to check that f is a convex function and for any $s \in \mathbb{R}$ and $c \in C$, $f(c + s\bar{x}) = -s$ (Exercise 4.1.16). It follows that

$$f'(0; y) = \begin{cases} -s & \text{if } y = rc + s\bar{x} \text{ for some } c \in C \text{ and } r, s \in \mathbb{R}, \\ +\infty & \text{otherwise.} \end{cases}$$

Now we show that $\partial f(0) = \emptyset$. Suppose on the contrary that $x^* \in \partial f(0)$. Since $\text{span}(C)$ is dense in X, for any $s \in \mathbb{R}$ we can find $r \in \mathbb{R}$ and $c \in C$ such that $rc + s\bar{x}$ is close to a unit vector so that

$$-\|x^*\| - 1 \leq \langle x^*, rc + s\bar{x}\rangle \leq f'(0; rc + s\bar{x}) = -s,$$

which is a contradiction.

Corollary 4.1.12 (Basic separation). *Let X be a Banach space, and suppose $C \subset E$ is a closed nonempty convex set, and suppose $x_0 \notin C$. Then there exists $\phi \in E$ such that*

$$\sup_C \phi < \langle \phi, x_0\rangle.$$

Proof. See the proof of Corollary 2.1.21. □

Corollary 4.1.13 (Weak closures). *Let X be a Banach space, and suppose C is a convex subset of X. Then C is weakly closed if and only if it is norm closed.*

Proof. Use Corollary 4.1.12 to express a norm closed convex set as an intersection of weakly closed half-spaces. □

Corollary 4.1.14. *Suppose $f : X \to (-\infty, +\infty]$ is a lsc convex function. Then f is weakly-lsc.*

Proof. The epigraph of f is a closed convex set (Proposition 4.1.1). By the previous corollary it is weakly closed, then Exercise 4.1.1(b) shows that f is weakly lsc. □

Corollary 4.1.15 (Separation theorem). *Let X be a Banach space, and suppose $C \subset X$ has nonempty interior and $x_0 \notin \text{int } C$. Then there exists $\phi \in X^* \setminus \{0\}$ so that*

$$\sup_C \phi \leq \phi(x_0) \quad \text{and} \quad \phi(x) < \phi(x_0) \text{ for all } x \in \text{int } C.$$

Proof. Let $\bar{x} \in \text{int } C$, and $\gamma_{C-\bar{x}}$ be the gauge functional of $C - \bar{x}$. Because C has a nonempty interior, $\gamma_{C-\bar{x}}$ is continuous (Exercise 4.1.30(d)) and so by the max formula (4.1.10), we let $\phi \in \partial \gamma_{C-\bar{x}}(x_0)$. Using the subdifferential inequality, we have

$$\phi(x) - \phi(x_0) \leq \gamma_{C-\bar{x}}(x) - \gamma_{C-\bar{x}}(x_0) \leq 0 \quad \text{for all } x \in C.$$

When $x \in \text{int } C$, $\gamma_{C-\bar{x}}(x) < \gamma_{C-\bar{x}}(x_0)$ so strict inequality is obtained in that case. □

4.1 Continuity and subdifferentials

Let X be a Banach space, and suppose C is a convex subset of X. We shall say $\phi \in X^* \setminus \{0\}$ is a *supporting linear functional* for C if there exists $\bar{x} \in C$ such that $\phi(\bar{x}) = \sup_C \phi$, and we say ϕ *supports* C at $\bar{x}$; in this case, $\bar{x}$ is a *support point* of C and the set of points $\{x : \phi(x) = \phi(\bar{x})\}$ is a *supporting hyperplane* for C. Observe that a supporting linear functional of B_X is *norm-attaining*, that is, there exists $\bar{x} \in B_X$ such that $\phi(\bar{x}) = \sup_{B_X} \phi = \|\phi\|$.

The following observations are straighforward consequences of the previous results.

Remark 4.1.16 (Supporting functionals). Let X be a Banach space.

(a) Suppose C is a convex subset of X with nonempty interior. Each boundary point $\bar{x} \in C$ is a support point of C.
(b) For each $x \in S_X$, there exists $\phi \in S_{X^*}$ such that $\phi(\bar{x}) = 1$.
(c) The *duality mapping* $J : X \to X^*$ defined by $J := \frac{1}{2}\partial\left(\|\cdot\|^2\right)$ is nonempty for each $x \in X$ and satisfies
$$J(x) = \{x^* \in X^* : \|x\|^2 = \|x^*\|^2 = \langle x, x^* \rangle\}.$$

Corollary 4.1.17 (Separation theorem). *Suppose A and B are convex subsets of a Banach space such that A has nonempty interior and $\mathrm{int}\, A \cap B = \emptyset$. Then there exists $\phi \in X^* \setminus \{0\}$ so that $\sup_A \phi \leq \inf_B \phi$.*

Proof. Let $C = A - B$, then C is convex, has nonempty interior and $0 \notin \mathrm{int}\, C$. According to Corollary 4.1.15, there exists $\phi \in X^* \setminus \{0\}$ so that $\sup_C \phi \leq \phi(0)$. Thus $\phi(a - b) \leq 0$ for all $a \in A$ and $b \in B$; this proves the result. □

In contrast to the finite-dimensional setting (Theorem 2.4.7), the condition in the previous result that one of the sets has nonempty interior cannot be removed; see Exercise 4.1.9.

Theorem 4.1.18 (Sandwich theorem). *Let X and Y be Banach spaces and let $T : X \to Y$ be a bounded linear mapping. Suppose that $f : X \to \mathbb{R}$, $g : Y \to \mathbb{R}$ are proper convex functions which together with T satisfy $f \geq -g \circ T$ and either*

$$0 \in \mathrm{core}(\mathrm{dom}\, g - T\,\mathrm{dom}\, f) \text{ and both } f \text{ and } g \text{ are lsc}, \quad (4.1.2)$$

or the condition

$$T\,\mathrm{dom}\, f \cap \mathrm{cont}\, g \neq \emptyset. \quad (4.1.3)$$

Then there is an affine function $\alpha : X \to \mathbb{R}$ of the form $\alpha(x) = \langle T^*y^*, x \rangle + r$ satisfying $f \geq \alpha \geq -g \circ T$. Moreover, for any $\bar{x}$ satisfying $f(\bar{x}) = (-g \circ T)(\bar{x})$, we have $-y^* \in \partial g(T\bar{x})$.

Proof. Define the function $h : Y \to [-\infty, +\infty]$ by

$$h(u) := \inf_{x \in X}\{f(x) + g(Tx + u)\}.$$

Then one can check h is a proper convex function. We next show that 0 is a point of continuity of h under condition (4.1.2). Indeed, in this case dom $h = $ dom $g - T$ dom f, and without loss of generality we assume $f(0) = g(0) = 0$. Then define

$$S := \{u \in Y : \text{ there exists } x \in B_X \text{ with } f(x) + g(Tx + u) \le 1\}.$$

Then S is convex, and we next show $0 \in \text{core } S$.

Let $y \in Y$. Then (4.1.2) implies there exists $t > 0$ such that $ty \in \text{dom } g - T \text{ dom } f$. Now choose $x \in \text{dom } f$ such that $Tx + ty \in \text{dom } g$, then

$$f(x) + g(Tx + ty) = k < \infty.$$

Let $m \ge \max\{\|x\|, |k|, 1\}$. Because f and g are convex with $f(0) = g(0) = 0$, we have

$$f\left(\frac{x}{m}\right) + g\left(T\left(\frac{x}{m}\right) + \frac{ty}{m}\right) \le 1.$$

Consequently, $ty/m \in S$, and it follows that $0 \in \text{core } S$. Because S is also convex, the Baire category theorem ensures that $0 \in \text{int } \overline{S}$ (Exercise 4.1.20). Thus, there exists $r > 0$ so that

$$0 \in rB_X \subset \overline{S} \subset S + \frac{r}{2}B_X.$$

Multiplying the previous inclusions by $1/2^i$ for $i = 1, 2, \ldots$ we obtain

$$\frac{r}{2^i}B_X \subset \frac{1}{2^i}S + \frac{r}{2^{i+1}}B_X.$$

It then follows that

$$\frac{r}{2}B_X \subset \frac{1}{2}S + \frac{1}{4}S + \cdots + \frac{1}{2^i} + \frac{r}{2^{i+1}}B_X.$$

Thus for any $u \in \frac{r}{2}B_X$, there exist $s_1, s_2, \ldots, s_i \in S$ such that

$$u \in \frac{1}{2}s_1 + \frac{1}{4}s_2 + \cdots + \frac{1}{2^i}s_i + \frac{r}{2^{i+1}}B_X.$$

Let $u_n = \sum_{i=1}^n 2^{-i}s_i$ it follows that $u_n \to u$. Because f and g are lsc, convex with $f(0) = 0 = g(0)$ it follows that $u \in S$. Consequently, $\frac{r}{2}B_X \subset S$ and so h is bounded above on $\frac{r}{2}B_X$, which means h is continuous at 0 as desired (Proposition 4.1.4).

In the event that (4.1.3) is valid, let $x_0 \in \text{dom } f$ be such that g is continuous at Tx_0. Let $r > 0$, be such that g is bounded above by K on the neighborhood $Tx_0 + rB_Y$. Then $h(u) \le f(x_0) + g(Tx_0 + u) \le f(x_0) + K$ for all $u \in rB_Y$. As before, h is continuous at 0.

4.1 Continuity and subdifferentials

In either case, according to the max formula (4.1.10), there is a subgradient $-\phi \in \partial h(0)$. Hence for all $u \in Y$ with $u = Tv$ where $v \in X$, and $x \in X$ we have

$$0 \leq h(0) \leq h(u) + \langle \phi, u \rangle$$
$$\leq f(x) + g(T(x+v)) + \langle \phi, Tv \rangle$$
$$= [f(x) - \langle T^*\phi, x \rangle] - [-g(T(x+v)) - \langle \phi, T(x+v) \rangle].$$

Then $a \leq b$ where

$$a = \inf_{x \in X}[f(x) - \langle T^*\phi, x \rangle] \text{ and } b = \sup_{v \in X}[-g(T(v)) - \langle \phi, T(v) \rangle].$$

Consequently for any $r \in [a, b]$, we have

$$f(x) \geq r + \langle T^*\phi, x \rangle \geq -(g \circ T)(x) \quad \text{for all } x \in X.$$

This proves the theorem. □

Even when $X = Y = \mathbb{R}$ and T is the identity operator in the sandwich theorem (4.1.18), it is easy to see that the conclusion can fail if one only assumes the domains of f and g have nonempty intersection (see Exercise 4.1.37 or Figure 2.15 in Section 2.3). The Fenchel duality theorem will be presented in Section 4.4 after conjugate functions are introduced. For now, we present some other useful consequences of the sandwich theorem.

Theorem 4.1.19 (Subdifferential sum rule). *Let X and Y be Banach spaces, and let $f : X \to (-\infty, +\infty]$ and $g : Y \to (-\infty, +\infty]$ be convex functions and let $T : X \to Y$ be a bounded linear mapping. Then at any point $x \in X$ we have the sum rule*

$$\partial(f + g \circ T)(x) \supset \partial f(x) + T^*(\partial g(Tx))$$

with equality if (4.1.2) or (4.1.3) hold.

Proof. The inclusion is straightforward, so we prove the reverse inclusion. Suppose $\phi \in \partial(f + g \circ T)(\bar{x})$. Because shifting by a constant does not change the subdifferential, we may assume without loss of generality that

$$x \mapsto f(x) + g(Tx) - \phi(x)$$

attains its minimum of 0 at $\bar{x}$. According to the sandwich theorem (4.1.18) there exists an affine function $A := \langle T^*y^*, \cdot \rangle + r$ with $-y^* \in \partial g(T\bar{x})$ such that

$$f(x) - \phi(x) \geq A(x) \geq -g(Ax).$$

Then equality is attained at $x = \bar{x}$. It is now a straightforward exercise to check that $\phi + T^*y^* \in \partial f(\bar{x})$. □

Corollary 4.1.20 (Compact separation). *Let C and K be closed convex subsets of a Banach space X. If K is compact and $C \cap K = \emptyset$, then there exists $\phi \in X^*$ such that $\sup_{x \in K} \phi(x) < \inf_{x \in C} \phi(x)$.*

Proof. See Exercise 4.1.34. □

Sometimes it is useful to separate epigraphs from points with affine functions. In this case, we do not require a continuity condition as in the sandwich theorem (4.1.18).

Theorem 4.1.21 (Epigraph-point separation). *Suppose $f : X \to (-\infty, +\infty]$ is a proper convex function that is lsc at some point in its domain.*

(a) Then f is bounded below on bounded sets.
(b) If f is lsc at x_0, and $\alpha < f(x_0)$ for some $x_0 \in X$ ($f(x_0) = +\infty$ is allowed), then there exists $\phi \in X^*$ such that $\phi(x) - \phi(x_0) \leq f(x) - \alpha$ for all $x \in X$.

Proof. First, suppose f is lsc at $x_0 \in \operatorname{dom} f$ and let $\alpha < f(x_0)$. The lower-semicontinuity of f implies that there exists $r > 0$ so that $f(x) \geq \alpha$ if $x \in B_r(x_0)$. Apply the sandwich theorem (4.1.18) to find $\phi \in X^*$ and $a \in \mathbb{R}$ so that

$$\alpha - \delta_{B_r(x_0)} \leq \phi(x) + a \leq f(x).$$

Then

$$\phi(x) - \phi(x_0) \leq f(x) - [\phi(x_0) + a] \leq f(x) - \alpha \quad \text{for all } x \in X.$$

This proves (b) in the case $x_0 \in \operatorname{dom} f$, and (a) also follows easily from this.

Now suppose $x_0 \notin \operatorname{dom} f$ and $\alpha < f(x_0)$, and fix some $y_0 \in \operatorname{dom} f$. Let $r > \|x_0 - y_0\|$, and by (a) we choose K so that $f(x) \geq K$ if $x \in B_r(x_0)$. Now choose $\delta \in (0, r)$ so that $f(x) > \alpha$ for $x \in B_\delta(x_0)$, and choose M so large that $\alpha - M\delta \leq K$. Then let $g(x) := -\alpha + M\|x - x_0\|$ if $x \in B_r(x_0)$, and let $g(x) := \infty$ otherwise. Since g is continuous at y_0, we can apply the sandwich theorem (4.1.18) as in the preceding paragraph to complete the proof. □

The following is a weak*-version of the separation theorem.

Theorem 4.1.22 (Weak*-separation). *Let X be a Banach space, and suppose C is a weak*-closed convex subset of X^*, and $\phi \in X^* \setminus C$. Then there exists $x_0 \in X$ such that $\sup_C x_0 < \phi(x_0)$.*

Proof. See Exercise 4.1.10. □

As a consequence of this, we have the following weak*-version of Theorem 4.1.21.

Proposition 4.1.23 (Weak*-epi separation). *Suppose $f : X^* \to (-\infty, +\infty]$ is a proper weak*-lsc convex function. Let $\alpha \in \mathbb{R}$ and $x_0^* \in \operatorname{cont} f$ be such that $\alpha < f(x_0^*)$, then there exists $x_0 \in X$ such that*

$$\langle x_0, x^* - x_0^* \rangle \leq f(x^*) - \alpha$$

4.1 Continuity and subdifferentials

for all $x^* \in X^*$.

Proof. Because $\operatorname{epi} f$ is a weak*-closed convex set and $(x_0^*, \alpha) \notin \operatorname{epi} f$, Theorem 4.1.22 ensures the existence of $(x_0, a) \in X \times \mathbb{R}$ so that

$$\langle (x_0, a), (x_0^*, \alpha) \rangle > \sup_{\operatorname{epi} f} (x_0, a) \text{ where } \langle (x, t), (x^*, r) \rangle := x^*(x) + tr.$$

Since $x_0 \in \operatorname{int} \operatorname{dom} f$, it follows that $a < 0$. By scaling we may assume $a = -1$. In this case,

$$\langle x_0, x_0^* - \alpha \rangle > \langle x_0, x^* \rangle - t \text{ for all } (x^*, t) \in \operatorname{epi} f.$$

In particular, letting $t = f(x^*)$ provides the desired inequality. □

Given a function $f : X \to (-\infty, +\infty]$ and $x_0 \in \operatorname{dom} f$ and $\varepsilon > 0$, the ε-*subdifferential* of f at x_0 is defined by

$$\partial_\varepsilon f(x_0) := \{\phi \in X^* : \phi(x) - \phi(x_0) \leq f(x) - f(x_0) + \varepsilon, \ x \in X\}. \quad (4.1.4)$$

When $x_0 \notin \operatorname{dom} f$, one defines $\partial_\varepsilon f(x_0) := \emptyset$. We next show that the domain of the ε-subdifferential always coincides with the domain of f when f is a proper lsc convex function. Also, this concept will be very useful in studying differentiability and exposed points as we will see in later sections.

Proposition 4.1.24. *Suppose f is a proper convex function, then $\partial_\varepsilon f(x) \neq \emptyset$ for all $x \in \operatorname{dom} f$ such that f is lsc at x and all $\varepsilon > 0$.*

Proof. Apply Theorem 4.1.21(b) with $\alpha = f(x_0) - \varepsilon$ for $x_0 \in \operatorname{dom} f$. □

Because subdifferentials are nonempty at points of continuity, we can now prove the following characterization of convex functions that are bounded on bounded sets.

Proposition 4.1.25. *Suppose $f : X \to (-\infty, +\infty]$ is proper and convex. Then the following are equivalent:*

(a) f is Lipschitz on bounded subsets of X;
(b) f is bounded on bounded subsets of X;
(c) f is bounded above on bounded subsets of X;
(d) ∂f maps bounded subsets of X into bounded nonempty subsets of X^*.

Proof. Clearly, (a) $\Rightarrow$ (b) $\Rightarrow$ (c); and (c) $\Rightarrow$ (a) follows from Lemma 4.1.3.
    (a) $\Rightarrow$ (d): Let D be a bounded subset of X, and let f have Lipschitz constant K on $D + B_X$. Let $x \in D$ then the max formula (4.1.10) ensures $\partial f(x) \neq \emptyset$, and we and let $\phi \in \partial f(x)$. Then $\phi(h) \leq f(x+h) - f(x) \leq K\|h\|$ for all $h \in B_X$. Therefore, $\|\phi\| \leq K$, and so $\partial f(D) \subset K B_{X^*}$.
    (d) $\Rightarrow$ (a): Suppose f is not Lipschitz on some bounded set D in X. Then for each $n \in \mathbb{N}$, choose $x_n, y_n \in D$ such that $f(y_n) - f(x_n) > n\|y_n - x_n\|$, and let $\phi_n \in \partial f(y_n)$. Then $\phi_n(x_n - y_n) \leq f(x_n) - f(y_n) < -n\|x_n - y_n\|$. Thus $\|\phi_n\| > n$ and so $\partial f(D)$ is not bounded. □

Perhaps, more importantly, we have the following local version of the previous result.

Proposition 4.1.26. *Let $f : X \to (-\infty, +\infty]$ be a proper convex function. Then f is continuous at x_0 if and only if ∂f is a nonempty on some open neighborhood U of x_0 and $\partial f(U)$ is a bounded subset of X^* (when f is additionally lsc these conditions occur if and only if $x_0 \in \operatorname{core} \operatorname{dom} f$). Moreover, ∂f is a nonempty weak*-compact convex subset of X^* at each point of continuity of f.*

Proof. Exercise 4.1.40. □

Exercises and further results

4.1.1 (Closed functions). ⋆

(a) Prove Proposition 4.1.1.
(b) Suppose $f : X \to (-\infty, +\infty]$ is a proper convex function. Show that f is weakly lsc on X if and only if its epigraph is weakly closed in $X \times \mathbb{R}$ if and only if its epigraph is norm closed in $X \times \mathbb{R}$ if and only if f is lsc.
(c) Suppose $f : X^* \to (-\infty, +\infty]$ is a proper convex function. Show that f is weak*-lsc on X^* if and only if its epigraph is weak*-closed in $X^* \times \mathbb{R}$ if and only if its lower level sets are weak*-closed.

Hint. For (b), suppose $x_\alpha \to x$ weakly but $\lim f(x_\alpha) = t < f(x)$. Then the epigraph of f is not weakly closed. Conversely, if (x_α, t_α) converges weakly to (x, t) where $(x_\alpha, t_\alpha) \in \operatorname{epi} f$, but $(x, t) \notin \operatorname{epi} f$, then

$$\liminf f(x_\alpha) \leq \liminf t_\alpha = t < f(x).$$

□

4.1.2 (Closures of functions). ⋆

(a) Prove Proposition 4.1.2.
(b) In the following three cases determine whether $(\operatorname{cl} f)(x) = \sup\{\alpha(x) : \alpha \leq f$ and $\alpha : X \to \mathbb{R}$ is a continuous affine function$\}$.

   (i) $f : \mathbb{R} \to [-\infty, +\infty]$ is convex.
   (ii) $f : X \to (-\infty, +\infty]$ is a proper convex function where X is infinite-dimensional.
   (iii) $f : X \to (-\infty, +\infty]$ is convex and $\operatorname{cl} f$ is proper.

(c) Provide an example of a proper convex function $f : X \to (-\infty, +\infty]$ such that $\operatorname{cl} f$ is not proper. Note that X must be infinite-dimensional (see Exercise 2.4.8).

Hint. (b) For (i), consider $f(0) := -\infty$ and $f(t) := +\infty$ otherwise. Show (ii) is false: consider f such that $f(x) := 0$ for all x in a dense convex subset of B_X and $f(x) = +\infty$ otherwise. Show (iii) is true by using the epi-point separation theorem (4.1.21).

4.1 Continuity and subdifferentials

(c) Let c_{00} be the nonclosed linear subspace of all finitely supported sequences in c_0. Consider, for example, $f : c_0 \to (-\infty, +\infty]$ for which $f(x) := \sum x_i$ if $x \in c_{00}$, and $f(x) := +\infty$ otherwise. Then $\operatorname{cl} f \equiv -\infty$. □

4.1.3.* Let f be a convex function on a Banach space. Show that for any $a \in \mathbb{R}$, the lower level set $f^{-1}(-\infty, a]$ is convex.

4.1.4.* Let X be a Banach space and let $f : X \to (-\infty, +\infty]$ be an extended-valued function. Verify that f is convex if and only if $\operatorname{epi} f$ is a convex subset of $X \times \mathbb{R}$.

4.1.5 (Stability properties).* (a) Show that the intersection of a family of arbitrary convex sets is convex.

(b) Conclude that $f(x) := \sup\{f_\alpha(x) : \alpha \in A\}$ is convex (and lsc) when $\{f_\alpha\}_{\alpha \in A}$ is a collection of convex (and lsc) functions.

(c) Suppose (f_α) is a net of convex functions on a Banach space X. Show that $\limsup_\alpha f_\alpha$ and $\lim f_\alpha$ (if it exists) are convex functions.

(d) Show by example that $\min\{f, g\}$ need not be convex when f and g are convex (even linear).

4.1.6 (Midpoint convexity). A function $f : X \to (-\infty, +\infty]$ is *midpoint convex* if $f(\frac{x+y}{2}) \leq \frac{1}{2} f(x) + \frac{1}{2} f(y)$ for all $x, y \in X$. Show that f is convex if it is midpoint convex and lsc. Provide an example of a measurable function $f : \mathbb{R} \to (-\infty, +\infty]$ that is midpoint convex but not convex. (Note that there is nothing very special about the midpoint.)

Hint. Let $f(t) := 0$ if t is rational and $f(t) := \infty$ otherwise. □

4.1.7 (Midpoint convexity, II). Show that a midpoint convex and measurable real-valued function is convex.

Hint. If f is measurable then it is bounded above on some interval. □

4.1.8.* Let C be a convex subset of a Banach space. Show that d_C and δ_C are convex functions.

4.1.9.* Show by example that separation may fail in c_0 and ℓ_p for $1 \leq p < \infty$ for a closed convex cone and a ray.

Hint. For c_0 let $A = c_0^+$, that is $A := \{(x_i) \in c_0 : x_i \geq 0, i \in \mathbb{N}\}$. Let B be the ray $\{tx - y : t \geq 0\}$ where $x = (4^{-i})$ and $y = (2^{-i})$. Check that $A \cap B = \emptyset$ and suppose $\phi \in \ell_1 \setminus \{0\}$ is bounded below on A, say $\phi = (s_i)$. Verify that $s_i \geq 0$ for all $i \in \mathbb{N}$ and $s_i > 0$ for at least one i and so $\phi(x) > 0$. Thus $\phi(tx - b) \to \infty$ as $t \to \infty$, so ϕ cannot separate A and B. Similarly for ℓ_p. □

4.1.10.*

(a) Suppose X is a vector space, and let $f, f_1, f_2, \ldots, f_n$ be linear functionals on X. If $\bigcap_{i=1}^n f_i^{-1}(0) \subset f^{-1}(0)$, then f is a linear combination of $f_1, f_2, \ldots, f_n$.
(b) Prove Theorem 4.1.22.

Hint. (a) See [199, Lemma 3.9].

(b) Suppose C is weak*-closed in X^* and that $\phi \in X^* \setminus C$. Find a weak*-open neighborhood U of the form

$$U = \{x^* \in X^* : |x^*(x_i)| < \varepsilon \text{ for } i = 1, \ldots, n\}$$

for some $x_1, \ldots, x_n \in X$ and $\varepsilon > 0$. Then $\phi \notin C + U$ and so by the separation theorem (4.1.17) there exists $F \in X^{**}$ so that

$$\sup_C F < \sup_{C+U} F \leq F(\phi).$$

Using (a) show that F is a linear combination of the x_i; see [199, Theorem 3.18] for full details. $\square$

4.1.11 (Separating family).★ Let $S \subset X^*$. Then S is a *separating family* for X (or S *separates points of X*) provided for any $x, y \in X$ with $x \neq y$ there exists $\phi \in S$ such that $\phi(x) \neq \phi(y)$. We shall say S *total* if for each $x \in X \setminus \{0\}$, there exists $\phi \in S$ such that $\phi(x) \neq 0$.

(a) Suppose $S \subset X^*$. Show that S is a separating family for X if and only if it is total.
(b) Suppose X is separable, show that there is a countable subset of X^* that separates points of X. (Hint: take a supporting functional of B_X at each point of a countable dense subset of S_X.)
(c) Show there is a countable separating family for $\ell_\infty(\mathbb{N})$ in its predual $\ell_1(\mathbb{N})$, or more generally, when X is separable there is a countable $S \subset X$ that is separating for X^*.

4.1.12 (Krein–Milman theorem). Suppose X is a Banach space. If $K \subset X$ is convex and weakly compact, then K is the closed convex hull of its extreme points. If $K \subset X^*$ is convex and weak*-compact, then K is the weak*-closed convex hull of its extreme points.

Hint. Suppose H is a supporting hyperplane for K. Show that H contains an extreme point of K. Then use a separation theorem to deduce the conclusion. See, for example, [199, Theorem 3.37] for more details. $\square$

4.1.13 (Goldstine's theorem).★ Show that the weak*-closure of B_X in X^{**} is $B_{X^{**}}$.

Hint. Alaoglu's theorem (4.1.6) ensures $B_{X^{**}}$ is weak*-compact. Thus, the weak*-closure of B_X must be contained in it. Should that containment be proper, then a contradiction would be obtained from the weak*-separation theorem (4.1.22). $\square$

4.1.14 (Bounded above but non-Lipschitz convex functions). (a) Let $\phi = (1, 1, 1, \ldots) \in \ell_\infty$ and define $f : \ell_1 \to (-\infty, +\infty]$ by

$$f(x) := \sup_{n \in \mathbb{N}} \{\phi(x) + n|x(n)| - n^2\}$$

where $x \in \ell_1$ and $x(n)$ denotes the n-th coordinate of x. Show that f is a continuous convex function. Let $C := \{x : \phi(x) + n|x(n)| - n^2 \leq -n \text{ for all } n \in \mathbb{N}\}$. Show that f

is bounded above on $C+B_{\ell_1}$, but that f is not Lipschitz on C (hence the boundedness assumption on C in Lemma 4.1.3 is necessary). (b) Suppose X is a Banach space and $f: X \to (-\infty, +\infty]$ is convex. Suppose C is convex and f is bounded (both above and below) on $C + \delta B_X$ for some $\delta > 0$. Show that f is Lipschitz on C.

Hint. For (a), f is a supremum of lsc convex functions and is real-valued. Verify that f is bounded above by 1 on $C + B_{\ell_1}$. To show that f is not Lipschitz on C, fix $m \in \mathbb{N}$, $m > 1$, and consider $\bar{x} = -me_m - m^2 e_{m^2}$. For $n \in \mathbb{N} \setminus \{m, m^2\}$, $\phi(\bar{x}) + n|x(n)| - n^2 = -m - m^2 - n^2 \leq -n$, while $\phi(\bar{x}) + m|x(m)| - m^2 = -m - m^2$ and $\phi(\bar{x}) + m^2|x(m^2)| - m^4 = -m - m^2$. Thus $\bar{x} \in C$ and $f(\bar{x}) = -m - m^2$. Now consider $\bar{y} := \bar{x} - e_m$, verify that $\bar{y} \in C$ and $f(\bar{y}) = -1 - m^2$ and hence f is not Lipschitz on C. Part (b) follows from the proof of Lemma 4.1.3. □

4.1.15. The Hilbert cube in ℓ_2 is defined by

$$H := \{x = (x_1, x_2, \dots) \in \ell_2 : |x_i| \leq 1/2^i, i = 1, 2, \dots\}.$$

Show that the Hilbert cube is a symmetric compact convex set of ℓ_2 satisfying core $H = \emptyset$ and $\overline{\text{span}(H)} = \ell_2$.

4.1.16.* Prove that the function f defined in Example 4.1.11 is convex and has the property that for any $s \in \mathbb{R}$ and $c \in C$, $f(c + s\bar{x}) = -s$.

4.1.17. With some additional work we can also construct a convex function whose subdifferential is empty on a dense subset of its domain. Let $X = \ell_2$ and H be the Hilbert cube defined in Exercise 4.1.15 and define $f: X \to (-\infty, +\infty]$ by

$$f(x) := \begin{cases} -\sum_{i=1}^{\infty} \sqrt{2^{-i} + x_i} & \text{if } x \in H, \\ +\infty & \text{otherwise.} \end{cases}$$

Show that f is lsc and $\partial f(x) = \emptyset$ for any $x \in H$ such that $x_i > -2^{-i}$ for infinitely many i.

Hint. See [349, Example 3.8]. □

4.1.18 (Domain of subdifferential).* Suppose the function $f: \mathbb{R}^2 \to (-\infty, +\infty]$ is defined by

$$f(x_1, x_2) := \begin{cases} \max\{1 - \sqrt{x_1}, |x_2|\} & \text{if } x_1 \geq 0, \\ +\infty & \text{otherwise.} \end{cases}$$

Prove that f is convex but that dom ∂f is not convex.

4.1.19 (Recognizing convex functions). Let X be a normed linear space, $U \subset X$ be an open convex set, and $f: U \to \mathbb{R}$.

(a) Suppose $\partial f(\bar{x}) \neq \emptyset$ for each $\bar{x} \in U$. Show that f is a convex function.
(b) Is the converse of (a) true when X is a Banach space?

Hint. (a) See the proof of Proposition 2.1.15. (b) The answer is yes when X is finite-dimensional. However, when X is an infinite-dimensional Banach space, the converse is true if f is also lsc. Otherwise the converse may fail; for example, consider a discontinuous linear functional on X. □

4.1.20 (Core versus interior). ⋆

(a) Supply the full details for the proof of Proposition 4.1.5.
(b) A convex set S in a normed linear space is said to be *absorbing* if $\bigcup_{t \geq 0} tS$ is the entire space. Suppose C is a closed convex subset of a Banach space X. Show that $0 \in \operatorname{core} C$ if and only if $0 \in \operatorname{int} C$ if and only if $\bigcup_{n=1}^{\infty} nC = X$. Hence, $0 \in \operatorname{core} C$ if and only if C is absorbing.
(c) Let $F \subset \mathbb{R}^2$ be defined by $F := \{(x, y) : y = 0 \text{ or } |y| \geq x^2\}$. Show that $0 \in \operatorname{core} F$ but $0 \notin \operatorname{int} F$.

4.1.21 (Example of a discontinuous linear functional).

(a) Let $\Lambda : c_0 \to \mathbb{R}$ be the linear functional $\Lambda(x) := \sum x_i$ where $x = (x_i) \in c_{00}$. Verify that Λ is a discontinuous linear functional on the normed space c_{00}.
(b) Extend Λ to a discontinuous linear functional on c_0, by extending an algebraic basis of c_{00} to c_0.

4.1.22 (Discontinuous linear functionals). ⋆ Let X be a normed space. Show that the following are equivalent

(a) X is finite-dimensional.
(b) Every linear function f is continuous on X.
(c) Every absorbing convex set has zero in its interior.

Hint. (c) ⇒ (b): $f^{-1}(-1, 1)$ is absorbing and convex and symmetric.
 (b) ⇒ (a): use the existence of an infinite linearly independent set $\{e_i\}$ to define a discontinuous everywhere finite linear functional with $f(e_i/\|e_i\|) = i$.
 (a) ⇒ (c) is obvious. □

4.1.23 (Convex series closed sets). A set $C \subset X$ is called *convex series closed* if $\bar{x} \in C$ whenever $\bar{x} = \sum_{i=1}^{\infty} \lambda_i x_i$ where $\lambda_i \geq 0$, $\sum_{i=1}^{\infty} \lambda_i = 1$, $x_i \in C$.

(a) Show that open and closed convex sets in a Banach space are convex series closed (this remains true for convex G_δ-sets).
(b) Show that every convex set in a Euclidean space is convex series closed.
(c) Let C be convex series closed in a Banach space. Show $\operatorname{int} C = \operatorname{int} \overline{C}$.
(d) (Open mapping theorem) Show that a continuous (or just closed) linear mapping T between Banach spaces is onto if and only if it is open.

Hint. For (c), see the proof that $0 \in \operatorname{int} S$ in the proof of the sandwich theorem (4.1.18). For (d), it is easy that an open map is onto, for the converse, use part (c), and show, for example, that $0 \in \operatorname{core} \overline{T(B_X)}$. For more information on convex series closed sets, see [121, p. 113ff]. □

4.1.24 (Convexity of p-norm). Recall that the space $\ell_p := \ell_p(\mathbb{N})$ is defined to be the vector space of all sequences $(x_k)_{k=1}^{\infty}$ for which $\sum_{k=1}^{\infty} |x_k|^p$ is finite. Show for

$1 \leq p < \infty$, that the p-norm defined on $\ell_p(\mathbb{N})$ by $\|x\|_p := \left(\sum_{k=1}^{\infty} |x_k|^p\right)^{1/p}$, is indeed a norm on the complete space ℓ_p.

Hint. Both methods suggested in Exercise 2.3.4 to show the function is a norm are applicable, and also apply to the Lebesgue integral analog on $L_p(\Omega, \mu)$ for a positive measure μ: $\|x\|_p^p := \int_\Omega |x(t)|^p \, \mu(dt)$. Completeness is left to the reader. □

4.1.25 (Reflexive Banach spaces). * Recall that a Banach space is *reflexive* if the canonical mapping of X into X^{**} is onto; see [199, p. 74]. Show that a Banach space X is reflexive if and only if B_X is weakly compact. Deduce that a Banach space is reflexive if and only if each closed bounded convex set is weakly compact.

A much deeper and extremely useful characterization of weakly compact sets is James' theorem which we state as follows, and refer the reader to [199, Theorem 3.55] and [182] for its proof.

Theorem 4.1.27 (James). *Let C be a closed convex subset of a Banach space X. Then C is weakly compact if and only if every $\phi \in X^*$ attains it supremum on C. Consequently, a Banach space X is reflexive if and only if every $\phi \in X^*$ attains its supremum on B_X.*

Hint. If $B_X = B_{X^{**}}$, then B_X is weak*-compact in X^{**}, and thus weakly compact in X. For the other direction, if B_X is weakly compact, then it is weak*-closed in X^{**}. Now use Goldstine's theorem (Exercise 4.1.13). For the remaining parts, recall that a closed convex set is weakly closed (Corollary 4.1.13). □

4.1.26 (Superreflexive Banach spaces). * Let X and Y be Banach spaces. We say Y is *finitely representable* in X if for every $\varepsilon > 0$ and every finite-dimensional subspace F of Y, there is a linear isomorphism T of F onto a subspace Z in X such that $\|T\| \|T^{-1}\| < 1 + \varepsilon$. A Banach space X is said to be *superreflexive* if every Banach space finitely representable in X is reflexive. Obviously, a superreflexive Banach space is reflexive, however, the following are more subtle to show.

(a) Every Banach space is finitely representable in c_0.
(b) Let X be a Banach space, then X^{**} is finitely representable in X.
(c) $(\sum \ell_\infty^n)_2$ is reflexive but not superreflexive.
(d) Show that the Banach spaces of Exercise 4.1.24 are superreflexive for $1 < p < \infty$.

Hint. See [199, p. 291–294] for these facts and more; note that part (b) follows from the stronger property known as the *principle of local reflexivity* – see [199, Theorem 9.15]. □

4.1.27 (Hilbert spaces). * An *inner product* on a real or complex vector space X is a scalar-valued function $\langle \cdot, \cdot \rangle$ on $X \times X$ such that:

(i) for every $y \in X$, the function $x \mapsto \langle x, y \rangle$ is linear;
(ii) $\overline{\langle x, y \rangle} = \langle y, x \rangle$ where the left-hand side denotes the complex conjugate;
(iii) $\langle x, x \rangle \geq 0$ for each $x \in X$;
(iv) $\langle x, x \rangle = 0$ if and only if $x = 0$.

A Banach space X is called a *Hilbert space* if there is an inner product $\langle \cdot, \cdot \rangle$ on X such that $\|x\| = \sqrt{\langle x, x \rangle}$.

(a) Show the parallelogram equality holds on a Hilbert space H, that is
$$\|x+y\|^2 + \|x-y\|^2 = 2\|x\|^2 + 2\|y\|^2 \quad \text{for all } x, y \in H.$$

(b) Show that $\langle x, y \rangle = \frac{1}{4}\left(\|x-y\|^2 - \|x-y\|^2\right)$ for all x, y in a Hilbert space over the real scalars.

Hint. See [199] for these and further properties of Hilbert spaces. □

4.1.28 (Lipschitz convex functions).* Suppose $f : X \to \mathbb{R}$ is convex and satisfies $f(x) \leq K\|x\|$ for all $x \in X$. Show that f has Lipschitz constant K.

Hint. For $\|h\| > 0$, $\dfrac{f(x+h)-f(x)}{\|h\|} \leq \limsup_{n \to \infty} \dfrac{f(x+nh)-f(x)}{n\|h\|} \leq K.$ □

4.1.29. Calculate the gauge function for $C := \operatorname{epi}(1/x) \cap \mathbb{R}_+^2$ and conclude that a gauge function is not necessarily lsc.

4.1.30.* Let C be a nonempty convex subset of a Banach space X and let γ_C be the gauge function of C.

(a) Show that γ_C is convex and when $0 \in C$ it is sublinear. Is it necessary to assume $0 \in C$?
(b) Show that if $x \in \operatorname{core} C$ then $\operatorname{dom} \gamma_{C-x} = X$.
(c) Suppose Λ is a discontinuous linear functional on X (Exercise 4.1.22) and let $C = \{x : |\Lambda(x)| \leq 1\}$. Show $0 \in \operatorname{core} C$ and γ_C is not continuous. Is $\overline{C} \subset \{x : \gamma_C(x) \leq 1\}$?
(d) Suppose $0 \in \operatorname{core} C$ and C is closed. Show that $C = \{x \in X : \gamma_C(x) \leq 1\}$, $\operatorname{int} C = \{x \in X : \gamma_C(x) < 1\}$ and that γ_C is Lipschitz.

Hint. No, for the last question in (c), otherwise Λ would be continuous. For (d), show $0 \in \operatorname{int} C$ and use Exercise 4.1.28. □

4.1.31. Let C_1 and C_2 be closed convex subsets of a Banach space X. Then $C_1 \subset C_2$ if and only if, for any $x^* \in X^*$, $\sigma_{C_1}(x^*) \leq \sigma_{C_2}(x^*)$. Thus, a closed convex set is characterized by its support function.

4.1.32. Interpret the sandwich theorem (4.1.18) geometrically in the case when T is the identity map.

4.1.33 (Normal cones). Let C be a closed convex subset of X. We define the *normal cone* of C at $\bar{x} \in C$ by $N_C(\bar{x}) := \partial \delta_C(\bar{x})$. Sometimes we will also use the notation $N(C; \bar{x})$. Show that $x^* \in N(C; x)$ if and only if $\langle x^*, y - x \rangle \leq 0$ for all $y \in C$.

4.1.34.* Prove Corollary 4.1.20.

4.1.35. Let X be a Banach space and let $f : X \to (-\infty, +\infty]$ be a convex function. Suppose that $\bar{x} \in \operatorname{core}(\operatorname{dom} f)$. Show that for any $d \in X$, $t \to g(d; t) := (f(\bar{x} + td) - f(\bar{x}))/t$ is a nondecreasing function.

4.1.36.* Supply any additional details needed to prove Proposition 4.1.23.

4.1 Continuity and subdifferentials

4.1.37. Provide an example of convex functions $f : \mathbb{R} \to (-\infty, +\infty]$ and $g : \mathbb{R} \to (-\infty, +\infty]$ such that $f \geq -g$, $\operatorname{dom} f \cap \operatorname{dom} g \neq \emptyset$ but there is no affine function $\alpha : \mathbb{R} \to \mathbb{R}$ such that $-g \leq \alpha \leq f$.

Hint. Let $f := -\sqrt{x}$ for $x \geq 0$ and $g := \delta_{(-\infty, 0]}$. □

4.1.38.* Prove Proposition 4.1.8.

4.1.39.* Prove the subdifferential of a convex function at a given point is a weak*-closed convex set.

4.1.40.* Prove Proposition 4.1.26.

4.1.41. Prove the following functions $x \in \mathbb{R} \mapsto f(x)$ are convex and calculate ∂f:

(a) $|x|$;

(b) $\delta_{\mathbb{R}_+}$;

(c) $\begin{cases} -\sqrt{x} & \text{if } x \geq 0, \\ \infty & \text{otherwise;} \end{cases}$

(d) $\begin{cases} 0 & \text{if } x < 0, \\ 1 & \text{if } x = 0, \\ \infty & \text{otherwise.} \end{cases}$

4.1.42. (Subgradients of norm) Calculate $\partial \|\cdot\|$. Generalize your result to an arbitrary sublinear function.

4.1.43 (Subgradients of maximum eigenvalue). Denote the largest eigenvalue of an N by N symmetric matrix by λ_1. Prove that $\partial \lambda_1(0)$ is the set of all N by N symmetric matrices with trace 1.

4.1.44 (Subdifferential of a max-function). Let X be a Banach space and suppose that $g_i : X \to (-\infty, +\infty]$, $i \in I$ are lsc convex functions where I is a finite set of integers. Let $g_M := \max\{g_i : i \in I\}$ and for each x, let $I(x) := \{i \in I : g_i(x) = g_M(x)\}$. Note that Exercise 2.2.6 holds in Banach space in the weak*-topology.

(a) Suppose $\bar{x} \in \operatorname{dom} g_M$. Show that

$$\overline{\operatorname{conv}}^{w^*} \left(\bigcup_{i \in I(\bar{x})} \partial g_i(\bar{x}) \right) \subset \partial g_M(\bar{x}).$$

(b) Suppose g_i is continuous at $\bar{x}$ for each $i \in I$, show that

$$\overline{\operatorname{conv}}^{w^*} \left(\bigcup_{i \in I(\bar{x})} \partial g_i(\bar{x}) \right) = \partial g_M(\bar{x}).$$

Can the weak*-closure be dropped on the left-hand side?

(c) Provide an example showing the inclusion in (a) may be proper if just one of the functions g_i is discontinuous at $\bar{x}$.

(d) Suppose

$$\operatorname{dom} g_j \cap \bigcap_{i \in I(\bar{x}) \setminus \{j\}} \operatorname{cont} g_i \neq \emptyset$$

for some index j in $I(\bar{x})$. Show the result of (b) actually remains correct if one uses the convention

$$0\, \partial g_i(\bar{x}) := \partial(0\, g_i)(\bar{x}) = N_{\operatorname{dom} g_i}(\bar{x}).$$

which changes things at most for the index j.

Hint. (a) It is clear that $\partial g_i(\bar{x}) \subset \partial g_M(\bar{x})$ for each $i \in I(x)$, and $\partial g_M(\bar{x})$ is a weak*-closed convex set. For (b), suppose $\phi \in \partial g_M(\bar{x}) \setminus C$ where $C := \overline{\operatorname{conv}}^{w^*}\left(\bigcup_{i \in I(\bar{x})} \partial g_i(\bar{x})\right)$. By the weak*-separation theorem (4.1.23), choose $h \in X$ such that $\phi(h) > \sup_C h$. According to the max formula (4.1.10) we have $g'_M(\bar{x}, h) \geq \phi(h) > g'_i(\bar{x}; h)$ for each $i \in I(x)$. Using this and the fact $g_i(\bar{x}) < g_M(\bar{x})$ for $i \in I \setminus I(x)$ deduce the contradiction: $g_M(\bar{x} + \delta h) > \max_i g_i(\bar{x} + \delta h)$ for some $\delta > 0$. In order to drop the weak*-closure, observe that what remains on the left-hand side is the convex hull of a finite union of weak*-compact convex sets. For (c), let $X = \mathbb{R}$, $g_1(t) := t$ and $g_2(t) := -\sqrt{t}$ for $t \geq 0$ and $g_2(t) := +\infty$ otherwise. Observe that $\partial g_1(0) \cup \partial g_2(0) = \{1\}$ and $\partial g_M(0) = (-\infty, 1]$. For (d), in the harder containment, one may assume 0 belongs to the set on the left and consider $\inf\{t: g_i(x) \leq t, i \in I(\bar{x})\}$. □

4.1.45 (Failure of convex calculus).

(a) Find convex functions $f, g: \mathbb{R} \to (-\infty, \infty]$ with

$$\partial f(0) + \partial g(0) \neq \partial(f + g)(0).$$

(b) Find a convex function $g: \mathbb{R}^2 \to (-\infty, \infty]$ and a linear map $A: \mathbb{R} \to \mathbb{R}^2$ with $A^* \partial g(0) \neq \partial(g \circ A)(0)$.

4.1.46 (DC functions). Let X be a Banach space and C be a nonempty convex subset of X. A function $f: C \to \mathbb{R}$ is a *DC function* if it can be written as the difference of two continuous convex functions.

(a) Prove that a DC function is locally Lipschitz and directionally differentiable at any x in the interior of its domain.
(b) Show that in a Hilbert space, $f(\cdot) = \frac{1}{2}\|\cdot\|^2 - \frac{1}{2}d_F^2(\cdot)$ is convex for any closed set F.
(c) For f as in (b), show that $f^*(x) = \|x\|^2/2$ if $x \in F$ and $f^*(x) = +\infty$ otherwise.

There is a wide array of work on DC functions. In this direction let us also note the broader paper [424] and then [148, 284] concerning the representation of Lipschitz functions as differences of convex functions on superreflexive and Hilbert spaces, see Theorem 5.1.25. The paper [425] shows that the infinite-dimensional version of Hartman's theorem (see Exercise 3.2.11 and [250, 422]) fails, but does provide some positive results in particular cases. In [259] it is shown that on each nonreflexive space there is a positive continuous convex function f such that $1/f$ is not DC which leads to a characterization of such spaces.

4.1 Continuity and subdifferentials

Hint. For (b) and then (c), observe that

$$f(x) = \frac{1}{2} \sup\{\|x\|^2 - \|x-y\|^2 \,:\, y \in F\}$$
$$= \sup\left\{\langle x, y\rangle - \frac{1}{2}\|y\|^2 \,:\, y \in F\right\},$$

where the second equality follows easily from the first. □

4.1.47. Show that for $x \in \mathbb{R}$, the function $x \mapsto \int_0^x t \sin(1/t)\,dt$ is continuously differentiable, hence locally Lipschitz, but is not locally DC in any neighbourhood of zero. Construct a similar everywhere twice differentiable example.

Hint. The integrand is not of bounded variation in any neighborhood of zero. □

4.1.48. Suppose f is a proper lsc convex function on a Banach space X. Show that

$$f'(x;h) = \lim_{\varepsilon \to 0^+} \sup\{\phi(h) \,:\, \phi \in \partial_\varepsilon f(x)\}$$

for any $x \in \mathrm{dom}\, f$.

Hint. See [349, Proposition 3.16]. □

4.1.49 (Pshenichnii–Rockafellar conditions). Let X be a Banach space, let C be a closed convex subset of X and let $f: X \mapsto (-\infty, +\infty]$ be a convex function. Suppose that $C \cap \mathrm{cont}\, f \neq \emptyset$ or $\mathrm{int}\, C \cap \mathrm{dom}\, f \neq \emptyset$ and f is bounded below on C. Show that there is an affine function $\alpha \leq f$ with $\inf_C f = \inf_C \alpha$. Moreover, $\bar{x}$ is a solution to

$$\text{minimize } f(x) \quad \text{subject to } x \in C \subset X \tag{4.1.5}$$

if and only if it satisfies $0 \in \partial f(\bar{x}) + N_C(\bar{x})$, where $N_C := \partial \delta_C(\cdot)$ is the *normal cone*.

Hint. Use the sandwich theorem (4.1.18) on f and $m - \delta_C$ where $m := \inf_C f$. Then apply the subdifferential sum rule (4.1.19) to $f + \delta_C$ at $\bar{x}$. □

4.1.50 (Constrained optimality). Let $C \subseteq X$ be a convex subset of a Banach space X and let $f: X \mapsto (-\infty, +\infty]$ be convex. Consider characterizing $\min\{f(x) \,:\, x \in C\}$. Although in most settings we can apply the Pshenichnii–Rockafellar conditions for optimality it is worth recording the following: $\bar{x}$ minimizes f over C if and only if $f'(\bar{x}; h) \geq 0$ for all $h \in \mathbb{R}_+(C - \bar{x})$.

4.1.51. Apply the Pshenichnii–Rockafellar conditions to the following two cases:

(a) C a single point $\{x^0\} \subset X$,
(b) C a polyhedron $\{x \,:\, Ax \leq b\}$, where $b \in \mathbb{R}^N = Y$.

4.1.52 (Hahn–Banach extension theorem for operators [64]). ** Let Y be a vector space and suppose Y is ordered by a convex cone S. We say (Y, S) has the *Hahn–Banach extension property* if given any vector spaces $M \subseteq X$ and a linear map $T: M \to Y$ and S-sublinear operator $p: X \to Y$ such that $Tx \leq_S p(x)$ for all x in M there is always a dominated linear extension $\overline{T}: M \to Y$ such that $\overline{T}|_M = T$ and $\overline{T}x \leq_S p(x)$ for all x in X. Recall that (Y, S) is *order-complete* if every set in Y with an upper bound has a supremum. (As discussed briefly in Exercise 2.4.32.)

(a) Follow the proof for $(\mathbb{R}, \mathbb{R}_+)$ to show that *if (Y,S) is order-complete then it has the Hahn–Banach extension property.*
(b) Show the converse.

Hint. Let $A \subset Y$ be set with an upper-bound and let B be the set of all upper bounds in Y for A. Let I index these sets with repetition as need be: $A := \{a(i) : i \in I\}$ and $B := \{b(i) : i \in I\}$. Let X be the functions of finite support on I and define a sublinear map by $p(x) := \sum_{i \in I} x(i)^+ b(i) - x(i)^- a(i)$. Take $M := \{x \in X : \sum_{i \in I} x(i) = 0\}$. Let T be a p-dominated extension of 0 from the hyperplane M and let $e(i)$ denote the i-th unit vector. Show that $c := T(e(i))$ is independent of $i \in I$ and is the least upper bound of A – so the ability to extend just from a hyperplane already forces order-completeness! □

(c) Suppose that $X := C(\Omega)$ with the sup norm is order-complete in the usual ordering induced by $S := C(\Omega)_+$. Show that for every Banach space Y such that (isomorphically) $Y \supseteq X$ there is a norm-one linear projection of Y onto X (or its isomorph). Such a Banach space X is called *injective* [266]. This is certainly the case for ℓ_∞.

Hint. Let e be the unit function in X. Define an order-sublinear operator $Q: Y \to X$ by $Q(y) := \|y\|e$ and let ι be injection of X into Y. Then $\iota(x) \leq_S Q(x)$ on X. Any linear dominated extension produces a norm-one projection. □

4.1.53 (A general sup formula). ** In [249, Theorem 4] the authors make a very careful study the subdifferential of the pointwise supremum

$$f(x) := \sup_{t \in T} f_t(x) \qquad (4.1.6)$$

of extended-real-valued convex functions $\{f_t : t \in T\}$ for $x \in X$ a locally convex space, and T a nonempty index set. For $\varepsilon > 0$, denote $T_\varepsilon(x) := \{t \in T : f_t(x) \geq f(x) - \varepsilon\}$ and $\mathcal{F}_z := \{F \subseteq X : F \text{ is linear, dim } F < \infty, z \in F\}$. In Banach space the result is:

Theorem 4.1.28. *Let X be a Banach space. Suppose convex functions f and $f_t, t \in T$ as above are given such that for all $x \in X$ (4.1.6) holds and $\operatorname{cl} f(x) = \sup_{t \in T} \operatorname{cl} f_t(x)$. Then for any $\alpha > 0$*

$$\partial f(x) = \bigcap_{L \in \mathcal{F}_\varepsilon, \varepsilon > 0} \operatorname{cl} \left\{ \operatorname{conv} \left(\bigcup_{t \in T_\varepsilon(x)} \partial_{\alpha\varepsilon} f_t(x) \right) + N_{L \cap \operatorname{dom} f}(x) \right\}$$

for every $x \in X$.

Note that Theorem 4.1.28 makes no additional assumptions on the index set or the underlying functions. Many simpler special cases arise: if X is Euclidean, if all functions are lsc, or if interiority conditions are imposed. The results are especially elegant when each f_t is continuous and affine.

(a) Deduce, in particular, for the support function $\sigma_A(x)$ of $A \subset X^*$ that

$$\partial \sigma_A(x) = \bigcap_{L \in \mathcal{F}_\varepsilon, \varepsilon > 0} \mathrm{cl}\left\{ \mathrm{conv}\,(A_\varepsilon) + \left(\overline{\mathrm{conv}}\left(L^\perp \cup A\right)\right)_\infty \cup \{x\}^\perp \right\}$$

for every $x \in X$. Here $A_\varepsilon := \{x^* \in A : \langle x, x^* \rangle > \sigma_A(x) - \varepsilon\}$, and C_∞ is the recession cone of C.

(b) Compare the compact maximum formula of Exercise 2.2.6.
(c) Recover the max-function result of Exercise 4.1.44.

Finally, convex calculus formulae such as (4.1.19) are then rederivable in a very natural and original fashion.

4.1.54 (Relaxation of sum rule [297]).** Let $\varepsilon > 0$ be given. Let f and g be convex functions on a Banach space and suppose that $x \in \mathrm{dom}\, f \cap \mathrm{dom}\, g$. Show that

$$\partial_\varepsilon (f+g)(x) \supseteq \bigcup \{\partial_\alpha f(x) + \partial_\beta g(x) : \alpha > 0, \beta > 0, \alpha + \beta \leq \varepsilon\}. \quad (4.1.7)$$

Determine an interiority condition for equality in (4.1.7).

4.2 Differentiability of convex functions

A function $f : X \to [-\infty, +\infty]$ is said to be *Fréchet differentiable* at $x_0 \in \mathrm{dom}\, f$ if there exists $\phi \in X^*$ such that for each $\varepsilon > 0$ there exists $\delta > 0$ so that

$$|f(x_0 + h) - f(x_0) - \phi(h)| \leq \varepsilon \|h\| \quad \text{if } \|h\| < \delta.$$

In this case, we say that ϕ is the *Fréchet derivative* of f at x_0. Notice that f is Fréchet differentiable at x_0 if and only if for some $f'(x_0) \in X^*$ the following limit exists uniformly for $h \in B_X$

$$\lim_{t \to 0} \frac{f(x_0 + th) - f(x_0)}{t} = \langle f'(x_0), h \rangle.$$

If the above limit exists for each $h \in S_X$, but not necessarily uniformly, then f is said to be *Gâteaux differentiable* at x_0. In other words, f is Gâteaux differentiable at $x_0 \in \mathrm{dom}\, f$ if for each $\varepsilon > 0$ and each $h \in S_X$, there exists $\delta > 0$ so that

$$|f(x_0 + th) - f(x_0) - \phi(th)| \leq \varepsilon t \quad \text{if } 0 \leq t < \delta \text{ and where } \phi \in X^*$$

In either of the above cases, we may write $\phi = f'(x_0)$ or $\phi = \nabla f(x_0)$.

A *bornology* on X is a family of bounded sets whose union is all of X, which is closed under reflection through the origin and under multiplication of positive scalars, and the union of any two members of the bornology is contained in some member of the bornology. We will denote a general bornology by β, but our attention will focus mainly on the following three bornologies: $\mathcal{F}$ the Gâteaux bornology of all finite sets; $\mathcal{W}$ the weak Hadamard bornology of weakly compact sets; and $\mathcal{B}$ the Fréchet bornology of all bounded sets. Because β contains all points in X, the topology on X^* of uniform convergence on β sets is a Hausdorff topology which we denote by τ_β.

Given a bornology β on X, we will say a real-valued function is β-*differentiable* at $x \in X$, if there is a $\phi \in X^*$ such that for each β-set S, the following limit exists uniformly for $h \in S$
$$\lim_{t \to 0} \frac{f(x+th) - f(x)}{t} = \phi(h).$$

In particular, f is *Gâteaux differentiable* at x if β is the Gâteaux bornology. Similarly for the *weak Hadamard* and *Fréchet* bornologies.

Observe that when f is Fréchet differentiable, then it is continuous, irrespective of whether it is convex. If f is Fréchet differentiable, then it is Gâteaux differentiable. When f is Gâteaux differentiable, then the derivative is unique. Further, as is standard terminology, we will say a norm on X is *Gâteaux* (resp *Fréchet*) *differentiable* if it is Gâteaux (resp. Fréchet) differentiable at every point of $X \setminus \{0\}$.

Example 4.2.1. There exists a function $f : \mathbb{R}^2 \to \mathbb{R}$ such that f is Gâteaux differentiable everywhere, but f is not continuous at $(0,0)$.

Proof. Let $f(x,y) := \frac{x^4 y}{x^8 + y^2}$ if $(x,y) \neq (0,0)$ and let $f(0,0) := 0$. Then f is C^∞-smooth on $\mathbb{R}^2 \setminus \{(0,0)\}$, one can check that the Gâteaux derivative of f is 0 at $(0,0)$, but f is not continuous at $(0,0)$. □

Proposition 4.2.2. *Suppose $f : X \to (-\infty, +\infty]$ is lsc convex and Gâteaux differentiable at x, then f is continuous at x.*

Proof. This follows because x is in the core of the domain of f (see Proposition 4.1.5). □

Actually, the existence of the limit forces the point to be in the interior of the domain, the lower-semicontinuity of the function was crucial in that respect. See Exercise 4.2.3 for a convex function on a Banach space that is Gâteaux differentiable at 0 but 0 is not in the interior of the domain of the function.

Example 4.2.3. On any infinite-dimensional Banach space there is a convex function that is Gâteaux differentiable at 0, lsc at 0, but discontinuous at 0.

Proof. Let ϕ be a discontinuous linear functional on X and let $f = \phi^2$. Then f is discontinuous, f is Gâteaux differentiable at 0 with $f'(0) = 0$. □

Fact 4.2.4. *Suppose $f : X \to (-\infty, +\infty]$ is convex. If f is Gâteaux differentiable at x_0, then $f'(x_0) \in \partial f(x_0)$.*

Proof. The expression in the limit $\langle f'(x_0), h \rangle = \lim_{t \to 0^+} \frac{f(x_0+th) - f(x_0)}{t}$ is nondecreasing as $t \to 0^+$ by the three slope inequality (2.1.1). Therefore
$$\langle f'(x_0), th \rangle \leq f(x_0 + th) - f(x_0) \quad \text{for all } h \in S_X \text{ and all } t \geq 0.$$

Consequently, $\langle f'(x_0), y \rangle \leq f(x_0 + y) - f(x_0)$ for all $y \in X$. □

Corollary 4.2.5 (Unique subgradients). *Suppose a convex function f is continuous at x_0. Then f is Gâteaux differentiable at x_0 if and only if $\partial f(x_0)$ is a singleton.*

Proof. Suppose f is Gâteaux differentiable at x_0. Then by Fact 4.2.4, $f'(x_0) \in \partial f(x_0)$. Suppose $\Lambda \in \partial f(x_0)$ and $\Lambda \neq f'(x_0)$, then $\Lambda(h) > \langle f'(x_0), h \rangle$ for some $h \in S_X$. The subdifferential inequality then implies the contradiction $f'(x_0; h) > \langle f'(x_0), h \rangle$. For the converse, suppose $\partial f(x_0) = \{\Lambda\}$. Then, the max formula (4.1.10) implies $f'(x_0; h) = \Lambda(h)$ and $f'(x_0; -h) = -\Lambda(h)$ for all $h \in S_X$. It then follows that $f'(x_0) = \Lambda$ as a Gâteaux derivative. □

The next example shows that the previous corollary may fail if the function is merely lsc.

Example 4.2.6. Let X be a separable infinite-dimensional Banach space, then there is a lsc convex function on X such that $\partial f(0)$ is a singleton, but f is not Gâteaux differentiable at 0.

Proof. Let $\{x_n\}_{n=1}^{\infty} \subset S_X$ be norm dense in X. Let

$$C := \overline{\text{conv}} \left(\left\{ \frac{1}{n} x_n \right\}_{n=1}^{\infty} \cup \{0\} \right).$$

Now let f be the indicator function of C. Then $0 \in \partial f(0)$. Suppose $\phi \in \partial f(0)$. Then $\phi(\frac{1}{n} x_n) \leq f(\frac{1}{n} x_n) - f(0) = 0$ for all n. Because the collection $\{x_n\}_{n=1}^{\infty}$ is norm dense in S_X, this means $\phi(x) \leq 0$ for all $x \in X$, i.e., $\phi = 0$. Therefore, $\partial f(0) = \{0\}$. Because C has empty interior, f is not continuous at 0, and according to Proposition 4.2.2 cannot be Gâteaux differentiable at 0. □

When f is a convex function that is continuous at x_0 one can describe the differentiability of convex functions without mentioning the derivative. This is because the max formula (4.1.10) ensures that there are candidates for $\nabla f(x_0)$ in the subdifferential.

Proposition 4.2.7 (Symmetric Fréchet differentiability). *Suppose $f : X \to (-\infty, +\infty]$ is a convex function that is continuous at x_0. Then the following are equivalent.*

(a) f is Fréchet differentiable at x_0.
(b) For each $\varepsilon > 0$, there exists $\delta > 0$ such that

$$f(x_0 + h) + f(x_0 - h) - 2f(x_0) \leq \varepsilon \|h\| \quad \text{whenever } \|h\| \leq \delta.$$

(c) For each $\varepsilon > 0$, there exists $\delta > 0$ such that

$$f(x_0 + h) + f(x_0 - h) - 2f(x_0) < \varepsilon \delta \quad \text{whenever } \|h\| \leq \delta.$$

Proof. (a) $\Rightarrow$ (b): Suppose f is Fréchet differentiable at x_0. Given $\varepsilon > 0$, choose $\delta > 0$ such that

$$|f(x_0 + h) - f(x_0) - \langle f'(x_0), h \rangle| \leq \frac{\varepsilon}{2}\|h\| \text{ whenever } \|h\| \leq \delta.$$

Using the previous inequality with h and $-h$ we obtain

$$f(x_0 + h) + f(x_0 - h) - 2f(x_0) = f(x_0 + h) - f(x_0) - \langle f'(x_0), h \rangle$$
$$+ f(x_0 - h) - f(x_0) - \langle f'(x_0), -h \rangle \leq \varepsilon \|h\|.$$

(b) $\Rightarrow$ (c) is straightforward, so we prove (c) $\Rightarrow$ (a) by contraposition. Suppose f is not Fréchet differentiable at x_0. According to the max formula (4.1.10), $\partial f(x_0)$ is not empty, so we let $\phi \in \partial f(x_0)$. Then for some $\varepsilon > 0$, there exist $u_n \in S_X$ and $t_n \to 0^+$ so that

$$f(x_0 + t_n u_n) - f(x_0) - \phi(t_n u_n) > \varepsilon t_n.$$

Let $h_n = t_n u_n$; now $\phi(h_n) \geq f(x_0) - f(x_0 - h_n)$ and so

$$f(x_0 + h_n) + f(x_0 - h_n) - 2f(x_0) > \varepsilon \|h_n\|$$

for all n. The convexity of f ensures that no $\delta > 0$ exists as needed for (c). □

We leave it to the reader to check that the following characterization of Gâteaux differentiability can be proved analogously.

Proposition 4.2.8 (Symmetric Gâteaux differentiability). *Suppose $f : X \to (-\infty, +\infty]$ is a convex function that is continuous at x_0. Then f is Gâteaux differentiable at x_0 if and only if for each $\varepsilon > 0$, and each $h \in S_X$ there exists $\delta > 0$ depending on ε and h such that*

$$f(x_0 + th) + f(x_0 - th) - 2f(x_0) \leq \varepsilon t \quad \text{whenever } 0 \leq t < \delta.$$

A nice application of Proposition 4.2.7 is the following result.

Proposition 4.2.9. *Suppose that f is continuous and convex on the nonempty open convex set V of a Banach space X. Then the set of points of Fréchet differentiability of f is a possibly empty G_δ.*

Proof. Let

$$G_n := \left\{ x \in V : \exists \delta > 0 \text{ such that } \sup_{\|h\| \leq \delta} f(x+h) + f(x-h) - 2f(x) < \frac{\delta}{n} \right\}.$$

It follows from Proposition 4.2.7 that $G = \bigcap_{n \in \mathbb{N}} G_n$ is the set of points of Fréchet differentiability of f. One need only check that G_n is open for each $n \in \mathbb{N}$, and this is a straightforward consequence of the local Lipschitz property of f. □

4.2 Differentiability of convex functions

The points of Gâteaux differentiability of a continuous convex function need not contain a G_δ, but on many Banach spaces they do; see for example Theorem 4.6.3 and Remark 4.6.4 in Section 4.6. The following is a version of Šmulian's theorem [404] that is an extremely useful test for differentiability.

Theorem 4.2.10 (Šmulian's theorem). *Suppose the convex function f is continuous at x_0. Then, the following are equivalent.*

(a) f is Fréchet differentiable at x_0.
(b) For each sequence $x_n \to x_0$ and $\phi \in \partial f(x_0)$, there exist $n_0 \in \mathbb{N}$ and $\phi_n \in \partial f(x_n)$ for $n \geq n_0$ such that $\phi_n \to \phi$.
(c) $\phi_n \to \phi$ whenever $\phi_n \in \partial f(x_n)$, $\phi \in \partial f(x_0)$ and $x_n \to x_0$.
(d) $(\phi_n - \Lambda_n) \to 0$ whenever $\phi_n \in \partial_{\varepsilon_n} f(x_n)$, $\Lambda_n \in \partial_{\varepsilon_n} f(y_n)$ and (x_n) and (y_n) converge to x_0 and $\varepsilon_n \to 0^+$.
(e) $\phi_n \to \phi$ whenever $\phi_n \in \partial_{\varepsilon_n} f(x_0)$, $\phi \in \partial f(x_0)$ and $\varepsilon_n \to 0^+$.

Proof. (a) $\Rightarrow$ (e): Suppose that (e) does not hold, then there exist $\varepsilon_n \to 0^+$, $\phi_n \in \partial_{\varepsilon_n} f(x_0)$, $\phi \in \partial f(x_0)$ and $\varepsilon > 0$ such that

$$\|\phi_n - \phi\| > \varepsilon \quad \text{for all } n.$$

Now choose $h_n \in S_X$ such that $(\phi_n - \phi)(h_n) \geq \varepsilon$ and let $t_n = 2\varepsilon_n/\varepsilon$. Then

$$\frac{t_n \varepsilon}{2} \leq t_n \varepsilon - \varepsilon_n \leq \phi_n(t_n h_n) - \phi(t_n h_n) - \varepsilon_n$$
$$\leq f(x_0 + t_n h_n) - f(x_0) - \phi(t_n h_n)$$

and so (a) does not hold.

(e) $\Rightarrow$ (d): Because f is continuous at x_0, we know that f has Lipschitz constant, say M, on $B_{2r}(x_0)$ for some $r > 0$ (Proposition 4.1.4). In the case $\|x_n - x_0\| \leq r$, one can check that $\|\phi_n\| \leq M + \varepsilon_n/r$. Consequently, $\phi_n(x_n) - \phi_n(x_0) \to 0$. Thus, for $y \in X$, one has

$$\phi_n(y) - \phi_n(x_0) = \phi_n(y) - \phi_n(x_n) + \phi_n(x_n) - \phi_n(x_0)$$
$$\leq f(y) - f(x_n) + \varepsilon_n + \phi_n(x_n) - \phi_n(x_0)$$
$$\leq f(y) - f(x_0) + \varepsilon_n + |f(x_n) - f(x_0)| + |\phi_n(x_0) - \phi_n(x_0)|.$$

Consequently, $\phi_n \in \partial_{\varepsilon'_n} f(x_0)$ where $\varepsilon'_n = \varepsilon_n + |f(x_n) - f(x_0)| + |\phi_n(x_n) - \phi_n(x_0)|$ and $\varepsilon'_n \to 0$. According to (e), $\phi_n \to \phi$. Similarly, $\Lambda_n \to \phi$ and so (d) holds.

(d) $\Rightarrow$ (c): Letting $\Lambda_n = \phi$ and $y_n = x_0$, we see that (c) follows directly from (d).

(c) $\Rightarrow$ (b): This follows because the continuity of f at x_0 implies f is continuous in a neighborhood of x_0 (Proposition 4.1.4), and so by the max formula (4.1.10), $\partial f(x_n)$ is eventually nonempty whenever $x_n \to x_0$.

(b) $\Rightarrow$ (a) Let us suppose (a) fails. Then there exist $t_n \downarrow 0$, $h_n \in B_X$ and $\varepsilon > 0$ such that

$$f(x_0 + t_n h_n) - f(x_0) - \phi(t_n h_n) > \varepsilon t_n \quad \text{where } \phi \in \partial f(x_0).$$

Let $\phi_n \in \partial f(x_0 + t_n h_n)$ (for sufficiently large n). Now,

$$\phi_n(t_n h_n) \geq f(x_0 + t_n h_n) - f(x_0) \geq \phi(t_n h_n) + \varepsilon t_n$$

and so $\phi_n \not\to \phi$. Thus (b) fails. □

For Gâteaux differentiability, norm convergence of the functionals is replaced with weak*-convergence.

Theorem 4.2.11 (Šmulian's theorem). *Suppose the convex function f is continuous at x_0. Then, the following are equivalent.*

(a) f is Gâteaux differentiable at x_0.
(b) For each sequence $x_n \to x$ and $\phi \in \partial f(x_0)$, there exist $n_0 \in \mathbb{N}$ and $\phi_n \in \partial f(x_n)$ for $n \geq n_0$ such that $\phi_n \to_{w^*} \phi$.
(c) $\phi_n \to_{w^*} \phi$ whenever $\phi_n \in \partial f(x_n)$, $\phi \in \partial f(x_0)$ and $x_n \to x_0$.
(d) $(\phi_n - \Lambda_n) \to_{w^*} 0$ whenever $\phi_n \in \partial_{\varepsilon_n} f(x_n)$, $\Lambda_n \in \partial_{\varepsilon_n} f(y_n)$ and (x_n) and (y_n) converge to x_0 and $\varepsilon_n \to 0^+$.
(e) $\phi_n \to_{w^*} \phi$ whenever $\phi_n \in \partial_{\varepsilon_n} f(x_0)$, $\phi \in \partial f(x_0)$ and $\varepsilon_n \to 0^+$.

A more general bornological version of Šmulian's theorem is given in Exercise 4.2.8. The following is an immediate consequence of two versions of Šmulian's theorem that were just presented. In Section 6.1 we will re-cast Šmulian's theorem in terms of selections of set-valued mappings.

Corollary 4.2.12 (Continuity of derivative mapping). *Suppose the convex function $f : U \to \mathbb{R}$ is continuous. Then f is Fréchet (resp. Gâteaux) differentiable on U if and only if $x \to f'(x)$ is norm-to-norm continuous (resp. norm-to-weak*-continuous) on U.*

It is often useful to know if the gauge of the lower level set of a differentiable convex function shares the same differentiability properties as the original function. This is usually done by applying the implicit function theorem; when the differentiability is merely Gâteaux but not C^1-smooth, a restriction to finite dimensions can be used because in finite dimensions, Gâteaux differentiability implies C^1-smoothness for convex functions. Below, we give an alternate argument for this type of situation.

Theorem 4.2.13 (Implicit function theorem for gauges). *Suppose $f : X \to (-\infty, +\infty]$ is a lsc convex function and that $C := \{x : f(x) \leq \alpha\}$ where $f(0) < \alpha$ and $0 \in \mathrm{int}\, C$. Suppose f is β-differentiable at all $x \in \mathrm{bdry}(C)$. Then γ_C, the gauge of C, is β-differentiable at all x where $\gamma_C(x) > 0$.*

Proof. First, fix $x_0 \in \mathrm{bdry}(C)$ and let $\phi = \frac{f'(x_0)}{\langle f'(x_0), x_0 \rangle}$. Then $\phi(x_0) = 1$, and so it is the natural candidate for being the β-derivative $\gamma_C'(x_0)$. This is precisely, what we will prove. In what follows, we let $H = \{h : \phi(h) = 0\}$.

We now show that $\phi \in \partial \gamma_C(x_0)$. Observe that if $\phi(u) = 1$, then $u = x_0 + h$ for some $h \in H$, and so $\langle f'(x_0), h \rangle = 0$. Now $f(u) \geq f(x_0) + \langle f'(x_0), h \rangle$ by the

4.2 Differentiability of convex functions

subgradient inequality. Consequently $f(u) \geq \alpha$ and so $\gamma_C(u) \geq 1 = \phi(u)$. Because γ_C is positively homogeneous, this shows $\gamma_C(x) \geq \phi(x)$ for all $x \in X$. Therefore,

$$\phi(x) - \phi(x_0) = \phi(x) - 1 \leq \gamma_C(x) - 1 = \gamma_C(x) - \gamma_C(x_0) \quad \text{for all } x \in X,$$

which shows that $\phi \in \partial \gamma_C(x_0)$. Moreover, ϕ is the only subgradient of γ_C at x_0. Indeed, suppose $\psi \in \partial \gamma_C(x_0)$. Then the subdifferential inequality implies $\psi(x_0) = 1$. Also, if $h \in H$, then $\psi(h) = \psi(x_0) - \psi(x_0 - h) \leq \gamma_C(x_0) - \gamma_C(x_0 - h) \leq 0$ which implies $\psi(h) = 0$ for all $h \in H$. Consequently, the kernels of ψ and ϕ are the same, and $\psi(x_0) = \phi(x_0) = 1$, and so $\phi = \psi$. Therefore $\partial \gamma_C(x_0) = \{\phi\}$. According to Corollary 4.2.5, γ_C is Gâteaux differentiable at x_0 with Gâteaux derivative ϕ.

Because both ϕ and γ_C are positively homogeneous, it follows that ϕ is the Gâteaux derivative of γ_C at λx_0 for each $\lambda > 0$. Finally, the derivative mapping $x \to f'(x)$ is norm to τ_β-continuous on bdry(C) (Exercise 4.2.8), and because $\gamma'_C(x) = \frac{f'(x)}{\langle f'(x), x \rangle}$ for all x in the boundary of C, it follows that $x \to \gamma'_C(x)$ is norm-to-τ_β continuous at all x where $\gamma_C(x) \neq 0$. According to Exercise 4.2.8, γ_C is β-differentiable at all x where $\gamma_C(x) \neq 0$. $\square$

The next two results provide a version of Šmulian's theorem for uniformly continuous derivatives.

Proposition 4.2.14 (Uniform Fréchet differentiability). *Suppose $f : X \to \mathbb{R}$ is a continuous convex function. Then the following are equivalent.*

(a) $x \to f'(x)$ is a uniformly continuous mapping on X, i.e. if $x_n, y_n \in X$ and $\|x_n - y_n\| \to 0$, then $\|f'(x_n) - f'(y_n)\| \to 0$.
(b) $\|\phi_n - \Lambda_n\| \to 0$ whenever $\phi_n \in \partial_{\varepsilon_n} f(x_n)$, $\Lambda_n \in \partial_{\varepsilon_n} f(y_n)$, $x_n, y_n \in X$, $\varepsilon_n \to 0^+$ and $\|x_n - y_n\| \to 0$.
(c) f is uniformly smooth; that is for each $\varepsilon > 0$, there exists $\delta > 0$ such that

$$f(x+h) + f(x-h) - 2f(x) \leq \varepsilon \|h\| \quad \text{whenever } \|h\| < \delta \text{ and } x \in X.$$

(d) f is uniformly Fréchet differentiable on X. That is,

$$\lim_{t \to 0} \frac{f(x+th) - f(x)}{t} = \langle f'(x), h \rangle$$

uniformly for $h \in S_X$ and $x \in X$.

Proof. (a) $\Rightarrow$ (d): Suppose that (d) does not hold. Then there exist $t_n \to 0$, $h_n \in S_X$ and $x_n \in X$ and $\varepsilon > 0$ such that

$$|f(x_n + t_n h_n) - f(x_n) - \langle f'(x_n), t_n h_n \rangle| \geq \varepsilon |t_n| \quad \text{for all } n \in \mathbb{N}.$$

According to the mean value theorem, there exist $0 < \lambda_n < 1$ such that

$$|\langle f'(x_n + \lambda_n t_n h_n), t_n h_n \rangle - \langle f'(x_n), t_n h_n \rangle| \geq \varepsilon |t_n| \quad \text{for } n \in \mathbb{N}.$$

Consequently, f' is not uniformly continuous.

(d) $\Rightarrow$ (c): Suppose (d) is true. Given $\varepsilon > 0$, we choose $\delta > 0$ so that $|f(x+ty) - f(x) - \langle f'(x), ty \rangle| \leq \frac{\varepsilon}{2}|t|$ for all $|t| \leq \delta$, $x \in X, y \in S_X$. Then,

$$f(x+ty) + f(x-ty) - 2f(x) \leq \varepsilon t \text{ for all } 0 \leq t < \delta, x \in X, y \in S_X.$$

Now, any h can be written $h = ty$ for some $y \in S_X$ with $\|h\| = t$ and so (c) holds.

(c) $\Rightarrow$ (b): Suppose (b) doesn't hold. Then there exist sequences (x_n), (y_n) with $\|x_n - y_n\| \to 0$, and $\varepsilon_n \to 0^+$ with $\phi_n \in \partial_{\varepsilon_n} f(x_n)$, $\Lambda_n \in \partial_{\varepsilon_n} f(y_n)$ where $\|\phi_n - \Lambda_n\| > \varepsilon$ for all n. Now choose h_n such that $\langle \phi_n - \Lambda_n, h_n \rangle > \varepsilon \|h_n\|$, $\|h_n\| \to 0$, $\|h_n\|\varepsilon > 4\varepsilon_n$, and $\|h_n\| \geq \|x_n - y_n\|$. Then

$$\frac{\varepsilon}{4}\left\|h_n + \frac{x_n - y_n}{2}\right\| < \frac{\varepsilon}{2}\|h_n\| < \varepsilon\|h_n\| - 2\varepsilon_n \leq (\phi_n - \Lambda_n)(h_n) - 2\varepsilon_n$$

$$\leq f(x_n + h_n) + f(y_n - h_n) - f(x_n) - f(y_n)$$

$$\leq f(x_n + h_n) + f(y_n - h_n) - 2f\left(\frac{x_n + y_n}{2}\right)$$

$$= f\left(\frac{x_n + y_n}{2} + \left[h_n + \frac{x_n - y_n}{2}\right]\right)$$

$$+ f\left(\frac{x_n + y_n}{2} - \left[h_n + \frac{x_n - y_n}{2}\right]\right) - 2f\left(\frac{x_n + y_n}{2}\right).$$

Thus condition (c) doesn't hold.

(b) $\Rightarrow$ (a): First, f is Fréchet differentiable by Šmulian's theorem (4.2.10). Thus condition (b) implies the uniform continuity of the derivative. □

From the proof of the previous result, we obtain a relevant bounded set version of the above.

Proposition 4.2.15 (Uniform smoothness on bounded sets). *Suppose $f : X \to \mathbb{R}$ is a continuous convex function. Then the following are equivalent.*

(a) $x \mapsto f'(x)$ *is a uniformly continuous mapping on bounded sets.*
(b) $\|\phi_n - \Lambda_n\| \to 0$ *whenever $\phi_n \in \partial_{\varepsilon_n} f(x_n)$, $\Lambda_n \in \partial_{\varepsilon_n} f(y_n)$, where (x_n) and (y_n) are bounded, $\varepsilon_n \to 0^+$ and $\|x_n - y_n\| \to 0$.*
(c) f *is uniformly smooth on bounded sets; that is, for each $K > 0$ and $\varepsilon > 0$, there exists $\delta > 0$ such that $f(x+h) + f(x-h) - 2f(x) \leq \varepsilon\|h\|$ whenever $\|x\| \leq K$ and $\|h\| \leq \delta$.*

Finally, we state a directional variant of Proposition 4.2.14.

Proposition 4.2.16 (Uniform Gâteaux differentiability). *Suppose $f : X \to \mathbb{R}$ is a continuous convex function. Then the following are equivalent.*

(a) *For each $h \in S_X$, $x \mapsto \langle f'(x), h \rangle$ is a uniformly continuous mapping on X, i.e. given $h \in S_X$, then $|\langle f'(x_n), h \rangle - \langle f'(y_n), h \rangle| \to 0$ whenever $x_n, y_n \in X$ and $\|x_n - y_n\| \to 0$.*

(b) $(\phi_n - \Lambda_n) \to 0$ weak* whenever $\phi_n \in \partial_{\varepsilon_n} f(x_n)$, $\Lambda_n \in \partial_{\varepsilon_n} f(y_n)$ $x_n, y_n \in X$, $\varepsilon_n \to 0$ and $\|x_n - y_n\| \to 0$.
(c) For each $\varepsilon > 0$ and $h \in S_X$, there exists $\delta > 0$ such that

$$f(x+th) + f(x-th) - 2f(x) \leq \varepsilon t \text{ whenever } 0 < t < \delta \text{ and } x \in X.$$

(d) f is uniformly Gâteaux differentiable on X. That means, for each $h \in S_X$

$$\lim_{t \to 0} \frac{f(x+th) - f(x)}{t} = \langle f'(x), h \rangle$$

uniformly for $x \in X$.

Proof. It is left for the reader to check that an appropriate directional modification of the proof of Proposition 4.2.14 works. □

Exercises and further results

4.2.1. (a) Let $C := \{x \in c_0 : |x_n| \leq n^{-2}\}$, and let f be the indicator function of C. As in Example 4.2.6, show that $\partial f(0) = \{0\}$, but show directly that f is not Gâteaux differentiable by showing $\lim_{k \to \infty} f(0 + \frac{1}{k}h) - f(0) = \infty$ where $h := (n^{-1}) \in c_0$.
(b) Let C be as in (a). Is d_C Gâteaux differentiable at 0?

4.2.2. Use the definition of Gâteaux differentiability to show that the function f in Example 4.2.3 satisfies $f'(0) = 0$. Is the function f lsc everywhere?
Hint. Let $h \in X$, Then $\phi(h) = \alpha$. Now

$$\lim_{t \to 0} [f(0+th) - f(0)]/t = \lim_{t \to 0} (t^2 \alpha^2)/t = 0.$$

No, f is not lsc, otherwise f would be continuous at 0. □

4.2.3. Let X be an infinite-dimensional Banach space. Find an example of a convex function $f : X \to [0, \infty]$ that is Gâteaux differentiable at 0, but 0 is not in the interior of the domain of f.
Hint. Modify Example 4.2.3. Take the function f given therein, and let $g(x) := f(x)$ if $f(x) \leq 1$ and $g(x) := \infty$ otherwise. □

4.2.4.* Let C be a closed convex subset of a Banach space X such that $0 \in C$. Prove that the following are equivalent.

(a) $\bigcup_{n=1}^{\infty} nC$ is norm dense in X.
(b) $\partial \delta_C(0) = \{0\}$.
(c) $\partial f(0) = \{0\}$ where $f := d_C(\cdot)$.

Hint. (a) $\Rightarrow$ (b): Let $D \subset X$ be dense in X such that $D \subset \bigcup_{n=1}^{\infty} nC$. Given $h \in D$, then $\frac{1}{n} h \in C$ for some $n \in \mathbb{N}$. Now

$$\phi\left(\frac{1}{n}h\right) - \phi(0) \leq \delta_C\left(\frac{1}{n}h\right) - \delta_C(0) = 0.$$

Therefore, $\phi(x) \leq 0$ for all x in some dense subset of X. Therefore $\phi = 0$.

(b) $\Rightarrow$ (c): $0 \in \partial f(0) \subset \partial \delta_C(0) = \{0\}$.

(c) $\Rightarrow$ (a): Proceed by contraposition. Suppose $\bigcup_{n=1}^{\infty} nC$ is not dense in X, then choose $h \in X$ and $\delta > 0$ such that $(h + \delta B_X) \cap (\bigcup_{n=1}^{\infty} nC) = \emptyset$. Then $\frac{1}{k}(h + \delta B_X) \cap (\bigcup_{n=1}^{\infty} nC) = \emptyset$ for all $k \in \mathbb{N}$, and so $d_C(\frac{1}{k}h) \geq \frac{1}{k}\delta$ for all $k \in \mathbb{N}$. Now conclude that $\partial f(0) \neq \{0\}$.

See [77, Proposition 2] for a different approach to a related result. □

4.2.5. Generalize Example 4.2.6 by proving that the following are equivalent for a Banach space X.

(a) X admits a closed convex set K with $0 \in K \setminus \text{int } K$ and $\bigcup nK$ is dense in X.
(b) There is a lsc proper convex $f : X \to (-\infty, +\infty]$ such that $\partial f(0)$ is a singleton, but f is not Gâteaux differentiable at 0.

Hint. (a) $\Rightarrow$ (b): Let K be as given, and let $f(x) = \delta_K(x)$, the indicator function of K. Then f is not continuous at 0 since 0 is not in the interior of the domain of f. However, $\partial f(0) = \{0\}$ by Exercise 4.2.4

(b) $\Rightarrow$ (a): Let f be a function as given in (b), and let $\phi \in \partial f(0)$. By replacing f with $f - \phi$, we may assume that $\partial f(0) = \{0\}$ and $f(x) \geq f(0) = 0$ for all $x \in X$. Now f is not continuous at 0, because f is not Gâteaux differentiable at 0. Let $C := \{x \in B_X : f(x) \leq 1\}$. Then $0 \in C$, but $0 \notin \text{int } C$. Let $C_1 := \{x \in B_X : \lim_{t \to 0^+} f(th)/t \leq 1\}$. One can check that C_1 is convex, and if $h \in C_1$, then $th \in C$ for some $t > 0$. Now let $C_2 = \{h \in C : \lim_{t \to 0^+} f(th)/t \leq 1\}$. Then C_2 is convex. Now the desired set is K, the norm closure of C_2. Check that $f(x) \geq d_K(x)$ if $x \in B_X \setminus K$ since $\lim_{t \to 0^+} f(th)/t > 1$ for $h \in B_X \setminus K$. Therefore, $0 \leq d_K(x) \leq f(x)$ for all $x \in B_X$. Therefore, when $g(x) = d_K(x)$, one has $\partial g(0) = \{0\}$ and so $\bigcup_{n=1}^{\infty} nK$ is norm dense in X by Exercise 4.2.4.

Note [77, p. 1126] asks whether every Banach space admits such a set K as in (a). □

4.2.6.* Suppose C is a closed convex set with empty interior in a Banach space.

(a) Prove that $f := d_C(\cdot)$ is not Fréchet differentiable at any $x \in C$.
(b) Find an example of f as in (a) where f is Gâteaux differentiable at some point(s) in C.

Hint. (a) Translate C so that $0 \in C$. Because $0 \in \partial f(0)$, it is a candidate for the derivative. Suppose f is Fréchet differentiable at 0. Then there exists $n \in \mathbb{N}$ so that $f(h) \leq \|h\|/4$ whenever $\|h\| \leq 1/n$. Now consider the closed convex set nC which also has empty interior. For each $h \in S_X$, deduce that there exists $x \in nC$ so that $\|x - h\| < 1/3$. Because nC has empty interior, we choose y_0 with $\|y_0\| < 1/3$ so that $y_0 \notin nC$. Use the basic separation theorem (4.1.12) to show that there exists $\phi \in S_{X^*}$ so that $\sup_{nC} \phi < \phi(y_0) < 1/3$. This contradicts the property that $d_{nC}(h) < 1/3$ for each $h \in S_X$.

(b) See Exercise 4.2.4. □

4.2 Differentiability of convex functions

4.2.7.* Prove Šmulian's theorem for Gâteaux differentiability (Theorem 4.2.11).

Hint. This follows from the following more general exercise. □

4.2.8 (Šmulian's theorem for β-differentiability).* Suppose the proper convex function f is continuous at x_0. Then, the following are equivalent.

(a) f is β-differentiable at x_0.
(b) For each sequence $x_n \to x$ and $\phi \in \partial f(x_0)$, there exist $n_0 \in \mathbb{N}$ and $\phi_n \in \partial f(x_n)$ for $n \geq n_0$ such that $\phi_n \to_{\tau_\beta} \phi$.
(c) $\phi_n \to_{\tau_\beta} \phi$ whenever $\phi_n \in \partial f(x_n)$, $\phi \in \partial f(x_0)$ and $x_n \to x_0$.
(d) $(\phi_n - \Lambda_n) \to_{\tau_\beta} 0$ whenever $\phi_n \in \partial_{\varepsilon_n} f(x_n)$, $\Lambda_n \in \partial_{\varepsilon_n} f(y_n)$, $x_n \to x_0$, $y_n \to x_0$ and $\varepsilon_n \to 0^+$.
(e) $\phi_n \to_{\tau_\beta} \phi$ whenever $\phi_n \in \partial_{\varepsilon_n} f(x_0)$, $\phi \in \partial f(x_0)$ and $\varepsilon_n \to 0^+$.
(f) Given any β-set W and any $\varepsilon > 0$, there exists $\delta > 0$ depending on ε and W such that $f(x + th) + f(x - th) - 2f(x_0) \leq t\varepsilon$ whenever $0 \leq t \leq \delta$.

Hint. (a) $\Rightarrow$ (f): Given W and $\varepsilon > 0$, by the definition of β-differentiability, choose $\delta > 0$ so that $|f(x_0 + th) - f(x_0) - f'(x_0)(th)| \leq \frac{\varepsilon}{2}|t|$ for all $|t| \leq \delta$ and $h \in W$. Now (e) follows directly from this.

(f) $\Rightarrow$ (e): Suppose that (e) does not hold, then there exist $\varepsilon_n \to 0^+$, $\phi_n \in \partial_{\varepsilon_n} f(x_0)$, $\phi \in \partial f(x_0)$ and $\varepsilon > 0$ such that

$$\sup_W |\phi_n - \phi| > \varepsilon \text{ for all } n \text{ and some } \beta\text{-set } W.$$

Now choose $h_n \in W$ such that $(\phi_n - \phi)(h_n) \geq \varepsilon$ and let $t_n = 2\varepsilon_n/\varepsilon$. Then

$$\frac{t_n \varepsilon}{2} \leq t_n \varepsilon - \varepsilon_n \leq \phi_n(t_n h_n) - \phi(t_n h_n) - \varepsilon_n$$
$$\leq f(x_0 + t_n h_n) - f(x_0) + f(x_0 - t_n h_n) - f(x_0)$$

and so (f) does not hold.

(e) $\Rightarrow$ (d): As in the proof of Theorem 4.2.10 (e) $\Rightarrow$ (d) we have $\phi_n, \Lambda_n \in \partial_{\varepsilon'_n}(x_0)$ where $\varepsilon'_n \to 0$. Thus $\phi_n \to_{\tau_\beta} \phi$ and $\Lambda_n \to_{\tau_\beta} \phi$ and so (d) follows.

(d) $\Rightarrow$ (c) and (c) $\Rightarrow$ (b) are straightforward. To prove (b) $\Rightarrow$ (a) we suppose f is not β-differentiable at x_0. Then there exist $t_n \downarrow 0$ and $h_n \in W$ where W is a β-set and $\varepsilon > 0$ such that

$$f(x_0 + t_n h_n) - f(x_0) - \phi(t_n h_n) \geq \varepsilon t_n \quad \text{where } \phi \in \partial f(x_0).$$

For sufficiently large n, there exist $\phi_n \in \partial f(x_0 + t_n h_n)$. Now,

$$\phi_n(t_n h_n) \geq f(x_0 + t_n h_n) - f(x_0) \geq \phi(t_n h_n) + \varepsilon t_n$$

and so $\phi_n \not\to_{\tau_\beta} \phi$. □

4.2.9.*

(a) Suppose $f : X^* \to \mathbb{R}$ is convex and weak*-lsc (hence continuous as a finite valued convex function). Is it true that $\partial f(x^*) \cap X \neq \emptyset$ for each $x^* \in X^*$?

(b) Suppose $f : X^* \to (-\infty, +\infty]$ is convex and weak*-lsc. Is it true that $\partial_\varepsilon f(x^*) \cap X \neq \emptyset$ for each $\varepsilon > 0$, and each $x^* \in \text{dom}(f)$?

Hint. (a) No, for $f := \|\cdot\|_1$ the usual norm on ℓ_1, consider $x^* := (2^{-i})$.
(b) Yes, use the weak* epi-separation theorem (4.1.23). □

4.2.10 (Šmulian's theorem β-differentiability weak*-lsc functions).* Suppose the weak*-lsc convex function $f : X^* \to (-\infty, +\infty]$ is continuous at ϕ_0. Then, the following are equivalent.

(a) f is β-differentiable at ϕ_0.
(b) $(x_n - y_n) \to_{\tau_\beta} 0$ whenever $x_n \in \partial_{\varepsilon_n} f(\phi_n) \cap X$, $y_n \in \partial_{\varepsilon_n} f(\Lambda_n) \cap X$, $\phi_n \to \phi_0$, $\Lambda_n \to \phi_0$ and $\varepsilon_n \to 0^+$.
(c) $x_n \to_{\tau_\beta} \Phi$ whenever $x_n \in \partial_{\varepsilon_n} f(\phi_0)$, $\Phi \in \partial f(\phi_0)$ and $\varepsilon_n \to 0^+$.

Hint. The point of this exercise is that one need only consider ε-subgradients from X. Clearly (a) $\Rightarrow$ (c) follows from Šmulian's theorem as in Exercise 4.2.8, and (c) $\Rightarrow$ (b) is proved as in the Exercise 4.2.8.

Now suppose f is not β-differentiable at ϕ_0. Then there is a β-set $W \subset X^*$ with $w_n \in W$ and $t_n \to 0^+$ and $\varepsilon > 0$ such that

$$f(\phi_0 + t_n w_n) + f(\phi_0 - t_n w_n) - 2f(\phi_0) \geq \varepsilon t_n.$$

Choose $x_n \in \partial_{\varepsilon_n} f(\phi_0 + t_n w_n)$ and $y_n \in \partial_{\varepsilon_n} f(\phi_0 - t_n w_n)$ where $\varepsilon_n = t_n \varepsilon/4$. Then

$$(x_n - y_n)(t_n w_n) \geq \varepsilon t_n/2$$

and so $(x_n - y_n)_n$ does not converge τ_β to 0. □

4.2.11.* (a) Suppose $f : X^* \to \mathbb{R}$ is weak*-lsc and convex, and that f is Fréchet differentiable at $x_0^* \in X^*$. Show that $f'(x_0^*) \in X$. What if f is only Gâteaux or weak Hadamard differentiable?
(b) Suppose X is weakly sequentially complete and $f : X^* \to (-\infty, +\infty]$ is weak*-lsc and convex. Suppose f is Gâteaux differentiable at $x_0^* \in X^*$. Show that $f'(x_0^*) \in X$.

Hint. (a) Let $\Phi = f'(x_0^*)$. Choose $x_n \in \partial_{1/n} f(x_0^*) \cap X$. By Šmulian's theorem (4.2.10), $x_n \to \Phi$ in norm, and so $\Phi \in X$. This need not be true if f is weak Hadamard differentiable, see hint for Exercise 4.2.9(a).

(b) Let Φ and x_n be as in the proof of (a). Then by Šmulian's theorem (4.2.11) $x_n \to_{w^*} \Phi$. Thus (x_n) is weakly Cauchy. Thus $x_n \to_w x$ for some $x \in X$. Necessarily, $x = \Phi$. □

4.2.12 (Monotone gradients and convexity).* Suppose $f : U \to \mathbb{R}$ is Gâteaux differentiable and U is an open convex subset of a Banach space. Show that f is convex if and only if $\langle \nabla f(x) - \nabla f(y), x - y \rangle \geq 0$ for all $x, y \in U$. Moreover, f is strictly convex if and only if the preceding inequality is strict.

Hint. See the proof of Theorem 2.2.6. □

4.2.13 (Gâteaux differentiability spaces). Suppose X is a Banach space such that each continuous convex function on X has a point of Gâteaux differentiability. Prove that B_{X^*} is weak*-sequentially compact.

Hint. We follow [199, Theorem 10.10]. Let $(\phi_n)_{n \geq 1} \subset B_{X^*}$, and let $A_n = \overline{(\phi_i)_{i \geq n}}^{w^*}$ and let $A = \cap_{n \geq 1} A_n$. Define $p : X \to \mathbb{R}$ by $p(x) = \sup\{\phi(x) : \phi \in A\}$. Then p is a continuous convex function, and therefore, is Gâteaux differentiable at some $x_0 \in X$. Now x_0 attains its supremum on A, and so $\phi_0(x_0) = p(x_0)$ for some $\phi_0 \in A$. Now, it is easy to verify that $\phi_0 \in \partial p(x_0)$. Now $\phi_0 \in A_n$ for each n, and so for each $j \in \mathbb{N}$, there is a ϕ_{n_j} with $n_j \geq n$ such that $|\phi_{n_j}(x_0) - \phi_0(x_0)| < \frac{1}{j}$, now one can easily verify $\phi_{n_j} \in \partial p_{\frac{1}{j}}(x_0)$. Thus by Šmulian's theorem (4.2.11), $\phi_{n_j} \to_{w^*} \phi_0$. □

4.3 Variational principles

We begin with Ekeland's variational principle.

Theorem 4.3.1 (Ekeland's variational principle). *Let (X, d) be a complete metric space and let $f : X \to (-\infty, +\infty]$ be a lsc function bounded from below. Suppose that $\varepsilon > 0$ and $z \in X$ satisfy*

$$f(z) < \inf_X f + \varepsilon.$$

Suppose $\lambda > 0$ is given, then there exists $y \in X$ such that

(a) $d(z, y) \leq \lambda$,
(b) $f(y) + (\varepsilon/\lambda)d(z, y) \leq f(z)$, *and*
(c) $f(x) + (\varepsilon/\lambda)d(x, y) > f(y)$, *for all $x \in X \setminus \{y\}$.*

Proof. We prove this in the case $\lambda = 1$, and note the general case follows by replacing $d(\cdot, \cdot)$ with $\lambda^{-1}d(\cdot, \cdot)$. Define a sequence (z_i) by induction starting with $z_0 := z$. Suppose that we have defined z_i. Set

$$S_i := \{x \in X : f(x) + \varepsilon d(x, z_i) \leq f(z_i)\}$$

and consider two possible cases: (i) $\inf_{S_i} f = f(z_i)$. Then we define $z_{i+1} := z_i$. (ii) $\inf_{S_i} f < f(z_i)$. We choose $z_{i+1} \in S_i$ such that

$$f(z_{i+1}) < \inf_{S_i} f + \frac{1}{2}[f(z_i) - \inf_{S_i} f] = \frac{1}{2}[f(z_i) + \inf_{S_i} f] < f(z_i). \quad (4.3.1)$$

We show that (z_i) is a Cauchy sequence. In fact, if (i) ever occurs, then z_i is stationary for i large. Otherwise,

$$\varepsilon d(z_i, z_{i+1}) \leq f(z_i) - f(z_{i+1}). \quad (4.3.2)$$

Adding (4.3.2) up from i to $j - 1 > i$ we have

$$\varepsilon d(z_i, z_j) \leq f(z_i) - f(z_j). \quad (4.3.3)$$

Observe that the sequence $(f(z_i))$ is decreasing and bounded from below by $\inf_X f$, and therefore convergent. We conclude from (4.3.3) that (z_i) is Cauchy. Let $y := \lim_{i \to \infty} z_i$. We show that y satisfies the conclusions of the theorem. Setting $i = 0$ in (4.3.3) we have

$$\varepsilon d(z, z_j) + f(z_j) \leq f(z). \tag{4.3.4}$$

Taking limits as $j \to \infty$ yields (b). Since $f(z) - f(y) \leq f(z) - \inf_X f < \varepsilon$, (a) follows from (b). It remains to show that y satisfies (c). Fixing i in (4.3.3) and taking limits as $j \to \infty$ yields $y \in S_i$. That is to say

$$y \in \bigcap_{i=1}^{\infty} S_i.$$

On the other hand, if $x \in \bigcap_{i=1}^{\infty} S_i$ then, for all $i = 1, 2, \ldots$,

$$\varepsilon d(x, z_{i+1}) \leq f(z_{i+1}) - f(x) \leq f(z_{i+1}) - \inf_{S_i} f. \tag{4.3.5}$$

It follows from (4.3.1) that $f(z_{i+1}) - \inf_{S_i} f \leq f(z_i) - f(z_{i+1})$, and therefore $\lim_i [f(z_{i+1}) - \inf_{S_i} f] = 0$. Taking limits in (4.3.5) as $i \to \infty$ we have $\varepsilon d(x, y) = 0$. It follows that

$$\bigcap_{i=1}^{\infty} S_i = \{y\}. \tag{4.3.6}$$

Notice that the sequence of sets (S_i) is nested, i.e. for any i, $S_{i+1} \subset S_i$. In fact, for any $x \in S_{i+1}$, $f(x) + \varepsilon d(x, z_{i+1}) \leq f(z_{i+1})$ and $z_{i+1} \in S_i$ yields

$$f(x) + \varepsilon d(x, z_i) \leq f(x) + \varepsilon d(x, z_{i+1}) + \varepsilon d(z_i, z_{i+1})$$
$$\leq f(z_{i+1}) + \varepsilon d(z_i, z_{i+1}) \leq f(z_i), \tag{4.3.7}$$

which implies that $x \in S_i$. Now, for any $x \neq y$, it follows from (4.3.6) that when i sufficiently large $x \notin S_i$. Thus, $f(x) + \varepsilon d(x, z_i) \geq f(z_i)$. Taking limits as $i \to \infty$ we conclude $f(x) + \varepsilon d(x, y) \geq f(y)$; now observe the inequality can be made strict for all $x \neq y$ because the argument works for any $\varepsilon > 0$ such that $f(z) < \inf_X f + \varepsilon$. $\square$

Theorem 4.3.2 (Brøndsted–Rockafellar). *Suppose f is a proper lsc convex function on the Banach space X. Then given any $x_0 \in \mathrm{dom} f$, $\varepsilon > 0$, $\lambda > 0$ and $x_0^* \in \partial_\varepsilon f(x_0)$, there exist $x \in \mathrm{dom} f$ and $x^* \in X^*$ such that*

$$x^* \in \partial f(x), \ \|x - x_0\| \leq \varepsilon/\lambda \text{ and } \|x^* - x_0^*\| \leq \lambda.$$

In particular, the domain of ∂f is dense in the domain of f.

Proof. First, $\langle x_0^*, x - x_0 \rangle \leq f(x) - f(x_0)$ for all $x \in X$. Now define the function g by $g(x) := f(x) - \langle x_0^*, x \rangle$ for $x \in X$. Then g is proper and lsc and has the same domain as f. Observe that $g(x_0) \leq \inf_X g + \varepsilon$, and so Ekeland's variational principle (4.3.1)

4.3 Variational principles

implies that there exists $y \in \mathrm{dom}\, f$ such that $\lambda \|y-x_0\| \leq \varepsilon$, $\lambda\|x-y\|+g(x) \geq g(y)$ for all $x \in X$. Now define the function h by $h(x) := \lambda\|x-y\|-g(y)$. Then $g(x) \geq -h(x)$ for all x and $g(y) = -h(y)$. According to the sandwich theorem (4.1.18) there is an affine separator $\alpha + \phi$ where $\phi \in X^*$ and $\alpha \in \mathbb{R}$ such that

$$-h(x) \leq \alpha + \phi(x) \leq g(x) \quad \text{for all } x \in X,$$

and moreover $\phi \in \partial g(y)$ and $\phi \in \partial h(y)$. Let $x := y$ and $x^* := \phi + x_0^*$. Then $\|\phi\| \leq \lambda$ and $x^* \in \partial f(x)$ as desired. □

Corollary 4.3.3. *Suppose f is a proper lsc convex function on the Banach space X. Given any $x_0 \in \mathrm{dom}\, f$ and $\varepsilon > 0$ then there is an $x \in \mathrm{dom}\, \partial f$ so that $|f(x) - f(x_0)| < \varepsilon$ and $\|x - x_0\| < \varepsilon$.*

Proof. Let $\varepsilon > 0$ and $x_0 \in \mathrm{dom}\, f$. Fix $x_0^* \in \partial_{\varepsilon/2} f(x_0)$ and let $K := \max\{\|x_0^*\|, 1\}$. According to the Brøndsted-Rockafellar theorem (4.3.2) there exist $x \in \mathrm{dom}\, f$ and $x^* \in X^*$ such that

$$x^* \in \partial f(x), \|x - x_0\| \leq \frac{\varepsilon}{2K} \text{ and } \|x^* - x_0^*\| \leq K.$$

Because $x_0^* \in \partial_{\varepsilon/2} f(x_0)$, we have

$$f(x) - f(x_0) \geq x_0^*(x - x_0) - \varepsilon/2 \geq -K(\varepsilon/2K) - \varepsilon/2 = -\varepsilon.$$

Also, $\|x^*\| \leq 2K$, and the subdifferential inequality implies

$$f(x_0) - f(x) \geq x^*(x_0 - x) \geq -2K\|x_0 - x\| = -\varepsilon.$$

Together, the previous two inequalities imply $|f(x) - f(x_0)| \leq \varepsilon$ as desired. □

We now present the Bishop-Phelps theorem in the classical setting of norm-attaining functionals. An important more general version is given in Exercise 4.3.6.

Theorem 4.3.4 (Bishop–Phelps). *Suppose X is a Banach space, then the set of functionals that attain their norms on B_X is dense in X^*.*

Proof. Given $\phi_0 \in X^*$, and $\varepsilon \in (0, 1)$, choose $x_0 \in S_X$ such that $\phi_0(x_0) > \|\phi_0\| - \varepsilon$. Consider the function f defined by $f(x) := \|\phi_0\|\|x\|$. Then $\phi_0 \in \partial_\varepsilon f$. According to the Brøndsted–Rockafellar theorem (4.3.2) with $\lambda = \sqrt{\varepsilon}$ there is an $x_1 \in X$ with $\|x_1 - x_0\| < \sqrt{\varepsilon}$ and $\|\phi - \phi_0\| < \sqrt{\varepsilon}$ such that $\phi \in \partial f(x_1)$. Then ϕ attains its norm at $x_1/\|x_1\|$. □

The following lemma provides a unified approach to several important classes of variational principles including smooth variational principles. Exercise 4.3.4 outlines how Lemma 4.3.5 can be used to obtain versions of Ekeland's variational principle and the Brøndsted–Rockafellar theorem. For this, we shall say a function $f : X \to (-\infty, +\infty]$ attains its *strong minimum* at x_0 if $f(x) > f(x_0)$ for all $x \neq x_0$ and $x_n \to x_0$ whenever $f(x_n) \to f(x_0)$.

Lemma 4.3.5. *Let X be a Banach space, and let $f : X \to (-\infty, +\infty]$ be lsc, bounded below with $\dom f \neq \emptyset$. Let $(Y, \|\cdot\|_Y)$ be a Banach space of bounded continuous real valued functions on X such that:*

(a) *for every $g \in Y$, $\|g\|_Y \geq \|g\|_\infty$;*
(b) *for every $g \in Y$ and every $u \in X$, $\|\tau_u g\|_Y = \|g\|_Y$ where $\tau_u g(x) = g(x+u)$ for each $x \in X$;*
(c) *for every $g \in Y$ and every $a > 0$, the function $h : X \to \mathbb{R}$ defined by $h(x) := g(ax)$ belongs to Y;*
(d) *there exists $b \in Y$ such that b has bounded nonempty support in X.*

Then the set of all $g \in Y$ such that $f + g$ attains its strong minimum on X is a dense G_δ-subset of Y.

Proof. Consider the sets

$$U_n = \{g \in Y : \text{there exists } x_0 \in X \text{ such that}$$
$$(f+g)(x_0) < \inf\{(f+g)(x) : x \in X \setminus B(x_0, 1/n)\}\}.$$

First, because $\|\cdot\|_Y \geq \|\cdot\|_\infty$, it follows that U_n is an open set in Y. Next we will show that U_n is dense in Y.

Let $g \in Y$ and $\varepsilon > 0$. We will find $h \in Y$ and $x_0 \in X$ such that $\|h\|_Y < \varepsilon$ and

$$(f+g+h)(x_0) < \inf\{(f+g+h)(x) : x \in X \setminus B(x_0, 1/n)\}.$$

Conditions (b) and (d) ensure that there is a bump function $b \in Y$ such that $b(0) \neq 0$. Replacing $b(x)$ with $\alpha_1 b(\alpha_2 x)$ using appropriately chosen $\alpha_1, \alpha_2 \in \mathbb{R}$ we assume that $b(0) > 0$, $\|b\|_Y < \varepsilon$ and $b(x) = 0$ whenever $\|x\| \geq 1/n$. Because $f + g$ is bounded below, we select $x_0 \in X$ such that

$$(f+g)(x_0) < \inf\{(f+g)(x) : x \in X\} + b(0).$$

Let $h(x) = -b(x - x_0)$. According to (b), $h \in Y$ and $\|h\|_Y < \varepsilon$. Additionally,

$$(f+g+h)(x_0) = (f+g)(x_0) - b(0) < \inf_X (f+g).$$

If $x \in X \setminus B(x_0, 1/n)$, then

$$(f+g+h)(x) = (f+g)(x) \geq \inf_X (f+g).$$

Therefore, $g + h \in U_n$ which in turn implies U_n is dense in Y. Consequently, $G = \bigcup_{n=1}^\infty U_n$ is a dense G_δ-subset of Y.

We show for $g \in G$ that $f + g$ attains its strong minimum on X. Indeed, for each $n \geq 1$, let $x_n \in X$ be such that

$$(f+g)(x_n) < \inf\{(f+g)(x) : x \in X \setminus B(x_n, 1/n)\}.$$

Now $x_p \in B(x_n, 1/n)$ for $p \geq n$. Otherwise, by the choice of x_n we have $(f+g)(x_p) > (f+g)(x_n)$. Then since $\|x_n - x_p\| \geq 1/n \geq 1/p$, by the choice of x_p we have $(f+g)(x_n) > (f+g)(x_p)$ which is a contradiction.

Thus (x_n) is a Cauchy sequence in X, and so it must converge to some $x_\infty \in X$. The lower-semicontinuity of f then implies

$$(f+g)(x_\infty) \leq \liminf (f+g)(x_n)$$
$$\leq \liminf \big[\inf\{(f+g)(x) : x \in X \setminus B(x_n, 1/n)\}\big]$$
$$\leq \inf\{(f+g)(x) : x \in X \setminus \{x_\infty\}\}.$$

Therefore, $f + g$ attains its minimum at x_∞, and we now show that this is a strong minimum. To this end, suppose $(f+g)(y_n) \to (f+g)(x_\infty)$ but that (y_n) does not converge to x_∞. By passing to a subsequence if necessary, we assume that $\|y_n - x_\infty\| \geq \varepsilon$ for all n and some $\varepsilon > 0$. It follows that there is an integer p such that $\|x_p - y_n\| > 1/p$ for all n. Consequently,

$$(f+g)(x_\infty) \leq (f+g)(x_p) < \inf\{(f+g)(x) : \|x - x_p\| > 1/p\}$$
$$\leq (f+g)(y_n) \text{ for all } n.$$

This contradiction shows that $(f + g)$ attains its strong minimum at x_0

To conclude that the set of $g \in Y$ such that $f + g$ attains its strong minimum on X is a dense G_δ-set (rather than residual as has just been shown), one need only observe that if $(f + g)$ attains its strong minimum at some $x_0 \in X$, then $g \in U_n$ for each n where U_n is as defined above. $\square$

Theorem 4.3.6 (Smooth variational principle). *Let X be a Banach space that admits a Lipschitz function with bounded nonempty support that is Fréchet differentiable (resp. Gâteaux differentiable). Then for every proper lsc bounded below function f defined on X and every $\varepsilon > 0$, there exists a function g which is Lipschitz and Fréchet differentiable (resp. Gâteaux differentiable) on X such that $\|g\|_\infty \leq \varepsilon$, $\|g'\|_\infty \leq \varepsilon$ and $f + g$ attains its strong minimum on X.*

Proof. Consider the space Y of functions that are bounded, Lipschitz and Fréchet differentiable (resp. Gâteaux differentiable) on X, where $\|\cdot\|_Y$ is defined by $\|g\|_Y = \|g\|_\infty + \|g'\|_\infty$. Then Y is a Banach spaces satisfying the conditions of Lemma 4.3.5. $\square$

Definition 4.3.7 (Fréchet subdifferential [121]). Let $f: X \to (-\infty, +\infty]$ be a proper lsc function on a Banach space X. We say f is *Fréchet-subdifferentiable* and x^* is a *Fréchet-subderivative* of f at x if $x \in \mathrm{dom} f$ and

$$\liminf_{\|h\| \to 0} \frac{f(x+h) - f(x) - \langle x^*, h \rangle}{\|h\|} \geq 0. \qquad (4.3.8)$$

The set of all Fréchet-subderivatives of f at x is denoted by $\partial_F f(x)$. This object is the *Fréchet subdifferential* of f at x. We set $\partial_F f(x) = \emptyset$ if $x \notin \mathrm{dom} f$.

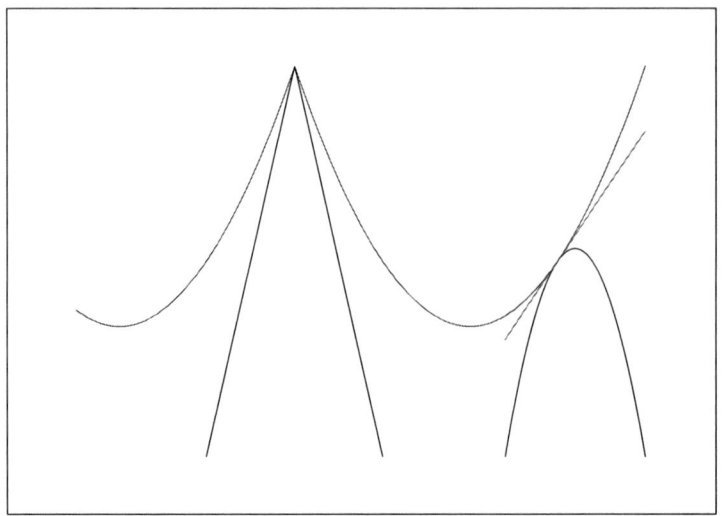

Figure 4.1 An ε-subdifferential at 0 and a Fréchet subdifferential at 3/2 for $(1-|x|)^2$.

Recall, that the *lower directional derivative* from above is defined by

$$f^-(x;h) := \liminf_{t\to 0^+} \frac{f(x+th)-f(x)}{t}. \qquad (4.3.9)$$

Fréchet subdifferentials are enormously useful and while much can be done with their Gâteaux analogs we shall not do so herein, see [122, 121]. We do record that $\partial_F f(x) \subseteq \partial_G f(x) := \{x : \langle x^*, h\rangle \le f^-(x;h) \text{ for all } h \in X\}$.

We say a Banach space is *Fréchet smooth* provided that it has an equivalent norm that is differentiable, indeed C^1-smooth, for all $x \ne 0$. The following 'approximate mean value theorem' encodes a great deal of variational information in terms of Fréchet subdifferentials. It will be used to good effect several times later in the book.

Theorem 4.3.8 (Limiting approximate mean value theorem [121]). *Let $f: X \to (-\infty,+\infty]$ be a lsc function on a Fréchet smooth Banach space X. Suppose $a \ne b \in X$ with $f(a) < \infty$ and that $-\infty < r \le \infty$ satisfies $r \le f(b)-f(a)$. Then there exist $c \in [a,b)$ and a sequence (x_i) with $(x_i, f(x_i)) \to (c, f(c))$ and $x_i^* \in \partial_F f(x_i)$ such that* (i) $\liminf_{i\to\infty}\langle x_i^*, c-x_i\rangle \ge 0$; (ii) $\liminf_{i\to\infty}\langle x_i^*, b-a\rangle \ge r$; *and* (iii) $f(c) \le f(a)+|r|$.

Exercises and further results

4.3.1. Show that Ekeland's variational principle holds if and only if the metric space (X, d) is complete.

Hint. The 'if' part was done; for the converse consider the function $f(x) := \lim_{i\to\infty} d(x_i, x)$ where (x_i) is a Cauchy sequence in X. For $\varepsilon \in (0,1)$, choose $y \in X$ such that $f(y) \le \varepsilon$ and $f(y) \le f(x)+\varepsilon d(x,y)$ for all $x \in X$. Show that $y = \lim_{i\to\infty} x_i$. □

4.3 Variational principles

4.3.2 (Banach's fixed point theorem). Let (X, d) be a complete metric space and let $\phi : X \to X$. Suppose there exists $0 < k < 1$ such that $d(\phi(x), \phi(y)) \leq k d(x, y)$ for all $x, y \in X$. The there is a unique *fixed point* $y \in X$ such that $\phi(y) = y$.

Hint. Let $f(x) := d(x, \phi(x))$. Apply Theorem 4.3.1 to f with $\lambda = 1$ and $0 < \varepsilon < 1-k$. Show that the $y \in X$ such that $f(x) + \varepsilon d(x, y) \geq f(y)$ for all $x \in X$ is the desired fixed point. □

Banach's theorem holds on any incomplete space with the fixed point property for continuous self-maps such as the set consisting of the diameters of the closed unit disk in the plane with rational slope. Thus, Ekeland's principle is in many ways a more refined tool.

4.3.3. ** (Diametral sets, approximate fixed points and nonexpansive mappings [400]) One of the most longstanding questions in geometric fixed point theory is whether a nonexpansive self-map T of a closed bounded convex subset C of a reflexive space X must have a fixed point. Here we sketch the convex geometry underlying much current research. We will have use for the *radius function*, also called the *farthest point function* defined by
$$r_C(x) := \sup_{y \in C} \|x - y\|.$$
Note that $r_C = r_{\overline{\text{conv}}C}$ is convex and continuous if C is bounded.

(a) (Approximate fixed points) A sequence (x_n) in C is an *approximate fixed point sequence* for T if $\|x_n - T(x_n)\| \to_n 0$. Apply the Banach fixed point theorem (Exercise 4.3.2) to show that approximate fixed points exist for nonexpansive mappings when C is closed, bounded and convex in a Banach space. Deduce in Euclidean space that T has fixed point.

Hint. Consider $T_n := (1 - 1/n) T + 1/n\, c_0$ for some $c_0 \in C$. □

(b) (Diametral sets) A general strategy is to appeal to weak-compactness and Zorn's lemma to show that T must have a minimal invariant weakly-compact convex subset C_0 and to look for conditions to force it to be singleton. (Note that $C_0 = \overline{\text{conv}} T(C_0)$ is forced by minimality.) Prove the following useful result:

Lemma 4.3.9 (Convex invariance). *Let C be a minimal invariant weakly compact convex set for a nonexpansive mapping mapping T on a Banach space. Suppose (i) $\psi : C \to \mathbb{R}$ is lsc and convex; and (ii) satisfies $\psi(T(x)) \leq \psi(x)$ for $x \in C$. Then ψ is constant on C.*

(c) Apply Lemma 4.3.9 to $\psi := r_C$ to prove:

Theorem 4.3.10 (Brodskii–Milman). *Let C be a minimal invariant weakly compact convex set for a nonexpansive T. Then for every $x \in C$ one has $r_C(x) \equiv \text{diam}(C)$.*

Such a set is called *diametral*. Show that the set $C := \overline{\text{conv}}_{n \in \mathbb{N}}\{e_n\}$ in $c_0(\mathbb{N})$ is weakly-compact convex and diametral (since $e_n \to_w 0$).

(d) Let C be minimal with (a_n) be an approximate fixed point sequence. Let (x_k) be an arbitrary subsequence of (a_n). Apply Lemma 4.3.9 to $\psi := \limsup_k \|x - x_k\|$

to prove a result of Goebel and Karlovitz, namely that $\lim_n \|x - a_n\| = \mathrm{diam}(C)$ for all $x \in C$. So nontrivial minimal sets are quite peculiar geometrically.

(e) (*Normal structure*) A space has *normal structure* if all diametral weakly compact convex sets are singleton. Show that every uniformly convex Banach space has normal structure (see Section 5.1 for some basic properties of uniformly convex spaces). Hence in a uniformly convex Banach space every nonexpansive self-map of a closed bounded convex subset has a fixed point. We thus recover classical results of Browder – the UC case – and of Kirk. Much more can be followed up in [400].

4.3.4. This exercise provides an alternate approach to Ekeland's variational principle and some of its consequences using Lemma 4.3.5.

(a) Use Lemma 4.3.5 to prove the following version of Ekeland's variational principle:

Theorem 4.3.11 (Ekeland's variational principle). *Let X be a Banach space and let $f : X \to (-\infty, +\infty]$ be lsc, bounded below with $\mathrm{dom} f \neq \emptyset$. Given $\varepsilon > 0$, there exists $x_0 \in \mathrm{dom} f$ such that for all $x \in X$ one has*

$$f(x) \geq f(x_0) - \varepsilon \|x - x_0\| \quad \text{and} \quad f(x_0) \leq \inf_X f + \varepsilon.$$

(b) Use Theorem 4.3.11 to prove Theorem 4.3.1 when X is assumed to be a Banach space, and the starting assumption on z is that $f(z) < \inf_X f + \varepsilon/4$.

(c) Use (b) do derive the conclusion of the Brøndsted–Rockafellar theorem under the stronger assumption $x_0^* \in \partial_{\varepsilon/4} f(x_0)$.

(d) Use (c) to derive the Bishop–Phelps theorem and Corollary 4.3.3.

Hint. (a) The Banach space of all bounded Lipschitzian functions on X equipped with the norm

$$\|g\|_Y := \|g\|_\infty + \sup \left\{ \frac{|g(x) - g(y)|}{\|x - y\|} : x \neq y \right\}$$

satisfies the conditions required in Lemma 4.3.5. Accordingly, there exist $g \in Y$ and $x_0 \in X$ such that $\|g\|_Y \leq \varepsilon/2$ and $f + g$ attains its minimum at x_0. Consequently,

$$f(x) \geq f(x_0) + g(x_0) - g(x) \geq f(x_0) - \varepsilon \|x - x_0\| \quad \text{for all } x \in X.$$

Additionally, because $\|g\|_Y \leq \varepsilon/2$, one has

$$f(x) \geq f(x_0) + g(x_0) - g(x) \geq f(x_0) - \varepsilon.$$

Thus we conclude that $f(x_0) \leq \inf_X f + \varepsilon.$

4.3 Variational principles

(b) Let $g(x) = f(x) + \frac{\lambda}{2}\|x - z\|$; then f and g have the same domain, and according to Theorem 4.3.11 there exists $y \in \text{dom } g$ such that

$$g(x) \geq g(y) - \frac{\lambda}{2}\|x - y\| \quad \text{and} \quad g(y) \leq \inf_X g + \frac{\varepsilon}{4}.$$

(c) and (d) follow from straightforward modifications of the given proofs. □

4.3.5. Let X be a Banach space, $x_0 \in S_X$, $\varepsilon > 0$ and $\phi_0 \in S_X$ with $\phi_0(x_0) > 1 - \varepsilon^2/2$. Use the Brøndsted–Rockafellar theorem (4.3.2) to show there exist $\bar{x} \in S_X$ and $\bar{\phi} \in S_{X^*}$ such that $\|x_0 - \bar{x}\| < \varepsilon$, $\|\phi_0 - \bar{\phi}\| < \varepsilon$ and $\bar{\phi}(\bar{x}) = 1$.

Hint. Note $\phi_0 \in \partial_{\varepsilon^2/2}\|x_0\|$, after applying the Brøndsted–Rockafellar theorem, find an appropriate $\bar{x} \in S_X$. A proof that does not use the Brøndsted–Rockafellar theorem may be found in [60]. □

4.3.6 (Another Bishop–Phelps theorem).* Suppose C is a nonempty closed convex subset of a Banach space X. Show that

(a) The support points of C are dense in its boundary.
(b) The support functionals of C are dense in the cone of all those functionals that are bounded above on C.

Hint. Let $f := \delta_C$. For (a), let $0 < \varepsilon < 1$ and $x_1 \in X \setminus C$ be such that $\|x_0 - x_1\| < \varepsilon$. By the basic separation theorem (4.1.12) find $x_0^* \in S_{X^*}$ such that $\sigma_C(x_0^*) < \langle x_0^*, x_1 \rangle$. Apply the Brøndsted–Rockafellar theorem (4.3.2) to f with $\lambda = \sqrt{\varepsilon}$. For (b), suppose $\sigma_C(x_0^*) < \infty$. Choose $x_0 \in C$ with $\langle x_0^*, x_0 \rangle > \sigma_C(x_0^*) - \varepsilon$ where $0 < \varepsilon < \|x_0^*\|^2$. Apply the Brøndsted–Rockafellar theorem (4.3.2) as in (a). See [350, Theorem 3.18] for further details. □

4.3.7. Note that for a convex function f one has $\partial f = \partial_F f = \partial_G f$. Show that for a concave function g one has rather $\{\nabla_F g\} = \partial_F g$ and $\{\nabla_G g\} = \partial_G g$. ($\nabla_F f(x)$ and $\nabla_G f(x)$ denote Fréchet and Gâteaux derivatives respectively).

4.3.8. ** Use Ekeland's variational principle (4.3.1) to show that a Banach space with a Fréchet differentiable bump function is an Asplund space.

Hint. See the paper [194]. □

4.3.9.* Show that if a Banach space has a Fréchet differentiable (resp. Gâteaux differentiable) norm, then it has a Lipschitz C^1-smooth (resp. Lipschitz Gâteaux differentiable) bump function.

Hint. Construct a smooth Lipschitz bump function from the norm by composing it with an appropriate real-valued function. □

4.3.10.* Show that if a Banach space has a Fréchet differentiable (resp. Gâteaux differentiable) bump function that is Lipschitz, then every continuous convex function on X is Fréchet (resp. Gâteaux) differentiable on a dense subset of its domain. Deduce the same respective conclusions when X has an equivalent Fréchet (resp. Gâteaux) differentiable norm.

Hint. For the first part use the smooth variational principle (4.3.6), deduce the second part from Exercise 4.3.9. □

4.3.11. Suppose f is a proper lsc convex function on X. Show that

$$f(x) = \sup\{\phi(x-y) + f(y) : \phi \in \partial f(y), y \in \text{dom}(\partial f)\}$$

for any $x \in \text{dom} f$.

Hint. One proof is similar to the proof of Corollary 4.3.3. □

4.3.12.** (a) (Supporting hyperplane lemma). Let X be a Banach space, and let $f : X \to (-\infty, +\infty]$ be a proper lsc convex function. Suppose that C is a closed bounded convex set so that the distance between $\text{epi} f$ and C is positive. Show that there exist $x_0 \in \text{dom} \partial f$ and $\phi_0 \in \partial f(x_0)$ so that C lies below the graph of

$$x \mapsto f(x_0) + \phi_0(x - x_0) - \varepsilon.$$

(b) Let f be a proper lsc convex function defined on a Banach space X. For each $y \in X^*$ with $f^*(y) < \infty$, show that there exists a sequence (y_n) in range(∂f) strongly convergent to y with $f^*(y) = \lim_{n \to \infty} f^*(y_n)$.

Hint. Part (a) was shown in [44, Lemma 4.10] using the comprehensive variational principle stated below in Theorem 4.3.12. Warning: we do not know of a simple direct proof using the separation techniques and variational principles proved from this chapter. Use the supporting hyperplane lemma of part (a) to deduce (b); see [20, Proposition 4.3] for further details. Note also the conjugate function f^* is introduced with basic properties in Section 4.4 below. □

The following comprehensive result from [63] is useful for showing many results on convex functions, of which Exercise 4.3.12 is just one such example.

Theorem 4.3.12 (Borwein's variational principle). *Suppose that f is a proper lsc convex function on a Banach space X, that $\varepsilon > 0$, $\beta \geq 0$ and that $x_0 \in \text{dom} f$. Suppose, further, that $x_0^* \in \partial f_\varepsilon(x_0)$. Then there exist points $x_\varepsilon \in \text{dom} f$ and $x_\varepsilon^* \in X^*$ such that*

(a) $x_\varepsilon^* \in \partial f(x_\varepsilon)$ and $\|x_\varepsilon - x_0\| \leq \sqrt{\varepsilon}$;
(b) $|f(x_\varepsilon) - f(x_0)| \leq \sqrt{\varepsilon}(\sqrt{\varepsilon} + 1/\beta)$ where by convention $1/0 = +\infty$;
(c) $\|x_\varepsilon^* - x_0^*\| \leq \sqrt{\varepsilon}(1 + \beta\|x_0^*\|)$;
(d) $|\langle x_\varepsilon^*, y \rangle - \langle x_0^*, y \rangle| \leq \sqrt{\varepsilon}(\|y\| + \beta|\langle x_0^*, y \rangle|)$ for all $y \in X$; and
(e) $x_\varepsilon^* \in \partial_{2\varepsilon} f(x_0)$.

The key results needed in the proof of Theorem 4.3.12 are Ekeland's variational principle, and the sum rule for subdifferentials (4.1.19); see [349, Chapter 3] for a proof and several applications illustrating the utility of this result. Indeed, a vigorous reader can provide the proof given those hints and the advice to apply Ekeland's variational principle to the function $g := f - x_0^*$ in the renorm $\|x\|_\beta := \|x\| + \beta|\langle x_0^*, x \rangle|$.

For now, we are content to mention a powerful subdifferential formula whose proof as given in [35] relies on careful application of Borwein's variational principle (4.3.12).

Theorem 4.3.13 (Subdifferential formula [35]). *Let E be a Banach space. Suppose int $\text{dom} f \neq \emptyset$ and $x \in E$. Define a set $S(x)$ in E^* by requiring $x^* \in S(x)$ if and only if there exist bounded nets (x_α) in $\text{int dom} f$ and (x_α^*) in X^* such that for every α, $x_\alpha^* \in \partial f(x_\alpha)$, $x_\alpha \to x$, $x_\alpha^* \to_{w^*} x^*$, and $f(x_\alpha) \to f(x)$. Let $N(x) = N_{\text{dom} f}(x)$. Then letting cl_{w^*} represent the weak*-closure:*

$$\partial f(x) = \text{cl}_{w*}\bigl(N(x) + \text{cl}_{w*} \text{conv } S(x)\bigr).$$

Furthermore, if $\text{dom } \nabla f$ is dense in $\text{dom } \partial f$, then define $G(x)$ by $y^* \in G(x)$ precisely when there exists a bounded net (y_α) in $\text{dom } \nabla f$ such that $(\nabla f(y_\alpha))$ is bounded, $y_\alpha \to x$, $\nabla f(y_\alpha) \to_{w^*} y^*$, $f(y_\alpha) \to f(x)$. In this case,

$$\partial f(x) = \text{cl}_{w*}\bigl(N(x) + \text{cl}_{w*} \text{conv } G(x)\bigr).$$

Note that the assumption on denseness in the 'Furthermore' part is always satisfied in a Gâteaux differentiability space (GDS) and thus in all separable and all Euclidean spaces. Moreover, when E is a GDS this construction can be performed with sequences because the dual ball is weak*-sequentially compact, see Exercise 4.2.13.

4.4 Conjugate functions and Fenchel duality

We begin this section by developing basic properties of the Fenchel conjugate of a functions. The section concludes with a sketch Fenchel duality theory which can be covered any time after the definition of conjugate functions and the basic Proposition 4.4.1.

4.4.1 Properties of conjugate functions

Let X be a Banach space. The *Fenchel conjugate* of a function $f : X \to [-\infty, +\infty]$ is the function $f^* : X^* \to [-\infty, +\infty]$ defined by

$$f^*(x^*) := \sup_{x \in X}\{\langle x^*, x\rangle - f(x)\}.$$

The function f^* is convex and if the domain of f is nonempty then f^* never takes the value $-\infty$. We can consider the conjugate of f^* called the *biconjugate* of f and denoted by f^{**}. This is a function on X^{**}.

We refer the reader to Table 2.1 on p. 45 for some basic examples of Fenchel conjugates, and to Table 2.2 on p. 46 for some properties on transformed conjugates.

The following is an elementary but important result concerning conjugate functions; part (a) is the famous Fenchel–Young inequality.

Proposition 4.4.1. *Let $f : X \to [-\infty, +\infty]$, $x \in X$, $x^* \in X^*$ and $\varepsilon > 0$. Then*

(a) $f(x) + f^*(x^*) \geq \langle x^*, x\rangle$, *and equality holds if and only if $x^* \in \partial f(x)$.*
(b) $x^* \in \partial_\varepsilon f(x)$ *if and only if $f(x) + f^*(x^*) \leq \langle x, x^*\rangle + \varepsilon$.*
(c) *If $f \leq g$, then $g^* \geq f^*$.*
(d) $f^{**}|_X \leq f$.

Proof. To prove (b), observe that $x^* \in \partial_\varepsilon f(x)$ if and only if $x^*(y) - x^*(x) \leq f(y) - f(x) + \varepsilon$ for all $y \in X$, if and only if $x^*(y) - f(y) \leq x^*(x) - f(x) + \varepsilon$ for all $y \in X$, if and only if $f^*(x^*) \leq x^*(x) - f(x) + \varepsilon$, if and only if $f(x) + f^*(x^*) \leq x^*(x) + \varepsilon$. This proves (b).

Now the definition of f^* implies $f^*(x^*) \geq x^*(x) - f(x)$ for all $x \in X$ and $x^* \in X^*$. Consequently, $f(x) + f^*(x^*) \geq x^*(x)$ for all $x \in X$, $x^* \in X$. Applying the proof of (b) with $\varepsilon = 0$ implies $f(x) + f^*(x^*) \leq x^*(x)$ if and only if $x^* \in \partial f(x)$. Thus (a) is true.

The definition of the conjugate immediately implies (c), while using the definition and then (a) we obtain $f^{**}(x) = \sup\{\langle x, x^* \rangle - f^*(x^*) : x^* \in X^*\} \leq f(x)$. □

We next explore how properties such as boundedness and continuity are related with a convex function and its conjugate. We begin by addressing when the second conjugate of a function is equal to the original function.

Proposition 4.4.2. *(a) Suppose $f : X \to (-\infty, +\infty]$ is convex and proper. Then $f^{**}(x) = f(x)$ at $x \in X$ if and only if f is lsc at x. In particular, f is lsc if and only if $f^{**}|_X = f$.*
(b) Suppose $f : X^* \to (-\infty, +\infty]$ is convex, proper and lsc. Then $(f^*|_X)^* = f$ if and only if f is weak-lsc.*

Proof. (a) First, if $f^{**}(x_0) = f(x_0)$, then f is lsc at x_0 because f^{**} is lsc and $f^{**} \leq f$ [Proposition 4.4.1(d)]. For the converse, let x_0 be a point where f is lsc. By Proposition 4.4.1(d), $f^{**}(x_0) \leq f(x_0)$ and so it suffices to show $f^{**}(x_0) \geq \alpha$ whenever $f(x_0) > \alpha$. So suppose $f(x_0) > \alpha$. Using the epi-separation theorem (4.1.21), choose $\phi \in X$ such that $\phi(x) - \phi(x_0) \leq f(x) - \alpha$ for all $x \in X$. Thus $\phi(x) - f(x) \leq \phi(x_0) - \alpha$ for all $x \in X$, and so $f^*(\phi) \leq \phi(x_0) - \alpha$. Thus $f^{**}(x_0) \geq \phi(x_0) - [\phi(x_0) - \alpha] = \alpha$.

(b) Is left as an exercise which can be similarly derived using the weak*-epi separation theorem (4.1.23). □

Proposition 4.4.2(a) remains valid when $f \equiv +\infty$ or $f \equiv -\infty$. However, some improper examples may fail. For example, let $f : X \to [-\infty, +\infty]$ be defined by $f(0) := -\infty$, $f(x) := +\infty$ if $x \neq 0$. Then $f^* \equiv +\infty$ so $f^{**} \equiv -\infty$. Examples such as this provide a compelling reason why some sources define the closure of a function that takes the value of $-\infty$ at some point, as the function identically equal to $-\infty$; see for example [369, p. 52].

Notationally, we let $\mathrm{cl} f$ or $\bar{f}$ denote the closure of f, and we let $\overline{\mathrm{conv}} f$ be the function whose epigraph is the closed convex hull of the epigraph of f.

Proposition 4.4.3. *Suppose $f : X \to [-\infty, +\infty]$. Then*

(a) $f^* = (\mathrm{cl} f)^* = (\overline{\mathrm{conv}} f)^*$.
(b) $f^{**}|_X = \overline{\mathrm{conv}} f$ if $\overline{\mathrm{conv}} f$ is proper.

Proof. (a) First $f^* \leq (\mathrm{cl} f)^* \leq (\overline{\mathrm{conv}} f)^*$ by Proposition 4.4.1(c). It remains to prove $f^* \geq (\overline{\mathrm{conv}} f)^*$. For this, let $\phi \in X^*$. If $f^*(\phi) = +\infty$ there is nothing to do, so we suppose $f^*(\phi) \leq \alpha$ for some $\alpha \in \mathbb{R}$. Then $\phi(x) - f(x) \leq \alpha$ for all $x \in X$. Let $g := \phi - \alpha$. Then $g \leq \overline{\mathrm{conv}} f$, consequently $(\overline{\mathrm{conv}} f)^* \leq g^*$ (Proposition 4.4.1(c)). Clearly, $g^*(\phi) = \alpha$ and so $(\overline{\mathrm{conv}} f)^*(\phi) \leq \alpha$ as desired.

4.4 Fenchel duality

(b) Suppose $\overline{\text{conv}} f$ is proper. Then $(\overline{\text{conv}} f)^{**}|_X = \overline{\text{conv}} f$ according to Proposition 4.4.2(a). By (a) of this proposition, we know $((\overline{\text{conv}} f)^*)^* = (f^*)^*$ from which the conclusion now follows. □

Fact 4.4.4. *Let $f : X \to (-\infty, +\infty]$ be a proper lsc convex function. Then*

(a) *f is bounded on bounded sets if and only if f^{**} is bounded on bounded sets in X^{**}.*
(b) *f is continuous at $x_0 \in X$ if and only if f^{**} is continuous at x_0.*
(c) *f is Fréchet differentiable at $x_0 \in X$ if and only if f^{**} is Fréchet differentiable at x_0.*

Proof. (a) $\Rightarrow$: Suppose $f \leq K$ on αB_X. Then $f^{**} \leq f \leq K$ on αB_X. Because f^{**} is weak*-lsc, $\{u \in X^{**} : f^{**}(x) \leq k\}$ is a weak*-closed subset of X^{**}, and it contains αB_X. According to Goldstine's theorem (Exercise 4.1.13), it contains $\alpha B_{X^{**}}$. Therefore, $f^{**} \leq K$ on $\alpha B_{X^{**}}$.
$\Leftarrow$: $f^{**}|_X = f$ is bounded on bounded sets.
The details of (b) and (c) are outlined in Exercise 4.4.22. □

The next result gives some useful information about the connection between the subdifferential of a function and its conjugate.

Proposition 4.4.5. *Let $f : X \to (-\infty, +\infty]$ be a function, and $x_0 \in \text{dom} f$.*

(a) *If $\phi \in \partial f(x_0)$, then $x_0 \in \partial f^*(\phi)$. Conversely, if additionally f is a convex function that is lsc at x_0 and $x_0 \in \partial f^*(\phi)$, then $\phi \in \partial f(x_0)$.*
(b) *Suppose $\phi \in \partial_\varepsilon f(x_0)$. Then $x_0 \in \partial_\varepsilon f^*(\phi)$. Conversely, if additionally f is a convex function that is lsc at x_0 and $x_0 \in \partial_\varepsilon f^*(\phi)$, then $\phi \in \partial_\varepsilon f(x_0)$.*

Proof. We first prove (b). Suppose $\phi \in \partial_\varepsilon f(x_0)$. Then $f^*(\phi) \leq \phi(x_0) - f(x_0) + \varepsilon$ [by Proposition 4.4.1(b)], now for $x^* \in X^*, f^*(x^*) \geq x^*(x_0) - f(x_0)$, therefore

$$f^*(x^*) - f^*(\phi) \geq [x^*(x_0) - f(x_0)] - [\phi(x_0) - f(x_0) + \varepsilon] = x^*(x_0) - \phi(x_0) - \varepsilon.$$

Consequently, $x_0 \in \partial_\varepsilon f^*(\phi)$. This proves the first part of (b). Now suppose that f is a convex function that is lsc at x_0. By Propositions 4.4.2(a) and 4.4.1(d), $f(x_0) = f^{**}(x_0)$ and $f^{**}|_X \leq f$ which together with $\phi \in \partial_\varepsilon f^{**}(x_0)$ imply $\phi_\varepsilon \in \partial_\varepsilon f(x_0)$. This proves (b); we notice that (a) follows by letting $\varepsilon = 0$ in the preceding. □

The following give useful criteria relating certain boundedness or Lipschitz properties of a function to its conjugate.

Proposition 4.4.6. *Suppose $f : X \to (-\infty, +\infty]$ is a proper lsc convex function. Then f is Lipschitz with Lipschitz constant $k \geq 0$ if and only if $\text{dom} f^* \subset k B_{X^*}$.*

Proof. $\Rightarrow$: Suppose $\Lambda \in X^*$, and $\|\Lambda\| > k$, then $\sup_{x \in X} \Lambda(x) - f(x) = +\infty$, and so $\Lambda \notin \text{dom} f^*$.
$\Leftarrow$: Suppose f does not satisfy a Lipschitz condition with constant k. Then there exist $x, y \in X$ such that $f(x) - f(y) > l\|x - y\|$ where $l > k$ (we allow $f(x) = +\infty$).

Then $f(x) > f(y) + l\|x - y\|$ and so the epi-separation theorem (4.1.21) ensures that there is a $\phi \in X^*$ such that $\phi(h) - \phi(x) \leq f(h) - [f(y) + l\|x - y\|]$ for all $h \in X$. In particular, $\phi \in \text{dom} f^*$, and $\|\phi\| \geq l$ because $\phi(y) - \phi(x) \leq -l\|x - y\|$. □

Remark 4.4.7. Observe that lower-semicontinuity is needed in Proposition 4.4.6 when X is infinite-dimensional. Indeed, consider $f : X \to (-\infty, +\infty]$ such that $f(x) := 0$ for all x in a dense subspace of X, and $f(x) := +\infty$ otherwise. Then $\text{dom}(f^*) = \{0\}$, but that f is not even lsc. However, in finite-dimensional spaces, Proposition 4.4.6 is valid for proper convex functions on $\mathbb{R}^n$ (see Exercise 2.3.13).

A function $f : X \to (-\infty, +\infty]$ is said to be *coercive* if $\lim_{\|x\| \to \infty} f(x) = +\infty$; if $\lim_{\|x\| \to \infty} \frac{f(x)}{\|x\|} = +\infty$ then f is said to be *supercoercive*; whereas f is called *cofinite* if its conjugate is defined everywhere on X^*. Some relations among these notions are developed in Exercise 4.4.23. An extremely useful fact is that supercoercive convex functions are related in a dual fashion to convex functions that are bounded on bounded sets.

Fact 4.4.8. *Suppose f is a proper convex function that is lsc at some point in its domain. Then the following are equivalent.*

(a) f is coercive.
(b) There exist $\alpha > 0$ and $\beta \in \mathbb{R}$ such that $f \geq \alpha \|\cdot\| + \beta$.
(c) $\liminf_{\|x\| \to \infty} f(x)/\|x\| > 0$.
(d) f has bounded lower level sets.

Proof. (a) $\Rightarrow$ (b): Suppose f is coercive. Assume first that $0 \in \text{dom} f$, and $f(0) = 0$. Then choose $r > 0$ such that $f(x) \geq 1$ if $\|x\| \geq r$. Now for $\|x\| \geq r$ we have

$$f\left(\frac{r}{\|x\|}x\right) \leq \frac{\|x\| - r}{\|x\|} f(0) + \frac{r}{\|x\|} f(x).$$

Therefore, $\frac{r}{\|x\|} f(x) \geq 1$ and so $f(x) \geq \frac{\|x\|}{r}$ if $\|x\| \geq r$. Because f is lsc at some point in its domain, there is an $M > 0$ such that $f(x) \geq -M$ on rB_X. Therefore, $f \geq \alpha \|\cdot\| + \beta$ where $\alpha = 1/r$ and $\beta = -M$. Consequently (b) holds for a translate of f, and it is easy to check that then holds for f as well with an appropriate adjustment to β.

Now (b) $\Rightarrow$ (c) $\Rightarrow$ (d) $\Rightarrow$ (a) follow directly from the definitions. □

The following fact thus relates the coercivity of f to a bound on its conjugate function in a neighborhood of the origin.

Fact 4.4.9. *Let $\alpha > 0$ and $\beta \in \mathbb{R}$. Then $f \geq \alpha \|\cdot\| + \beta$ if and only if $f^* \leq -\beta$ on αB_{X^*}.*

Proof. Observe that $f \geq \alpha \|\cdot\| + \beta$ if and only if $f^* \leq (\alpha \|\cdot\| + \beta)^*$ if and only if $f^* \leq \delta_{\alpha B_{X^*}} - \beta$. □

4.4 Fenchel duality

Because a convex function is continuous at a point if and only if it is bounded above on a neighborhood of that point (Proposition 4.1.4), the previous two facts immediate provide:

Theorem 4.4.10 (Moreau–Rockafellar). *Let $f : X \to (-\infty, +\infty]$ be proper, convex and lsc at some point in its domain. Then f is coercive if and only if f^* is continuous at 0.*

In fact, more generally, one has the following result.

Corollary 4.4.11 (Moreau–Rockafellar). *Let $f : X \to (-\infty, +\infty]$ be convex, proper and lsc, and let $\phi \in X^*$. Then $f - \phi$ is coercive if and only if f^* is continuous at ϕ.*

Proof. According to Theorem 4.4.10, $f - \phi$ is coercive if and only if $(f - \phi)^*$ is continuous at 0. Now $(f - \phi)^*(x^*) = f^*(x^* + \phi)$ for $x^* \in X^*$, thus $(f - \phi)^*$ is continuous at 0 if and only if f^* is continuous at ϕ. □

Theorem 4.4.12 (Moreau–Rockafellar dual [325]). *Let $f : X \to (-\infty, +\infty]$ be a lsc convex proper function. Then f is continuous at 0 if and only if f^* has weak*-compact lower level sets.*

Proof. Observe that f is continuous at 0 if and only if f^{**} is continuous at 0 (Fact 4.4.4(b)) if and only if f^* is coercive (Theorem 4.4.10) if and only if f^* has bounded lower level sets (Fact 4.4.8) if and only if f^* has weak*-compact lower level sets (by Alaoglu's theorem 4.1.6 and Exercise 4.1.1(c) which implies the lower level sets of f^* are weak*-closed). □

Theorem 4.4.13 (Conjugates of supercoercive functions). *Suppose $f : X \to (-\infty, +\infty]$ is a lsc proper convex function. Then*

(a) f is supercoercive if and only if f^* is bounded on bounded sets.
(b) f is bounded on bounded sets if and only if f^* is supercoercive.

Proof. (a) $\Rightarrow$: Given any $\alpha > 0$, there exists M such that $f(x) \geq \alpha \|x\|$ if $\|x\| \geq M$. Now there exists $\beta \geq 0$ such that $f(x) \geq -\beta$ if $\|x\| \leq M$. Therefore $f \geq \alpha \| \cdot \| + (-\beta)$. Thus, Fact 4.4.9 implies $f^* \leq \beta$ on αB_{X^*}.

$\Leftarrow$: Let $\alpha > 0$. Now there exists K such that $f^* \leq K$ on αB_{X^*}. Then $f \geq \alpha \| \cdot \| - K$ and so $\liminf_{\|x\| \to \infty} \frac{f(x)}{\|x\|} \geq \alpha$.

(b) According to (a), f^* is supercoercive if and only if f^{**} is bounded on bounded sets which according to Fact 4.4.4(b) occurs if and only if f is bounded on bounded sets. □

Next we introduce infimal convolutions. Some of their many applications include smoothing techniques and approximation.

Definition 4.4.14. Let f and g be proper extended real-valued functions on a normed linear space X. The *infimal convolution* of f and g is defined by

$$(f \square g)(x) := \inf_{y \in X} f(y) + g(x - y).$$

Geometrically, the infimal convolution of f and g is the largest extended real-valued function whose epigraph contains the sum of epigraphs of f and g, consequently it is a convex function; see Exercise 2.3.14 for some further facts about infimal convolutions that are also valid in Banach spaces. The following is a useful result concerning the conjugate of the infimal convolution.

Lemma 4.4.15. *Let X be a normed linear space and let f and g be proper functions on X. Then $(f \square g)^* = f^* + g^*$.*

Proof. Let $\phi \in X^*$. Using the definitions and then properties of infima's and suprema's one obtains

$$(f \square g)^*(\phi) = \sup_{x \in X} \phi(x) - (f \square g)(x)$$

$$= \sup_{x \in X} \phi(x) - \inf_{y \in X}[f(y) + g(x - y)]$$

$$= \sup_{x \in X, y \in Y} \sup \phi(y) - f(y) + \phi(x - y) - g(x - y)$$

$$= \sup_{y \in X} \phi(y) - f(y) + \sup_{v \in X} \phi(v) - g(v)$$

$$= f^*(\phi) + g^*(\phi). \qquad \square$$

The following are basic approximation facts.

Lemma 4.4.16. *Suppose $f : X \to \mathbb{R}$ is a convex function bounded on bounded sets. If $f_n \leq f$ for each n, and $f_n^* \to f^*$ uniformly on bounded subsets of the domain of f^*, then $f_n \to f$ uniformly on bounded subsets of X.*

Proof. Let $D \subset X$ be bounded and let $\varepsilon > 0$. Then $\partial f(D)$ is bounded because f is bounded on bounded sets (Proposition 4.1.25). Choose $N \in \mathbb{N}$ such that $f_n^*(\phi) \leq f^*(\phi) + \varepsilon$ for each $\phi \in \partial f(D)$, and $n \geq N$. Now let $x \in D$, and let $\phi \in \partial f(x)$. Then for each $n \geq N$, one has

$$f^*(\phi) = \phi(x) - f(x) \geq f_n^*(\phi) - \varepsilon \geq \phi(x) - f_n(x) - \varepsilon.$$

Therefore, $f(x) - \varepsilon \leq f_n(x) \leq f(x)$ for all $x \in D$, and all $n \geq N$. $\qquad \square$

Corollary 4.4.17 (Yosida approximation). *Let $f : X \to \mathbb{R}$ be convex and bounded on bounded sets. Then both $f \square n\|\cdot\|^2$ and $f \square n\|\cdot\|$ converge uniformly to f on bounded sets.*

Proof. This follows from Lemmas 4.4.15 and 4.4.16. Indeed,

$$(f \square n\|\cdot\|^2)^* = f^* + \frac{1}{2n}\|\cdot\|_*^2 \quad \text{and} \quad (f \square n\|\cdot\|)^* = f^* + h_n^*,$$

where $\|\cdot\|_*$ denotes the dual norm and $h_n^*(\phi) := 0$ if $\|\phi\| \leq n$, and $h_n^*(\phi) := \infty$ otherwise. $\qquad \square$

4.4.2 Fenchel duality theory

Conjugate functions are ubiquitous in optimization. Our next result is phrased in terms of convex programming problems. The formulation is in many aspects similar to the duality theory in linear programming.

The Fenchel duality theorem can be viewed as a dual representation of the sandwich theorem (4.1.18).

Theorem 4.4.18 (Fenchel duality). *Let X and Y be Banach spaces, let $f: X \to (-\infty, +\infty]$ and $g: Y \to (-\infty, +\infty]$ be convex functions and let $T: X \to Y$ be a bounded linear map. Define the primal and dual values $p, d \in [-\infty, +\infty]$ by the Fenchel problems*

$$p := \inf_{x \in X} \{f(x) + g(Tx)\}$$
$$d := \sup_{x^* \in Y^*} \{-f^*(T^*x^*) - g^*(-x^*)\}. \quad (4.4.1)$$

Then these values satisfy the weak duality *inequality $p \geq d$. Suppose further that f, g and T satisfy either*

$$0 \in \mathrm{core}(\mathrm{dom}\, g - T\, \mathrm{dom} f) \quad (4.4.2)$$

and both f and g are lsc, or the condition

$$T\, \mathrm{dom} f \cap \mathrm{cont}\, g \neq \emptyset. \quad (4.4.3)$$

Then $p = d$, and the supremum in the dual problem (4.4.1) is attained if finite.

Proof. The weak duality is left as Exercise 4.4.14. For the strong duality, follow the proof given in finite-dimensional spaces (Theorem 2.3.4), noting that the continuity of h given therein on infinite-dimensional spaces is established in the proof of the sandwich theorem (4.1.18). $\square$

To relate Fenchel duality and convex programming with linear constraints, we let g be the indicator function of a point, which gives the following particularly elegant and useful corollary.

Corollary 4.4.19. (Fenchel duality for linear constraints) *Given any function $f: X \to (-\infty, +\infty]$, any bounded linear map $T: X \to Y$, and any element b of Y, the weak duality inequality*

$$\inf_{x \in X} \{f(x) : Tx = b\} \geq \sup_{x^* \in Y} \{\langle b, x^* \rangle - f^*(T^*x^*)\}$$

holds. If f is lsc and convex and b belongs to $\mathrm{core}(T\, \mathrm{dom} f)$ then equality holds, and the supremum is attained when finite.

Proof. Exercise 4.4.16. $\square$

Fenchel duality can be used to conveniently calculate polar cones. Recall that for a set K in a Banach space X, the *(negative) polar cone* of K is the convex cone

$$K^- := \{x^* \in X^* : \langle x^*, x \rangle \leq 0, \text{ for all } x \in K\}.$$

Analogously, $K^+ = \{x^* \in X^* : \langle x^*, x \rangle \geq 0, \text{ for all } x \in K\}$ is the *(positive) polar cone*. The cone K^{--} is called the *bipolar* – sometimes in the second dual and sometimes in the predual, X. Here, we take it in X. An important example of the polar cone is the normal cone to a convex set $C \subset X$ at a point $x \in C$, since $N_C(x) = (C - x)^-$ (see Exercise 4.1.33).

The following calculus for polar cones is a direct consequence of the Fenchel duality theorem (4.4.18).

Corollary 4.4.20. *Let X and Y be Banach spaces, let $K \subset X$ and $H \subset Y$ be cones and let $T: X \to Y$ be a bounded linear map. Then*

$$K^- + T^*H^- \subset (K \cap T^{-1}H)^-.$$

Equality holds if H and K are closed and convex and satisfy $H - TK = Y$.

Proof. Observe that for any cone K, we have $K^- = \partial \delta_K(0)$. The result follows directly from Theorem 4.4.18. □

Exercises and further results

4.4.1. Many important convex functions f on a reflexive Banach space equal their biconjugate f^{**}. Such functions thus occur as natural pairs, f and f^*. Table 2.1 on p. 45 shows some elegant examples on $\mathbb{R}$, and Table 2.2 on p. 46 describes some simple transformations of these examples. Check the calculation of f^* and check $f = f^{**}$ for functions in Table 2.1. Verify the formulas in Table 2.2.

4.4.2.* Suppose $f: X \to \mathbb{R}$ is a proper lsc function. Suppose f^* is Fréchet differentiable at $\phi_0 \in X^*$. Suppose $x_0 = \nabla(f^*)(\phi_0)$. Show that $x_0 \in X$, $f^{**}(x_0) = f(x_0)$ and consequently $\phi_0 \in \partial f(x_0)$.

Hint. That $x_0 \in X$ is from Exercise 4.2.11. For notational purposes, let $\hat{f} = f^{**}|_X$. Since $\hat{f}(x_0) \leq f(x_0)$, we will show $\hat{f}(x_0) \geq f(x_0)$. Now $x_0 \in \partial f^*(\phi)$. Also, f^* is Fréchet differentiable at ϕ and so f^* is continuous at ϕ. Now choose $x_n \in X$ such that

$$\phi(x_n) - f(x_n) \geq f^*(\phi) - \varepsilon_n \text{ where } \varepsilon_n \to 0^+.$$

Then $\phi(x_n) - \hat{f}(x_n) \geq f^*(\phi) - \varepsilon_n$ and so $x_n \in \partial_{\varepsilon_n} f^*(\phi)$ for all n. According to Šmulian's theorem (4.2.10) $x_n \to x_0$. In particular, $\phi(x_n) \to \phi(x_0)$. Therefore,

$$\hat{f}(x) = \phi(x) - f^*(\phi)$$
$$= \lim_{n \to \infty} \phi(x) - [\phi(x_n) - f(x_n)]$$
$$= \lim_{n \to \infty} f(x_n).$$

4.4 Fenchel duality

Moreover $\liminf_{n\to\infty} f(x_n) \geq f(x_0)$ because f is lsc. Therefore, $\hat{f}(x_0) \geq f(x_0)$ as desired. It now follows immediately that $\phi_0 \in \partial f(x_0)$. □

4.4.3.* Suppose $f : X \to (-\infty, +\infty]$ is such that f^{**} is proper. Show that f is convex if f^* is Fréchet differentiable at all $x \in \text{dom}(\partial f^*)$.

Hint. Use Exercise 4.4.2 and the dense graph consequence of the Brøndsted-Rockafellar theorem (Corollary 4.3.3); see the proof of Theorem 4.5.1. □

4.4.4. Calculate the conjugate and biconjugate of the function

$$f(x_1, x_2) := \begin{cases} \dfrac{x_1^2}{2x_2} + x_2 \log x_2 - x_2 & \text{if } x_2 > 0, \\ 0 & \text{if } x_1 = x_2 = 0, \\ \infty & \text{otherwise.} \end{cases}$$

4.4.5. Let X be a Hilbert space, and $f : X \to [-\infty, +\infty]$ be proper. Show that $f = f^*$ or $f(x) \geq f^*(x)$ for all x if and only if $f = \|\cdot\|^2/2$ where $\|\cdot\|$ is the inner product norm.

4.4.6 (Maximum entropy example). (a) Let $a^0, a^1, \ldots, a^N \in X$. Prove the function

$$g(z) := \inf_{x \in \mathbb{R}^{N+1}} \left\{ \sum_{n=0}^{N} \exp^*(x_n) : \sum_{n=0}^{N} x_n = 1, \sum_{n=0}^{N} x_n a^n = z \right\}$$

is convex.

(b) For any point y in $\mathbb{R}^{N+1}$, prove

$$g^*(y) = \sup_{x \in \mathbb{R}^{N+1}} \left\{ \sum_{n=0}^{N} (x_n \langle a^n, y \rangle - \exp^*(x_n)) : \sum_{n=0}^{N} x_n = 1 \right\}.$$

(c) Deduce the conjugacy formula

$$g^*(y) = 1 + \log \left(\sum_{n=0}^{N} \exp \langle a^n, y \rangle \right).$$

(d) Compute the conjugate of the function of $x \in \mathbb{R}^{N+1}$,

$$\begin{cases} \sum_{n=0}^{N} \exp^*(x_n) & \text{if } \sum_{n=0}^{N} x_n = 1, \\ \infty & \text{otherwise.} \end{cases}$$

4.4.7 (Conjugate of indicator function). Let X be a reflexive Banach space and let C be a closed convex subset of X. Show that $\delta_C^* = \sigma_C$ and $\delta_C^{**} = \delta_C$.

4.4.8 (Kernel average of convex functions [39]).** Let f_1, f_2 and g be proper lsc convex functions on X. Define $P(\lambda_1, f_1, \lambda_2, f_2, g): X \to [-\infty, +\infty]$ at $x \in X$ by

$$P(\lambda_1, f_1, \lambda_2, f_2, g) := \inf_{\lambda_1 y_1 + \lambda_2 y_2 = x} \{\lambda_1 f_1(y_1) + \lambda_2 f_2(y_2) + \lambda_1 \lambda_2 g(y_1 - y_2)\}$$

$$= \inf_{x = z_1 + z_2} \left\{ \lambda_1 f_1\left(\frac{z_1}{\lambda_1}\right) + \lambda_2 f\left(\frac{z_2}{\lambda_2}\right) + \lambda_1 \lambda_2 g\left(\frac{z_1}{\lambda_1} - \frac{z_2}{\lambda_2}\right) \right\}.$$

This is called the *g-average* of f_1 and f_2.

With appropriate choices of g, f_1, and f_2, show how to recover the: (a) arithmetic average; (b) epigraphical average; and (c) infimal convolution operation from the g-average. See [39] for this, and for the development of conjugacy, subdifferentiability and several other properties of these averaging operations.

4.4.9. Let $K \subset X$ be a closed convex cone. Show that both d_K and δ_K are convex functions and, for any $x \in X$,

$$\partial d_K(x) \subset \partial \delta_K(0) \cap B_{X^*},$$

and

$$\partial \delta_K(x) \subset \partial \delta_K(0).$$

4.4.10.† We consider an *objective function* p_N involved in the coupon collection problem given by

$$p_N(q) := \sum_{\sigma \in S_N} \left(\prod_{i=1}^{N} \frac{q_{\sigma(i)}}{\sum_{j=i}^{N} q_{\sigma(j)}} \right) \left(\sum_{i=1}^{N} \frac{1}{\sum_{j=i}^{N} q_{\sigma(j)}} \right),$$

summed over *all N!* permutations; so a typical term is

$$\left(\prod_{i=1}^{N} \frac{q_i}{\sum_{j=i}^{N} q_j} \right) \left(\sum_{i=1}^{N} \frac{1}{\sum_{j=i}^{n} q_j} \right).$$

For example, with $N = 3$ this is

$$q_1 q_2 q_3 \left(\frac{1}{q_1 + q_2 + q_3} \right) \left(\frac{1}{q_2 + q_3} \right) \left(\frac{1}{q_3} \right) \left(\frac{1}{q_1 + q_2 + q_3} + \frac{1}{q_2 + q_3} + \frac{1}{q_3} \right).$$

Show that p_N is *convex* on the positive orthant. Furthermore show that $1/p_N$ is concave. This is the base case of Example 1.3.9.

Hint.

(a) Establish

$$p_N(x_1, \ldots, x_N) = \int_0^1 \left(1 - \prod_{n=1}^{N} (1 - t^{x_n}) \right) \frac{dt}{t}. \qquad (4.4.4)$$

(b) Use
$$1 - e^{-tx_n} = x_n \int_0^t e^{-x_n y_n} \, dy_n,$$

to establish

$$1 - \prod_{n=1}^N (1 - e^{-tx_n}) = \left(\prod_{n=1}^N x_n\right)\left(\int_{\mathbb{R}_+^N} e^{-\langle x, y\rangle} \, dy - \int_{S_t^N} e^{-\langle x, y\rangle} \, dy\right),$$

where
$$S_t^N = \{y \in \mathbb{R}_+^N : 0 < y_n \le t \text{ for } n = 1, \ldots, N\}.$$

(c) Derive

$$\int_0^\infty \left(1 - \prod_{n=1}^N (1 - e^{-tx_n})\right) dt = \left(\prod_{n=1}^N x_n\right) \int_0^\infty dt \int_{\mathbb{R}_+^N \setminus S_t^N} e^{-\langle x, y\rangle} \, dy$$

$$= \left(\prod_{n=1}^N x_n\right) \int_0^\infty dt \int_{\mathbb{R}_+^N} e^{-\langle x, y\rangle} \chi_t(y) \, dy,$$

where
$$\chi_t(y) = \begin{cases} 1 & \text{if } \max(y_1, \ldots, y_N) > t, \\ 0 & \text{otherwise.} \end{cases}$$

(d) Show that the integral in (c) can be expressed as the *joint expectation* of Poisson distributions. Explicitly, if $x = (x_1, \ldots, x_N)$ is a point in the positive orthant $\mathbb{R}_+^N$, then

$$\int_0^\infty \left(1 - \prod_{n=1}^N (1 - e^{-tx_n})\right) dt = \left(\prod_{n=1}^N x_i\right) \int_{\mathbb{R}_+^N} e^{-\langle x, y\rangle} \max(y_1, \ldots, y_N) \, dy.$$

(e) Deduce that

$$p_N(x_1, \ldots, x_N) = \int_{\mathbb{R}_+^N} e^{-(y_1 + \cdots + y_N)} \max\left(\frac{y_1}{x_1}, \ldots, \frac{y_N}{x_N}\right) dy,$$

and hence that p_N is positive, decreasing and convex, as is the integrand.

(f) To derive the stronger result that $1/p_N$ is concave, let

$$h(a, b) := \frac{2ab}{a+b}.$$

Then h is concave and show that the concavity of $1/p_N$ is equivalent to

$$p_N\left(\frac{x + x'}{2}\right) \le h(p_N(x), p_N(x')) \quad \text{for all } x, x' \in \mathbb{R}_+^N. \tag{4.4.5}$$

(g) Recover the full case of Example 1.3.9.

The history of this problem and additional details can be found in Borwein, Bailey and Girgensohn [74, p. 36]. This book and its sister volume by Borwein and Bailey [73] also discuss how to use methods of experimental mathematics to gain insights on this and other related problems. □

4.4.11. As in the proof of the sandwich theorem (4.1.18), define
$$h(u) := \inf_{x \in X} \{f(x) + g(Tx + u)\}.$$
Prove that
$$\text{dom } h = \text{dom } g - T \text{ dom } f.$$

4.4.12 (Normals to an intersection). Let C_1 and C_2 be two convex subsets of X and let $x \in C_1 \cap C_2$. Suppose that C_1 and C_2 are closed and $0 \in \text{core}(C_1 - C_2)$ or $C_1 \cap \text{int } C_2 \neq \emptyset$. Show that
$$N_{C_1 \cap C_2}(x) = N_{C_1}(x) + N_{C_2}(x).$$

4.4.13. Let $K(x^*, \varepsilon) := \{x \in X : \varepsilon \|x^*\| \|x\| \leq \langle x^*, x \rangle\}$ be a Bishop–Phelps cone. Show that
$$N(K(x^*, \varepsilon); 0) = \partial \delta_{K(x^*,\varepsilon)}(0) \subset \bigcup_{r \geq 0} rB_{\varepsilon\|x^*\|}(-x^*).$$

4.4.14.* Prove the weak Fenchel duality in Theorem 4.4.18.

Hint. This follows immediately from the Fenchel-Young inequality (Proposition 4.4.1(a)). □

4.4.15. Let X be a reflexive Banach space. Suppose that $A: X \to X^*$ is a bounded linear operator, C a convex subset of X and D a nonempty closed bounded convex subset of X^*. Show that
$$\inf_{x \in C} \sup_{y \in D} \langle y, Ax \rangle = \max_{y \in D} \inf_{x \in C} \langle y, Ax \rangle.$$

Hint: Apply the Fenchel duality theorem (4.4.18) to $f = \delta_C$ and $g = \delta_D^*$.

4.4.16 (Fenchel duality for linear constraints).* Prove Corollary 4.4.19. Deduce duality theorems for the following separable problems.
$$\inf \left\{ \sum_{n=1}^{N} p(x_n) : Ax = b \right\},$$
where the map $A: \mathbb{R}^N \to \mathbb{R}^M$ is linear, $b \in \mathbb{R}^M$, and the function $p: \mathbb{R} \to (-\infty, +\infty]$ is convex, defined as follows:

(a) (Nearest points in polyhedra) $p(t) = t^2/2$ with domain $\mathbb{R}_+$.
(b) (Analytic center) $p(t) = -\log t$ with domain int $\mathbb{R}_+$.
(c) (Maximum entropy) $p = \exp^*$. What happens if the objective function is replaced by $\sum_{n=1}^{N} p_n(x_n)$?

4.4 Fenchel duality

4.4.17 (Symmetric Fenchel duality). Let X be a Banach space. For functions $f, g \colon X \to [-\infty, +\infty]$, define the *concave conjugate* $g_* \colon X \to [-\infty, +\infty]$ by

$$g_*(x^*) := \inf_{x \in X} \{\langle x^*, x \rangle - g(x)\}.$$

Prove

$$\inf(f - g) \geq \sup(g_* - f^*),$$

with equality if f is lsc and convex, g is usc and concave, and

$$0 \in \operatorname{core}(\operatorname{dom} f - \operatorname{dom}(-g)),$$

or f is convex, g is concave and

$$\operatorname{dom} f \cap \operatorname{cont} g \neq \emptyset.$$

4.4.18. Let X be a Banach space and let $K \subset X$ be a cone. Show that $\delta_{K^-} = \delta_K^*$, and therefore $\delta_{K^{--}} = \delta_K^{**}$.

4.4.19 (Sum of closed cones). Let X be a Banach space (see Exercise 2.4.25).

(a) Prove that any cones $H, K \subset X$ satisfy $(H + K)^- = H^- \cap K^-$.
(b) Deduce that if H and K are closed convex cones then they satisfy $(H \cap K)^- = \operatorname{cl}(H^- + K^-)$. In $\mathbb{R}^3$, define sets

$$H := \{x \colon x_1^2 + x_2^2 \leq x_3^2, x_3 \leq 0\} \text{ and}$$

$$K := \{x \colon x_2 = -x_3\}.$$

(c) Prove H and K are closed convex cones.
(d) Calculate the polar cones H^-, K^- and $(H \cap K)^-$.
(e) Prove $(1, 1, 1) \in (H \cap K)^- \setminus (H^- + K^-)$, and deduce that the sum of two closed convex cones is not necessarily closed.

4.4.20 (Coercivity and minimization).* Let X be a Banach space.

(a) Show that every proper coercive lsc convex function attains its minimum on X if and only if X is reflexive. In particular, when $f \colon X \to (-\infty, +\infty]$ is lsc coercive and convex on a reflexive space X, conclude that $0 \in \operatorname{range} \partial f$.
(b) Suppose X is reflexive and $f \colon X \to (-\infty, +\infty]$ is proper, supercoercive lsc and convex. Show that $\operatorname{range} \partial f = X^*$.
(c) Is (b) true if X is not reflexive?

Hint. (a) Suppose X is reflexive, then the properness and coercivity of f implies that $\{x \colon f(x) \leq M\}$ is nonempty and bounded for some $M > 0$. This a weakly compact set hence f attains its minimum on this set. Outside of the set $f(x) > M$, and so that minimum is an absolute minimum for f.

Conversely, suppose X is not reflexive. According to James' theorem (4.1.27) there is a functional $\phi \in S_{X^*}$ that does not attain its norm on B_X. Then $f := \phi + \delta_{B_X}$ does not attain its minimum on X.

(b) For each $\phi \in X^*$, $f - \phi$ is supercoercive. Now $f - \phi$ attains its minimum at $\bar{x}$ for some $\bar{x} \in X$. Then $\phi \in \partial f(\bar{x})$.

(c) No: for example let f be defined by $f(x) := (\sup_{n \in \mathbb{N}} |x(n) - 1|)^2 + \sum_{n=1}^{\infty} \frac{1}{2^n} |x(n) - 1|^2$ where $x = (x(n)) \in c_0$. Then f is a continuous and supercoercive convex function that does not attain its minimum (observe that $x_n = (1, 1, \ldots, 1, 0, \ldots)$ is a minimizing sequence) and so $0 \notin \text{range} \, \partial f$. More generally, in any nonreflexive space, consider $f := \|\cdot\|^2$. Using James' theorem (4.1.27), one can show that the subdifferential map is not onto. □

4.4.21 (Infimal convolutions and approximation). Suppose that f is convex, and $f_n \leq f$ for all $n \in \mathbb{N}$.

(a) If f is Lipschitz convex on X, show that $f_n \to f$ uniformly on X provided that $f_n^* \to f^*$ uniformly on bounded subsets of the domain of f^*.
(b) Suppose $f : X \to (-\infty, +\infty]$ is a proper lsc convex function. If $f_n^* \to f^*$ pointwise on the domain of f^*, show that $f_n(x) \to f(x)$ for each $x \in X$.
(c) Conclude that $f \square n \|\cdot\|^2$ converges uniformly (resp. pointwise) to f provided that f is Lipschitz (resp. lsc proper) and convex.

Hint. (a) Repeat the proof of Lemma 4.4.16 noting the domain of f^* is bounded.

(b) Let $x_0 \in X$. Given $\varepsilon > 0$, and any number $\alpha < f(x_0)$ it suffices to show that there exists N such that $f_n(x_0) > \alpha - \varepsilon$ for all $n \geq N$. Now choose $\phi \in X^*$ such that $\phi(x) - \phi(x_0) \leq f(x) - \alpha$ for all $x \in X$. Then $f^*(\phi) \leq \phi(x_0) - \alpha$ and so $\phi \in \text{dom} \, f^*$. Now choose $N \in \mathbb{N}$ so that $f_n^*(\phi) \leq f^*(\phi) - \varepsilon$ for $n \geq N$. Now, for all $n \geq N$,

$$\phi(x_0) - \alpha \geq f^*(\phi) \geq f_n^*(\phi) - \varepsilon \geq \phi(x_0) - f_n(x_0) - \varepsilon.$$

Therefore, $f_n(x_0) \geq \alpha - \varepsilon$. □

4.4.22.* Prove Fact 4.4.4 (b), (c).

Hint. (b) Suppose f is continuous at x_0, then there exist $\delta > 0$ and $M > 0$ so that $f(x) \leq M$ for $x \in x_0 + \delta B_X$ (Proposition 4.1.4). Now suppose $x^{**} \in X^{**}$ and $\|x^{**} - x_0\| \leq \delta$. According to Goldstine's theorem (Exercise 4.1.13) there is a net $(x_\alpha) \subset \delta B_X$ with $x_\alpha \to_{w^*} (x^{**} - x_0)$. The weak*-lower-semicontinuity of f^{**} implies that

$$f^{**}(x^{**}) \leq \liminf_\alpha f^{**}(x_0 + x_\alpha) = \liminf_\alpha f(x_0 + x_\alpha) \leq M.$$

Thus f^{**} is bounded above on a neighborhood of x^{**}, and thus it is continuous at x^{**} (Proposition 4.1.4).

Conversely, if f^{**} is continuous at $x_0 \in X$, then so is f because $f^{**}|_X = f$ (Proposition 4.4.2(a)).

(c) Suppose f is Fréchet differentiable at x_0. Let $\varepsilon > 0$, the according to Proposition 4.2.7 there exists $\delta > 0$ so that

$$f(x_0 + h) + f(x_0 - h) - 2f(x_0) \leq \varepsilon \|h\| \text{ if } \|h\| < \delta.$$

Now suppose $h \in X^{**}$ and $\|h\| < \delta$. Then there exist $h_\alpha \in X$ with $\|h_\alpha\| = \|h\| < \delta$ and $h_\alpha \to_{w^*} h$ by Goldstine's theorem (Exercise 4.1.13). Using the weak*-lower-semicontinuity of f^{**} and the fact $f^{**}|_X = f$ (Proposition 4.4.2(a)) we have

$$f^{**}(x_0 + h) + f^{**}(x_0 - h) - 2f^{**}(x_0)$$
$$\leq \liminf_\alpha f^{**}(x_0 + h_\alpha) + f^{**}(x_0 - h_\alpha) - 2f^{**}(x_0)$$
$$= \liminf_\alpha f(x_0 + h_\alpha) + f(x_0 - h_\alpha) - 2f(x_0) \leq \varepsilon \|h\|.$$

Applying Proposition 4.2.7 we conclude that f^{**} is Fréchet differentiable at x_0. The converse, as in (b), follows from Proposition 4.4.2(a). □

4.4.23 (Coercive versus cofinite functions). *

(a) Let f be a lsc proper convex function. Consider the following conditions:
  (i) f is supercoercive;
  (ii) $f - y^*$ is coercive for every $y^* \in X^*$;
  (iii) $\mathrm{dom}\, f^* = X^*$.

Show that (i) $\Rightarrow$ (ii) $\Leftrightarrow$ (iii). If X is finite-dimensional, show that (ii) $\Rightarrow$ (i).

(b) Provide an example showing that (ii) need not imply (i) when $X = \ell_2$.

(c) Given any normed linear space X, find a continuous function f such that $\lim_{\|x\| \to \infty} f(x)/\|x\| = 0$ but f has bounded lower level sets.

Hint. For (a), see [35, Theorem 3.4]. For (b), define the conjugate function f^* by $f^*(x) := \|x\|^2 + \sum_{n=1}^\infty (x_i)^{2n}$, f^* is a continuous convex function and supercoercive. However, f^* is not bounded on $2B_{\ell_2}$, since $f(2e_n) = 2^{2n}$. Therefore, $f = f^{**}$ cannot be supercoercive. For (c), consider $f := \sqrt{\|\cdot\|}$. □

4.4.24 (Risk function duality).† Let X be a linear space of measurable functions on Ω, containing the constants, in the a.e. pointwise ordering. Following Artzner *et al.* [385], an extended real-valued function ρ is a *coherent risk function* if it satisfies:

A1. ρ is convex.
A2. ρ is monotone nondecreasing.
A3. ρ is translation equivariant: $\rho(x + a) = \rho(x) + a$ for all real a.
A4. ρ is positively homogeneous.

Denote $\mathcal{A} := \mathrm{dom}\, \rho^*$. Show that (A2) holds if and only if $\mathcal{A}$ consists only of nonnegative measures while (A3) holds if and only if $\mathcal{A}$ consists only for measures with $\mu(\Omega) = 1$. Finally, it is obvious that (A4) coincides with being able to write $\rho(x) = \sup_{\sigma \in \mathcal{A}} \int_\Omega x \, d\sigma$.

186   *Convex functions on Banach spaces*

Let $c > 0$, $p \geq 1$ and a probability measure μ be given. Determine under what conditions on μ, c and p the following are coherent risk functions.

(a) (Mean-deviation) $\rho(x) := \int_\Omega x \, d\mu + c \|x - \int_\Omega x \, d\mu\|_p$.
(b) (Mean-semideviation) $\rho(x) := \int_\Omega x \, d\mu + c \|x - \int_\Omega x^+ \, d\mu\|_p$.
(c) (Conditional value at risk, Rockafellar–Uryasev). Fix $p = 1, \varepsilon_1, \varepsilon_2 > 0$. Set $\tau := \varepsilon_2/(\varepsilon_1 + \varepsilon_2)$. Let G be the cumulative distribution function with respect to μ, and define $\rho(x) := (1 - \varepsilon_1) \int_\Omega x \, d\mu + \varepsilon_1 \mathrm{CVaR}_\tau(x)$ where

$$\mathrm{CVaR}_\tau(x) := \inf_{z \in \mathbb{R}} \left\{ z + \int_{-\infty}^{\infty} (x-z)^+ \, dG \right\}.$$

In each case sublinearity and continuity of ρ are clear, but only the third is always a coherent risk function.

4.4.25 (von Neumann minimax theorem). Show that the minimax theorem of Exercise 2.4.21 remains valid with the same proof if in (b)(i) 'bounded' is replaced by 'compact' – which is equivalent in the Euclidean case.

4.4.26 (The Krein–Šmulian theorem [379]). ** The normed space case of the Krein–Šmulian or Banach–Dieudonné theorem asserts that *a convex set C in a dual Banach space X^* is weak*-closed as soon as its intersection with all closed balls is*. We sketch a proof. Let $C_n := C \cap (n) B_{X^*}$.

(a) It suffices to show that if $0 \notin C$ then 0 can be separated from C.
(b) Inductively, there exist sequences $(x_n)_{n=1}^\infty$, $(y_n)_{n=1}^\infty$ in X such that

$$\|y_n\| < 1/n, \quad x_{n+1} \in \mathrm{conv}\{x_n, y_n\} \quad \text{and} \quad \min_{c \in C_{n+1}} \langle x_{n+1}, c \rangle > 1.$$

Hint. The inductive step for $n-1$, implies that $\min_{c \in C_{n+1}} \langle x_n, c \rangle > 1$. Let $D_n := \mathrm{conv}\{x_n, \{y : \|y\| < 1/n\}\}$. Consider any $c \in C_{n+1}$. If $\|c\| \leq n$ set $x := x_n$; otherwise there is a point y with $\|y\| < 1/n$, $\langle y, c \rangle > 1$. In any event

$$\sup_{x \in D_n} \min_{c \in C_{n+1}} \langle x, c \rangle = \min_{c \in C_{n+1}} \sup_{x \in D_n} \langle x, c \rangle > 1$$

where the equality follows from the convex minimax theorem of Exercise 4.4.25 since C_n is weak*-compact. Thus the inductive step holds for n.   □

(c) Observe that for all $n > 0$ we have $x_n \in \overline{\mathrm{conv}}\{0, y_1, y_2, \ldots, y_n, \ldots\}$ which is a norm-compact convex set. Any norm cluster point of $(x_n)_{n=1}^\infty$ separates C from 0 in the weak*-topology.

4.5 Čebyšev sets and proximality

4.5.1 A sufficient condition for convexity of functions

This subsection develops a nice result for checking the convexity of a function using the smoothness of its conjugate which is of independent interest, and which is useful in the study of Čebyšev sets in Hilbert spaces. Given a bornology β on X^*, we use

4.5 Čebyšev sets and proximality

τ_β to denote the topology on X^{**} of uniform convergence on β-sets. We will say $f : X \to (-\infty, +\infty]$ is *sequentially τ_β-lsc* if for every sequence $(x_n) \subset X \subset X^{**}$, and $x \in X$, $\liminf f(x_n) \geq f(x)$ whenever $x_n \to_{\tau_\beta} x$. Notice, that the τ_β-topology restricted to X is at least as strong as the weak topology on X.

Theorem 4.5.1. *Suppose* $f : X \to (-\infty, +\infty]$ *is such that* f^{**} *is proper. If* f^* *is β-differentiable at all* $x^* \in \mathrm{dom}(\partial f^*)$ *and* f *is sequentially τ_β-lsc, then* f *is convex.*

Proof. First, $f^{**} \leq f$, so it suffices to show that $f(x) \leq f^{**}(x)$ for all $x \in X$. For notational purposes, let $\hat{f} = f^{**}|_X$. If $x \notin \mathrm{dom}\,\hat{f}$ then $\hat{f}(x) = \infty$, so $\hat{f}(x) \geq f(x)$. So let $x \in \mathrm{dom}\,\hat{f}$. We first handle the case $x \in \mathrm{dom}(\partial\hat{f})$. Indeed, for such x, let $\phi \in \partial\hat{f}(x)$. Then $x \in \partial f^*(\phi)$. Also, f^* is lsc everywhere and β-differentiable at ϕ and so f^* is continuous at ϕ. Now choose $x_n \in X$ such that

$$\phi(x_n) - f(x_n) \geq f^*(\phi) - \varepsilon_n \text{ where } \varepsilon_n \to 0^+.$$

Then $\phi(x_n) - \hat{f}(x_n) \geq f^*(\phi) - \varepsilon_n$ and so $x_n \in \partial_{\varepsilon_n} f^*(\phi)$ for all n. According to Šmulian's theorem (Exercise 4.2.10) $x_n \to_{\tau_\beta} x$. In particular, $\phi(x_n) \to \phi(x)$. Therefore,

$$\hat{f}(x) = \phi(x) - f^*(\phi)$$
$$= \lim_{n \to \infty} \phi(x) - [\phi(x_n) - f(x_n)]$$
$$= \lim_{n \to \infty} f(x_n).$$

Now f is sequentially τ_β-lsc, and so $\liminf_{n \to \infty} f(x_n) \geq f(x)$. Therefore, $\hat{f}(x) \geq f(x)$ when $x \in \mathrm{dom}(\partial\hat{f})$.

Now suppose $x \in \mathrm{dom}\,\hat{f} \setminus \mathrm{dom}(\partial\hat{f})$. The dense graph consequence of the Brøndsted–Rockafellar theorem (Corollary 4.3.3) now asserts that there exists a sequence $x_n \to x$ such that $x_n \in \mathrm{dom}(\partial\hat{f})$ and $|\hat{f}(x_n) - \hat{f}(x)| \to 0$. Consequently,

$$f(x) \leq \liminf f(x_n) = \liminf \hat{f}(x_n) = \hat{f}(x).$$

Thus, for any $x \in X$, $f(x) \leq \hat{f}(x)$, and so $f(x)$ is convex. $\square$

The important special cases of this are recorded as follows.

Corollary 4.5.2. *Suppose* $f : X \to (-\infty, +\infty]$ *is such that* f^{**} *is proper.*

(a) *Suppose* f^* *is Fréchet differentiable at all* $x^* \in \mathrm{dom}(\partial f^*)$ *and* f *is lsc. Then* f *is convex.*
(b) *Suppose* f^* *is Gâteaux differentiable at all* $x^* \in \mathrm{dom}(\partial f^*)$ *and* f *is sequentially weakly lsc. Then* f *is convex.*

Figure 4.2 neatly illustrates the 'failure' of Corollary 4.5.2 in one dimension.

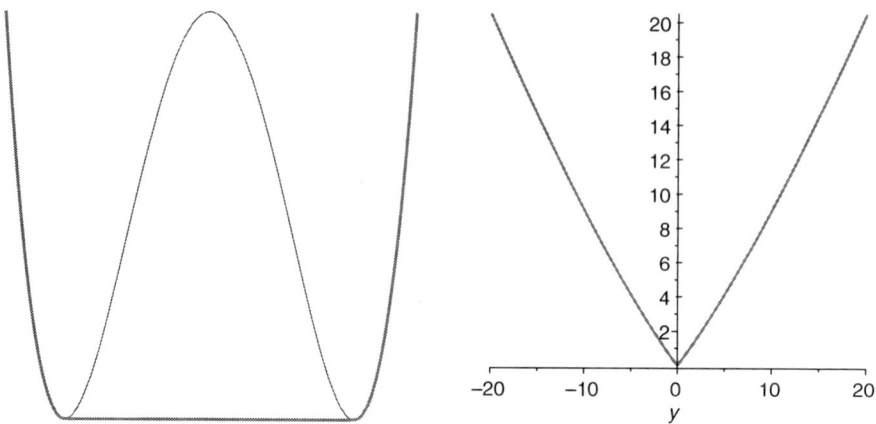

Figure 4.2 $(1 - 2x^2)^2$ with second conjugate (L) and conjugate (R).

4.5.2 Čebyšev sets and proximality

Let C be a nonempty subset of a normed space. We define the *nearest point mapping* by

$$P_C(x) := \{v \in C : \|v - x\| = d_C(x)\}.$$

A set C is said to be a *Čebyšev set* if $P_C(x)$ is a singleton for every $x \in X$. If $P_C(x) \neq \emptyset$ for every $x \in X$, then C is said to be *proximal*; the term *proximinal* is also used. A norm $\|\cdot\|$ is said to be *strictly convex* if $\|x + y\| < 2$ whenever $\|x\| = \|y\| = 1$ and $x \neq y$.

Fact 4.5.3. *Let C be a nonempty closed convex subset of a Banach space X.*

(a) If X is reflexive, then $P_C(x) \neq \emptyset$ for each $x \in X$.
(b) If X is strictly convex, then $P_C(x)$ is either empty or a singleton for each $x \in X$.

In particular, every closed convex set in a strictly convex reflexive Banach space is Čebyšev.

Proof. See Exercise 4.5.2. □

The following is deeper because of its connection to James' theorem (4.1.27).

Theorem 4.5.4. *A Banach space X is reflexive if and only if every closed convex nonempty subset of X is proximal if and only if for every nonempty closed convex subset C of X, $P_C(x) \neq \emptyset$ for at least one $x \notin C$.*

Proof. (Outline) Let X be reflexive and let C be a closed convex subset of X. For $\bar{x} \in X$, define $f(\cdot) := \|\cdot - \bar{x}\| + \delta_C$. Then f is coercive and convex, and therefore attains its minimum on C (Exercise 4.4.20). Thus C is proximal.

Suppose X is not reflexive, then James' theorem (4.1.27) ensures the existence of $\phi \in S_{X^*}$ such that $\phi(x) < 1$ for all $x \in B_X$. Let $C := \{x \in X : \phi(x) = 0\}$; it is left as Exercise 4.5.3 to show that $P_C(x) = \emptyset$ for all $x \notin C$. □

4.5 Čebyšev sets and proximality

A norm is said to have the *Kadec–Klee property* if the weak and norm topologies agree on its unit sphere (this is also known as the *Kadec property*). Before focusing our attention on Čebyšev sets in Hilbert spaces, let us mention the following theorem.

Theorem 4.5.5 (Lau–Konjagin). *Every closed set A in a Banach space densely (equivalently generically) admits nearest points if and only if the norm has the Kadec–Klee property and the space is reflexive.*

A proof of the preceding theorem is outlined in [72] and the full proof can be found in [80]. Theorem 4.5.5 implies in particular that every norm-closed set in a Hilbert space has 'a lot of' nearest points – but not necessarily for a sufficiently odd set in a sufficiently perverse renorm as discussed further in Exercise 8.4.6. Somewhat relatedly, in Exercise 4.5.8 we describe a remarkable result of Odell and Schlumprecht on *distortion* of the norm on a Hilbert space which shows how perverse an equivalent renorm can be at least from some vantage points.

This leads us to mention a fundamental question of Klee's [277] in 1961: Is every Čebyšev set in a Hilbert space convex? At this stage, it is known that every weakly closed Čebyšev set in a Hilbert space is convex; we will present a proof of this in what follows. To begin, we establish a nice duality formula.

Fact 4.5.6. *Let C be a closed nonempty subset of a Hilbert space. Let $f := \frac{1}{2}\|\cdot\|^2 + \delta_C$. Then $d_C^2 = \|\cdot\|^2 - 2f^*$.*

Proof. Given f as defined, we compute

$$f^*(y) = \sup\left\{\langle x,y\rangle - \frac{1}{2}\|x\|^2 : x \in C\right\}$$

$$= \sup\left\{\langle x,y\rangle - \frac{1}{2}\langle x,x\rangle : x \in C\right\}$$

$$= \sup\left\{\frac{1}{2}\langle y,y\rangle + \langle x,y\rangle - \frac{1}{2}\langle y,y\rangle - \frac{1}{2}\langle x,x\rangle : x \in C\right\}$$

$$= \frac{1}{2}\langle y,y\rangle + \sup\left\{-\frac{1}{2}\langle x,x\rangle + \langle x,y\rangle - \frac{1}{2}\langle y,y\rangle : x \in C\right\}$$

$$= \frac{1}{2}\langle y,y\rangle - \frac{1}{2}\inf\{\langle x,x\rangle - 2\langle x,y\rangle + \langle y,y\rangle : x \in C\}$$

$$= \frac{1}{2}\langle y,y\rangle - \frac{1}{2}d_C^2(y).$$

Therefore,

$$d_C^2 = \|\cdot\|^2 - 2f^* \tag{4.5.1}$$

as desired. □

We next characterize closed convex sets via the differentiability of the distance function in Hilbert spaces.

Theorem 4.5.7. *Let X be a Hilbert space and let C be a nonempty closed subset of X. Then the following are equivalent.*

(a) *C is convex.*
(b) *d_C^2 is Fréchet differentiable.*
(c) *d_C^2 is Gâteaux differentiable.*

Proof. (a) $\Rightarrow$ (c): If C is convex, then d_C^2 is Fréchet differentiable by Exercise 4.5.6.
(c) $\Rightarrow$ (b): Let $f := \frac{1}{2}\|\cdot\|^2 + \delta_C$. Then $f^* = (\|\cdot\|^2 - d_C^2)/2$ and so f^* is Gâteaux differentiable. Thus the derivatives of f^* and $\|\cdot\|^2$ are both norm-to-weak continuous. Consequently the derivative of d_C^2 is norm-to-weak continuous. By the Kadec–Klee property of the Hilbert norm, the derivative of d_C^2 is norm-to-norm continuous (see Exercise 4.5.4). Consequently, so is the derivative of f^*, and so f^* is Fréchet differentiable according to Šmulian's theorem (4.2.10). Therefore, d_C^2 is also Fréchet differentiable.
(b) $\Rightarrow$ (a): Suppose d_C^2 is Fréchet differentiable. With f as in the previous part, we conclude that f^* is Fréchet differentiable. By Corollary 4.5.2(a), we know f is convex. Therefore, $C = \mathrm{dom} f$ is convex. $\square$

Proposition 4.5.8. *Suppose X is a reflexive Banach space, and let C be a weakly closed Čebyšev subset of X. Then $x \mapsto P_C(x)$ is norm-to-weak continuous. If, moreover, the norm on X has the Kadec–Klee property, then $x \mapsto P_C(x)$ is norm-to-norm continuous.*

Proof. Let $x \in X$ and suppose $x_n \to x$. Then $\|x_n - P_C(x_n)\| \to \|x - P_C(x)\|$. According to the Eberlein–Šmulian theorem (see e.g. [199, p. 85]), there is a subsequence $(P_C(x_{n_k}))_{k \in \mathbb{N}}$ that converges weakly to some $\bar{x} \in X$. Then $\bar{x} \in C$ because C is weakly closed. Now

$$\|x - P_C(x)\| = \lim_{k \to \infty} \|x_{n_k} - P_C(x_{n_k})\| \geq \|x - \bar{x}\|.$$

Because $\bar{x} \in C$, we conclude that $\bar{x} = P_C(x)$. Thus $P_C(x_{n_k}) \to_w P_C(x)$. Standard arguments now imply $x \mapsto P_C(x)$ is norm-to-weak continuous.
Now suppose the norm on X has the Kadec–Klee property. Then in the previous paragraph, we have $x_{n_k} - P_C(x_{n_k}) \to_w x - P_C(x)$, and hence $x_{n_k} - P_C(x_{n_k}) \to x - P_C(x)$ which implies $P_C(x_{n_k}) \to P_C(x)$ from which we deduce the norm-to-norm continuity. $\square$

Theorem 4.5.9. *Let X be a Hilbert space and suppose C is a nonempty weakly closed subset of X. Then the following are equivalent.*

(a) *C is convex.*
(b) *C is a Čebyšev set.*
(c) *d_C^2 is Fréchet differentiable.*
(d) *d_C^2 is Gâteaux differentiable.*

4.5 Čebyšev sets and proximality

Proof. The implication (a) ⇒ (b) follows from Proposition 4.5.3. We next prove (b) ⇒ (c). For this, again consider $f := \frac{1}{2}\|\cdot\|^2 + \delta_C$. We first show that

$$\partial f^*(x) = \{P_C(x)\}, \quad \text{for all } x \in X. \tag{4.5.2}$$

Indeed, for $x \in X$, (4.5.1) implies

$$f^*(x) = \frac{1}{2}\|x\|^2 - \frac{1}{2}\|x - P_C(x)\|^2 = \langle x, P_C(x)\rangle - \frac{1}{2}\|P_C(x)\|^2$$
$$= \langle x, P_C(x)\rangle - f(P_C(x)).$$

Consequently, $P_C(x) \in \partial f^*(x)$ for $x \in X$. Now suppose $y \in \partial f^*(x)$, and define $x_n = x + \frac{1}{n}(y - P_C(x))$. Then $x_n \to x$, and hence $P_C(x_n) \to P_C(x)$ by Proposition 4.5.8. Using the subdifferential inequality, we have

$$0 \leq \langle x_n - x, P_C(x_n) - y\rangle = \frac{1}{n}\langle y - P_C(x), P_C(x_n) - y\rangle.$$

This now implies:

$$0 \leq \lim_{n \to \infty} \langle y - P_C(x), P_C(x_n) - y\rangle = -\|y - P_C(x)\|^2.$$

Consequently, $y = P_C(x)$ and so (4.5.2) is established. Now f^* is continuous, and Proposition 4.5.8 the ensures that the mapping $x \mapsto P_C(x)$ is norm-to-norm continuous. Consequently, Šmulian's theorem (4.2.10) implies that f^* and d_C^2 are Fréchet differentiable.

The implication (c) ⇒ (d) is trivial. Finally, (d) ⇒ (a) follows from Corollary 4.5.2(b) because f is weakly lsc (use the fact that C is weakly closed). □

Exercises and further results

4.5.1.* Suppose $f : X \to (-\infty, +\infty]$ is such that f^{**} is proper. Suppose f^* is Fréchet differentiable at $x^* \in X^*$, and f is lsc and proper. Show that $f(x) = f^{**}(x)$ where $x = \nabla f^*(x^*)$.
Hint. First, show that $x \in X$ (see Exercise 4.2.11). Then examine the proof of Theorem 4.5.1. Another proof is given in [205, Lemma 3]. □

4.5.2.* Prove Fact 4.5.3.

4.5.3.* Let X be a nonreflexive Banach space, and let $\phi \in S_{X^*}$ be a functional that does not attain its norm on B_X whose existence is ensured by James' theorem (4.1.27). Let $C := \{x \in X : \phi(x) = 0\}$. Show that if $\phi(x) \neq 0$, then $P_C(x) = \emptyset$.

4.5.4.* Let A be a nonempty closed subset of a Banach space X whose dual norm is weak*-Kadec (i.e. the weak* and norm topologies agree on its sphere). Suppose d_A^2 is Gâteaux differentiable with derivative that is norm-to-weak*-continuous. Show that the derivative is norm-to-norm continuous.
Hint. Let $x \in X$, and let $f := d_A$. Show that $\|f'(x_n)\| \to \|f'(x)\|$ whenever $x_n \to x$, and then use the weak*-Kadec property. □

4.5.5.* Let X be a reflexive Banach space. Show that an equivalent norm on X is Gâteaux differentiable if and only if its dual norm is strictly convex.

Hint. When the dual norm is strictly convex show for each $x \neq 0$ that $\partial \|x\|$ is a singleton. Conversely, suppose $x, y \in S_{X^*}$ are such that $\|x + y\| = 2$. Because of reflexivity, we can choose $\phi \in S_X$ so that $\phi(x+y) = 2$. Thus $x, y \in \partial \|\phi\|$. By Gâteaux differentiability, $x = y$. □

4.5.6. Let H be a Hilbert space with inner product norm $\|\cdot\|$. Suppose $C \subset H$ is closed and convex. Without using Theorem 4.5.9, show that d_C is Fréchet differentiable at each $x \notin C$; conclude that d_C^2 is Fréchet differentiable everywhere.

Hint. Let $x \notin C$. Because H is reflexive, there exists $y \in C$ such that $\|y - x\| = d_C(x)$. Choose $\phi_n \in \partial_{\varepsilon_n} d_C(x)$ where $\varepsilon_n \to 0^+$, and let $\phi \in \partial d_C(x)$. Verify that $\|\phi_n\| \to 1$ and $\|\phi\| = 1$ and $\|\phi_n + \phi\| \to 2$. By the parallelogram law, $\|\phi_n - \phi\| \to 0$. Therefore d_C is Fréchet differentiable at x by Šmulian's theorem (4.2.10). See Exercise 5.3.11 for stronger more general results. □

4.5.7 (Making a hole in the space [230]).* Let E be a nonreflexive Banach space and let $x^* \in S_{X^*}$ be a non-norm-attaining functional on the space whose existence is ensured by James' theorem (4.1.27). Define

$$\Phi(x) := 1 + x^*(x) + \max\{2(\|x\| - 1), 0\}.$$

(a) Show that Φ is convex, continuous and coercive, but that $0 = \inf_{x \in B_X} \Phi(x)$ is not attained. Moreover, $\Phi(B_X) = \Phi(S_X) = (0, 2)$.

(b) Fix $a > 0$. We now construct an interesting mapping T_a. Select a sequence $(x_n)_{n=0}^\infty$ on the unit sphere with $x^*(x_n) = 2^{-n}$ and form a curve γ by connecting the points by segments with x_0 connected to ∞ by a radial ray. For $x \in X$ and $y := \gamma(\Phi(x)/2)$ there is a unique point $z := T_a(x)$ on the ray extending $[x, y]$ with $\Phi(z) = \Phi(x) + a$.

(c) Show that T_a maps X onto $\{x : \Phi(x) > a\}$, is invertible and that T_a and T_a^{-1} are locally Lipschitz.

(d) Fix $a = 2$ and let $p_2(x) := \inf\{t > 0 : \Phi(tx) \geq 2\}$ be the Minkowski gauge. Define

$$T_B(x) := p_2(T_2(x)) \frac{T_2(x)}{\|T_2(x)\|}.$$

Show that T_B removes the unit ball $B = B_X$ from the range.

(e) Show that

$$T(x) := \left(1 - \frac{1}{\|T_B(x)\|}\right) T_B(x),$$

maps X onto $X \setminus \{0\}$.

(f) Finally, show that T_0 is a fixed-point free locally Lipschitz involution ($T_0^2 = I$). It is an open question as to whether a fixed-point free uniformly continuous involution exists.

4.5.8 (The distortion of Hilbert space [334, 335]).* A Banach space $(X, \|\cdot\|)$ is said to be *distortable* if there exist an equivalent norm $|\cdot|$ on X and a $\lambda > 1$ such that,

4.5 Čebyšev sets and proximality

for all infinite-dimensional subspaces $Y \subseteq X$, $\sup\{|y|/|x| : x, y \in Y, \|x\| = \|y\| = 1\} > \lambda$. If this holds for all $\lambda > 1$ then the space is *arbitrarily distortable*. Odell and Schlumprecht prove that a separable infinite-dimensional Hilbert space is arbitrarily distortable. Adding an earlier result of V. Milman, they prove that any space not containing an isomorphic copy of ℓ_1 or c_0 is distortable. (R. C. James proved ℓ_1 and c_0 are not distortable.)

Distortability of the Hilbert space ℓ_2 is equivalent to the existence of two separated sets in the sphere of ℓ_2, each of which intersect every infinite-dimensional closed subspace of ℓ_2 [334]. Indeed [335] yields a sequence of (*asymptotically orthogonal*) subsets $(C_i)_{i=1}^\infty$ of the unit sphere of ℓ_2 such that (a) each set C_i intersects each infinite-dimensional closed subspace of ℓ_2, and (b) $\sup\{|\langle x, y\rangle| : x \in C_i, y \in C_j\} \to 0$ as $\min\{i, j\} \to \infty$.

The next exercise examines in greater generality the Euclidean argument given in Section 3.5 as part of our discussion of the Čebyšev problem.

4.5.9. ** Let X be a normed space. As in Section 3.5, a subset $C \subset X$ is said to be *approximately convex* if for any norm ball $D \subset X$, there is a norm ball $D' \supset D$ disjoint from C with arbitrarily large radius.

(a) Show that every convex set is approximately convex.
(b) Show that every approximately convex set in a Banach space is convex if and only if the dual norm is strictly convex.
(c) Suppose C is a Čebyšev set in X and suppose $x \mapsto P_C(x)$ is norm-to-norm continuous. If the norm on X is such that its dual is strictly convex, show that C is convex.
(d) Conclude that every weakly closed Čebyšev set is convex in a reflexive Banach space whose norm is Gâteaux differentiable and has the Kadec–Klee property.

Hint. (a) follows from the separation theorem (4.1.15).

(b) is a result of Vlasov's [428, p. 776] discussed in [229, Theorem 4, p. 244]. The direction used in (c) is proved as in Exercise 3.5.2.

(c) First show for $x \notin C$,

$$\limsup_{y \to x} \frac{d_C(y) - d_C(x)}{\|y - x\|} = 1.$$

This will require a mean value theorem, and the continuity of the metric projection. Now for $\alpha > d_C(x)$, choose real numbers $\sigma \in (0, 1)$ and ρ satisfying

$$\frac{\alpha - d_C(x)}{\sigma} < \rho < \alpha - \beta.$$

Now apply Ekeland's variational principle (Theorem 4.3.1) to the function $-d_C + \delta_{x+\rho B_X}$ to show there is a point $v \in X$ that satisfies both:

$$d_C(x) + \sigma \|x - v\| \leq d_C(v)$$
$$d_C(z) - \sigma \|z - v\| \leq d_C(v) \quad \text{for all } z \in x + \rho B_X.$$

Conclude that $\|x - v\| = \rho$, and then show that C is approximately convex, and then use (b).

(d) Use Proposition 4.5.8 for the continuity property of P_C. Because the norm on X is Gâteaux differentiable, the dual norm on X^* is strictly convex, and therefore we may apply (c). □

4.5.10 (Čebyšev suns). Another concept of interest in the study of Čebyšev sets is that of a *sun*. A Čebyšev set S of a Hilbert space is called a *sun* if for $x \in S$ every point on the ray $P_S(x) + t(x - P_S(x))$ where $t \geq 0$ has nearest point $P_S(x)$. Show that every Čebyšev sun is approximately convex.

4.5.11 (Convexity of Čebyšev suns [175]).** For a Čebyšev set C of a Hilbert space X, show that the following are equivalent.

(a) C is convex.
(b) C is a sun (as defined in Exercise 4.5.10).
(c) $x \mapsto P_C(x)$ is a nonexpansive mapping.

Hint. A proof based on the Brouwer fixed point theorem is given in [175] and sketched in [72, Proposition 5]. Note that Exercises 4.5.10 and 4.5.9 show the equivalence of (a) and (b). The equivalent of (a) and (c) is from [346], see also [229, Theorem 6, p. 247]. □

4.5.12 (Bregman–Čebyšev sets).** In [41] the authors provide an analysis in Euclidean space that simultaneously recovers Theorem 4.5.9 for the Euclidean norm and the Kullback–Leibler divergence. More precisely, let f be convex and let $C \subset \operatorname{int} \operatorname{dom} f$ be a closed set. Try to apply the ideas of Section 7.6 on zone consistency and of this Section 4.5 to prove that:

> A Bregman–Čebyšev set C – the associated Bregman projection is everywhere singleton – is convex provided the function f inducing the Bregman distance is Legendre and supercoercive.

4.6 Small sets and differentiability

We have already seen several results that deal with the size of the set of points of (non) differentiability of continuous convex functions on finite-dimensional spaces. In this section, we will give an overview of some types of small sets in Banach spaces, then we will discuss first-order differentiability of convex functions and locally Lipschitz functions in relation to these notions. The section concludes with some observations on second-order differentiability of convex functions.

4.6.1 Small sets

This section outlines five basic types of small sets. These classes of sets will be closed under translation, countable unions and inclusion.

A. Countable sets. This is a familiar notion of smallness, and we have already seen that when the domain of a convex function is an open interval in $\mathbb{R}$, there are at most

4.6 Small sets and differentiability

countably many points of nondifferentiability (Theorem 2.1.2(d)). However, by considering the function $f(x,y) = |x|$, this result is no longer valid in $\mathbb{R}^2$. Therefore, the notion of countability is too small for differentiability results concerning continuous convex function beyond $\mathbb{R}$.

B. First category sets. This is a very powerful and widely used notion of smallness. Let X be metric space. A subset S of X is *nowhere dense* if $\overline{S}$ has empty interior. A subset F of X is of *first category* if it is a union of countably many nowhere dense sets. A set is said to be of the *second category* if it is not of the first category. The space X is called a *Baire space* if for any set F of first category, the complement $T \setminus F$ is everywhere dense. The complement of a set of first category is said to be *residual* or *fat*. A first-category set is sometimes also said to be *meager* or *thin*. A set containing a dense G_δ is said to be *generic*.

On finite-dimensional spaces, the set of points of differentiability of a convex function is a dense G_δ (Corollary 2.5.2), and in infinite dimensions the set of points of Fréchet differentiability of a continuous convex function is a possibly empty G_δ-set (Proposition 4.2.9). Moreover, there are infinite-dimensional spaces where even a norm can be nowhere Fréchet or indeed Gâteaux differentiable (Exercise 4.6.7). Nevertheless, there are wide classes of spaces where the points of nondifferentiability of a continuous convex function are of the first category.

In addition to its usefulness in questions of Fréchet differentiability of convex functions, another attractive aspect of the notion of category is that the dimension of the Banach space is not an issue in its definition. However, on the negative side, recall that Remark 2.5.5 shows the points of a nondifferentiability of a Lipschitz function on $\mathbb{R}$ may be of the second category and as a consequence Exercise 2.6.13 shows the points where a continuous convex function on the real line fails to be twice differentiable can be a set of the second category. Hence another notion of smallness is needed in this venue.

C. Null sets. We have already seen the almost everywhere differentiability of continuous convex functions on $\mathbb{R}^n$ (Theorem 2.5.1) and its more difficult extension of Rademacher's theorem (2.5.4) for locally Lipschitz functions; additionally, Alexandrov's theorem (2.6.4) asserts the continuous convex functions on $\mathbb{R}^n$ have second-order Taylor expansions almost everywhere. In infinite-dimensional spaces, care must be exercised in defining classes of null sets because there is no analog of the *Lebesgue measure* on an infinite-dimensional Banach space. However, classes of *null sets* can be defined to possess the following properties:

(N1) They are closed under inclusion, translation and countable unions.
(N2) No nonempty open set is null.
(N3) They coincide with the class of Lebesgue null sets in finite-dimensional spaces.

By a *Borel measure* on a topological space X we mean any measure defined on $\mathcal{B}(X)$ the Borel subsets of X, which is the σ-algebra generated by the open sets. By a *Radon measure* μ on X we mean any Borel measure on X which satisfies: (i) $\mu(K) < \infty$ for each compact $K \subset X$; (ii) $\mu(A) = \sup\{\mu(K) : K \subset A, K \text{ compact}\}$ for each

$A \in \mathcal{B}(X)$. We will say that a Borel subset N of a Banach space X is *Haar null* if there exists a (not necessarily unique) Radon probability measure p on X such that

$$p(x + N) = 0 \quad \text{for each } x \in X.$$

In such a case, we shall call the measure p a *test-measure* for N. More generally we say that an arbitrary subset $N \subset X$ is *Haar-null* if it is contained in a Haar-null Borel set.

The following lists some of the basic properties of Haar-null sets.

Proposition 4.6.1. *Let X be a Banach space. Then the following are true.*

(a) *Every subset of a Haar-null set in X is Haar-null.*
(b) *If A is Haar-null, so is $x + A$ for every $x \in X$.*
(c) *If A is Haar-null, then there exists a test-measure for A with compact support.*
(d) *If A is Haar-null, then $X \setminus A$ is dense in X.*
(e) *If $\{A_n\}_{n \in \mathbb{N}}$ are Haar-null sets, then so is $\bigcup_{n=1}^{\infty} A_n$.*

Proof. The proofs (a), (b) and (c) are left as an exercise. To prove (d) it is sufficient to show that there are no nonempty open Haar-null sets. For this, let U be a nonempty open subset of X and suppose by way of contradiction that there exists a test measure p for U on X. Let A denote the support of p. As A is separable, for some $x_0 \in X$, $(x_0 + U) \cap A \neq \emptyset$. Thus, $p(x_0 + U) \geq p((x_0 + U) \cap A) > 0$ which contradicts the fact that p is a test measure for U.

(e) First observe that without loss of generality we may assume that each set A_j is Borel. For each $j \in \mathbb{N}$, let p_j be a test measure for A_j on G. Let H be the smallest closed subspace of X that contains the support of each p_j. Since the support of each p_j is separable (see [105, Theorem 2.1(a)]), it is not too difficult to see that H is separable. Next let p_j^* denote the restriction of p_j to H. It follows from [152, Theorem 1] that there exists a Radon probability measure p^* on H that is a test measure for each set of the form $\bigcup\{B_j : j \in \mathbb{N}\}$, provided that $B_j \in \mathcal{B}(H)$ and p_j^* is a test measure for B_j. Let p be the extension of p^* to X. We claim that p is a test measure for $\bigcup\{A_j : j \in \mathbb{N}\}$. To prove this, we must show that for each $x \in X$, $p(x + \bigcup\{A_j : j \in \mathbb{N}\}) = 0$. Now fix $x \in X$. Then

$$p\left(x + \bigcup_{j=1}^{\infty} A_j\right) = p^*\left(\left(x + \bigcup_{j=1}^{\infty} A_j\right) \cap H\right) = p^*\left(\bigcup_{j=1}^{\infty}(x + A_j) \cap H\right).$$

However, each p_j^* is a test measure for $(x+A_j) \cap H$ since p_j is a test measure for A_j and $h + ((x+A_j) \cap H) = ((h+x) + A_j) \cap H \subset (h+x) + A_j$ for each $h \in H$. Therefore, p^* is a test measure for $\bigcup\{(x+A_j) \cap H : j \in \mathbb{N}\}$, and so $p^*(\bigcup\{(x+A_j) \cap H : j \in \mathbb{N}\}) = 0$, which implies that $p(x + \bigcup\{A_j : j \in \mathbb{N}\}) = 0$. □

Proposition 4.6.2. *Suppose C is a closed convex set and $d \in X \setminus \{0\}$ are such that C contains no line segment in the direction d. Then C is Haar null.*

Proof. Indeed, $\mu(S) := \lambda\{t \in [0,1] : td \in S\}$ where λ is the Lebesgue measure on $\mathbb{R}$ defines a Borel probability measure μ such that $\mu(x + C) = 0$ for all $x \in X$. □

From this, it follows that compact sets in infinite-dimensional Banach spaces are Haar null.

Another important class of *negligible* sets (the term is used in various contexts) is defined follows. For each $y \in X \setminus \{0\}$, let $\mathcal{A}(y)$ denote the family of all Borel sets A in X which intersect each line parallel to y in a set of one-dimensional Lebesgue measure 0. If $\{x_n\}$ is an at most countable collection in $Z \setminus \{0\}$, we let $\mathcal{A}(\{x_n\})$ denote the collection of all Borel sets A that can be written as $A = \bigcup A_n$, where $A_n \in \mathcal{A}(x_n)$ for every n. A set is *Aronszajn null* if for each nonzero sequence $\{x_n\}$ whose linear span is dense in X, A can be decomposed into a union of Borel sets $\{A_n\}$ such that $A_n \in \mathcal{A}(x_n)$ for every n.

It is known that an Aronszajn null set is Haar null, but there are Haar null sets that are not Aronszajn null. In fact, there is another well known class, the *Gaussian null* sets, which we will not define here, and moreover, they are known to coincide with the Aronszajn null sets; see [52] for further information.

D. Porosity. A subset S of a Banach space X is called *porous* if there is a number $\lambda \in (0, 1)$ such that for every $x \in S$ and every $\delta > 0$ there is a $y \in X$ such that $0 < \|y - x\| < \delta$ and $S \cap B_r(y) = \emptyset$ where $r = \lambda\|y - x\|$. If S is a countable union of porous sets, then we will say that S is σ-*porous*. The complement of a σ-porous set is said to be a *staunch set*.

E. Angle small sets. Let X be a Banach space, for $x^* \in X^*$ and $\alpha \in (0, 1)$ consider the cone
$$K(x^*, \alpha) := \{x \in X : \alpha\|x\|\|x^*\| \leq \langle x^*, x \rangle\}.$$

For a fixed $\alpha \in (0, 1)$, a subset S of X is said to be α-*cone meager* if for every $x \in S$ and $\varepsilon > 0$ there exists $z \in B_\varepsilon(x)$ and $x^* \in X^* \setminus \{0\}$ such that
$$S \cap [z + \text{int } K(x^*, \alpha)] = \emptyset.$$

The set S is said to be *angle small* if for every $\alpha \in (0, 1)$ it can be expressed as a countable union of α-cone meager sets.

It is easy to see from the definition that both σ-porous and angle small sets are of the first category.

4.6.2 First-order differentiability

A large amount of study has been directed at determining the extent to which Theorem 2.5.1 on the almost everywhere differentiability of convex functions on $\mathbb{R}^n$, Rademacher's theorem (2.5.4) and Alexandrov's theorem (2.6.4) extend to more general Banach spaces. Because these concepts may involve mappings between Banach spaces, we include the following definitions. Let X and Y be Banach spaces, and let U be an open subset of X. A function $f : U \to Y$ is said to be Fréchet differentiable at

$\bar{x} \in U$ if given $\varepsilon > 0$, there exists $\delta > 0$ and a continuous linear mapping $T : X \to Y$ so that

$$\|f(\bar{x} + h) - f(\bar{x}) - Th\| \leq \varepsilon \|h\| \quad \text{whenever } \|h\| < \delta.$$

If the limit exists directionally, that is for each $\varepsilon > 0$ and $h \in S_X$, there exists $\delta > 0$ (depending on ε and h) so that

$$\|f(\bar{x} + th) - f(\bar{x}) - T(th)\| \leq \varepsilon t \quad \text{whenever } 0 < t < \delta,$$

then f is said to be Gâteaux differentiable at $\bar{x}$. In both cases, we will write T as $f'(x)$.

We begin with a classical result of Mazur.

Theorem 4.6.3 (Mazur [311]). *Suppose X is a separable Banach space and f is a continuous convex function defined on an open convex subset U of X. Then the set of points at which f is Gâteaux differentiable is a dense G_δ-subset of U.*

Proof. We provide a sketch of the proof, which like Theorem 2.5.1 reduces to a one-dimensional argument. First, let $\{x_n\}_{n\in\mathbb{N}}$ be dense in S_X. For each $n, m \in \mathbb{N}$, let

$$A_{n,m} := \{x \in U : \text{ there exist } x^*, y^* \in \partial f(x) \text{ with } \langle x^* - y^*, x \rangle \geq 1/m\}.$$

Because f is Gâteaux differentiable at x if and only if $\partial f(x)$ is a singleton (Corollary 4.2.5), it follows that f fails to be Gâteaux differentiable at x if and only if $x \notin \bigcup_{n,m} A_{n,m}$. Because X it separable, it follows that bounded subsets of X^* are metrizable in the weak*-topology, this along with Alaoglu's theorem (4.1.6) can be used to show $A_{n,m}$ is closed. One can then use Theorem 2.1.2(d) to show that $U \setminus A_{n,m}$ is dense in U. □

Remark 4.6.4 (Extensions of Mazur's theorem). (a) Mazur's theorem has numerous extensions. A Banach space X is called a *weak Asplund space* if every continuous convex function defined on a nonempty convex open subset of X is Gâteaux differentiable on a dense G_δ-subset of its domain. It was shown in [355] a Banach space with a Gâteaux differentiable norm is a weak Asplund space (a weaker result is given in Exercise 4.6.6); in particular, weakly compactly generated (WCG) Banach spaces are weak Asplund spaces; see [180] for renorming of WCG spaces.

(b) Using a smooth variational principle, it is easy to check that a Banach space with a Gâteaux differentiable norm is a *Gâteaux differentiability space*, that is, every continuous convex function defined on a nonempty convex open subset of is Gâteaux differentiable on a dense subset of its domain. However, the process of obtaining a dense G_δ-subset as in (a) is nontrivial and often uses a topological result called a *Banach–Mazur game*; see [355] and [350, p. 69ff]. Further, Moors and Somasundaram [322] show that a Gâteaux differentiability space need not be a weak Asplund space, thus solving a long-standing open question. However, it had been known earlier that in contrast to the Fréchet differentiability case, a continuous convex function could be Gâteaux differentiable on a dense set that

is not residual [158]; further information on examples like this is provided in Exercise 6.1.10.
(c) The classical example that the usual norm on ℓ_1 is nowhere Fréchet differentiable (see Exercise 4.6.7(a)) shows that the Fréchet analog of Mazur's theorem is not valid.
(d) Mazur's theorem does not extend to ℓ_∞; see Exercise 4.6.7(b).

Next, we state a Gâteaux differentiability version of Rademacher's theorem (2.5.4) for separable Banach spaces. This result uses the concept of a Banach space with the *Radon–Nikodým property* (we will study some properties of this class of spaces in Section 6.6); for now, let us mention that this class includes reflexive Banach spaces and separable dual spaces.

Theorem 4.6.5. *Let X be a separable Banach space and let Y be a space with the Radon–Nikodým property. Let f be a Lipschitz function from an open subset U of X into Y. Then the set of points in U at which f is not Gâteaux differentiable is Aronszajn null.*

A Banach space X is said to be an *Asplund space* if the points of differentiability of every continuous convex function $f : X \to \mathbb{R}$ are a dense (automatically G_δ) subset of X; see Exercise 4.6.11 for a common alternate formulation of the definition of Asplund spaces. As noted in Remark 4.6.4(d), not every separable Banach space is an Asplund space, however, those spaces with separable duals are Asplund spaces:

Theorem 4.6.6. *Suppose that the Banach space X has a separable dual and f is a continuous convex function on X. Then the points at which f fails to be Fréchet differentiable on X is an angle small set.*

Proof. According to Šmulian's theorem (4.2.10), it suffices to show that the set

$$A := \{x \in \mathrm{dom}(\partial f) : \lim_{\delta \to 0^+} \mathrm{diam}\, \partial f[B(x,\delta)] > 0\}$$

is angle small. First, we write $A = \bigcup_{n \in \mathbb{N}} A_n$ where

$$A_n := \{x \in \mathrm{dom}(\partial f) : \lim_{\delta \to 0^+} \mathrm{diam}\, \partial f[B(x,\delta)] > 1/n\}.$$

Let $\{x_k^*\}_{k=1}^\infty$ be a dense subset in X^* and suppose $0 < \alpha < 1$. Now let

$$A_{n,k} := \{x \in A_n : \mathrm{dist}(x_k^*, \partial f(x)) < \alpha/4n\}.$$

To complete the proof, it suffices to show that each $A_{n,k}$ is α-cone meager. For this, suppose that $x \in A_{n,k}$ and $\varepsilon > 0$. Because $x \in A_n$, there exist $0 < \delta < \varepsilon$ and $z_1, z_2 \in B(x,\delta)$ and $z_i^* \in \partial f(z_i)$ so that $\|z_1^* - z_2^*\| > 1/n$. Consequently, for any $x^* \in \partial f(x)$, one of $\|z_i^* - x^*\| > 1/2n$. Because $\mathrm{dist}(x_k^*, \partial f(x)) < \alpha/4n$, we can choose $x^* \in \partial f(x)$ such that $\|x_k^* - x^*\| < \alpha/4n$. From this, one sees that there are points $z \in B(x, \varepsilon)$ and $z^* \in \partial f(z)$ such that

$$\|z^* - x_k^*\| \geq \|z^* - x^*\| - \|x_k^* - x^*\| > 1/2n - \alpha/4n > 1/4n.$$

Next we show that $A_{n,k} \cap (z + \text{int } K(z^* - x_k^*, \alpha)) = \emptyset$. That is,

$$A_{n,k} \cap \{y \in X : \langle z^* - x_k^*, y - z \rangle > \alpha \|z^* - x_k^*\| \|y - z\|\} = \emptyset.$$

Now, if $y \in \text{dom}(\partial f)$ and $\langle z^* - x_k^*, y - z \rangle > \alpha \|z^* - x_k^*\| \|y - z\|$ and if $y^* \in \partial f(y)$, then

$$\begin{aligned}
\langle y^* - x_k^*, y - z \rangle &= \langle y^* - z^*, y - z \rangle + \langle z^* - x_k^*, y - z \rangle \\
&\geq \langle z^* - x_k^*, y - z \rangle > \alpha \|z^* - x_k^*\| \|y - z\| \\
&\geq \frac{\alpha}{4n} \|y - z\|.
\end{aligned}$$

Consequently, $\|y^* - x_k^*\| \geq \alpha/4n$ and so $y \notin A_{n,k}$. □

It will be shown in Corollary 6.6.10 that a Banach space X is an Asplund space if and only if each of its separable subspaces has a separable dual. The case of Fréchet differentiability of locally Lipschitz functions is highly nontrivial, we state the following result of Preiss.

Theorem 4.6.7 (Preiss [354]). *Each locally Lipschitz real-valued function on an Asplund space is Fréchet differentiable at the points of a dense set.*

However, there is no guarantee that the set of points where the function is not differentiable is small in any of the senses we have discussed. Indeed, a Lipschitz function may fail to be differentiable at a set of second category (Remark 2.5.5). Next, we mention examples on various Asplund spaces showing that the points of nondifferentiability need not be a null set in the various notions introduced here. The following is from [107].

Example 4.6.8. Let $C = c_0^+$, i.e. $C = \{(x_i) \in c_0 : x_i \geq 0 \text{ for all } i \in \mathbb{N}\}$. Then C is not Haar null. Moreover, C has empty interior, so d_C is a Lipschitz convex function that fails to be Fréchet differentiable at any $x \in C$.

Proof. This is outlined in Exercise 4.6.10. □

Using the following theorem from [309] whose proof we omit, one can create examples of the previous sort on any nonreflexive (Asplund) space.

Theorem 4.6.9. *A separable Banach space is not reflexive if and only if there is a closed convex set C with empty interior in X which contains a translate of any compact set in E. In particular, C is not Haar null.*

The next example sharpens the previous observations by showing that the points of nondifferentiability may fail to lie in the more restrictive class of Aronszajn null sets.

Example 4.6.10. On any separable space there is a Lipschitz convex function whose points of nondifferentiability is not Aronszajn null.

Proof. (Outline) Let X be a separable Banach space and let μ be a finite Borel measure on X. Because μ is regular, there exists an increasing sequence of compact convex sets (C_n) such that $\mu(X \setminus \bigcup_{n=1}^{\infty} C_n) = 0$. Now let $f_n := d_{C_n}$, and let $f := \sum_{n=1}^{\infty} 2^{-n} f_n$. Then f is a Lipschitz convex function and fails to be Fréchet differentiable at the points of $\bigcup_{n=1}^{\infty} C_n$ by Exercise 4.6.8. See [107, p. 53] for a proof that this set is not Aronszajn null. □

Finally, we state without proof a more striking example showing the points of differentiability of a convex function can be small, even on a separable Hilbert space. The reader is referred to [52, Example 6.46] for the construction and proof of this example in the ℓ_2 case.

Example 4.6.11. There is a continuous convex function $f : \ell_2 \to \mathbb{R}$ such that set in which f is Fréchet differentiable is an Aronszajn null set. In fact, there is such a norm on any separable superreflexive space.

We mention one final result concerning sets where a continuous convex function on a separable Banach space can fail to be differentiable. For this, a subset S of X is said to be a *hypersurface* if it is the graph of a function defined on a hyperplane of X which is the difference of two Lipschitz convex functions. The following theorem is from Zajíček's work [442]; see also [52, Theorem 4.20].

Theorem 4.6.12. *Let D be a subset of a separable Banach space X. Then there exists a continuous convex function on X which is nowhere differentiable on D if and only if D is contained in a countable union of hypersurfaces.*

4.6.3 Second-order differentiability on Banach spaces

We will say a real-valued function $f : U \to \mathbb{R}$ is twice Gâteaux differentiable (resp. twice Fréchet differentiable) at $\bar{x} \in U$ if there is an open neighborhood V of $\bar{x}$ such that f is Gâteaux differentiable (resp. Fréchet differentiable) on V, and the mapping $f'(x) : V \to X^*$ is Gâteaux differentiable (resp. Fréchet differentiable). As in the Euclidean space case, we denote the second derivative of f at $\bar{x}$ by $f''(\bar{x})$ or $\nabla^2 f(\bar{x})$, and it is a continuous linear mapping from X to X^*, that is $\nabla^2 f(\bar{x}) \in \mathcal{L}(X; \mathcal{L}(X; \mathbb{R}))$. The definitions above for both the Fréchet and Gâteaux cases, imply the following limit exists in the norm sense for each $h \in X$

$$\nabla^2 f(\bar{x})(h) = \lim_{t \to 0} \frac{\nabla f(\bar{x} + th) - \nabla f(\bar{x})}{t}.$$

If the above limit is only known to exist in the weak*-sense, we will say that f has a *weak*-second-order Gâteaux* (or *Fréchet*) *derivative* at $\bar{x}$. In any of the above cases, we may let $f''(\bar{x})(h, k) := \langle h, \nabla^2 f(\bar{x})(k) \rangle$. Thus, $f''(\bar{x})$ identifies with a continuous bilinear mapping from $X \times X \to \mathbb{R}$, that is $f''(\bar{x}) \in \mathcal{L}_2(X; \mathbb{R})$; see [144, Chapter 1, §1.9] for a formal development of the natural isometry of $\mathcal{L}_2(X; \mathbb{R})$ with

$\mathcal{L}(X; \mathcal{L}(X; \mathbb{R}))$ and more generally. With this notation, and the definitions just given, one can check that

$$f''(\bar{x})(h,k) = \lim_{t \to 0} \frac{\langle \nabla f(\bar{x}+tk), h\rangle - \langle \nabla f(\bar{x}), h\rangle}{t},$$

when $f''(\bar{x})$ exists in any four senses above.

We leave it as Exercise 4.6.13 for the reader to verify that $\nabla^2 f(\bar{x})$ is symmetric in each of the above cases. As in the finite-dimensional case, the second derivative can be used to check convexity.

Theorem 4.6.13 (Second-order derivatives and convex functions). *Let U be an open convex subset of a Banach space X, and suppose that $f : U \to \mathbb{R}$ has a weak*-second-order Gâteaux derivative at each $x \in U$. Then f is convex if and only if $\nabla^2 f(\bar{x})(h,h) \geq 0$ for all $\bar{x} \in U$, and $h \in X$. If $\nabla^2 f(\bar{x})(h,h) > 0$ for all $\bar{x} \in U$, $h \in X$, then f is strictly convex.*

Proof. It is left for the reader to modify the finite-dimensional version (Theorem 2.2.8) to the Banach space setting. □

We now turn our attention to the generalized second derivatives and Taylor expansions in Banach spaces.

Let X be a Banach space, U a nonempty open convex set and $f : U \to \mathbb{R}$ be a continuous convex function. Then f is said to have a *weak second-order Taylor expansion* at $x \in U$ if there exists $y^* \in \partial f(x)$ and a continuous linear operator $A : X \to X^*$ such that f has a representation of the form

$$f(x+th) = f(x) + t\langle y^*, h\rangle + \frac{t^2}{2}\langle Ah, h\rangle + o(t^2) \quad (t \to 0), \tag{4.6.1}$$

for each $h \in X$; if more arduously,

$$f(x+h) = f(x) + \langle y^*, h\rangle + \frac{1}{2}\langle Ah, h\rangle + o(\|h\|^2) \quad (\|h\| \to 0) \tag{4.6.2}$$

for $h \in X$, we say that f has a *strong second-order Taylor expansion* at x.

As in the finite-dimensional case, Taylor expansions will be related to generalized second derivatives. However, the notions of strong and weak Taylor expansions are different in the infinite-dimensional setting:

Example 4.6.14. *Strong and weak second-order Taylor expansion are distinct concepts in infinite-dimensional spaces.*

Proof. See [107, Example 1, p. 50]. □

Let X be a Banach space, U a nonempty open convex set and $f : U \to \mathbb{R}$ be a continuous convex function. Then ∂f is said to have:

4.6 Small sets and differentiability

(a) a generalized *weak*-Gâteaux derivative* (resp. *weak*-Fréchet derivative*) at $x \in U$ if there exists a bounded linear operator $T : X \to X^*$ such that

$$\lim_{t \to 0} \frac{y_t^* - y^*}{t} = Th$$

in the weak*-sense for any fixed $h \in X$ and all $y_t^* \in \partial f(x+th), y^* \in \partial f(x)$ (resp. uniformly for all $h \in B_X$, $y_t^* \in \partial f(x+th)$ and $y^* \in \partial f(x)$);

(b) a generalized *Gâteaux derivative* (resp. *Fréchet derivative*) at x if (a) holds with the respectively limits in the sense of the dual norm.

Given these notions, [107, Theorem 3.1] establishes the following result.

Theorem 4.6.15. *Let X be a Banach space, U a nonempty open convex set, $f : U \to \mathbb{R}$ be a continuous convex function and $x \in U$. Then:*

(a) f has a weak second-order Taylor expansion at x if and only if ∂f has a generalized weak*-Gâteaux derivative at x.

(b) f has a strong second-order Taylor expansion at x if and only if ∂f has a generalized Fréchet derivative at x.

In both cases the linear mapping $T : X \to X^*$ is the same.

Next we examine whether there are infinite-dimensional spaces on which continuous convex functions have generalized second-order derivatives, except at possibly a *small set*; we already saw in Exercise 2.6.13 that even on $\mathbb{R}$ a continuous convex function may have second-order derivatives at only a set of first category. In other words, does Alexandrov's theorem (2.6.4) work in infinite-dimensional Banach spaces for any of the notions of null sets introduced here? The following observation, along with examples for first-order Fréchet differentiability show that it does not.

Proposition 4.6.16. *Let U be a nonempty open convex subset of a Banach space X, $f : U \to \mathbb{R}$ be a continuous convex function, and $x \in U$. If f has a weak second-order Taylor expansion at x, then f is Fréchet differentiable at x.*

Proof. For each $h \in S_X$, there exists $C_h > 0$, $\delta_h > 0$ so that

$$f(x+th) - f(x) - \langle \phi, th \rangle \le C_h t^2, \quad 0 \le t \le \delta_h.$$

Now let $F_{n,m} := \{h \in B_X : f(x+th) - f(x) - \langle \phi, x \rangle \le nt^2, 0 \le t \le 1/m\}$. Then $\bigcup F_m = B_X$. According to the Baire category theorem, there exists an open $U \subset B_X$, and $n_0, m_0 \in \mathbb{N}$ so that

$$f(x+th) - f(x) - \langle \phi, th \rangle \le n_0 t^2, \quad \text{whenever } 0 \le t \le 1/m_0, \ h \in U.$$

Now use the convexity of f to show there exists $\delta > 0$, and $K > 0$ so that

$$f(x+th) - f(x) - \langle \phi, th \rangle \leq Kt^2, \text{ whenever } 0 \leq t \leq \delta, \ h \in B_X. \qquad (4.6.3)$$

In particular, this shows f is Fréchet differentiable at x. □

In fact, (4.6.3) is a property stronger than the Fréchet differentiability of f at x_0. Indeed, it is a condition that is equivalent to f being what is called *Lipschitz smooth* at x_0 (see Exercise 4.6.14), that is there exist $\delta > 0$ and $C > 0$ so that

$$\|\phi_x - \nabla f(x_0)\| \leq C\|x - x_0\| \text{ whenever } \|x - x_0\| < \delta, \text{ and } \phi_x \in \partial f(x). \qquad (4.6.4)$$

Lipschitz smoothness and related notions will be discussed in more detail in Section 5.5. For now, our main interest in the previous proposition is that a point where a continuous convex function has a generalized second-order derivative, is automatically a point of Fréchet differentiability. Consequently, we have the following:

Remark 4.6.17 (Failure of Alexandrov's theorem in infinite dimensions). (a) Let $C := \{f \in L_2[0,1] : |f| \leq 1 \text{ a.e.}\}$. According to [216, Section 5], the nearest point mapping $P_C : H \to C$ is nowhere Fréchet differentiable. Moreover, P_C is the Fréchet derivative of the convex function

$$f(x) = \frac{1}{2}\|x\|^2 - \frac{1}{2}\|x - P_Cx\|^2; \qquad (4.6.5)$$

see [216] for details. Consequently, the function $f : L_2[0,1] \to \mathbb{R}$ is nowhere twice Fréchet differentiable.
(b) Let C be the positive cone in $\ell_2(\Gamma)$ where $|\Gamma| > \aleph_0$. Then the nearest point mapping P_C is nowhere Gâteaux differentiable, therefore, $f : \ell_2(\Gamma) \to \mathbb{R}$ as in (4.6.5) is nowhere twice Gâteaux differentiable.
(c) Let $C = c_0^+$, then d_C fails to have a weak second-order Taylor expansion at each point of the set C which is not Haar null; see Example 4.6.8.
(d) In any separable nonreflexive Banach space, there is a closed convex set C that is not Haar null. Thus d_C fails to have a weak second-order Taylor expansion at each $x \in C$; see Theorem 4.6.9.
(e) On any separable Banach space, there is a Lipschitz convex function that fails to have a second-order Taylor expansion at each point of a set that is not Aronszajn null set; see Example 4.6.10.
(f) There is a continuous convex function $f : \ell_2 \to \mathbb{R}$ such that the points where f has a weak second-order Taylor expansion is Aronszajn null; see Example 4.6.11. In fact, such an example exists on any separable superreflexive space. In particular, the set of points where f fails to have a weak second-order Taylor expansion is not Haar null.

Exercises and further results

4.6.1. Give an example of a set in $[0,1]$ that is first category but has measure one.

Hint. Find open sets O_n containing the rationals with measure less than $1/n$. Then look at the complement of $G := \bigcap O_n$. □

4.6.2.* Prove Proposition 4.6.1 (a), (b) and (c).

4.6.3 (Angle small sets). (a) Show that any angle small set is of first category. (b) Show that a subset of $\mathbb{R}$ is angle small if and only if it is countable.

Hint. For (b) show that an α-cone meager subset of $\mathbb{R}$ can contain at most two points. □

4.6.4 (Porosity). (a) Show that the Cantor set is porous, and conclude that there are porous sets that are not angle small. (b) Is an angle small set always σ-porous. (c) Show that a σ-porous set in Euclidean space has outer measure zero.

Hint. See [443] for more information on porous and σ-porous sets. □

4.6.5.* Fill in the details to the proof of Mazur's theorem.

Hint. See [350, p. 12] for the full details. □

4.6.6 (Generalizing Mazur). Show that if a Banach space X has a norm whose dual norm is strictly convex, then every continuous convex function on X is generically Gâteaux differentiable. Note: a norm is strictly convex if $\|\frac{x+y}{2}\| < \frac{1}{2}\|x\| + \frac{1}{2}\|y\|$ whenever $\|x\| = \|y\| > 0$, and it is easy to renorm a separable Banach space so its dual norm is strictly convex (Proposition 5.1.11).

Hint. Suppose f is a continuous convex function, and define $\psi(x) := \inf\{\|x^*\| : x^* \in \partial f(x)\}$ where $\|\cdot\|$ is a strictly convex dual norm. Show that ψ is lsc and then conclude that ψ is continuous at each point of a residual subset, say G. Use the strict convexity of the dual norm to show that at any x there is at most one $x^* \in \phi(x)$ with $\|x^*\| = \psi(x)$. Now suppose $x \in G$ and $\partial f(x)$ is not a singleton, then there exists $y^* \in \partial f(x)$ with $\|y^*\| > \psi(x)$. Choose h with $\|h\| = 1$ so that $y^*(h) > \psi(x)$. Now let $x_n := x + \frac{1}{n}y$. Use the subdifferential inequality to show that $\psi(x_n) \geq y^*(h)$ for all n. This contradicts the continuity of ψ at x. □

4.6.7.* (a) Show the usual norm $\|\cdot\|_1$ is nowhere Fréchet differentiable on ℓ_1. Conclude that ℓ_1 has no equivalent Fréchet differentiable norm.
(b) Show that $f(x) := \limsup_{n\to\infty} |x_n|$ is nowhere Gâteaux differentiable on ℓ_∞. Conclude that there is a norm on ℓ_∞ that is nowhere Gâteaux differentiable.

Hint. (a) Let $x := (x_n) \in \ell_1$. Let $\delta_n \to 0^+$ where $\delta_n \geq |x_n|$. Choose points $h^n \in \ell_1$ where $h^n = 3\delta_n e_n$. Then $\|h^n\| \to 0$, but

$$\|x + h^n\|_1 + \|x - h^n\|_1 - 2\|x\|_1 \geq 4\delta_n > \|h_n\|$$

and so $\|\cdot\|_1$ is not Fréchet differentiable at x. Use a smooth variational principle to conclude there is no equivalent Fréchet differentiable norm on ℓ_1. □

4.6.8.* Show that the function f in the proof of Example 4.6.10 fails to be Fréchet differentiable at each point of $\bigcup C_n$. Can the same be said about the Gâteaux differentiability of f?

Hint. First d_{C_n} is not Fréchet differentiable at any point of C_n; see Exercise 4.2.6. Then the same is true of the sum of kd_{C_n} for any $k > 0$ with any convex function. For

4.6.9. Let C be the positive cone in c_0. For any compact set $K \subset c_0$, show that a translate of K is contained in C. Hence this is a concrete case of Theorem 4.6.9. (See also Exercise 4.6.16.)

4.6.10.★ Verify Example 4.6.8.

Hint. We follow [107, Example 2, p. 53]. Use the regularity of Borel measures to deduce that c_0^+ is not Haar null because Exercise 4.6.9 shows it contains a translate of each compact subset of c_0. It is easy to show that c_0^+ has empty interior. Then apply Exercise 4.2.6 to deduce the distance function $d_{c_0^+}$ is not Fréchet differentiable at any points of c_0^+. □

4.6.11.★ Show that X is an Asplund space if and only if for every open convex set $U \subset X$, and every continuous convex function $f : U \to \mathbb{R}$, f is Fréchet differentiable on a dense G_δ-subset of U.

Hint. Use local Lipschitz property of f, i.e. given $x \in U$, there exists $V \subset U$ such that $x \in V$, and $f|_V$ is Lipschitz. Now extend f to a Lipschitz convex function $\tilde{f}$ on X that agrees with f on V. Use fact that X is Asplund to deduce that $\tilde{f}$ is Fréchet differentiable at some point in V. Then note that points of Fréchet differentiability are always G_δ. □

4.6.12 (Locally residual implies residual). Recall that a subset R is *residual* if it contains as a subset the intersection of countably many dense open sets. Show that a subset R of X is residual if, and only if, it is *locally residual* (i.e. for each $x \in X$ there exists an open neighbourhood U of x such that $R \cap U$ is residual in U). Actually something less than this is sufficient: it suffices to be be locally residual at the points of a dense subset of X.

Hint. See, for example [201, Proposition 1.3.2, p. 16]. □

4.6.13.★ Let X be a Banach space, U an open convex subset of X and $f : U \to \mathbb{R}$ convex. Suppose f has a weak* second-order Gâteaux derivative at $\bar{x} \in U$. Show that $f''(\bar{x})$ is symmetric.

Hint. For any $h, k \in X$, let $g(s, t) = f(\bar{x} + sh + tk)$. By the symmetry result in finite dimensions (Theorem 2.6.1), deduce $g_{st}(0) = g_{ts}(0)$, and conclude that $f''(\bar{x})(h, k) = f''(\bar{x})(k, h)$. □

4.6.14 (Lipschitz smooth points). Show that the condition in (4.6.3) is equivalent to the defining condition (4.6.4) for a point of Lipschitz smoothness.

Hint. See Theorem 5.5.3 and its hint for a more general result. □

4.6.15.★★ Prove Theorem 4.6.15.

Hint. Full details can be found in [107, Theorem 3.1]. Can the equivalence in finite dimensions be used to produce another proof, at least for some parts of the theorem?
□

4.6.16 (Further results on boundaries of closed convex sets).★★ We collect some striking results relating to Theorem 4.6.9.

(a) It is shown in [309] that a separable Banach space X is nonreflexive if and only if there is a convex closed subset Q of X with empty interior which contains translates of all compact sets $K \subset X$ (Q may also be required to be bounded and K a compact subset of the unit ball).

(b) Consider the space $C(K)$ of continuous functions on a compact set K containing a nonisolated point p in K. Then Q can be the subset of K whose members attain their minima at p. In particular the nonnegative cone in c_0 is not Haar null but has empty interior.

(c) Determine the differentiability properties of d_Q for such a convex Q.

(d) In [308] the following is shown. Let X be a reflexive Banach space, and let $C \subset X$ be a closed, convex and bounded set with empty interior. Then, for every $\delta > 0$, there is a nonempty finite set $F \subset X$ with an arbitrarily small diameter, such that C contains at most $\delta \cdot |F|$ points of any translation of F.

(e) Deduce that the boundary of a closed convex subset of a separable reflexive Banach space is Haar null if and only if it has empty interior.

Hint. All Borel measures on a separable space are *tight*:

$$\mu(C) = \sup\{\mu(K) : K \subset C, K \text{ compact}\}.$$

□

(f) Deduce that this property characterizes reflexive spaces amongst separable spaces.

4.6.17.** Suppose that X is a separable Banach space that admits a differentiable norm whose derivative is (locally) Lipschitz on the unit sphere. Show that X admits a twice Gâteaux differentiable norm whose first derivative is (locally) Lipschitz on the unit sphere.

Hint. This requires significant computations which we omit. Let $f_0(x) := \|x\|^2$, and then let

$$f_{n+1}(x) := \int_{\mathbb{R}^{n+1}} f_0\left(x - \sum_{i=0}^{n} t_i h_i\right) \phi_0(t_0)\phi_1(t_1)\cdots\phi_n(t_n)\, dt_0\, dt_1 \cdots dt_n$$

for $n = 0, 1, 2, \ldots$, where $\phi_0 : \mathbb{R} \to \mathbb{R}$ is a C^∞-function, $\phi_0 \geq 0$, ϕ_0 is even, $\phi_0(t) = 0$ if $|t| > 1/2$ and $\int_{-\infty}^{+\infty} \phi_0 = 1$; $\phi_n(t) = 2^n \phi_0(2^n t)$ for $t \in \mathbb{R}$ and $\{h_i\}_{n=1}^{\infty}$ is dense in the unit sphere of X. Then each f_n is a continuous convex function, and the crucial step is that (f_n) converges uniformly on bounded sets to a continuous convex function f that is twice Gâteaux differentiable with locally Lipschitz first derivative, in fact

$$f'(x)(h) = \lim_n \int_{\mathbb{R}^{n+1}} f_0'\left(x - \sum_{i=0}^{n} t_i h_i\right)(h) \cdot \Pi_{i=0}^{n}\phi_i(t_i)dt_0 \cdots dt_n \text{ for } x, h \in X.$$

and then show

$$f''(x)(h, h_i) = \lim_n \int_{\mathbb{R}^{n+1}} f_0'\left(x - \sum_{i=0}^n t_i h_i\right)(h)\phi_0(t_0) \cdots \phi_i'(t_i) \cdots \phi_n(t_n)\, dt_0\, dt_1 \cdots dt_n$$

Use the density of $\{h_i\}_{i=1}^\infty$ in S_X and the local Lipschitz property of f' to obtain $f''(x)(h, k)$ is a bounded symmetric bilinear map. Finally, an implicit function theorem is used on an appropriate lower level set of f to obtain the equivalent norm. See [203] for the full details.

Note: an analog of this exercise for *bump functions*, that is, functions with bounded nonempty support is given in [206, Theorem 6]. □

We conclude this section by stating some interesting related results.

Remark 4.6.18 (A staunch Rademacher theorem). Recall that the complement of a σ-porous set is said to be *staunch*. In [55] Bessis and Clarke recover Rademacher's theorem (2.5.4) as a byproduct of a study of the staunchness of sets on which the partial Dini-subdifferentials exist in $\mathbb{R}^n$ and produce the Dini-subdifferential of a Lipschitz function – and much more. In particular, they show that if the gradient exists on a staunch set then the points of Fréchet differentiability are staunch.

Remark 4.6.19 (Convex functions on small sets [84]). Let X be a Banach space, C a G_δ-subset of X, and $f : C \to \mathbb{R}$ a locally Lipschitz convex function. One may show that if X is a space of class (S) as defined in [84, 201] (respectively, an Asplund space), then f is Gâteaux (respectively, Fréchet) differentiable on a dense G_δ-subset of C. The results unify earlier ones which assume that C has nonsupport points or that C is not contained in a closed hyperplane. Indeed the nonsupport points $N(C)$ are a G_δ-subset of X when X is separable. Class (S) spaces include all WCG spaces and much more.

By contrast, in [306] it is shown that there is a convex locally Lipschitz function on a closed subset C of ℓ_2 which is Fréchet differentiable on a dense subset D of the nonsupport points $N(C)$ but every selection for ∂f is discontinuous at each point on D.

Remark 4.6.20 (Fréchet points of Lipschitz functions). In stark distinction to the behavior of convex functions, a very recent result Kurka [281] shows that the sets of Fréchet subdifferentiability of Lipschitz functions on a Banach space are always Borel sets if and only if the space is reflexive.

5
Duality between smoothness and strict convexity

[Banach] would spend most of his days in cafés, not only in the company of others but also by himself. He liked the noise and the music. They did not prevent him from concentrating and thinking. There were cases when, after the cafés closed for the night, he would walk over to the railway station where the cafeteria was open around the clock. There, over a glass of beer, he would think about his problems. (Andrzej Turowicz)[1]

5.1 Renorming: an overview

This section presents some classical results concerning norms on Banach spaces. We focus on some constructions on separable spaces as well as basic duality results for smoothness and convexity. Material in later sections concerning the duality between exposed points and smooth points, and between convexity and smoothness for convex functions is motivated by the corresponding duality in the classical case of norms. A familiarity with the more concrete case of norms can make the general convex function case much more transparent.

For a subset $A \subset X$, the *polar* of A denoted by A° is the subset of X^* defined by

$$A^\circ := \{x^* \in X^* : \langle a, x^* \rangle \leq 1 \text{ for all } a \in A\}.$$

Given a subset $B \subset X^*$, the *prepolar* of B denoted by $B_\circ$ is the subset of X defined by

$$B_\circ := \{x \in X : \langle x, b^* \rangle \leq 1 \text{ for all } b \in B\}.$$

The following is a basic fundamental result concerning polars of sets in Banach spaces.

Theorem 5.1.1 (Bipolar theorem). *Let X be a Banach space, and let $A \subset X$ and $B \subset X^*$. Then*

(a) $(A^\circ)_\circ$ *is the closed balanced convex hull of A.*
(b) $(B_\circ)^\circ$ *is the weak*-closed balanced convex hull of B.*

Proof. See Exercise 5.1.1. □

The bipolar theorem can be used to recognize dual norms.

---
[1] Turowicz, another Lvov mathematician, quoted in R. Kaluza, *The Life of Stefan Banach*, Boston, 1996.

Fact 5.1.2. *Let $\|\cdot\|$ be a norm on X^* that is equivalent to the dual norm of the original norm on X. Then the following are equivalent:*

(a) *$\|\cdot\|$ is a dual norm on X^*;*
(b) *$\|\cdot\|$ is weak*-lsc;*
(c) *the unit ball B of $\|\cdot\|$ is weak*-closed.*

Proof. (a) $\Rightarrow$ (b): Suppose $\|\cdot\|$ is a dual norm. Then $\|x^*\| = \sup\{\langle x, x^*\rangle : \|x\| \leq 1\}$. As a supremum of weak*-lsc functions, $\|\cdot\|$ is weak*-lsc.

(b) $\Rightarrow$ (c) is immediate since $B = \{x^* \in X^* : \|x^*\| \leq 1\}$ is the lower level set of a weak*-lsc function.

(c) $\Rightarrow$ (a): According to the bipolar theorem (5.1.1), $B = (B_\circ)^\circ$ and so $\|\cdot\|$ is the dual norm of the norm that is the gauge, $\gamma_{B_\circ}$. $\square$

Recall that Gâteaux and Fréchet differentiability for convex functions were introduced in Section 4.2. A norm $\|\cdot\|$ is said to be *Gâteaux differentiable* (resp. *Fréchet differentiable*) if $\|\cdot\|$ is Gâteaux differentiable (resp. Fréchet differentiable) as a function on $X \setminus \{0\}$. Because of the positive homogeneity of norms, this is equivalent to $\|\cdot\|$ being Gâteaux differentiable (resp. Fréchet differentiable) on S_X. Also, because derivatives of convex functions are subgradients, one can readily verify when a norm $\|\cdot\|$ is differentiable at $x \in X \setminus \{0\}$, its derivative will be the unique linear functional $\phi \in S_{X^*}$ satisfying $\phi(x) = \|x\|$.

Proposition 5.1.3 (Symmetric differentiability of norms). *Let $(X, \|\cdot\|)$ be a Banach space.*

(a) *The norm $\|\cdot\|$ is Fréchet differentiable at $x \in X \setminus \{0\}$ if and only if for each $\varepsilon > 0$, there exists $\delta > 0$ so that*

$$\|x + h\| + \|x - h\| - 2\|x\| \leq \varepsilon \|h\| \text{ whenever } \|h\| \leq \delta.$$

(b) *The norm $\|\cdot\|$ is Gâteaux differentiable at $x \in X \setminus \{0\}$ if and only if for each $\varepsilon > 0$ and $h \in S_X$, there exists $\delta > 0$ so that*

$$\|x + th\| + \|x - th\| - 2\|x\| \leq \varepsilon |t| \text{ whenever } |t| \leq \delta.$$

Proof. Part (a) is a restatement of Proposition 4.2.7, whilst part (b) is a restatement of Proposition 4.2.8. $\square$

Proposition 5.1.4 (Šmulian). *Let $(X, \|\cdot\|)$ be a normed space, and $x \in S_X$. Then the following are equivalent.*

(a) *$\|\cdot\|$ is Fréchet differentiable at x.*
(b) *$(\phi_n - \Lambda_n) \to 0$ whenever $\phi_n(x_n) \to 1$, $\Lambda_n(y_n) \to 1$ and $x_n \to x$, $y_n \to x$ and $\phi_n, \Lambda_n \in B_{X^*}$.*
(c) *$(\phi_n - \phi) \to 0$ whenever $\phi_n, \phi \in B_{X^*}$ and $\phi_n(x) \to \phi(x) = 1$.*

5.1 Renorming: an overview

Proof. (This is similar to convex function case in Theorem 4.2.10.) (a) ⇒ (c): Suppose (c) is not true. Let $\phi \in S_{X^*}$ be a supporting linear functional of x. Then there exist $\phi_n \in B_X$ such that $\phi_n(x) \to 1$, and $\varepsilon > 0$ and $h_n \in B_{X^*}$ such that $(\phi - \phi_n)(h_n) \geq \varepsilon$. Now choose $t_n \to 0^+$ such that $1 - \phi_n(x) \leq \frac{t_n \varepsilon}{2}$. Then

$$\|x + t_n h_n\| + \|x - t_n h_n\| - 2\|x\| \geq \phi_n(x + t_n h_n) + \phi(x - t_n h_n) - 2\|x\|$$
$$\geq \phi_n(x) + \phi(x) + t_n(\phi_n - \phi)(h_n) - 2$$
$$\geq t_n \varepsilon + \phi_n(x) - 1 \geq \frac{t_n \varepsilon}{2}.$$

Consequently, $\|\cdot\|$ is not Fréchet differentiable at x by Proposition 5.1.3(a).

(c) ⇒ (b): Let $\Lambda_n, \phi_n, x_n, y_n$ be as in (b). Choose a supporting functional $\phi \in S_{X^*}$ with $\phi(x) = 1$. Now $\phi_n(x) \to 1$ and $\Lambda_n(x) \to 1$ since $x_n \to x$ and $y_n \to x$. According to (c) $\phi_n \to \phi$ and $\Lambda_n \to \phi$ and so $(\phi_n - \Lambda_n) \to 0$ as desired.

(b) ⇒ (a): Suppose $\|\cdot\|$ is not Fréchet differentiable at x. Then for $\phi \in S_{X^*}$ with $\phi(x) = 1$, there exist $t_n \to 0^+$, $h_n \in S_X$ and $\varepsilon > 0$ such that

$$\|x + t_n h_n\| - \|x\| - \phi(t_n h_n) > \varepsilon t_n$$

for all n. Choose $\phi_n \in S_{X^*}$ so that $\phi_n(x + t_n h_n) = \|x + t_n h_n\|$. The previous inequality implies $\phi_n(t_n h_n) - \phi(t_n h_n) > \varepsilon t_n$ and so $\phi_n \not\to \phi$ which shows (b) is not true. □

The analogous version for Gâteaux differentiability is listed as follows.

Proposition 5.1.5 (Šmulian). *Let $(X, \|\cdot\|)$ be a normed space, and $x \in S_X$. Then the following are equivalent.*

(a) $\|\cdot\|$ *is Gâteaux differentiable at x.*
(b) $(\phi_n - \Lambda_n) \to_{w^*} 0$ *whenever* $\phi_n(x_n) \to 1$, $\Lambda_n(y_n) \to 1$ *and* $x_n \to x$, $y_n \to x$ *and* $\phi_n, \Lambda_n \in B_{X^*}$.
(c) $(\phi_n - \phi) \to_{w^*} 0$ *whenever* $\phi_n, \phi \in B_{X^*}$, $\phi_n(x) \to \phi(x) = 1$.

Proof. See Exercise 5.1.2. □

Let C be a closed convex subset of a Banach space X. A point $x \in C$ is an *extreme point* of C, if $C \setminus \{x\}$ is convex. We say that $x \in C$ is an *exposed point* of C if there is a $\phi \in X^*$ such that $\phi(x) = \sup_{u \in C} \phi(u)$ and $\phi(x) > \phi(u)$ for all $u \in C \setminus \{x\}$; alternatively, we say ϕ *exposes* $x \in C$. We say $x \in C$ is a *strongly exposed point* of C, if there is a $\phi \in X^*$ such that $\phi(x) = \sup_{u \in C} \phi(u)$ and $x_n \to x$ whenever $x_n \in C$ and $\phi(x_n) \to \phi(x)$; alternatively, we say ϕ *strongly exposes* $x \in C$.

Proposition 5.1.6. *Let $(X, \|\cdot\|)$ be a Banach space, $x_0 \in S_X$ and $\phi_0 \in S_{X^*}$. Then x_0 exposes (resp. strongly exposes) $\phi_0 \in B_{X^*}$ if and only if $\|\cdot\|$ is Gâteaux (resp. Fréchet) differentiable at x_0 with $\nabla \|x_0\| = \phi_0$.*

Proof. We prove only the Gâteaux case; the other case is similar. (⇒) Suppose $(\phi_n) \subset B_{X^*}$ satisfies $\phi_n(x) \to 1$. Suppose that $\phi_n \not\to_{w^*} \phi_0$. According to Alaoglu's theorem (4.1.6), there is a subnet $\phi_{n_\alpha} \to_{w^*} \bar{\phi} \in B_{X^*}$ where $\bar{\phi} \neq \phi_0$. Now $\bar{\phi}(x_0) < 1$

since x_0 exposes B_{X^*} at ϕ_0. This contradiction shows $\phi_n \to_{w^*} \phi_0$. According to Proposition 5.1.5, $\|\cdot\|$ is differentiable at x_0.

The converse follows immediately from Proposition 5.1.5. □

Corollary 5.1.7. *The norm $\|\cdot\|$ on X is Gâteaux differentiable at $x_0 \in S_X$ if and only if there exists a unique support functional $\phi \in S_{X^*}$.*

Proof. Suppose $\|\cdot\|$ is Gâteaux differentiable at $x_0 \in S_X$. Then by Proposition 5.1.6, x_0 exposes $\phi \in B_{X^*}$ where $\phi = \nabla\|x_0\|$. Hence ϕ must be the unique support functional of x_0. Conversely, suppose $\phi \in S_{X^*}$ is a unique support functional for $x_0 \in S_X$, it follows that x_0 exposes $\phi \in B_{X^*}$, and so $\|\cdot\|$ is differentiable at x_0 with $\phi = \nabla\|x_0\|$. □

A norm $\|\cdot\|$ on X is said to be *strictly convex* (or *rotund*) if $x = y$ whenever $\|x+y\| = 2$, and $\|x\| = \|y\| = 1$. The usual norm on a Hilbert space is the classic example of a strictly convex norm.

Fact 5.1.8. *Suppose $(X, \|\cdot\|)$ is a normed space. Then:*

(a) $2\|x\|^2 + 2\|y\|^2 - \|x+y\|^2 \geq 0$ *for all* $x, y \in X$.
(b) *If* $2\|x_n\|^2 + 2\|y_n\|^2 - \|x_n + y_n\|^2 \to 0$, *then* $(\|x_n\| - \|y_n\|) \to 0$.

Proof. Both (a) and (b) follow from the inequality

$$2\|x\|^2 + 2\|y\|^2 - \|x+y\|^2 \geq 2\|x\|^2 + 2\|y\|^2 - (\|x\| + \|y\|)^2$$
$$= (\|x\| - \|y\|)^2$$

which is a consequence of the parallelogram law in $\mathbb{R}$. □

The next fact provides some basic reformulations of strictly convex norms.

Fact 5.1.9. *Let $(X, \|\cdot\|)$ be a normed space. Then the following are equivalent.*

(a) $\|\cdot\|$ *is strictly convex.*
(b) *Every $x \in S_X$ is an extreme point of B_X.*
(c) *If $x, y \in X$ satisfy $2\|x\|^2 + 2\|y\|^2 - \|x+y\|^2 = 0$, then $x = y$.*
(d) *If $x, y \in X \setminus \{0\}$ satisfy $\|x+y\| = \|x\| + \|y\|$, then $x = \lambda y$ for some $\lambda > 0$.*

Proof. The equivalence of (a) and (b) is a straightforward consequence of the definitions.

(a) $\Rightarrow$ (d): Suppose $\|x+y\| = \|x\| + \|y\|$ for some $x, y \in X \setminus \{0\}$. We may assume $0 < \|x\| \leq \|y\|$. Then using the triangle inequality

$$\left\|\frac{x}{\|x\|} + \frac{y}{\|y\|}\right\| \geq \left\|\frac{x}{\|x\|} + \frac{y}{\|x\|}\right\| - \left\|\frac{y}{\|x\|} - \frac{y}{\|y\|}\right\|$$
$$= (1/\|x\|)\|x+y\| - \|y\|(1/\|x\| - 1/\|y\|)$$
$$= (1/\|x\|)(\|x\| + \|y\|) - \|y\|(1/\|x\| - 1/\|y\|) = 2.$$

Therefore, $\left\|x/\|x\| + y/\|y\|\right\| = 2$ and so $x/\|x\| = y/\|y\|$.

(d) $\Rightarrow$ (c): Suppose that $2\|x\|^2 + 2\|y\|^2 - \|x+y\|^2 = 0$. If $x = 0$, it is clear that $y = 0$ and vice versa. So we may now assume $x, y \in X \setminus \{0\}$. Thus,

$$0 = 2\|x\|^2 + 2\|y\|^2 - \|x+y\|^2 \geq 2\|x\|^2 + 2\|y\|^2 - (\|x\| + \|y\|)^2$$
$$= (\|x\| - \|y\|)^2 \geq 0.$$

Therefore, $\|x+y\| = \|x\| + \|y\|$ and $\|x\| = \|y\|$, and so with the help of (d) we obtain that $x = y$.

(c) $\Rightarrow$ (a) is easy to verify. $\square$

Proposition 5.1.10. *Suppose* $T : X \to Y$ *is a one-to-one continuous linear operator. If* $\|\cdot\|_Y$ *is a strictly convex norm on* Y, *then:*

(a) $\|\cdot\|_1$ *defined by* $\|x\|_1 := \|x\| + \|Tx\|_Y$ *is a strictly convex norm on* X.
(b) $\|\cdot\|_2$ *defined by* $\|x\|_2^2 := \|x\|^2 + \|Tx\|_Y^2$ *is a strictly convex norm on* X.

Proof. We prove only (b), and leave (a) as Exercise 5.1.4. Suppose $\|u\|_2^2 + \|v\|_2^2 - \|u+v\|_2^2 = 0$, then $\|Tu\|_Y^2 + \|Tv\|_Y^2 - \|Tu+Tv\|_Y^2 = 0$ and so $Tu = Tv$ by the strict convexity of $\|\cdot\|_Y$ (Fact 5.1.9). Because T is one-to-one, $u = v$ and so $\|\cdot\|$ is strictly convex by Fact 5.1.9. $\square$

Proposition 5.1.11. *Suppose* X *is separable, then* X *admits an equivalent norm whose dual is strictly convex.*

Proof. Let $\{x_n\}_{n=1}^\infty$ be dense in S_X. Consider the linear operator $T : X^* \to \ell_2$ defined by $Tf := \{2^{-n}f(x_n)\}$ for $f \in X^*$. Then T is one-to-one. Now define $\|\cdot\|$ by $\|f\|^2 := \|f\|^2 + \|Tf\|_2^2$. Using Fact 5.1.2 one can check this is an equivalent dual norm on X^*. The strict convexity of $\|\cdot\|$ follows from Proposition 5.1.10(b). $\square$

Let $(X, \|\cdot\|)$ be a normed space. The norm $\|\cdot\|$ is said to be *locally uniformly convex* (LUC) if $\|x_n - x\| \to 0$ whenever $x_n, x \in B_X$ are such that $\|x_n + x\| \to 2$. Such norms are also called *locally uniformly rotund*.

Fact 5.1.12. *Let* $(X, \|\cdot\|)$ *be a normed space. Then the following are equivalent.*

(a) $\|\cdot\|$ *is locally uniformly convex.*
(b) $\|x_n - x\| \to 0$ *whenever* $x_n, x \in X$ *are such that* $2\|x\|^2 + 2\|x_n\|^2 - \|x_n+x\|^2 \to 0$.

Proof. (a) $\Rightarrow$ (b): Suppose $2\|x\|^2 + 2\|x_n\|^2 - \|x_n+x\|^2 \to 0$, then $(\|x_n\| - \|x\|) \to 0$ (Fact 5.1.8). In the event $x = 0$, the conclusion is trivial. So we may suppose $x_n, x \in X \setminus \{0\}$. Now let $u = x/\|x\|$ and $u_n = x_n/\|x_n\|$. Then $u, u_n \in S_X$ and $\|u_n + u\| \to 2$. Therefore, $\|u_n - u\| \to 0$. Consequently, $\|x_n - x\| \to 0$. The implication (b) $\Rightarrow$ (a) is easy to check. $\square$

Proposition 5.1.13. *Suppose* $(X, \|\cdot\|)$ *is a normed space.*

(a) *If the dual norm is strictly convex, then* $\|\cdot\|$ *is Gâteaux differentiable.*
(b) *If the dual norm is locally uniformly convex, then* $\|\cdot\|$ *is Fréchet differentiable.*

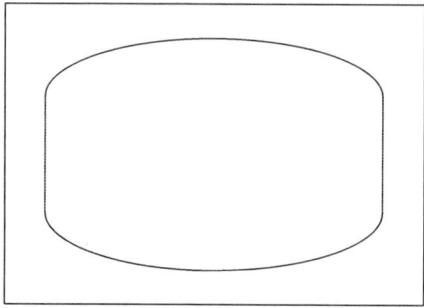 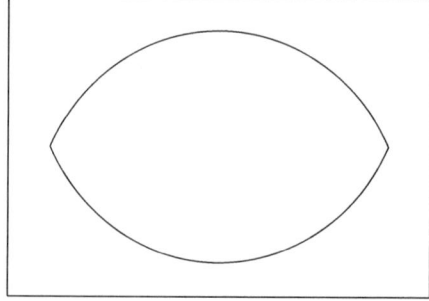

Figure 5.1 Smooth and rotund balls.

Proof. (a) Let $x \in S_X$ and choose a supporting functional $f \in S_{X^*}$ so that $f(x) = 1$ (Remark 4.1.16). Suppose $g \in S_{X^*}$ satisfies $g(x) = 1$. Then, $\|f+g\| \geq (f+g)(x) = 2$. Because $\|\cdot\|$ is strictly convex, then $g = f$. According to Corollary 5.1.7, $\|\cdot\|$ is Gâteaux differentiable at x.

(b) This part is left as Exercise 5.1.6. □

Theorem 5.1.14 (Kadec). *(a) Suppose X is separable. Then X admits an equivalent locally uniformly convex norm.*

(b) Suppose X^* is separable. Then X admits an equivalent norm whose dual is locally uniformly convex.

Proof. (a) Let X be a separable Banach space. Let $\{x_n\}_{n=1}^\infty \subset S_X$ be dense in S_X, and let $\{f_n\}_{n=1}^\infty \subset S_{X^*}$ be a separating family for X (see Exercise 4.1.11). For $n \in \mathbb{N}$, put $F_n = \operatorname{span}\{x_1, \ldots, x_n\}$ and note that $d_{F_n}(x) \to 0$ for each $x \in X$ as $n \to \infty$ where the distance is measured in the original norm. Define another norm $\|\cdot\|$ on X by

$$\|x\|^2 := \|x\|^2 + \sum_{n=1}^\infty 2^{-n} d_{F_n}(x)^2 + \sum_{n=1}^\infty 2^{-n} f_n^2(x),$$

where $\|\cdot\|$ is the original norm on X. Because $d_{F_i}(\cdot)$ is a positive homogeneous subadditive function for each i, $\|\cdot\|$ is an equivalent norm on X.

We now show that $\|\cdot\|$ is locally uniformly convex. Indeed, assume that $x_k, x \in X$ are such that

$$\lim_{k \to \infty} (2\|x\|^2 + 2\|x_k\|^2 - \|x + x_k\|^2) = 0. \tag{5.1.1}$$

Each of the expressions in the following are nonnegative (use parallelogram inequality as in Fact 5.1.8), thus it follows from the above equation that

$$\lim_{k \to \infty} [2\|x\|^2 + 2\|x_k\|^2 - \|x + x_k\|^2] = 0,$$

$$\lim_{k \to \infty} [2d_{F_n}^2(x) + 2d_{F_n}^2(x_k) - d_{F_n}^2(x + x_k)] = 0 \text{ for every } n,$$

$$\lim_{k \to \infty} [2f_n^2(x) + 2f_n^2(x_k) - f_n^2(x + x_k)] = 0 \text{ for every } n.$$

5.1 Renorming: an overview

As a consequence of the previous limits (reason as in Fact 5.1.8) we obtain

$$\lim_{k\to\infty} \|x_k\| = \|x\|, \tag{5.1.2}$$

$$\lim_{k\to\infty} d_{F_n}(x_k) = d_{F_n}(x) \text{ for every } n, \tag{5.1.3}$$

$$\lim_{k\to\infty} f_n(x_k) = f_n(x) \text{ for every } n. \tag{5.1.4}$$

Now, $\overline{\{x_k\}_{k=1}^\infty \cup \{x\}}$ is norm compact (see Exercise 5.1.9). Therefore, the topology of pointwise convergence induced by $\{f_n\}_{n=1}^\infty$ is equivalent to the norm topology on $\overline{\{x_k\}_{k=1}^\infty \cup \{x\}}$ (because two Hausdorff topologies agree on a set that is compact in the stronger topology). Thus by (5.1.4) we have $\|x_k - x\| \to 0$ which completes the proof.

(b) Follow the argument as above, choosing $\{x_n\}_{n=1}^\infty$ norm dense in S_{X^*} and $\{f_n\}_{n=1}^\infty$ norm dense in S_X (and hence separating on X^*). Given the original dual norm $\|\cdot\|$ on X^*, all the terms in the definition of $\|\cdot\|$ are weak*-lsc. Therefore, $\|\cdot\|$ will be an equivalent dual norm on X^*. □

We now mention some important consequences of the results just presented.

Corollary 5.1.15. *Suppose X^* is separable. Then:*

(a) X admits an equivalent Fréchet differentiable norm; and
(b) every continuous convex function on X is Fréchet differentiable on a dense (automatically) G_δ-set.

Proof. Part (a) follows from Theorem 5.1.14(b) and Proposition 5.1.13(b). Part (b) follows from part (a) and a smooth variational principle (Exercise 4.3.10). □

Corollary 5.1.16. *Let X be a separable Banach space. Then the following are equivalent.*

(a) X^* is separable.
(b) X admits an equivalent norm whose dual is locally uniformly convex.
(c) X admits an equivalent Fréchet differentiable norm.

Proof. (a) $\Rightarrow$ (b) $\Rightarrow$ (c) were shown in Theorem 5.1.14 and Proposition 5.1.13. The implication (c) $\Rightarrow$ (a) is left as Exercise 5.1.10. □

The conditions in Corollary 5.1.16 remain equivalent in certain classes of nonseparable Banach spaces, as we shall outline later in this section, but first, we introduce and present some properties of uniformly convex norms.

Let $(X, \|\cdot\|)$ be a Banach space. The *modulus of convexity* of $\|\cdot\|$ is defined by

$$\delta_X(\varepsilon) := \inf\left\{ 1 - \left\|\frac{x+y}{2}\right\| : x,y \in B_X, \|x-y\| \geq \varepsilon \right\}, \quad \text{for } \varepsilon \in [0,2].$$

The norm $\|\cdot\|$ is called *uniformly convex* (UC) if $\delta_X(\varepsilon) > 0$ for all $\varepsilon \in (0,2]$, and the space $(X, \|\cdot\|)$ is called a *uniformly convex space*. The term *uniformly rotund* is often used in place of uniformly convex.

Fact 5.1.17. *Let $(X, \|\cdot\|)$ be a Banach space. Then the following are equivalent.*

(a) X is uniformly convex.
(b) $\lim_{n\to\infty} \|x_n - y_n\| = 0$ whenever $x_n, y_n \in X$, $n \in \mathbb{N}$, $(x_n)_{n=1}^\infty$ is bounded and $\lim_{n\to\infty}(2\|x_n\|^2 + 2\|y_n\|^2 - \|x_n + y_n\|^2) = 0$.
(c) $\lim_{n\to\infty} \|x_n - y_n\| = 0$ whenever $x_n, y_n \in B_X$, $n \in \mathbb{N}$ and $\lim_{n\to\infty} \|x_n + y_n\| = 2$.

Proof. Observe that (c) is a sequential reformulation of the definition, so it is equivalent to (a).

(c) $\Rightarrow$ (b): Suppose $(2\|y_n\|^2 + 2\|x_n\|^2 - \|y_n + x_n\|^2) \to 0$, then $(\|y_n\| - \|x_n\|) \to 0$ (by Fact 5.1.8). Thus when $\|x_n\| \to 0$ or $\|y_n\| \to 0$, the conclusion is trivial. So we may suppose $\|x_n\| \geq \varepsilon$ and $\|y_n\| \geq \varepsilon$ for some $\varepsilon > 0$. Now let $u_n = x_n/\|x_n\|$ and $v_n = y_n/\|y_n\|$. It follows that $u_n, v_n \in S_X$ and $\|u_n + v_n\| \to 2$. Therefore, $\|u_n - v_n\| \to 0$. Consequently, $\|x_n - y_n\| \to 0$. The implication (b) $\Rightarrow$ (a) is trivial. $\square$

Let $(X, \|\cdot\|)$ be a Banach space. For $\tau > 0$ we define the *modulus of smoothness* of $\|\cdot\|$ by

$$\rho_X(\tau) := \sup\left\{\frac{\|x + \tau h\| + \|x - \tau h\| - 2}{2} : \|x\| = \|h\| = 1\right\}.$$

The norm $\|\cdot\|$ is said to be *uniformly smooth* if $\lim_{\tau \to 0^+} \frac{\rho_X(\tau)}{\tau} = 0$, and the space $(X, \|\cdot\|)$ is said to be *uniformly smooth*.

Proposition 5.1.18. *Let $(X, \|\cdot\|)$ be a Banach space. Then the following are equivalent.*

(a) X is uniformly smooth.
(b) For each $\varepsilon > 0$, there exists $\delta > 0$ such that for all $x \in S_X$ and $y \in \delta B_X$,

$$\|x + y\| + \|x - y\| - 2 \leq \varepsilon \|y\|.$$

(c) The norm is uniformly Fréchet differentiable, that is, the limit

$$\lim_{t \to 0} \frac{\|x + th\| - \|x\|}{t} = \|x\|'(h) \text{ exist uniformly for } x, h \in S_X.$$

(d) The norm $\|\cdot\|$ is Fréchet differentiable with uniformly continuous derivative on S_X.

Proof. The proof is left as Exercise 5.1.18. $\square$

Theorem 5.1.19. *Let X be a Banach space, then $\|\cdot\|$ is uniformly smooth if and only if its dual norm is uniformly convex.*

Proof. We will prove each implication by contraposition. First, suppose the dual norm is not uniformly convex. Then there are $\phi_n, \Lambda_n \in B_{X^*}$ and $\varepsilon > 0$ such that $\|\phi_n + \Lambda_n\| \to 2$ but $\|\phi_n - \Lambda_n\| > 2\varepsilon$ for each n. Now choose $x_n \in S_X$ such that

$(\phi_n + \Lambda_n)(x_n) \to 2$ and let $t_n \to 0^+$ be such that $(\phi_n + \Lambda_n)(x_n) \geq 2 - \varepsilon t_n$. Finally, choose $h_n \in S_X$ so that $(\phi_n - \Lambda_n)(h_n) \geq 2\varepsilon$. Then

$$\|x_n + t_n h_n\| + \|x_n - t_n h_n\| - 2 \geq \phi_n(x_n + t_n h_n) + \Lambda_n(x_n - t_n h_n) - 2$$
$$\geq 2 - \varepsilon_n t_n + t_n(\phi_n - \Lambda_n)(h_n) - 2 \geq \varepsilon t_n.$$

Therefore, $\|\cdot\|$ is not uniformly smooth.

Conversely, suppose that $\|\cdot\|$ is not uniformly smooth. Then there exist $(h_n) \subset X$, $(x_n) \subset S_X$ and $\varepsilon > 0$ such that $\|h_n\| \to 0^+$ and

$$\|x_n + h_n\| + \|x_n - h_n\| - 2 \geq \varepsilon \|h_n\|.$$

Now choose $\phi_n, \Lambda_n \in S_{X^*}$ so that $\phi(x_n + h_n) = \|x_n + h_n\|$ and $\Lambda_n(x_n - h_n) = \|x_n - h_n\|$. Then

$$\|\phi_n + \Lambda_n\| + (\phi_n - \Lambda_n)(h_n) \geq (\phi_n + \Lambda_n)(x_n) + (\phi_n - \Lambda_n)(h_n)$$
$$= \phi_n(x_n + h_n) + \Lambda_n(x_n - h_n) \geq 2 + \varepsilon \|h_n\|.$$

This inequality now implies $\|\phi_n + \Lambda_n\| \to 2$, and $\|\phi_n - \Lambda_n\| \geq \varepsilon$. Therefore, the dual norm is not uniformly convex. □

Theorem 5.1.20 (Milman–Pettis). *Suppose X admits a uniformly convex norm, then X is reflexive.*

Proof. Let $\Phi \in B_{X^{**}}$. Applying Goldstine's theorem (Exercise 4.1.13), we choose a net $(x_\alpha) \subset B_X$ such that $x_\alpha \to_{w^*} \Phi$. Then $\|x_\alpha + x_\beta\| \to 2$ and so $\|x_\alpha - x_\beta\| \to 0$. Thus (x_α) is norm convergent, and so $\Phi \in X$. □

As a consequence of the previous two results, we obtain

Corollary 5.1.21. *(a) Suppose X is uniformly convex or uniformly smooth, then X is reflexive.*
(b) A norm on X is uniformly convex if and only if its dual norm is uniformly smooth.

The duality relationships involving uniform convexity and uniform smoothness can be more precisely deduced from the following fundamental relationship.

Proposition 5.1.22. *Let $(X, \|\cdot\|)$ be a Banach space, and let $\delta_X(\cdot)$ be the modulus of convexity of $\|\cdot\|$ and let $\rho_{X^*}(\cdot)$ be the modulus of smoothness of the dual norm. Then for every $\tau > 0$,*

$$\rho_{X^*}(\tau) = \sup\left\{\tau \frac{\varepsilon}{2} - \delta_X(\varepsilon) : 0 < \varepsilon \leq 2\right\}.$$

Similarly, let $\rho_X(\cdot)$ be the modulus of smoothness of $\|\cdot\|$ and $\delta_{X^*}(\cdot)$ be the modulus of convexity of the dual norm. Then, for every $\tau > 0$,

$$\rho_X(\tau) = \sup\left\{\tau \frac{\varepsilon}{2} - \delta_{X^*}(\varepsilon) : 0 < \varepsilon \leq 2\right\}.$$

Proof. The proof follows along the same lines as the case for convex functions in Theorem 5.4.1 given later; see, for example, [199, Lemma 9.9]. □

A norm on a Banach space X is said to have *modulus of convexity of power type p* if there exists $C > 0$ so that $\delta(\varepsilon) \geq C\varepsilon^p$ for $0 \leq \varepsilon \leq 2$. A norm on a Banach space X is said to have *modulus of smoothness of power type q* if there exists $k > 0$ so that $\rho(\tau) \leq k\tau^q$ for all $\tau > 0$.

Corollary 5.1.23. *Let $p \geq 2$, and let $1/p + 1/q = 1$. Then a norm on X has modulus of convexity of power type p if and only if its dual norm has modulus of smoothness of power type q.*

Proof. This is left as Exercise 5.1.21. □

Next, we state the following fundamental renorming theorem.

Theorem 5.1.24 (James–Enflo). *A Banach space is superreflexive if and only if it admits an equivalent uniformly convex norm if and only if it admits an equivalent uniformly smooth norm.*

Proof. See [199, Theorem 9.18] for the full proof and see also [197]. A proof that X is superreflexive when it admits a uniformly convex or uniformly smooth norm is sketched in Exercise 5.1.20. □

We mention an important and deep theorem of Pisier's [353] whose proof is beyond the scope of this book: Each superreflexive Banach space can be renormed to have a norm with modulus of convexity of power type p for some some $p \geq 2$.

5.1.1 Approximation of functions

The existence of nice norms on Banach spaces is often closely connected with nice approximations. For example, we highlight the following which completely characterizes superreflexive spaces in terms of Lasry–Lions type approximations [284]. Some further results and references to smooth approximations are given in Exercise 5.1.33.

Theorem 5.1.25. *A Banach space X is superreflexive if and only if every real-valued Lipschitz function on X can be approximated uniformly on bounded sets by difference of convex functions which are Lipschitz on bounded sets.*

Proof. This is from [148], we prove the only if part following [52, Theorem 4.21, p. 94]. By the James–Enflo theorem (5.1.24), we can assume the norm on X is uniformly convex. Let $f : X \to \mathbb{R}$ be Lipschitz, with Lipschitz constant 1. Define

$$f_n(x) := \inf\{f(y) + n(2\|x\|^2 + 2\|y\|^2 - \|x+y\|^2) : y \in Y\}.$$

Then represent $f_n(x) = h_n(x) - g_n(x)$ where $h_n(x) = 2n^2\|x\|^2$ and

$$g_n(x) := \sup\{n\|x+y\|^2 - 2n\|y\|^2 - f(y) : y \in E\}.$$

5.1 Renorming: an overview

It is left as to Exercise 5.1.29 to verify that g_n and h_n are convex, and f_n, g_n and h_n are Lipschitz on bounded sets, and $f_n(x) \leq f(x)$ for all x. We will show that (f_n) converges uniformly to f on the set rB_X for each $r \geq 1$. For this, consider the set

$$A_n := \{y : f(y) + n(2\|x\|^2 + 2\|y\|^2 - \|x+y\|^2) \leq f(x)\}.$$

Since $f_n(x) \leq f(x)$, we need only consider $y \in A_n$, so fix such a y. According to the triangle inequality, and because f is Lipschitz with Lipschitz constant 1,

$$n(\|x\| - \|y\|)^2 \leq n(2\|x\|^2 + 2\|y\|^2 - \|x+y\|^2) \tag{5.1.5}$$
$$\leq f(x) - f(y) \leq \|x - y\|.$$

Thus it suffices to show $\sup\{\|x - y\| : y \in A_n\} \to 0$ uniformly on rB_X.

According to (5.1.5), if $r \geq 1$ and $n \geq 3$, then $\|y\| \leq 2r$ for each $y \in A_n$; this follows because if $\|y\| \geq 2r$, then the left-hand side is bigger than $3(\|y\| - r)^2$ while the right-hand side is at most $\|y\| + r$. Consequently, $|\|x\| - \|y\|| \leq \sqrt{3r/n}$. Now let $z = h\|x\|/\|y\|$ then $\|z\| = \|x\|$ and $\|z - y\| \leq |\|x\| - \|y\||$. Using the second and last terms of (5.1.5), one can show (Exercise 5.1.29(b))

$$\|x + z\|^2 \geq 4\|x\|^2 - C(r)/\sqrt{n}. \tag{5.1.6}$$

According to the definition of the modulus of convexity,

$$\|x - z\| \leq \|x\|\delta^{-1}\left(\frac{C_1(r)}{\sqrt{n}\|x\|^2}\right).$$

Separating the cases $\|x\| \geq n^{-1/8}$ and $\|x\| \leq n^{-1/8}$, we obtain, respectively, the two estimates

$$\|x - y\| \leq \|x - z\| + \|z - y\| \leq r\delta^{-1}(C_1(r)n^{-1/4}) + (3r/n)^{1/2},$$
$$\|x - y\| \leq 2\|x\| + |\|x\| - \|y\|| \leq 2n^{-1/8} + (3r/n)^{1/2}.$$

Consequently, $\sup\{\|x - y\| : y \in A_n\} \to 0$ uniformly on rB_X as desired.
For the converse, we refer the reader to [52, p. 95]. □

Notice in the above that in the case of a Hilbert space, the parallelogram identity simplifies the expression for f_n to

$$f_n(x) = \inf\{f(y) + n\|x - y\|^2 : y \in Y\}.$$

5.1.2 Further remarks on renorming

We now state some generalizations of the local uniformly convex renormings of separable spaces. First, a Banach space is *weakly compactly generated* (WCG) if it contains a weakly compact set whose span is norm dense. The reader can find the following two theorems and much more in [180]

Theorem 5.1.26. *(a) Suppose X is a WCG space, then X admits an equivalent locally uniformly convex norm, and X^* admits an equivalent dual strictly convex norm.*
(b) Suppose X^* is a WCG space, then X admits an equivalent locally uniformly convex norm, and X^* admits an equivalent dual locally uniformly convex norm.

Remark 5.1.27 (Spaces without strictly convex norms [171, 167]). Let Γ be uncountable. M.M. Day was the first to show that there is no equivalent strictly convex norm on $\ell_c^\infty(\Gamma)$, the Banach subspace of $\ell^\infty(\Gamma)$ of all elements of with countable support. In consequence there is none on $\ell^\infty(\Gamma)$.

The following is a generalization of Corollary 5.1.16.

Theorem 5.1.28. *Let X be a weakly compactly generated Banach space. Then the following are equivalent.*

(a) X admits an equivalent norm whose dual is locally uniformly convex.
(b) X admits an equivalent Fréchet differentiable norm.
(c) X is an Asplund space.

Proof. The implications (a) $\Rightarrow$ (b) $\Rightarrow$ (c) are valid in general Banach spaces following along the same lines of reasoning as in the proof of Corollary 5.1.15. The more difficult implication (c) $\Rightarrow$ (a) can be found for example in [180, Corollay VII.1.13]. □

In general the implications in the previous theorem cannot be reversed in a strong sense as cited in Remark 5.1.30. First, we describe the $C(K)$ spaces that are Asplund spaces. A topological space is said to be *scattered* if it contains no nonempty perfect subset, or equivalently if every nonempty subset has a (relatively) isolated point. With this terminology we state:

Theorem 5.1.29 (Asplund $C(K)$ spaces). *Let K be a compact Hausdorff space. Then the following are equivalent.*

(a) $C(K)$ is an Asplund space.
(b) K is scattered.
(c) $C(K)$ contains no isometric copy of $C[0, 1]$.
(d) $C(K)$ contains no isomorphic copy of ℓ_1.

Remark 5.1.30. (a) There are $C(K)$ Asplund spaces that admit no equivalent Gâteaux differentiable norm.
(b) The space $C[0, \omega_1]$ admits an equivalent Fréchet differentiable norm, but it cannot be renormed with an equivalent norm whose dual is strictly convex.

Proof. Part (a) is Haydon's result from [252] and (b) was shown by Talagrand in [411]. □

5.1 Renorming: an overview

Remark 5.1.31. Let X be a Banach space.

(a) If X admits a norm whose dual norm is Fréchet differentiable, then X is reflexive.
(b) Suppose X is a reflexive Banach space. Then X can be renormed so that the norm on X is Fréchet differentiable, but its dual norm on X^* is not locally uniformly convex.

Proof. Part (a) is due to Šmulian [404] and is left as Exercise 5.1.17, while (b) can be found in [441, Theorem 2.3]. □

Next, we mention an averaging theorem concerning the renorming of spaces whose norms and their duals simultaneously possess certain rotundity properties.

Theorem 5.1.32 (Asplund averaging theorem). *Let X be a Banach space.*

(a) *Suppose X admits an equivalent locally uniformly convex norm, and X^* admits an equivalent dual locally uniformly convex norm. Then X admits a locally uniformly convex norm whose dual is locally uniformly convex.*
(b) *The analog of (a) remains valid with strict convexity replacing local uniform convexity in either or both X and X^*.*
(c) *Suppose X admits a uniformly convex norm. Then X (and X^*) admit norms that are simultaneous uniformly convex and uniformly smooth.*

Proof. We refer the reader to [204] for a Baire category proof of a stronger result. For (c), one needs the James–Enflo theorem (5.1.24) that a Banach space is superreflexive if and only if admits a uniformly convex norm. Exercise 5.1.24 outlines a simple proof of a weaker results combining differentiability and rotundity properties. □

Corollary 5.1.33. *Suppose X^* is a WCG space, then X admits an equivalent locally uniformly convex norm and Fréchet differentiable norm.*

Finally, we mention some classical results on *higher-order derivatives* of norms on L_p-spaces. For this, n-th derivatives are identified as continuous n-linear mappings, i.e. $f^{(n)} \in \mathcal{L}_n(X; \mathbb{R})$. For example $f : U \to \mathbb{R}$ is *three times Fréchet differentiable at $\bar{x} \in U$*, if it is twice Fréchet differentiable on an open neighborhood V of $\bar{x}$, and the mapping $x \mapsto f''(x)$ is a Fréchet differentiable mapping from X to $\mathcal{L}_2(X; \mathbb{R})$ where $\mathcal{L}_2(X; \mathbb{R})$ represents the continuous bilinear mappings from X to $\mathbb{R}$, and $f^{(3)}(\bar{x}) \in \mathcal{L}_3(X; \mathbb{R})$ the continuous 3-linear mappings from X to $\mathbb{R}$. Higher-order Fréchet derivatives will always be symmetric, so we say that f is k-times Fréchet differentiable at $x \in X$ if for the continuous symmetric form of the $f^{(k-1)}$-th Fréchet derivative of f defined on a neighborhood U of x, one has

$$\lim_{t \to 0} \frac{f^{(k-1)}(x + th_k)(h_1, \ldots, h_{k-1}) - f^{(k-1)}(x)(h_1, \ldots, h_{k-1})}{t}$$

exists uniformly on $h_1, \ldots, h_k$ in B_X and is a continuous symmetric k-linear form in the variables $h_1, h_2, \ldots, h_k$. Properties of continuity, Lipschitzness and the like for

$f^{(k)}$ are meant in the context of the k-linear forms involved. For example, $f^{(k)}$ is said to be *Lipschitz* on X if there exists $M > 0$ so that

$$\|f^{(k)}(x) - f^{(k)}(y)\| := \sup\{|(f^{(k)}(x) - f^{(k)}(y))(h_1, h_2, \ldots, h_k)| :$$
$$h_1, h_2, \ldots, h_k \in B_X\}$$
$$\leq M\|x - y\| \quad \text{for } x, y \in X.$$

A real-valued function f whose k-th Fréchet derivative is continuous on an open set $U \subset X$ is said to be C^k-*smooth on* U and this is denoted by $f \in C^k(U)$. If f has continuous k-th Fréchet derivatives for all $k \in \mathbb{N}$ on U, then f is said to be C^∞-*smooth on* U. When $U = X$, we simply say f is C^k-smooth or C^∞-smooth as appropriate. See [144, Chapter 1] for a careful development of differentiability on Banach spaces; let us note that in [144] and many similar references 'differentiability' refers to what we have called Fréchet differentiability.

Theorem 5.1.34 (Smoothness on L_p-spaces). *(a) If p is an even integer, then $\|\cdot\|^p$ is C^∞-smooth on L_p, $(\|\cdot\|^p)^{(p)}$ is constant and $(\|\cdot\|^p)^{(p+1)}$ is zero.*
(b) If p is an odd integer, then $\|\cdot\|^p$ is C^{p-1}-smooth and $(\|\cdot\|^p)^{(p-1)}$ is Lipschitz on L_p.
(c) If p is not an integer, then $\|\cdot\|^p$ is $C^{\lfloor p \rfloor}$-smooth and $(\|\cdot\|^p)^{\lfloor p \rfloor}$ is $(p - \lfloor p \rfloor)$-Hölder smooth on L_p.

Proof. We refer the reader to [180, Theorem V.1.1, p. 184] for a proof of this important classical theorem. □

Exercises and further results

5.1.1.* Prove the bipolar theorem (5.1.1).

Hint. The bipolar theorem holds generally in locally convex spaces, see for example [160, p. 127] for more details. □

5.1.2.* Prove Šmulian's theorem (5.1.5) for Gâteaux differentiability of norms.

5.1.3.* Deduce the norm cases of Šmulian's theorem (Theorems 5.1.4 and 5.1.5) from the respective cases for convex functions.

Hint. For $\phi \in S_{X^*}$, $\phi \in \partial_\varepsilon \|x\|$ if and only if $\phi(x) \geq 1 - \varepsilon$. □

5.1.4.* Prove Proposition 5.1.10(a).

5.1.5.* Suppose $\|\cdot\|$ is a dual norm on X^*. Show that $\|\cdot\|$ is Fréchet differentiable at $\phi \in S_{X^*}$ if and only if $\phi(x) = 1$ for some $x \in S_X$ and $\|x_n - x\| \to 0$ whenever $(x_n) \subset B_X$ is such that $\phi(x_n) \to 1$.

Hint. Mimic the proof of Šmulian's theorem (5.1.4), making sure to verify $\nabla \|\phi\| \in X$. See Exercise 4.2.10 for a more general result. □

5.1.6.* Prove Proposition 5.1.13(b)

Hint. Use Šmulian's theorem (5.1.4) as appropriate. □

5.1 Renorming: an overview

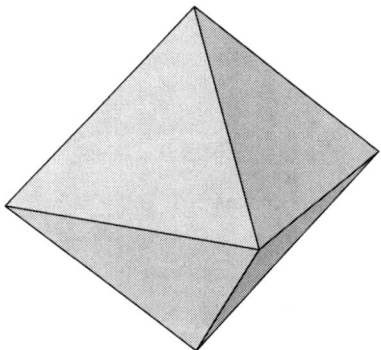

Figure 5.2 The 1-ball in three dimensions.

5.1.7. The ball for three-dimensional ℓ_1 is shown in Figure 5.2. Observe that the ball is smooth at points $(x_1, x_2, x_3) \in \mathbb{R}^3$ where $x_i \neq 0$, $i = 1, 2, 3$. Prove the following concerning the usual norm $\|\cdot\|_1$ on ℓ_1.

(a) $\|\cdot\|_1$ is Gâteaux differentiable at $x := (x_i)$ on the unit sphere of $\ell_1(\mathbb{N})$ if and only if $x_i \neq 0$ for all i.
(b) $\|\cdot\|_1$ is nowhere Fréchet differentiable on $\ell_1(\mathbb{N})$.
(c) $\|\cdot\|_1$ is nowhere Gâteaux differentiable on $\ell_1(\Gamma)$ if Γ is uncountable.

Hint. For (a), show that these are the points that have unique support functionals. For (b), consider only the points $x = (x_i) \in S_{\ell_1}$ of Gâteaux differentiability, let $y^n \in S_{\ell_\infty}$ where $y_i^n := \text{sign}(x_i)$, $i = 1, \ldots, n$, and $y_i = 0$ otherwise. Then $y^n(x) = 1 - \varepsilon_n$ where $\varepsilon_n := \sum_{i=n+1}^{\infty} |x_i|$. Now use Šmulian's theorem (5.1.4). Deduce (c) from the argument for (a). $\square$

5.1.8. Describe the points $(x_i) \in c_0$ at which its usual norm is Fréchet differentiable. Verify directly that this is a dense G_δ (in fact open) subset of c_0.

5.1.9.* Use (5.1.2) and (5.1.3) to show the set $\overline{\{x_k\}_{k=1}^\infty \cup \{x\}}$ in the proof of Theorem 5.1.14 is compact.

Hint. Use that bounded balls in finite-dimensional subspaces are compact to show for each $\varepsilon > 0$, the set has a finite ε-net. $\square$

5.1.10.* Suppose X is separable and admits a Fréchet differentiable norm, show that X^* is separable.

Hint. Let $\{x_n\}_{n=1}^\infty \subset S_X$ be dense. For each n, choose $f_n \in S_{X^*}$ such that $f_n(x_n) = 1$ (Remark 4.1.16). According to the Bishop–Phelps theorem (4.3.4), it suffices to show that $\overline{\{f_n\}_{n=1}^\infty}$ contains the norm-attaining functionals in S_{X^*} (this will show S_{X^*} is separable). Now let $f \in S_X$ be a norm-attaining functionals say $f(x) = 1$. Choose $x_{n_k} \to x$. Now $f_{n_k}(x_{n_k}) = 1$, and so $f_{n_k}(x) \to 1$. According to Šmulian's theorem (5.1.4), $f_{n_k} \to f$. $\square$

5.1.11. Recall that a norm is $\|\cdot\|$ on X is said to have the *Kadec property*, if the weak and norm topologies agree on S_X. We will say a norm on X^* is *weak*-Kadec* if the norm and weak*-topologies agree on its sphere (this is sometimes called the weak*-Kadec–Klee property). Suppose X has an equivalent norm whose dual norm is weak*-Kadec. Show that X is an Asplund space.

We remark that Raja [360] has shown the stronger result that if X^* has an equivalent (dual) weak*-Kadec norm, then X^* has an equivalent dual locally uniformly convex norm.

Hint. Use the fact that X is an Asplund space if and only each of its separable subspaces has a separable dual (Corollary 6.6.10). Consider the separable case first, and show X^* separable. Indeed, if X is separable, then B_{X^*} is weak*-separable, so we fix a countable set $\{f_n\} \subset B_{X^*}$ that is weak*-dense in B_{X^*}. Now fix $f \in S_{X^*}$. Then find a subnet $f_{n_\alpha} \to_{w^*} f$. By the weak*-Kadec property, $\|f_{n_\alpha} - f\| \to 0$. Conclude S_{X^*} is separable, and hence X^* is as well.

In general, if X is not separable, consider a separable subspace Y of X, and show that the restricted norm on Y has weak*-Kadec dual norm (see e.g. [111, Proposition 1.4] for the straightforward details). Then Y^* is separable, and so X is an Asplund space. □

5.1.12. Suppose a norm on X^* has the weak*-Kadec property. Show that it is automatically a dual norm.

Hint. It suffices to show that the unit ball with respect to this norm is weak*-closed. Suppose $\|x_\alpha\| \leq 1$ and $x_\alpha \to_{w^*} x$ where $\|x\| = K$. It suffices to show that $K \leq 1$. So suppose it is not. Consider the line through x_α and x. Since this line passes through the interior of the ball of radius K, it must intersect its sphere in two places. One of those points is x, call the other x'_α (draw the picture). Now $x'_\alpha - x = \lambda_\alpha(x_\alpha - x)$ where $1 < \lambda_\alpha \leq 2K/(K-1)$. Then $(x'_\alpha - x) \to_{w^*} 0$ and so by the weak*-Kadec property $\|x'_\alpha - x\| \to 0$, but then $\|x_\alpha - x\| \to 0$ and $\|x\| < K$, a contradiction. □

5.1.13. (a) Suppose that X is a Banach space and its second dual norm is weak*-Kadec. Prove that X is reflexive.
(b) Suppose that X has an equivalent norm whose dual is Fréchet differentiable, show that X is reflexive.
(c) Does the existence of a Fréchet differentiable convex conjugate function on X^* imply that X is reflexive? See also Exercise 5.1.28 below.

Hint. (a) Modify the Lindenstrauss–Tzafriri proof given for the Milman–Pettis theorem (5.1.20).

(b) Let $\Phi \in S_{X^{**}}$ be a norm-attaining functional, say $\Phi(\phi) = 1$ where $\phi \in S_{X^*}$. Take $(x_n) \subset S_X$ such that $\phi(x_n) \to 1$. By Šmulian's theorem (5.1.4) $\|x_n - \Phi\| \to 0$. Thus $\Phi \in X$. According to the Bishop–Phelps theorem (4.3.4), $\bar{S}_X = S_{X^{**}}$, and this means $X = X^{**}$.

(c) No, for trivial reasons, for example consider any Banach space and $f := \delta_{\{0\}}$ the indicator function of 0. Then f^* is the zero function on X^*. □

5.1.14. Consider ℓ_p with $1 < p < 2$. Let $C = \{x \in \ell_p : \sum |x_i|^p \leq 1\}$, that is $C = B_{\ell_p}$. Show that $C \subset \ell_2$ is closed, convex, bounded (and hence weakly compact), and that C has extreme points that are not support points.

Hint. Let $\phi \in \ell_p^* \setminus \ell_2$. Then ϕ attains its supremum on C, at say $\bar{x} \in C$ because C is a weakly compact set in ℓ_p. Moreover, ϕ is unique up to nonzero scalar multiplication by the differentiability of the norm on ℓ_p on its sphere. Also, $\bar{x}$ is an extreme point of C since $\|\cdot\|_p$ is strictly convex. Now suppose $\Lambda \in \ell_2^* = \ell_2$ supports $\bar{x} \in C \subset \ell_2$, say $\sup_{x \in C} \Lambda = \langle \Lambda, x \rangle$. Then Λ is bounded on B_{ℓ_p} and so $\Lambda \in \ell_p^*$. Therefore, Λ is a nonzero multiple of ϕ by the uniqueness mentioned above. This is a contradiction, because $\Lambda \in \ell_2$ while $\phi \notin \ell_2$. Therefore, we conclude that $\bar{x}$ is not a support point of C. □

5.1.15. Let X be a finite-dimensional Banach space, and $\|\cdot\|$ an equivalent norm on X.
(a) Show $\|\cdot\|$ is strictly convex if and only if it is uniformly convex.
(b) Show $\|\cdot\|$ is Gâteaux differentiable if and only if it is uniformly smooth.
(c) Show $x_0 \in B_X$ is strongly exposed if and only if it is exposed.

5.1.16.* Let X be a Banach space with equivalent norms $\|\cdot\|_1$ and $\|\cdot\|_2$. Let $\|\cdot\|$ be defined by $\|\cdot\|^2 := \|\cdot\|_1^2 + \|\cdot\|_2^2$. Suppose $\|\cdot\|_1$ is strictly (resp. locally uniformly, uniformly) convex, show that $\|\cdot\|$ is an equivalent strictly (resp. locally uniformly, uniformly) convex.
Hint. Apply Fact 5.1.9(c), Fact 5.1.12(b) and Fact 5.1.17(b) respectively. □

5.1.17. Suppose X is a reflexive Banach space. Show that a norm $\|\cdot\|$ on X is Fréchet differentiable and locally uniformly convex if and only if its dual norm is Fréchet differentiable and locally uniformly convex. (Compare to Remark 5.1.31.)
Hint. Suppose the dual norm to $\|\cdot\|$ is Fréchet differentiable and locally uniformly convex. Then $\|\cdot\|$ is Fréchet differentiable. Now suppose $x_n, x \in S_X$ are such that $\|x_n + x\| \to 2$. Choose $\phi \in S_{X^*}$ so that $\phi(x) = 1$ and $\phi_n \in S_{X^*}$ so that $\phi_n(x_n + x) = \|x_n + x\|$. Then $\phi_n(x) \to 1$. Because $\|\cdot\|$ is Fréchet differentiable, Šmulian's theorem (5.1.4) implies, $\phi_n \to \phi$. Therefore, $\phi(x_n) \to 1$. Because the dual norm is Fréchet differentiable at ϕ and $\phi(x) = 1$, Šmulian's theorem (5.1.4) implies $x_n \to x$ as desired. □

5.1.18.* Prove Proposition 5.1.18.
Hint. See the proof of Proposition 4.2.14. For further details, see, for example, [199, Fact 9.7]. □

5.1.19.* Use Proposition 5.1.22 to deduce the duality relations between uniform convexity and uniform smoothness.
Hint. See [199, Theorem 9.10], for example, for a proof of this. □

5.1.20.* Suppose the norm on X is uniformly convex (resp. uniformly smooth) and Y is finitely representable in X. Show that Y admits an equivalent uniformly convex (resp. uniformly smooth) norm. Deduce that such an X is superreflexive.
Hint. See Exercise 4.1.26 for facts concerning finite representability; see also [199, Proposition 9.16] for the first part. Deduce superreflexivity from the first part and the Milman–Pettis theorem (5.1.20). This shows the easier part of the James–Enflo theorem (5.1.24). □

5.1.21.* Use Proposition 5.1.22 to prove Corollary 5.1.23.

5.1.22.* This considers when the sequences in Fact 5.1.17(b) can be unbounded; a more general result will be given in Theorem 5.4.6 below. Let X be a Banach space, show that the following are equivalent.

(a) $\|x_n - y_n\| \to 0$ whenever $2\|x_n\|^2 + 2\|y_n\|^2 - \|x_n + y_n\|^2 \to 0$.
(b) $\|\cdot\|$ is uniformly convex with modulus of convexity of power type 2.

Hint. (a) $\Rightarrow$ (b): Suppose $\|\cdot\|$ does not have modulus of convexity of power type 2. Then there exist $u_n, v_n \in S_X$ such that $\|u_n - v_n\| \geq \frac{1}{n}$ while $\|u_n + v_n\| \geq 2 - \frac{\varepsilon_n}{n^2}$ where $\varepsilon_n \to 0^+$. Now let $x_n = nu_n$, $y_n = nv_n$; then

$$\|x_n - y_n\| \geq 1, \quad \text{and} \quad \|x_n + y_n\| \geq 2n - \frac{\varepsilon_n}{n}.$$

Consequently,

$$2\|x_n\|^2 + 2\|y_n\|^2 - \|x_n + y_n\|^2 \leq 2n^2 + 2n^2 - \left(2n - \frac{\varepsilon_n}{n}\right)^2$$

$$= 4\varepsilon_n - \frac{\varepsilon_n^2}{n^2},$$

and so $2\|x_n\|^2 + 2\|y_n\|^2 - \|x_n + y_n\|^2 \to 0$ which means (a) fails.

(b) $\Rightarrow$ (a): Suppose $\|\cdot\|$ has modulus of convexity of power type 2. Now suppose that $\|x_n - y_n\| \geq \delta$ but $2\|x_n\|^2 + 2\|y_n\|^2 - \|x_n + y_n\|^2 \to 0$. Then $(\|x_n\| - \|y_n\|)^2 \to 0$. Let $\|x_n\| = \alpha_n$. Because $\|\cdot\|$ is uniformly convex, we know $\alpha_n \to \infty$. Thus, by replacing y_n with $\alpha_n \frac{y_n}{\|y_n\|}$, we have

$$\liminf \|x_n - y_n\| \geq \delta \quad \text{and} \quad \|y_n\| = \|x_n\| = \alpha_n.$$

Thus we may assume $\|x_n - y_n\| > \eta$ for all n where $\eta = \frac{\delta}{2}$. Now let $u_n = \frac{x_n}{\alpha_n}$ and $v_n = \frac{y_n}{\alpha_n}$. Then $\|u_n - v_n\| \geq \frac{\eta}{\alpha_n}$, and because the modulus of convexity of $\|\cdot\|$ is of power type 2, there is a $C > 0$ so that $\|u_n + v_n\| \leq 2 - C\left(\frac{\eta}{\alpha_n}\right)^2$. Therefore, $\|x_n + y_n\| \leq 2\alpha_n - \frac{C\eta^2}{\alpha_n}$. Consequently, we compute

$$2\|x_n\|^2 + 2\|y_n\|^2 - \|x_n + y_n\|^2 \geq 2\alpha_n^2 + 2\alpha_n^2 - \left(2\alpha_n - \frac{C\eta^2}{\alpha_n}\right)^2$$

$$= 4C\eta^2 - \frac{C^2\eta^4}{\alpha_n^2} \not\to 0.$$

This contradiction completes the proof. □

5.1.23.* Let $1 < p \leq 2$. Show that a norm $\|\cdot\|$ has modulus of smoothness of power type p if and only if there exists $C > 0$ so that $\|J(x) - J(y)\| \leq C\|x - y\|^{p-1}$ for all $x, y \in S_X$, where J is the duality mapping.

Hint. The similar proof for the analogous result for convex functions is given in Exercise 5.4.13; for further details see [180, Lemma IV.5.1]. □

5.1 Renorming: an overview

5.1.24 (Asplund averaging). This exercise outlines a simple approach of John and Zizler [268] to the following averaging results.

(a) Suppose X admits an equivalent norm that is locally uniformly convex (resp. strictly convex), and X^* admits a dual norm that is locally uniformly convex. Show that X admits a norm that is locally uniformly convex (resp. strictly convex) and Fréchet differentiable.

(b) Suppose X admits a strictly convex norm (resp. locally uniformly convex) and X^* admits a dual norm that is strictly convex (resp. locally uniformly convex), then X admits a norm that is strictly convex and Gâteaux differentiable (resp. strictly convex and Fréchet differentiable).

(c) According to the James–Enflo theorem (5.1.24) X admits a uniformly convex norm if and only if X^* admits a uniformly convex norm. With the help of this theorem, show that if X admits a uniformly convex norm, then X admits a uniformly convex norm that is uniformly smooth.

Hint. Let $\|\cdot\|$ be locally uniformly convex on X and let $\|\cdot\|_1$ be a norm whose dual $\|\cdot\|_1^*$ is locally uniformly convex on X^*. Consider $\|\cdot\|_n$ on X whose dual is defined by $\|\phi\|_n^* := \sqrt{\|\phi\|^2 + \frac{1}{n}\|\phi\|_1^2}$. Then $\|\cdot\|_n$ is Fréchet differentiable and the sequence $(\|\cdot\|_n)$ converges uniformly to $\|\cdot\|$ on bounded sets. Now let $\|\cdot\|$ on X be defined by $\|x\|^2 := \sum_{k=1}^\infty \frac{1}{2^n}\|x\|_n^2$. Then $\|\cdot\|$ is Fréchet differentiable, and

$$2\|x_n\|^2 + 2\|x\|^2 - \|x + x_n\|^2 \to 0 \Rightarrow 2\|x_n\|_k^2 + 2\|x\|_k^2 - \|x + x_n\|_k^2 \to 0 \text{ for all } k.$$

Because of uniform convergence the latter implies $2\|x\|^2 + 2\|x_n\|^2 - \|x+x_n\|^2 \to 0$ which implies $\|x - x_n\| \to 0$ and hence $\|x - x_n\| \to 0$.

The other cases are analogous. □

5.1.25.* Suppose X is a separable or more generally a WCG Banach space. Show that X admits a norm that is simultaneously locally uniformly convex and Gâteaux differentiable.

5.1.26.* Construct an equivalent norm on ℓ_2 that is strictly convex but there is a point on S_X which is not strongly exposed. Conclude that ℓ_2 has an equivalent norm that is Gâteaux differentiable but fails to be Fréchet differentiable at some point of S_X.

Hint. Define $\|\cdot\|$ by

$$\|x\| := \max\left\{\frac{1}{2}\|x\|, |x_1|\right\} + \sqrt{\sum_{n=1}^\infty \frac{x_i^2}{2^i}}.$$

Verify that $\|\cdot\|$ is strictly convex, and that $e_1/\|e_1\|$ is supported by e_1, but not strongly exposed in the unit ball of $\|\cdot\|$. Use duality for the differentiability assertion. □

5.1.27 (Constructing smooth norms).* Suppose X is a Banach space and $f : X \to \mathbb{R}$ is a coercive, continuous convex function.

(a) Let β be a bornology on X. Suppose that f is β-differentiable, then X admits an equivalent β-differentiable norm. Moreover, if X is a dual space and f is weak*-lsc, show the norm constructed is a dual norm.

(b) Suppose f' is uniformly continuous (resp. satisfies an α-Hölder condition where $0 < \alpha \le 1$) on bounded subsets of X. Show that X admits an equivalent uniformly smooth norm (resp. norm whose derivative satisfies an α-Hölder condition on its sphere).

Hint. Replace f with $\frac{1}{2}[f(x) + f(-x)]$. Check that this symmetric f has: the same smoothness properties as the original f, and a minimum at 0, and so one may shift f by a constant and assume $f(0) = 0$. The norm $\|\cdot\|$ with unit ball $B = \{x : f(x) \le 1\}$ will have the desired properties. According to the Implicit function theorem (4.2.13) and its proof $\|x\|' = \frac{f'(x)}{\langle f'(x), x \rangle}$ for $\|x\| = 1$.

For (a), use the fact that $\langle f'(x), x \rangle \ge 1$ (since $f(0) = 0$ and $f(x) = 1$), it follows that the derivative of $\|\cdot\|$ is τ_β-continuous on its sphere if the derivative of f is. The assertion concerning dual norms follows when B is weak*-closed.

For (b), compute, for example

$$\|\|x\|' - \|y\|'\| = \left\| \frac{f'(x)}{\langle f'(x), x \rangle} - \frac{f'(y)}{\langle f'(y), y \rangle} \right\|$$

$$= \left\| \frac{f'(x)}{\langle f'(x), x \rangle} - \frac{f'(x)}{\langle f'(y), y \rangle} + \frac{f'(x)}{\langle f'(y), y \rangle} - \frac{f'(y)}{\langle f'(y), y \rangle} \right\|$$

$$\le \frac{\|f'(x)\|}{|\langle f'(x), x \rangle \langle f'(y), y \rangle|} |\langle f'(x), x \rangle - \langle f'(y), y \rangle|$$

$$+ \frac{1}{|\langle f'(y), y \rangle|} \|f'(x) - f'(y)\|.$$

and deduce the appropriate uniformity for $\|\cdot\|'$. Note also a dual approach to this exercise is given in Theorem 5.4.3. $\square$

5.1.28. Suppose X is a Banach space. Show that X is reflexive if and only if there exists a function $f : X^* \to \mathbb{R}$ that is coercive, weak*-lsc, convex and Fréchet differentiable. Formulate and prove a local version, i.e. where f is Fréchet differentiable on a neighborhood, of x, and $f(\cdot) - \nabla f(x)$ satisfies an appropriate growth condition.

Hint. Use Exercise 5.1.27, Remark 5.1.31 and the fact a reflexive Banach space admits a norm whose dual is Fréchet differentiable, as follows for example, from Corollary 5.1.33. $\square$

5.1.29.*

(a) Prove that g_n and h_n are convex, and f_n, g_n and h_n are Lipschitz on bounded sets, and $f_n(x) \le f(x)$ for all x, where f_n, g_n and h_n are from the proof of Theorem 5.1.25.

(b) Verify (5.1.6) in the proof of Theorem 5.1.25.

5.1.30 (Uniformly Gâteaux differentiable norms).** A (Gâteaux differentiable) norm $\|\cdot\|$ on a Banach X is said to be *uniformly Gâteaux differentiable* if for each $h \in S_X$

the limit

$$\lim_{t \to 0} \frac{\|x + th\| - \|x\|}{t} = \nabla \|x\|(h)$$

is uniform of $x \in S_X$. Show that the following are equivalent.

(a) $\|\cdot\|$ is uniformly Gâteaux differentiable.
(b) For each $h \in S_X$ and $\varepsilon > 0$ there exists $\delta > 0$ such that

$$\|x + th\| + \|x - th\| - 2\|x\| \leq \varepsilon t \text{ for all } 0 < t < \delta, \, x \in S_X.$$

(c) The norm $\|\cdot\|$ is Gâteaux differentiable and for each $h \in S_X$ the mapping $x \mapsto \nabla\|x\|(h)$ is uniformly continuous on S_X.

Hint. Modify the proof of Proposition 4.2.14 appropriately for directions $h \in S_X$. For further information on spaces that admit uniformly Gâteaux differentiable norms, see, for example, [199, p. 395ff]. □

5.1.31 (Weak*-uniformly convex norms).** A norm $\|\cdot\|$ on a Banach space X is said to be *weakly uniformly convex* if $(x_n - y_n) \to 0$ weakly whenever $x_n, y_n \in B_X$ are such that $\|x_n + y_n\| \to 2$. Analogously, a dual norm $\|\cdot\|$ on X^* is said to be *weak*-uniformly convex* if $(x_n - y_n) \to 0$ weak* whenever $x_n, y_n \in B_X^*$ are such that $\|x_n + y_n\| \to 2$. Show the following claims.

(a) A norm is uniformly Gâteaux differentiable if and only if its dual norm is weak*-uniformly convex.
(b) A norm is weakly uniformly convex if and only if its dual norm is uniformly Gâteaux differentiable.
(c) Suppose X is separable. Show that X^* admits an equivalent dual weak*-uniformly convex norm.
(d) Suppose X^* is separable. Show that X admits an equivalent weakly uniformly convex norm.

Hint. Both (a) and (b) are directional versions of the duality between uniform convexity and uniform smoothness (modify the proof of Theorem 5.1.19 or see [180, p. 63] for further details). For (c), let $\{x_n\}_{n=1}^\infty$ be dense in S_X, define the norm $\|\cdot\|$ on X^* by $\|f\|^2 := \|f\|^2 + \sum_{n=1}^\infty 2^{-n} f^2(x_n)$. Suppose that $\|f_k\| \leq 1$, $\|g_k\| \leq 1$ and $\|f_k + g_k\| \to 2$. As in the proof of Theorem 5.1.14, conclude, for each $n \in \mathbb{N}$ that $(f_k - g_k)(x_n) \to 0$ as $k \to \infty$. The proof of (d) is similar to (c): define $\|\cdot\|$ on X by $\|x\|^2 := \|x\|^2 + \sum_{n=1}^\infty 2^{-n} f_n^2(x)$ where $\{f_n\}_{n=1}^\infty$ is dense in S_{X^*}. □

5.1.32 (Extension of rotund norms [412]).** Suppose X admits an equivalent uniformly convex norm and Y is a closed subspace of X. Show that any equivalent uniformly convex norm on Y can be extended to an equivalent uniformly convex norm on X. Analogous statements are valid for strict convexity, local uniform convexity and several other forms of rotundity (see [412, Theorem 1.1]).

Hint. Let $|\cdot|_Y$ be a given equivalent uniformly convex norm on Y. Extend this to an equivalent norm $|\cdot|$ on X. Let $\|\cdot\|$ be a uniformly convex norm on X such that $\|\cdot\| \leq \frac{1}{\sqrt{2}}|\cdot|$. Define a function $p(\cdot)$ on $B_X(|\cdot|)$ by

$$p^2(x) := |x|^2 + q(x) + d_Y^2(x) \quad \text{where } q(x) := d_Y^2(x)e^{\|x\|^2}.$$

Show that $q(\cdot)$ is convex on $\{x : \|x\| \leq 1\}$ by showing $(r,s) \mapsto r^2 e^{s^2}$ is convex on $(0, 1/\sqrt{2}) \times (0, 1/\sqrt{2})$. (Indeed, the Hessian is positive definite if $|s| < 1/\sqrt{2}$ and r is arbitrary, see also Exercise 2.4.30.)

Let $B := \{x : p(x) \leq 1\}$. Show that B is the unit ball of an equivalent uniformly convex norm on X that extends the norm $|\cdot|_Y$. See [412] for full details. □

5.1.33 (Smooth approximation). ⋆⋆ A vast amount of research has been devoted to the question as to when continuous functions on a Banach space can be approximated uniformly by smooth functions (for various types of smoothness). Seminal results include Kurzweil's paper [282] on analytic approximations and Bonic and Frampton's paper on smooth approximation [61]. Many such results depend on the structure of the space and the existence of a smooth norm or perhaps only a smooth bump function on the space. See [180, Chapter 8] and the references therein for a broader account of this topic. Since the publication of the book [180] there have been some important developments on analytic approximations, in this direction we refer the reader to [149, 202, 176, 177, 223]. We should also mention that ensuring the approximating functions are not only smooth but also Lipschitz when the original function is Lipschitz is quite subtle; see, for example, [224, 225].

On the other hand, there are results that only assume the existence of a nice norm on X to obtain smooth approximations. In this direction, one has

Theorem 5.1.35 (Frontisi [221]). *Suppose X admits a locally uniformly convex norm, and every Lipschitz convex function on X can be approximated by C^k-smooth functions. Then every continuous function on X can be approximated uniformly by C^k-smooth functions.*

The following two exercises provide an application of Theorem 5.1.35.

(a) Suppose that X^* admits an equivalent dual locally uniformly convex norm (resp. strictly convex norm). Show that every convex function bounded on bounded sets of X can be approximated by C^1-smooth (resp. Gâteaux differentiable) convex functions that are bounded on bounded sets.
(b) Conclude that if X admits a locally uniformly convex norm, and X^* admits a dual locally uniformly convex norm, then every continuous function on X can be approximated uniformly on X by C^1-smooth functions.
    As an application of Theorem 5.1.25 show that:
(c) Every Lipschitz function on a superreflexive space can be approximated uniformly on bounded sets by functions whose derivatives are uniformly continuous on bounded sets. Conversely, note that a Banach space is superreflexive if every Lipschitz function can be so approximated. This follows because from such

an approximate one can construct a bump function with uniformly continuous derivative, according to Exercise 5.5.7 below, the space is superreflexive.

Hint. Observe that (a) and (b) become almost trivial if one uses a stronger theorem from [222, Theorem 4.2] that shows in Theorem 5.1.35, one need only approximate *norms* by uniformly on bounded sets by C^k-smooth functions. For (a), suppose the norm $\|\cdot\|$ is such that its dual is locally uniformly convex. Given a convex function f bounded on bounded sets, consider $f_n = f \square n \|\cdot\|^2$. For (c), do the same but with a norm that is uniformly smooth. □

5.1.34 (Rehilbertizing, [248]).** Let us call a bounded linear operator on a Hilbert space X *strongly positive definite* (s-pd) if for some $c > 0$ one has $\langle Ax, x \rangle \geq c\|x\|^2$ for all $x \in X$ (i.e. it is linear, monotone and coercivity). (a) Show that the space X is Euclidean if and only if such operators coincide with positive definite ones. To study s-pd operators in general, we shall consider a second inner product

$$[x, y] = [x, y]_H := \langle Hx, x \rangle$$

induced by a strongly positive definite operator H. (b) Show that every norm-equivalent inner product is induced by some s-pd H.

Let $A^{[*]}$ be the adjoint of A in the induced inner-product. (c) Show that (i) $A^{[*]} = H^{-1}AH$ for all $A \in B(X)$; (ii) $A^{[*]} = A^*$ if and only if HA and A^*H are both self-adjoint in $\langle \cdot, \cdot \rangle$; and (iii) when $AH = HA$ then $A = A^*$ if and only if $A = A^{[*]}$. Let $\rho(A) := \sup\{|\lambda| : \lambda \in \sigma(A)\}$ denote the *spectral radius* of A: $\sigma(A)$ denotes the spectrum of A, and $\rho(A) = \lim_{n \to \infty} \|A^n\|^{1/n}$ (in the original operator norm) [28].

Theorem 5.1.36 (Stein). *Suppose $H = H^*$ and B are bounded operators on a Hilbert space X and that $T_B(H) := H - B^*HB$ is strongly positive definite Then H is strongly positive definite if and only if $\rho(B) < 1$.*

The original proof considered point, continuous, and residual spectra of B separately. Renorming, provides a more transparent proof. It is worthwhile to consider a proof in the Euclidean case.

Hint. [$\rho(B) < 1 \Rightarrow H$ p-sd]. Note first for $n \in \mathbb{N}$, that

$$T_B(H) + B^*T_B(H)B + \cdots + B^{*n}T_B(H)B^n = T_{B^n}(H). \quad (5.1.7)$$

As $\rho(B) < 1$, there is $n \in \mathbb{N}$ with $\|B^{nk}\| \to 0$ as $k \to \infty$. Since $T_{B^{nk}}(H)$ is s-pd, H is positive semidefinite. Confirm H is also invertible and so s-pd.
[$\rho(B) < 1 \Leftarrow H$ p-sd]. Note $[x, y] = \langle x, Hy \rangle = \langle Hy, x \rangle$. Fix $x \in S_X$, the original sphere. Then $0 < c \leq \langle x, Hx \rangle - \langle Bx, HBx \rangle = [x, x] - [Bx, Bx] \Leftrightarrow [Bx, x]_H/[x, x]_H \leq 1 - c/\langle x, Hx \rangle \leq 1 - c/\|H\|$. By homogeneity, $\|B\|_H < 1$ (the induced operator norm). Each such norm dominates $\rho(B)$. □

5.2 Exposed points of convex functions

Let $f : X \to [-\infty, +\infty)$. Suppose $x_0 \in \operatorname{dom} f$ and $f(x_0) > f(x)$ for all $x \in X \setminus \{x_0\}$, then we say f attains its *strict maximum* at x_0; a *strict minimum* can be defined analogously when $f : X \to (-\infty, +\infty]$. Now suppose $f : X \to [-\infty, +\infty)$. Then f is said to attain a *strong maximum* at $x_0 \in \operatorname{dom} f$, if $f(x_0) \geq f(x)$ for all $x \in C$, and $x_n \to x_0$ whenever $f(x_n) \to f(x_0)$; a *strong minimum* can be defined analogously when $f : X \to (-\infty, +\infty]$.

Example 5.2.1. Suppose C is a closed convex subset of a Banach space X and $\phi \in X^*$ strongly exposes $x_0 \in C$. Then $\phi - \delta_C$ attains its strong maximum at x_0 and $\delta_C - \phi$ attains its strong minimum at x_0.

We will say that $x_0 \in \operatorname{dom} f$ is an *exposed point* of f if $(x_0, f(x_0))$ is an exposed point of $\operatorname{epi} f$. Notice that $(x_0, f(x_0))$ is exposed by a functional of the form $(\phi, -1) \in X^* \times \mathbb{R}$. Indeed, suppose the functional $(\phi, r_0) \in X^* \times \mathbb{R}$ attains its *strict maximum* at $(x_0, f(x_0)) \in \operatorname{epi} f$. Then $r_0 < 0$ because

$$(\phi, r_0)(x_0, f(x_0)) := \phi(x_0) + r_0 f(x_0) > (\phi, r_0)(x_0, t) \quad \text{when } t > f(x_0).$$

Now it is easy to check that $(\phi_0, -1)$ where $\phi_0 = \frac{1}{|r_0|}\phi$ exposes $(x_0, f(x_0))$ in $\operatorname{epi} f$.

Figure 5.3 shows an example of an extreme point of a function that is not exposed.

Proposition 5.2.2. *For a function $f : X \to (-\infty, +\infty]$, the following are equivalent.*

(a) *x_0 is an exposed point of f, where $\operatorname{epi} f$ is exposed by $(\phi_0, -1)$ at $(x_0, f(x_0))$.*
(b) *$\phi_0 - f$ attains a strict maximum at x_0.*

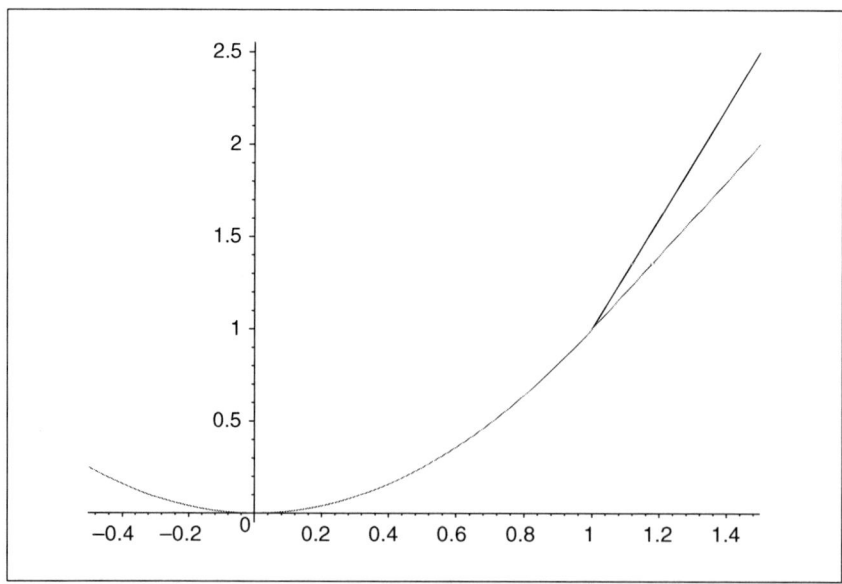

Figure 5.3 Convex functions with $(1, 1)$ exposed (upper), and extreme but not exposed (lower) in the epigraph.

5.2 Exposed points of convex functions

(c) $f - \phi_0$ attains a strict minimum at x_0.
(d) $\phi_0 \in \partial f(x_0)$ and $\phi_0 \notin \partial f(x)$ if $x \neq x_0$ (succinctly $(\partial f)^{-1}(\phi_0) = \{x_0\}$).

Proof. (a) $\Rightarrow$ (b): Now $(\phi_0, -1)(x_0, f(x_0)) > (\phi_0, -1)(x, f(x))$ and so $\phi_0(x_0) - f(x_0) > \phi_0(x) - f(x)$ for all $x \in \text{dom} f$. Consequently, (b) is true.

(b) $\Rightarrow$ (a): Suppose that $\phi_0 - f$ attains a strict maximum at x_0. Let $(x, t) \in \text{epi} f$ with $(x, t) \neq (x_0, f(x_0))$. If $x \neq x_0$, then $(\phi_0, -1)(x, t) = \phi_0(x) - t \leq \phi_0(x) - f(x) < \phi(x_0) - f(x_0) = (\phi_0, -1)(x_0, f(x_0))$. If $x = x_0$, then $t > f(x_0)$, and so clearly $(\phi_0, -1)(x_0, t) < (\phi_0, -1)(x_0, f(x_0))$.

It clear that (b) and (c) are equivalent, and the equivalence of (b) and (d) follows because, in general, $(\phi - f)$ attains its maximum at x if and only if $\phi \in \partial f(x)$. Thus $(\phi_0 - f)$ attains a strict maximum at x_0 if and only if $(\partial f)^{-1}(\phi_0) = \{x_0\}$. □

We will say that $x_0 \in \text{dom} f$ is a *strongly exposed point* of f if $(x_0, f(x_0))$ is a strongly exposed point of the epigraph of f.

Theorem 5.2.3. *For a proper function $f : X \to (-\infty, +\infty]$, the following conditions (a) through (d) are equivalent.*

(a) x_0 is a strongly exposed point of f, where $\text{epi} f$ is exposed by $(\phi_0, -1)$ at $(x_0, f(x_0))$.
(b) $\phi_0 - f$ attains a strong maximum at x_0.
(c) $f - \phi_0$ attains a strong minimum at x_0.
(d) $\phi_0 \in \partial f(x_0)$ and if $\phi_0 \in \partial_{\varepsilon_n} f(x_n)$ where $\varepsilon_n \to 0^+$, then $\|x_n - x\| \to 0$.

If, moreover, f is lsc with bounded domain or if f is lsc convex, then each of the above conditions is equivalent to:

(e) f^* is Fréchet differentiable at ϕ_0 with $\nabla f^*(\phi_0) = x_0$.

Proof. The proof of the equivalence of (a) and (b) is similar to the argument given in Proposition 5.2.2; see Exercise 5.2.7. The equivalence of (b) and (c) is straightforward, and their equivalence with (d) can be derived easily from the following observation. Given $\phi_0 \in \partial f(x_0)$ observe that

$\phi_0 \in \partial_{\varepsilon_n} f(x_n)$ with $\varepsilon_n \to 0^+$

$\Leftrightarrow \phi_0(x_0) - \phi_0(x_n) \leq f(x_0) - f(x_n) + \varepsilon_n$ where $\varepsilon_n \to 0^+$
$\Leftrightarrow (\phi_0 - f)(x_0) \leq (\phi - f)(x_n) + \varepsilon_n$ where $\varepsilon_n \to 0^+$
$\Leftrightarrow (\phi_0 - f)(x_n) \to (\phi_0 - f)(x_0)$.

(e) $\Rightarrow$ (d): Suppose additionally f is lsc, and that (e) holds with $x_0 = \nabla f^*(\phi_0)$. According to Exercise 4.4.2, $\phi_0 \in \partial f(x_0)$. Now suppose $\phi_0 \in \partial_{\varepsilon_n} f(x_n)$ where $\varepsilon_n \to 0^+$. Proposition 4.4.5(b) implies $x_n \in \partial_{\varepsilon_n} f^*(\phi_0)$. Šmulian's theorem (4.2.10) then shows $\|x_n - x_0\| \to 0$ as desired.

(c) $\Rightarrow$ (e): Suppose additionally f has bounded domain with the diameter of the domain not exceeding M where $M > 0$. Let f, ϕ_0 and x_0 be as in (c). Let $0 < r \leq M$ be given; using (c), we choose $\varepsilon > 0$ so that $(f - \phi_0)(x_0 + h) \geq \varepsilon$ if $\|h\| \geq r/2$. Now let $g(\cdot) = \frac{\varepsilon}{M} d_C(\cdot) + (f - \phi_0)(x_0)$ where $C = \{x : \|x - x_0\| \leq r/2\}$. Then g is a continuous convex function such that $g \leq f - \phi_0$, and $g(x_0) = (f - \phi_0)(x_0)$ and $f^{**} - \phi_0 \geq g$; moreover $g(x) \geq r\varepsilon/(2M)$ whenever $\|x - x_0\| \geq r$, and so

$$(f^{**} - \phi_0)(x) \geq (f^{**} - \phi_0)(x_0) + \frac{r\varepsilon}{2M} \text{ whenever } \|x - x_0\| \geq r.$$

Because $0 < r \leq M$ was arbitrary, this shows $f^{**} - \phi_0$ attains its strong minimum at x_0.

Now, $x_0 \in \partial f^*(\phi_0)$ and we suppose $x_n \in \partial_{\varepsilon_n} f^*(\phi_0)$ where $\varepsilon_n \to 0^+$. Then $(f^{**} - \phi_0)(x_n) \to (f^{**} - \phi_0)(x_0)$ and consequently $\|x_n - x_0\| \to 0$. According to Šmulian's theorem from Exercise 4.2.10, f^* is Fréchet differentiable at ϕ_0 with $\nabla f^*(\phi_0) = x_0$.

The equivalence of (a) and (e) in the case f is a proper lsc convex function is left as Exercise 5.2.2. □

Exercise 5.2.8 illustrates that (d) and (e) of the previous theorem are logically independent without additional assumptions on f. We next examine the natural duality between smoothness and exposed points.

Proposition 5.2.4. *Let $f : X \to (-\infty, +\infty]$ be a lsc convex function with $x \in \text{dom } f$.*

(a) *Then f^* is strongly exposed by $x \in X$ at $\phi \in X^*$ if and only if f is Fréchet differentiable at x with $f'(x) = \phi$.*

(b) *Suppose additionally f is continuous at x. Then f^* is exposed by $x \in X$ at $\phi \in X^*$ if and only if f is Gâteaux differentiable at x with $f'(x) = \phi$.*

Proof. We prove only part (a); the similar proof of (b) is left as Exercise 5.2.10. First, suppose f^* is strongly exposed by $x \in X$ at $\phi \in X^*$. Then f is continuous at x by Exercise 5.2.6; also $\phi \in \partial f(x)$. Now suppose $\phi_n \in \partial_{\varepsilon_n} f(x)$ where $\varepsilon_n \to 0^+$. According to Proposition 4.4.5(b), $x \in \partial_{\varepsilon_n} f^*(\phi_n)$. Now Theorem 5.2.3 implies $\|\phi_n - \phi\| \to 0$. Finally, Šmulian's theorem (4.2.10) implies f is Fréchet differentiable at x.

Conversely, suppose f is Fréchet differentiable at x with $f'(x) = \phi$. Then $\phi \in \partial f(x)$ and so Proposition 4.4.5(a) implies that $x \in \partial f^*(\phi)$. Now suppose $x \in \partial_{\varepsilon_n} f^*(\phi_n)$ where $\varepsilon_n \to 0^+$. Then $\phi_n \in \partial_{\varepsilon_n} f(x)$. According to Šmulian's theorem (4.2.10), $\|\phi_n - \phi\| \to 0$. Therefore, Theorem 5.2.3 implies f^* is strongly exposed at ϕ by x. □

Exercises and further results

5.2.1 (Exposed points in Euclidean space). Suppose E is a Euclidean space, and $f : E \to (-\infty, +\infty]$ is a proper lsc convex function. Show that x_0 is an exposed point of f if and only if it is strongly exposed. Find an example of a proper convex function g on $\mathbb{R}^2$ that has an exposed point that is not strongly exposed.

Hint. Suppose x_0 is an exposed point of f. Choose $\phi \in \partial f(x_0)$ so that $f - \phi$ attains its strict minimum at x_0. Now suppose $(f - \phi)(x_n) \to (f - \phi)(x_0)$ but $\|x_n - x_0\| \not\to 0$. Using the convexity of $f - \phi$ and by appropriately choosing $y_n := \lambda_n x_n + (1 - \lambda_n) x_0$, there is a bounded sequence (y_n) such that $(f - \phi)(y_n) \to (f - \phi)(x_0)$ but

$\|y_n - x_0\| \nrightarrow 0$, and by passing to a subsequence we may assume $y_n \to \bar{y}$. Using the lower semicontinuity of $f - \phi$ we arrive at the contradiction $(f - \phi)(\bar{y}) = (f - \phi)(x_0)$. One such example is $g(x, y) := x^2$ when $x > 0$, $g(0, 0) := 0$ and $g(x, y) := +\infty$ otherwise. Show that g is exposed at $(0, 0)$ by ϕ where ϕ is the zero-functional, but $(g - \phi)(n^{-1}, 1) \to (g - \phi)(0, 0)$. □

5.2.2 (Tikhonov well-posed).* A function f is said to be *Tikhonov well-posed* if every minimizing sequence converges in norm to the unique minimizer. Equivalently, this says f attains its strong minimum. Suppose f is a proper lsc convex function.

(a) Show that f is Tikhonov well-posed with minimum at $\bar{x}$ if and only if f^* has Fréchet derivative $\bar{x}$ at 0.

(b) Deduce that (a) through (e) are equivalent in Theorem 5.2.3 when f is assumed to be a proper lsc convex function.

(c) Show that f is Fréchet differentiable at 0 with $\nabla f(0) = x^*$ if and only if f^* is Tikhonov well-posed with minimum at x^*.

Hint. For (a), use Šmulian's theorem as in the proof of Theorem 5.2.3, and use the fact $f^{**}|_X = f$. Observe that (c) follows from Proposition 5.2.4. □

5.2.3.* Let f be a proper lsc, convex function such that f^* is Fréchet differentiable on $\text{dom}(\partial f^*)$. Suppose that $x_n \to \bar{x}$ weakly, $f(x_n) \to f(\bar{x})$ and $\partial f(\bar{x}) \neq \emptyset$. Show that $\|x_n - \bar{x}\| \to 0$. Is it necessary to assume $x_n \to \bar{x}$ weakly?

Hint. This is from [94, Lemma 2.8]. Show there exists $\phi \in \partial f(\bar{x})$ that strongly exposes f; the weak convergence ensures $\phi(x_n) \to \phi(\bar{x})$. Yes, it is necessary to assume $x_n \to \bar{x}$ weakly, otherwise one could consider $f(x) = \|x\|^2$ on $X = \ell_2$, and $(e_n) \subset S_X$. Then $f(e_n) = 1$, but (e_n) is not norm convergent. □

5.2.4 (Exposed points). (a) Suppose x_0 exposes f^* at ϕ_0 and f is lsc, convex and proper. Suppose $f^*(\phi_n) - \langle x_0, \phi_n \rangle \to f^*(\phi_0) - \langle x_0, \phi_0 \rangle$; show that $\phi_n \in \partial f_{\varepsilon_n}(x_0)$ where $\varepsilon_n \to 0$.

(b) Find an example of a Lipschitz convex function $f : \mathbb{R} \to \mathbb{R}$ such that the epigraph of f has an extreme point which is not an exposed point.

Hint. For (a), observe first, $x_0 \in \partial f^*(\phi_0)$, and therefore, $\phi_0 \in \partial f(x_0)$, which implies $f^*(\phi_0) - \langle x_0, \phi_0 \rangle = -f(x_0)$. Consequently, $f^*(\phi_n) - \langle x_0, \phi_n \rangle \to -f(x_0)$. In other words, $\phi_n(x) - f(x) - \phi_n(x_0) \leq -f(x_0) + \varepsilon_n$ for all $x \in X$ where $\varepsilon_n \to 0^+$. Thus $\phi_n \in \partial_{\varepsilon_n} f(x_0)$.

For (b), consider the function $f(t) := t^2$ if $|t| \leq 1$, and $f(t) := 2|t| - 1$ if $|t| \geq 1$; see Figure 5.3. □

5.2.5 (Exposed points). Suppose $f^* : X^* \to (-\infty, +\infty]$ is a proper, weak*-lsc convex function that is exposed at $\phi_0 \in X^*$ by $x_0 \in X$ and that

$$f^*(\phi_n) - \langle x_0, \phi_n \rangle \to f^*(\phi_0) - \langle x_0, \phi_0 \rangle.$$

(a) Show that $\phi_n \to_{w^*} \phi_0$ whenever $(\phi_n)_{n=1}^\infty$ is bounded.
(b) Find an example where $(\phi_n)_{n=1}^\infty$ is unbounded and f^* is Lipschitz.

(c) Is $(\phi_n)_{n=1}^{\infty}$ bounded if f^* is the conjugate of a lsc convex function f that is continuous at x_0?

Hint. (a) Suppose $\phi_n \not\to_{w^*} \phi_0$. Because $(\phi_n)_{n=1}^{\infty}$ is bounded, it then has a weak*-convergent subnet (ϕ_{n_α}) that converges to $\bar\phi \neq \phi_0$. Now, by the weak*-lower semicontinuity of $f^*(\cdot) - \langle x_0, \cdot \rangle$, we have that

$$f^*(\phi_0) - \langle x_0, \phi_0 \rangle = \limsup_n f^*(\phi_n) - \langle x_0, \phi_n \rangle$$
$$\geq \liminf_\alpha f^*(\phi_{n_\alpha}) - \langle x_0, \phi_{n_\alpha} \rangle$$
$$\geq f^*(\bar\phi) - \langle x_0, \bar\phi \rangle.$$

(b) Define $f^* : \ell_1 \to \mathbb{R}$ where $f((x_i)) := \sum 2^{-i} |x_i|$. Then f^* is exposed by $0 \in c_0$, but $f^*(ne_n) - \langle 0, ne_n \rangle = \frac{1}{n}$ and $ne_n \not\to_{w^*} 0$.
(c) Yes: $\phi_n \in \partial_{\varepsilon_n} f(x_0)$ where $\varepsilon_n \to 0$ by Exercise 5.2.4(a). Hence (ϕ_n) must be bounded because f is continuous at x_0. □

5.2.6.* Let $f : X \to (-\infty, +\infty]$ be a lsc proper convex function.

(a) Suppose $\phi \in X^*$ strongly exposes f at x_0. Show that $f - \phi$ is coercive.
(b) Suppose x_0 strongly exposes f^* at ϕ, show that f is continuous on a neighborhood of x_0.

Hint. (a) There exists $\delta > 0$ such that $(f - \phi)(u) \geq (f - \phi)(x) + \delta$ whenever $\|u - x\| = 1$. The convexity of $(f - \phi)$ now implies $(f - \phi)(u) > (f - \phi)(x) + n\delta$ whenever $\|x - u\| \geq n$ where $n \in \mathbb{N}$. Consequently, if $\|u\| \to \infty$, $\|u - x\| \to \infty$, and so $(f - \phi)(u) \to \infty$.
 (b) By part (a), $f^* - x_0$ is coercive, and by a Moreau–Rockafellar result (Corollary 4.4.11) f^{**} is continuous at x_0. Deduce the conclusion because $f^{**}|_X = f$. □

5.2.7.* Prove the equivalence of (a) and (b) in Theorem 5.2.3.

Hint. (a) $\Rightarrow$ (b): Suppose $(x_0, f(x_0))$ is strongly exposed by $(\phi_0, -1)$, and that $(\phi_0 - f)(x_n) \to (\phi_0 - f)(x_0)$. Then

$$(\phi_0, -1)(x_n, f(x_n)) \to (\phi_0, -1)(x_0, f(x_0))$$

and (a) implies $\|(x_n, f(x_n)) - (x_0, f(x_0))\| \to 0$ which implies $\|x_n - x_0\| \to 0$.
 (b) $\Rightarrow$ (a): Suppose that $\phi_0 - f$ has a strong maximum at x_0. Then by Proposition 5.2.2, $(\phi_0, -1)$ exposes epi$f$ at $(x_0, f(x_0))$. Now if $(x_n, t_n) \in$ epi$f$ and $(\phi_0, -1)(x_n, t_n) \to (\phi_0, -1)(x_0, f(x_0))$, then $(\phi_0, -1)(x_n, f(x_n)) \to (\phi_0, -1)(x_0, f(x_0))$ since $f(x_n) \leq t_n$ for all n. Therefore,

$$(\phi_0 - f)(x_n) \to (\phi_0 - f)(x_0). \quad (5.2.1)$$

Now, (b) implies that $\|x_n - x_0\| \to 0$. Therefore $\phi_0(x_n) \to \phi_0(x_0)$; this with (5.2.1) implies $f(x_n) \to f(x_0)$. Therefore, $\|(x_n, f(x_n)) - (x_0, f(x_0))\| \to 0$ as desired. □

5.2.8.* Show that neither of Theorem 5.2.3(c) or Theorem 5.2.3(e) generally implies the other.

Hint. Let $f(t) = \min\{|t|, 1\}$. Then $f - 0$ attains a strong minimum at 0, but $f^* = \delta_{\{0\}}$ is not Fréchet differentiable at 0. Let $g(t) = |t|$ if $t \neq 0$ and $g(0) = +\infty$. Then $g^* = \delta_{[-1,1]}$ so $\nabla g^*(0) = 0$, but $f - 0$ does not attain its strong minimum at 0. □

5.2.9. (a) Does $f(x) := \|x\|$ have any exposed points as a convex function?
(b) Show that $x_0 \in S_X$ is an exposed (resp. strongly exposed) point of B_X if and only if x_0 is an exposed (resp. strongly exposed) point of the convex function $f(x) = \|x\|^2$.

Hint. (a) For any norm, 0 is the only exposed point of $f(x) = \|x\|$.
(b) Observe that $\phi \in S_{X^*}$ strongly exposes x_0 if and only if $\phi(x_0) = 1$ and $x_n \to x_0$ whenever $\|x_n\| \to 1$ and $\phi(x_n) \to 1$.
Then $\phi \in \partial f(x_0)$ if and only if $\|\phi/2\| = 1$ and $\phi(x_0)/2 = 1$, and $(\phi - f)(x_n) \to (\phi - f)(x_0)$ implies $\|x_n\| \to 1$. Thus $(\phi - f)(x_n) \to (\phi - f)(x_0)$ implies $\phi(x_n) \to \phi(x_0)$ and $\|x_n - x_0\| \to 0$. □

5.2.10.* Prove Proposition 5.2.4(b).

Hint. Suppose f^* is exposed at ϕ by $x \in X$. Then $x \in \partial f^*(\phi)$ and so Proposition 4.4.5(a) implies $\phi \in \partial f(x)$. Now Proposition 5.2.2 implies $\partial f(x) = \{\phi\}$ and so f is Gâteaux differentiable at x according to Corollary 4.2.5.
Conversely, suppose f is Gâteaux differentiable at x with $f'(x) = \phi$. Then $\partial f(x) = \{\phi\}$. Then $x \in \partial f^*(\phi)$ according to Proposition 4.4.5(a), and moreover, $x \notin \partial f^*(\Lambda)$ for $\Lambda \neq \phi$ (or else $\Lambda \in \partial f(x)$). Therefore, Proposition 5.2.2 implies that f^* is exposed by x at ϕ. □

5.2.11. This is a function version of Exercise 5.1.14. Let $1 < p < 2$, and let $f : \ell_2 \to (-\infty, +\infty]$ be defined by

$$f(x) := \begin{cases} \sum_{i=1}^{\infty} |x_i|^p & \text{if } x = (x_i) \in \ell_p; \\ \infty & \text{otherwise.} \end{cases}$$

Show that f is a lsc strictly convex function and thus every point in dom f is an extreme point of f. However, there are points in the *relative interior* of the domain of f where the subdifferential is empty.

Hint. Let $g(x) := \|x\|^p$ where $x \in \ell_p$. Then g is a continuous supercoercive convex function on a reflexive space, therefore its subdifferential is onto ℓ_p^*. Thus we find $\bar{x} \in \ell_p$ so that $\phi = \partial g(\bar{x})$ (note that g is differentiable) and $\phi \in \ell_p^* \setminus \ell_2$. Now $\bar{x}$ is in the relative interior of the domain of f, if $\Lambda \in \partial f(\bar{x})$, then $\Lambda|_{\ell_p} = \phi$, but this cannot happen since $\phi \notin \ell_2$. □

5.2.12 (Perturbed minimization principles). Let X be a Banach space, and let $C \subset X$ be a closed bounded convex set. Show that the following are equivalent.

(a) Every weak*-lsc convex function $f : X^* \to \mathbb{R}$ such that $f \leq \sigma_C$ is Fréchet differentiable on a dense G_δ-subset of X^*.

(b) Given any proper lsc bounded below function $f : C \to (-\infty, +\infty]$ and $\varepsilon > 0$, there exist $\phi \in \varepsilon B_{X^*}$ and $x_0 \in C$ such that $f - \phi$ attains its strong minimum at x_0.

Hint. (a) $\Rightarrow$ (b): Suppose $f : C \to \mathbb{R}$ is bounded below on C. Then there exists $a \in \mathbb{R}$ so that $f + a \geq \delta_C$. Consequently, $f^* - a = (f + a)^* \leq \delta_C^* \leq \sigma_C$. Given $\varepsilon > 0$, there exists $\phi \in \varepsilon B_{X^*}$ so that $f^* - a$ and hence f^* is Fréchet differentiable at ϕ. Conclude that $f - \phi$ attains its strong minimum at x_0.

(b) $\Rightarrow$ (a): Take any weak*-lsc convex $g \leq \sigma_C$. Let $f = g^*|_X$. Then $f \geq \delta_C$, and $f^* = g$. Now let $\Lambda \in X^*$ be arbitrary, then $f + \Lambda$ is bounded below on C, so $f + \Lambda$ is strongly exposed by some $\phi \in \varepsilon B_{X^*}$. This implies $(f + \Lambda)^*$ is Fréchet differentiable at ϕ. But $(f + \Lambda)^*(\cdot) = g^*(\cdot - \Lambda)$, and so g^* is Fréchet differentiable at $\Lambda + \phi$. Conclude that the points of differentiability of f is a dense (automatically) G_δ-set. □

5.3 Strictly convex functions

Recall a proper convex function $f : X \to (-\infty, +\infty]$ is *strictly convex* if

$$f(\lambda x + (1 - \lambda)y) < \lambda f(x) + (1 - \lambda)f(y)$$

whenever $x \neq y$, $0 < \lambda < 1$, $x, y \in \mathrm{dom}\, f$.

Notice that some confusion may arise, because even when $\|\cdot\|$ is a strictly convex norm, the function $f(\cdot) := \|\cdot\|$ is not a strictly convex function. Thus one must be careful to note the context in which the adjective strictly convex is used, the same will be true for locally uniformly convex, and uniform convex functions as defined later in this section. Figure 7.1 illustrates the nuances of strict convexity reasonably well.

Fact 5.3.1. *For a proper convex function $f : X \to (-\infty, +\infty]$, the following are equivalent.*

(a) *f is strictly convex.*
(b) *$f\left(\frac{x+y}{2}\right) < \frac{1}{2}f(x) + \frac{1}{2}f(y)$ for all distinct $x, y \in \mathrm{dom}\, f$.*

Proof. Clearly, (a) implies (b). The converse can be proved by observing when f is not strictly convex, then its graph contains a line segment by the three-slope inequality (Fact 2.1.1) and so then (b) will fail for some x and y. □

Example 5.3.2. Recall that $t \mapsto t^2$ is strictly convex on $\mathbb{R}$ because its derivative is strictly increasing. Let $f : \mathbb{R} \to (-\infty, +\infty]$ be defined by $f(t) = t^2$ if t is rational and $f(t) = +\infty$ otherwise. Then

$$f\left(\frac{t+s}{2}\right) < \frac{1}{2}f(t) + \frac{1}{2}f(s) \text{ for all } s \neq t,\ s,t \in \mathrm{dom}\, f.$$

but f is not convex, and hence not strictly convex.

5.3 Strictly convex functions

We will say a convex function f is *strictly convex* at $x_0 \in \text{dom} f$ if $f(\frac{y+x_0}{2}) < \frac{1}{2}f(x_0) + \frac{1}{2}f(y)$ for all $y \in \text{dom} f \setminus \{x_0\}$.

Fact 5.3.3. *Suppose $f : X \to (-\infty, +\infty]$ is convex, lsc and proper. Then f is strictly convex if and only if $(x_0, f(x_0))$ is an extreme point of $\text{epi} f$ whenever $x_0 \in \text{dom} f$.*

Proof. For the 'only if' implication, suppose f is strictly convex and $x_0 \in \text{dom} f$. Suppose that $(x_0, f(x_0)) = \lambda(x, t) + (1-\lambda)(y, s)$ where $0 < \lambda < 1$ and $x, y \in \text{dom} f$, $f(x) \leq t$, and $f(y) \leq s$. If $x = y = x_0$, then $s = t = f(x_0)$. Thus we suppose $x \neq x_0$ and $y \neq x_0$. According to the strict convexity of f, we deduce that $f(x_0) < \lambda t + (1-\lambda)s$ which is a contradiction. For the converse, suppose $x, y \in \text{dom} f$, $x \neq y$, and that $x_0 = \lambda x + (1-\lambda)y$ where $0 < \lambda < 1$. Given that $(x_0, f(x_0))$ is an extreme point of $\text{epi} f$, we deduce that $f(x_0) < \lambda f(x) + (1-\lambda) f(y)$. Therefore f is strictly convex. □

A convex function f is said to be *locally uniformly convex* at $x_0 \in \text{dom} f$ if

$$\|x_n - x_0\| \to 0 \text{ whenever } \frac{1}{2}f(x_n) + \frac{1}{2}f(x_0) - f\left(\frac{x_n + x_0}{2}\right) \to 0, \ x_n \in \text{dom} f.$$

We will say a convex function $f : X \to (-\infty, +\infty]$ is *locally uniformly convex* if it is locally uniformly convex at each $x \in \text{dom} f$. Observe that a locally uniformly convex function is strictly convex.

Proposition 5.3.4. *Suppose $f : X \to (-\infty, +\infty]$ is a proper lsc function and that $\phi \in \partial f(x_0)$.*

(a) If f is strictly convex at x_0, then ϕ exposes f at x_0.
(b) If f is locally uniformly convex at x_0, then ϕ strongly exposes f at x_0.

Proof. (a) Suppose $x \in \text{dom} f$ and $x \neq x_0$. Write $x = x_0 + h$, then $f(x_0 + \frac{1}{2}h) < \frac{1}{2}f(x_0) + \frac{1}{2}f(x_0 + h)$. Now $\frac{1}{2}\phi(h) = \phi(x_0 + \frac{1}{2}h) - \phi(x_0) < \frac{1}{2}f(x_0 + h) - \frac{1}{2}f(x_0)$ and so $\phi(h) < f(x_0 + h) - f(x_0)$, that is $\phi(x) - \phi(x_0) < f(x) - f(x_0)$.
(b) Suppose (x_n) is such that $f(x_n) - \phi(x_n) \to f(x_0) - \phi(x_0)$. Now observe

$$0 \leq \frac{1}{2}[f(x_n) + f(x_0)] - f\left(\frac{x_n + x_0}{2}\right)$$

$$\leq f(x_n) - \frac{1}{2}\phi(x_n) + \frac{1}{2}\phi(x_0) - \left[f(x_0) + \frac{1}{2}\phi(x_n) - \frac{1}{2}\phi(x_0)\right]$$

now the right-hand side of this goes to 0, so since f is locally uniformly convex at x_0, we have that $\|x_n - x_0\| \to 0$. □

The converses to the preceding proposition do not hold in the following sense. Let $f(\cdot) := \|\cdot\|$. Then $\phi = 0$ strongly exposes f at $x = 0$. However, f is not strictly convex at $x = 0$.

Proposition 5.3.5. *Suppose X is a Banach space and $f : X \to \mathbb{R}$ is a continuous convex function. Then the following are equivalent.*
(a) *f is strictly convex.*
(b) *Every $x_0 \in X$ is an exposed point of f.*
(c) *For every $x_0 \in X$, $(x_0, f(x_0))$ is an extreme point of $\operatorname{epi} f$.*

Proof. First, note that (a) $\Rightarrow$ (b) follows from Proposition 5.3.4 because $\operatorname{dom}(\partial f) = X$. Also, (c) $\Rightarrow$ (a) follows from Fact 5.3.3.

(b) $\Rightarrow$ (c): Suppose $x\, y \in X$ and let $x_0 := \lambda x + (1-\lambda) y$ where $0 < \lambda < 1$. Choose $\phi \in \partial f(x_0)$ such that $f - \phi$ attains a strict minimum at x_0. Then

$$f(x_0) < \lambda (f - \phi)(x) + (1-\lambda)(f - \phi)(y) + \phi(x_0)$$
$$= \lambda f(x) + (1-\lambda) f(y),$$

and so $(x_0, f(x_0))$ is an extreme point of $\operatorname{epi} f$. □

Proposition 5.3.6. *Suppose $f : X \to (-\infty, +\infty]$ is convex lsc and proper.*

(a) *If f^* is strictly convex, then f is Gâteaux differentiable at each point in the interior of its domain.*
(b) *If f^* is locally uniformly convex, then f is Fréchet differentiable at each point in the interior of its domain.*

Proof. Suppose $x_0 \in \operatorname{int}(\operatorname{dom} f)$. Then f is continuous at x_0 and so $\partial f(x_0) \neq \emptyset$, say $\phi \in \partial f(x_0)$. Thus $\phi \in \operatorname{dom}(f^*)$, and $x_0 \in \partial f^*(\phi)$. Because f^* is strictly convex (locally uniformly convex), Proposition 5.3.4 implies x_0 exposes (strongly exposes) f^* at ϕ and so f is Gâteaux (Fréchet) differentiable at x_0 by Proposition 5.2.4. □

We now present a rather satisfactory duality theorem for reflexive spaces.

Theorem 5.3.7. *Suppose X is a reflexive Banach space and $f : X \to \mathbb{R}$ is a continuous cofinite convex function. Then:*

(a) *f is Gâteaux differentiable if and only if f^* is strictly convex.*
(b) *The following are equivalent:*

  (i) *f is Fréchet differentiable;*
  (ii) *f^* is strongly exposed at each $x^* \in X^*$ by each subgradient in $\partial f(x^*)$;*
  (iii) *f^* is strongly exposed at each $x^* \in X^*$.*

Proof. (a) The 'if' implication follows immediately from Proposition 5.3.6. For the converse, let $x_0^* \in X^*$. Because f^* is continuous, we fix $x_0 \in X$ such that $x_0 \in \partial f^*(x_0^*)$. Now $x_0^* \in \partial f(x_0)$, and thus Proposition 5.2.4 ensures that x_0 exposes f^* at x_0^*. Thus every $x^* \in X^*$ is an exposed point of f^*, and so f^* is strictly convex according to Proposition 5.3.5.

(b) (i) $\Rightarrow$ (ii): Suppose f is Fréchet differentiable and $x_0^* \in X^*$. Choose any $x_0 \in \partial f^*(x_0^*)$. Because $x_0^* \in \partial f(x_0)$, Proposition 5.2.4 implies f^* is strongly exposed at x_0^* by x_0.

(ii) $\Rightarrow$ (iii) follows because $\partial f^*(x^*)$ is nonempty at each $x^* \in X^*$. We now prove (iii) $\Rightarrow$ (i): Suppose that f^* is strongly exposed at each $x^* \in X^*$. Now fix $x_0 \in X$ and suppose $\phi \in \partial f(x_0)$ and $\phi_n \in \partial_{\varepsilon_n} f(x_0)$ where $\varepsilon_n \to 0^+$. Then $x_0 \in \partial f^*(\phi)$, $\phi_n \to_{w^*} \phi$ (by Šmulian's theorem (4.2.11) since f is Gâteaux differentiable by part (a) and Proposition 5.3.5) and $f^*(\phi_n) \to f^*(\phi)$. Now let y strongly expose f^* at ϕ. Because $\phi_n \to \phi$ weakly, it now follows that $\limsup (f^* - y)(\phi_n) \leq (f^* - y)(\phi)$. Consequently, $\|\phi_n - \phi\| \to 0$. According to Smulian's theorem (4.2.10), f is Fréchet differentiable at x_0. $\square$

A proper convex function $f : X \to (-\infty, +\infty]$ is said to be *uniformly convex on bounded sets* if $\|x_n - y_n\| \to 0$ whenever $(x_n), (y_n)$ are bounded sequences in $\mathrm{dom}\, f$ such that

$$\frac{1}{2} f(x_n) + \frac{1}{2} f(y_n) - f\left(\frac{x_n + y_n}{2}\right) \to 0.$$

If $\|x_n - y_n\| \to 0$ whenever $x_n, y_n \in \mathrm{dom}\, f$ are such that

$$\frac{1}{2} f(x_n) + \frac{1}{2} f(y_n) - f\left(\frac{x_n + y_n}{2}\right) \to 0,$$

then f is said to be *uniformly convex*. We do not require $\mathrm{dom}\, f$ to have nonempty interior. Thus, for example, if $\mathrm{dom}\, f$ is a singleton, then f is trivially uniformly convex.

We leave the proof of the following observation as an exercise.

Proposition 5.3.8. *Let f and g be proper convex functions, if f is strictly convex (resp. locally uniformly convex, uniformly convex), then $f + g$ is strictly convex (resp. locally uniformly convex, uniformly convex) when it is proper.*

Intuitively, the previous proposition holds because adding a convex function to another only makes the function 'more convex'. Also, let us note:

Remark 5.3.9. (a) Let X be finite-dimensional and let $f : X \to \mathbb{R}$ be continuous and convex. Then f is strictly convex if and only if f is uniformly convex on bounded sets.
(b) See Exercise 5.3.9 for a lsc strictly convex function $f : \mathbb{R}^2 \to (-\infty, +\infty]$ that is not uniformly convex on bounded sets.
(c) Let X be finite-dimensional, and let $f : X \to \mathbb{R}$ be continuous and convex. Then f is Gâteaux differentiable if and only if f is uniformly smooth on bounded sets.

Proof. (a) Suppose f is strictly convex, but not uniformly convex on bounded sets. Then there are bounded sequences (x_n) and (y_n) such that $\|x_n - y_n\| \geq \varepsilon$ for all n and some $\varepsilon > 0$, yet

$$\frac{1}{2} f(x_n) + \frac{1}{2} f(y_n) - f\left(\frac{x_n + y_n}{2}\right) \to 0.$$

Because of compactness, there are subsequences $(x_{n_k}) \to \bar{x}$ and $(y_{n_k}) \to \bar{y}$. The continuity of f then implies

$$\frac{1}{2}f(\bar{x}) + \frac{1}{2}f(\bar{y}) - f\left(\frac{\bar{x}+\bar{y}}{2}\right) = 0.$$

This contradicts the strict convexity of f. The other direction is trivial. Similarly, (c) follows from a compactness argument and Theorem 2.2.2. □

The function, $f(t) := |t|^p$ for $p > 1$, is differentiable and strictly convex (for example $f''(t) > 0$ for $t > 0$). Thus part (a) of the next example is immediate from the previous observation. While (c) can be obtained rather easily with the help of the mean value theorem.

Example 5.3.10. Consider $f : \mathbb{R} \to \mathbb{R}$ defined by $f(t) := |t|^p$ where $p \geq 1$. Then

(a) f is uniformly convex and uniformly smooth on bounded sets for $p > 1$.
(b) f is uniformly convex if and only if $p \geq 2$.
(c) f is uniformly smooth if and only if $1 < p \leq 2$.

Proof. See Exercise 5.3.6(b). □

The next example builds certain convex functions from norms.

Example 5.3.11. Let $p > 1$ and X be a Banach space and suppose its norm $\|\cdot\|$ is strictly (resp. locally uniformly, uniformly) convex. Then the function $f := \|\cdot\|^p$ is strictly convex (resp. locally uniformly convex, uniformly convex on bounded sets).

Proof. First, suppose $\|\cdot\|$ is uniformly convex. Now suppose $(x_n), (y_n) \subset X$ are bounded sequences such that

$$\frac{1}{2}\|x_n\|^p + \frac{1}{2}\|y_n\|^p - \left\|\frac{x_n+y_n}{2}\right\|^p \to 0. \tag{5.3.1}$$

The uniform convexity of the function $g = |\cdot|^p$ on bounded subsets of $\mathbb{R}$ then implies $(\|x_n\| - \|y_n\|) \to 0$, and thus (5.3.1) ensures $(\|(x_n + y_n)/2\| - \|x_n\|) \to 0$. The uniform convexity of $\|\cdot\|$ implies $\|x_n - y_n\| \to 0$ as desired (for this, boundedness of the sequences is essential). The other proofs are similar: for example, in the local uniformly convex case, replace the sequence (y_n) with a single point x. □

It is worth noting that determining the uniform convexity of $\|\cdot\|^p$ is actually quite delicate, and depends on the modulus of convexity of the norm as will be seen later in Theorem 5.4.6. We now turn from examples to duality results.

Proposition 5.3.12. *Suppose $f : X \to \mathbb{R}$ is continuous convex. If f^* is uniformly convex on bounded sets, then f' is uniformly continuous on sets D such that $\partial f(D)$ is bounded.*

Proof. Suppose $\partial f(D)$ is bounded, (x_n) and (y_n) are sequences in D such that $\|x_n - y_n\| \to 0$. First, f is Fréchet differentiable because f^* is locally uniformly convex

5.3 Strictly convex functions

(Proposition 5.3.6). Now let $\phi_n := f'(x_n)$ and $\Lambda_n := f'(y_n)$. Then (ϕ_n) and (Λ_n) are bounded and $f^*(\phi_n) = \phi_n(x_n) - f(x_n)$ and $f^*(\Lambda_n) = \Lambda_n(y_n) - f(y_n)$ and note that $f^*(\frac{\phi_n+\Lambda_n}{2}) \geq \langle \frac{\phi_n+\Lambda_n}{2}, \frac{x_n+y_n}{2}\rangle - f(\frac{x_n+y_n}{2})$. Therefore,

$$0 \leq \frac{1}{2}f^*(\phi_n) + \frac{1}{2}f^*(\Lambda_n) - f^*\left(\frac{\phi_n + \Lambda_n}{2}\right)$$

$$\leq \frac{1}{2}[\phi_n(x_n) - f(x_n)] + \frac{1}{2}[\Lambda_n(y_n) - f(y_n)]$$

$$- \left[\left\langle \frac{\phi_n + \Lambda_n}{2}, \frac{x_n + y_n}{2}\right\rangle - f\left(\frac{x_n + y_n}{2}\right)\right]$$

$$= \frac{1}{2}\phi_n\left(\frac{x_n - y_n}{2}\right) + \frac{1}{2}\Lambda_n\left(\frac{y_n - x_n}{2}\right) + f\left(\frac{x_n + y_n}{2}\right)$$

$$- \frac{1}{2}f(x_n) - \frac{1}{2}f(y_n)$$

$$\leq \frac{1}{2}\phi_n\left(\frac{x_n - y_n}{2}\right) + \frac{1}{2}\Lambda_n\left(\frac{y_n - x_n}{2}\right).$$

The last term goes to 0 because $\|x_n - y_n\| \to 0$ and the functionals are uniformly bounded. Because f^* is uniformly convex on bounded sets, $\|\phi_n - \Lambda_n\| \to 0$. □

Proposition 5.3.13. *Suppose $f : X \to \mathbb{R}$ is convex and has uniformly continuous derivative. Then f^* is uniformly convex.*

Proof. Suppose $(\phi_n), (\Lambda_n) \subset \operatorname{dom} f^*$ are such that

$$\frac{1}{2}f^*(\phi_n) + \frac{1}{2}f^*(\Lambda_n) - f^*\left(\frac{\phi_n + \Lambda_n}{2}\right) \to 0.$$

Given $\varepsilon_n \to 0^+$, choose $x_n \in X$ such that $f^*(\frac{\phi_n+\Lambda_n}{2}) = (\frac{\phi_n+\Lambda_n}{2})(x_n) - f(x_n) + \varepsilon_n$. Then

$$\frac{1}{2}[f^*(\phi_n) - (\phi_n(x_n) - f(x_n))] + \frac{1}{2}[f^*(\Lambda_n) - (\Lambda_n(x_n) - f(x_n))] - \varepsilon_n \to 0.$$

Now each of the preceding terms in brackets is nonnegative, consequently, they both must go to 0 since their sum goes to 0. It follows that $\phi_n \in \partial_{\tilde{\varepsilon}_n} f(x_n)$ and $\Lambda_n \in \partial_{\tilde{\varepsilon}_n} f(x_n)$ where $\tilde{\varepsilon}_n \to 0^+$. The uniform smoothness of f implies $\|\phi_n - \Lambda_n\| \to 0$ (Proposition 4.2.14). Thus f^* is uniformly convex. □

We are now ready for a duality theorem concerning uniform convexity and uniform smoothness on bounded sets.

Theorem 5.3.14. *Suppose $f : X \to (-\infty, +\infty]$ is a convex function.*

(a) *f is supercoercive and uniformly smooth on bounded sets if and only if f^* is supercoercive, bounded and uniformly convex on bounded sets.*
(b) *f is supercoercive, bounded and uniformly convex on bounded sets if and only if f^* is supercoercive and uniformly smooth on bounded sets.*

Proof. (a) Suppose f is supercoercive and uniformly smooth on bounded sets. Then f is also bounded on bounded sets because its derivative uniformly continuous on bounded sets (Proposition 4.2.15). Now f^* is supercoercive and bounded on bounded sets by Theorem 4.4.13. Suppose sequences (ϕ_n) and (Λ_n) as in the proof of Proposition 5.3.13 are bounded, the supercoercivity of f then forces the sequence (x_n) therein to be bounded, therefore the uniform smoothness of f on bounded sets ensures that $\|\phi_n - \Lambda_n\| \to 0$ (Proposition 4.2.15).

Conversely, because f^* is supercoercive and bounded on bounded sets, Theorem 4.4.13 ensures f is supercoercive and bounded on bounded sets. Therefore, ∂f is bounded on bounded sets. According to Proposition 5.3.12, $x \to f'(x)$ is uniformly continuous on bounded sets.

(b) This can be deduced from (a) because the existence of such a function f on a Banach space implies it is reflexive (Exercise 5.3.3). □

The above proof uses supercoercivity in a very crucial fashion. The following remark illustrates that this is a necessary restriction.

Remark 5.3.15. (a) Consider $f(t) := e^t - t$, then f is coercive, uniformly convex on bounded subsets of $\mathbb{R}$, but f^* is not everywhere defined, moreover f^* is not uniformly smooth bounded sets.
(b) In Exercise 5.3.10, an example of a function $f : \mathbb{R}^2 \to \mathbb{R}$ is given so that f is not strictly convex, while f^* is globally Lipschitz and uniformly smooth on bounded sets.
(c) Together (a) and (b) show that we cannot remove supercoercivity from Theorem 5.3.14 and still have duality between uniform smoothness on bounded sets and uniform convexity on bounded sets.

Thus far we almost have the duality between uniformly convex and uniformly smooth functions. The key to completing this is that uniformly convex functions are automatically supercoercive as we will see shortly.

For a proper convex function $f : X \to (-\infty, +\infty]$, we define the *modulus of convexity*, $\delta_f : [0, \infty] \to [0, \infty]$, of f by

$$\delta_f(\varepsilon) := \inf\left\{\frac{1}{2}f(x) + \frac{1}{2}f(y) - f\left(\frac{x+y}{2}\right) \; : \; \|x - y\| \geq \varepsilon, \, x, y \in \mathrm{dom} f\right\}.$$

As usual, we define the infimum over the empty set as $+\infty$. Observe that f is uniformly convex if and only if $\delta_f(\varepsilon) > 0$ for each $\varepsilon > 0$.

Fact 5.3.16. *Suppose $f : X \to (-\infty, +\infty]$ is a uniformly convex function. Then:*

(a) $\delta_f(2\varepsilon) \geq 4\delta_f(\varepsilon)$ *for all $\varepsilon > 0$. Consequently,* $\liminf_{\varepsilon \to \infty} \dfrac{\delta_f(\varepsilon)}{\varepsilon^2} > 0$; *and*

(b) $\liminf_{\|x\| \to \infty} \dfrac{f(x)}{\|x\|^2} > 0.$

5.3 Strictly convex functions

Proof. (a) Let $\varepsilon > 0$ and let $\eta > 0$. If $\delta_f(2\varepsilon) = +\infty$, there is nothing further to do, otherwise, choose $x, y \in X$ such that $\|x - y\| \geq 2\varepsilon$, and

$$f\left(\frac{x+y}{2}\right) > \frac{1}{2}f(x) + \frac{1}{2}f(y) - \delta_f(2\varepsilon) - \eta. \tag{5.3.2}$$

By shifting f, we may and do assume $x = 0$. Now

$$f\left(\frac{0+y}{2}\right) \leq \frac{1}{2}f\left(\frac{y}{4}\right) + \frac{1}{2}f\left(\frac{3}{4}y\right) - \delta_f(\varepsilon)$$

$$\leq \frac{1}{4}f(0) + \frac{1}{4}f\left(\frac{y}{2}\right) + \frac{1}{4}f\left(\frac{y}{2}\right) + \frac{1}{4}f(y) - 2\delta_f(\varepsilon)$$

$$\leq \frac{1}{4}f(0) + \frac{1}{4}f(y) + \frac{1}{4}f(0) + \frac{1}{4}f(y) - 2\delta_f(\varepsilon) - \frac{1}{2}\delta_f(2\varepsilon).$$

Comparing this with (5.3.2) noting that η is arbitrarily small, yields $\delta_f(2\varepsilon) \geq 4\delta_f(\varepsilon)$. Consequently, by induction, one can show $\delta_f(2^n) \geq (2^n)^2 \delta_f(1)$, from which it follows that $\liminf_{\varepsilon \to \infty} \delta_f(\varepsilon) \geq \frac{1}{4}\delta_f(1)\varepsilon^2$.

(b) Without loss of generality, we may shift f so that $0 \in \operatorname{dom} f$. Now let $\phi \in \partial_1 f(0)$, and then let $g(x) = f(x) - \phi(x) + 1$. Then g is uniformly convex and $g \geq 0$. Moreover, if $\|x\| = \varepsilon$, then $\frac{1}{2}g(x) + \frac{1}{2}g(0) - g(x/2) \geq \frac{1}{4}\delta_g(1)\varepsilon^2$, and so $g(x) \geq \frac{1}{4}\delta_g(1)\varepsilon^2 - g(0)$ hence $\liminf_{\|x\| \to \infty} \frac{g(x)}{\|x\|^2} > 0$, and so the same is true for f. □

Theorem 5.3.17. *Suppose $f : X \to (-\infty, +\infty]$ is a proper lsc convex function.*

(a) f is uniformly smooth if and only if f^* is uniformly convex.
(b) f is uniformly convex if and only if f^* is uniformly smooth.

Proof. We prove (b) first and then deduce (a). Suppose f^* is uniformly smooth, then f^{**} is uniformly convex by Proposition 5.3.13. Thus $f = f^{**}|_X$ is also uniformly convex. Conversely, suppose f is uniformly convex. Then f is supercoercive by Fact 5.3.16(b), so f^* is continuous everywhere. Suppose by way of contradiction that f^* is not uniformly smooth. Then there exist $(\phi_n), (\Lambda_n) \subset X^*$ where $\|\Lambda_n\| \to 0^+$ and $\varepsilon > 0$ so that

$$\frac{1}{2}f^*(\phi_n + \Lambda_n) + \frac{1}{2}f^*(\phi_n - \Lambda_n) - f^*(\phi_n) \geq 2\varepsilon\|\Lambda_n\|.$$

Choose $x_n, y_n \in X$ so that $f^*(\phi_n + \Lambda_n) \leq (\phi_n + \Lambda_n)(x_n) - f(x_n) + \varepsilon\|\Lambda_n\|$ and $f^*(\phi_n - \Lambda_n) \leq (\phi_n - \Lambda_n)(y_n) - f(y_n) + \varepsilon\|\Lambda_n\|$. Hence one can estimate

$$\frac{1}{2}\Lambda_n(x_n - y_n) + \frac{1}{2}\phi_n(x_n + y_n) - \frac{1}{2}f(x_n) - \frac{1}{2}f(y_n) - f^*(\phi_n) \geq \varepsilon\|\Lambda_n\|.$$

This with $\phi_n\left(\frac{x_n+y_n}{2}\right) - f\left(\frac{x_n+y_n}{2}\right) - f^*(\phi_n) \leq 0$ implies

$$\frac{1}{2}\Lambda_n(x_n - y_n) + f\left(\frac{x_n + y_n}{2}\right) - \frac{1}{2}f(x_n) - \frac{1}{2}f(y_n) \geq \varepsilon\|\Lambda_n\|. \tag{5.3.3}$$

Thus $\Lambda_n(x_n - y_n) \geq 2\varepsilon \|\Lambda_n\|$ for all n, and hence $\|x_n - y_n\| \geq \varepsilon$ for all n. This with (5.3.3) implies $|\Lambda_n(x_n - y_n)| \geq \delta_f(\|x_n - y_n\|) \geq \delta_f(\varepsilon) > 0$. Because $\|\Lambda_n\| \to 0$, this implies $\|x_n - y_n\| \to \infty$, but then Fact 5.3.16(a) implies $\delta_f(\|x_n - y_n\|)/\|x_n - y_n\| \to \infty$ which is impossible since $\|\Lambda_n\| \to 0$. This proves (b).

To prove (a), suppose f is uniformly smooth, then f^* is uniformly convex according to Proposition 5.3.13. Conversely, if f^* is uniformly convex, then f^{**} is uniformly smooth by part (b), and hence so is $f^{**}|_X$. □

In contrast to the Milman–Pettis theorem (5.1.20), Banach spaces with uniformly convex or uniformly smooth functions are not necessarily reflexive, for example, any constant function is uniformly smooth and the indicator function of a point is trivially uniformly convex. We close this section with the following interesting consequence of Fact 5.3.16(b).

Corollary 5.3.18. *Suppose $f : X \to (-\infty, +\infty]$ is a proper lsc uniformly convex function. Then f attains its strong minimum on X, and $\partial f(X) = X^*$.*

Proof. Because f is supercoercive (Fact 5.3.16(b)) and lsc, it is bounded below. Thus we let $(x_n) \subset \operatorname{dom} f$ be a sequence such that $f(x_n) \to \inf_X f$. Then

$$\lim_{m,n \to \infty} \frac{1}{2} f(x_n) + \frac{1}{2} f(x_m) - f\left(\frac{x_n + x_m}{2}\right) \to 0$$

and so by the uniform convexity of f, $\lim_{m,n} \|x_n - x_m\| \to 0$. Therefore, (x_n) is Cauchy and converges to some $\bar{x} \in X$. By the lower-semicontinuity of f, $f(\bar{x}) = \inf_X f$.

Now let $\phi \in X^*$. Then $f - \phi$ is uniformly convex, and so it attains its minimum at some $\bar{x}$. Thus $f(x) - \phi(x) \geq f(\bar{x}) - \phi(\bar{x})$, and so $\phi \in \partial f(\bar{x})$. □

Exercises and further results

5.3.1. Verify that the definition of locally uniformly convex functions is not changed if we assume the sequences $(x_n)_{n=1}^\infty \subset \operatorname{dom} f$ are bounded.

5.3.2. Suppose X is a Banach space, $1 < p < \infty$ and $\|\cdot\|$ is a uniformly smooth norm on X. Show that the function $f = \|\cdot\|^p$ is uniformly smooth on bounded sets. *Hint.* This can be done directly, or as a dual to Example 5.3.11. □

5.3.3.* Suppose X is a Banach space. Show the following are equivalent.

(a) There is a proper lsc locally uniformly convex (resp. uniformly convex on bounded sets) function $f : X \to (-\infty, +\infty]$ that is continuous at some point.
(b) X admits an equivalent locally uniformly convex (resp. uniformly convex) norm.
(c) There is a continuous supercoercive $f : X \to (-\infty, +\infty]$ locally uniformly convex (resp. uniformly convex on bounded sets) function.

Conclude that a Banach space is superreflexive if and only if it admits a lsc uniformly convex function whose domain has nonempty interior. Show, additionally, in the locally uniformly convex case, that the norm constructed in (b) is a dual norm when X is a dual space and the function is weak*-lsc.

Hint. (a) ⇒ (b): Without loss, assume f is continuous at 0, and let $\phi \in \partial f(0)$. Exercise 5.2.6 implies $f - \phi$ is coercive and symmetrize this new function. We provide the details in the uniformly convex case only. Now let $g(x) := \frac{f(x)+f(-x)}{2}$ and let $r > 0$ be such that $B_{2r} \subset \text{dom} f$. Now for $\|h\| = r$ we have

$$\frac{1}{2}g(h) + \frac{1}{2}g(0) - g\left(\frac{h}{2}\right) \geq \delta_g(r) > 0.$$

Thus $g(h) \geq 2\delta_g(r)$ for all h such that $\|h\| = r$. Let consider the norm $\|\cdot\|$ whose unit ball is $B = \{x : g(x) \leq \delta_g(r)\}$. Then g is Lipschitz on B. Suppose $\|x_n\| = \|y_n\| = 1$ and

$$\frac{1}{2}\|x_n\| + \frac{1}{2}\|y_n\| - \left\|\frac{x_n + y_n}{2}\right\| \to 0.$$

Then $g(\frac{x_n+y_n}{2}) \to \delta_g(r)$ since g is Lipschitz on B, consequently $\frac{1}{2}g(x_n) + \frac{1}{2}g(y_n) - g(\frac{x_n+y_n}{2}) \to 0$ and so $\|x_n - y_n\| \to 0$ and hence $\|x_n - y_n\| \to 0$.
For (b) ⇒ (c), take $f = \|\cdot\|^2$ where $\|\cdot\|$ is guaranteed by (b). □

5.3.4. Suppose $f : X \dashrightarrow \mathbb{R}$ is a coercive uniformly smooth function.

(a) Use the implicit function theorem for gauges (4.2.13) to show that X has an equivalent uniformly smooth norm.
(b) Show that f^2 is supercoercive and uniformly smooth on bounded sets. Deduce from duality and Exercise 5.3.3 that X has an equivalent uniformly smooth norm.

5.3.5 (Constructing uniformly convex functions). For this exercise, we will say a real function ϕ is *uniformly increasing* if for each $\delta > 0$, there exists $\varepsilon > 0$ so that $\phi(s) + \varepsilon \leq \phi(t)$ whenever $s, t \in \text{dom}\,\phi$ and $s + \delta \leq t$.

(a) Let $f : X \to (-\infty, +\infty]$ be a uniformly convex function, and suppose ϕ is a convex uniformly increasing function whose domain contains the range of f. Then $\phi \circ f$ is uniformly convex.
(b) Let $p \geq 1$, and consider $\phi : [0, +\infty) \to \mathbb{R}$ defined by $\phi(t) = t^p$. Show that ϕ is a uniformly increasing convex function.
(c) Let H be a Hilbert space with inner product norm $\|\cdot\|$, and let $p \geq 2$. Show that $\|\cdot\|^p$ is a uniformly convex function.
(d) Suppose $(X, \|\cdot\|)$ is a Banach space, and $\|\cdot\|$ has modulus of convexity of power type 2. Show that the function $\|\cdot\|^p$ is uniformly convex for $p \geq 2$. See Theorem 5.4.6 for a more general and precise result.
(e) Show $e^{\|\cdot\|^p}$ is uniformly convex when $p \geq 2$, and $\|\cdot\|$ is as in (c) or (d). See Exercise 5.4.7 for related constructions of uniformly convex functions.

Hint. (a) Let $x, y \in \text{dom} f$. Suppose $\|x - y\| = \varepsilon$ where $\varepsilon > 0$, and choose $\eta > 0$ so that $\phi(s) + \eta \leq \phi(t)$ wherever $s + \delta_f(\varepsilon) \leq t$. Then

$$\phi\left(f\left(\frac{x+y}{2}\right)\right) \leq \phi\left(\frac{1}{2}f(x) + \frac{1}{2}f(y) - \delta_f(\varepsilon)\right)$$

$$\leq \phi\left(\frac{1}{2}f(x) + \frac{1}{2}f(y)\right) - \eta$$

$$\leq \frac{1}{2}\phi(f(x)) + \frac{1}{2}\phi(f(y)) - \eta.$$

(b) Given $\delta > 0$, show that $s^p + \varepsilon \leq t^p$ where $\varepsilon = \min\{p\delta^p/2^{p-1}, 2^{-1}\delta^p\}$ or any other valid estimate. Suppose $s + \delta < t$. In the case $s \leq \delta/2$, $t^p - s^p \geq \delta^p - \delta^p/2^p \geq \delta^p/2$. In the case $s > \delta/2$, $t^p - s^p \geq ps^{p-1}(t-s) \geq p\delta^p/2^{p-1}$.

(c) The parallelogram law implies $f = \|\cdot\|^2$ is uniformly convex. By parts (a) and (b), $f^p = \|\cdot\|^{2p}$ is uniformly convex for each $p \geq 1$.

(d) Use Exercise 5.1.22 to conclude $\|\cdot\|^2$ is uniformly convex and proceed as in (c).

(e) Note that $t \mapsto e^t$ is uniformly increasing on $[0, \infty)$. □

5.3.6. ⋆

(a) Prove Proposition 5.3.8.
(b) Verify Example 5.3.10.
(c) Prove Remark 5.3.9(c).

Hint. (b) To verify Example 5.3.10 parts (b) and (c), note that $f = |\cdot|$ is neither uniformly convex nor uniformly smooth. Use the mean value theorem to show that $f = |\cdot|^p$ is uniformly smooth if and only if $1 < p \leq 2$. The statement on uniform convexity can be derived dually. An alternative proof of the uniform convexity statement is as follows. Exercise 5.3.5(c) ensures f is uniformly convex for $p \geq 2$, while Fact 5.3.16(b) ensures f is not uniformly convex for $p < 2$. □

5.3.7. Suppose $f : X \to (-\infty, +\infty]$ is a lsc convex function. If $f : X \to \mathbb{R}$ is a locally uniformly convex function, must f be coercive? What if f is uniformly convex on bounded sets?

Hint. No: once again $f : \mathbb{R} \to \mathbb{R}$ where $f(t) = e^t$ provides a counterexample. Compare to Fact 5.3.16(b). □

5.3.8. Suppose $f : X \to (-\infty, +\infty]$ is a lsc convex function. Suppose that f^* is strictly convex at all points of $\text{dom}(\partial f^*)$. Show that f is Gâteaux differentiable at all points in the interior of its domain.

5.3.9 (Strictly versus uniformly convex functions). Define $f : \mathbb{R}^2 \to (-\infty, +\infty]$ by

$$f(x, y) := \begin{cases} -\sqrt[4]{xy} + \frac{1}{x} + \max\left\{\frac{1}{y}, 1\right\} & \text{if } x > 0 \text{ and } y > 0; \\ +\infty & \text{otherwise.} \end{cases}$$

Verify that f is a lsc convex function that is continuous in the extended real-valued sense and that f is strictly convex, but that f is not uniformly convex on bounded subsets of its domain.

5.3 Strictly convex functions

Hint. Check that $-\sqrt[4]{xy}$ is strictly convex for $x > 0, y > 0$ by computing its Hessian. Now, f is strictly convex because it is a sum of a strictly convex function with other convex functions. Consider, for example the points $(\frac{1}{n}, 2)$ and $(\frac{1}{n}, 4)$ to see that f is not uniformly convex. □

5.3.10. Let $f(x, y) := \begin{cases} -\sqrt[4]{xy} & \text{if } 0 \leq x \leq 1 \text{ and } 0 \leq y \leq 1; \\ \infty & \text{otherwise.} \end{cases}$

(a) Show that f is not strictly convex but that it is strictly convex on $\text{dom}(\partial f)$.
(b) Show that f^* is Lipschitz and Gâteaux differentiable everywhere on $\mathbb{R}^2$.
(c) Conclude that the derivative of f^* is uniformly continuous on bounded sets.
(d) Conclude that the derivative of f^* is not uniformly continuous everywhere.

Hint. (a) By computing the Hessian, one can see that $-\sqrt[4]{xy}$ is strictly convex if $x > 0$ and $y > 0$. It is not strictly convex when $x = 0$ or $y = 0$. Notice also, $\text{dom}(\partial f)(x, y) = \emptyset$ if $x = 0$ or $y = 0$.

(b) f^* is Lipschitz because the domain of f is bounded. f^* is Gâteaux differentiable by Exercise 5.3.8.

(c) Follows from Remark 5.3.9.

(d) f is not uniformly convex, therefore f^* cannot be uniformly smooth. □

5.3.11 (Differentiability of distance functions). Suppose X is a Banach space with closed convex subset C.

(a) Suppose the dual norm on X^* is strictly convex. Show that d_C^2 is Gâteaux differentiable.
(b) Suppose the dual norm on X^* is locally uniformly convex. Show that d_C^2 is Fréchet differentiable.
(c) Suppose the dual norm on X^* is uniformly convex. Show that d_C^2 has a uniformly continuous derivative on bounded sets. It is worth noting that d_C^2 may not have a globally uniformly continuous derivative. For example, Theorem 5.4.5 below shows that even if $C = \{0\}$ and $X = \ell_p$ with its usual norm for $1 < p < 2$, d_C^2 will not be uniformly smooth.

Hint. One approach is as follows: f is the infimal convolution of $\|\cdot\|^2$ with the indicator function of C. The conjugate of this function is the sum of the support function and one-fourth the dual norm squared. The later function is strictly convex (resp. locally uniformly convex, uniformly convex on bounded sets), hence so is the sum. Now use duality results. □

5.3.12. Suppose f and g are proper lsc convex functions on a Banach space X such that $f \square g$ is continuous. Show that $f \square g$ is Gâteaux (resp. Fréchet, uniformly) smooth provided f^* is strictly (resp. locally uniformly, uniformly) convex. Provide an example where $f \square g$ can fail to be Gâteaux differentiable even when f is Fréchet differentiable.

Hint. Use Lemma 4.4.15 with Proposition 5.3.8 and appropriate duality results. The desired example can be found in [419, Remark 2.5(a)]. □

5.3.13 (Directional uniform convexity).** Given a Banach space X with a topology τ, a convex function $f : X \to (-\infty, +\infty]$ is said to be τ-*uniformly convex* if $(x_n - y_n) \to_\tau 0$ whenever $(x_n), (y_n)$ are sequences in dom f such that

$$\frac{1}{2}f(x_n) + \frac{1}{2}f(y_n) - f\left(\frac{x_n + y_n}{2}\right) \to 0.$$

When τ is the weak topology on X (resp. weak* topology when X is a dual space), we say f is *weakly uniformly convex* (resp. *weak*-uniformly convex*).

(a) Suppose that $f : X \to \mathbb{R}$ is continuous convex function such that $x \mapsto f'(x)$ is norm-to-weak*-uniformly continuous. Show that f^* is weak*-uniformly convex.
(b) Suppose $f : X \to \mathbb{R}$ is continuous convex. Suppose f^* is weak*-uniformly convex. Show that $x \mapsto f'(x)$ is norm-to-weak*-uniformly continuous on sets D such that $\partial f(D)$ is bounded.
(c) Suppose $f : X \to \mathbb{R}$ is a Lipschitz convex function. Show that $x \mapsto f'(x)$ is uniformly Gâteaux differentiable if and only if f^* is weak*-uniformly convex.

Hint. Mimic the proofs given for the analogous cases for uniformly convex and uniformly smooth functions. □

5.3.14. Suppose f is a proper lsc convex function on a Banach space X.

(a) Suppose f^* is a supercoercive strictly convex function. Show that f is bounded on bounded sets and Gâteaux differentiable.
(b) Suppose f^* is supercoercive and Gâteaux differentiable everywhere. Show that f is strictly convex.
(c) Suppose f^* is Gâteaux differentiable everywhere. Show that f is strictly convex on the interior of its domain. In particular, if $X = \mathbb{R}$, f is strictly convex.
(d) Suppose X is reflexive. Show that $f : X \to \mathbb{R}$ is strictly convex, supercoercive and bounded on bounded sets if and only if f^* is Gâteaux differentiable, supercoercive and bounded on bounded sets.

Hint. Use appropiate duality theorems to deduce (a). For (b), f is bounded on bounded sets, hence continuous. Let $x \in X$, and take $\phi \in \partial f(x)$. We claim that ϕ exposes f at x. Indeed, if not, then there exist $y \neq x$ such that $f(y) = f(x) + \phi(y) - \phi(x)$. From this, $\phi \in \partial f(y)$. Now $x, y \in \partial f^*(\phi)$ which implies f^* is not Gâteaux differentiable at ϕ. To prove (c), one can mimic the proof of (b), since $\partial f(x) \neq \emptyset$ if $x \in \text{int}(\text{dom} f)$. □

5.3.15. Suppose X is a Banach space and $f : X \to (-\infty, +\infty]$ is a lsc convex function. Show that the following are equivalent.
(a) f^* is supercoercive, bounded on bounded sets and Fréchet differentiable.
(b) X is reflexive and f is supercoercive, bounded on bounded sets and f is strongly exposed at each $x \in X$.

Hint. (a) $\Rightarrow$ (b): Because f^* is supercoercive and bounded on bounded sets, so is f (Theorem 4.4.13). The reflexivity of X follows from Exercise 5.1.28. Let $x \in X$;

5.3 Strictly convex functions

choose $\phi \in \partial f(x)$ because f is continuous at x. By Theorem 5.2.3, f is strongly exposed at x by ϕ since f^* is Fréchet differentiable at ϕ.

(b) $\Rightarrow$ (a): Because f is supercoercive and bounded on bounded sets, so is f^*. Use Theorem 5.3.7 to deduce that f^* is Fréchet differentiable. □

5.3.16. Suppose X is a Banach space. Show that X is reflexive if and only if it admits a continuous convex cofinite function f such that f and f^* are both Fréchet differentiable.

Hint. If X is reflexive consider $f := \frac{1}{2}\|\cdot\|^2$ where $\|\cdot\|$ is an equivalent locally uniformly convex and Fréchet differentiable norm on X. For the converse, use the Moreau-Rockafellar dual theorem (4.4.12) to deduce that f^* is coercive, and then apply Exercise 5.1.28 to deduce X is reflexive. □

5.3.17. Suppose X is a reflexive Banach space and $f : X \to (-\infty, +\infty]$ is a lsc proper function.
(a) Suppose f is continuous, cofinite and f and f^* are both Fréchet differentiable. Show that f and f^* are both locally uniformly convex.
(b) Show that f is supercoercive, bounded on bounded sets, locally uniformly convex and Fréchet differentiable if and only if f^* is.
(c) Find an example of a supercoercive Fréchet differentiable function f such that f^* is not locally uniformly convex.

Hint. (a) We will show that f is locally uniformly convex. For this, suppose that $(x_n)_{n=1}^\infty$ is bounded (see Exercise 5.3.1) and

$$\frac{1}{2}f(x) + \frac{1}{2}f(x_n) - f\left(\frac{x + x_n}{2}\right) \to 0.$$

Let $\phi \in \partial f(x)$ and $\phi_n \in \partial f(\frac{x+x_n}{2})$. Then

$$\frac{1}{2}f(x) + \frac{1}{2}f(x_n) - \left[\phi_n\left(\frac{x + x_n}{2}\right) - f^*(\phi_n)\right] \to 0 \text{ and so}$$

$$\frac{1}{2}[f(x) - \phi_n(x)] + \frac{1}{2}[f(x_n) - \phi_n(x_n)] + f^*(\phi_n) \to 0.$$

Now both $f(x) - \phi_n(x) \geq -f^*(\phi_n)$ and $f(x_n) - \phi_n(x_n) \geq -f^*(\phi_n)$, so we conclude that

$$\phi_n(x) - f(x) - f^*(\phi_n) \to 0 \text{ and } \phi_n(x_n) - f(x_n) - f^*(\phi_n) \to 0. \quad (5.3.4)$$

Thus $f(x) + f^*(\phi_n) \leq \phi_n(x) + \varepsilon_n$ where $\varepsilon_n \to 0^+$. According to Proposition 4.4.1(b), $\phi_n \in \partial_{\varepsilon_n} f(x)$. Because f is Fréchet differentiable at x, Šmulian's theorem (4.2.10) implies $\|\phi_n - \phi\| \to 0$. Because $(x_n)_{n=1}^\infty$ is bounded, this implies $\phi_n(x_n) - \phi(x_n) \to 0$. From (5.3.4), we also have $\phi(x_n) - f(x_n) \geq f^*(\phi) - \varepsilon_n$ where $\varepsilon_n \to 0^+$. Again, Proposition 4.4.1(b) implies $\phi \in \partial_{\varepsilon_n} f(x_n)$ which in turn implies $x_n \in \partial_{\varepsilon_n} f^*(\phi)$ (Proposition 4.4.5). Because f^* is Fréchet differentiable at ϕ, Šmulian's theorem (4.2.10) implies $\|x_n - x\| \to 0$ as desired. Because $f^{**} = f$, the argument implies f^* is locally uniformly convex as well.

Part (b) follows from (a) because f is supercoercive and bounded on bounded sets if and only if f^* is by Theorem 4.4.13 and then f is Fréchet differentiable when f^* is locally uniformly convex by Proposition 5.3.6. For part (c), consider $f := \|\cdot\|^2$ where $\|\cdot\|$ is a Fréchet differentiable norm whose dual norm is not locally uniformly convex as cited in Remark 5.1.31(b). □

5.3.18 (Limitations on convexity/smoothness duality). Let X be a Banach space, and suppose f is a lsc proper convex function on X. Provide examples for the following assertions.

(a) f Fréchet differentiable and supercoercive does not imply f^* is strictly convex.
(b) f locally uniformly convex and supercoercive does not imply f^* is Gâteaux differentiable.
(c) f^* Fréchet differentiable and coercive does not imply f is strictly convex.
(d) f^* locally uniformly convex, Fréchet differentiable and coercive does not imply f is Fréchet differentiable.

Hint. For (a), consider, $f(x) := \|x\|^2$ where $\|\cdot\|$ is Fréchet differentiable on $C[0, \omega_1]$; see Remark 5.1.30(b). For (b) let $f(x) := \|x\|^2$ where $\|\cdot\|$ is locally uniformly convex on ℓ_1. For (c), let $g(x, y) := f(x - 1/2, y - 1/2)$ where f is the function from Exercise 5.3.10. Compute that $g^*(x, y) = f^*(x, y) + \langle (x, y), (1/2, 1/2) \rangle$. Therefore, g^* has uniformly continuous derivative on bounded sets because f^* has such a derivative. However, f is not strictly convex, and so g is not strictly convex. However, g^* is coercive because g is continuous at $(0, 0)$ by the Moreau–Rockafellar theorem (4.4.10). For (d), let $f^*(t) := e^t - t$. □

5.4 Moduli of smoothness and rotundity

This section will look at finer properties of uniformly convex and uniformly smooth functions. For a convex function $f : X \to \mathbb{R}$, we define the *modulus of smoothness* of f, for $\tau \geq 0$, by

$$\rho_f(\tau) := \sup\left\{\frac{1}{2}f(x + \tau y) + \frac{1}{2}f(x - \tau y) - f(x) : x \in X, y \in S_X\right\}.$$

Notice that f must be defined everywhere in order for its modulus of smoothness to be finite. Also, a continuous convex function f is *uniformly smooth* if $\lim_{\tau \to 0^+} \rho_f(\tau)/\tau = 0$; see Proposition 4.2.14 for characterizations of uniformly smooth convex functions.

Theorem 5.4.1. *Suppose $f : X \to (-\infty, +\infty]$ is a proper lsc convex function.*

(a) *Suppose that $f : X \to \mathbb{R}$. Then $\rho_f(\tau) = \sup\left\{\tau\frac{\varepsilon}{2} - \delta_{f^*}(\varepsilon) : 0 \leq \varepsilon\right\}$.*
(b) *Suppose that $f^* : X^* \to \mathbb{R}$. Then $\rho_{f^*}(\tau) = \sup\left\{\tau\frac{\varepsilon}{2} - \delta_f(\varepsilon) : 0 \leq \varepsilon\right\}$.*

Proof. The proofs of (a) and (b) are similar, so we will prove only (b), and leave the proof of (a) to Exercise 5.4.9. We will first show that $\delta_f(\varepsilon) + \rho_{f^*}(\tau) \geq \tau\frac{\varepsilon}{2}$ for

5.4 Moduli of smoothness and rotundity

$\varepsilon \geq 0$ and $\tau > 0$. Let $\eta > 0$. When $\delta_f(\varepsilon)$ is finite, we choose $x, y \in \operatorname{dom} f$ such that $\|x - y\| \geq \varepsilon$ and

$$\frac{1}{2}f(x) + \frac{1}{2}f(y) - f\left(\frac{x+y}{2}\right) \leq \delta_f(\varepsilon) + \eta.$$

Now let $\phi \in \partial_\eta f(\frac{x+y}{2})$ and $\Lambda \in \partial\|x - y\|$. Then

$$\rho_{f^*}(\tau) \geq \frac{1}{2}f^*(\phi + \tau\Lambda) + \frac{1}{2}f^*(\phi - \tau\Lambda) - f^*(\phi)$$

$$\geq \frac{1}{2}[(\phi + \tau\Lambda)(x) - f(x)] + \frac{1}{2}[(\phi - \tau\Lambda)(y) - f(y)]$$

$$- \left[\phi\left(\frac{x+y}{2}\right) - f\left(\frac{x+y}{2}\right) + \eta\right]$$

$$= \tau\Lambda\left(\frac{x-y}{2}\right) + f\left(\frac{x+y}{2}\right) - \frac{1}{2}f(x) - \frac{1}{2}f(y) - \eta$$

$$\geq \tau\left\|\frac{x-y}{2}\right\| - \delta_f(\varepsilon) - 2\eta.$$

Because $\eta > 0$ was arbitrary, $\rho_{f^*}(\tau) + \delta_f(\varepsilon) \geq \tau\frac{\varepsilon}{2}$.

On the other hand, for any $\tau \geq 0$ and any $\eta > 0$, given $\Lambda, \phi \in X^*$, we can choose $x, y \in X$ so that

$$\frac{1}{2}f^*(\phi + \tau\Lambda) + \frac{1}{2}f^*(\phi - \tau\Lambda) - f^*(\phi) \leq$$

$$\leq \frac{1}{2}[(\phi + \tau\Lambda)(x) - f(x) + (\phi - \tau\Lambda)(y) - f(y)]$$

$$- \left[\phi\left(\frac{x+y}{2}\right) - f\left(\frac{x+y}{2}\right)\right] + \eta$$

$$= f\left(\frac{x+y}{2}\right) - \frac{1}{2}f(x) - \frac{1}{2}f(x) + \frac{1}{2}\tau\Lambda(x-y) + \eta$$

$$\leq \frac{1}{2}\tau\varepsilon - \delta_f(\varepsilon) + \eta.$$

where $\varepsilon = \|x - y\|$. Therefore, $\rho_{f^*}(\tau) \leq \sup\{\tau\frac{\varepsilon}{2} - \delta_f(\varepsilon) : 0 \leq \varepsilon\}$ for each $\tau > 0$. □

One can derive the duality between uniformly convex and uniformly smooth convex functions using the previous result; see Exercise 5.4.9. We say a proper lsc convex function f has *modulus of convexity of power type p*, if there is a $C > 0$ such that $\delta_f(\varepsilon) \geq C\varepsilon^p$ for all $\varepsilon \geq 0$. Also, f is said to have *modulus of smoothness of power type p* if there exists $C > 0$ such that $\rho_f(\tau) \leq C\tau^p$ for $\tau > 0$.

Theorem 5.4.2. Suppose $f : X \to (-\infty, +\infty]$ is a proper lsc convex function, and $p > 1$.

(a) f has modulus of convexity of power type p if and only if f^* has modulus of smoothness of power type q where $1/p + 1/q = 1$;

(b) f^* has modulus of convexity of power type p if and only if f has modulus of smoothness of power type q where $1/p + 1/q = 1$.

Proof. See Exercise 5.4.10. Also note that Exercise 5.4.6(d) confirms the case $p < 2$ only occurs when the domain of the uniformly convex function is a singleton. □

The next theorem shows norms with moduli of convexity of power type can be constructed naturally from functions with such moduli.

Theorem 5.4.3. *Suppose X is a Banach space and $f : X \to (-\infty, +\infty]$ is a proper lsc convex function with* int dom $f \neq \emptyset$. *If f has modulus of convexity of power type $p \geq 2$ (on bounded sets), then X admits an equivalent with norm modulus of convexity of power type p.*

Proof. First, shift f so that $0 \in$ int dom f. Now let $\phi \in \partial f(0)$, then $f - \phi$ has same convexity properties of f, and $f(0) = 0$ is the minimum of f. Now let $g(x) := \frac{f(x)+f(-x)}{2} + \|x\|^2$. Let $r > 0$ be such that $B_{2r} \subset$ int dom g and g is Lipschitz on B_r. Then g has the same convexity properties of f, $g(0) = 0$, g is continuous on B_{2r}, and $g(x) \geq r^2$ if $\|x\| \geq r$. Define an equivalent norm $\|\cdot\|$ on X as the norm whose ball is B where $B := \{x : g(x) \leq r^2\}$.

Let $\alpha > 0$ and $\beta > 0$ denote constants such that $\|\cdot\| \geq \alpha \|\cdot\|$ and $\|\cdot\| \geq \beta \|\cdot\|$, and let $K > 0$ be the Lipschitz constant of g on B. Let $C > 0$ be such that $\delta_g(\varepsilon) \geq C\varepsilon^p$ for $\varepsilon > 0$. Let $\|x\| = \|y\| = 1$, and suppose $\|x - y\| \geq \varepsilon$. Then $\|x - y\| \geq \beta\varepsilon$ and $g(x) = g(y) = r^2$; this implies

$$g\left(\frac{x+y}{2}\right) \leq r^2 - C(\beta\varepsilon)^p.$$

The Lipschitz property of g now implies $\|\frac{x+y}{2} - u\| \geq \frac{C}{K}(\beta\varepsilon)^p$ for any u with $g(u) = r^2$, that is $\|u\| = 1$. This then implies

$$\left\|\frac{x+y}{2}\right\| \leq 1 - \alpha\frac{C}{K}\beta^p\varepsilon^p,$$

which shows $\|\cdot\|$ has modulus of convexity of power type p. □

5.4.1 Spaces with nontrivial uniformly convex functions

We have see that nontrivial uniformly convex functions can only exist on superreflexive spaces. A natural question is then whether uniformly convex functions are easy to construct on such spaces. This subsection will demonstrate for $2 \leq p < \infty$ that $f := \|\cdot\|^p$ is uniformly convex if and only if the norm $\|\cdot\|$ has modulus of convexity of power type p.

Lemma 5.4.4. *Let $0 < r \leq 1$, then $|t^r - s^r| \leq |t - s|^r$ for all $s, t \in [0, \infty)$.*

Proof. First, for $x \geq 0$, $(1 + x)^r \leq 1 + x^r$ (see [410, Example 4.20]). Setting $x = (t - s)/s$ with $t \geq s > 0$, and then multiplying by s^r, we get $t^r \leq s^r + (t - s)^r$. The conclusion follows from this. □

5.4 Moduli of smoothness and rotundity

Theorem 5.4.5. *For $1 < q \leq 2$, the following are equivalent in a Banach space $(X, \|\cdot\|)$.*

(a) The norm $\|\cdot\|$ has modulus of smoothness of power type q.
(b) The derivative of $f := \|\cdot\|^q$ satisfies a $(q-1)$-Hölder condition.
(c) The function $f := \|\cdot\|^q$ has modulus of smoothness of power type q.
(d) The function $f := \|\cdot\|^q$ is uniformly smooth.

Proof. (a) $\Rightarrow$ (b): Assume that $\|\cdot\|$ has modulus of smoothness of power type q. Given $x \in X \setminus \{0\}$, let ϕ_x denote a support functional of x. According to Exercise 5.1.23, $\|\cdot\|$ has a (Fréchet) derivative satisfying a $(q-1)$-Hölder-condition on its sphere; this implies that each $x \neq 0$ has a unique support functional, and there exists $C > 0$ such that

$$\|\phi_x - \phi_y\| \leq C \|x - y\|^{q-1} \text{ for all } x, y \in S_X. \tag{5.4.1}$$

Let $f(\cdot) := \|\cdot\|^q$. Then $f'(0) = 0$, and $f'(x) = q \|x\|^{q-1} \phi_x$ for $x \neq 0$. Thus if $x = 0$ or $y = 0$, then $\|f'(x) - f'(y)\| \leq q \|x - y\|^{q-1}$. Let $x, y \in X \setminus \{0\}$. Then

$$f'(x) - f'(y) = q \|x\|^{q-1} \phi_x - q \|y\|^{q-1} \phi_y$$
$$= q \|x\|^{q-1} (\phi_x - \phi_y) + \left(q \|x\|^{q-1} - q \|y\|^{q-1}\right) \phi_y. \tag{5.4.2}$$

Using Lemma 5.4.4 we also compute

$$\left| q \|x\|^{q-1} - q \|y\|^{q-1} \right| \leq q \left| \|x\| - \|y\| \right|^{q-1} \leq q \|x - y\|^{q-1}. \tag{5.4.3}$$

We now work on an estimate for $q \|x\|^{q-1} (\phi_x - \phi_y)$. We may and do assume that $0 < \|y\| \leq \|x\|$. Suppose first $\|y\| \leq \|x\|/2$. Then

$$q \|x\|^{q-1} \|\phi_x - \phi_y\| \leq 2q \|x\|^{q-1} \leq q 2^q \|x - y\|^{q-1}. \tag{5.4.4}$$

Suppose $\|y\| \geq \|x\|/2$, and let $x' = \lambda x$ where $\lambda = \|y\|/\|x\|$, so that $\|x'\| = \|y\|$. Then

$$\|x' - y\| \leq \|x' - x\| + \|x - y\| = \|x\| - \|y\| + \|x - y\| \leq 2 \|x - y\|. \tag{5.4.5}$$

Now let $\alpha = \|y\|$. Observe that ϕ_x and ϕ_y are also support functionals for $\alpha^{-1} x'$ and $\alpha^{-1} y$ respectively. Applying (5.4.1), the fact that $\|x\| \leq 2\alpha$, and (5.4.5) we obtain

$$\|\phi_x - \phi_y\| \leq C \|\alpha^{-1} x' - \alpha^{-1} y\|^{q-1} \leq \frac{C}{\alpha^{q-1}} \|x' - y\|^{q-1}$$
$$\leq \frac{C 2^{q-1}}{\|x\|^{q-1}} (2 \|x - y\|)^{q-1} = \frac{C 4^{q-1}}{\|x\|^{q-1}} \|x - y\|^{q-1}.$$

Consequently, $q \|x\|^{q-1} \|\phi_x - \phi_y\| \leq C 4^{q-1} q \|x-y\|^{q-1}$. This inequality and (5.4.4) show there exists $K > 0$ such that

$$q \|x\|^{q-1} \|\phi_x - \phi_y\| \leq K \|x-y\|^{q-1} \quad \text{for all } x, y \in X \setminus \{0\}. \tag{5.4.6}$$

Combining (5.4.2), (5.4.3) and (5.4.6) shows that f' satisfies a $(q-1)$-Hölder-condition.

(b) $\Rightarrow$ (c) follows from Exercise 5.4.13 and (c) $\Rightarrow$ (d) is trivial, so we prove (d) $\Rightarrow$ (a). Suppose $\|\cdot\|$ does not have modulus of smoothness of power type q. Then using Exercise 5.1.23 there are $x_n, y_n \in S_X$ such that $\|x_n - y_n\| \to 0$ while

$$\|\phi_{x_n} - \phi_{y_n}\| \geq n \|x_n - y_n\|^{q-1}.$$

Let $\delta_n = \|x_n - y_n\|$ and define $u_n = \frac{1}{\delta_n \sqrt{n}} x_n$ and $v_n = \frac{1}{\delta_n \sqrt{n}} y_n$. Then $\|u_n - v_n\| = \frac{1}{\sqrt{n}} \to 0$. However

$$\|f'(u_n) - f'(v_n)\| = \left\| q \|u_n\|^{q-1} \phi_{u_n} - q \|v_n\|^{q-1} \phi_{v_n} \right\|$$
$$= \left\| q \|u_n\|^{q-1} \phi_{x_n} - q \|v_n\|^{q-1} \phi_{y_n} \right\|$$
$$= \frac{q}{\delta_n^{q-1} n^{\frac{q-1}{2}}} \|\phi_{x_n} - \phi_{y_n}\|$$
$$\geq \frac{q}{\delta_n^{q-1} n^{\frac{q-1}{2}}} \left(n \delta_n^{q-1} \right) = q n^{\frac{3-q}{2}} \to \infty.$$

Consequently, f' is not uniformly continuous, and so Proposition 4.2.14 shows that that $f(\cdot) = \|\cdot\|^q$ is not a uniformly smooth function. $\square$

The duality between uniform smoothness and uniform convexity enables us to derive the dual version of Theorem 5.4.5 for uniformly convex functions.

Theorem 5.4.6. *Let $(X, \|\cdot\|)$ be a Banach space, and let $2 \leq p < \infty$. Then the following are equivalent.*

(a) *The norm $\|\cdot\|$ on X has modulus of convexity of power type p.*
(b) *The function $f := \|\cdot\|^p$ has modulus of convexity of power type p.*
(c) *The function $f := \|\cdot\|^p$ is uniformly convex.*

Proof. (a) $\Rightarrow$ (b): Let us assume that $\|\cdot\|$ has modulus of convexity of power type p, then the modulus of smoothness of the dual norm on X^*, which we denote in this proof as $\|\cdot\|_*$, is of power type q where $\frac{1}{p} + \frac{1}{q} = 1$; see Exercise 5.1.21. By Theorem 5.4.5 the function $g := \frac{1}{q} \|\cdot\|_*^q$ has modulus of smoothness of power type q. The Fenchel conjugate of g is $g^* = \frac{1}{p} \|\cdot\|^p$. Thus g^* – and hence $\|\cdot\|^p$ – has modulus of convexity of power type p according to Theorem 5.4.2.

(b) $\Rightarrow$ (c) is trivial, so we prove (c) $\Rightarrow$ (a). Indeed, assuming that $f(\cdot) = \|\cdot\|^p$ is a uniformly convex function, then Theorem 5.3.17 shows that f^* (and hence $\|\cdot\|_*^q$)

is a uniformly smooth function. According to Theorem 5.4.5, $\|\cdot\|_*$ has modulus of smoothness of power type q. Therefore $\|\cdot\|$ has modulus of convexity of power type p; see Exercise 5.1.21. □

Example 5.4.7. Let $f := \|\cdot\|_p^r$ where $\|\cdot\|_p$ is the usual norm on ℓ_p for $p > 1$. Then f is uniformly convex if and only if $r \geq \max\{p, 2\}$.

Proof. This follows from Theorem 5.4.6, because the norm $\|\cdot\|_p$ has modulus convexity of power type $\max\{p, 2\}$ (see [180]). □

5.4.2 Growth rates of uniformly convex functions and renorming

We now try to more explicitly construct a uniformly convex norm whose modulus of convexity is related to the growth rate of a given uniformly convex function on the Banach space. We begin with some preliminary results.

Lemma 5.4.8. *Let $\|\cdot\|$ be a norm on a Banach space X. Suppose $\|x\| = \|y\|$, and $\|x - y\| \geq \delta$ where $0 < \delta \leq 2\|x\|$. Then $\inf_{t \geq 0} \|x - ty\| \geq \delta/2$.*

Proof. Assume that $\|x - t_0 y\| < \delta/2$ for some $t_0 \geq 0$. Then $|1 - t_0| \|y\| < \delta/2$ and so

$$\|x - y\| \leq \|x - t_0 y\| + |1 - t_0| \|y\| < \delta,$$

which is a contradiction. □

The next lemma will be used later to estimate the modulus of convexity of a norm constructed by using lower level sets of a symmetric uniformly convex function.

Lemma 5.4.9. *Let $\{\|\cdot\|_n\}_{n \geq N}$ be a family of norms on $(X, \|\cdot\|)$ satisfying*

$$\frac{1}{2^{n+1}} \|\cdot\| \leq \|\cdot\|_n \leq \frac{1}{2^n} \|\cdot\| \quad \text{for } n \geq N. \tag{5.4.7}$$

For each $n \geq N$, suppose there exists $d_n > 0$ so that

$$\left\| \frac{x+y}{2} \right\|_n \leq 1 - d_n, \quad \text{whenever } \|x\|_n = \|y\|_n = 1 \text{ and } \|x - y\| \geq 1.$$

Then there exist an equivalent norm $|\cdot|$ on X and $M \in \mathbb{N}$ so that the modulus of convexity of the norm $|\cdot|$ satisfies

$$\delta_{|\cdot|}(t) \geq \frac{d_n}{n^2} \quad \text{whenever } \frac{1}{2^{n-M-1}} \leq t \leq 2 \text{ and } n \geq M.$$

Proof. Choose $M \geq \max\{4, N\}$, and define $|\cdot|$ by

$$|\cdot| := \sum_{m=M}^{\infty} \frac{2^{m+1}}{m^2} \|\cdot\|_m.$$

Observe that, $|\cdot| \leq \sum_{m=M}^{\infty} \frac{2^{m+1}}{m^2 2^m} \|\cdot\| \leq \sum_{m=4}^{\infty} \frac{2}{m^2} \|\cdot\| \leq \|\cdot\|$; and then

$$\frac{1}{2^M} \|\cdot\| \leq \frac{1}{M^2} \|\cdot\| \leq \frac{2^{M+1}}{M^2} \|\cdot\|_M \leq |\cdot| \leq \|\cdot\|. \tag{5.4.8}$$

Now suppose that $|x| = |y| = 1$ and $|x - y| \geq \frac{1}{2^{n-M-1}}$ where $n \geq M$ is fixed. Because $|x| = |y| = 1$, it follows from (5.4.8) that

$$1 \leq \|x\| \leq 2^M \quad \text{and} \quad 1 \leq \|y\| \leq 2^M. \tag{5.4.9}$$

We assume, without loss of generality, $\|x\|_n \leq \|y\|_n$. Now let us denote $a = \|x\|_n^{-1}$ and $b = \|y\|_n^{-1}$. It follows from (5.4.7) and (5.4.9) that $2^{n-M} \leq b \leq a \leq 2^{n+1}$, which in turn implies $|ax - ay| \geq 2$.

According to Lemma 5.4.8, $|ax - by| \geq 1$, and hence $\|ax - by\| \geq 1$. Thus we compute

$$\left\|\frac{ax + ay}{2}\right\|_n \leq \left\|\frac{ax + by}{2}\right\|_n + \frac{1}{2}(a - b)\|y\|_n$$

$$\leq \frac{1}{2}\|ax\|_n + \frac{1}{2}\|by\|_n + \frac{1}{2}(a - b)\|y\|_n - d_n$$

$$= \frac{a}{2}(\|x\|_n + \|y\|_n) - d_n.$$

This inequality implies

$$\left\|\frac{x + y}{2}\right\|_n \leq \frac{1}{2}\|x\|_n + \frac{1}{2}\|y\|_n - \frac{d_n}{a}. \tag{5.4.10}$$

Thus, using (5.4.10), and the triangle inequality for $\|\cdot\|_j$ when $j \neq n$, and then that $a \leq 2^{n+1}$ we obtain

$$\left|\frac{x + y}{2}\right| \leq \sum_{j=M}^{\infty} \frac{2^{j+1}}{2j^2} \|x\|_j + \sum_{j=M}^{\infty} \frac{2^{j+1}}{2j^2} \|y\|_j - \frac{2^{n+1}d_n}{n^2 a} \leq 1 - \frac{d_n}{n^2},$$

which finishes the proof. $\square$

Theorem 5.4.10. *Let $(X, \|\cdot\|)$ be a Banach space and let $F : [0, +\infty) \to [0, +\infty)$ be a continuous convex function satisfying $F(0) = 0$. Suppose $f : X \to \mathbb{R}$ is a continuous uniformly convex function satisfying $f(x) \leq F(\|x\|)$ for all $x \in X$. Then there is an equivalent norm $|\cdot|$ on X such that given any $\gamma > 0$, there are constants $\alpha > 0$ and $\beta > 0$ so that*

$$\delta_{|\cdot|}(t) \geq \frac{\alpha}{F(\beta t^{-1})} t^\gamma \quad \text{for } 0 < t \leq 2.$$

Proof. As before, we may and do assume f is centrally symmetric; notice that this new f will still be bounded above by $F(\|\cdot\|)$. Because $F(0) = 0$, the convexity of F ensures that $F(\lambda t) \leq \lambda F(t)$ for $0 \leq \lambda \leq 1$; in particular F is nondecreasing on $[0, +\infty)$ because it is nonnegative there. Consequently, $F(\|\cdot\|)$ is a convex function.

According to Fact 5.3.16, there exists $N > 0$ so that $f(x) \geq 0$ if $\|x\| \geq N$. Now replace f with $[f(\cdot) + F(\|\cdot\|)]/2$; then we have

$$F\left(\frac{\|x\|}{2}\right) \leq \frac{1}{2}F(\|x\|) \leq f(x) \leq F(\|x\|) \text{ whenever } \|x\| \geq N. \tag{5.4.11}$$

For $n \geq N$, let $\|\cdot\|_n$ be norm whose unit ball is $B_n := \{x : f(x) \leq F(2^n)\}$. It follows from (5.4.11) that if $\|x\|_n = 1$, then $2^n \leq \|x\| \leq 2^{n+1}$. Consequently,

$$\frac{1}{2^{n+1}}\|\cdot\| \leq \|\cdot\|_n \leq \frac{1}{2^n}\|\cdot\|.$$

Let $M_n := \sup\{f'_+(u; v) : \|u\|_n = 1, \|v\| = 1\}$. If $\|u\|_n = 1$, and $\|v\| = 1$, then $\|u\| \leq 2^{n+1}$ and we compute

$$f'_+(u; v) \leq \frac{f(u + 2^{n+1}v) - f(u)}{2^{n+1}} \leq \frac{F(2 \cdot 2^{n+1}) - 0}{2^{n+1}} = \frac{F(2^{n+2})}{2^{n+1}}. \tag{5.4.12}$$

It follows that $M_n \leq 2^{-(n+1)}F(2^{n+2})$.

Now suppose $\|x\|_n = \|y\|_n = 1$, and $\|x - y\| \geq 1$. Letting δ_f denote the modulus of convexity of f with respect to $\|\cdot\|$, the uniform convexity of f ensures $\delta_f(1) > 0$. Then denoting $z = \frac{x+y}{2}$ and $z' = z/\|z\|_n$ we obtain $f(x) = f(y) = f(z') = F(2^n)$, and so

$$\delta_f(1) \leq \frac{1}{2}f(x) + \frac{1}{2}f(y) - f\left(\frac{x+y}{2}\right) = f(z') - f(z) \leq f'_+(z', z' - z)$$

$$= \|z' - z\| f'_+\left(z', \frac{z' - z}{\|z' - z\|}\right) \leq M_n \|z' - z\|. \tag{5.4.13}$$

Consequently, using $\|\cdot\|_n \geq \frac{1}{2^{n+1}}\|\cdot\|$, (5.4.13) and then the bound on M_n, we obtain

$$\left\|\frac{x+y}{2}\right\|_n = 1 - \|z' - z\|_n \leq 1 - \|z' - z\|\frac{1}{2^{n+1}} \leq 1 - \frac{\delta_f(1)}{M_n} \cdot \frac{1}{2^{n+1}}$$

$$\leq 1 - \frac{\delta_f(1)}{F(2^{n+2})}. \tag{5.4.14}$$

Applying Lemma 5.4.9, we find an equivalent norm $|\cdot|$ and $M \geq N$ such that

$$\delta_{|\cdot|}(t) \geq \frac{\delta_f(1)}{F(2^{n+2})} \cdot \frac{1}{n^2}, \text{ whenever } n \geq M \text{ and } \frac{1}{2^{n-M-1}} \leq t \leq 2.$$

Given $\gamma > 0$, we fix $n_0 \geq M$ so large that $n^{-2} \geq (2^{-n})^\gamma$ for all $n \geq n_0$. Then

$$\delta_{|\cdot|}(t) \geq \frac{\delta_f(1)}{F(2^{n+2})} \cdot \left(\frac{1}{2^n}\right)^\gamma \text{ whenever } n \geq n_0 \text{ and } \frac{1}{2^{n-M-1}} \leq t \leq 2.$$

Let $\alpha := \delta_f(1) \min\left\{\left(\frac{1}{2^{n_0+1}}\right)^\gamma, \left(\frac{1}{2^{M+2}}\right)^\gamma\right\}$ and $\beta := \max\{2^{n_0+3}, 2^{M+4}\}$. The previous inequality and along with the fact F is nondecreasing, ensure that for $\frac{1}{2^{n_0-M-1}} \leq t \leq 2$

we have

$$\delta_{|\cdot|}(t) \geq \frac{\delta_f(1)}{F(2^{n_0+2})}\left(\frac{1}{2^{n_0}}\right)^\gamma \geq \frac{\delta_f(1)}{F(2^{n_0+3}t^{-1})}\left(\frac{t}{2^{n_0+1}}\right)^\gamma \geq \frac{\alpha t^\gamma}{F(\beta t^{-1})};$$

and for $\frac{1}{2^{n-M-1}} \leq t \leq \frac{1}{2^{n-M-2}}$ where $n \geq n_0 + 1$, we have

$$\delta_{|\cdot|}(t) \geq \frac{\delta_f(1)}{F(2^{n+2})}\left(\frac{1}{2^n}\right)^\gamma \geq \frac{\delta_f(1)}{F(2^{M+4}t^{-1})}\left(\frac{t}{2^{M+2}}\right)^\gamma \geq \frac{\alpha t^\gamma}{F(\beta t^{-1})}.$$

Altogether, $\delta_{|\cdot|}(t) \geq \frac{\alpha}{F(\beta t^{-1})}t^\gamma$ for $0 < t \leq 2$, as desired. □

Corollary 5.4.11. *Let $(X, \|\cdot\|)$ be a Banach space, and let $p \geq 2$. Suppose $f : X \to \mathbb{R}$ is a continuous uniformly convex function such that $f(x) \leq \|x\|^p$ for all $x \in X$. Then for any $r > p$, X admits an equivalent norm with modulus of convexity of power type r.*

Proof. Apply the previous theorem with $F(t) := t^p$. □

In the case $p = 2$ we will prove the following sharp result.

Theorem 5.4.12. *Let $(X, \|\cdot\|)$ be a Banach space. Then there is a continuous uniformly convex function $f : X \to \mathbb{R}$ satisfying $f(\cdot) \leq \|\cdot\|^2$ if and only if X admits an equivalent norm with modulus of convexity of power type 2.*

Before proving this theorem, we will present a preliminary lemma.

Lemma 5.4.13. *Let X be a Banach space. Suppose $\{\|\cdot\|_n\}_{n \in \mathbb{N}}$ are norms on $(X, \|\cdot\|)$ so that*

$$K\|\cdot\| \leq \|\cdot\|_n \leq \|\cdot\|, \qquad (5.4.15)$$

for some $K > 0$ and all $n \in \mathbb{N}$. Then, there exists an equivalent norm $|\cdot|$ such that

$$\delta_{|\cdot|}(t) \geq \liminf \delta_{\|\cdot\|_n}(t), \quad \text{for} \quad 0 < t < 2.$$

Proof. Let us consider a free (nonprincipal) ultrafilter $\mathcal{U}$ on $\mathbb{N}$. Then $\lim_\mathcal{U} \|x\|_n$ exists for each $x \in X$, where $\lim_\mathcal{U} \|x\|_n = L$ means for each $\varepsilon > 0$, there exists $A \in \mathcal{U}$ such that $|\|x\|_n - L| < \varepsilon$ for all $n \in A$. Now define $|\cdot| : X \to [0, +\infty)$ by

$$|x| := \lim_\mathcal{U} \|x\|_n, \quad \text{for all} \quad x \in X.$$

The definition of $|\cdot|$ together with (5.4.15) ensure $|\cdot|$ is an equivalent norm on X.

We will proceed by reductio ad absurdum. Assume there exists $t \in (0, 2)$ such that $\delta_{|\cdot|}(t) < \liminf \delta_{\|\cdot\|_n}(t)$. Since $\delta_{|\cdot|}$ is continuous – see [246] – there exists $t' \in (t, 2)$ such that $\delta_{|\cdot|}(t') < \liminf \delta_{\|\cdot\|_n}(t)$. Then, there exist $x, y \in X$ and a constant $a > 0$ such that $|x| = |y| = 1$, $|x - y| \geq t'$ and $1 - |(x+y)/2| < a < \liminf \delta_{\|\cdot\|_n}(t)$. For this x and y, let $x_n = x/\|x\|_n$ and $y_n = y/\|y\|_n$. By the definition of $|\cdot|$, there exists $A \in \mathcal{U}$ such that $\|x_m - y_m\|_m \geq t$ and $1 - \|(x_m + y_m)/2\|_m < a$ for all $m \in A$.

Therefore $\delta_{\|\cdot\|_m}(t) < a < \liminf \delta_{\|\cdot\|_n}(t)$ for all $m \in A$, which yields a contradiction, since $\mathcal{U}$ is free and then A is infinite. □

Proof. (Theorem 5.4.12) First, if X admits an equivalent norm that has modulus of convexity of power type 2, then it has such a norm $|\cdot|$ satisfying $|\cdot| \leq \|\cdot\|$. According to Theorem 5.4.6, $f(\cdot):=|\cdot|^2$ is uniformly convex as desired.

Conversely, suppose $f : X, \to \mathbb{R}$ is a uniformly convex function such that $f(\cdot) \leq \|\cdot\|^2$. Proceeding as in Theorem 5.4.10 when $F(t):=t^2$ we obtain norms $\{\|\cdot\|_n\}_{n\geq N}$ satisfying $\frac{1}{2^{n+1}} \|\cdot\| \leq \|\cdot\|_n \leq \frac{1}{2^n} \|\cdot\|$ and then (5.4.14) becomes

$$\left\|\frac{x+y}{2}\right\|_n \leq 1 - \frac{\delta_f(1)}{16} \left(\frac{1}{2^n}\right)^2, \text{ whenever } \|x\|_n = \|y\|_n = 1 \text{ and } \|x-y\|_n \geq \frac{1}{2^n}.$$

The previous inequality implies

$$\delta_{\|\cdot\|_n}(2^{-n}) \geq \frac{\delta_f(1)}{16}(2^{-n})^2.$$

According to [213, Corollary 11] there is a universal constant $L > 0$ such that

$$\frac{\delta_{|\cdot|_n}(2^{-n})}{(2^{-n})^2} \leq 4L\frac{\delta_{|\cdot|_n}(\eta)}{\eta^2} \text{ for } 2^{-n} \leq \eta \leq 2.$$

Let $R:=\frac{\delta_f(1)}{64L}$; then the previous two inequalities imply

$$\delta_{|\cdot|_n}(t) \geq Rt^2 \text{ for } 2^{-n} \leq t \leq 2. \tag{5.4.16}$$

For each $n \geq N$, let us consider the new norm $|\cdot|_n := 2^n \|\cdot\|_n$. These new norms satisfy $\frac{1}{2}\|\cdot\| \leq |\cdot|_n \leq \|\cdot\|$ and $\delta_{|\cdot|_n}(\cdot) = \delta_{\|\cdot\|_n}(\cdot)$. Applying Lemma 5.4.13 and then (5.4.16) we obtain a norm $|\cdot|$ satisfying

$$\delta_{|\cdot|}(t) \geq \liminf_{n\to\infty} \delta_{|\cdot|_n}(t) = \liminf_{n\to\infty} \delta_{\|\cdot\|_n}(t) \geq Rt^2 \text{ for } 0 < t \leq 2,$$

which finishes the proof. □

Exercises and further results

5.4.1. (a) Consider the convex function on the real line $t \mapsto |t|^p$. Show that this function is uniformly smooth with modulus of smoothness of power type p when $1 < p \leq 2$. Use duality to conclude that this function is uniformly convex with modulus of convexity of power type p when $2 \leq p < \infty$.
(b) Let $b > 1$. Define $f : \mathbb{R} \to (-\infty, +\infty]$ by $f(t) := +\infty$ if $t < 0$, and $f(t) := b^t$ if $t \geq 0$. Compute f^* and show that f^* has modulus of smoothness of power type q for any $p \in (1, 2]$. Deduce that f has modulus of convexity of power type p for any $p \geq 2$.

Hint. (a) One can use Lemma 5.4.4 and Exercise 5.4.13 for the modulus of smoothness assertion. □

5.4.2. (a) Suppose a function f on $\mathbb{R}$ satisfies $f^{(n)} \geq \alpha > 0$ on $[a, \infty)$ where $n \geq 2$ is a fixed integer, and that $f^{(k)} \geq 0$ on $[a, \infty)$ for $k \in \{2, \ldots, n+1\}$. Define the function g by $g(x) := f(x)$ for $x \geq a$ and $g(x) := +\infty$ for $x < a$. Show that g is uniformly convex with modulus of convexity of power type n.
(b) Let $b > 1$ and $g(x) := b^x$ for $x \geq 0$, and $g(x) := +\infty$ otherwise. Use (a) to show that g is uniformly convex with modulus of convexity of power type p for any $p \geq 2$.
(c) Let $p \geq 2$ and $g(x) := x^p$ for $x \geq 0$, and $g(x) := +\infty$ otherwise. Show that g is uniformly convex with modulus of convexity of power type p.

Hint. Use Taylor's theorem for (a) and (c). □

5.4.3. Let $f : X \to \mathbb{R}$ be a continuous convex function, and let $p > 0$. Show that f has modulus of convexity of power type p if and only if there exists $C > 0$ such that

$$f(\bar{x} + h) \geq f(\bar{x}) + \phi(h) + C\|h\|^p \quad \text{whenever } \bar{x}, h \in X, \ \phi \in \partial f(\bar{x}).$$

5.4.4. Suppose f and g are proper lsc convex functions.
(a) Suppose f is uniformly convex, show that $\delta_f \leq \delta_h$ when $h := f + g$ is proper.
(b) Suppose f is uniformly smooth, show that $\rho_h \leq \rho_f$ when $h := f \square g$ is proper.

In particular, modulus of convexity of power type is preserved by a proper sum, and modulus of smoothness of power type is preserved by a proper infimal convolution.

Hint. Part (a) is straightforward. According to Lemma 4.4.15, $h^* = f^* + g^*$. Then part (b) follows from (a) and Theorem 5.4.1(a).

5.4.5. (a) Suppose $f : \mathbb{R} \to \mathbb{R}$ is a C^2-smooth function. Show that f is affine (i.e. $f'' = 0$ everywhere) if and only if f satisfies an α-Hölder condition for some $\alpha > 1$.
(b) Use the fundamental theorem of calculus to show that if $f : \mathbb{R} \to \mathbb{R}$ is C^1-smooth and f' satisfies an α-Hölder condition for some $\alpha > 1$, then $f'' = 0$ everywhere, and thus f is affine.

Hint. (a) Clearly an affine function satisfies any such α-Hölder condition. Suppose $f''(x_0) \neq 0$ for some $x_0 \in \mathbb{R}$. By replacing f with $-f$ as necessary, there is an open interval I containing x_0 and $\varepsilon > 0$ so that $f'' > \varepsilon$ on I. Then $|f'(t) - f'(s)| \geq \varepsilon|t - s|$ for all $s, t \in I$, as $|s - t| \to 0^+$ this will contradict the α-Hölder condition. □

5.4.6. If a norm has modulus of convexity of power type p, it follows that $p \geq 2$. Dually, if a norm has modulus of convexity of power type q, it follows that $q \leq 2$. See [180, p. 154,157].

(a) Verify that $\delta_{\{0\}}$ has modulus of convexity of power type p for any $p > 0$.
(b) Verify that any constant function has modulus of smoothness of power type q for any $q > 0$.
(c) Show that a convex function can have modulus of smoothness of power type $q > 2$ if and only if it is affine.
(d) Show that a lsc uniformly convex function can have modulus of convexity of power type $p < 2$ if and only if its domain is a singleton.
(e) Construct a continuously uniformly convex function whose modulus of convexity is not of power type p for any $p \geq 2$.

5.4 Moduli of smoothness and rotundity

Hint. (c) Use Exercise 5.4.5(b) and Exercise 5.4.13. Check that (d) is dual to (c). For (e), let $\|\cdot\|$ on $\mathbb{R}^2$ be a norm that does not satisfy a modulus of convexity of power type p for any $p \geq 2$ (see [245]). Let $\|\cdot\|$ be the usual norm on $\mathbb{R}^2$. Check that $f := \|\cdot\|^2 + \max\{\|\cdot\|^2 - 2, 0\}$ is one such function. □

5.4.7 (More constructions of uniformly convex functions). (a) Let $p \geq 2$, and suppose $f : [0, +\infty) \to [0, +\infty]$ is a uniformly convex function with modulus of convexity of power type p that satisfies $f'(t) \geq Kt^{p-1}$ for some $K > 0$ and for all $t \geq 0$. Suppose $\|\cdot\|$ is an equivalent norm on a Banach space with modulus of convexity of power type p. Show that $h := f \circ \|\cdot\|$ is a uniformly convex function with modulus of convexity of power type p.

(b) Suppose f, h and $\|\cdot\|$ are as in (a), except that the modulus of convexity of f is not known to be of power type p. Show that h is uniformly convex.

(c) Examples on ℓ_p spaces where $1 < p < \infty$. Let $\|\cdot\|_p$ be the canonical norm on ℓ_p.

  (i) If $1 < p \leq 2$, show that $f := \|\cdot\|_p^r$ is uniformly convex for any $r \geq 2$. However, f is uniformly convex on bounded sets, but not uniformly convex if $1 < r < 2$.

  (ii) If $p > 2$, show that $f := \|\cdot\|_p^r$ is uniformly convex for any $r \geq p$. However, f is uniformly convex on bounded sets, but is not uniformly convex for $1 < r < p$.

  (iii) Let $b > 1$. Show that $f := b^{\|\cdot\|_p}$ is uniformly convex with modulus of convexity of power type $r = \max\{2, p\}$ for any $1 < p < \infty$.

Hint. (a) Let $K > 0$ be chosen so that $f'(t) \geq Kt^{p-1}$ for all $t \geq 0$ and let $C > 0$ be chosen so that $\delta_f(\varepsilon) \geq C\varepsilon^p$ for all $\varepsilon > 0$. Suppose $x, y \in X$, and $\|x - y\| \geq \delta$ where $\delta > 0$ is fixed. We may assume $\|y\| \leq \|x\|$.

In the case $\|y\| + \delta/2 \leq \|x\|$, we get

$$\frac{1}{2}f(\|x\|) + \frac{1}{2}f(\|y\|) - f\left(\frac{\|x+y\|}{2}\right) \geq C\left(\frac{\delta}{2}\right)^p. \qquad (5.4.17)$$

In the case $\|y\| + \delta/2 > \|x\|$, let $a = \|y\|$ and $\tilde{x} = x/\|x\|$, $\tilde{y} = y/\|y\|$. Then $\|y - a\tilde{x}\| > \delta/2$. Consequently, $\|\tilde{y} - \tilde{x}\| > \frac{\delta}{2a}$. Then

$$\left\|\frac{\tilde{x}+\tilde{y}}{2}\right\| \leq 1 - C_1\left(\frac{\delta}{2a}\right)^p \text{ and so } \left\|\frac{x+y}{2}\right\| \leq \frac{1}{2}\|x\| + \frac{1}{2}\|y\| - C_1 a\left(\frac{\delta}{2a}\right)^p \quad (5.4.18)$$

where $C_1 > 0$ is a fixed constant for the modulus of convexity of the norm. In the case, $C_1 a \left(\frac{\delta}{2a}\right)^p \geq a/2 \geq \delta/8$ (since $\|x\| + \|y\| \geq \delta$, and so $\|y\| \geq \delta/4$) use properties of f to show

$$f\left(\left\|\frac{x+y}{2}\right\|\right) \leq \frac{1}{2}f(\|x\|) + \frac{1}{2}f(\|y\|) - C\left(\frac{\delta}{8}\right)^p.$$

In the other case, the right hand side of the second inequality in (5.4.18) above is clearly $\geq a/2$. Now use the fact $f'(t) \geq K(a/2)^{p-1}$ when $t \geq a/2$. Consequently,

$$f\left(\left\|\frac{x+y}{2}\right\|\right) \leq f\left(\frac{1}{2}\|x\| + \frac{1}{2}\|y\|\right) - C_1 a \left(\frac{\delta}{2a}\right)^p \cdot K\left(\frac{a}{2}\right)^{p-1}$$

$$\leq \frac{1}{2}f(\|x\|) + \frac{1}{2}f(\|y\|) - C_1 K \left(\frac{\delta}{4}\right)^p.$$

Putting all this together one can deduce that $f \circ \|\cdot\|$ has modulus of convexity of power type p as desired.

For (b), notice that the modulus of convexity of $\|\cdot\|$ in conjunction with the derivative of f is crucial. However, uniform convexity of f will be enough to deduce uniform convexity of the composition.

For (c), use the known moduli of ℓ_p norms. That is, if $1 < p \leq 2$, $\|\cdot\|_p$ has modulus of convexity of power type 2. If $p > 2$, then $\|\cdot\|_p$ has modulus of convexity of power type p but not less, and trivially a norm with modulus of convexity of power type p also satisfies power type r when $r \geq p$. Thus, deduce part (iii) using part (a) and Exercise 5.4.2(b). □

5.4.8.* Let X be a Banach space. Use the results from this section along the James–Enflo theorem (5.1.24) and Pisier's theorem stated on p. 218 to show the following are equivalent.

(a) X admits an equivalent uniformly convex norm.
(b) There is a function $f : X \to \mathbb{R}$ that is supercoercive, bounded and uniformly convex on bounded sets.
(c) There is a lsc uniformly convex $f : X \to (-\infty, +\infty]$ whose domain has nonempty interior.
(d) X is superreflexive (hence both X and X^* admit uniformly convex and uniformly smooth norms).
(e) There is a function $h : X \to \mathbb{R}$ that is supercoercive and uniformly smooth on bounded sets.
(f) There is an equivalent norm $\|\cdot\|$ and $p \geq 2$ so that $f(\cdot) := \|\cdot\|^p$ is a uniformly convex function on X.

Hint. (a) $\Rightarrow$ (b): Let $\|\cdot\|$ be a uniformly convex norm on X, and let $f(x) = \|x\|^2$. (b) $\Rightarrow$ (c): Consider f plus the indicator function of the unit ball.

(c) $\Rightarrow$ (a): See the hint from Exercise 5.3.3.

By the James–Enflo theorem (5.1.24), (a) and (d) are equivalent, consequently, X admits a uniformly smooth norm $|\cdot|$ and so $h(x) = |x|^2$ is an appropriate function for (e). Now (e) implies X^* has a function as in (b), which then implies X^* and hence X is superreflexive.

The equivalence of (a) and (f) requires Pisier's theorem [353] and Theorem 5.4.6.
□

5.4.9 (Uniform convexity duality).* (a) Supply the details for the proof of Theorem 5.3.17(a) using Theorem 5.4.1(a).

5.4 Moduli of smoothness and rotundity

(b) Prove Theorem 5.4.1(a).

Hint. (a) Suppose f is uniformly convex, then $\operatorname{dom} f^* = X^*$. Let $\varepsilon_0 > 0$. Find $\eta > 0$ so that $\tau \frac{\varepsilon}{2} - \delta_f(\varepsilon) \leq 0$ when $0 \leq \tau \leq \eta$ and $\varepsilon > \varepsilon_0$. Therefore, by Theorem 5.4.1(b), $\rho_{f^*}(\tau) \leq \tau \frac{\varepsilon_0}{2}$ if $0 \leq \tau < \eta$. Consequently, f^* is uniformly smooth.

Conversely, suppose f is not uniformly convex. If $\operatorname{dom} f^* \neq X$, there is nothing further to do. Otherwise, we fix $\varepsilon_0 > 0$ such that $\delta_f(\varepsilon_0) = 0$. Then for each $\tau > 0$, by Theorem 5.4.1 (b), $\rho_{f^*}(\tau) \geq \tau \frac{\varepsilon_0}{2}$ and so f^* is not uniformly smooth.

(b) Let $\eta > 0$ and $\varepsilon > 0$. In the event $\delta_{f^*}(\varepsilon) = \infty$, clearly $\rho_f(\tau) + \delta_{f^*}(\varepsilon) \geq \frac{\tau\varepsilon}{2}$. Otherwise, choose $\phi, \Lambda \in \operatorname{dom} f^*$ such that $\|\phi - \Lambda\| \geq \varepsilon$ and

$$\frac{1}{2}f^*(\phi) + \frac{1}{2}f^*(\Lambda) - f\left(\frac{\phi + \Lambda}{2}\right) \leq \delta_{f^*}(\varepsilon) + \eta.$$

Now choose $x \in \partial_n f(\frac{\phi+\Lambda}{2})$ and $y \in S_X$ such that $y \in \partial_\eta \|\lambda - \phi\|$. Then,

$$\rho_f(t) \geq \frac{1}{2}f(x + ty) + \frac{1}{2}f(x - ty) - f(x)$$

$$\geq \frac{1}{2}[f(x + ty) - \phi(x + ty)] + \frac{1}{2}[f(x - ty) - \Lambda(x - ty)]$$

$$+ \left\langle \frac{\phi + \Lambda}{2}, x \right\rangle - f(x) + \frac{1}{2}t(\phi - \Lambda)(y)$$

$$\geq f^*\left(\frac{\phi + \Lambda}{2}\right) - \eta - \frac{1}{2}f^*(\phi) - \frac{1}{2}f^*(\Lambda) + \frac{t}{2}\varepsilon - \eta$$

$$\geq -\delta_{f^*}(\varepsilon) + \frac{t\varepsilon}{2} - 3\eta.$$

Therefore, $\rho_f(\tau) + \delta_{f^*}(\varepsilon) \geq \frac{\tau\varepsilon}{2}$ for $\tau > 0$, $\varepsilon > 0$.

On the other hand, when $\phi \in \partial f(x + ty)$ and $\Lambda \in \partial f(x - ty)$ where $y \in S_X$ and $t \geq 0$, we compute

$$\frac{1}{2}f(x + ty) + \frac{1}{2}f(x - ty) - f(x) \leq$$

$$\leq \frac{1}{2}[\phi(x + ty) - f^*(\phi) + \Lambda(x - ty) - f^*(\Lambda)]$$

$$- \left[\left\langle x, \frac{\phi + \Lambda}{2}\right\rangle - f^*\left(\frac{\phi + \Lambda}{2}\right)\right]$$

$$= f^*\left(\frac{\phi + \Lambda}{2}\right) - \frac{1}{2}f^*(\phi) - \frac{1}{2}f^*(\Lambda) + \frac{1}{2}t(\phi - \Lambda)(y)$$

$$\leq -\delta_{f^*}(\varepsilon) + \frac{1}{2}\varepsilon t,$$

where $\varepsilon = \|\phi - \Lambda\|$. Thus $\rho_f(\tau) \leq \sup\{\tau\frac{\varepsilon}{2} - \delta_{f^*}(\varepsilon) : 0 \leq \varepsilon\}$. □

5.4.10.* Use Theorem 5.4.1 to prove Theorem 5.4.2.

5.4.11 (Moduli of convexity). Define the *gauge of uniform convexity* by

$$p_f(\varepsilon) := \inf \left\{ \frac{(1-\lambda)f(x) + \lambda f(y) - f((1-\lambda)x + \lambda y)}{\lambda(1-\lambda)} : \right.$$

$$\left. 0 < \lambda < 1, \|x-y\| = \varepsilon, x, y \in \operatorname{dom} f \right\}.$$

(a) Show that $2\delta_f(\varepsilon) \leq p_f(\varepsilon) \leq 4\delta_f(\varepsilon)$.
(b) Conclude that the definition of uniform convexity used here is equivalent that used in Zălinescu's book [445].
(c) Prove that $p_f(ct) \geq c^2 p_f(t)$ for all $c \geq 1$ and all $t \geq 0$.

Hint. (a) See [444, p. 347]. For $0 < \lambda \leq \frac{1}{2}$, and $\|x-y\| = \varepsilon$ with $x, y \in \operatorname{dom} f$, one computes

$$f(\lambda x + (1-\lambda)y) = f\left(2\lambda \left(\frac{x+y}{2}\right) + (1-2\lambda)y\right)$$

$$\leq 2\lambda f\left(\frac{x+y}{2}\right) + (1-2\lambda)f(y)$$

$$\leq \lambda f(x) + \lambda f(y) - 2\lambda \delta_f(\varepsilon) + (1-2\lambda)f(y)$$

$$\leq \lambda f(x) + (1-\lambda)f(y) - (2\lambda)(1-\lambda)\delta_f(\varepsilon).$$

Therefore, $2\lambda(1-\lambda)\delta_f(\varepsilon) \leq \lambda f(x) + (1-\lambda)f(y) - f(\lambda x + (1-\lambda)y)$, or, in other words,

$$2\delta_f(\varepsilon) \leq \frac{\lambda f(x) + (1-\lambda)f(y) - f(\lambda x + (1-\lambda)y)}{\lambda(1-\lambda)} \leq p_f(\varepsilon).$$

Where we note above that there is no restriction assuming $0 < \lambda \leq \frac{1}{2}$ since the roles of x and y can be reversed. On the other hand, letting $\lambda = \frac{1}{2}$ in the definition of $p_f(\varepsilon)$, one sees that $p_f(\varepsilon) \leq 4\delta_f(\varepsilon)$.
(b) A proper convex function $f : X \to (-\infty, +\infty]$ is uniformly convex if and only if $\delta_f(\varepsilon) > 0$ for each $\varepsilon > 0$ if and only if $p_f(\varepsilon) > 0$ for each $\varepsilon > 0$.
(c) See [445, pp. 203–204]. □

5.4.12 (Banach limits). A *Banach limit* is a translation invariant linear functional Λ on ℓ_∞ such that $\liminf_k x_k \leq \Lambda((x_k)) \leq \limsup_k x_k$. Thus Λ agrees with the limit on c_0.

(a) Apply the Hahn–Banach extension theorem to the linear function $\lambda(x) := \lim_n \left(\sum_{i=1}^n x_i\right)/n \leq \limsup_k x_k$, on the space of Cesàro convergent sequences to deduce the existence of Banach limits.
(b) Let $K > 0$ and let $\|\cdot\|_n \leq K\|\cdot\|$ for each $n \in \mathbb{N}$ be given. Let L be a positive linear functional on ℓ_∞.
   (i) Show that $L(\|\cdot\|_n)$ and $\|\cdot\|$ are equivalent norms with $(L(\|\cdot\|_n))^* \leq L(\|\cdot\|_n^*)$. Under what conditions are the later two norms equal?

(ii) In the case that L is a Banach limit show that the modulus of convexity of $L(\|\cdot\|_n)$ exceeds the lim inf of the moduli of convexity of the sequence of norms.

5.4.13.★ Let $f: X \to \mathbb{R}$ be a continuous convex function. Show that f has modulus of smoothness of power type p where $p > 1$ if and only if there exists $C > 0$ so that $\|f'(x) - f'(y)\| \le C\|x-y\|^{p-1}$ for all $x, y \in X$ if and only if f^* has modulus of convexity of power type q where $1/p + 1/q = 1$.

Hint. The final 'if and only if' follows directly from Theorem 5.4.2. Note also the case $p > 2$ occurs only when f is affine. A standard proof of the first 'if and only if' is as follows.

Suppose f' satisfies the given Hölder condition. Then given $x \in X$, $y \in S_X$ and $\tau > 0$, using the mean value theorem we have $c_1, c_2 \in [0,1]$ such that

$$\frac{1}{2}f(x+\tau y) + \frac{1}{2}f(x-\tau y) - f(x) = \frac{1}{2}[f(x+\tau y) - f(x)]$$
$$- \frac{1}{2}[f(x) - f(x-\tau y)]$$
$$= \frac{1}{2}[f'(x+c_1\tau y)(\tau y) - f'(x-c_1\tau y)(\tau y)]$$
$$\le \frac{1}{2}K(2\tau)^{p-1}(\tau) \le K\tau^p.$$

Conversely, suppose f has modulus of smoothness of power type p. Fix $x, y \in X$, and choose $h \in S_X$ arbitrary and let $\tau := \|x-y\|$. Then

$$f'(x)(\tau h) - f'(y)(\tau h) \le f(x+\tau h) - f(x) - f'(y)(\tau h)$$
$$= f(x+\tau h) - f(y) - f'(y)(x+\tau h - y)$$
$$\quad + f(y) - f(x) + f'(y)(x-y)$$
$$\le f(x+\tau h) - f(y) - f'(y)(x+\tau h - y)$$
$$= f(x+\tau h) - f(y) + f'(y)(y - x - \tau h)$$
$$\le f(x+\tau h) - f(y) + f(2y - x - \tau h) - f(y)$$
$$= f(y + (x+\tau h - y)) + f(y - (x+\tau h - y)) - 2f(y)$$
$$\le 2C(2\tau)^p = 2^{1+p}C\tau\|x-y\|^{p-1}.$$

Thus $\|f'(x) - f'(y)\| \le 2^{p+1}C\|x-y\|^{p-1}$. □

5.5 Lipschitz smoothness

Let C be a nonempty bounded subset of a Banach space X, let $x \in C$, and let $\phi \in X^*$. Following [200], we say C is *Lipschitz exposed* at x by ϕ if there exists $c > 0$ such that, for all $y \in C$,

$$\|x - y\|^2 \leq c\langle \phi, x - y \rangle. \tag{5.5.1}$$

In this case, ϕ *Lipschitz exposes* C at x. Clearly, this condition will never be satisfied for an unbounded set. Thus, for any set $C \subset X$, we say C is *Lipschitz exposed* at $x \in C$ by $\phi \in X^*$, if ϕ strongly exposes x and if there exists $\delta > 0$ and $c > 0$ so that (5.5.1) is satisfied for all $y \in C$ with $\|y - x\| \leq \delta$. It is left as Exercise 5.5.1 to check that this latter definition coincides with the former when C is bounded.

Consider a proper function $f : X \to (-\infty, +\infty]$. We will say that $x_0 \in \operatorname{dom} f$ is a *Lipschitz exposed point* of f if there exists $\phi \in X^*$ such that $f - \phi$ attains its strong minimum at x_0 and there exist $\delta > 0$ and $c > 0$ such that

$$\|x - x_0\|^2 \leq c[(f - \phi)(x) - (f - \phi)(x_0)] \text{ whenever } x \in \operatorname{dom} f, \|x - x_0\| \leq \delta. \tag{5.5.2}$$

In this case we say that ϕ *Lipschitz exposes* f at x_0. It is left as Exercise 5.5.2 to verify this is equivalent to $(x_0, f(x_0))$ being a Lipschitz exposed in $\operatorname{epi} f$ by the functional $(\phi, -1)$. If $\|x - x_0\|^2$ is replaced with $\|x - x_0\|^{1+\frac{1}{\alpha}}$ where $0 < \alpha \leq 1$ in (5.5.2), we will say f is α-*Hölder exposed* by ϕ at x_0; of course, when $\alpha = 1$, this is the same as Lipschitz exposedness.

Proposition 5.5.1. *For a proper function $f : X \to (-\infty, +\infty]$, the following are equivalent.*
(a) f is α-Hölder exposed by ϕ at x_0.
(b) There exist $\phi \in \partial f(x_0)$, $\delta > 0$, $\eta > 0$ and $c > 0$ such that $(f - \phi)(y) \geq (f - \phi)(x_0) + \eta$ whenever $\|x - y\| \geq \delta$, and moreover for $\varepsilon > 0$

$$\|x - x_0\|^{1+\frac{1}{\alpha}} \leq c\varepsilon \text{ whenever } \phi \in \partial_\varepsilon f(x) \text{ where } x \in \operatorname{dom} f, \|x - x_0\| < \delta.$$

Proof. (a) $\Rightarrow$ (b): Let $\delta > 0$ be as in the definition of α-Hölder exposedness. Because ϕ strongly exposes f at x_0, there exists $\eta > 0$ such that $(f - \phi)(y) \geq (f - \phi)(x_0) + \eta$ whenever $\|x - x_0\| \geq \delta$. Suppose $\phi \in \partial f(x_0)$ and $\phi \in \partial_\varepsilon f(x)$ where $\|x - x_0\| \leq \delta$, $x \in \operatorname{dom} f$ and $\varepsilon > 0$. Then

$$\phi(x_0) - \phi(x) \leq f(x_0) - f(x) + \varepsilon,$$

and so $(f - \phi)(x) - (f - \phi)(x_0) \leq \varepsilon$. According to (a), $\|x - x_0\|^{1+\frac{1}{\alpha}} \leq c\varepsilon$.

(b) $\Rightarrow$ (a): Let $\varepsilon := (f - \phi)(x) - (f - \phi)(x_0)$ where $\phi \in \partial f(x_0), x \in \operatorname{dom} f$ and $\|x - x_0\| \leq \delta$. Then

$$f(x_0) - f(x) + \varepsilon = \phi(x_0) - \phi(x),$$

and so for any $z \in X$,

$$\phi(z) - \phi(x) = \phi(z) - \phi(x_0) + \phi(x_0) - \phi(x)$$
$$\leq f(z) - f(x_0) + f(x_0) - f(x) + \varepsilon = f(z) - f(x) + \varepsilon.$$

5.5 Lipschitz smoothness

This shows $\phi \in \partial_\varepsilon f(x)$ and so

$$\|x - x_0\|^{1+\frac{1}{\alpha}} \leq c\varepsilon, \text{ thus } \|x - x_0\|^{1+\frac{1}{\alpha}} \leq c[(f - \phi)(x) - (f - \phi)(x_0)]$$

as desired. □

We next introduce smoothness notions which will be relevant in discussions on duality. A convex function $f : X \to (-\infty, +\infty]$ is said to be α-Hölder smooth at x where $0 < \alpha \leq 1$ if there exists $\delta > 0$ such that $\partial f(y) \neq \emptyset$ for $\|y - x\| < \delta$, and if there exists $c > 0$ so that

$$\|\Lambda - \phi\| \leq c\|y - x\|^\alpha \text{ whenever } \|y - x\| < \delta, \Lambda \in \partial f(y), \phi \in \partial f(x).$$

Again, the case when $\alpha = 1$ is of particular importance, and in that event we say f is *Lipschitz smooth* at x. Notice that these definitions imply the subdifferential is locally bounded at x and so f is continuous at x, and then Šmulian's theorem (4.2.10) implies f is Fréchet differentiable at x.

Example 5.5.2. (a) If $f : X \to \mathbb{R}$ is twice differentiable at a point x_0, then x_0 is a point of Lipschitz smoothness of f (see Proposition 4.6.16 and the paragraph following it). In particular, the functions $f(t) := e^t$ and $g(t) := |t^p|$ with $p > 2$ are not uniformly smooth, yet each $t \in \mathbb{R}$ is a point of Lipschitz smoothness for these functions.
(b) Let $f(t) := |t|^{1+\alpha}$ for $0 < \alpha \leq 1$. Then each point of $\mathbb{R}$ is a point of α-Hölder smoothness.
(c) There are nonreflexive Banach spaces such as c_0 that admit C^∞-smooth norms (see [180, Theorem V.1.5]), and hence coercive C^∞-smooth convex functions; yet because it is not superreflexive, c_0 does not admit a uniformly smooth coercive convex function.

For a convex function $f : X \to (-\infty, +\infty]$, we define the *modulus of smoothness* of f at $x_0 \in \text{dom} f$ for $\tau \geq 0$ by

$$\rho_f(x_0, \tau) := \sup \left\{ \frac{1}{2}f(x_0 + \tau y) + \frac{1}{2}f(x_0 - \tau y) - f(x) : y \in S_X \right\}.$$

A useful characterization of Hölder smooth points is the following local version of Exercise 5.4.13.

Theorem 5.5.3 (Lipschitz/Hölder smooth points). *Let $f : X \to (-\infty, +\infty]$ be a convex function, and let $0 < \alpha \leq 1$. Then the following are equivalent.*
(a) f is α-Hölder smooth at x.
(b) There exist $c > 0$ and $\delta > 0$ and $\phi \in X^*$ such that

$$|f(x+h) - f(x) - \langle \phi, h \rangle| \leq c\|h\|^{1+\alpha} \text{ whenever } \|h\| < \delta.$$

(c) f is continuous at x and there exist $c > 0$ and $\delta > 0$ such that

$$f(x+h) + f(x-h) - 2f(x) \leq c\|h\|^{1+\alpha} \text{ whenever } \|h\| < \delta;$$

in other words, $\rho_f(x_0, t) \leq \frac{c}{2}t^{1+\alpha}$ whenever $0 \leq t < \delta$.

Proof. (a) $\Rightarrow$ (b): Let $\delta > 0$ and $c > 0$ be as given in the definition of α-Hölder smoothness. Let $\|h\| < \delta$, $\Lambda \in \partial f(x+h)$ and $\phi \in \partial f(x)$. Then

$$0 \leq f(x+h) - f(x) - \langle \phi, h \rangle \leq \langle \Lambda, h \rangle - \langle \phi, h \rangle$$
$$\leq \|\Lambda - \phi\| \|h\| \leq c\|h\|^\alpha \|h\| = c\|h\|^{1+\alpha}.$$

(b) $\Rightarrow$ (c): Suppose that (b) is valid and $\|h\| < \delta$. Then

$$f(x+h) - f(x) + f(x-h) - f(x) = f(x+h) - f(x) - \phi(h)$$
$$+ f(x-h) - f(x) - \phi(-h)$$
$$\leq 2c\|h\|^{1+\alpha}.$$

(c) $\Rightarrow$ (a): Let $\|y - x\| < \delta/2$, $\Lambda \in \partial f(y)$ and $\phi = \nabla f(x)$ (notice f is Fréchet differentiable at x by Proposition 4.2.7). Let $\varepsilon > 0$ and choose $h \in X$ such that $\|h\| = \|y-x\|$ and $\Lambda(h) - \phi(h) \geq \|\Lambda - \phi\|\|h\| - \varepsilon$. Now,

$$\Lambda(h) - \phi(h) \leq f(y+h) - f(y) - \phi(h)$$
$$= f(y+h) - f(x) + f(x) - f(y) - \phi(h)$$
$$\leq f(y+h) - f(x) + \phi(x-y) - \phi(h)$$
$$= f(y+h) - f(x) + \phi(x-y-h)$$
$$\leq f(x + (y-x+h)) - f(x) + f(x - (y-x+h)) - f(x)$$
$$\leq c\|y - x + h\|^{1+\alpha} \leq 4c\|h\|\|y-x\|^\alpha.$$

Because $\varepsilon > 0$ was arbitrary $\|\Lambda - \phi\| \leq 4c\|y-x\|^\alpha$. □

Unlike points of Fréchet smoothness, the points of Lipschitz smoothness of a continuous convex function need not contain a dense G_δ-set.

Example 5.5.4. There is a continuous convex function $f : \mathbb{R} \to \mathbb{R}$ such that the Lipschitz smooth points of f do not form a generic set.

Proof. Choose open sets $G_n \subset \mathbb{R}$ such that $\mu(G_n) < 1/n$ and $\mathbb{Q} \subset G_n$ where μ denotes the Lebesgue measure. Then let $G := \bigcap_{n=1}^\infty G_n$. Now $\mu(G) = 0$, thus [133, Theorem 7.6, p. 288] ensures there is a continuous strictly increasing function g such that $g'(t) = \infty$ for all $t \in G$. Now $f(x) := \int_0^x g(t)\,dt$ is the desired convex function. □

Proposition 5.5.5. *Suppose $f : X \to (-\infty, +\infty]$ is a lsc proper convex function and $x_0 \in \operatorname{dom} f$. Then f^* is α-Hölder exposed at ϕ_0 by $x_0 \in X$ if and only if x_0 is a point of α-Hölder smoothness of f where $\nabla f(x_0) = \phi_0$.*

Proof. $\Rightarrow$: Since an α-Hölder exposed point is strongly exposed, f is Fréchet differentiable at x_0 according to Proposition 5.2.4. Thus we can choose $\eta > 0$ so that

$\|x - x_0\| \leq \eta$ implies $\partial f(x) \neq \emptyset$ and if $\phi \in \partial f(x)$ and $\phi_0 \in \partial f(x_0)$ we have $\|\phi - \phi_0\| < \delta$ where $\delta > 0$ is as in (b) of Proposition 5.5.1.

Now let $\|x - x_0\| \leq \eta$ and let $\phi \in \partial f(x)$ and $\phi_0 \in \partial f(x_0)$. Then

$$\|\phi - \phi_0\|^{1+\frac{1}{\alpha}} \leq c[(f^* - x_0)(\phi) - (f^* - x_0)(\phi_0)]$$

$$= c[f^*(\phi) - \phi(x_0) - f^*(\phi_0) + \phi_0(x_0)]$$

$$= c[\phi(x) - f(x) - \phi(x_0) - \phi_0(x_0) + f(x_0) + \phi_0(x_0)]$$

$$= c[\phi(x) - \phi(x_0) + f(x_0) - f(x)]$$

$$\leq c[\phi(x) - \phi(x_0) + \phi_0(x_0) - \phi_0(x)]$$

$$= c[\phi(x - x_0) - \phi_0(x - x_0)]$$

$$\leq c\|\phi - \phi_0\|\|x - x_0\|.$$

Therefore, $\|\phi - \phi_0\| \leq c^\alpha \|x - x_0\|^\alpha$ as desired.

$\Leftarrow$: Let $\phi := \nabla f(x_0)$ and let $c > 0$ and $\delta > 0$ be such that

$$f(x_0 + h) - f(x_0) - \phi(h) \leq c\|h\|^{1+\alpha} \quad \text{whenever } \|h\| \leq \delta.$$

Suppose $\varepsilon > 0$, $\|\Lambda - \phi\| < \delta$, and $x_0 \in \partial_\varepsilon f^*(\Lambda)$. Then $\Lambda \in \partial_\varepsilon f(x_0)$ (Proposition 4.4.5(b)) and consequently

$$\Lambda(x_0 + h) - \Lambda(x_0) - \phi(h) \leq c\|h\|^{1+\alpha} + \varepsilon.$$

This implies $\langle \Lambda - \phi, h \rangle \leq c\|h\|^{1+\alpha} + \varepsilon$. In the case $\varepsilon < \delta^{1+\alpha}$, we take the supremum over h with $\|h\| = \varepsilon^{\frac{1}{1+\alpha}}$. This implies $\|\Lambda - \phi\| \leq (c+1)\varepsilon^{\frac{\alpha}{1+\alpha}}$. Raising both sides to the $(1 + 1/\alpha)$-power implies $\|\Lambda - \phi\|^{1+\frac{1}{\alpha}} \leq (c+1)^{1+\frac{1}{\alpha}}\varepsilon$. In the case $\varepsilon \geq \delta^{1+\alpha}$ it follows that $\|\phi - \Lambda\|^{1+\frac{1}{\alpha}} \leq C\varepsilon$ where $C = \delta^{\frac{1}{\alpha} - \alpha}$. The result now follows from Proposition 5.5.1. $\square$

Corollary 5.5.6. *Suppose $f : X \to (-\infty, +\infty]$ is a proper lsc function. Then f is α-Hölder exposed by ϕ at x_0 if f^* is α-Hölder smooth at ϕ. The converse holds, if additionally, $\mathrm{dom} f$ is bounded or f is convex.*

Proof. Suppose f^* is α-Hölder smooth at ϕ with $\nabla f^*(\phi) = x_0$ which is necessarily in X by Exercise 4.2.11. Then f^{**} is α-Hölder exposed by ϕ at x_0. Moreover, because f is lsc $f(x_0) = f^{**}(x_0)$ and $f \geq f^{**}$ it follows that f is α-Hölder exposed by ϕ at x_0.

For the converse, we now suppose $\mathrm{dom} f$ is bounded. According to Exercise 5.5.4, we choose $c > 0$ so that

$$f(x) \geq f(x_0) + \phi(x - x_0) + c\|x - x_0\|^{1+\frac{1}{\alpha}} \quad \text{for all } x \in X.$$

Consequently, $f^{**}(x^{**}) \geq f(x_0) + \phi(x^{**} - x_0) + c\|x^{**} - x_0\|^{1+\frac{1}{\alpha}}$ for all x^{**} in X^{**} where $\|\cdot\|$ now represents the canonical second dual norm on X^{**}. Thus f^{**} is α-Hölder exposed by ϕ at x_0. According to Proposition 5.5.5, f^* is α-Hölder smooth at ϕ with $\nabla f^*(\phi) = x_0$ as desired. The proof of the converse in the case f is lsc and convex can be argued similarly using Exercise 5.5.4(b). □

Given $x_0 \in \operatorname{dom} f$, we define the *modulus of convexity* of f at x_0 by

$$\delta_f(x_0, \varepsilon) := \inf\left\{ \frac{1}{2}f(x) + \frac{1}{2}f(x_0) - f\left(\frac{x_0+x}{2}\right) : \|x - x_0\| = \varepsilon, x \in \operatorname{dom} f \right\}.$$

Proposition 5.5.7. *Suppose f is a lsc convex function and $\alpha \geq 1$. If $\delta_f(x_0, \varepsilon) \geq C\varepsilon^{1+\frac{1}{\alpha}}$ for some $C > 0$ and $0 \leq \varepsilon \leq 1$ at x_0, then f is α-Hölder exposed at x_0 by any $\phi \in \partial f(x_0)$ when $\partial f(x_0)$ is not empty.*

Proof. Let $\phi_0 \in \partial f(x_0)$. Then for $\|x - x_0\| \leq 1$, we have

$$f(x_0) + \phi\left(\frac{x - x_0}{2}\right) \leq f\left(\frac{x_0+x}{2}\right) \leq \frac{1}{2}f(x_0) + \frac{1}{2}f(x) - C\|x_0 - x\|^{1+\frac{1}{\alpha}}.$$

Therefore

$$C\|x_0 - x\|^{1+\frac{1}{\alpha}} \leq \frac{1}{2}f(x) - \phi\left(\frac{x}{2}\right) - \frac{1}{2}f(x_0) + \phi\left(\frac{x_0}{2}\right).$$

Consequently, $2C\|x_0 - x\|^{1+\frac{1}{\alpha}} \leq (f - \phi)(x) - (f - \phi)(x_0)$ for $\|x - x_0\| \leq 1$. Because f is convex, $f - \phi$ attains a strong minimum at x_0, and so f is α-Hölder exposed by ϕ at x_0. □

Exercises and further results

5.5.1.* Prove that the two definitions for Lipschitz exposed points at the beginning of this section agree for bounded nonempty sets.

Hint. Clearly the definition from [200] implies the point is strongly exposed, so one direction is clear. For the other, suppose C is a bounded set of diameter less than M. Let $\delta > 0$, $c > 0$ be as specified. Because x is strongly exposed, there exists $\varepsilon > 0$ and such that $\|y - x\| \leq \delta$ if $y \in C$ and $\phi(y) \geq \phi(x) - \varepsilon$. Choose $k > 0$ so large that $M^2 \leq k\varepsilon$. Now let $c_M := \max\{c, k\}$, and verify

$$\|y - x\|^2 \leq c_M[\phi(x) - \phi(y)] \text{ whenever } y \in C$$

to finish the exercise. □

5.5.2.* Show that x_0 is a Lipschitz exposed point of f with corresponding ϕ as in the definition if and only if $(x_0, f(x_0))$ is Lipschitz exposed in $\operatorname{epi} f$ by $(\phi, -1)$.

5.5.3.* Following the literature (see [141]) we say a function f is *uniformly convex at the point* x_0 if $\delta_f(x_0, \varepsilon) > 0$ for each $\varepsilon > 0$. Show that f is uniformly convex at x_0 if and only if it is locally uniformly convex at x_0 (as we defined on p. 239).

5.5.4.* Suppose $f : X \to (-\infty, +\infty]$ is a proper function, and that $\phi \in X^*$ α-Hölder exposes f at x_0.

(a) If dom f is a bounded set, show that there exists $c > 0$ so that

$$f(x) \geq f(x_0) + \phi(x - x_0) + c\|x - x_0\|^{1+\frac{1}{\alpha}} \text{ for all } x \in X. \tag{5.5.3}$$

(b) Suppose f is lsc and convex. Show that there is a continuous convex function $g : \mathbb{R} \to [0, +\infty)$ such that $g(t) := k|t|^{1+\frac{1}{\alpha}}$ whenever $|t| \leq \delta$ and $g(t) := c|t|$ for $|t| > \delta$ where $\delta > 0$, $k > 0$ and $c > 0$ so that

$$f(x) \geq f(x_0) + \phi(x - x_0) + g(\|x - x_0\|) \text{ for all } x \in X. \tag{5.5.4}$$

(c) Deduce that if either dom f is bounded or f is lsc and convex, then ϕ also α-Hölder exposes $f^{**} : X^{**} \to (-\infty, +\infty]$ at x_0.

5.5.5 (Perturbed minimization principles). This exercise provides a local variant for functions of some duality results from [200]. Let X be a Banach space, and let $C \subset X$ be a closed bounded convex set. Show that the following are equivalent.

(a) Every weak*-lsc convex function $g : X^* \to \mathbb{R}$ such that $g \leq \sigma_C$ has a dense set of α-Hölder smooth points.
(b) Given any proper lsc bounded below function $f : C \to (-\infty, +\infty]$ and $\varepsilon > 0$, there exists $\phi \in \varepsilon B_{X^*}$, $x_0 \in C$ and $k > 0$ such that

$$f(x_0) + \langle \phi, x - x_0 \rangle + k\|x - x_0\|^{1+\frac{1}{\alpha}} \leq f(x)$$

for all $x \in C$, that is f is α-Hölder exposed by ϕ at x_0.

Hint. This is similar to Exercise 5.2.12. (a) $\Rightarrow$ (b): Suppose $f : C \to \mathbb{R}$ is bounded below on C. Then there exists $a \in \mathbb{R}$ so that $f + a \geq \sigma_C$. Consequently, $f^* - a = (f + a)^* \leq \delta_C^* \leq \sigma_C$. Given $\varepsilon > 0$, there exists $\phi \in \varepsilon B_{X^*}$ so that $f^* - a$ and hence f^* is α-Hölder smooth at ϕ. Now apply Corollary 5.5.6 to get the conclusion.
(b) $\Rightarrow$ (a): Take any weak*-lsc convex $g \leq \sigma_C$. Let $f := g^*|_X$. Then $f \geq \delta_C$, and $f^* = g$. Now let $\Lambda \in X^*$ be arbitrary, then $f + \Lambda$ is bounded below on C, so $f + \Lambda$ is α-Hölder exposed by some $\phi \in \varepsilon B_{X^*}$. This implies $(f + \Lambda)^*$ is α-Hölder smooth at ϕ. But $(f + \Lambda)^*(\cdot) = g^*(\cdot - \Lambda)$, and so g^* is α-Hölder smooth at $\Lambda + \phi$. □

5.5.6 (Isomorphic characterizations of Hilbert spaces).*** There are many characterizations of Hilbert spaces using smoothness and rotundity that can be derived from Kwapien's theorem [283] that states a Banach space is isomorphic to a Hilbert space if it has *type 2* and *cotype 2*. For example, if X admits equivalent norms, one which has modulus of smoothness of power type 2, and another which has modulus of convexity of power type 2, then X is isomorphic to a Hilbert space. An important work by Meshkov [314] showed X is isomorphic to a Hilbert space if both X and X^* have bump functions that are C^2-smooth. In [200, Theorem 3.3] it is established that X is isomorphic to a Hilbert space if X admits a bump function with locally Lipschitz derivative and X^* is a *Lipschitz determined* space, that is, a space on which each continuous convex function is Lipschitz smooth at a dense set of points. Building upon results from [178], in [420] it is shown a Banach space is isomorphic to a Hilbert space if X and X^* have continuous bump functions that are everywhere twice Gâteaux differentiable (the definition therein stipulated the second derivative must be symmetric). A good survey of these types of results using an elementary approach is given in [205].

It is well-known that the usual norms on ℓ_p for $1 < p \leq 2$ have modulus of rotundity of power type 2, while the usual norms on ℓ_q for $2 \leq q < \infty$ have modulus of smoothness of power type 2, but it is only when $p = 2$ that the norm has both such properties [180]; see also Theorem 5.1.34. The following two isomorphic characterizations of Hilbert spaces can be derived in an elementary fashion.

(a) Let X be a Banach space and suppose there exists $x \in S_X$ such that $\|\cdot\|$ is twice Gâteaux differentiable at x and B_X is Lipschitz exposed at x (by $\nabla\|x\|$). Show that X is isomorphic to a Hilbert space.
(b) Let X be a Banach space such that $\|\cdot\|^2$ has a strong second-order Taylor expansion at 0. Show that X is isomorphic to a Hilbert space.

Hint. (a) Use the Lipschitz exposedness of X to show there exists $c > 0$ such that

$$\|h\|^2 \leq c\nabla^2\|h\|(h,h) \quad \text{for all } h \in H$$

where $H := \{y \in X : \langle \nabla\|x\|, y\rangle = 0\}$. Because the second derivative is symmetric, $h \mapsto \sqrt{\nabla^2\|h\|(h,h)}$ defines an inner product norm on H; see [200, Proposition 3.1] for the full details. For (b), note $\nabla\|\cdot\|^2 = 0$ when $x = 0$, then the bilinear form in the Taylor expansion of $\|\cdot\|^2$ at 0 can be used to create an inner product norm on X; see [180, p. 184]. □

5.5.7 (Fabian, Whitfield, Zizler [203]).** Show that if X admits a Lipschitz smooth (resp. Hölder smooth, uniformly smooth) bump function, then X admits an equivalent Lipschitz smooth (resp. Hölder smooth, uniformly smooth) norm. In particular, X is superreflexive in any of these cases.

Hint. For the Lipschitz case, other cases are similar. Let $\phi : X \to \mathbb{R}$ be differentiable and symmetric with ϕ' Lipschitz, $\phi \leq 0$, $\inf \phi = -1 = \phi(0)$ and $\mathrm{supp}\, \phi \subset \frac{1}{2}B_X$. Define

$$\psi(x) := \inf\left\{\sum_{i=1}^n \alpha_i \phi(x_i) : \sum \alpha_i x_i = x, \alpha_i \geq 0, \sum \alpha_i = 1, x_i \in B_X, n \in \mathbb{N}\right\}.$$

Let $G:=\{x \in B_X : \psi(x) < -1/2\}$. Show ψ is convex on G and use the symmetric definition of differentiability to show ψ' satisfies a Lipschitz condition on G. Let $Q:=\{x \in X : \psi(x) \leq -3/4\}$. Note that Q is a convex set with nonempty interior. Use the implicit function theorem for gauges to show that γ_Q has Lipschitz derivative on the set $\{x : Q(x) = 1\}$. An equivalent Lipschitz smooth norm is then $\|x\|:=\gamma_Q(x)+\gamma_Q(-x)$. □

5.5.8 (Tang [413]).** Suppose X admits a uniformly Gâteaux differentiable bump function. Show that X admits an equivalent uniformly Gâteaux differentiable norm.

5.5 Lipschitz smoothness

Hint. One can show that such a bump function is automatically bounded [206, p. 699] and hence Lipschitz [313, Remark 2.1]. By composing the bump function with an appropriate real function, we may then suppose $b : X \to [0, 1]$ is a uniformly Gâteaux differentiable bump function such that $b(x) = 1$ if $\|x\| \leq 1/r$ and $b(x) = 0$ if $\|x\| \geq 1/2$ where $r > 2$. Let $\phi(x) := \sum_{n=1}^{\infty} r^n[1 - b(r^{-n}x)]$. Establish that

$$\phi(0) = 0, \quad \max\{0, (\|x\| - r)/r\} \leq \phi(x) \leq r^2(\|x\| + 1) \text{ for all } x \in X \quad (5.5.5)$$

and ϕ has uniform directional modulus of smoothness; see [313, Lemma 2.2] for more details.

Let ψ be defined by

$$\psi(x) := \inf \left\{ \sum_{i=1}^{n} \alpha_i \phi(x_i) : x = \sum_{i=1}^{n} \alpha_i x_i, \alpha_i \geq 0, \sum_{i=1}^{n} \alpha_i = 1, n \in \mathbb{N} \right\}.$$

Show that ψ is a Lipschitz convex uniformly Gâteaux differentiable function that satisfies the same bounds as ϕ in (5.5.5). Now ψ^* has bounded domain because ψ is Lipschitz, moreover $0 \in \text{int dom } \psi^*$, and ψ^* is weak*-uniformly rotund. Symmetrize ψ^* and use an appropriate lower level set to create an equivalent dual weak*-uniformly rotund norm on X^*. For further details, see [413]. For a very nice alternate approach, see [206, Theorem 3]. □

6
Further analytic topics

Harald Bohr is reported to have remarked "Most analysts spend half their time hunting through the literature for inequalities they want to use, but cannot prove." (D. J. H. Garling)[1]

6.1 Multifunctions and monotone operators

This section gives a brief account of some properties of set-valued mappings which we will typically call *multifunctions*. Particular attention will be given to the relation between multifunctions and subdifferentials of convex functions.

Let X be a Banach space, and let $T : X \to 2^{X^*}$ be a multifunction from X into subsets of X^*. We will say T is a *monotone operator* provided the following perhaps unnatural inequality holds:

$$\langle x^* - y^*, x - y \rangle \geq 0$$

whenever $x, y \in X$ and $x^* \in T(x), y^* \in T(y)$. It is not required that $T(x)$ be nonempty. The *domain* (or *effective domain*) $D(T)$ of T is the collection of $x \in X$ such that $T(x)$ is nonempty. This notion of monotonicity forms the basis for all of Chapter 9. A natural example of a monotone operator is the subdifferential of a convex function; this was outlined for convex functions on Euclidean spaces in Exercise 2.2.21, and the reader can easily check that the argument extends to general normed spaces.

Let X and Y be Hausdorff topological spaces and suppose that $T : X \to 2^Y$ is a multifunction from X into the subsets of Y. If S is a subset of X, we define $T(S) = \bigcup_{x \in A} T(x)$. The mapping T is said to be *upper-semicontinuous at the point* $x \in X$ if for each open set V in Y containing $T(x)$ there is an open neighborhood U of x such that $T(U) \subset V$. We will say that T is *upper-semicontinuous* (USC) on a set $S \subset X$, if it is upper-semicontinuous at each $x \in S$.

Proposition 6.1.1. *Suppose* $f : X \to (-\infty, +\infty]$ *is a convex function that is continuous at* $x_0 \in X$. *Then the map* $x \mapsto \partial f(x)$ *is norm-to-weak* upper-semicontinuous at* x_0.

---

[1] On p. 575 of Garling's very positive review of Michael Steele's *The Cauchy Schwarz Master Class* in the *American Mathematical Monthly*, June–July 2005, 575–579.

Proof. Let W be a weak*-open subset of X^* that contains $\partial f(x_0)$. Now suppose by way of contradiction that there exists $x_n \to x_0$ and $x_n^* \in \partial f(x_n) \setminus W$. Now by the local boundedness property of the subdifferential (Proposition 4.1.26), there exists $K > 0$ so that $x_n^* \in KB_{X^*}$ for all n. Now let x^* be a weak*-cluster point of (x_n^*). Then one can check that $x^* \in \partial f(x) \setminus W$ which is a contradiction. □

The following result is a natural generalization of Theorem 4.6.6.

Theorem 6.1.2. *Suppose that the Banach space X has a separable dual and that $T : X \to 2^{X^*}$ is monotone. Then there exists an angle small set $S \subset D(T)$ such that T is single-valued and norm-to-norm upper-semicontinuous at each point of $D(T) \setminus S$.*

Proof. Replace ∂f with T in the proof of Theorem 4.6.6. □

If T is a set-valued map, and ϕ is a single-valued mapping on $D(T)$ such that $\phi(x) \in T(x)$ for each $x \in D(T)$, then ϕ is said to be a *selection* for T.

Proposition 6.1.3. *Let β be a bornology on the Banach space X. Suppose $T : X \to 2^{X^*}$ is a multifunction such that $T(x) \neq \emptyset$ for all x in some neighborhood of x_0. Then the following are equivalent.*

(a) *$T(x_0)$ is a singleton and T is norm-to-τ_β upper-semicontinuous at x_0.*
(b) *Each selection for T is norm-to-τ_β continuous at x_0.*

Proof. (a) $\Rightarrow$ (b): Let U be an open neighborhood of x_0 such that $T(x)$ is nonempty for each $x \in U$, and let ϕ be any selection for T. Let V be a τ_β-open set containing $\phi(x_0)$. Because $\phi(x_0) = T(x_0)$, we can use the upper-semicontinuity of T to choose U_1 an open subset of U such that $T(x) \in V$ for all $x \in U_1$. Consequently, $\phi(x) \in V$ for all $x \in U_1$, and so ϕ is norm-to-τ_β continuous at x_0.

(b) $\Rightarrow$ (a): It is easy to check that (b) implies $T(x_0)$ is a singleton since τ_β is a Hausdorff topology. Let us suppose that T is not norm-to-τ_β upper-semicontinuous at x_0. Then for some τ_β-open set V containing $T(x_0)$, we can find $x_n \to x_0$ such that $T(x_n) \not\subset V$. Eventually, $T(x_n)$ is not empty, so there exist n_0 and $\phi_n \in T(x_n)$ such that $\phi_n \notin V$ for all $n \geq n_0$. Hence, any selection ϕ for T such that $\phi(x_n) = \phi_n$, we have that ϕ is not norm-to-τ_β continuous. □

The following proposition is a straightforward consequence of Šmulian's characterizations of differentiability given in Theorems 4.2.10 and 4.2.11 and Proposition 6.1.3.

Proposition 6.1.4. *Suppose $f : X \to (-\infty, +\infty]$ is convex, and continuous at x_0. Then the following are equivalent.*

(a) *f is Gâteaux differentiable (resp. Fréchet differentiable) at x_0.*
(b) *There is a selection for the subdifferential map ∂f which is norm-to-weak* (resp. norm-to-norm) continuous at x_0.*
(c) *Every selection for the subdifferential map ∂f is norm-to-weak* (resp. norm-to-norm) continuous at x_0.*
(d) *$\partial f(x_0)$ is a singleton and $x \mapsto \partial f(x)$ is norm-to-weak* (resp. norm-to-norm) upper-semicontinuous at x_0.*

Proof. Exercise 6.1.5. □

A subset G of $X \times X^*$ is said to be *monotone* provided $\langle x^* - y^*, x - y \rangle \geq 0$ whenever $(x, x^*), (y, y^*) \in G$. Let $T : X \to 2^{X^*}$ be a monotone operator, then it is easy to see that its *graph* is a monotone set where its graph is denoted by $\operatorname{graph}(T)$ (or sometimes $G(T)$) and defined by $\operatorname{graph}(T) := \{(x, x^*) \in X \times X^* : x^* \in T(x)\}$. We will say T is a *maximal monotone operator* provided its graph is maximal in the family of monotone sets in $X \times X^*$ ordered by inclusion.

Theorem 6.1.5 (Rockafellar). *Suppose $f : X \to (-\infty, +\infty]$ is a lsc proper convex function. Then ∂f is a maximal monotone operator.*

Proof. We follow the nice argument given in [5]. It is straightforward to establish that ∂f is monotone; see Exercise 2.2.21. To show that it is maximal, suppose $(x_0, v_0) \in X \times X^*$ is such that

$$\langle x - x_0, v - v_0 \rangle \geq 0,$$

holds true whenever $v \in \partial f(x)$. We will show that $v_0 \in \partial f(x_0)$. For this, define $f_0 : X \to (-\infty, +\infty]$ by

$$f_0(x) := f(x + x_0) - \langle x, v_0 \rangle. \tag{6.1.1}$$

Let $g(x) := \frac{1}{2}\|x\|^2$ and apply the Fenchel duality theorem (4.4.18) to f and g to find $u \in X^*$ such that

$$\inf_{x \in X} \left\{ f_0(x) + \frac{1}{2}\|x\|^2 \right\} = -f_0^*(u) - \frac{1}{2}\|u\|^2.$$

Because f_0 is lsc, proper and convex, both sides of the previous equation are finite. Consequently, we rewrite the equation as

$$\inf_{x \in X} \left\{ f_0(x) + \frac{1}{2}\|x\|^2 \right\} + f_0^*(u) + \frac{1}{2}\|u\|^2 = 0 \tag{6.1.2}$$

Now there exists a minimizing sequence $(y_n) \subset X$ such that

$$\frac{1}{n^2} \geq f_0(y_n) + \frac{1}{2}\|y_n\|^2 + f_0^*(u) + \frac{1}{2}\|u\|^2$$

$$\geq \langle u, y_n \rangle + \frac{1}{2}\|y_n\|^2 + \frac{1}{2}\|u\|^2$$

$$\geq \frac{1}{2}(\|y_n\| - \|u\|)^2 \geq 0, \tag{6.1.3}$$

where the second inequality follows from the Fenchel-Young inequality (Proposition 4.4.1(a)). Using the previous equation we obtain

$$f_0(y_n) + f_0^*(u) - \langle u, y_n \rangle \leq \frac{1}{n^2}.$$

This implies $u \in \partial_{1/n^2} f_0(y_n)$ for each n. According to the Brøndsted-Rockafellar theorem (4.3.2), there are sequences $(z_n) \subset X$ and $(w_n) \subset X^*$ such that

$$w_n \in \partial f_0(z_n), \quad \|w_n - u\| \leq 1/n \text{ and } \|z_n - y_n\| \leq 1/n. \tag{6.1.4}$$

It now follows that

$$\langle z_n, w_n \rangle \geq 0. \tag{6.1.5}$$

According to (6.1.3), one has

$$\|y_n\| \to \|u\|, \quad \langle y_n, u \rangle \to -\|u\|^2, \text{ as } n \to \infty, \tag{6.1.6}$$

which combined with (6.1.4) and (6.1.5) yields $u = 0$. Therefore, $y_n \to 0$. Because f_0 is lsc, $x = 0$ minimizes $f_0(x) + \frac{1}{2}\|x\|^2$ and, using (6.1.2) we have

$$f_0(0) + f_0^*(0) = 0.$$

This implies $0 \in \partial f_0(0)$ which in turn implies $u \in \partial f(x_0)$ as desired. $\square$

Let x_0 be in the closure of $D(T)$ where T is a multifunction from a topological space X to a normed space Y. We will say that T is *locally bounded* at $x_0 \in X$ if there exists an open neighborhood U of x_0 so that $T(U)$ is a bounded subset of Y. The following result thus generalizes the local boundedness property of the subdifferential (Proposition 4.1.26).

Theorem 6.1.6. *Let X be a Banach space, and let $T : X \to 2^{X^*}$ be a monotone operator. If $x \in$ core dom T, then T is locally bounded at x.*

Proof. See Exercise 6.1.6. $\square$

Consider an upper-semicontinuous (USC) set-valued mapping T from a topological space X into the subset of a topological (resp. linear topological) space Y. Then T is called a *usco* (resp. *cusco*) on X if for each $x \in X$, $T(x)$ is nonempty, (resp. convex) and compact. These names arise because the mappings are (*c*onvex) *u*pper-*s*emicontinuous *c*ompact valued. Figure 6.1 illustrates the behavior of a cusco in $\mathbb{R}^2$.

An usco (cusco) mapping T from a topological space X to subsets of a topological (linear topological) space Y is called a *minimal usco* (*minimal cusco*) if its graph does

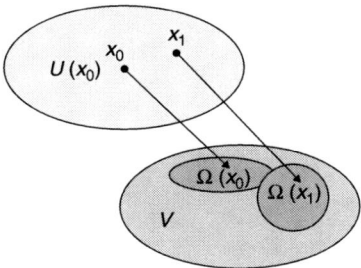

Figure 6.1 A cusco with $\Omega(x_0) \subset V$.

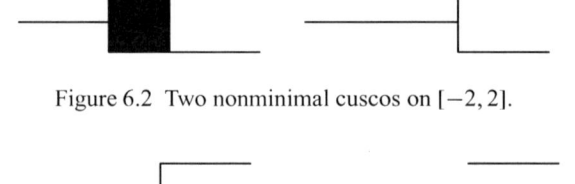

Figure 6.2 Two nonminimal cuscos on $[-2, 2]$.

Figure 6.3 A minimal cusco and smaller minimal usco on $[-2, 2]$.

not strictly contain the graph of any other usco (cusco) mapping X into the subsets of Y. Figure 6.2 shows two nonminimal cuscos. Figure 6.3 shows a minimal cusco and a smaller minimal usco inside the cuscos of Figure 6.2.

The notions of monotone and maximal monotone operators can be extended naturally to subsets of Banach spaces. With all of this in mind, the next result shows that minimal weak*-cusco's are natural generalizations of maximal monotone operators.

Theorem 6.1.7. *Let U be an open set in a Banach space X, and let $T : X \to 2^{X^*}$ be a maximal monotone operator with $U \subset D(T)$. Then T_U is a minimal weak*-cusco.*

The proof of this theorem will use the following lemma.

Lemma 6.1.8. *Let X be a Banach space and suppose that $D \subset X$ is open. If $T : D \to 2^{X^*}$ is monotone and norm-to-weak* upper-semicontinuous and $T(x)$ is nonempty, convex and weak*-closed for all $x \in D$, then T is maximal monotone in D.*

Proof. We need only show that if $(y, y^*) \in D \times X^*$ satisfies
$$\langle y^* - x^*, y - x \rangle \geq 0 \ \text{ for all } x \in D, x^* \in T(x), \tag{6.1.7}$$
then $y^* \in T(y)$. So suppose such a $y^* \notin T(y)$, by the weak*-separation theorem (4.1.22) there exists $z \in X$ such that $T(y) \subset W$ where $W = \{z^* \in X^* : \langle z^*, z \rangle < \langle y^*, z \rangle\}$. Because W is weak*-open and T is norm-to-weak* upper-semicontinuous, there exists a neighborhood U of y in D such that $T(U) \subset W$. For $t > 0$ sufficiently small, $y + tz \in U$, and consequently $T(y+tz) \subset W$. For any $u^* \in T(y+tz)$, equation (6.1.7) implies that
$$0 \leq \langle y^* - u^*, y - (y + tz) \rangle = -t \langle y^* - u^*, z \rangle,$$
and consequently $\langle u^*, z \rangle \geq \langle y^*, z \rangle$. From this, one sees that $u^* \notin W$ which is a contradiction. □

Proof. We now turn to the proof of Theorem 6.1.7. First, Exercise 6.1.3 implies that T_U is a weak*-cusco, so it remains to show it is minimal. Indeed, if $F : U \to 2^{X^*}$ is a weak*-cusco such that $G(F) \subset G(T_U)$, then according to Lemma 6.1.8 F is maximal monotone. Consequently, $F = T_U$ as desired. □

Multifunctions can be an effective tool for classifying Banach spaces, and we state the following result as an example of this. First, recall that an *Asplund* space is a

6.1 Multifunctions and monotone operators

Banach space on which every continuous convex function is Fréchet differentiable on a dense subset.

Theorem 6.1.9. *A Banach space X is Asplund if and only if every locally bounded minimal weak*-cusco from a Baire space into X^* is generically single-valued and norm-continuous.*

Of course, the 'if' part of this theorem implies that X is Asplund if every continuous convex function is generically Fréchet differentiable on X. The 'only if' part states that a more general phenomenon occurs in Asplund spaces. We refer the reader to [68, 104] for a proof of this result, and related work. We will provide some other characterizations of Asplund spaces in Section 6.6.

6.1.1 Selections and fixed points in Banach space

We now examine briefly the Banach space forms of the Euclidean material of Section 3.4. The Banach space form of Theorem 3.4.1 is due to Fan.

Theorem 6.1.10 (Kakutani–Fan [208]). *If $\emptyset \neq C \subset X$ is a compact convex subset of a Banach space, then every cusco $\Omega : C \to C$ has a fixed point.*

This uses the Banach space version of Theorem 3.4.3 with Brouwer's theorem replaced by Schauder's fixed point theorem (Exercise 6.1.11).

Theorem 6.1.11 (Cellina approximate selection theorem [21]). *Let X, Y be Banach spaces with $K \subset X$ compact and convex. Suppose $\Omega : K \to Y$ is an usco. For any $\varepsilon > 0$ there is a continuous map $f : K \to Y$ such that*

$$d_{G(\Omega)}(x, f(x)) < \varepsilon \text{ for all points } x \text{ in } K, \tag{6.1.8}$$

and with range $f \subseteq \overline{\text{conv}}$ range Ω.

The distance d_G is defined analogously as in Theorem 6.1.11 and the proof is largely unchanged (Exercise 6.1.13).

A topological space is said to be *paracompact* if every open cover of the topological space has a locally finite open refinement. Any compact or metrizable space is paracompact. Equivalently, for every open cover of a paracompact space K, there exists a (continuous) partition of unity on K subordinate to that cover. While this is some work to show in generality, in the metrizable case the Euclidean proof we used needs little change. The full version of Theorem 3.4.4 follows. It is well described in [260].

Theorem 6.1.12 (Michael selection theorem [315]). *Let K be a paracompact space and let Y be a Banach space. Suppose $\Omega : K \to Y$ is LSC with nonempty closed convex images. Then given any point $(\bar{x}, \bar{y})$ in $G(\Omega)$, there is a continuous selection f of Ω satisfying $f(\bar{x}) = \bar{y}$.*

Conversely, fix a topological space K. Suppose every LSC multifunction with nonempty convex closed values mapping K to an arbitrary Banach space, admits a continuous selection. Then K is paracompact.

In the monotone case we rarely have LSC mappings and we settle for finding selections which are generically continuous in some topology.

Exercises and further results

6.1.1. Let X be a Banach space and suppose $f : X \to \mathbb{R}$ is a function. Define the multifunction $T_f : X \to 2^\mathbb{R}$ by $T_f(x) := \{t \in \mathbb{R} : t \geq f(x)\}$. Show that T_f is USC if and only if f is lsc.

Notwithstanding this example, one should see Exercise 6.2.10 below that illustrates the rationale for the term upper-semicontinuous in terms of lim sup as defined for sets.

Hint. Suppose f is lsc. Let $U \subset \mathbb{R}$ be open. Either $V := \{x : T_f(x) \subset U\}$ is empty, or $\bar{x} \in V$ for some $\bar{x} \in X$. Then $(a, +\infty) \subset U$ where $a < f(\bar{x})$, and use the lower-semicontinuity of f to find $\delta > 0$ so that $B_\delta(\bar{x}) \subset V$.

Conversely, let $\bar{x} \in X$, and let $\varepsilon > 0$. Let $U := \{t : t > f(\bar{x}) - \varepsilon\}$. Use the fact that $V = \{x : T_f(x) \subset U\}$ is open to conclude f is lsc. □

6.1.2 (Nonexpansive extensions). ** There is a tight relationship between nonexpansive mappings and monotone operators in Hilbert spaces:

Lemma 6.1.13. *Let H be a Hilbert space. Suppose P and T are multifunctions from H to 2^H whose graphs are related by the condition $(x, y) \in \text{graph } P$ if and only if $(v, w) \in \text{graph } T$ where $x = w + v$ and $y = w - v$. Then $\text{dom } P = \text{range}(T + I)$. Moreover, P is nonexpansive (and single-valued) if and only if T is monotone.*

The Kirszbraun–Valentine theorem on the existence of nonexpansive extensions to all of Hilbert space of nonexpansive mappings follows easily.

Theorem 6.1.14. [Kirszbraun–Valentine] *Let H be a Hilbert space with D a subset of H. Suppose $P : D \to H$ is a nonexpansive mapping. Then there exists a nonexpansive mapping $\widehat{P} : H \to H$ with $\text{dom } P = H$ and $\widehat{P}|_D = P$.*

Hint. Associate to P a monotone T as in Lemma 6.1.13 and extend T to a maximal $\widehat{T}$. Define $\widehat{P}$ from $\widehat{T}$ using Lemma 6.1.13 again. Theorem 9.3.5 shows $\text{dom}(\widehat{P}) = \text{range}(\widehat{T} + I) = H$. Finally, check that $\widehat{P}$ is an extension of P. □

This result is highly specific to Hilbert space. For example, if the preceding result holds in a strictly convex space X with dimension exceeding one, then X is a Hilbert space [387].

6.1.3.* Suppose that X is a Banach space and $T : X \to 2^{X^*}$ is maximal monotone. Suppose that $D \subset D(T)$ is a nonempty open subset, show that $T|_D$ is a weak*-cusco.

Hint. Use the local boundedness of T (Theorem 6.1.6); see [121, Exercise 5.1.17] for further details. □

6.1.4. Suppose $f : X \to (-\infty, +\infty]$ is a convex function that is continuous at x_0. Show that $\partial f(x_0)$ is a singleton if and only each (resp. some) selection for ∂f is norm-to-weak* continuous at x_0. Generalize this to maximal monotone operators.

Hint. We discuss the convex case only. There are many reasons for this, for example, $\partial f(x_0)$ is a singleton if and only if f is Gâteaux differentiable at x_0. So this follows from Šmulian's theorems. This exercise also follows directly by combining Proposition 6.1.3 and 6.1.1. □

6.1.5.* (a) Prove Proposition 6.1.4.
(b) State and prove Proposition 6.1.4 in the more general norm-to-τ_β setting.

Hint. For (b) replace, e.g. Gâteaux differentiability and weak*-topology with β-differentiability and the τ_β-topology in the statement. The use the β-differentiability version of Šmulian's theorem (Exercise 4.2.8) along with Proposition 6.1.3 to prove the result. □

6.1.6.* Prove Theorem 6.1.6.

Hint. Following the argument as in [121, Theorem 5.1.8], let $x^* \in T(x)$ and replace T with $y \mapsto T(y+x) - x^*$; thus we may assume $x = 0$ and $0 \in T(0)$. Define $f : X \to (-\infty, +\infty]$ by

$$f(x) := \sup\{\langle y^*, x - y\rangle : y \in \text{dom } T, \|y\| \le 1 \text{ and } y^* \in T(y)\}.$$

Then f is lsc and convex as a supremum of such functions. Use the monotonicity of T to show $f(0) = 0$. Then use the monotonicity of T and that fact $0 \in \text{core dom } T$ to show $0 \in \text{core dom } f$. Thus f is bounded on a neighborhood of 0. Then show T is bounded on a neighborhood of 0. □

6.1.7. Let $X := \ell_2$. Show there is a maximal monotone operator $T : X \to X$ such that T is everywhere norm-to-weak continuous, but fails to be norm-to-norm continuous at some point in X.

Hint. Exercise 5.1.26 shows there is a norm, say $\|\cdot\|$ on X that is everywhere Gâteaux differentiable on S_X, but fails to be Fréchet differentiable at some point of S_X. Let $f(\cdot) := \|\cdot\|^2$, and consider $T := \partial f$. □

6.1.8.* Let X and Y be Hausdorff spaces and suppose $T : X \to 2^Y$ is an usco (resp. cusco). Show that there is a minimal usco (resp. cusco) contained in T.

Proof. Use Zorn's lemma. See, for example [121, Proposition 5.1.23]. □

6.1.9. Let X be a Banach space whose dual norm is strictly convex. Show that every minimal weak*-cusco from a Baire space into X^* is generically single-valued.

Hint. See [120, Theorem 2.10]; this is a generalization of Exercise 4.6.6, the proof is similar where here one considers the weak*-cusco in place of the subdifferential mapping. □

6.1.10 (Nongenericity of Gâteaux differentiability points [320, 321]).** Herein all topological spaces are assumed to be at least completely regular, K is a compact Hausdorff space and $M, \Phi : (C(K), \|\cdot\|_\infty) \to \mathbb{R}$ are defined by $M(f) := \sup\{f(k) : k \in K\}$, and $\Phi(f) := \{k \in K : f(k) = M(f)\}$.

(a) Φ is a minimal usco that is also an open mapping.
(b) M is a continuous convex function on $(C(K), \|\cdot\|_\infty)$ and is Gâteaux differentiable at $f \in C(K)$ if and only if $\Phi(f)$ is a singleton.
(c) One can use the prior part to show that if K has a dense set composed of G_δ-points then M is Gâteaux differentiable at the points of a dense subset of $(C(K), \|\cdot\|_\infty)$. This leads to:

Theorem 6.1.15. *If M is Gâteaux differentiable at the points of a dense G_δ-subset, G, in $(C(K), \|\cdot\|_\infty)$ then K contains a dense G_δ-subset X that is (completely) metrizable.*

Hint. Define $F : K \to 2^G$ by,

$$F(k) := \{f \in C(K) : k \in \Phi(f)\} \cap G.$$

Then F is densely defined and has closed graph relative to $K \times G$. Moreover, F is *demi-lower-semicontinuous* [320] and [320] shows there exists a dense G_δ-subset X of K and a continuous selection $\sigma : X \to G$ of $F|_X$. Since $\Phi|_G$ is single-valued (and so may be viewed as a continuous mapping) and $\Phi|_G(\sigma(x)) = \{x\}$ for all $x \in X$, σ^{-1} is also continuous. Hence X is homeomorphic to $\sigma(X) \subseteq G$; which is metrizable with respect to the usually metric on $(C(K), \|\cdot\|_\infty)$. The fact that X is completely metrizable follows from the fact that a metric space P is completely metrizable (by some metric) if, and only if, it is Čech-complete [201]. □

(d) Let S be a first countable Baire space that does not contain a dense completely metrizable subset (e.g. the *Sorgenfrey line*). Then for Čech compactification $K := \beta(S)$, M is densely Gâteaux differentiable but not on any dense G_δ-subset of $(C(K), \|\cdot\|_\infty)$.
(e) An explicit example of a continuous convex function that is densely differentiable but not on any second category set is given in [321].

6.1.11 (Schauder fixed point theorem).* Let $K \subset X$ be a compact convex subset of a Banach space X and let $f : K \to K$ be norm-continuous. Show that f has a fixed point in K.

Hint. Since K is compact we can and do suppose that X is separable and has a strictly convex norm. We also suppose $0 \in K$. For each finite-dimensional subspace $F \subset X$ let $P_F : K \to K \cap F$ be the continuous metric projection onto $K \cap F$. Define $f_F : K \cap F \to K \cap F$ via $f_F(k) := P_F(f(k))$. Then each f_F has a fixed point. Now take limits. □

6.1.12.* Prove Theorem 6.1.10.
6.1.13.* Prove Theorem 6.1.11.

6.2 Epigraphical convergence: an introduction

We begin by illustrating that pointwise convergence does not necessarily preserve minimizers of functions, nor is it necessarily preserved through conjugation (for this, given a function h, we use the terminology $\operatorname{argmin} h := \{x : h(x) = \inf h\}$). These deficiencies can be overcome, in many settings, using notions of epigraphical convergence as we shall see throughout this section.

Example 6.2.1. (a) Consider Lipschitz convex functions $f_n : \mathbb{R} \to \mathbb{R}$ defined by $f_n(t) := n|t - 1/n| - 1$ for $n \in N$. Then (f_n) converges pointwise to $f := \delta_{\{0\}}$, and $\operatorname{argmin} f_n$ is a subset of the compact set $[0, 1]$ for each n, however

$$\liminf_{n \to \infty} \inf_{\mathbb{R}} f_n = -1 < 0 = \inf_{\mathbb{R}} f.$$

(b) Consider functions $g_n : \mathbb{R} \to \mathbb{R}$ defined by $g_n(t) := t/n$ for $n \in N$. Then $g_n \to 0$ uniformly on bounded sets, but $g_n^* = \delta_{\{1/n\}}$ and $g^* = \delta_{\{0\}}$ and so (g_n^*) does not even converge pointwise to g^*.

In addition to addressing issues raised by the previous example, limits of sets, or *epigraphical convergence* as applied to epigraphs of functions, is an important tool in the study of multifunctions as seen in works such as [22]. In our limited coverage our primary focus will be directly on limits of epigraphs of functions. These epi-limits will be natural tools for which to analyze approximations of functions when minimizers are of primary concern, and they also provide a setting with which to study generalized derivatives of functions as outlined in Exercise 6.2.27.

This section introduces the following notions of convergence: Kuratowski–Painlevé, Mosco, Attouch–Wets, Wijsman, and slice convergence. All of these forms of convergence are equivalent for lsc convex functions on $\mathbb{R}^n$ (see Theorem 6.2.13). These notions are often introduced in the language of sets, and then the respective definitions apply naturally to epigraphs of functions. We begin with Kuratowski–Painlevé convergence because of its relation with lower and upper limits of sets.

Definition 6.2.2. Let Y be a Hausdorff topological space and let (F_i) be a sequence of subsets of Y. The sequential *lower* and *upper* limits of (F_i) are defined by

$$\liminf_{i \to \infty} F_i := \left\{ \lim_{i \to \infty} y_i : y_i \in F_i \text{ for all } i = 1, 2, \ldots \right\}$$

and

$$\limsup_{i \to \infty} F_i := \left\{ \lim_{k \to \infty} y_{i_k} : y_{i_k} \in F_{i_k} \text{ for some } i_k \to \infty \right\}.$$

When the two limits are equal, that is, if $F = \liminf_{i \to \infty} F_i = \limsup_{i \to \infty} F_i$ we say (F_i) converges *Kuratowski–Painlevé* to F.

In a metric space both the sequential lower and upper limits are closed. However, this is not true in general (Exercise 6.2.2). It is also easy to check that these lower and upper limits in metric spaces can be described by using the distance between a set and a point.

Lemma 6.2.3. *Let X be a metric space and let (A_i) be a sequence of subsets in X. Then*

$$\liminf_{i \to \infty} A_i = \left\{ x \in X : \limsup_{i \to \infty} d_{A_i}(x) = 0 \right\}$$

and

$$\limsup_{i \to \infty} A_i = \left\{ x \in X : \liminf_{i \to \infty} d_{A_i}(x) = 0 \right\}.$$

Proof. Exercise 6.2.3. □

Because our primary focus is on functions, we will say a sequence of functions $f_n : X \to [-\infty, +\infty]$ converges *Kuratowski–Painlevé* to the lsc function $f : X \to [-\infty, +\infty]$ if the epigraphs of f_n converge Kuratowski–Painlevé to the epigraph of f in $X \times \mathbb{R}$ endowed with any equivalent product norm. This is often referred to as *epi-convergence*, and it is easy to see that this convergence is invariant under equivalent renormings. In fact, this can be very nicely formulated for functions as follows.

Lemma 6.2.4. *Let X be a metric space and let $f, f_1, f_2, f_3, \ldots$ be lsc functions mapping X to $(-\infty, +\infty]$. Then (f_i) converges Kuratowski–Painlevé to f if and only if at each point $x \in X$ one has*

$$\liminf_{i \to \infty} f_i(x_i) \geq f(x) \text{ for every sequence } x_i \to x \tag{6.2.1}$$

and

$$\limsup_{i \to \infty} f_i(x_i) \leq f(x) \text{ for some sequence } x_i \to x. \tag{6.2.2}$$

Proof. Exercise 6.2.4. □

Epigraphical convergence of functions has useful applications in minimization problems as we now illustrate.

Theorem 6.2.5. *Let X be a metric space and let $f_i : X \mapsto (-\infty, +\infty]$ be a sequence of lsc functions that converges Kuratowski–Painlevé to f. Then*

$$\limsup_{i \to \infty} (\operatorname{argmin} f_i) \subset \operatorname{argmin} f. \tag{6.2.3}$$

If, moreover, there is a compact set $K \subset X$ such that $\operatorname{argmin} f_i \subset K$ and $\operatorname{argmin} f_i \neq \emptyset$ for all i, then

$$\lim_{i \to \infty} (\inf f_i) = \inf f, \tag{6.2.4}$$

6.2 Epigraphical convergence: an introduction

Proof. For the first part, we may assume $\limsup(\operatorname{argmin} f_i) \neq \emptyset$ otherwise there is nothing to do. Let $\bar{x} \in \limsup(\operatorname{argmin} f_i)$. Then there exist $x_{i_j} \in \operatorname{argmin} f_{i_j}$ so that $x_{i_j} \to \bar{x}$. It follows from Lemma 6.2.4, $f(\bar{x}) \leq \liminf_j f_{i_j}(x_{i_j}) \leq \limsup_i (\operatorname{argmin} f_i)$. On the other hand, given any $x \in X$ there exists a sequence $z_k \to x$ so that $\limsup f_k(z_k) \leq f(x)$. Therefore,

$$f(\bar{x}) \leq \limsup(\operatorname{argmin} f_i) \leq \limsup f_k(z_k) = f(x).$$

This shows $\bar{x} \in \operatorname{argmin} f$ as desired.

For the 'moreover' part, given $\alpha > \inf f$, we may choose $\bar{x} \in X$ so that $f(\bar{x}) < \alpha$. According to Lemma 6.2.4, there exists a sequence $x_i \to \bar{x}$ so that

$$f(\bar{x}) \geq \limsup f_i(y_i) \geq \limsup(\inf f_i).$$

Consequently, $\inf f \geq \limsup(\inf f_i)$. On the other hand, let $y_i \in \operatorname{argmin} f_i$, using compactness and Lemma 6.2.4, there is a convergent subsequence, say $y_{i_k} \to \bar{y}$ such that

$$f(\bar{y}) \leq \liminf_{k \to \infty} f_{i_k}(y_{i_k}) = \liminf(\inf f_i) \leq \limsup(\inf f_i) \leq \inf f,$$

which completes the proof. □

In Exercise 6.2.7 it is shown that the 'moreover' part of the previous theorem can fail in the absence of compactness. Because of this, and because bounded sets in reflexive Banach spaces are relatively weakly compact, the following notion of convergence has been widely used in reflexive Banach spaces. We will focus on closed convex subsets of Banach spaces when considering further notions of set convergence.

Definition 6.2.6. [Mosco convergence] Let X be a Banach space and let C_n, C be closed convex subsets of X. We will say (C_n) *converges Mosco* to C if each of the following conditions hold.

(M1) For each $x \in C$ there is a sequence (x_n) converging in norm to x such that $x_n \in C_n$ for each $n \in \mathbb{N}$.
(M2) Whenever $n_1 < n_2 < n_3 < \ldots$, and $x_{n_k} \in C_{n_k}$ is such that $x_{n_k} \to_w x$, then $x \in C$.

Given a sequence of proper lsc convex functions (f_n) on a Banach space X, we will say (f_n) converges *Mosco* to the lsc convex function f if $(\operatorname{epi} f_n)$ converges Mosco to $\operatorname{epi} f$ as subsets of $X \times \mathbb{R}$.

It is clear that Mosco and Kuratowski–Painlevé convergence are equivalent in finite-dimensional spaces. Because our primary interest here is, as we noted, with functions, we state the following reformulation of Mosco convergence for functions as follows.

Fact 6.2.7. *Let X be a Banach space, and let (f_n) be a sequence of proper lsc convex functions on X. Then (f_n) converges Mosco to the lsc convex function f if and only if both of the following two conditions are true.*

(M1) For each $x \in X$, there exist $x_n \to x$ such that $\limsup f_n(x_n) \leq f(x)$.
(M2) $\liminf_{n\to\infty} f_n(x_n) \geq f(x)$ whenever $x_n \to_w x$.

The proof of this fact is left as Exercise 6.2.6. Because lsc convex functions are automatically weakly-lsc, Mosco convergence is nicely suited for preserving bounded sets of minimizers in reflexive Banach spaces, more precisely we have the following.

Theorem 6.2.8. *Let X be a Banach space and let (f_n) be a sequence of proper lsc convex functions that converges Mosco to the proper lsc convex function f. Then*

$$\limsup_{i\to\infty}(\operatorname{argmin} f_i) \subset \operatorname{argmin} f \qquad (6.2.5)$$

where the limsup is considered in the weak topology. If, moreover, there is a weakly compact set $K \subset X$ such that $\operatorname{argmin} f_i \subset K$ and $\operatorname{argmin} f_i \neq \emptyset$ for all i, then

$$\lim_{i\to\infty}(\inf f_i) = \inf f. \qquad (6.2.6)$$

Proof. Modify the proof of Theorem 6.2.5 as appropriate. □

Some other properties of minimizers can be found in Exercise 6.2.7. In addition to its nice properties concerning minimization problems, Mosco convergence is preserved through conjugation on reflexive Banach spaces as detailed in the next theorem. Also the various forms of epigraphical convergence we discuss in this section are implied by uniform convergence on bounded sets, however, none of the forms of epigraphical convergence we discuss imply or are implied by pointwise convergence; see Theorem 6.2.14 below.

Theorem 6.2.9. *Let X be a reflexive Banach space, and let $f : X \to (-\infty, +\infty]$ be a proper lsc convex function. Then the sequence of proper lsc convex functions (f_n) converges Mosco to f if and only if (f_n^*) converges Mosco to f^*.*

Proof. Suppose (f_n) converges Mosco to f on X. Let $\phi \in X^*$ and $\phi_n \to_w \phi$, and let $\alpha < f^*(\phi)$. Now choose $x \in X$ so that $\phi(x) - f(x) > \alpha$. According to Fact 6.2.7 there exists a sequence $x_n \to x$ such that $f_n(x_n) \to f(x)$. Consequently, $\phi_n(x_n) \to \phi(x)$ and so

$$\liminf_{n\to\infty} f_n^*(\phi_n) \geq \liminf_{n\to\infty} \phi_n(x_n) - f_n(x_n)$$
$$= \liminf_{n\to\infty} \phi(x_n) - f(x_n) = \phi(x) - f(x) > \alpha.$$

It now remains to show given $\phi \in X^*$ that there is a sequence $(\phi_n) \to \phi$ such that $\limsup_{n\to\infty} f_n^*(\phi_n) \leq f^*(\phi)$. If $f^*(\phi) = \infty$, there is nothing further to do, otherwise choose $x_n \in X$ such that

$$\phi(x_n) - f(x_n) \geq f^*(\phi) - \frac{1}{n}.$$

Let $\alpha := f^*(\phi)$, then $\phi(x) - \alpha \leq f(x)$ for all $x \in X$. Consider the affine function $L_n(x) := \phi(x) - \alpha - 1/n$. Then $L_n(x_n) \geq f(x_n) - 2/n$. Let

$$W_n := \overline{\text{conv}}(\{\pm x_1, \pm x_2, \ldots, \pm x_n\} + nB_X).$$

Then W_n is a closed convex weakly compact set. Using Fact 6.2.7 and the Eberlein–Šmulian theorem, we also can choose n_k such that

$$f_n(x) \geq f(x) - \frac{1}{k} \quad \text{for } x \in W_k,\ n \geq n_k.$$

Consider $g_n = L_n - \delta_{W_n}$. According to the sandwich theorem (4.1.18), there is an affine function A_n say $A_n = \phi_n - \alpha_n$ such that $f_n \geq A_n \geq g_n$. Accordingly,

$$\liminf_{n \to \infty}(A_n - L_n) \geq 0$$

uniformly on bounded sets. Again, using Fact 6.2.7, we can choose $x_{k,j}$ such that $\|x_{k,j} - x_k\| \leq \frac{1}{k}$ and $f_j(x_{k,j}) \leq f(x_k) + 1/k$ for $j \geq n_k$. Now for $j \geq n_k$

$$L_j(x_{k,j}) \geq L_j(x_k) - \|\phi\| \|x_{k,j} - x_k\|$$

$$\geq f(x_k) - \frac{2}{k} - \|\phi\| \|x_{k,j} - x_k\|$$

$$\geq f_j(x_{k,j}) - \frac{3}{k} - \|\phi\| \|x_{k,j} - x_k\|$$

$$\geq A_{n_k}(x_{k,j}) - \frac{3}{k} - \|\phi\| \|x_{k,j} - x_k\|.$$

According to Exercise 6.2.1 we conclude that

$$\|\phi_k - \phi\| \to 0 \quad \text{and} \quad \alpha_k \to \alpha.$$

Then $f_k^*(\phi_k) \leq \alpha_k$ and $\limsup_{k \to \infty} f_k^*(\phi_k) \leq f^*(\phi)$ as desired. The converse can be deduced using biconjugates. □

It is important to note that the previous theorem is valid only in reflexive spaces, because if Mosco convergence of sequences of proper lsc convex functions is preserved through conjugation, then the space is reflexive; see [46, Theorem 3.3]. Thus, we now introduce a form of convergence that is stronger than Mosco convergence in infinite-dimensional spaces. Let X be a normed linear space and let A and B be nonempty, closed and bounded subsets of X. Then the *Hausdorff distance* D_H between A and B is defined by $D_H(A,B) := \max\{e(A,B), e(B,A)\}$ where *excess functional* is defined by $e(A,B) := \sup\{d_B(a) : a \in A\}$. Let $A, A_1, A_2, A_3, \ldots$ be closed subsets of X, we will say (A_n) converges *Attouch–Wets* to A if $d_{A_n}(\cdot)$ converges uniformly to $d_A(\cdot)$ on bounded subsets of X.

We will say a sequence of lsc functions f_n converges *Attouch–Wets* to the lsc function f on X if (epi f_n) converges Attouch–Wets to epi f in $X \times \mathbb{R}$ endowed with an equivalent product norm. It is easy to see that Attouch–Wets convergence is invariant under renorming, and that it is implied by uniform convergence on bounded sets. One can intuitively visualize Attouch–Wets convergence as being uniform convergence of epigraphs on bounded sets. While Attouch–Wets convergence is weaker than uniform convergence of the functions on bounded sets, for certain nice functions the two notions are equivalent (Exercise 6.2.18). Additionally, Attouch–Wets convergence has the advantage of being preserved under conjugation as seen in the following important theorem that is stated without proof.

Theorem 6.2.10. *Let X be a Banach space, and let $f, f_1, f_2, f_3, \ldots$ be proper lsc convex functions on X. Then (f_n) converges Attouch–Wets to f if and only if (f_n^*) converges Attouch–Wets to f^*.*

For a proof of this theorem, one can consult [44, Theorem 7.2.11] or [297, Theorem 9.1.4]. The next theorem, which again, we only state, shows the precise connection between Attouch–Wets convergence of convex functions, and convergence of their regularizations.

Theorem 6.2.11. *Suppose X is a normed linear space and let f, f_n be proper lsc convex functions. Then (f_n) converges Attouch–Wets to f if and only if $f_n \square \lambda \|\cdot\|$ converges uniformly on bounded sets to $f \square \lambda \|\cdot\|$ for all sufficiently large λ.*

One can consult [45, Theorem 4.3] for a proof of the preceding theorem. The final two notions of set convergence that we will introduce are as follows. Let X be a normed space, and let (C_α) and C be nonempty closed convex sets in X. The net (C_α) is said to *converge Wijsman* to C if $d_{C_\alpha}(x) \to d_C(x)$ for every $x \in X$. If $d(W, C_\alpha) \to d(W, C)$ for every closed bounded convex subset W of X, then (C_α) is said to *converge slice* to C. Again, these notions apply to lsc convex functions when we apply them to epigraphs under a product norm on $X \times \mathbb{R}$.

It is evident that slice convergence is formally stronger than Wijsman convergence, and that both of the previous definitions depend on the norm of X (or $X \times \mathbb{R}$ for epigraphs). However, slice convergence is invariant under renorming while Wijsman convergence is not, that and more is captured in the following nice geometric theorem.

Theorem 6.2.12. *Let X be a normed space, and let C_α, C be closed convex sets in X. Then (C_α) converges slice to C if and only if (C_α) converges Wijsman to C with respect to every equivalent norm on X. In particular, slice convergence in invariant under renorming.*

Proof. See Exercise 6.2.23. □

The usefulness of slice convergence stems from the fact it agrees with Mosco convergence in reflexive Banach spaces, and carries many of the nice features of Mosco convergence to the nonreflexive setting. Exercise 6.2.7 illustrates its application to minimization problems. Roughly speaking, Mosco convergence is the weakest form of epigraphical convergence that extends the key nice properties of epiconvergence in finite-dimensional spaces to reflexive spaces, and then slice convergence is the weakest form of epigraphical convergence that preserves these properties in general normed spaces. With all of these notions now defined, let us observe

6.2 Epigraphical convergence: an introduction

Theorem 6.2.13. *Let X be a finite-dimensional normed space. Then Kuratowski–Painlevé, Mosco, Attouch–Wets, Wijsman and slice convergence are equivalent for sequences of closed convex sets in X.*

Proof. See Exercise 6.2.15. □

Theorem 6.2.14. *Let X be a normed space and let $f, f_n : X \to (-\infty, +\infty]$ be a sequence of lsc proper convex functions. Consider the following statements*

(a) (f_n) *converges uniformly on bounded sets to f.*
(b) (f_n) *converges Attouch–Wets to f.*
(c) (f_n) *converges slice to f.*
(d) (f_n) *converges Mosco to f.*
(e) (f_n) *converges Wijsman to f.*
(f) (f_n) *converges Kuratowski–Painlevé to f.*
(g) (f_n) *converges pointwise to f.*

Then (a) $\Rightarrow$ (b) $\Rightarrow$ (c) $\Rightarrow$ (d), (c) $\Rightarrow$ (e), (d) $\Rightarrow$ (f), (e) $\Rightarrow$ (f), (a) $\Rightarrow$ (g), but none of these implications reverse in general. Moreover, in general, (d) $\not\Rightarrow$ (e), (e) $\not\Rightarrow$ (d), (b) $\not\Rightarrow$ (g) and (g) $\not\Rightarrow$ (f).

Proof. Many of these relations and more are developed in Exercises 6.2.5, 6.2.11, 6.2.16, 6.2.12, 6.2.14, 6.2.7, and the remaining are left for Exercise 6.2.17. □

Historically, Wijsman convergence is important because it was the mode of epigraphical convergence that was shown to be preserved under conjugation in finite-dimensional spaces. However, outside of finite-dimensional spaces it fails to preserve bounded miminizing sequences (Exercise 6.2.7) and it is not generally preserved under conjugation (Exercise 6.2.13). Nevertheless, these are far from reasons to dismiss Wijsman convergence, and in fact, in ℓ_p spaces with $1 < p < \infty$ under their usual norms, it is equivalent to (the much harder to check) slice convergence as follows from the next general theorem.

Theorem 6.2.15. *Let X be a Banach space. Then Wijsman and slice convergence coincide for nets of closed convex sets if and only if the dual norm on X^* is weak*-Kadec.*

The proof of this theorem can be found in [111], however, it is easy to verify the 'only if' portion of this theorem (see Exercise 6.2.22). Moreover, for reflexive Banach spaces a proof is suggested in Exercise 6.2.24.

Exercises and further results

6.2.1.* Suppose $\alpha_n = \Lambda_n - a_n$ and $\beta_n = \phi_n - b_n$ are continuous affine functions on X (i.e., $\Lambda_n, \phi_n \in X^*$ and $a_n, b_n \in \mathbb{R}$ for each n) such that:
(a) $\liminf_{n\to\infty}(\beta_n - \alpha_n) \geq 0$ uniformly on bounded sets; and

(b) there are sequences $(x_n) \subset X$ and $\varepsilon_n \to 0^+$ such that $\beta_n(x_k) < \alpha_n(x_k) + \varepsilon_k$ for all $k \le n$.

Show that $(\Lambda_n - \phi_n) \to 0$ in X^* and $(c_n - k_n) \to 0$ in $\mathbb{R}$. Conclude that $(\alpha_n - \beta_n) \to 0$ uniformly on bounded sets.

Hint. Let $\varepsilon > 0$ and then choose ε_k such that $\varepsilon_k < \varepsilon/2$. Now let $n_0 \ge k$ be such that

$$\beta_n(x) \ge \alpha_n(x) - \varepsilon/2 \text{ if } x \in x_k + B_X, \ n \ge n_0.$$

Because $n_0 \ge k$, we also obtain

$$\alpha_n(x_k) > \beta_n(x_k) - \varepsilon/2 \text{ if } n \ge n_0.$$

Now for $h \in B_X$, and $n \ge n_0$ we have

$$\begin{aligned}(\Lambda_n - \phi_n)(h) &= (\alpha_n - \beta_n)(h) \\ &= (\alpha_n - \beta_n)(x_k + h) - (\alpha_n - \beta_n)(x_k) \\ &\le \frac{\varepsilon}{2} + \frac{\varepsilon}{2} = \varepsilon.\end{aligned}$$

Therefore, $\|\Lambda_n - \phi_n\| \le \varepsilon$ for $n \ge n_0$. Thus $(\Lambda_n - \phi_n) \to 0$ as desired.

Now fix x_k, by what was just shown, we know that $(\phi_n - \Lambda_n)(x_k) \to 0$, and for all sufficiently large n,

$$-\varepsilon_k < (\beta_n - \alpha_n)(x_k) < \varepsilon_k.$$

Consequently, $\lim_{n \to \infty} |a_n - b_n| \le \varepsilon_k$ for each k. $\square$

6.2.2.* Let X be a metric space, and let (A_i) be a sequence of subsets of X. Show that $\limsup A_i$ and $\liminf A_i$ are both closed. Show that this may fail if X is merely a Hausdorff topological space.

6.2.3.* Prove Lemma 6.2.3.

6.2.4.* Prove Lemma 6.2.4.

6.2.5 (Examples concerning epigraphical and pointwise convergence).*

(a) Find an example of a sequence of Lipschitz convex functions on ℓ_2 that converges Mosco but not pointwise to 0.
(b) Find a sequence of equi-Lipschitz convex functions on ℓ_2 that converges pointwise but not Mosco to 0.
(c) Suppose (f_n) is a sequence of equicontinuous convex functions on a Banach space. Suppose (f_n) converges Kuratowski–Painlevé to the lsc proper convex function f. Show that (f_n) converges pointwise to f.
(d) Suppose $f, f_1, f_2, \ldots$ are lsc proper convex functions on a Banach space such that $f_n \ge f$ for all n, and (f_n) converges pointwise to f. Show that (f_n) converges Mosco (in fact slice) to f.
(e) Find a sequence of proper lsc convex functions (f_n) on $\mathbb{R}$ that converges Attouch–Wets but not pointwise to a proper lsc convex function f.

Hint. (a) Let $f := 0$ and define f_n by $f_n(\cdot) := n^2 |\langle e_n, \cdot \rangle|$. For $x := (1/n) \in \ell_2$ we have $f(x) = 0$ while $f_n(x) = n$ so $f_n(x) \to \infty$. Clearly (M2) is satisfied because $f_n \ge 0$ for each n. For any $u := (u_n) \in \ell_2$, define the sequence $(x_n) \subset \ell_2$ by $x_n(j) := u_j$ if

$j \neq n$, and $x_n(j) := 0$ if $j = n$. Then $(x_n) \to u$ and $f_n(x_n) = 0$ for each n which shows $f_n(x_n) \to f_n(u)$. Thus (M1) is satisfied.

(b) Let $f := 0$ and $f_n := \langle e_n, \cdot \rangle$. Then $f_n \to 0$ pointwise because $e_n \to_w 0$. However, given $x_n := -e_n$, we have $x_n \to_w 0$, but $f_n(x_n) = -1$. Then $\liminf_{n \to \infty} f_n(x_n) < f(0)$ and so (f_n) does not converge Mosco to f.

(c) Let $x_0 \in X$, if $f(x_0) = \infty$, there is nothing to do since $\liminf f_n(x_0) \geq f(x_0)$. Otherwise, select $(x_n) \to x_0$ so that $f_n(x_n) \to f(x)$. Choose $\delta > 0$ so that $|f_n(y-x_0)| < \varepsilon/2$ if $|y - x_0| < \delta$ for all n, and choose n_0 so that $|f_n(x_n) - f_n(x_0)| < \varepsilon/2$ for $n \geq n_0$. Then for $n \geq n_0$, we have

$$|f_n(x_0) - f(x_0)| \leq |f_n(x_0) - f_n(x_n)| + |f_n(x_n) - f(x_0)| < \varepsilon/2 + \varepsilon/2 = \varepsilon.$$

(d) (M1) follows immediately from pointwise convergence. For (M2), suppose $x_n \to_w x$, then $\liminf_{n \to \infty} f_n(x_n) \geq \liminf_{n \to \infty} f(x_n) \geq f(x)$. For slice convergence, use pointwise convergence to show $\limsup d(W, \operatorname{epi} f_n) \leq d(W, \operatorname{epi} f)$ for any set $W \subset X \times \mathbb{R}$. Because $\operatorname{epi} f_n \subset \operatorname{epi} f$ for all n, it is clear that $\liminf d(W, \operatorname{epi} f_n) \geq d(W, \operatorname{epi} f)$.

(e) Consider $f_n := \delta_{[0, 1-1/n]}$ and $f := \delta_{[0,1]}$. Then $f_n(1) = \infty$ for all n, while $f(1) = 0$. □

6.2.6.* Prove Fact 6.2.7.

6.2.7 (Epigraphical convergence and minimizing sequences).* Assume X is a Banach space and all functions are proper lsc convex functions on X.

(a) Suppose (f_n) converges Kuratowski–Painlevé to f. Verify that

$$\limsup_{n \to \infty} \left(\inf_X f_n \right) \leq \inf_X f.$$

(b) Suppose (f_n) converges Mosco to f and (x_n) is a sequence converging weakly to x such that $f_n(x_n) \leq \inf_X f_n + \varepsilon_n$ where $\varepsilon_n \to 0^+$. Show that $x \in \operatorname{argmin} f$ and $f(x) = \lim_{n \to \infty}(\inf_X f_n)$.

(c) Suppose X is reflexive, (f_n) converges Mosco to f, $f_n(x_n) = \inf_X f_n$ and (x_n) is a bounded sequence. Show that $\inf_X f = \lim_{n \to \infty} f_n(x_n)$ and $\operatorname{argmin} f \neq \emptyset$.

(d) Suppose X is not reflexive. Let $(x_n) \subset S_X$ be a sequence with no weakly convergent subsequence. Let $f_n(x) := -t$ when $x = tx_n$ for $0 \leq t \leq 1$ and $f_n(x) := +\infty$ otherwise. Show that (f_n) converges Mosco to $\delta_{\{0\}}$ but $\lim (\inf_X f_n) < \inf_X \delta_{\{0\}}$. See also [46, Theorem 5.1] for further results.

(e) Consider the Lipschitz convex functions $f_n : \mathbb{R} \to \mathbb{R}$ defined by $f_n(t) := -t/n$ if $t \leq n^2$ and $f_n(t) := -n$ if $t > n^2$. Verify that (f_n) converges Mosco to 0, but $\limsup_{n \to \infty} f_n(n^2) = -\infty$ while $\inf_{\mathbb{R}} f = 0$. Similarly, let $g_n(t) := -t/n$; verify (g_n) converges Mosco to 0. Deduce the 'moreover' part of Theorem 6.2.5 may fail when the argmins are empty or not contained in a compact set.

(f) Suppose (f_n) converges slice to f and there exists a bounded sequence (x_n) such that $\liminf_{n \to \infty} f_n(x_n) = \liminf_{n \to \infty}(\inf_X f_n)$. Show that

$$\inf_X f = \lim_{n \to \infty} \left(\inf_X f_n \right)$$

(g) Let $X = c_0$ endowed with its usual norm, and consider the functions f_n defined by $f_n(x) := \max\{\langle e_n, x\rangle, -1\}$ and let $f := 0$. Show that (f_n) converges pointwise and Wijsman to f with respect to ℓ_1-product norm defined on $X \times \mathbb{R}$. In contrast to (f), observe that $\inf_X f_n = -1$ is attained on B_X, but $\lim (\inf_X f_n) < \inf_X f$. Does (f_n) converges Wijsman to f with respect to the ℓ_∞-product norm on $X \times \mathbb{R}$?

Hint. (a) This follows from (6.2.2) in Lemma 6.2.4. For (b), observe that

$$f(x) \leq \liminf_{n \to \infty} f_n(x_n) \leq \liminf_{n \to \infty} (\inf_X f_n + \varepsilon_n)$$

$$= \liminf_{n \to \infty} (\inf_X f_n) \leq \limsup_{n \to \infty} \left(\inf_X f_n\right) \leq \inf_X f.$$

Deduce (c) from (b) and the Eberlein–Šmulian theorem. In (e), notice that the sequence converges uniformly on bounded sets.

(f) By part (a) it suffices to show that $\inf f \leq \liminf_{n \to \infty} (\inf f_n)$. Assuming the contrary, we suppose there exists a number α such that

$$\liminf_{n \to \infty} (\inf f_n) < \alpha < \inf f.$$

Let $C = \{(x, \alpha) : \|x\| \leq M\}$ where $M > 0$ is chosen so that $\|x_n\| \leq M$ for all n. Then $d(C, \text{epi} f) \geq \inf f - \alpha$ (using the ℓ_1-product norm) while $\liminf_{n \to \infty} d(C, \text{epi} f_n) = 0$ which contradicts slice convergence.

(g) Observe that (f_n) converges pointwise to f. Now let $(x, t) \in X \times \mathbb{R}$. If $t \geq 0$, then $d_{\text{epi} f}((x, t)) = 0$ and since (f_n) converges pointwise to 0, $d_{\text{epi} f_n}((x, t)) \leq |f_n(x)| \to 0$. If $t < 0$, $d_{\text{epi} f}((x, t)) = |t|$. If n_0 is such that $f_n(x) > t$ for $n > n_0$ show that $d_{\text{epi} f_n}((x, t)) = |f_n(x) - t|$ for $n > n_0$. For the last question, consider $d_{\text{epi} f}(0, -1) = 1$ in the ℓ_∞ product norm. However, $(-e_n/2, -1/2) \in \text{epi} f_n$ for each n, and thus $d_{\text{epi} f_n}(0, -1) \leq 1/2$. □

6.2.8. Use $f_n(t) := \begin{cases} a_n & \text{if } t = x_n \\ \infty & \text{if } t \neq x_n \end{cases}$ and $f(t) := \begin{cases} a & \text{if } t = \bar{x} \\ \infty & \text{if } t \neq \bar{x} \end{cases}$ to show that any form of convergence that reasonably preserves minimizers on weakly compact sets implies Mosco convergence.

6.2.9. Let X be a Banach space, show that the following are equivalent.

(a) There is a sequence $(\phi_n) \in S_{X^*}$ that converges weak* to 0.
(b) There is a sequence of equi-Lipschitz convex functions (f_n) on X that converges Wijsman to 0 with respect to the ℓ_1-product norm on $X \times \mathbb{R}$, such that

$$\liminf \left(\inf_{B_X} f_n\right) < 0.$$

Hint. (a) ⇒ (b): Let $f_n := \phi_n$. (b) ⇒ (a): Since (f_n) converges pointwise to 0 by Exercise 6.2.5(c), we may and do assume $f_n(0) = 0$ for all n, and we choose $x_n \in B_X$ such that $\liminf f_n(x_n) < \alpha < 0$. Let $\phi_n \in \partial f_n(0)$. Then $\limsup \|\phi_n\| \geq |\alpha|$. However, $(\phi_n) \to_{w^*} 0$, otherwise there is an $h \in X$ such that $\limsup \phi_n(h) > 0$, and the subdifferential inequality would yield $\limsup f_n(h) > 0$ which is a contradiction. □

6.2.10.* Let X and Y be Banach spaces and suppose $T : X \to Y$ is a multifunction. This exercise shows some relation of the upper-semicontinuity of T to the lim sup of sets as given in Definition 6.2.2.

(a) Suppose Y is finite-dimensional and T is locally bounded at $\bar{x}$. Show that T is norm-to-norm upper-semicontinuous at $\bar{x}$ if $\limsup Tx_n \subset T\bar{x}$ in the norm topology on Y whenever $\|x_n - \bar{x}\| \to 0$.
(b) Suppose Y is reflexive and T is locally bounded at $\bar{x}$. Show that T is norm-to-weak upper-semicontinuous at $\bar{x}$ if $\limsup Tx_n \subset T\bar{x}$ in the weak topology on Y whenever $\|x_n - \bar{x}\| \to 0$.

Hint. For (b), suppose T is not norm-to-weak upper-semicontinuous at $\bar{x}$. Then there is a weak open neighborhood W of $T\bar{x}$ and $x_n \to \bar{x}$ such that $Tx_n \not\subset W$, and by the local boundedness property we may assume $Tx_n \subset MB_Y$ for some $M > 0$ and all n. Now choose $y_n \in Tx_n \setminus W$. By the Eberlein–Šmulian theorem (see e.g., [199, p. 85]), there is a subsequence (y_{n_k}) that is weakly convergent, say to $\bar{y}$. Now $\bar{y} \notin W$. This is a contradiction because $\limsup Tx_n \not\subset T\bar{x}$ in the weak topology on Y. □

6.2.11.* Let X be a Banach space, show that the following are equivalent.

(a) There is a sequence $(\phi_n) \subset S_{X^*}$ such that $(\phi_n) \to_{w^*} 0$.
(b) There is a sequence of 1-Lipschitz convex functions converging Mosco but not Attouch-Wets to 0.
(c) There is a sequence of nonnegative 1-Lipschitz convex functions converging slice but not Attouch-Wets to 0.

Thus, according to the Josefson–Nissenzweig theorem (8.2.1), Mosco, slice and Attouch–Wets convergence agree if and only if a space is finite-dimensional.

Hint. (a) $\Rightarrow$ (c): Consider $X \times \mathbb{R}$ endowed with the ℓ_1-product norm. Let $(\phi_n) \subset S_{X^*}$ be such that $(\phi_n) \to_{w^*} 0$. Define $f_n := \max\{\phi_n, 0\}$. Then $f_n \to 0$ pointwise and so it follows easily that (f_n) converges slice to 0 (Exercise 6.2.5(d)). However, consider $x_n \in 2B_X$ such that $\phi_n(x_n) = 1$. Then $(x_n, 0) \in \mathrm{epi} f$, but $d_{\mathrm{epi} f_n}((x_n, 0)) \geq |f(x_n) - 0| = 1$. Thus (f_n) does not converge Attouch–Wets to f. Observe (c) $\Rightarrow$ (b) is immediate because slice convergence implies Mosco convergence. To show (b) implies (a), suppose (b) is not true. Because Mosco convergence implies pointwise convergence for equi-Lipschitz functions (Exercise 6.2.5(c)), we deduce there exist a bounded sequence (x_n) and $\varepsilon > 0$ so that $f_n(x_n) > \varepsilon$ for infinitely many n (if $f_n(y_n) < -\varepsilon$ for infinitely many n, use the convexity and fact $f_n \to 0$ pointwise to find the x_n). For such n, let $C_n := \{x : f_n(x) \leq \varepsilon\}$. By the separation theorem (4.1.17), choose $\phi_n \in S_{X^*}$ so that $\phi_n(x_n) > \sup_C \phi_n$. Use the fact $f_n \to 0$ pointwise to deduce that $\phi_n \to_{w^*} 0$. □

6.2.12.* Show that Wijsman convergence implies Kuratowski–Painlevé convergence, but not conversely in infinite-dimensional spaces.

Hint. Let (C_n) be a sequence of closed convex sets converging Wijsman to the closed convex set C. If $x \in C$, then $d_{C_n}(x) \to d_C(x) = 0$. Consequently, there exists $x_n \in C_n$ so that $x_n \to x$. On the other hand, suppose $x_{n_k} \in C_{n,k}$ and $x_{n_k} \to \bar{x}$. Then $d_{C_{n,k}}(\bar{x}) \to 0$, and so $\bar{x} \in C$ as desired.

Let $X = c_0$, and consider $X \times \mathbb{R}$ in the ℓ_∞ norm, then consider the sequence (f_n) given in the proof of Exercise 6.2.7(g). Then (f_n) does not converge Wijsman to 0 in the ℓ_∞-product norm on $c_0 \times \mathbb{R}$. However, it converges Wijsman in the ℓ_1-product norm to 0, thus by the previous paragraph it converges Kuratowski–Painlevé in $X \times \mathbb{R}$. Now use the Josefson–Nissenzweig theorem (8.2.1) to produce an analogous example in any infinite-dimensional Banach space. □

6.2.13. Let X be an infinite-dimensional Banach space. Show there is a sequence of 1-Lipschitz convex functions (f_n) converging Wijsman to $f := 0$ with respect to the ℓ_1-product norm, but (f_n^*) fails to converge Kuratowski–Painlevé to $\delta_{\{0\}} = f^*$.

Hint. Use the Josefson–Nissenzweig theorem (8.2.1) to choose $(\phi_n)_{n=1}^\infty \subset S_{X^*}$ such that $\phi_n \to_{w^*} 0$. Let $f_n := \phi_n$ and $f := 0$. Check that (f_n) converges Wijsman to f with respect to the ℓ_1-product norm on $X \times \mathbb{R}$. However, $f_n^* = \delta_{\{\phi_n\}}$ and $f^* = \delta_{\{0\}}$. There is no sequence $(x_n^*) \subset X^*$ such that $x_n^* \to 0$ and $f_n^*(x_n^*) \to f^*(0)$. Thus (f_n^*) does not converge Kuratowski–Painlevé to f^*. □

6.2.14.* (a) Show that slice convergence implies Mosco convergence for sequences of closed convex sets in arbitrary normed spaces.
(b) Let X be a reflexive Banach space. Show that Mosco and slice convergence are equivalent for sequences of closed convex sets.

Hint. (a) Slice convergence implies Mosco convergence in arbitrary normed spaces. Indeed, suppose (C_n) converges slice to C. Let $x \in C$, then $d_C(x) = 0$ and so $d_{C_n}(x) \to 0$ from which we conclude that there exist $x_n \in C_n$ such that $x_n \to x$. Now suppose $x_{n_k} \in C_{n_k}$ is such that $x_{n_k} \to_w x$. Then (x_{n_k}) is a bounded sequence contained in MB_X for some $M > 0$. Suppose $x \notin C$, then choose $\phi \in X^*$ so that $\phi(x) > \alpha > \sup_C \phi$. Consider $W = MB_X \cap \{u \in X : \phi(u) \geq \alpha\}$. Then eventually $x_{n_k} \in W$ and so $\lim d(W, C_{n_k}) = 0$ while $d(W, C) > 0$ which contradicts slice convergence.

(b) Suppose (C_n) converges Mosco to C in the reflexive space X. If $x \in C$, then there exist $x_n \in C_n$ with $x_n \to x$. Thus $\limsup d(W, C_n) \leq d(W, C)$ for any set $W \in X$. Now suppose there is a closed bounded convex set W such that $\liminf d(W, C_n) < d(W, C)$. Use weak compactness, and weak lower-semicontinuity of the norm to get a contradiction. □

6.2.15.* Prove Theorem 6.2.13.

Hint. By a compactness argument, slice and Wijsman convergence coincide in finite-dimensional spaces. Also, Mosco and Kuratowski–Painlevé convergence are equivalent in finite-dimensional spaces. Combining this with Exercise 6.2.14 shows that all four of these notions are equivalent in finite-dimensional spaces. Finally, Attouch–Wets convergence is the strongest of these notions in general, so we now show that it is implied by Wijsman convergence in finite-dimensional spaces. Indeed, if (A_n) converges Wijsman to A, then $d_{A_n}(\cdot) \to d_A(\cdot)$ pointwise, and thus uniformly on bounded sets because they are equi-Lipschitz. □

6.2.16.* Suppose X is a Banach space that is not reflexive. Show that there is a sequence of proper lsc convex functions on X that converges Mosco but not Wijsman.

Hint. Let H be a nonreflexive hyperplane of X, choose a sequence $(h_n) \in H$ with $\|h_n\| = 1/2$ so that (h_n) has no weakly convergent subsequence. Now let $\|x_0\| = 1$ and consider $x_n := h_n + x_0$. Now (x_n) has no weakly convergent subsequence. Let $C_n := [0, x_n]$, and let $f_n := \delta_{C_n}$ and $f = \delta_{\{0\}}$. Then (f_n) converges Mosco to f, but not Wijsman in most any reasonable product norm. Consider distance of $(x_0, 0)$ to the epigraphs. □

6.2.17.* Prove Theorem 6.2.14.

Hint. Note that many of the relations or failure thereof can be found in Exercises 6.2.5, 6.2.11, 6.2.16, 6.2.12, 6.2.14, 6.2.7. Then establish the remaining statements. □

6.2.18. Suppose f is convex and bounded on bounded sets and that (f_n) is a sequence of proper lsc convex functions. Suppose (f_n) converges Attouch–Wets to f. Show that (f_n) converges uniformly to f on bounded sets.

Hint. Suppose (f_n) does not converge to f uniformly on bounded sets. Then there exists $\varepsilon > 0$ and a bounded sequence (x_n) such that one of the following is true.
  (i) $f_n(x_n) + \varepsilon < f(x_n)$ for infinitely many n.
  (ii) $f(x_n) + \varepsilon < f_n(x_n)$ for infinitely many n.

If (i) occurs, then Attouch–Wets convergence fails (use Lipschitz property of f to show that the excess of $\mathrm{epi}\, f_n$ over f is uniformly large).

In the case of (ii), suppose Attouch–Wets convergence doesn't fail. Then there are $(y_n, s_n) \in \mathrm{epi}\, f_n$ such that $\|y_n - x_n\| \leq \varepsilon_n$ and $|s_n - f(x_n)| < \varepsilon_n$ where $\varepsilon_n \to 0$. Now let $t_n = f(x_n) + \varepsilon$ and use the sandwich theorem (4.1.18) to find $\phi_n \in X^*$, $\alpha_n \in \mathbb{R}$ so that $\phi_n - \alpha_n$ separates (x_n, t_n) and $\mathrm{epi}\, f_n$ for infinitely many n. Now, passing to a subsequence

$$\frac{\phi_n(x_n - y_n)}{\|x_n - y_n\|} \geq \frac{t_n - s_n}{\varepsilon_n} = \frac{\varepsilon - \varepsilon_n}{\varepsilon_n} = M_n.$$

where $M_n \to \infty$. Choose h_n such that $\|h_n\| = 1$ and $\phi_n(h_n) \geq M_n - 1$ and let $z_n = x_n + h_n$. Then $f(z) \leq f(x_n) + 2M$ where M is a Lipschitz constant for f on $\{x_n\}_{n=1}^{\infty} + 2B_X$ and $\|z - z_n\| \leq 1/2$. However, $f_n(z) \geq f(x_n) + \phi(z) \geq f(x_n) + M_n - 1 - M_n/2 \geq f(x_n) + M_n/2 - 1$ for $\|z - z_n\| \leq 1/2$. Hence $d_{\mathrm{epi}\, f_n}((x_n, f(x_n))) > 1/2$ for large n. □

6.2.19. The definition of Attouch–Wets convergence applies naturally to nonconvex functions; see [44] for further details. Let (X, d) be a metric space and let (f_n) be a sequence of lsc real-valued functions on X such that f is Lipschitz on bounded subsets of X and (f_n) is eventually equi-Lipschitz on bounded subsets of X. Suppose that (f_n) converges Attouch–Wets to f, show that (f_n) converges uniformly on bounded sets to f.

Hint. See [44, Proposition 7.1.3]. □

6.2.20. Suppose $f, f_1, f_2, \ldots$ are real-valued functions on X such that (f_n) converges uniformly to f on X. Show that $\mathrm{dom}\, f_n^* = \mathrm{dom}\, f^*$ for all large n, and (f_n^*) converges uniformly to f^* on the domain of f^*.

Hint. Let $\varepsilon > 0$, and choose $n_0 \in \mathbb{N}$ so that $|f_n(x) - f(x)| \leq \varepsilon$ for all $n \geq n_0$. Let $\phi \in X^*$, then

$$\phi(x) - f_n(x) - \varepsilon \leq \phi(x) - f(x) \leq \phi(x) - f_n(x) + \varepsilon$$

for all $n \geq n_0$. Consequently, if $f^*(\phi) < \infty$, then $|f_n^*(\phi) - f^*(\phi)| \leq \varepsilon$ for all $n \geq n_0$. If $f^*(\phi) = \infty$, then $f_n^*(\phi) = \infty$ for all $n \geq n_0$. □

6.2.21. Suppose f is convex, supercoercive and bounded on bounded sets and that (f_n) is a sequence of proper lsc convex functions on X. Show that (f_n) converges to f uniformly on bounded sets if and only if (f_n^*) converges uniformly to f^* on bounded sets.

Hint. Suppose (f_n) converges to f uniformly on bounded sets, then (f_n) converges Attouch–Wets to f. According to Theorem 6.2.10, (f_n^*) converges Attouch–Wets to f^*. Now f^* is bounded on bounded sets, so according to Exercise 6.2.18, (f_n^*) converges uniformly on bounded set to f^*.

Now f^* is supercoercive and bounded on bounded sets. Hence so is f^{**}. The argument from the previous paragraph shows that (f_n^{**}) converges to f^{**} uniformly on bounded sets, hence the same is true for their restrictions to X, and these restrictions are f_n and f according to Proposition 4.4.2. □

6.2.22. Suppose the dual norm on X^* does not have the weak*-Kadec property (sequentially). Then we can choose norm-attaining $\phi_n, \phi \in S_{X^*}$ such that $\phi_n \to_{w^*} \phi$ while $\|\phi_n - \phi\| > \varepsilon$ for some $\varepsilon > 0$. Let $C_n := \{x \in X : \phi_n(x) = 1\}$ and $C := \{x \in X : \phi(x) = 1\}$. Show that (C_n) converges Wijsman but not slice to C.

Hint. Use the Bishop–Phelps theorem (4.3.4) to find the appropriate norm–attaining functionals. □

6.2.23.* Let $(X, \|\cdot\|)$ be a normed space and let C_α, C be closed convex subsets of X. Show that (C_α) converges slice to C if and only if (C_α) converges Wijsman to C with respect to every equivalent norm on X. In particular slice convergence is invariant under renorming.

Hint. First show that Wijsman convergence implies

$$\limsup_\alpha d(W, C_\alpha) \leq d(W, C)$$

for all closed bounded sets W, and each equivalent norm on X. For the second step, show that if (C_α) converges slice to C, then $\liminf d_{C_\alpha}(x) \geq d_C(x)$ for each $x \in X$, and with respect to each equivalent norm on X. For the third step, show that if slice convergence fails, then there is a norm with respect to which Wijsman convergence fails.

To verify the first step, let $\varepsilon > 0$ and let $\|\cdot\|$ be an equivalent norm on X. Fix $x \in W$ and $c \in C$ so that $\|x - c\| \leq d_{|\cdot|}(W, C) + \varepsilon$. Because (C_α) converges Wijsman to C with respect to some norm on X, there exist $c_\alpha \in C_\alpha$ such that $c_\alpha \to c$. Now

$$\limsup_\alpha d_{|\cdot|}(W, C_\alpha) \leq \limsup \|x - c_\alpha\| \leq d_{|\cdot|}(W, C) + \varepsilon.$$

6.2 Epigraphical convergence: an introduction   299

For the second step, let $x \in X \setminus C$ and let $r = d_C(x) - \varepsilon$ where the distance is measured with respect to $\|\cdot\|$. Now let $W = x + rB$ where B is the unit ball with respect to $\|\cdot\|$. Now $d(W, C) > 0$, and so $\lim_\alpha d_{C_\alpha}(x) > 0$ which implies that $C_\alpha \cap W = \emptyset$ eventually, and so
$$\liminf_\alpha d_{C_\alpha}(x) \geq d_C(x).$$
This shows the second step; moreover, the first two steps together prove that slice convergence implies Wijsman convergence with respect to every equivalent norm on X.

Conversely, if convergence is not slice, choose W closed bounded and convex for which $d(W, C_\alpha) \not\to d(W, C)$. If Wijsman fails with respect to the given norm, we are done. So, suppose not. Then by part (a) we know that $\limsup d(W, C_\alpha) \leq d(W, C)$. Suppose $W + rB$ meets a subnet of (C_α), but $W + (r + \varepsilon)B \cap C = \emptyset$. Without loss of generality, assume $0 \in W$. Separating these sets, say $W + rB \subset \{x : \phi(x) \leq k\}$ where $k > 0$, but $C \subset \{x : \phi(x) \geq k + \delta\}$. Let $U = \{x : |\phi(x)| \leq k, \|x\| \leq M\}$ where M is so large that $W + rB_X \subset U$. Conclude that Wijsman convergence fails with respect to the norm whose unit ball is U. See [44, Theorem 2.4.5] for further details. □

6.2.24. A net of nonempty closed convex sets (C_λ) is said to converge *Mosco* to a nonempty closed convex set C, if (i) for every $x \in C$, there is a net $x_\lambda \to x$ where $x_\lambda \in C_\lambda$ for all λ; and (ii) for every weakly compact set W such that $W \cap C_{\lambda_\alpha} \neq \emptyset$ for a subnet, $W \cap C \neq \emptyset$. Prove that Wijsman convergence implies Mosco convergence for nets of closed convex sets in a Banach space X if and only if Mackey and weak*-convergence agree on the sphere of its dual ball.

Hint. Suppose Mackey and weak*-convergence do not coincide. Find norm-attaining functionals $\phi_\alpha \in S_{X^*}$ such that $(\phi_\alpha) \to_{w^*} \phi$ but (ϕ_α) does not converge Mackey to ϕ. Let $C_\alpha := \phi_\alpha^{-1}(1)$ and $C := \phi^{-1}(1)$. Show that (C_α) converges Wijsman but not Mosco to C.

Conversely, suppose Mackey and weak*-convergence coincide on the dual sphere. Suppose also that there is a net of closed convex sets (C_α) that converges Wijsman but not Mosco to C. Then there is a weakly compact set W with $x_{\alpha_\lambda} \in C_{\alpha_\lambda} \cap W$ for some subnet, but $W \cap C = \emptyset$. By passing to a further subnet, assume (x_{λ_α}) converges weakly to $\bar{x}$. Then $\bar{x} \in W$ and so $d_C(\bar{x}) > 0$. By Wijsman convergence $\lim d_{C_\alpha}(\bar{x}) = d_C(\bar{x})$. Let $r_\alpha := d_{C_\alpha}(\bar{x})$. Choose $\phi_\alpha \in S_X^*$ so that ϕ_α separates C_α and the interior of $x_\alpha + r_\alpha B_X$. Now (ϕ_α) has a subnet weak*-convergent to ϕ; let $x_n \in C$ be such that $\|\bar{x} - x_n\| \to d_C(\bar{x})$. Choose nets $(x_{n,\alpha})_\alpha \to x_n$, $x_{n,\alpha} \in C_\alpha$, now $\phi_\alpha(x_{n,\alpha} - \bar{x}) \geq r_\alpha$ and one can thus show $\phi(x_n - \bar{x}) \geq r$ for all n. Thus $\phi \in S_{X^*}$; however (ϕ_α) does not converge uniformly on weakly compact sets to ϕ because $\phi(x_{\alpha_\lambda} - \bar{x}) \to 0$ while $\phi_{\alpha_\lambda}(x_{\alpha_\lambda} - \bar{x}) \geq r_\alpha$. □

6.2.25. Show that Wijsman and Mosco convergence coincide for sequences of closed convex sets in a Banach space X if and only if X is reflexive and the weak and norm topologies coincide on its dual sphere.

Hint. Look at solutions to Exercises 6.2.16 and 6.2.24. A proof of this result that integrates several other equivalent conditions can be found in [81]. □

6.2.26 (Stability under addition). Suppose $f, f_1, f_2, \ldots, g, g_1, g_2, \ldots$ are all proper lsc extended real-valued functions on a Banach space X. Show that $(f_n + g_n)$ converges Mosco and pointwise to $f + g$ if (f_n) and (g_n) converge both Mosco and pointwise to f and g respectively.

Hint. Let $x_n = x$ for each n. Then $x_n \to x$ and $\lim(f_n + g_n)(x) = f(x) + g(x)$. Also, suppose $x_n \to x$ weakly, then

$$\liminf_{n \to \infty} (f_n + g_n)(x_n) \geq \liminf_{n \to \infty} f_n(x_n) + \liminf_{n \to \infty} g_n(x_n)$$
$$\geq f(x) + g(x).$$

which completes the proof of Mosco convergence. □

6.2.27 (Generalized derivatives).** This outlines some results of Rockafellar's from [377] which generalize some of the second-order results from Section 2.6. Let $f : \mathbb{R}^n \to (-\infty, +\infty]$ be a proper convex function. If

$$f(x + h) = f(x) + f'(x; h) + o(\|h\|)$$

where $x \in \operatorname{int} \operatorname{dom} f$, then f is said to be *semidifferentiable* at x. If the limit

$$\lim_{h' \to h, t \downarrow 0} \frac{f(x + th') - f(x) - f'(x; h')}{\frac{1}{2} t^2}$$

converges uniformly on bounded sets to some continuous function $d^2 f(x)(\cdot)$, one then has

$$f(x + h) = f(x) + f'(x; h) + \frac{1}{2} d^2 f(x)(h) + o(\|h\|^2)$$

and f is said to be *twice semidifferentiable* at x. Now, consider the 'limit'

$$\lim_{w' \to w, t \downarrow 0} \frac{\partial f(x + tw') - \nabla f(x)}{t} = \lim_{w' \to w, t \downarrow 0} \Delta_t[\partial f](x)(w'), \quad (6.2.7)$$

where $x \in \operatorname{dom} \nabla f$. If the set limit in (6.2.7) exists and is nonempty in the sense of the liminf of sets for every w, then ∂f is said to be *semidifferentiable* at x, if this is true for *almost every* w, then ∂f is said to be *almost semidifferentiable*. With this, we state

Theorem 6.2.16 (Rockafellar [377]). *Let $x \in \operatorname{dom} \nabla f$. Then f is twice semidifferentiable at x if and only if ∂f is almost semidifferentiable at x.*

The reader can consult the paper [377] for full details and more information on other notions of generalized derivatives that use set-convergence on $\mathbb{R}^n$. A more comprehensive treatment on this subject can be found in the monograph [378]. See also [107, Section 6] for an approach to generalized second-order differentiability in infinite-dimensional spaces. Also, along these lines, we should mention the following theorem, because it (and its predecessors by Attouch [18, 19]) which related epi-convergence of functions with convergence of their subdifferentials have played an important role in this subject; see, for example, [374, 375, 377].

Theorem 6.2.17 (Attouch–Beer [20]). *Let X be a Banach space, and let $f, f_1, f_2, \ldots$ be proper lsc convex functions on X. Then (f_n) converges slice to f if and only if the following two conditions are satisfied*
(a) *$\partial f = \lim \partial f_n$ in the Kuratowski–Painlevé sense; and*
(b) *there exist $z \in \partial f(u)$ and for each n, $z_n \in \partial f_n(u_n)$ for which $(u, f(u), z) = \lim_{n \to \infty} (u_n, f(u_n), z_n)$.*

Hint. See [20, Theorem 4.2]. See also [297, Theorem 9.3.1] for a proof in the simpler reflexive space setting. □

6.3 Convex integral functionals

6.3.1 The autonomous case

This section will look at integral functionals of the form

$$I_\phi(x) := \int_S \phi(x(s)) \, d\mu(s)$$

where (S, μ) is a finite, complete measure space, and $x \in L_1(S, \mu)$, and $\phi : \mathbb{R} \to (-\infty, +\infty]$ is convex. There are much more general approaches to convex integral functionals (where there is dependence on s not only $x(s)$ as is discussed in the notes at the end of this section). This section will present a fairly naive approach that will capture results suitable for the development of strongly rotund functions in Section 6.4. Throughout this section, $\phi : \mathbb{R} \to (-\infty, +\infty]$ will be a convex proper lsc function.

Proposition 6.3.1. *Suppose $\phi : \mathbb{R} \to (-\infty, +\infty]$ is proper, convex and lsc, then I_ϕ is proper, lsc and convex on $L_1(S, \mu)$.*

Proof. Suppose $x_n \to_{L_1} x$, but $\liminf_{n \to \infty} I_\phi(x_n) < I_\phi(x)$. Then we fix $\varepsilon > 0$ and a subsequence (x_{n_k}) such that $I_\phi(x_{n_k}) < I_\phi(x) - \varepsilon$ for all $k \in \mathbb{N}$. Now passing to a further subsequence, we may assume $x_{n_k} \to x$ a.e. (since mean convergence implies convergence in measure). Suppose, for now, that $\phi \geq 0$, then $\phi \circ x_{n_k} \geq 0$, and the lower-semicontinuity of ϕ ensures that $\liminf_{k \to \infty} \phi(x_{n_k}(s)) \geq \phi(x(s))$ a.e. Thus Fatou's lemma implies

$$\liminf_k \int \phi(x_{n_k}(s)) \, d\mu(s) \geq \int \phi(x(s)) \, d\mu(s).$$

The remaining case of the proof of the lower-semicontinuity of I_ϕ is left as Exercise 6.3.1, as are the other parts of the proposition. □

In this section, in a couple of places we will need the following (deeper) characterization of weak compactness. For this, recall that a subset $W \subset L_1(S, \mu)$ is said to be *uniformly integrable* if for each $\varepsilon > 0$, there is a $\delta > 0$, so that for each measurable subset K of S with $\mu(K) < \delta$, one has

$$\int_K |x(s)| \, d\mu(s) < \varepsilon \text{ for each } x \in W.$$

Theorem 6.3.2 (Dunford–Pettis criterion). *A subset W of $L_1(S, \mu)$ is relatively weakly compact if and only if it is bounded and uniformly integrable.*

Proposition 6.3.3. *Suppose $\phi : \mathbb{R} \to (-\infty, +\infty]$ is supercoercive. Then I_ϕ has weakly compact lower level sets.*

Proof. Let $C := \{x \in L_1 : I_\phi(x) \leq k\}$ for some $k \in \mathbb{R}$. Because ϕ is bounded below and S is finite measure space, there is a bound M_1 such that $\int_E \phi(x(s))\,d\mu(s) \leq M_1$ for any measurable subset $E \subset S$ and $x \in C$.

Now choose $M_2 > 0$ so that $M_1/M_2 < \varepsilon/2$. The supercoercivity of ϕ implies there exists $\alpha > 0$ so that $\phi(t) \geq M_2|t|$ for $|t| \geq \alpha$. For a fixed $x \in C$, consider the two measurable sets

$$S_{x,1} := \{s : |x(s)| \leq \alpha\} \quad \text{and} \quad S_{x,2} := \{s : |x(s)| > \alpha\}.$$

Now observe that for any measurable $K \subset S$, with $\mu(K) < \varepsilon/2\alpha$, and $x \in C$,

$$\int_K |x(s)|\,d\mu(s) = \int_{K \cap S_{x,1}} |x(s)|\,d\mu(s) + \int_{K \cap S_{x,2}} |x(s)|\,d\mu(s)$$

$$\leq \alpha\mu(K) + \frac{1}{M_2} \int_{S_{x,2}} \phi(x(s))\,d\mu(s)$$

$$\leq \alpha\mu(K) + M_1/M_2 < \varepsilon.$$

Thus C is uniformly integrable, we leave it as an elementary analysis exercise to show C is bounded. The result then follows from the Dunford–Pettis theorem (6.3.2). □

Recall that a *simple function* is a function of the form $y = \sum_{k=1}^n a_k \chi_{S_k}$ where $a_k \in \mathbb{R}$, and since a_k may be 0, we may assume $\bigcup_{k=1}^n S_k = S$, and we may assume the S_k's are disjoint. The following beautiful conjugacy result is now accessible.

Theorem 6.3.4. *Let $\phi : \mathbb{R} \to (-\infty, +\infty]$ be a closed, proper convex function. Then $(I_\phi)^* = I_{\phi^*}$.*

Proof. For $y \in L_\infty(S, \mu)$ observe that

$$(I_\phi)^*(y) = \sup_{x \in L_1} \left\{ \langle x, y \rangle - \int_S \phi(x(s))\,d\mu(s) \right\}$$

$$= \sup_{x \in L_1} \int_S [x(s)y(s) - \phi(x(s))]\,d\mu(s)$$

$$\leq \int_S \phi^*(y(s))\,d\mu(s) = I_{\phi^*}(y).$$

For the reverse inequality, first consider the case where y is a simple function, say $y = \sum_{k=1}^n a_k \chi_{S'_k}$ with the S'_ks disjoint, and their union is all of S. Suppose also

6.3 Convex integral functionals

that ϕ^* is finite-valued (hence continuous); then we may choose $t_k \in \mathbb{R}$ such that $a_k t_k - \phi(t_k) = \phi^*(a_k)$ and let $x := \sum_{k=1}^n t_k \chi_{S_k}$. Then

$$I_{\phi^*}(y) = \sum_{k=1}^n \int_{S_k} \phi^*(a_k) \, d\mu(s) = \sum_{k=1}^n \phi^*(a_k)\mu(S_k)$$

$$= \sum_{k=1}^n (a_k t_k - \phi(t_k))\mu(S_k) = \int x(s)y(s) - \phi(x(s)) \, d\mu(s) \leq (I_\phi)^*(y).$$

Consequently, $I_{\phi^*}(y) \leq (I_\phi)^*(y)$ whenever $y \in L_\infty(S, \mu)$ is a simple function.

Still assume that ϕ^* is everywhere finite. Then ϕ^* is continuous and bounded on bounded sets, so the same is true of I_{ϕ^*} since S is a finite measure space, and thus $(I_\phi)^*$ is bounded above on bounded sets. Thus both I_{ϕ^*} and $(I_\phi)^*$ are continuous and are equal on a norm dense set (the collection of simple functions) in $L_\infty(S)$, which implies their equality everywhere.

In the case ϕ^* is not everywhere finite, consider the infimal convolution $f_n^* := \phi^* \square n|\cdot|$. Then the reader can check that $f_n^* \leq \phi^*$ and $f_n^* \to \phi^*$ pointwise on $\mathbb{R}$ (Exercise 6.3.5). Observe that $\psi(t) := \phi^*(t) + at + b > 0$ for all t for appropriate real numbers a, b the same is true of $f_n^*(t) + at + b$ where $n > |a|$. Now use Fatou's lemma to conclude that $I_{f_n^*}(y) \to I_{\phi^*}(y)$ for all $y \in L_\infty(S, \mu)$. Finally, $f_n \geq \phi$, and so $(I_\phi)^*(y) \geq (I_{f_n})^*(y)$ for all $y \in L_\infty(S, \mu)$ and all n. Altogether this implies the theorem is true. □

Lemma 6.3.5. *Suppose ϕ is not supercoercive (if and only if ϕ^* is not finite everywhere) and suppose S has a collection of measurable subsets $\{A_n\}_{n=1}^\infty$ such that $\mu(A_n) \to 0^+$, then I_ϕ does not have weakly compact level sets.*

Proof. We consider the case where there is $M > 0$ so that

$$\limsup_{t \to \infty} \frac{|\phi(t)|}{t} \leq M.$$

Now let $\alpha_n := \mu(A_n)$ and $x_n := \frac{1}{\alpha_n} \chi_{A_n}$. Then

$$\int_{A_n} x_n(s) \, d\mu(s) = 1$$

for all n, and so $\{x_n\}_{n=1}^\infty$ is not uniformly integrable. However,

$$\int_S |\phi(x_n(s))| \, d\mu(s) \leq \int_S M|x_n(s)| \, d\mu(s) = M.$$

Therefore, by the Dunford–Pettis theorem (6.3.2), the level set $\{x : I_\phi(x) \leq M\}$ is not relatively weakly compact. The other case is similar. □

Exercises and further results

6.3.1.* Complete the proof of Proposition 6.3.1.

Hint. For lower-semicontinuity, use the separation theorem (4.1.17) to find $a, b \in \mathbb{R}$ so that $\psi(t) := \phi(t) + at + b \geq 0$. Apply the proof already given to ψ and use L_1 convergence along with the fact the measure space is finite. Given that $\phi(t_0) < \infty$, it follows that $I_\phi(x_0) < \infty$ when $x_0 = t_0 \chi_S$. Also, given that $\phi(t) > at + b$, then $I_\phi(x) > -|a|\|x\|_{L_1} - |b|\mu(S) > -\infty$. The convexity of I_ϕ follows directly from properties of integrals and the convexity of ϕ. □

The following exercises illustrate some of the problems that may occur if we relax conditions on ϕ or S. Hence additional restrictions on ϕ will be needed in the general case which we reference in some further results stated at the end of this section.

6.3.2. Consider $S := [0, 1]$ with the usual Lebesgue measure. Let $\phi : \mathbb{R} \to (-\infty, +\infty]$ be such that $\phi(t) := 0$ for $t < 1$, $\phi(1) := 1$ and $\phi(t) := \infty$ for $t > 1$. Show that I_ϕ is not lsc.

Hint. Consider $x_n := (1 - 1/n)\chi_{[0,1]}$ and $x := \chi_{[0,1]}$. □

6.3.3.* Consider $S = \mathbb{R}$ with μ the usual Lebesgue measure, and let $\phi(t) := -\sqrt{t}$ if $t \geq 0$, and $\phi(t) := +\infty$ otherwise. Show that I_ϕ is not lsc on $L_1(\mathbb{R})$. Furthermore, I_ϕ is not bounded below and it takes the value $-\infty$ as an improper integral.

Hint. Let $f_n := \frac{1}{n^2}\chi_{[n,2n]}$. Then $\|f_n\|_{L_1} \to 0$. Now $I_\phi(0) = 0$, but

$$I_\phi(f_n) = \int_n^{2n} -\frac{1}{n} = -1,$$

and so I_ϕ is not lsc. Also, if $x := \frac{1}{t^2}\chi_{[1,\infty)}$, then $I_\phi(x) = -\infty$ as an improper integral. □

6.3.4.* Consider $L_1(\mathbb{R})$ with $\phi(t) := \max\{t^2 - 1, 0\}$. Then ϕ is supercoercive, but the lower level set $\{x \in L_1(\mathbb{R}) : I_\phi(x) \leq 0\}$ is not bounded.

Hint. Let $f_n := \chi_{[0,n]}$. Then $\|f_n\|_1 = n$ and $I_\phi(f_n) = 0$ for all n. □

6.3.5.* Verify the assertions concerning f_n^* in the proof of Theorem 6.3.4.

6.3.6 (Lebesgue–Radon–Nikodým theorem [389]). ** The theorem has many proofs. What follows is a proof-sketch based on viewing the problem via an abstract convex quadratic program. We wish to show that *if μ and ν are probability measures on a measure space (Ω, Σ) then one can find a μ-null set $N \in \Sigma$ and a measurable function g such that for all $E \in \Sigma$*

$$\nu(E) = \nu(E \cap N) + \int_E g \, d\mu. \tag{6.3.1}$$

Note that for $\nu \ll \mu$ this is the classical Radon–Nikodým theorem for which a neat approach due to von Neumann applies the Riesz representation theorem to the continuous linear function $f \mapsto \int_\Omega f d\mu$ on $L_2(\mu + \nu)$.

(a) Let $Q(f) := \frac{1}{2} \int_\Omega f^2 d\mu - \int_\Omega f dv$. For $n \in \mathbb{N}$, show the minimum of

$$v_n := \inf\{Q(f) : 0 \leq f \leq n\},$$

is attained by some $g_n \in L_1(\mu + v)$.

Hint. Q is convex and Lipschitz on order intervals of $L_1(\mu+v)$ which are weakly compact by the Dunford–Pettis criterion. □

(b) Without loss, assume g_n is everywhere finite. Set $C_n := \{\omega : g_n(\omega) < n\}$ and $N_n := \{\omega : g_n(\omega) = n\}$. Use Exercise 4.1.50 to show that $v(E) = \int_E g_n \, d\mu$, for $E \subseteq C_n$ and $v(E) \geq \int_E g_n \, d\mu = nv(E)$ for $E \subseteq N_n$. (Exercise 4.1.49 only applies directly when $v \ll \mu$.)

(c) Set $A_n := D_n \setminus (C_1 \cup C_2 \cup \cdots \cup C_{n-1})$ and $N := \bigcap_{n \in \mathbb{N}} N_n$. Deduce that $g := \sum_{n=1}^\infty \chi_{A_n} g_n$ and N are as required.

(d) Extend the result to σ-finite measures.

Further results. One can consider more general *integral functionals*, such as

$$I(x) := \int_S h(x(s), s) \, d\sigma(s), \qquad (6.3.2)$$

where $I : \mathcal{X} \to [-\infty, +\infty]$, $\mathcal{X}$ is some space of functions $x : S \to X$, and X is a real linear space. First, one needs to address whether such a functional is well-defined. Following [371, p. 7] we shall say that I is well-defined if for each $x \in \mathcal{X}$ the function $s \mapsto h(x(s), s)$ is measurable. If there exists a real-valued function $\alpha : S \to \mathbb{R}$ which is summable with respect to σ and satisfies $\alpha(s) \geq h(x(s), x)$ almost everywhere, the integral (6.3.2) has an unambiguous value either finite or $-\infty$ which we assign to $I(x)$. Otherwise, we define $I(x)$ to be ∞.

In most applications, X has topological structure, hence we can use the *Borel sets* (members of the algebra generated by open sets) for notions of measurability. As such, the integrand $h : X \times S \to [-\infty, +\infty]$ is said to be *measurable* on $X \times S$ if h is measurable with respect to the σ-algebra on $X \times S$ generated by the sets $B \times A$ where B is a Borel set and A is measurable in S. In this case, $h(x(s), s)$ is measurable in s whenever $x(s)$ is measurable in s, since the latter measurability implies that of the transformation $s \to (x(s), s)$, such functionals are called *normal integrands*. Then [371, Theorem 3] is as follows.

Theorem 6.3.6. *The integral functional I given in (6.3.2) is well defined on $\mathcal{X}$ in the above sense if, relative to the Borel structure on X generated by some topology, the integrand $h : X \times S \to [-\infty, +\infty]$ is measurable and the functions $x : S \to X$ are all measurable. If, in addition, h is convex in the X argument, then I is a convex function on $\mathcal{X}$.*

With the convexity of the integral functional I established in this more general setting, it is then interesting to look at its conjugate, subdifferentiability, and other natural properties. Under appropriate dual pairings, results such as

$$I^*(v) = \int_S h^*(v(s), s) \, d\sigma(s)$$

and $v \in \partial I(x)$ if and only if $v(s) \in \partial h(x(s), s)$ for almost every s can be derived. We refer the reader to [371, p. 58ff] for a development and precise statements of results of this nature.

More precisely, a *normal integrand* is a proper extended real-valued function on $\mathbb{R}^N \times \Omega$, where $(\Omega, \mathcal{M}, \mu)$ is (for simplicity) a complete σ-finite measure space such that (a) $\phi(\cdot, s)$ is lsc for each s in Ω and (b) the epigraph multifunction $E_\phi(s) := \{(x, \alpha) : \phi(x, s) \leq \alpha\}$ is a so-called measurable multifunction which when ϕ is closed is equivalent to the measurability of the epigraph of ϕ in the product space. Full details are given in [372]. For us it suffices to know that ϕ and $-\phi$ are normal if and only if ϕ is measurable in S and continuous in x. A special case of more general results of Rockafellar is:

Theorem 6.3.7 (Theorems 3C and 3H, [372]). *For $1 \leq p < \infty$ let $X = L_p(\Omega, \mathbb{R}^N)$. Suppose that ϕ is a normal integrand on $\Omega \times \mathbb{R}^N$ and that there exists $x \in L_p(\Omega, \mathbb{R}^N)$ and $y \in L_p(\Omega, \mathbb{R}^N)$ ($1/p + 1/q = 1$) with $I_\phi(x) < \infty$ and $I_\phi^*(y) < \infty$. Then I_ϕ^{**} is the largest lsc functional majorized by I_ϕ. In particular I_ϕ is lsc if and only if $\phi(\cdot, s)$ is convex for almost all s in Ω. In this latter case,*

$$I_\phi^* = I_{\phi^*}$$

and if $p > 1$ also

$$I_\phi^{**} = I_{\phi^{**}}.$$

There is a somewhat more delicate result when $p = 1$.

6.3.7. Under the hypotheses of Theorem 6.3.7, show that $\lambda \in \partial I_\phi(x)$ if and only if $\lambda(s) \in \partial \phi(x(s), s)$ for almost all s in Ω, the subgradient being taken with respect to the first variable.

6.4 Strongly rotund functions

Following [94], we will say a function $f : X \to (-\infty, +\infty]$ has the *Kadec property* if for each $\bar{x} \in \mathrm{dom}\, f$ one has $x_n \to \bar{x}$ in norm whenever $f(x_n) \to f(\bar{x})$ and $x_n \to \bar{x}$ weakly. A proper lsc function $f : X \to (-\infty, +\infty]$ is said to be *strongly rotund* if f is strictly convex on its domain, has weakly compact lower level sets, and has the Kadec property.

Throughout this section, we suppose that (S, μ) is a complete finite measure space (with nonzero μ), and $\phi : \mathbb{R} \to (-\infty, +\infty]$ is a proper, closed, convex function. We denote the interior of the domain of ϕ by (α, β) where $-\infty \leq \alpha < \beta \leq \infty$.

6.4 Strongly rotund functions

Since ϕ is a *normal convex integrand*, one can define the proper weakly lsc functional (see Proposition 6.3.1)

$$I_\phi : L_1(S, \mu) \to (-\infty, +\infty] \text{ by } I_\phi(x) = \int_S \phi(x(s)) \, d\mu(s).$$

Then the conjugate is $I_\phi^* : L_\infty(S, \mu) \to (-\infty, +\infty]$ is given by $I_\phi^* = I_{\phi^*}$ where ϕ^* is the conjugate of ϕ; see Theorem 6.3.4.

Lemma 6.4.1. *I_ϕ is strictly convex on its domain if and only if ϕ is strictly convex on its domain.*

Proof. See Exercise 6.4.2. □

Lemma 6.4.2. *I_{ϕ^*} is Fréchet differentiable everywhere on $L_\infty(S, \mu)$ if and only if ϕ^* is differentiable everywhere on $\mathbb{R}$.*

Proof. Suppose ϕ^* is differentiable on $\mathbb{R}$, then it is continuously differentiable. Given any $y \in L_\infty(S, \mu)$, pick m and M in $\mathbb{R}$ with $m \leq y(s) \leq M$ almost everywhere. Because $\nabla\phi^*$ is uniformly continuous on $[m-1, M+1]$, for almost every s, given any $\varepsilon > 0$, there is a $\delta > 0$ such that $|\nabla\phi^*(y(s) + v) - \nabla\phi^*(y(s))| < \varepsilon$ whenever $|v| < \delta$. According to the mean value theorem, for some $v' \in (-\delta, \delta)$, we have

$$|\phi^*(y(s) + v) - \phi^*(y(s)) - v(\nabla\phi^*(y(s)))| = |v||\nabla\phi^*(y(s) + v') - \nabla\phi^*(y(s))|$$
$$\leq \varepsilon|v|,$$

thus, if $\|h\|_\infty \leq \delta$, one has

$$\left| I_{\phi^*}(y + h) - I_{\phi^*}(y) - \int_S h(s) \nabla\phi^*(y(s)) \, d\mu(s) \right| \leq \varepsilon \|h\|_\infty.$$

This shows that I_{ϕ^*} has Fréchet derivative $\nabla I_{\phi^*}(y) = \nabla\phi^*(y(\cdot))$ at y.
The converse is left as an exercise by considering constant functions. □

Lemma 6.4.3. *If $x \in L_\infty(S, \mu)$ with ess inf $x > \alpha$ and ess sup $x < \beta$, then $\partial I_\phi(x) \neq \emptyset$.*

Proof. See Exercise 6.4.3. □

For any $m \in \mathbb{N}$, define for a fixed $x \in L_1(S, \mu)$,

$$S_m := \left\{ s \in S : (-m) \vee \left(\alpha + \frac{1}{m}\right) \leq x(s) \leq \left(\beta - \frac{1}{m}\right) \wedge m \right\}.$$

Lemma 6.4.4. *If $\alpha < x(s) < \beta$ almost everywhere, then $\mu(S_m^c) \downarrow 0$ as $m \to \infty$.*

Proof. The sets S_m are nested and increasing with

$$\bigcup_{m=1}^{\infty} S_m = \{s : \alpha < x(s) < \beta\}.$$

Consequently, $\mu(S_m) \uparrow \mu(S)$ as $m \to \infty$. □

For any measurable subset T of S, we denote the restrictions of μ and x to T by $\mu|_T$ and x_T, and $I_\phi^T : L_1(T, \mu_T) \to (-\infty, +\infty]$ by $I_\phi^T(z) := \int_T \phi(z(s))\,d\mu(s)$. With this notation

Lemma 6.4.5. *Suppose $x_n \to x$ weakly in $L_1(S, \mu)$ and $I_\phi(x_n) \to I_\phi(x) < \infty$. Then for any measurable subset T of S, $x_n|_T \to x|_T$ weakly in $L_1(T, \mu|_T)$ and $I_\phi^T(x_n|_T) \to I_\phi^T(x|_T) < \infty$.*

Proof. The weak convergence of $x_n|_A$ to x_A follows by integrating with respect to functions zero on A^c for any measurable $A \subset S$. Because I_ϕ^T and $I_\phi^{T^c}$ are both weakly lsc, it then follows that

$$\liminf I_\phi^T(x_n|_T) \geq I_\phi^T(x|_T) \quad \text{and} \quad \liminf I_\phi^{T^c}(x_n|_{T^c}) \geq I_\phi^{T^c}(x|_{T^c}).$$

On the other hand, because $I_\phi(z) = I_\phi^T(z|_T) + I_\phi^{T^c}(z|_{T^c})$ for any z, one has

$$\limsup I_\phi^T(x_n|_T) = \limsup(I_\phi(x_n) - I_\phi^{T^c}(x_n|_{T^c}))$$
$$= I_\phi(x) - \liminf I_\phi^{T^c}(x_n|_{T^c})$$
$$\leq I_\phi(x) - I_\phi^{T^c}(x|_{T^c}) = I_\phi^T(x|_T),$$

from which the result follows. □

Lemma 6.4.6. *Suppose ϕ^* is differentiable everywhere on $\mathbb{R}$. If $x_n \to x$ weakly in $L_1(S, \mu)$, $I_\phi(x_n) \to I_\phi(x) < \infty$ and $\alpha < x(s) < \beta$ almost everywhere, then it follows that $\|x_n - x\|_1 \to 0$.*

Proof. Because $x_n \to x$ weakly, it follows that $\{x_n\}_{n=1}^\infty \cup \{x\}$ is weakly compact in $L_1(S, \mu)$. Let $\varepsilon > 0$ be fixed. According to the Dunford–Pettis theorem (6.3.2), there exists $\delta > 0$ so that if $\mu(T) \leq \delta$, then $\int_T |x_n(s)|\,d\mu(s) < \varepsilon$ for all n, and $\int_T |x(s)|\,d\mu(s) < \varepsilon$. By Lemma 6.4.4, there is an m with $\mu(S_m^c) \leq \delta$, and so $\int_{S_m^c} |x_n(s) - x(s)|\,d\mu(s) < 2\varepsilon$ for all n.

Now, Lemma 6.4.5 implies as $n \to \infty$, $x_n|_{S_m} \to x|_{S_m}$ weakly in $L_1(S_m, \mu|_{S_m})$ and $I_\phi^{S_m}(x_n|_{S_m}) \to I_\phi^{S_m}(x|_{S_m}) < \infty$, and certainly $x|_{S_m} \in L_\infty(S_m, \mu|_{S_m})$ with

$$\alpha < \operatorname{ess\,inf} x|_{S_m} \quad \text{and} \quad \operatorname{ess\,sup} x|_{S_m} < \beta.$$

6.4 Strongly rotund functions

Thus by Lemma 6.4.3, $\partial I_\phi^{S_m}(x|_{S_m}) \neq \emptyset$, so noting that I_{ϕ^*} is Fréchet differentiable by Lemma 6.4.2 we can apply Exercise 6.4.1 (on $L_1(S_m, \mu|_{S_m})$) to deduce that $\int_{S_m} |x_n(s) - x(s)| \, d\mu(s) \to 0$ as $n \to \infty$.

Finally, for all n one has

$$\|x_n - x\|_1 = \int_{S_m} |x_n(s) - x(s)| \, d\mu(s) + \int_{S_m^c} |x_n(s) - x(s)| \, d\mu(s)$$

$$< \int_{S_m} |x_n(s) - x(s)| \, d\mu(s) + 2\varepsilon.$$

This implies $\limsup \|x_n - x\|_1 \leq 2\varepsilon$, and thus the result follows. □

Theorem 6.4.7. *Suppose ϕ^* is differentiable everywhere on $\mathbb{R}$. If $x_n \to x$ weakly in $L_1(S, \mu)$ and $I_\phi(x_n) \to I_\phi(x) < \infty$, then $\|x_n - x\|_1 \to 0$.*

Proof. Because $I_\phi(x) < \infty$, $\alpha \leq x(s) \leq \beta$ almost everywhere and for all n sufficiently large $\alpha \leq x_n(s) \leq \beta$ almost everywhere. Define

$$S^\gamma := \{s \in S : \alpha < x(s) < \beta\}, \text{ and } S^\alpha := \{x \in S : x(s) = \alpha\}$$

$$S^\beta := \{s \in S : x(s) = \beta\}.$$

According to Lemma 6.4.5, $x_n|_{S^\gamma} \to x|_{S^\gamma}$ weakly in $L_1(S^\gamma, \mu|_{S^\gamma})$, and

$$I_\phi^{S^\gamma}(x_n|_{S^\gamma}) \to I_\phi^{S^\gamma}(x|_{S^\gamma}) < \infty,$$

so applying Lemma 6.4.6, $\int_{S^\gamma} |x_n(s) - x(s)| \, d\mu(s) \to 0$. But now for all n sufficient large,

$$\|x_n - x\|_1 = \int_{S^\gamma} |x_n(s) - x(s)| \, d\mu(s) + \int_{S^\alpha} |x_n(s) - x(s)| \, d\mu(s)$$

$$+ \int_{S^\beta} |x_n(s) - x(s)| \, d\mu(s)$$

$$= \int_{S^\gamma} |x_n(s) - x(s)| \, d\mu(s) + \int_{S^\alpha} (x_n(s) - \alpha) \, d\mu(s)$$

$$+ \int_{S^\beta} (\beta - x_n(s)) \, d\mu(s)$$

$$\to 0,$$

as $n \to \infty$. □

We are now ready for the central result of this section.

Theorem 6.4.8. *If ϕ^* is differentiable everywhere on $\mathbb{R}$, then I_ϕ is strongly rotund on $L_1(S, \mu)$. The converse is also true if (S, μ) is not purely atomic.*

Proof. Suppose ϕ^* is differentiable everywhere on $\mathbb{R}$. Then Exercise 5.3.14(c), implies ϕ is strictly convex on its domain. Moreover, because dom $\phi^* = \mathbb{R}$, Exercise 4.4.23 implies ϕ is supercoercive. Consequently I_ϕ is strictly convex on its domain (Lemma 6.4.1) and I_ϕ has weakly compact lower level sets (Proposition 6.3.3). Finally, Theorem 6.4.7 establishes the Kadec property.

Conversely, the strict convexity of ϕ follows from the strict convexity of I_ϕ by Lemma 6.4.1. Because I_ϕ has compact lower level sets, we deduce that ϕ^* is everywhere finite by Lemma 6.3.5. Because $\phi^{**} = \phi$ is strictly convex, Proposition 5.3.6(a) implies ϕ^* is differentiable. □

Exercises and further results

6.4.1.* Suppose $f : X \to (-\infty, +\infty]$ is lsc and f^* is Fréchet differentiable on the domain of ∂f^*. Suppose $x_n \to \bar{x}$ weakly and $f(x_n) \to f(\bar{x})$ and $\partial f(\bar{x}) \neq \emptyset$. Show that $\|x_n - \bar{x}\| \to 0$.

Hint. Let $\phi \in \partial f(\bar{x})$. Then $\bar{x} = \nabla f^*(\phi)$. Now $\langle x_n, \phi \rangle - f(x_n) \to \langle \bar{x}, \phi \rangle - f(\bar{x})$ and so $\|x_n - \bar{x}\| \to 0$. (Use Šmulian's theorem (4.2.10) or characterizations of strongly exposed points.) □

6.4.2.* Prove Lemma 6.4.1.

6.4.3.* Prove Lemma 6.4.3.

Hint. With x as in Lemma 6.4.3, let $\psi = \phi'_+ \circ x$, show that $\psi \in L_\infty(S)$; use the fact that ϕ'_+ is nondecreasing and bounded above and below on [ess inf x, ess sup x], see Theorem 2.1.2. Then show $\psi \in \partial I_\phi(x)$. This follows because $\phi'_+(a)(b-a) \leq \phi(b) - \phi(a)$ for any $\alpha < a < \beta$, therefore,

$$\psi(y) - \psi(x) = \int_S \psi(s)[y(s) - x(s)] \, d\mu(s) = \int_S \phi'_+(x(s))[y(s) - x(s)] \, d\mu(s)$$

$$\leq \int_S \phi(y(s)) - \phi(x(s)) \, d\mu(s) = I_\phi(y) - I_\phi(x).$$

For an alternate proof, see [94, Lemma 3.3]. □

6.4.4. Suppose X is a Banach space. Show that there exists an everywhere finite strongly rotund function on X if and only if X is reflexive.

Hint. For one direction, use the Baire category theorem to show X has a weakly compact set with nonempty interior. For the other consider $f(x) := \|x\|^2$ where $\|\cdot\|$ is a locally uniformly convex Fréchet differentiable norm on the reflexive space X; see [94, Corollary 4.3]. □

6.4.5. Let $\phi(t):=t \log t$, show that $I_\phi(x) = \int_S x(s) \log(x(s)) \, d\mu(s)$ is strongly rotund on $L_1(S, \mu)$ when S is a finite measure space. Hence (minus) the Boltzmann–Shannon entropy is strongly rotund.

6.4.6 (L_1-norm nonattainment). Show that

$$\inf \left\{ \int_0^1 |x(t)| \, dt : \int_0^1 t x(t) dt = 1, x(t) \geq 0 \text{ a.e.} \right\}$$

is not attained – $\|\cdot\|_1$ is very far from strongly rotund. Note that the problem is unchanged if the objective is replaced by the linear functional $\int_0^1 x(t)\,dt$.

6.4.7 (Strongly rotund functions and optimal value problems).★★ Let X be a Banach space, $f : X \to (-\infty, +\infty]$ a proper convex function with weakly compact lower level sets, and suppose $C_1, C_2, \ldots$ and C_∞ are closed convex subsets of X. Consider the optimization problems

$$(P_n) \quad \inf\{f(x) : x \in C_n\}, \quad n = 1, 2, \ldots, \infty,$$

and let $V(P_n) \in [-\infty, +\infty]$ denote the value of each problem. Consider the conditions (M1), (M2) from the definition of Mosco convergence (6.2.6) applied to C_n, C_∞, and

$$C_\infty \subset \bigcup_{m=1}^\infty \bigcap_{n=m}^\infty C_n. \tag{6.4.1}$$

(a) Suppose (M2) holds. Show that $\liminf_{n\to\infty} V(P_n) \geq V(P_\infty)$ (finite or infinite).
(b) Suppose (6.4.1) holds. Show that $\limsup_{n\to\infty} V(P_n) \leq V(P_\infty)$.
(c) Suppose (M2) holds, $V(P_n) \to V(P_\infty)$, and (P_∞) has a unique optimum, say x_∞. Show that for any sequence of optimum solutions x_n for (P_n), $x_n \to x_\infty$ weakly.
(d) Suppose (C_n) converges Mosco to C, and that $\mathrm{int}(\mathrm{dom} f) \cap C_\infty \neq \emptyset$. Show that $V(P_n) \to V(P_\infty) < \infty$. Furthermore, if f^* is Fréchet differentiable on the domain of ∂f^*, show that (P_n) and (P_∞) have unique optimal solutions x_n and x_∞, respectively for all n sufficiently large, and $\|x_n - x_\infty\| \to 0$.
(e) Suppose (M2) and (6.4.1) hold. Show that $V(P_n) \to V(P_\infty)$ (finite or infinite). Furthermore, if $V(P_\infty) < \infty$ and f is strongly rotund, show that (P_n) and (P_∞) have unique optimal solutions x_n and x_∞, respectively (for all n sufficiently large), and $\|x_n - x_\infty\| \to 0$.

Hint. See [94, pp. 147–149]. □

6.4.8 (Burg entropy nonattainment).★★ Letting $\phi(t) := -\log t$, it is easy to show that generally $I_\phi(x) = -\int_S \log(x(s))\,d\mu(s)$ is is not strongly rotund on $L_1(S, \mu)$ when S is a finite measure space. Hence (minus) the Burg entropy is not strongly rotund.

Indeed, suppose S is the cube in three dimensions and consider the simple spectral estimation problem

$$v(\alpha) := \inf \left\{ -\int_S \log x(t)\,dV : \int_S x(t)\,dV = 1, \right.$$
$$\left. \int_S x(t) \cos(2\pi u)\,dV = \int_S x(t) \cos(2\pi v)\,dV = \int_S x(t) \cos(2\pi w)\,dV = \alpha \right\}$$

where $t := (u, v, w)$ and $dV := du dv dw$. One can show that for $0 \leq \alpha < 1$ the value $v(\alpha)$ is finite but that there is an value $\bar{\alpha} \approx 0.34053\ldots$ such that for $\alpha > \bar{\alpha}$ the value is not attained and for $\alpha \leq \bar{\alpha}$ it is attained. Nothing changes in the qualitative appearance of the problem. Moreover the value is related to Watson integrals for so-called face centered cubic lattices [93]. More of the corresponding mathematical physics is outlined in [74, pp. 120-121].

6.4.9. ** A lovely companion result linking weak and norm convergence was given by Visintin in [427]. The main result of the paper is the following:

Theorem 6.4.9. *Let Ω be endowed with a σ-finite, complete measure and let $u_n \to u$ weakly in $L_1(\Omega)^N$. If $u(x)$ is an extremal point of the closed convex hull of $\{u_n(x) : n = 1, 2, \cdots\}$ a.e. in Ω, then $u_n \to u$ strongly in $L_1(\Omega)^N$.*

The phenomenon described in Theorem 6.4.9 is well known in the calculus of variations where, as in Theorem 6.4.8, a weakly convergent minimizing sequence is usually strongly convergent. The result is applied to partial differential equations including a nonlinear generalization of the Stefan problem.

Use Visintin's theorem to show that all weakly convergent minimizing sequences for

$$\min_{x \in C} I_\phi(x)$$

are norm convergent if ϕ is closed and strictly convex while C is closed and convex in $L_1(\Omega)^N$.

6.5 Trace class convex spectral functions

The useful patterns observed in Section 3.2 extend to some central classes of infinite-dimensional functions. We sketch the situation. Details are to be found in [109] and [121, §7.3]. Let $\ell_2^{\mathbb{C}}$ be the complex Hilbert sequence space, and consider the bounded self-adjoint operators on $\ell_2^{\mathbb{C}}$, denoted by B_{sa}. An operator $T \in B_{sa}$ is *positive* (denoted $T \geq 0$) if $\langle Tx, x \rangle \geq 0$ for all $x \in \ell_2^{\mathbb{C}}$. Each $T \in B_{sa}$ induces a unique positive operator $|T| = (T^*T)^{\frac{1}{2}} \in B_{sa}$ (see Exercise 6.5.1) and thence $T^{\pm} := (|T| \pm -T)/2$.

For $T \geq 0$ in B_{sa} the (possibly infinite) *trace* is given by

$$\text{tr}(T) := \sum_{i=1}^{\infty} \langle Te^i, e^i \rangle, \qquad (6.5.2)$$

which is independent of the orthonormal basis $\{e^i\}$ used. The trace is extended to B_{sa} using $\text{tr}(T) := \text{tr}(T^+) - \text{tr}(T^-)$. Let $\mathcal{U}$ denote the unitary operators on $\ell_2^{\mathbb{C}}$ (i.e. $U^* = U^{-1}$). Denote by B_0 the space of compact, self-adjoint operators, and by B_1 those with finite trace. The following is well known.

Theorem 6.5.1. *For all $T \in B_0$ there exists $U \in \mathcal{U}$ with $T = U^* \text{diag}(\lambda(T)) U$.*

6.5 Trace class convex spectral functions

Theorem 6.5.1 and the fact that $\operatorname{tr}(ST) = \operatorname{tr}(TS)$ makes proving *Lidskii's Theorem* easy for self-adjoint operators:

$$\operatorname{tr}(T) = \sum_{i=1}^{\infty} \lambda_i(T)$$

where $(\lambda_i(T))$ is any *spectral sequence* for T: any sequence of eigenvalues of T (with multiplicity), as in Euclidean space.

The self-adjoint *Hilbert–Schmidt* operators B_2 are those $T \in B_{sa}$ with $T^2 = T^*T \in B_1$. As sets $B_1 \subset B_2 \subset B_0 \subset B_{sa}$, The *Schatten p-spaces* $B_p \subset B_0$ are defined for $p \in [1, \infty)$ by placing $T \in B_p$ if $\|T\|_p := (\operatorname{tr}(|T|^p))^{1/p} < \infty$. When T is self-adjoint

$$\|T\|_p := \left(\operatorname{tr}(|T|^p)\right)^{1/p} = \left(\sum_{i=1}^{\infty} |\lambda_i(T)|^p\right)^{1/p}.$$

For $1 < p, q < \infty$ with $1/p + 1/q = 1$, B_p and B_q are paired, and the sesquilinear form $\langle S, T \rangle := \operatorname{tr}(ST)$ implements the duality on $B_p \times B_q$. We likewise pair B_0 with B_1. For each $x \in \ell_2^{\mathbb{C}}$ define the operator $x \odot x \in B_1$ by $(x \odot x)y = \langle x, y \rangle x$, and for $x \in \ell_\infty$ we define the operator $\operatorname{diag} x \in B_{sa}$ pointwise by

$$\operatorname{diag} x := \sum_{i=1}^{\infty} x_i(e^i \odot e^i).$$

For $1 \leq p < \infty$, if $x \in \ell_p$ then $\operatorname{diag} x \in B_p$ and $\|\operatorname{diag} x\|_p = \|x\|_p$. If $x \in c_0$, then $\operatorname{diag} x \in B_0$ and $\|\operatorname{diag} x\| = \|x\|_\infty$. This motivates:

Definition 6.5.2. (Spectral sequence space) *For $1 \leq p < \infty$ we say ℓ_p is the spectral sequence space for B_p and c_0 the spectral sequence space for B_0.*

Definition 6.5.3. (Paired Banach spaces) *We say that V and W are paired Banach spaces $V \times W$, if $V = \ell_p$ and $W = \ell_q$ and $1 \leq p, q \leq \infty$ satisfy $p^{-1} + q^{-1} = 1$ or $V = c_0$ and $W = \ell_1$ or vice versa. We denote the norms on V and W by $\|\cdot\|_V$ and $\|\cdot\|_W$ respectively.*

Similarly, we say $\mathcal{V}$ and $\mathcal{W}$ are paired Banach spaces $\mathcal{V} \times \mathcal{W}$ where $\mathcal{V} = B_p$ and $\mathcal{W} = B_q$ or where $\mathcal{V} = B_0$ (with the operator norm) and $\mathcal{W} = B_1$ (or vice versa). We denote the norms on $\mathcal{V}$ and $\mathcal{W}$ by $\|\cdot\|_\mathcal{V}$ and $\|\cdot\|_\mathcal{W}$ respectively.

We always take V, W to be the spectral sequence space for the operator space $\mathcal{V}, \mathcal{W}$. Thus, fixing $V \times W$ fixes $\mathcal{V} \times \mathcal{W}$ and vice versa.

Unitarily and rearrangement invariant functions. Operators S and T in B_{sa} are *unitarily equivalent* if there is a $U \in \mathcal{U}$ such that $U^*TU = S$. We say $\phi: \mathcal{V} \to (-\infty, +\infty]$ is *unitarily invariant* if $\phi(U^*TU) = \phi(T)$ for all $T \in \mathcal{V}, U \in \mathcal{U}$.

In Section 3.2 we showed that a unitarily invariant function ϕ can be represented as $\phi = f \circ \lambda$ where $f : \mathbb{R}^n \to (-\infty, +\infty]$ is a rearrangement invariant function and $\lambda: S^n \to \mathbb{R}^n$ is the spectral mapping. We now sketch an analogous result for unitarily invariant functions on $\mathcal{V}$. A function $f: V \to (-\infty, +\infty]$ is *rearrangement invariant* if $f(x_\pi) = f(x)$ for any rearrangement π. The definition of a spectral mapping is less straightforward.

We need to mimic arranging components of a vector in $\mathbb{R}^n$ in lexicographic order. Let $I_>(x) := \{i : x_i > 0\}, I_=(x) := \{i : x_i = 0\}$, and $I_<(x) := \{i : x_i < 0\}$. The mapping $\Phi: V \to V$ is defined for $x \in V$ as follows. We pick the largest positive component of x, next a 0 component, then the most negative component and so on. If any of the sets $I_>(x), I_=(x)$ or $I_<(x)$ is exhausted we skip the corresponding step. The outcome is $\Phi(x)$. We summarize useful properties of Φ [121, §7.3].

Proposition 6.5.4. *For each $x \in \ell_2$ there is a permutation π with $(\Phi(x))_i = x_{\pi(i)}$ for all $i \in \mathbb{N}$, and for $x, y \in \ell_2$ we have $\Phi(x) = \Phi(y)$ if and only if there exists a permutation π with $y_i = x_{\pi(i)}$ for all $i \in \mathbb{N}$. Moreover, $\Phi^2 = \Phi$ and $f: \ell_2 \to (-\infty, +\infty]$ is rearrangement invariant if and only if $f = f \circ \Phi$.*

Proof. Exercise 6.5.2. □

We now define the *eigenvalue mapping* $\lambda: V \to V$ as follows. For any $T \in V$ let $\mu(T)$ be any spectral sequence of T. Then $\lambda(T) := \Phi(\mu(T))$, gives us a canonical spectral sequence for each compact self-adjoint operator T.

Proposition 6.5.5. *The mapping λ is unitarily invariant, and $\Phi = \lambda \circ \mathrm{diag}$.*

Proof. Exercise 6.5.3. □

Moreover, λ and diag act as inverses in the following sense.

Proposition 6.5.6 (Inverses). *For $x \in \ell_2$, $(\lambda \circ \mathrm{diag})(x)$ is a rearrangement of x, and for $T \in V$, $(\mathrm{diag} \circ \lambda)(T)$ is unitarily equivalent to T.*

Proof. Exercise 6.5.4. □

For any rearrangement invariant $f: V \to (-\infty, +\infty]$ we have that $f \circ \lambda$ is unitarily invariant, and for any unitarily invariant $\phi: V \to (-\infty, +\infty]$ that $f \circ \mathrm{diag}$ is rearrangement invariant.

Theorem 6.5.7 (Unitary invariance). *Let $\phi: V \to (-\infty, +\infty]$. The following are equivalent: (a) ϕ is unitarily invariant; (b) $\phi = \phi \circ \mathrm{diag} \circ \lambda$; (c) $\phi = f \circ \lambda$ for some rearrangement invariant $f: V \to (-\infty, +\infty]$. If (c) holds then $f = \phi \circ \mathrm{diag}$.*

Proof. Exercise 6.5.5. □

Symmetrically we have:

Theorem 6.5.8 (Rearrangement invariance). *Let $f: V \to (-\infty, +\infty]$. The following are equivalent: (a) f is rearrangement invariant; (b) $f = f \circ \lambda \circ \mathrm{diag}$; (c) $f = \phi \circ \mathrm{diag}$ for some unitarily invariant $\phi: V \to (-\infty, +\infty]$. If (c) holds then $\phi = f \circ \lambda$.*

Proof. Exercise 6.5.6. □

We now have the following useful formulas: $\|\mathrm{diag}(x)\|_V = \|x\|_V$, for all $x \in V$ and $\|\lambda(T)\|_V = \|T\|_V$, for all $T \in V$ (see Exercise 6.5.7).

6.5 Trace class convex spectral functions

Conjugacy of unitarily invariant functions. Proofs of the following results are in [109] and [121, §7.3]. Note their fidelity to those of Section 3.2.

Theorem 6.5.9 (Conjugacy). *Let* $\phi\colon \mathcal{V} \to (-\infty, +\infty]$ *be unitarily invariant. Then* $\phi^* \circ \mathrm{diag} = (\phi \circ \mathrm{diag})^*$.

Corollary 6.5.10 (Convexity). *Let* $\phi\colon \mathcal{V} \to (-\infty, +\infty]$ *be unitarily invariant. Then* ϕ *is proper, convex, and lsc if and only if* $\phi \circ \mathrm{diag}$ *is proper, convex, and lsc.*

Theorem 6.5.11. *Let* $f\colon \mathcal{V} \to (-\infty, +\infty]$ *be rearrangement invariant. Then* $(f \circ \lambda)^* = f^* \circ \lambda$.

Theorem 6.5.12. *Let* $\phi\colon \mathcal{V} \to (-\infty, +\infty]$ *be unitarily invariant. Then* $\phi = f \circ \lambda$ *for a rearrangement invariant* $f\colon \mathcal{V} \to (-\infty, +\infty]$. *Furthermore* $\phi^* = f^* \circ \lambda$.

Let $f\colon \mathcal{V} \to (-\infty, +\infty]$ *be rearrangement invariant. Then* $f = \phi \circ \mathrm{diag}$ *for a unitarily invariant* $\phi\colon \mathcal{V} \to (-\infty, +\infty]$. *Furthermore* $f^* = \phi^* \circ \mathrm{diag}$.

Subdifferentials of unitarily invariant functions. Key to the analysis is the following nontrivial result:

Theorem 6.5.13 (Commutativity). *Let* $f\colon \mathcal{V} \to (-\infty, +\infty]$ *be unitarily invariant lsc and convex. If* $T \in \partial f(S)$ *for* $(S, T) \in \mathcal{V} \times \mathcal{W}$ *then* $TS = ST$.

Theorem 6.5.14 (Convex subgradient). *Let* $f\colon \mathcal{V} \to (-\infty, +\infty]$ *be a unitarily invariant lsc convex function. For* $(S, T) \in \mathcal{V} \times \mathcal{W}$ *we have* $T \in \partial f(S)$ *if and only if there exist* $U \in \mathcal{U}$ *and* $(x, y) \in \mathcal{V} \times \mathcal{W}$ *with* $S = U^*(\mathrm{diag}\, x)U$, $T = U^*(\mathrm{diag}\, y)U$ *and* $y \in \partial(f \circ \mathrm{diag})(x)$.

The final result shows that differentiability of a convex unitarily invariant function f is again characterized by that of $f \circ \mathrm{diag}$.

Theorem 6.5.15. *Let* $f\colon \mathcal{V} \to (-\infty, +\infty]$ *be a unitarily invariant lsc convex function. Then* f *is Gâteaux differentiable at* $A \in \mathcal{V}$ *if and only if* $f \circ \mathrm{diag}$ *is Gâteaux differentiable at* $\lambda(A) \in V$.

We illustrate these spectral results with a couple of examples.

Example 6.5.16. The standard norm on ℓ_p for $1 < p < \infty$ is Gâteaux differentiable away from zero. Immediately by Theorem 6.5.15, so is the Schatten norm of B_p since $\lambda(A) = 0$ implies $A = 0$. Theorem 6.5.18 below extends this to Fréchet differentiability.

We next consider the Calderón norm.

Example 6.5.17. For $1 < p < \infty$ and q satisfying $1/p + 1/q = 1$ the symmetric norm defined by

$$|||x|||_p := \sup_n \left(n^{-\frac{1}{q}} \sum_{i=1}^{n} \lambda_i(\mathrm{diag}\, x) \right)$$

is the *Calderón norm* on ℓ_p. Clearly, this is rearrangement invariant and continuous on the sequence space, so $|||T|||_{\mathcal{V},p} := |||\lambda(T)|||_p$ induces a unitarily invariant function on B_p. Now apply Theorem 6.5.12 to get (as with Schatten p-norms) the very pleasant dual norm formula

$$(|||\cdot|||_{\mathcal{V},p})^* = (|||\cdot|||_p \circ \lambda)^* = (|||\cdot|||_p)^* \circ \lambda = |||\cdot|||_{\mathcal{V},q}.$$

A generalization can be found in Exercise 6.5.9.

The Fréchet case. With a good deal more work, Fréchet differentiability of spectral operators is similarly induced from rearrangement invariant functions. This requires a carefully constructed spectral map $\widehat{\lambda}(T)$ described in [101, §2]. The key nonexpansivity property given in [101, Theorem 2.3] is that

$$\|\widehat{\lambda}(T) - \widehat{\lambda}(S)\|_p \leq \|T - S\|_p$$

for $S, T \in B_p$ (or in B_0). By contrast λ is not even always continuous. This leads to:

Theorem 6.5.18 (Fréchet differentiability [101]). *Suppose that $V := \ell_p$ for $1 < p < \infty$ or $V := c_0$. Let ϕ be a rearrangement invariant proper closed convex function on V. Then $\phi \circ \widehat{\lambda}$ is norm-continuous or Fréchet differentiable at $T \in \mathcal{V}$ if and only if ϕ is norm-continuous or Fréchet differentiable at $\widehat{\lambda}(T)$.*

Moreover, when ϕ is insensitive to zero-eigenvalues, $\phi \circ \widehat{\lambda} = \phi \circ \lambda$.

Exercises and further results

We have seen that the central results of Section 3.2 extend neatly to compact self-adjoint operators.

6.5.1 (Square-root iteration).** Show that each positive semidefinite symmetric and bounded linear operator T on a Hilbert space has a unique positive semidefinite square root $T^{1/2}$. Show, in addition, that $T^{1/2}$ commutes with all bounded linear operators that commute with T.

Hint. Consider the iteration of Exercise 3.2.4. In the compact setting the proof in Exercise 3.2.4 does not change significantly; but the iteration converges as required in full generality [28, §23.1]. □

6.5.2.* Prove Proposition 6.5.4.
6.5.3.* Prove Proposition 6.5.5.
6.5.4.* Prove Proposition 6.5.6.
6.5.5.* Prove Theorem 6.5.7.
6.5.6.* Prove Theorem 6.5.8.
6.5.7. Let V and $\mathcal{V}$ be as in Definition 6.5.3. Prove that

$$\|\operatorname{diag}(x)\|_{\mathcal{V}} = \|x\|_V, \text{ for all } x \in V, \qquad \|\lambda(T)\|_V = \|T\|_{\mathcal{V}}, \text{ for all } T \in \mathcal{V}.$$

Our next exercise revisits the k-th largest eigenvalue; but now of a positive self-adjoint operator.

6.5.8 (*k*-th largest eigenvalue).** Consider $0 < S \in B_p$. Denote the k-th largest eigenvalue of S by $\mu_k(S)$. Then $\mu_k = \phi_k \circ \lambda$ where $\phi_k : \ell_2 \to \mathbb{R}$ is defined to be the k-th largest component of x. Show μ_k is locally Lipschitz on B_p.

Hint. As in the matrix case express ϕ_k as the difference of two convex continuous permutation invariant functions. Then σ_k is continuous, permutation invariant and convex and $\phi_k = \sigma_k - \sigma_{k-1}$.

Thus, ϕ_k is a locally Lipschitz permutation invariant function on ℓ_p, and therefore μ_k is a unitarily invariant locally Lipschitz function on B_p: μ_k is DC with continuous finite components by Theorem 6.5.18. □

6.5.9 (Calderón norms).** For any x of c_0, rearrange the components of $(|x_i|)$ into decreasing order to obtain a new element $\bar{x}$ of c_0.

(a) Show that for a fixed decreasing sequence $t_i \to 0$ ($t_i \geq 0$, not all zero)

$$x \in V \to \sup_{n \in \mathbb{N}} \left\{ \frac{\sum_{i=1}^n \bar{x}_i}{\sum_{i=1}^n t_i} \right\},$$

and

$$y \in W \to \sum_{i=1}^\infty t_i \bar{y}_i$$

are a pair of dual norms;
(b) Show their compositions with λ are dual norms on the spaces V and W.
(c) if $t_i := i^{1/q} - (1-i)^{1/q}$ $1 < q < \infty$ the pair are classical *Calderón norms*.

6.6 Deeper support structure

Let A be a subset of a Banach space X, recall that the *support function of A*, denoted by σ_A is defined on X^* by $\sigma_A(\phi) = \sup\{\phi(x) : x \in A\}$, where $\phi \in X^*$.

Lemma 6.6.1. *Let C be a closed convex bounded subset of X. Then $x_0 \in C$ is strongly exposed by $\phi \in X^*$ if and only if σ_C is Fréchet differentiable at ϕ, and $\sigma'_C(\phi) = x_0$.*

Proof. Suppose $x_0 \in C$ is strongly exposed by $\phi \in X^*$. Then $\phi(x_0) = \sup_C \phi$, and this implies $x_0 \in \partial \sigma_C(\phi)$ (Exercise 6.6.11). Now suppose $x_n \in \partial_{\varepsilon_n} \sigma_C(\phi)$ where $\varepsilon_n \to 0^+$. Using the separation theorem (4.1.17) we deduce that $x_n \in C$. Now for any $\Lambda \in X^*$,

$$\langle x_n, \Lambda - \phi \rangle \leq \sigma_C(\Lambda) - \sigma_C(\phi) + \varepsilon_n \text{ which implies}$$
$$\langle x_n, \Lambda - \phi \rangle \leq \sigma_C(\Lambda) - \phi(x_0) + \varepsilon_n.$$

Because this is true for $\Lambda = 0$, the previous inequality implies that $\phi(x_n) \to \phi(x_0)$. Because x_0 is strongly exposed by ϕ, this means $\|x_n - x_0\| \to 0$, and so the result follows from Šmulian's theorem (4.2.10).

Conversely, suppose σ_C is Fréchet differentiable at ϕ. Then $\partial \sigma_C(\phi) = \{x_0\}$ for some $x_0 \in X$ (see Exercise 4.2.11). Then it is straightforward to verify that $x_0 \in C$, and

$\sigma_C(\phi) = \phi(x_0)$ (Exercise 6.6.11). Suppose that $x_n \in C$ is such that $\phi(x_n) \to \phi(x_0)$, then $x_n \in \partial_{\varepsilon_n}\sigma_C(\phi)$ where $\varepsilon_n = \phi(x_0) - \phi(x_n)$, and then Šmulian's theorem (4.2.10) implies that $\|x_n - x_0\| \to 0$ which shows that ϕ strongly exposes C at x_0. □

Proposition 6.6.2. *Let C be a closed bounded convex subset of a Banach space X. Then the following are equivalent.*

(a) *The strongly exposing functionals of C form a dense G_δ-subset of X^* and C is the closed convex hull of its strongly exposed points.*
(b) *σ_C is Fréchet differentiable on a dense G_δ-subset of X^*.*

Proof. (a) ⇒ (b): This is an immediate consequence of Lemma 6.6.1.
(b) ⇒ (a): Lemma 6.6.1 shows that if σ_C is Fréchet differentiable at $\phi \in X^*$, then $\sigma'_C(\phi) = x_0$ where $x_0 \in C$ and ϕ strongly exposes x_0 and these are the only strongly exposing functionals of C. Thus the strongly exposing functionals of C form a dense G_δ-subset in X. Now suppose C_1 is the closed convex hull of the strongly exposed points of C and $C_1 \neq C$. Then there is an $x_0 \in C \setminus C_1$. Because G, the set of strongly exposing functionals of C is dense in X^*, the separation theorem (4.1.17) ensures that there is a $\phi \in G$ such that $\phi(x_0) > \sigma_{C_1}(\phi)$ which is a contradiction with the fact that ϕ is a strongly exposing functional of C. Hence C is the closed convex hull of its strongly exposed points. □

6.6.1 Asplund spaces

Recall that a Banach space X is an *Asplund space* provided each continuous convex function on X is Fréchet differentiable on a dense G_δ-subset of X.

Lemma 6.6.3. *If X is an Asplund space, then every nonempty bounded subset of X^* admits weak*-slices of arbitrarily small diameter.*

Proof. Let A be a nonempty and bounded subset of X^*, define the sublinear functional $p(x) := \sigma_A(x) = \sup\{x^*(x) : x^* \in A\}$. Then $\sigma_A(x) \leq M\|x\|$ where $M > 0$ is chosen so that $A \subset MB_{X^*}$. Now suppose every weak*-slice of A has diameter greater than some $\varepsilon > 0$. Then given $x \in X$, the slice $S(x, A, \varepsilon/3n)$ has diameter greater than ε. Thus we choose $x_n^*, y_n^* \in S(x, A, \varepsilon/n)$ such that $\|x_n^* - y_n^*\| > \varepsilon$, it simple to check $x_n^*, y_n^* \in \partial_{\varepsilon/n} p(x)$ and so by Šmulian's theorem (4.2.10) p is not Fréchet differentiable at x. □

Theorem 6.6.4. *A separable Banach space X is an Asplund space if and only if X^* is separable.*

Proof. If X^* is separable, then X is an Asplund space by Theorem 4.6.6.
Conversely, suppose X^* is not separable. Then neither is B_{X^*} and so there is an uncountable ε-net, say $(x_\alpha^*) \subset B_{X^*}$ for some $\varepsilon > 0$. That is $\|x_\alpha^* - x_\beta^*\| > \varepsilon$ if $\alpha \neq \beta$. Because X is separable it follows that B_{X^*} is metrizable in its weak*-topology. Thus A has at most countably many points which are not weak*-condensation points. Now

6.6 Deeper support structure

let B denote the set of the weak*-condensation points of A. Consequently any weak*-slice of B must contain at least two distinct points from (x_α^*) and so its diameter is at least ε. □

Lemma 6.6.5. *Suppose X is a Banach space, and $f : X \to \mathbb{R}$ is a continuous convex function. Let $A \subset X^*$ be such that every bounded subset of A is weak*-denotable. Let G_n be the set of $x \in X$ such that there is an open neighborhood V of x such that $\operatorname{diam}(\partial f(V) \cap A) < 1/n$ where by convention we let the diameter of the empty set be 0. Then G_n is a dense open subset of X.*

Proof. Clearly, G_n is open by definition, and now we will show it is dense. Let $x \in X$ and let U be any open neighborhood of x. Because f is continuous, by replacing U with a smaller neighborhood if necessary, we may and do assume that $\partial f(U)$ is bounded. Now let $D := \partial f(U) \cap A$. If D is empty there is nothing further to do as $x \in G_n$. Otherwise, by hypothesis, let

$$S := \{\phi \in D : \phi(z) > \alpha\}$$

be a slice of D whose diameter is less than $1/n$. Suppose $\phi \in S$. Then $\phi \in \partial f(x_0)$ for some $x_0 \in U$ and $\phi(z) > \alpha$. Now fix $r > 0$ so that $x_1 := x_0 + rz \in U$. Let $\Lambda \in \partial f(x_1)$. Then

$$\Lambda(x_1 - x_0) \geq f(x_1) - f(x_0) \geq \phi(x_1 - x_0)$$

which implies $r\langle \Lambda - \phi, z\rangle \geq 0$. Consequently, $\Lambda(z) \geq \phi(z) > \alpha$. Because the subdifferential map is norm-to-weak* upper-semicontinuous (Proposition 6.1.1) it follows that there exists $\delta > 0$ such that $B_\delta(x_1) \subset U$, and $\partial f(y) \subset \{x^* \in X^* : x^*(z) > \alpha\}$ for any $y \in B_\delta(x_1)$. Now, $\partial f(B_\delta(x_1)) \cap A$ is a possibly empty subset of S that has diameter less than $1/n$. Thus, $x_1 \in G_n \cap U$, and the density of G_n has been established. □

Theorem 6.6.6. *A Banach space X is an Asplund space if and only if every nonempty bounded subset of X^* admits weak*-slices of arbitrarily small diameter.*

Proof. ⇒: This follows from Lemma 6.6.3.

⇐: Suppose $f : X \to \mathbb{R}$ is a continuous convex function. Let G_n be as in Lemma 6.6.5 where we take A therein to be X^*. Now let $x \in G$ where $G := \bigcap G_n$. Let (x_k) be a sequence converging to x. Let $\phi_n \in \partial f(x_k)$; Now fix k_n so that $x_k \in G_n$ for $k \geq k_n$. Because we are applying Lemma 6.6.5 with $A = X^*$, we conclude that $\|\phi_k - \phi\| < 1/n$ for $k \geq k_n$, and so $\phi_n \to \phi$. According to Šmulian's theorem (4.2.10), f is Fréchet differentiable at each x in the dense G_δ-set G. □

Corollary 6.6.7. *Let X be a Banach space, then the following are equivalent.*

(a) *X is an Asplund space.*
(b) *Given any convex open subset U of X, and any continuous convex function $f : U \to \mathbb{R}$, f is Fréchet differentiable on a dense G_δ-subset of U.*
(c) *Every equivalent norm on X is Fréchet differentiable at least one point.*

Proof. (a) $\Rightarrow$ (b) is Exercise 4.6.11, while (b) $\Rightarrow$ (c) is obvious.

Let us now prove (c) $\Rightarrow$ (a) using a contrapositive argument. Suppose X is not an Asplund space, then there is a bounded nonempty subset A of X^* and an $\varepsilon > 0$ so that every weak*-slice of A has diameter greater than ε. Let $C := A \cup -A$, and let B be the weak*-closed convex hull of $C + B_{X^*}$. It is left as Exercise 6.6.8 to verify that B is the ball of an equivalent dual norm on X^*, and that the predual norm on X is nowhere Fréchet differentiable. □

Proposition 6.6.8. *A closed subspace of an Asplund space is an Asplund space.*

Proof. Let Y be a closed subspace of an Asplund space X. According to Theorem 6.6.6 it suffices to show that every bounded nonempty subset A of $Y^* = X^*/Y^\perp$ has weak*-slices of arbitrarily small diameter. Without loss of generality we may assume that A is weak*-compact and convex. The quotient map $Q : X^* \to Y^*$ is norm one, onto and weak*-to-weak* continuous. Let $\varepsilon > 0$. Because Q is an open map, $Q(B_{X^*})$ contains a neighborhood of the origin in Y^*. Because A is bounded, there is a $\lambda > 0$, so that $Q(\lambda B_{X^*}) = \lambda Q(B_{X^*}) \supset A$. According to Zorn's lemma there exists a minimal (under inclusion) set with these properties. Let C_1 be such a minimal set. Because X is an Asplund space, there is a weak*-slice $S := S(x, C_1, \alpha)$ of C_1 of diameter less than ε. Because S is relatively weak*-open, the set $A_1 := Q(C_1 \setminus S)$ is a weak*-compact convex set, which according to the minimality of C_1 is properly contained in A. If $x_1^*, x_2^* \in A \setminus A_1$, there exists $y_1^*, y_2^* \in S$ such that $Q(y_i^*) = x_i^*$ and

$$\|x_1^* - x_2^*\| = \|Q(y_1^* - y_2^*)\| \leq \|y_1^* - y_2^*\| < \varepsilon.$$

Thus $\operatorname{diam}(A \setminus A_1) \leq \varepsilon$. The weak*-separation theorem (4.1.22) implies there exists a weak*-slice of A which misses A_1, and hence has diameter at most ε, and we are done. □

The following outlines a separable reduction argument that allows us to characterize nonseparable Asplund spaces via their separable subspaces.

Theorem 6.6.9. *A Banach space X is an Asplund space if every separable closed subspace Y of X is an Asplund space.*

Proof. We sketch the details as given in [350, Theorem 2.14, p. 23]. Suppose f is continuous and convex on the nonempty open convex subset D of X and suppose that the set G of points $x \in D$ where f is Fréchet differentiable is not dense in D. We will construct a separable subspace Y of X such that $Y \cap D \neq \emptyset$, and the points of Fréchet differentiability of $f|_Y$ are not dense in $Y \cap D$. For each n, let

$$G_n(f) := \left\{ x \in D : \sup_{\|h\|=1} \frac{f(x+\delta h) + f(x-\delta h) - 2f(x)}{\delta} < \frac{1}{n}, \text{ for some } \delta > 0 \right\}.$$

Now $G := \bigcap G_n(f)$, and thus for some $m \in \mathbb{N}$, $G_m(f)$ is not dense in D. Thus we let U be a nonempty open subset of $D \setminus G_m(f)$. Next, we construct an increasing sequence Y_k of separable subspaces of X. First, fix $x_1 \in U$. It follows that there exists a sequence $(h_{1,j}) \subset S_X$ such that for all $\delta > 0$,

$$\sup_j \frac{f(x_1 + \delta h_{1,j}) + f(x_1 - \delta h_{1,j}) - 2f(x_1)}{\delta} \geq \frac{1}{2m}.$$

Let Y_1 be the closed linear span of x_1 and $\{h_{1,j}\}_{j=1}^\infty$. Then Y_1 is separable, and $x_1 \in Y_1 \cap U$. At this stage, $f|_{Y_1}$ fails to be differentiable at $x_1 \in Y_1$, and thus $x_1 \notin G_{2m}(f|_{Y_1})$. Now if $Y_1 \subset Y_2 \subset \ldots \subset Y_n$ have been chosen, one can choose a countable dense subset $\{x_{n,i}\}_{i=1}^\infty$ in $Y_n \cap U$, and then a countable set of directions $\{h_{n,k,i}\} \subset S_X$ to ensure that $x_{n,i} \notin G_{2m}(f|_{Y_{n+1}})$ where Y_{n+1} is the closed linear span of Y_n and $\{h_{n,k,i}\}$. Letting $Y := \bigcup Y_n$ one can then show that $f|_Y$ is not Fréchet differentiable at any point of $Y \cap U$. It is left to the reader to more rigorously fill in the details. □

Putting Theorem 6.6.4, Proposition 6.6.8 and Theorem 6.6.9 together, we obtain

Corollary 6.6.10. *A Banach space is an Asplund space if and only if every separable subspace has a separable dual if and only if each of its subspaces is an Asplund space.*

6.6.2 Radon–Nikodým and Krein–Milman properties

A nonempty subset of a Banach space is said to be *dentable* if it admits slices of arbitrarily small diameter. A Banach space X is said to have the *Radon–Nikodým property* (RNP) if every bounded nonempty subset of X is dentable. Further we shall say a nonempty bounded subset A of X has the *Radon–Nikodým property* if all of its nonempty subsets are dentable.

First, we state some basic facts about dentability and the RNP.

Fact 6.6.11. *Let X be a Banach space and suppose A is a nonempty subset of X.*

(a) *If A has a strongly exposed point, then A is dentable.*
(b) *If A is not dentable, then neither are $\mathrm{conv}(A)$, $\overline{A}$, $A \cup (-A)$, $A + B_X$.*
(c) *A has the RNP if and only if $\overline{\mathrm{conv}}(A)$ has the RNP if and only if each closed convex subset of $\overline{\mathrm{conv}}(A)$ has the RNP.*

In particular, a closed bounded convex set C has the RNP if each of its closed convex subsets has a strongly exposed point, and a Banach space has the RNP if and only if each of its bounded closed convex subsets has the RNP.

Proof. See Exercise 6.6.10 □

A Banach space X is said to be a *weak*-Asplund space* if every continuous weak*-lsc convex function on X^* is Fréchet differentiable on a dense G_δ-set.

Theorem 6.6.12. *Suppose X is a Banach space and C is a closed bounded convex subset of X. Then C has the RNP if and only if for each proper lsc bounded below function $f : C \to (-\infty, +\infty]$, f^* is Fréchet differentiable on a dense G_δ-subset of X^*.*

Proof. ⇐: Let $D \subset C$ be closed and convex, then $\delta_D^* = \sigma_D$ is Fréchet differentiable on a dense G_δ-subset of X^*, and so D has a strongly exposed point by Lemma 6.6.1. According to Fact 6.6.11, C has the RNP.

⇒: Let $K > 0$ be such that $C \subset KB_X$. Because f is bounded below, there exists $b \in \mathbb{R}$ so that $f \geq \delta_C + b$. Therefore, $f^* \leq \delta_C^* - b$ and so f^* has Lipschitz constant K on X^*. Now let G_n be the set of $\phi \in X^*$ for which there is an open neighborhood V of ϕ so that $\partial f^*(V) \cap C$ has diameter at most $1/n$. According to Lemma 6.6.5, G_n is a dense open subset of X^*. Also, let $\hat{f} := f^{**}|_X$. Then $\operatorname{dom}\hat{f} \subset C$ and $(\hat{f})^* = f^*$. Let $\phi \in \bigcap_{n=1}^\infty G_n$ and suppose f^* is not Fréchet differentiable at ϕ. Then there exist $y_n^* \in S_{X^*}$, $\varepsilon > 0$ and $t_n \to 0^+$ so that

$$f^*(\phi + t_n y_n^*) + f^*(\phi - t_n y_n^*) - 2f^*(\phi) \geq \varepsilon t_n.$$

The Brøndsted–Rockafellar theorem (4.3.2) implies $\overline{\operatorname{range} \partial \hat{f}} \supset \operatorname{dom} f^*$ and so we choose $\phi_n, \psi_n \in \operatorname{range} \partial \hat{f}$, say $\phi_n \in \partial\hat{f}(x_n)$ and $\psi_n \in \partial\hat{f}(y_n)$ so that

$$\|\phi_n - (\phi + t_n y_n^*)\| \leq \frac{\varepsilon t_n}{8K} \qquad \|\psi_n - (\phi - t_n y_n^*)\| \leq \frac{\varepsilon t_n}{8K}.$$

Because f^* has Lipschitz constant K, this implies

$$f^*(\phi_n) + f^*(\psi_n) - 2f^*(\phi) \geq \frac{3\varepsilon t_n}{4}.$$

Then the Fenchel–Young equality (Proposition 4.4.1(a)) implies

$$\phi_n(x_n) - f(x_n) + \phi_n(y_n) - f(y_n) - 2f^*(\phi) \geq \frac{3\varepsilon t_n}{4}.$$

Now using the fact that $x_n, y_n \in KB_X$, this implies

$$(\phi + t_n y_n^*)(x_n) - f(x_n) + (\phi - t_n y_n^*)(y_n) - 2f^*(\phi) \geq \frac{\varepsilon t_n}{2}.$$

This last inequality along with the fact $\phi(x_n) - f(x_n) \leq f^*(\phi)$ and $\phi(y_n) - f(y_n) \leq f^*(\phi)$ implies $\|x_n - y_n\| \geq \varepsilon/2$. Because $(\phi_n) \to \phi$ and $(\psi_n) \to \phi$ this contradicts that $\phi \in G_n$ when $n > 2/\varepsilon$. □

We now list several corollaries of the preceding result.

Corollary 6.6.13. *Let C be a closed bounded convex subset of a Banach space X. Then C has the RNP if and only if every closed convex subset of C is the closed convex hull of its strongly exposed points and its strongly exposing functionals form a dense G_δ-subset of X^*.*

Proof. Suppose C has the RNP, and let D be a closed convex subset of C. Then $\sigma_D = \delta_D^*$, and so Theorem 6.6.12 implies that $\sigma_D : X^* \to \mathbb{R}$ is Fréchet differentiable on a dense G_δ-subset of X^*. According to Proposition 6.6.2, D is the closed convex hull of its strongly exposed points. The converse follows easily from Fact 6.6.11. □

6.6 Deeper support structure

Theorem 6.6.14. *For a Banach space X, the following are equivalent.*

(a) *X has the RNP.*
(b) *Every weak*-lsc Lipschitz convex function on X^* is Fréchet differentiable on a dense G_δ-set.*
(c) *X^* is a weak*-Asplund space.*
(d) *For every closed convex subset C of X, the strongly exposing functionals of C are a dense G_δ-subset in X^*, and/or C is the closed convex hull of its strongly exposed points.*

Proof. (a) $\Rightarrow$ (b): Let $f : X^* \to \mathbb{R}$ have Lipschitz constant K and be weak*-lsc. Then $\hat{f} = (f^*|_X)^*$ by Proposition 4.4.2(b), and $\text{dom}(f^*) \subset KB_{X^{**}}$ according to Proposition 4.4.6. Consequently, $\text{dom}\hat{f} \subset KB_X$, and $\hat{f}$ is bounded below because it is a lsc convex function with bounded domain. According to the Theorem 6.6.12, $f = (\hat{f})^*$ is Fréchet differentiable on a dense G_δ-subset of X^*.

For the remaining implications, observe that (b) $\Rightarrow$ (c) is similar to Exercise 4.6.11, while (c) $\Rightarrow$ (d) follows from Corollary 6.6.13, and (d) $\Rightarrow$ (a) is in Fact 6.6.11. □

Before presenting versions of Stegall's variational principle, we present a simple general variational principle.

Proposition 6.6.15. *Suppose that X is a Banach space and $f : X \to (-\infty, +\infty]$ is a proper lsc function such that f^* is Fréchet differentiable at $\phi \in X^*$, then*

(a) *$(f^*)'(\phi) = x_0$ where $x_0 \in \text{dom} f$, and*
(b) *$(f - \phi)$ attains its strong minimum at x_0.*

Proof. First, Exercise 4.4.2, implies $(f^*)'(x_0^*) = x_0 \in X$ and $f^{**}(x_0) = f(x_0)$, and the Fenchel–Young equality (Proposition 4.4.1(a)) ensures that $f^{**}(x_0) < \infty$. This shows (a), and moreover implies that $f^*(\phi) = \phi(x_0) - f(x_0)$. Now suppose $(f - \phi)(x_n) \leq (f - \phi)(x_0) + \varepsilon_n$ where $\varepsilon_n \to 0^+$. Then $(\phi - f)(x_n) \geq f^*(\phi) - \varepsilon$ and so

$$\langle y^* - \phi, x_n \rangle = \langle y^* - f, x_n \rangle - \langle \phi - f, x_n \rangle$$
$$\leq f^*(y^*) - (f^*(\phi) + \varepsilon_n).$$

Therefore, $x_n \in \partial_{\varepsilon_n} f^*(\phi)$. Because f^* is Fréchet differentiable at ϕ, Šmulian's theorem (4.2.10) implies $\|x_n - x_0\| \to 0$ as desired. □

Corollary 6.6.16. *Suppose that X is a Banach space with the RNP and that $f : X \to (-\infty, +\infty]$ is a lsc function for which there exist $a > 0$ and $b \in \mathbb{R}$ such that $f(x) \geq a\|x\| + b$ for all $x \in X$. Then there is an $\varepsilon > 0$ so that the set*

$$\{x^* \in \varepsilon B_{X^*} : f - x^* \text{ attains its strong minimum on } X\}$$

is residual in εB_{X^*}.

Proof. The growth condition implies that f^* is bounded on a neighborhood of the origin (Fact 4.4.9) and hence is continuous on a neighborhood εB_{X^*} of the origin

(Proposition 4.1.4). Thus f^* is Fréchet differentiable on a dense G_δ-subset G of εB_{X^*}. By Proposition 6.6.15, $f - x^*$ attains its strong minimum at $(f^*)'(x^*) \in \operatorname{dom} f$ for each $x^* \in G$. □

Corollary 6.6.17 (Stegall's variational principle). *Suppose $C \subset X$ is a nonempty closed bounded convex set with the RNP, and suppose that $f : C \to \mathbb{R}$ is a lsc function on C that is bounded below. Then the set*

$$S = \{x^* \in X^* : f - x^* \text{ attains its strong minimum on } X\}$$

is residual in X^*.

Proof. According to Theorem 6.6.12, f^* is Fréchet differentiable on a dense G_δ-subset of X^*. Hence, like the previous corollary the result follows from Proposition 6.6.15. □

We close this section by mentioning some related results without proof.

Theorem 6.6.18. *A Banach space X is an Asplund space if and only if X^* has the RNP.*

We shall not prove this theorem, but let us note that Theorem 6.6.6 immediately implies X^* has the RNP if X is an Asplund space. A nice proof of the converse is given in [350, pp. 80–82].

A Banach space X is said to have the *Krein–Milman property* (KMP) if every closed convex subset of X is the closed convex hull of its extreme points. Because strongly exposed points are extreme points, it is clear that a Banach space with the RNP has the KMP. However, the converse is still not resolved. Nevertheless, there are partial results, for example if X^* is a dual space with the KMP, then it has the RNP (see [126, Theorem 4.4.1]).

Exercises and further results

6.6.1. Use Stegall's variational principle to give an alternate proof of the following. Suppose that $C \subset X$ is a nonempty bounded closed convex set with the RNP. Then C is the closed convex hull of its strongly exposed points. Moreover, the functionals which strongly expose points of C form a dense G_δ-subset of X^*.

Hint. For the second assertion, let

$$G_n := \{x^* \in X^* : \operatorname{diam} S(x^*, C, \alpha) < 1/n \text{ for some } \alpha > 0\}.$$

Show that G_n is open (easy exercise) and Stegall's variational principle easily shows that G_n is dense. Then show $x^* \in \bigcap G_n$ if and only if x^* strongly exposes C. For the first assertion, let D be the closed convex hull of the strongly exposed points of C. If $D \neq C$, use the second assertion with the separation theorem (4.1.17) to derive a contradiction. □

6.6 Deeper support structure

6.6.2. Use Stegall's variational principle (6.6.17) to derive the 'only if' implication in Theorem 6.6.12.

Hint. See Exercise 5.2.12. □

The next exercise exhibits an ingenious variational proof from [207] of a classical result originally proven with more technology.

6.6.3 (Pitt's theorem). Suppose $1 \leq p < q < \infty$. Show that every bounded linear operator from $\ell_q \to \ell_p$ is compact.

Hint. Let T be such a bounded linear operator. Apply Corollary 6.6.16 to $f(x) := \|x\|_q^q - \|Tx\|_p^p$ to find $x \in \ell_q$ and $x^* \in \ell_q^*$ such that

$$f(x+h) - f(x) - \langle x^*, h \rangle \geq 0 \quad h \in \ell_q.$$

This implies $f(x+h) + f(x-h) - 2f(x) \geq 0$ for all $h \in \ell_q$. Thus for all $h \in \ell_q$,

$$\|x+h\|_q^q + \|x-h\|_q^q - 2\|x\|_q^q \geq \|T(x+h)\|_p^p + \|T(x-h)\|_p^p - 2\|Tx\|_p^p.$$

Let (x_i) be a bounded sequence in ℓ_q, which by passing to a subsequence may be assumed to converge weakly to some $y \in \ell_q$. Show that $\|Tx_i - Ty\| \to 0$ as $i \to \infty$; see [121, Theorem 6.3.13] for full details. □

6.6.4.** Using a result of Moors [319] that shows (f) ⇒ (a) in Theorem 6.6.19, prove Theorem 6.6.19.

Hint. Clearly (b) ⇒ (c). One can show (c) ⇒ (a) by contraposition, i.e. if X fails the RNP, then X has a closed bounded set that is not dentable. Consequently, X has an equivalent norm whose unit ball is not dentable. Then the dual norm is not Fréchet differentiable anywhere. Thus (a) through (d) are equivalent. Finally, it is clear that (d) ⇒ (e) ⇒ (f), while, as mentioned, the implication (f) ⇒ (a) can be found in [319]. □

Theorem 6.6.19. *Let X be a Banach space. The following are equivalent.*

(a) X has the RNP.
(b) X^* is a weak*-Asplund space.
(c) Every equivalent dual norm on X^* is Fréchet differentiable at at least one point.
(d) Every nonempty closed bounded convex subset C of X is the closed convex hull of its strongly exposed points, and the strongly exposing functional form a dense G_δ-subset in X^*.
(e) The support functionals for every nonempty closed bounded convex set in X form a residual set in X^*.
(f) The norm-attaining functionals for each equivalent norm on X form a residual set in X^*.

6.6.5.* Verify that $x_n^*, y_n^* \in \partial_{\varepsilon/n} p(x)$ for each n in the proof of Lemma 6.6.3.

Proof. First $x_n^*(x) > p(x) - \varepsilon$ by definition of a slice. Also, $x_n^*(y) \leq p(y)$ for all $y \in X$ by the definition of p. Therefore, $x_n^*(y) - x_n^*(x) < p(y) - p(x) + \varepsilon$ for all $y \in X$. Similarly, for y_n^*. □

6.6.6. Let X be a nonseparable Banach space, show that B_X contains an uncountable ε-net for some $\varepsilon > 0$.

Hint. Suppose not, then consider $\{x_{n,j}\}_{j=1}^{\infty}$ a countable collection of maximal $1/n$-nets in B_X. This is a countable dense subset of B_X which is a contradiction. ☐

6.6.7 (Tang [414]). ⋆⋆ Suppose C is a weak*-closed convex subset of X^*. Every subset of C is weak* dentable if and only if every continuous convex function $f : X \to \mathbb{R}$ such that there exist $a > 0$ and $b \in \mathbb{R}$ with $f \le a\sigma_C + b$ is Fréchet differentiable on a dense G_δ-subset of X. For further related characterizations related to this, see Tang's paper [414] on Asplund functions.

Hint. $\Leftarrow$: Let D be a weak*-closed convex subset of A. Then $\sigma_D \le \sigma_C$ and so it is Fréchet differentiable on a dense subset of X. Conclude that D has strongly exposed points.

$\Rightarrow$: Suppose $f \le a\sigma_C + b$. Then f is Fréchet differentiable on a dense G_δ-subset of X if and only if $\frac{1}{a}(f - b)$ is. Thus we may assume $a = 1$ and $b = 0$. Now $f \le \sigma_C$ implies $f^*(\phi) \ge \sigma_C^*(\phi) = \delta_C(\phi)$. Consequently, $\mathrm{dom} f^* \subset C$, and so $\mathrm{dom}\, \partial f \subset C$. Now apply Lemma 6.6.5 with $A = C$ as in the proof of Theorem 6.6.6. ☐

6.6.8.⋆ Complete the details for the proof of Corollary 6.6.7.

Hint. For (c) $\Rightarrow$ (a), check that the weak*-slices of B have diameter greater than ε. Now B has no strongly exposed points, and so the predual norm has no point of Fréchet differentiability. ☐

6.6.9. Let A be a nonempty set in a Banach space. A point $x_0 \in A$ is called a *denting point* of A if for every $\varepsilon > 0$, there is a slice S of A containing x_0 of diameter less than ε. Show that x_0 is a denting point of A if and only if for every $\varepsilon > 0$, $x_0 \notin \overline{\mathrm{conv}}(A \setminus B_\varepsilon(x_0))$.

6.6.10.⋆ Prove Fact 6.6.11.

6.6.11.⋆ Let X be a Banach space.

(a) Suppose that A is a closed bounded subset of X, and $x_0 \in A$ is such that $\sigma_A(\phi) = \phi(x_0)$. Show that $x_0 \in \partial \sigma_A(\phi)$.
(b) Let C be a nonempty closed convex bounded set. Suppose $x_0 \in \partial \sigma_C(\phi)$. Show that $x_0 \in C$, and $\phi(x_0) = \sigma_C(\phi)$.
(c) Let C be a nonempty closed bounded convex set, and suppose $y \in \partial_\varepsilon \sigma_C(\phi)$ for some $\varepsilon > 0$, show that $y \in C$.

Hint. (a) Let $\Lambda \in X^*$, then $\langle x_0, \Lambda - \phi \rangle \le \sigma_A(\Lambda) - \phi(x_0) = \sigma_A(\Lambda) - \sigma_A(\phi)$ as desired.

(b) By (c), we know that $x_0 \in C$. The subdifferential inequality implies

$$\langle x_0, 0 - \phi \rangle \le \sigma_C(0) - \sigma_C(\phi)$$

and so $\phi(x_0) \ge \sigma_C(\phi)$, and since $x_0 \in C$, this means $\phi(x_0) = \sigma_C(\phi)$.

(c) Suppose $y \notin C$. According to the separation theorem (4.1.17), we can find $\Lambda \in X^*$ so that $\Lambda(y) > \sup_C \Lambda + 3\varepsilon$. Also, we choose $z \in C$ such that $\phi(z) > \sigma_C(\phi) - \varepsilon$.

Then
$$\langle y, \Lambda - \phi \rangle > \sigma_C(\Lambda) - \sigma_C(\phi) + 2\varepsilon$$
which contradicts that $y \in \partial_\varepsilon \sigma_C(\phi)$. □

6.6.12 (Anti-proximinal norms [88]).** Two equivalent norms $\|\cdot\|$ and $\|\cdot\|$ are said to form an *anti-proximinal pair* when neither the problem
$$\inf\{\|x - z\| : \|x\| \leq 1\}$$
nor the problem
$$\inf\{\|x - z\| : \|x\| \leq 1\}$$
is ever achieved except trivially – when z lies in the feasible set. Each such norm is called a companion to the other.

(a) Show that this mutual nonattainment is equivalent to openness of the sum $\{x : \|x\| \leq 1\} + \{x : \|x\| \leq 1\}$. Thus we exhibit two closed convex bodies whose sum is open.
(b) Show than no such pair can exist in a Banach space with the RNP.
(c) Find a companion norm in c_0, endowed with the usual norm $\|\cdot\|_\infty$.

Hint. Find a bounded linear mapping T on c_0 whose adjoint sends nonzero support functionals in ℓ_1 to nonsupport functionals and consider $\|x\| := \|Tx\|_\infty$. □

6.6.13. Show that the closed unit ball of c_0 under its usual norm does not have any extreme points. Show likewise that the closed unit ball in $L_1[0, 1]$ in its usual norm has no extreme points while the closed unit ball for $C[0, 1]$ in its usual norm has only two extreme points. In particular, conclude that none of $c_0, L_1[0, 1]$ or $C[0, 1]$ have the RNP.

6.6.14. Let X be a Banach space.

(a) Suppose the norm on X is locally uniformly convex, show that the dual norm is Fréchet differentiable on a dense subset of X^*. Conclude that the strongly exposing functionals of B_X form a dense G_δ-set in X^*.
(b) Give an example of a closed convex bounded subset C of a Banach space X such that the strongly exposing functionals of C forms a dense G_δ-subset of X^*, yet C fails the RNP.

Hint. For (a), use the local uniform convexity of the norm along with the dual version of Šmulian's theorem (Exercise 5.1.5) to conclude that the dual norm is Fréchet differentiable at every norm attaining functional. Then apply the Bishop–Phelps theorem (4.3.4). For (b), use that fact that the separable space c_0 admits an equivalent locally uniformly convex norm and Exercise 6.6.13. □

6.6.15 (Special case of Theorem 6.6.12). Let C be a closed bounded convex subset of X. If C has the RNP, then σ_C is Fréchet differentiable on a dense G_δ-subset of X^*. Mimic the proof of Theorem 6.6.12 to prove this using the Bishop–Phelps theorem (Exercise 4.3.6) instead of the Brøndsted–Rockafellar theorem (4.3.2).

Hint. Since C is bounded, assume $\|\cdot\|$ is an equivalent norm on X so that $C \subset B_X$. Let G_n be the set of $\phi \in X^*$ such that there exists an open neighborhood U of ϕ such that the diameter of $\partial \sigma_C(U) \cap X$ is less than $1/n$. According to Lemma 6.6.5, G_n is a dense open set in X^* for each $n \in \mathbb{N}$. Now suppose $\phi \in \bigcap G_n$, and $\|\phi\| \leq 1$, but that σ_C is not Fréchet differentiable at ϕ. Then there exist $t_n \to 0^+$, y_n^* such that $\|y_n^*\| = 1$ and

$$\sigma_C(\phi + t_n y_n^*) + \sigma_C(\phi - t_n y_n^*) - 2\sigma_C(\phi) \geq \varepsilon t_n.$$

Using the Bishop–Phelps theorem (Exercise 4.3.6), we choose ψ_n and ϕ_n that attain their suprema on C such that

$$\|\phi_n - (\phi + t_n y_n^*)\| \leq \frac{\varepsilon}{8} t_n \quad \text{and} \quad \|\psi_n - (\phi - t_n y_n^*)\| \leq \frac{\varepsilon}{8} t_n.$$

Then let $x_n, y_n \in S_X$ be points such that $\phi_n(x_n) = \psi_n(y_n) = 1$. Then

$$\phi_n(x_n) + \psi_n(y_n) - 2\sigma_C(\phi) \geq \frac{3}{4} \varepsilon t_n, \quad \text{which implies}$$

$$(\phi + t_n y_n^*)(x_n) + (\phi - t_n y_n^*)(y_n) - 2\sigma_C(\phi) \geq \frac{\varepsilon t_n}{2}.$$

Then $t_n y_n^*(x_n - y_n) \geq \frac{\varepsilon t_n}{2}$ and so $\|x_n - y_n\| \geq \frac{\varepsilon}{2}$. Now $(\phi_n) \to \phi$ and $(\psi_n) \to \phi$ therefore $\phi \notin G_n$ when $\frac{1}{n} < \frac{\varepsilon}{2}$, which is a contradiction. $\square$

6.6.16 (Smooth variational principles on RNP spaces).** Suppose X is a Banach space with the RNP, and that X has a C^k-smooth bump function b. Given any proper bounded below lsc function $f : X \to (-\infty, +\infty]$, show that there is a C^k-smooth function g and $x_0 \in X$ so that $f(x_0) = g(x_0)$ and $f \geq g$ (i.e. X admits a C^k-smooth variational principle).

Hint. Let $\phi := f + b^{-2}$ and apply Corollary 6.6.16; see [205, Theorem 19] for further details. $\square$

6.6.17 (Smooth variational principles and superreflexivity).** Suppose $k > 1$. Show that X admits a C^k-smooth variational principle if and only if X is superreflexive and admits a C^k-smooth bump function.

Combined with Exercise 6.6.16, this shows that a Banach space with the RNP that admits a C^2-smooth bump function is superreflexive. Stronger results in this direction can be found in [312, Proposition 2.3] which built upon [178] to show that the continuous bump function on a space with the RNP need only have a Gâteaux derivative that satisfies a directional Hölder condition at each point, in order for the space to be superreflexive.

Hint. For necessity, apply the C^k-smooth variational principle to $1/\|\cdot\|$ to obtain a C^k-smooth bump. To obtain superreflexivity, it is enough to show every separable subspace Y of X is superreflexive. For that, let $|\cdot|$ be an equivalent locally uniformly convex norm on Y. Then apply the variational principle to f defined by $f(x):=1/|x|$ if $x \in Y$, and $f(x):=\infty$ otherwise. Use the local uniform convexity of $|\cdot|$ to then construct a bump function on Y with Lipschitz derivative, and hence use Exercise 5.5.7 to conclude Y is superreflexive. The converse follows from Exercise 6.6.16. For more details, see [205, Theorem 20]. □

6.7 Convex functions on normed lattices

This section assumes the reader is familiar with normed lattices; notions not explained here can be found in [386]. We write the *lattice operations* as $x \vee y$ and $x \wedge y$ for the supremum and infimum of x and y respectively. As usual, we denote $x^+ := x \vee 0$ and $x^- := (-x)^+$, and $|x| := x \vee (-x)$.

In this section, use $\partial^a f$ to denote the algebraic subdifferential of f, that is elements from the algebraic dual X' that satisfy the subdifferential inequality. The proof of the next two simple facts are left as Exercise 6.7.2.

Fact 6.7.1. *Let X be a topological vector space, and $f : X \to (-\infty, +\infty]$ a positively homogeneous convex function. Then for $\phi \in X^*$, $\phi \in \partial f(\bar{x})$ if and only if $\phi \in \partial f(0)$ and $\phi(\bar{x}) = f(\bar{x})$.*

Fact 6.7.2. *Let X be a normed lattice. Define $g : X \to \mathbb{R}$ by $g(x) = \|x\|$. For $\bar{x} \neq 0$,*

$$\partial g(\bar{x}) = \{\phi \in X^* : \|\phi\| = 1, \phi(\bar{x}) = \|\bar{x}\|\}, \quad \partial g(0) = B_{X^*}.$$

Proposition 6.7.3. *Let X be a normed lattice. Then $x \mapsto \|x^+\|$ is a continuous convex function.*

Proof. For $x_1, x_2 \in X$ and $0 \leq \lambda \leq 1$ we have $\lambda x_1^+ \geq \lambda x_1, 0$ and $(1-\lambda)x_2^+ \geq (1-\lambda)x_2, 0$. Adding these we obtain $\lambda x_1^+ + (1-\lambda)x_2^+ \geq (\lambda x_1 + (1-\lambda)x_2)^+$. Because X is a normed lattice, this yields

$$\|(\lambda x_1 + (1-\lambda)x_2)^+\| \leq \|\lambda x_1^+ + (1-\lambda)x_2^+\|$$
$$\leq \lambda \|x_1^+\| + (1-\lambda)\|x_2^+\|,$$

which establishes the desired convexity. Continuity follows from the continuity of lattice operations; see [386, Section II.5.2]. □

The previous proposition and the max formula (4.1.10) yield

Corollary 6.7.4. *Let X be a normed lattice. Define $f : X \to \mathbb{R}$ by $f(x) = \|x^+\|$. Then $\partial f(x) \neq \emptyset$ for all $x \in X$.*

Proposition 6.7.5. *Let X be a normed lattice. Define $f : X \to \mathbb{R}$ by $f(x) = \|x^+\|$. For $\bar{x} \notin -X_+$,*

$$\partial f(\bar{x}) = \{\phi \in X^* : \phi \geq 0, \|\phi\| = 1, \phi(\bar{x}^-) = 0, \phi(\bar{x}^+) = \|\bar{x}^+\|\}.$$

For $\bar{x} \leq 0$,

$$\partial f(\bar{x}) = \{\phi \in X^* : \phi \geq 0, \|\phi\| \leq 1, \phi(\bar{x}) = 0\}.$$

Proof. The positive homogeneity of f along with Fact 6.7.1 establish that $\phi \in \partial f(\bar{x})$ if and only if $\phi \in \partial f(0)$ if and only if $\phi(\bar{x}) = f(\bar{x})$. Now $\phi \in \partial f(0)$ if and only if $\phi(x) \leq \|x^+\|$, for all $x \in X$; this is equivalent to $\phi \geq 0$ and $\|\phi\| \leq 1$. Accordingly, $\phi \in \partial f(\bar{x})$ if and only if $\phi \geq 0$, $\|\phi\| \leq 1$ and $\phi(\bar{x}) = \|\bar{x}^+\|$. This implies $\|\bar{x}^+\| = \phi(\bar{x}^+ - \bar{x}^-) \leq \phi(\bar{x}^+) \leq \|\phi\| \|\bar{x}^+\| \leq \|\bar{x}^+\|$, and so the result follows. □

The following makes this more precise when x is not negative.

Proposition 6.7.6. *Let X be a normed lattice. Define $f, g : X \to \mathbb{R}$ by $f(x) = \|x^+\|$, $g(x) = \|x\|$ for all $x \in X$. Then for $\bar{x} \notin -X_+$,*

$$\partial f(\bar{x}) = \partial g(\bar{x}^+) \cap \{\phi \in X^* : \phi \geq 0, \phi(\bar{x}^-) = 0\}.$$

In particular, for $\bar{x} \notin -X_+$, if the norm $\|\cdot\|$ is differentiable at $\bar{x}^+$, then f is differentiable at $\bar{x}$ with the same derivative.

Proof. The formula for $\partial f(\bar{x})$ follows from Fact 6.7.2 and Proposition 6.7.5. Also, if g is differentiable at $\bar{x}^+$, then $\partial g(\bar{x}^+)$ is a singleton. Now, Corollary 6.7.4 ensures $\partial f(\bar{x}) \neq \emptyset$, and thus $\partial f(\bar{x}) = \{\nabla g(\bar{x}^+)\}$. □

Example 6.7.7. As before, let $f(x) := \|x^+\|$, for $x \in X$.

(a) Let $X = L_p(T, \mu)$. For $x \neq 0$, $\nabla \|x\|_p = \|x\|_p^{1-p} |x|^{p-2} x$ (Exercise 6.7.3). According to Proposition 6.7.6, f is differentiable for $x \notin -X_+$, and $\nabla f(x) = \|x^+\|_p^{1-p} (x^+)^{p-1}$.

(b) Let $X = L_1(T, \mu)$. From Proposition 6.7.5 it follows that for any x and $\phi \in L_\infty(T, \mu)$, $\phi \in \partial f(x)$ if and only if

$$\phi = \begin{cases} 1 & \text{a.e., where } x(t) > 0; \\ 0 & \text{a.e., where } x(t) < 0; \\ \in [0, 1] & \text{a.e., where } x(t) = 0. \end{cases}$$

Thus f is differentiable at x if and only if $|x(t)| > 0$ a.e., and in this case, $\nabla f(x) = \chi_{\{t : x(t) > 0\}}$.

(c) Let $X = L_\infty(T, \mu)$ with the weak*-topology. According to Proposition 6.7.5, for $\phi \in L_1(T, \mu)$ and $x \notin -X_+$, $\phi \in \partial f(x)$ if and only if, $\int_T \phi d\mu = 1$, $\phi = 0$ a.e. where $x(t) < 0$ and $\int_T \phi x^+ d\mu = \operatorname{ess\,sup} x^+$, which is equivalent to $\phi \geq 0$, $\phi|_{T_x^c} = 0$ and $\int_{T_x} \phi d\mu = 1$, where $T_x = \{t \in T : x(t) = \operatorname{ess\,sup} x\}$. It follows that if $x(t) < \operatorname{ess\,sup} x$ a.e. then $\partial f(x) = \emptyset$. Also, f is never differentiable at x if (T, μ) is nonatomic.

(d) Let $X = C(T)$ where T is a compact Hausdorff space. For $x \in X$, define $T_x = \{t \in T : x(t) = \max_\tau x(\tau)\}$. According to Proposition 6.7.5, for $-x \in X_+$,

6.7 Convex functions on normed lattices

using the Riesz representation theorem,

$$\partial f(x) = \{\phi \in M(T) : \phi \geq 0, \phi(T) = 1, \mathrm{supp}\,\phi \subset T_x\},$$

(in other words, probability measures supported on T_x). Thus for $x \notin -X_+$, f is differentiable at x if and only if T_x is a singleton, $\{t_0\}$ say, in which case $\nabla f(x) = \delta_{t_0}$, a unit mass concentrated at t_0.

Let us note that a subspace I of X is an *ideal* if $y \in I$ whenever $x \in I$ and $|y| \leq |x|$. An ideal I of X is called a *band* if $A \subset I$ and $\sup A = x \in X$ implies $x \in I$. We denote the collection of bands in X by X^b.

Proposition 6.7.8. *Let (X, Y) be a dual pair, with X a vector lattice and Y partially ordered by $(X_+)^+$. Suppose $y_0 \in Y$, $y_0 \geq 0$, and define $h : X \to \mathbb{R}$ by $h(x) := \langle x^+, y_0 \rangle$. Then h is a positively homogeneous convex function such that, for any $x \in X$,*

$$\partial h(x) = \{y \in Y : 0 \leq y \leq y_0, \langle x^-, y \rangle = 0, \langle x^+, y_0 - y \rangle = 0\}.$$

If Y is an ideal in X^b, then $\partial h^a(x) = \partial h(x)$.

Proof. Clearly, h is positively homogeneous. Now let $x_1, x_2 \in X$ and $0 \leq \lambda \leq 1$. Then $\lambda x_1^+ \geq \lambda x_1^+$ and $(1-\lambda)x_2^+ \geq (1-\lambda)x_2, 0$, so

$$(\lambda x_1 + (1-\lambda)x_2)^+ \leq \lambda x_1^+ + (1-\lambda)x_2^+$$

and then, because $y_0 \geq 0$, it follows that

$$\langle (\lambda x_1 + (1-\lambda)x_2)^+, y_0 \rangle \leq \lambda \langle x_1^+, y_0 \rangle + (1-\lambda)\langle x_2^+, y_0 \rangle.$$

This shows h is convex. Now $y \in \partial h(0)$ if and only if $\langle x, y \rangle \leq \langle x^+, y_0 \rangle$ for all $x \in X$; this is equivalent to $y \in [0, y_0]$. According to Fact 6.7.1, $y \in \partial h(x)$ if and only if $y \in [0, y_0]$ and $\langle x, y \rangle = \langle x^+, y_0 \rangle$. But then

$$\langle x^+, y_0 \rangle \geq \langle x^+, y \rangle \geq \langle x, y \rangle = \langle x^+, y_0 \rangle.$$

Thus there is equality throughout and the result follows.

When $\phi \in X'$, $\phi \in \partial^a h(0)$ if and only if $\phi(x) \leq \langle x^+, y_0 \rangle$ for all $x \in X$. This is equivalent to $\phi \geq 0$ and $y_0 - \phi \geq 0$. Thus $\phi \in X^b$ and $0 \leq \phi \leq y_0$, for $\phi \in Y$ because Y is an ideal. Therefore, $\partial^a h(0) = \partial h(0)$, and consequently $\partial^a h(x) = \partial h(x)$ for all $x \in X$. □

Proposition 6.7.9. *Let X be a normed lattice, $\Lambda \in X^*$, $\Lambda \geq 0$, and define $h : X \to \mathbb{R}$ by $h(x) := \Lambda(x^+)$. Then h is a continuous convex function, and in particular, $\partial h(x) \neq \emptyset$ for all $x \in X$.*

Proof. The convexity follows from Proposition 6.7.8, and the continuity follows from properties of normed lattices as in [386, II.5.2]. The subdifferentiability then follows from the max formula (4.1.10). □

6.7.1 Hilbert lattices

A *Hilbert lattice* is a Banach lattice in which the norm induces an inner product. It is well known that a Hilbert lattice is isomorphic to $L_2(S,\mu)$ in the usual ordering for some measure space (S,Ω,μ) [386]. Thus, one is in fact looking at spaces like $L_2([0,1])$, $\ell_2(\mathbb{N})$, $\ell_2(\mathbb{R})$ and their products.

Since convex functions on Hilbert space are so central in applications, we describe some of the striking order theoretic results characterizing Hilbert lattices within Banach lattices.

Theorem 6.7.10 (Hilbert lattice characterizations). *Let H be a vector lattice Hilbert space with ordering cone S. Then the following are equivalent.*

(a) *H is a Hilbert lattice.*
(b) *For x, y in S, $\|x \wedge y\|^2 \leq \langle x, y \rangle \leq \|x \vee y\| \|x \wedge y\|$.*
(c) *For x, y in H, $x \wedge y = 0 \Leftrightarrow \langle x, y \rangle = 0, x \in S, y \in S$.*
(d) *$S = S^+$.*

Proof. (a) $\Rightarrow$ (b): It suffices to verify that

$$\langle x \wedge y, x \vee y \rangle = \langle x, y \rangle \tag{6.7.1}$$

whenever $x, y \geq 0$. Then, as the norm is monotone and $x \vee y \geq x \wedge y \geq_S 0$ we derive (b). Since the norm is absolute, we have that $x \vee y + x \wedge y = x + y$ and $\|x \vee y - x \wedge y\| = \|x - y\|$. By the parallelogram law we therefore have that

$$\|x \vee y\|^2 + \|x \wedge y\|^2 \pm 2\langle x \wedge y, x \vee y \rangle = \|x\|^2 + \|y\|^2 \pm 2\langle x, y \rangle.$$

Subtraction and cancellation yields the asserted result (6.7.1).

The implication (b) $\Rightarrow$ (c) is immediate.

(a) $\Rightarrow$ (d): Since $S \subset S^+$ ((a) implies (b)) it suffices to show the converse. Fix $x \in S^+$. Since $\langle x^+, x^- \rangle = 0$ ((a) implies (c)) we have

$$\|x^+\|^2 + \|x^-\|^2 = \langle x, x \rangle \leq \langle x^+, x \rangle \leq \langle x^+, x^+ \rangle$$

since $x^+ \leq_S x$. Hence $x^- = 0$ and $x = x^+ \in S$. Thus we have shown that all properties hold in a Hilbert lattice. The converse is left as Exercise 6.7.5. $\square$

The self-duality of the semidefinite matrices (Proposition 2.4.9) shows how strongly this result relies on vector lattice structure.

Example 6.7.11. The Sobolev space $W^{1,2}([0,1])$ of all absolutely continuous functions with square-integrable derivatives on $[0,1]$, endowed with the a.e. pointwise order and normed by

$$\|x\|^2 := \|x'\|_2^2 + |x(0)|^2$$

provides an example of a Hilbert space which is a vector lattice but not a Banach lattice: the norm preserves absolute values but is not order-preserving on the cone [119].

6.7 Convex functions on normed lattices

Another useful connection between order and geometry is due to Nemeth. We state it without proof.

Theorem 6.7.12 (Projection characterization [330]). *Suppose K is a closed generating convex cone in a real Hilbert space H, and that P_K is the metric projection onto K. Then H ordered by K is a Banach lattice if and only if P_K is isotone ($y \geq_K x$ implies $P_K(y) \geq_K P_K(x)$), and subadditive ($P_K(x) + P_K(y) \geq_K P_K(x+y)$, for all $x, y \in H$).*

Finally, we quote a pretty lattice variant on the classical Pythagorean characterization of Hilbert space.

Theorem 6.7.13 (Pythagorean lattice characterization [196]). *Let E be a Banach lattice with norm $\|\cdot\|$ and ordering cone K. Then E is a Hilbert lattice if and only if*

$$\frac{1}{2}\|x+y\|^2 + \frac{1}{2}\|x-y\|^2 = \|x\|^2 + \|y\|^2$$

for all x, y in K.

Exercises and further results

6.7.1 (The semidefinite order is not a lattice order). Show the Loewner-order is not a lattice order.

Hint. Suppose there is a matrix Z in S^2 satisfying

$$W \geq \begin{bmatrix} 1 & 0 \\ 0 & 0 \end{bmatrix} \text{ and } W \geq \begin{bmatrix} 0 & 0 \\ 0 & 1 \end{bmatrix} \Leftrightarrow W \geq Z,$$

in the semidefinite order. By considering diagonal matrices W, prove $Z = \begin{bmatrix} 1 & a \\ a & 1 \end{bmatrix}$ for some real a. Now using $W = I$, prove $Z = I$. Finally derive a contradiction by considering $W := \frac{2}{3}\begin{bmatrix} 2 & 1 \\ 1 & 2 \end{bmatrix}$.

It is illuminating to plot this situation. □

6.7.2.* Prove Fact 6.7.1 and Fact 6.7.2.

6.7.3. Let $X := L_p(T, \mu)$. For $x \neq 0$, show that $\nabla \|x\|_p = \|x\|_p^{1-p}|x|^{p-2}x$.

Hint. See [260, p. 170]. □

6.7.4. Let H be a Hilbert lattice ordered by S. Show that $P_S(x) = x^+$.

Hint. Using the characterization in Exercise 2.3.17(c) it suffices to show that $\langle x - x^+, s - x^+ \rangle \leq 0$ for all $s \in S$. Since S is a cone this means showing $\langle x^+, x^- \rangle = 0$ and $x^- \in S^+$. But $S \subset S^+$ by (a) implies (c) of Theorem 6.7.10.

6.7.5.* Finish the proof of Theorem 6.7.10.

Hint. To show (b) implies (a) notice the norm is monotone, and $x^+ \wedge x^- = 0$ implies the norm is absolute. For (c) implies (b), fix $x, y \geq 0$. Let $z := x \vee y$, $w := x \wedge y$. Then $x + y = w + z$ and $(x - z) \wedge (y - z) = 0$. By the parallelogram law $\langle x, y \rangle = \langle w, z \rangle \leq \|z\| \|w\|$. For (d) implies (c), suppose $x, y \geq_S 0$. Let $z := x \wedge y \geq_S 0$.

Suppose $\langle x, y \rangle = 0$. As $S \subset S^+$, $0 = \langle x, y \rangle \geq \langle z, y \rangle \geq \langle z, z \rangle \geq 0$ and so $x \wedge y = 0$. Conversely, if $z = 0$ then $x + y = x \vee y = |x - y|$ one may derive (6.7.1), after first showing the norm is absolute when $S^+ \subset S$ [119]. □

6.7.6 (New norms from old). ** In addition to methods already discussed there are several accessible ways of generating useful new norms from lattice norms. Two notable constructions are the following.

(a) (Substitution norms [172]). We call a Banach function space X on an index set S *full* if $|y(s)| \leq |x(s)|$, for all $s \in S$ and $x \in X$ implies $y \in X$; this holds in a Banach lattice sequence space such as $c_0(\mathbb{N})$ or $\ell_p(\mathbb{N})$ for $1 \leq p \leq \infty$. Let X be a full function space and for each s in S, let $(X_s, \|\cdot\|_s)$ be a normed space. Consider

$$P_X(X_s) := \{x : \psi(s) = \|x(s)\|_s \in X\}.$$

Show that $\|\|x\|\| := \|\psi\|$ defines a norm on $P_X(X_s)$. Show that $P_X(X_s)$ so normed is complete if and only if each component space is.

(b) Suppose each X_s is complete. For $1 \leq p \leq \infty$ and $1/q + 1/p = 1$, define $P_p(X_s) := P_{\ell_p(S)}(X_s)$ and let $m(S)$ denote the Radon measures on S in the variation norm. Show that (i) $P_1(X_s^*) = P_{c_0(S)}(X_s)^*$; (ii) $P_1(X_s)^* = P_{m(S)}(X_s)$; and (iii) $P_q(X_s^*) = P_p(X_s)^*$.

In particular for $1 < p < \infty$ the space $P_p(X_s)$ is reflexive if and only if each component space is.

(c) (Recursive norms on ℓ_1 [188]). Two fairly general constructions on ℓ_1 are as follows.

(i) For $0 < \gamma_n \leq \gamma_{n+1} \leq 1$ the norm $\|x\|_\gamma := \sup_n \gamma_n \sum_{k=n}^\infty \|x_n\|$ is an equivalent norm on ℓ^1.

(ii) Let $p := (p_n)_{n=1}^\infty$ be a sequence of numbers decreasing to 1 and let $1/q_n + 1/p_n = 1$. Let $x = (x_n)_{n=1}^\infty$ be a real sequence. Let $S(x) := (x_2, x_3, \ldots, x_m, \ldots)$ denote the unilateral shift. Define a sequence on seminorms $v_n(p, x)$ by $v_1(p, x) := |x_1|$ and

$$v_{n+1}(p, x) := \left(|x_1|^{p_1} + v_n(Sp, Sx)^{p_1}\right)^{1/p_1}.$$

Show that $v_{n+1}(p, \cdot)$ increases pointwise to a seminorm v_p which is equivalent to the 1-norm if and only if its conjugate v_q is equivalent to the supremum norm on c_0. In particular, show this is the case if $p_n := 2^n/(2^n - 1)$. Note the analogy to a continued fraction or radical in this construction.

6.7.7 (Lattice renorms). ** A striking result is that a Banach lattice has a locally uniformly rotund (LUR) renorm if and only if order intervals in the lattice are weakly compact [169]. (Equivalently the lattice is order continuous [169].) Deduce that, given an order interval $[0, x]$ in such a lattice, the end points are strongly exposed by positive continuous linear functionals.

6.7 Convex functions on normed lattices

Likewise a Banach lattice is reflexive if and only if it has an equivalent lattice renorm in which both the norm and the dual norm are simultaneously Fréchet and LUR [226].

6.7.8 (A weak lattice-like property [51]). ⋆⋆ Let X be a Banach space ordered by a closed convex cone K. The order is *directed* if given $x_1, x_2 \in X$ there exists $z \in X$ with $z \geq_K x_1$ and $z \geq_K x_2$. Clearly every vector lattice order is directed. An element $\lambda \neq 0$ of a convex cone is called an *extreme direction* if whenever $\lambda = \lambda_1 + \lambda_2$ for $k_1, k_2 \in K$ then actually $\lambda_1, \lambda_2 \in \mathbb{R}_+ \lambda$.

(a) Check that if $\operatorname{int} K \neq \emptyset$ then K is directed.
(b) Show that if $\operatorname{int} K \neq \emptyset$ then $\operatorname{ext} K^+ \neq \emptyset$.
(c) Suppose that $\lambda \in \operatorname{ext} K^+$ and that $x_1, x_2 \in X$ satisfy $\langle \lambda, x_1 \rangle \leq 0$ and $\langle \lambda, x_2 \rangle \leq 0$. Show that for every $\varepsilon > 0$ there exists $z_\varepsilon \in X$ with $x_1 \leq_K z_\varepsilon, x_2 \leq_K z_\varepsilon$ and $\langle \lambda, z_\varepsilon \rangle \leq \varepsilon$.

Hint. Consider the abstract linear program

$$-\infty < \mu := \inf\{\langle \lambda, z \rangle : z \in X, z \geq_K x_1, z \geq_K x_2\}.$$

Compute the dual linear program and justify the lack of a duality gap. Use the extremal nature of λ to deduce that $\mu \leq 0$ [51]. □

(d) Use the point $\lambda := (1, 1)$ in $(\mathbb{R}_+^2)^+$ and $x_1 := (1, -1), x_2 := (-1, 1)$ to show the necessity for something like extremality in (c).
(e) Show that for $1 \leq p < \infty$ the Banach lattice cone in $L_p([0, 1])$, with respect to Lebesgue measure, has no extreme directions.

6.7.9 (Cone quasiconvexity [51]). ⋆⋆ Let X, Z be Banach spaces with X ordered by a closed convex cone K. A function $F \colon Z \to (X, K)$ is quasiconvex with respect to a cone K if, for all x in X, all level sets $\{z \in Z : F(z) \leq_K x\}$ are convex. Clearly all cone convex functions as in Exercise 2.4.28 – which extends to Banach space – are cone quasiconvex. Show the following:

(a) Suppose that K is directed and K^+ is the weak*-closed convex hull of its extreme directions. Then F is K-quasiconvex if and only if $\lambda \circ F$ is (scalarly) quasiconvex for every $\lambda \in \operatorname{ext} K^+$.
(b) K^+ is the weak*-closed convex hull of its extreme directions if $\operatorname{int} K \neq \emptyset$. Thus (a) applies whenever K has nonempty interior.
(c) The result in (a) applies in the Banach lattice $\ell_p(\mathbb{N})$ ($1 \leq p < \infty$) whose cone has empty interior: F is ℓ_p^+-quasiconvex if and only if each coordinate is.
(d) Yet, the sum of two quasiconvex functions on $\mathbb{R}$ is usually not.

Hint. (a) Apply Exercise 6.7.8 (c). For (b) Suppose $e \in \operatorname{int} K \neq \emptyset$. Show that $C := \{\mu \in K^+ : \langle \mu, e \rangle = 1\}$ is a weak*-compact base for the cone and use the Krein–Milman theorem applied to C. (d) Consider the decomposition of any function of bounded variation into the sum of an increasing plus a decreasing function. □

6.7.10 (Continuous Hahn–Banach extension theorem for operators [63]). ⋆⋆ Show that a Banach lattice Y is order-complete if and only if it has the *continuous* Hahn–Banach extension property for Banach spaces. (That is, in the definition of Exercise 4.1.52 we require all spaces to be Banach and all mappings to be continuous.)

Hint. Adjust the construction of p from Exercise 4.1.52 so p is defined and continuous on $\ell_1(I)$. Note one can assume A, B lie in an order interval. □

6.7.11 (The convex geometry of choice [48, 65]).** Let X be a Banach space ordered by a closed convex cone K. It is proven in [65] for Zermelo–Frankel set theory (ZF) with the axiom of choice (AC) that every weakly compact set C admits a point, c^*, which is simultaneously extreme and *maximal* or *Pareto efficient*: $c \geq_K c^*$, $c \in C \Rightarrow c^* \geq_K c$. (When $K = 0$ this is just a version of the Krein–Milman theorem.) Remarkably, the converse is true. Indeed:

Theorem 6.7.14. *The following are equivalent in Zermelo–Frankel:*

(a) *The axiom of choice obtains.*
(b) *The closed unit ball of every dual Banach lattice has an extreme point.*
(c) *The closed unit ball of every dual Banach lattice has a minimal point in the dual lattice ordering.*

Hint. We outline (c) implies (a) [65]. The idea behind (b) implies (a) is similar [48]. We need to construct a choice function for a family of sets $\{A_i : i \in I\}$ which may be assumed disjoint subsets of some set U. Let $A := \cup_{i \in I} A_i$ and let X be the normed space defined by

$$X := \{x \in \mathbb{R}^A : \|x\| := \sum_{i \in I} \sup_{a \in A_i} |x(a)| < \infty\}.$$

Then, in the language of Exercise 6.7.6 (a), X is the substitution space of the spaces $X_i := \ell_\infty(A_i)$ in $\ell_1(I)$ and has dual

$$X^* := \{x^* * = (x_i^*) : \|x^*\| := \sup_{i \in I} \|x_i^*\|_i < \infty, x_i^* \in X_i^*\}.$$

Since each X_i is a Banach lattice, X^* is a dual Banach lattice with positive cone

$$K := \{x^* \in X^* : x_i^*(x_i) \geq 0 \text{ for all } x_i \in \ell_\infty(A_i)_+\}.$$

We now build a (multiple) choice function as follows. Let g be a maximal element of the ball. By maximality $g = |g| \in K$. Set $\alpha_i := \sup_{A_i} g$ and let

$$M(i) := \{a_i \in A_i : g(a_i) > \alpha_i/2\}.$$

We will show each $M(i)$ is nonempty and finite from which (a) follows.
(i) To show $M(i) \neq \emptyset$, it suffices to show $\alpha_i > 0$. Else $g(A_i) = 0$. Fix $a_i \in A_i$ and let g' agree with g except for $j = i$ where $g'_i = \delta_{a_i}$ (the Kronecker δ). Then $g \leq_K g' \in B_{X^*}$ which violates maximality. (ii) $M(i)$ is finite since $2 \geq 2g(\delta_{a_i}) \geq 2 \sum_{a_i \in M(i)} g(a_i) \geq \alpha_i \#(M(i))$, while $\alpha_i > 0$. □

Recall that weak*-compactness of the dual ball follows from the *Boolean prime ideal theorem* and so does not require the full axiom of choice – but combining the conclusion of the Krein–Milman theorem does! In the same vein, a result of Luxemburg is that the validity of the Hahn–Banach theorem is equivalent to Luxemburg's even weaker axiom (AL): *Every family with the finite intersection property of weak*-closed convex subsets of the dual ball of a Banach space has a common point.*

7
Barriers and Legendre functions

Considerable obstacles generally present themselves to the beginner, in studying the elements of Solid Geometry, from the practice which has hitherto uniformly prevailed in this country, of never submitting to the eye of the student, the figures on whose properties he is reasoning, but of drawing perspective representations of them upon a plane. ...I hope that I shall never be obliged to have recourse to a perspective drawing of any figure whose parts are not in the same plane. (Augustus De Morgan)[1]

7.1 Essential smoothness and essential strict convexity

This chapter is dedicated to the study of convex functions whose smoothness and curvature properties are preserved by Fenchel conjugation.

Definition 7.1.1. We will say a proper convex lower-semicontinuous function $f : \mathbb{R}^N \to (-\infty, +\infty]$ is:

(a) *essentially smooth in the classical sense* if it is differentiable on $\operatorname{int} \operatorname{dom} f \neq \emptyset$, and $\|\nabla f(x_n)\| \to \infty$ whenever $x_n \to x \in \operatorname{bdry} \operatorname{dom} f$;
(b) *essentially strictly convex in the classical sense*, if it is strictly convex on every convex subset of $\operatorname{dom} \partial f$;
(c) *Legendre in the classical sense*, if it is both essentially smooth and essentially strictly convex in the classical sense.

The duality theory for these classical functions is presented in [369, Section 26]. The qualification *in the classical sense* is used to distinguish the definition on $\mathbb{R}^n$ from the alternate definition for general Banach spaces given below which allows the classical duality results to extend to reflexive Banach spaces as shown in [35], which we follow.

Definition 7.1.2. A proper lsc convex function $f : X \to (-\infty, +\infty]$ is said to be

(a) *essentially smooth* if ∂f is both locally bounded and single-valued on its domain;

---

[1] Augustus de Morgan, 1806–1871, First President of the London Mathematical Society, quoted in Adrian Rice, "What Makes a Great Mathematics Teacher?" *American Mathematical Monthly*, June–July 1999, p. 540.

(b) *essentially strictly convex*, if $(\partial f)^{-1}$ is locally bounded on its domain and f is strictly convex on every convex subset of dom ∂f;
(c) *Legendre*, if it is both essentially smooth and essentially strictly convex.

In finite dimensions, where as usual the situation is both best understood and most satisfactory, the local boundedness conditions above are superfluous, and consequently the definitions above are compatible in Euclidean spaces as will be shown in Theorem 7.3.8.

Exercises and further results

7.1.1 (Essential vs. strict convexity). Confirm that

$$\max\{(x-2)^2 + y^2 - 1, -(xy)^{1/4}\},$$

drawn on p. 4, is essentially strictly convex with a nonconvex subgradient domain and is not strictly convex. Compute and plot its essentially smooth conjugate.

7.1.2 (An alternating convex inequality [134]). Suppose that f is convex on $[0, b]$, with $b \geq a_1 \geq a_2 \geq \cdots \geq a_n \geq 0$ while $0 \leq h_1 \leq h_2 \leq \cdots \leq h_n$. Show that

$$\sum_{k=1}^{n}(-1)^{k-1} h_k f(a_k) \geq f\left(\sum_{k=1}^{n}(-1)^{k-1} h_k a_k\right).$$

7.2 Preliminary local boundedness results

This section develops rudimentary results that will be used in the next section to develop the duality between essential strict convexity and essential smoothness.

Fact 7.2.1. *Suppose f is a proper lsc supercoercive convex function. Then:*

(a) $\phi_n \in \partial f(x_n)$ *and* $\|x_n\| \to \infty$, *imply* $\|\phi_n\| \to \infty$;
(b) $(\partial f)^{-1}$ *is bounded on bounded sets;*
(c) *if, additionally, f is strictly convex, then f is essentially strictly convex.*

Proof. Fix $x_0 \in \text{dom} f$. Because $\phi_n \in \partial f(x_n)$, it follows that $\|\phi_n\| \geq |f(x_n) - f(x_0)|/\|x_n\|$; thus $\|\phi_n\| \to \infty$ by supercoercivity which proves (a). Now (b) is a consequence of (a), while (c) is a consequence of (b) and the definitions involved. □

The next few facts consider boundedness properties of subdifferentials. Related results for Euclidean spaces were outlined in Exercises 2.1.23, 2.2.23 and Exercise 2.4.7.

Lemma 7.2.2. *Suppose $f : X \to (-\infty, +\infty]$ is a proper lsc convex function. Then ∂f fails to be locally bounded at each $x \in \text{bdry } \partial f$.*

Proof. Suppose ∂f is locally bounded at $x \in \overline{\text{dom } \partial f}$. Then choose $x_n \to x$ and $\phi_n \in \partial f(x_n)$ and $\|\phi_n\| \leq K$. Now for some subnet $\phi_{n_\alpha} \to_{w^*} \phi$, and one can check that $\phi \in \partial f(x)$. Therefore $x \in \text{dom } \partial f$.

Now suppose $x \in$ bdry ∂f, then there exists a neighborhood U of x with $\partial f(U)$ bounded. According to the Bishop–Phelps theorem (Exercise 4.3.6), there exists $z \in U \cap \overline{\text{dom } \partial f}$ and $w^* \in X^* \setminus 0$ such that $w^*(z) = \sup \langle w^*, \text{dom } \partial f \rangle$. Thus, ∂f is locally bounded at z and so by the first paragraph of this proof, $z \in \text{dom } \partial f$. Now let $z^* \in \partial f(z)$. Then, for $\lambda > 0$,

$$\langle z^* + \lambda w^*, x - z \rangle \leq f(x) - f(z) + \lambda w^*(x - z) \leq f(x) - f(z)$$

for all $x \in \text{dom} f$. Therefore, $z^* + \lambda w^* \in \partial f(z)$ for all $\lambda \neq 0$ and so $\partial f(z)$ is unbounded. This contradiction completes the proof. □

Corollary 7.2.3. *Suppose $f : X \to (-\infty, +\infty]$ is a proper lsc convex function and $x \in \text{dom } \partial f$. Then $x \in \text{int dom} f$ if and only if ∂f is locally bounded at x.*

Proof. Suppose $x \in \text{int dom} f$. Then ∂f is locally bounded at x by Proposition 4.1.26. The converse follows from Lemma 7.2.2. □

Lemma 7.2.4. *Suppose int dom$f \neq \emptyset$ and $x \in \text{dom } \partial f \setminus \text{int dom} f$, then $\partial f(x)$ is unbounded.*

Proof. Let $x \in \text{dom } \partial f \setminus \text{int dom} f$. Because dom f is convex with nonempty interior, according to the separation theorem (4.1.15) there exists $w^* \in X^* \setminus 0$ such that $\langle w^*, x \rangle = \sup \langle w^*, \text{dom} f \rangle$. As in the proof of Lemma 7.2.2, $x^* + \lambda w^* \in \partial f(x)$ if $x^* \in \partial f(x)$ and $\lambda > 0$. □

The previous result extends naturally to maximal monotone operators as the reader can find in [350, p. 30].

Exercises and further results

7.2.1 (Asymptotically well-behaved convex functions [23]). ** A proper closed convex function on $\mathbb{R}^n$ is said have *good asymptotical behavior* if for all sequences $(x_k), (y_k)$ such that $y_k \in \partial f(x_k)$ and $\lim_{k \to \infty} y_k = 0$ one has $\lim_{k \to \infty} f(x_k) = \inf\{f(x) : x \in \mathbb{R}^n\}$. For the following, suppose f, g are proper lsc convex functions on $\mathbb{R}^n$.

(a) Show that if f is inf-compact (i.e. for each λ, the set $\{x : f(x) \leq \lambda\}$ is bounded), then it has good asymptotical behavior.
(b) Suppose $\{x : f(x) \leq \lambda\}$ lies in the relative interior of dom f for all $\lambda > \inf f$. Show that f has good asymptotical behavior.
(c) Consider the following example of Rockafellar [369]: $f : \mathbb{R}^2 \to [0, +\infty)$ defined by $f(x, y) := y^2/x$ if $x > 0$; $f(0, 0) := 0$, and $f(x, y) := +\infty$ otherwise. Show that f does not have good asymptotic behavior.
(d) Let f and g be proper lsc convex functions on $\mathbb{R}^n$, and let $h := f \square g$. Suppose further that f is inf-compact and for each $x \in \mathbb{R}^n$ there exists $y \in \mathbb{R}^n$ such that $h(x) = f(x - y) + g(y)$. Show that g has good asymptotical behavior if and only if h has good asymptotical behavior.
(e) Does an essentially smooth convex function necessarily have good asymptotical behavior?

Hint. For (c), consider the sequence $x_k = (k^2, k)$. For (d), use the fact that $\partial h(x) = \partial f(x-y) \cap \partial g(y)$ when $h(x) = f(x-y) + g(x)$. See [23, Proposition 2.9] for more details. For (e), find an appropriate convex function $g : \mathbb{R}^2 \to [0, +\infty)$ such that $g(x, y) = 0$ if $x \geq 1$ and that $f + g$ is essentially smooth. See [23] for applications and further properties of asymptotically well-behaved convex functions. □

7.2.2. Suppose $f : X \to \mathbb{R}$ is continuous strictly convex and X is reflexive. (a) Is f^* Gâteaux differentiable everywhere on X^*? (b) What if $X = \mathbb{R}^n$ and f is coercive? (c) What if f is supercoercive?

Hint. Both (a) and (b) fail even when $X = \mathbb{R}$. Indeed, we can take $f(t) := e^t - t$ which is strictly convex and coercive, but f^* is not defined everywhere.
(c) Because $(f^*)^* = f$ is strictly convex, it follows that f^* is Gâteaux differentiable on the interior of its domain (see Proposition 5.3.6). Thus, if f is strictly convex and supercoercive, then f^* is bounded on bounded sets (see Theorem 4.4.13), and so f^* is Gâteaux differentiable everywhere. □

7.2.3. Verify the following relations.

(a) If f is continuous and Gâteaux differentiable everywhere, then f is essentially smooth.

Find examples on finite-dimensional spaces for (b), (c), and (d).
(b) f continuous and strictly convex $\not\Rightarrow f^*$ Gâteaux differentiable everywhere.
(c) f Gâteaux differentiable everywhere $\not\Rightarrow f^*$ strictly convex on its domain.
(d) f essentially strictly convex $\not\Rightarrow f$ is strictly convex.
(e) In finite dimensions, f strictly convex $\Rightarrow f$ essentially strictly convex; however, produce an example showing this is not true in ℓ_2.

Hint. (a) is easy. For (b) see the previous exercise. For (c) see Exercise 5.3.10. (d) the conjugate f^* of f in (c) is such and example. (e) See Example 7.3.9 below. □

7.2.4 (Three classic functions). We revisit the sometimes subtle difference between strict and essential strict convexity (see the example before Theorem 23.5 and examples before Theorem 26.3 in [369]). Define on $\mathbb{R}^2$:

$$f_1(r, s) := \begin{cases} \max\{1 - r^{1/2}, |s|\} & \text{if } r \geq 0; \\ \infty & \text{otherwise.} \end{cases}$$

$$f_2(r, s) := \begin{cases} s^2/(2r) - 2s^{1/2} & \text{if } r > 0 \text{ and } s \geq 0; \\ 0 & \text{if } r = s = 0; \\ \infty & \text{otherwise.} \end{cases}$$

$$f_3(r,s) := \begin{cases} s^2/(2r) + s^2 & \text{if } r > 0 \text{ and } s \geq 0; \\ 0 & \text{if } r = s = 0; \\ \infty & \text{otherwise.} \end{cases}$$

Confirm that

(a) dom ∂f_1 is not convex, and f_1 is not strictly convex on int dom f. Clearly, f_1 is not essentially strictly convex;
(b) f_2 is not strictly convex. However, dom ∂f_2 is convex, and f_2 is essentially strictly convex;
(a) dom f_3 = dom ∂f_3 is convex, f_3 is strictly convex on int dom f_3, but f_3 is not essentially strictly convex.

Finally, f_4 below, given in Exercise 7.1.1 and drawn in Chapter 1, is perhaps more borderline than any of the functions above:

$$f_4(r,s) := \begin{cases} \max\{(r-2)^2 + s^2 - 1, -(rs)^{1/4}\} & \text{if } r \geq 0 \text{ and } s \geq 0; \\ \infty & \text{otherwise.} \end{cases}$$

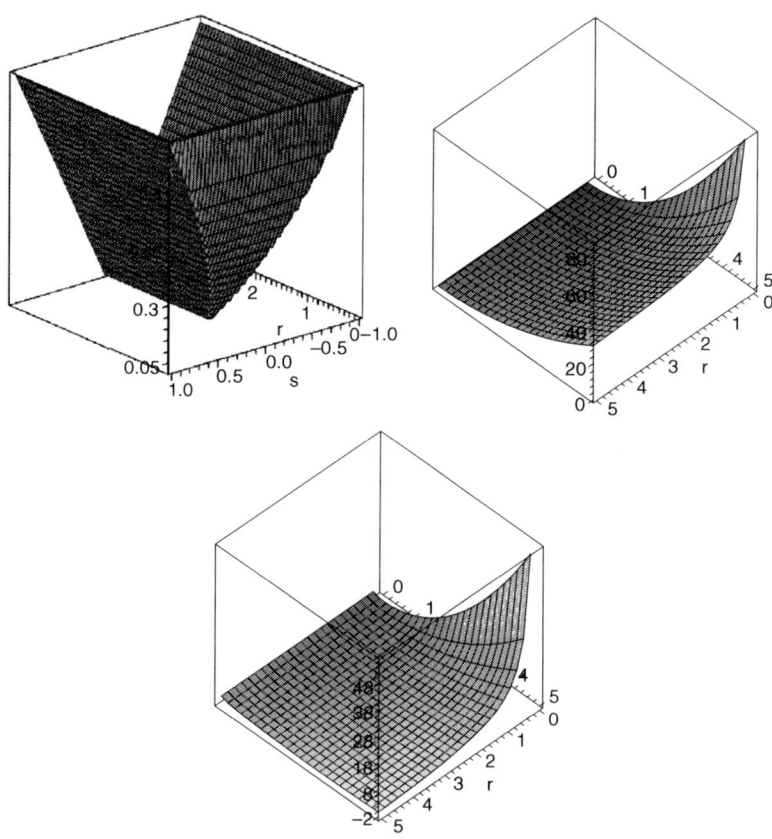

Figure 7.1 The functions f_1, f_2, f_3.

Then f_4 is not strictly convex, dom ∂f_4 is not convex, yet f_4 is essentially strictly convex! (Note that the conjugates of $f_1, \ldots, f_4$ are worth computing with respect to essential smoothness.)

Figure 7.1 illustrates the first three functions of Exercise 7.2.4.

7.2.5. Suppose $y \in \operatorname{int dom} f$ and $x \in \operatorname{dom} \partial f$. Choose $M > 0, r > 0$ such that $f(u) \leq f(x) + M$ for all $u \in y + rB_X$, and so that $\|\phi\| \leq M$ for some $\phi \in \partial f(x)$. Show that

(a) For $0 < \lambda < 1$, let $U_\lambda := (1 - \lambda)x + \lambda(y + rB_X)$. Show that $f(x) - \lambda M \leq f(u) \leq f(x) + \lambda M$ for all $u \in U_\lambda$.
(b) Let $u_\lambda = (1 - \lambda)x + \lambda y$ for $0 < \lambda < 1$. Show that $\partial f(u_\lambda) \neq \emptyset$, and $\partial f(u_\lambda) \subset \frac{M}{r} B_{X^*}$.

7.3 Legendre functions

This section highlights the satisfactory duality theory that exists for Legendre functions in reflexive spaces, and also shows the compatibility of the classical and general definitions in finite-dimensional spaces.

Lemma 7.3.1. *Suppose $f : X \to (-\infty, +\infty]$ is a lsc proper convex function. Then:*

(a) *∂f is single-valued on its domain if and only if f^* is strictly convex on line segments in* range ∂f.
(b) *For all $x, y \in X$, $x \neq y \Rightarrow \partial f(x) \cap \partial f(y) = \emptyset \Leftrightarrow f$ is strictly convex on line segments in* dom ∂f.

Proof. (a) $\Rightarrow$: We proceed by way of contradiction. Thus we assume there exist $y_1^*, y_2^* \in \operatorname{range} \partial f$ such that $y_1^* \neq y_2^*$, $[y_1^*, y_2^*] \subset \operatorname{range} \partial f$ and $\lambda_1, \lambda_2 \in (0, 1)$ with $\lambda_1 + \lambda_2 = 1$ and $f^*(\lambda_1 y_1^* + \lambda_2 y_2^*) = \lambda_1 f^*(y_1^*) + \lambda_2 f^*(y_2^*)$. Now let $y^* = \lambda_1 y_1^* + \lambda_2 y_2^*$. Then there exists $x \in X$ such that $y^* \in \partial f(x)$. Hence

$$0 = f(x) + f^*(y^*) - \langle y^*, x\rangle = \sum_{i=1}^{2} \lambda_i(f(x) + f^*(y^*) - \langle y_i^*, x\rangle) \geq 0.$$

It follows that both y_1^* and y_2^* belong to $\partial f(x)$, which is a contradiction.

$\Leftarrow$: Choose $y_1^*, y_2^* \in \partial f(x)$. Then $f(x) + f^*(y_i^*) = \langle y_i^*, x\rangle$, for $i = 1, 2$. For all nonnegative reals λ_1, λ_2 that add up to 1, we have:

$$f(x) + \lambda_1 f^*(y_1^*) + \lambda_2 f^*(y_2^*) = \langle \lambda_1 y_1^* + \lambda_2 y_2^*, x\rangle$$
$$\leq f(x) + f^*(\lambda_1 y_1^* + \lambda_2 y_2^*)$$
$$\leq f(x) + \lambda_1 f^*(y_1^*) + \lambda_2 f^*(y_2^*).$$

Therefore equality holds throughout. It follows that $x \in \partial f^*([y_1^*, y_2^*])$ and that $f^*|_{[y_1^*, y_2^*]}$ is affine. Consequently, $y_1^* = y_2^*$.
(b) is proved analogously. □

Theorem 7.3.2. *Suppose X is a reflexive Banach space. Then f is essentially smooth if and only if f^* is essentially strictly convex.*

Proof. This follows from Lemma 7.3.1 and the fact $(\partial f)^{-1} = \partial f^*$ in reflexive spaces. □

Exercise 5.3.10 provides a function $f : \mathbb{R}^2 \to (-\infty, +\infty]$ such that f^* is smooth, but $(f^*)^* = f$ is not strictly convex. So the theorem above highlights a nice tight duality for the 'essential' case. See also Exercise 7.2.2 for more information when f is strictly convex.

The following corollary is immediate from the previous theorem and the definition of Legendre functions.

Corollary 7.3.3. *Suppose X is reflexive. Then f is Legendre if and only if f^* is.*

Exercise 7.3.5 shows the previous corollary can fail outside of reflexive Banach spaces.

Theorem 7.3.4. *Suppose X is a Banach space and $f : X \to (-\infty, +\infty]$ is a proper lsc convex function. The following are equivalent.*

(a) *f is essentially smooth.*
(b) *$\operatorname{int} \operatorname{dom} f \neq \emptyset$ and ∂f is single-valued on its domain.*
(c) *$\operatorname{dom} \partial f = \operatorname{int} \operatorname{dom} f \neq \emptyset$ and ∂f is single-valued on its domain.*
(d) *$\operatorname{int} \operatorname{dom} f \neq \emptyset$, f is Gâteaux differentiable on $\operatorname{int} \operatorname{dom} f$ and $\|\nabla f(x_n)\| \to \infty$ for every sequence (x_n) in $\operatorname{int} \operatorname{dom} f$ converging to some point in $\operatorname{bdry} \operatorname{dom} f$.*

Proof. (a) $\Rightarrow$ (b): By the Brøndsted–Rockafellar theorem (4.3.2), $\operatorname{dom} \partial f \neq \emptyset$. Thus we fix $x \in \operatorname{dom} \partial f$. By assumption, ∂f is locally bounded at x. According to Corollary 7.2.3, $x \in \operatorname{int} \operatorname{dom} f$, and so $\operatorname{int} \operatorname{dom} f \neq \emptyset$.

(b) $\Rightarrow$ (c): First, $\operatorname{int} \operatorname{dom} f \subset \operatorname{dom} \partial f$ (Proposition 4.1.26). Suppose $x \in \operatorname{dom}(\partial f)$. According to Lemma 7.2.4, x cannot be a boundary point of $\operatorname{dom} f$. Hence $x \in \operatorname{int} \operatorname{dom} f$ and thus $\operatorname{dom} \partial f = \operatorname{int} \operatorname{dom} f$.

(c) $\Rightarrow$ (d): f is Gâteaux differentiable where it is continuous and ∂f is single-valued (Corollary 4.2.5). Let $x \in \operatorname{bdry} f$ and $(x_n) \subset \operatorname{int} \operatorname{dom} f$ be such that $x_n \to x$. Suppose by way of contradiction that $\liminf_n \|\nabla f(x_n)\| < \infty$. Pass to a subnet (x_α) of (x_n) such that $\nabla f(x_\alpha) \to_{w^*} x^*$. Because $x_n \to x$, it is easy to check that $x^* \in \partial f(x)$. Thus $x \in (\operatorname{dom} \partial f) \cap (\operatorname{bdry} \operatorname{dom} f)$ which contradicts (c).

(d) $\Rightarrow$ (a): The Gâteaux differentiability of f implies single-valuedness of ∂f and Proposition 4.1.26 guarantees the local boundedness of ∂f on $\operatorname{int} \operatorname{dom} f$. Thus it suffices to show that $\partial f(x) = \emptyset$ for all $x \in \operatorname{bdry} \operatorname{dom} f$. So, suppose $\partial f(x) \neq \emptyset$ for some $x \in \operatorname{bdry} \operatorname{dom} f$. Applying Exercise 7.2.5, the sequence $x_n = u_{1/n}$ converges to x, and $\limsup_n \|\nabla f(x_n)\| < \infty$. □

Lemma 7.3.5. *Suppose X is a Banach space and $f : X \to (-\infty, +\infty]$ is a proper lsc convex function. Suppose both $\operatorname{dom} \partial f$ and $\operatorname{dom} f^*$ are open. Then f is essentially strictly convex if and only if f is strictly convex on $\operatorname{int} \operatorname{dom} f$.*

7.3 Legendre functions

Proof. First, int dom $f^* \subset$ dom ∂f^* (Proposition 4.1.26). By the openness of dom f^* and Corollary 7.2.3, we deduce that ∂f^* is locally bounded on its domain. In particular, $(\partial f)^{-1}$ is locally bounded on its domain. Again int dom $f \subset$ dom ∂f, but the latter set is open and so both sets are equal, and convex. The equivalence now follows easily. □

The next results lists some properties of essentially strictly convex functions on reflexive spaces.

Theorem 7.3.6. *Suppose X is reflexive and f is essentially strictly convex. Then:*

(a) *For all $x, y \in X$, $x \neq y \Rightarrow \partial f(x) \cap \partial f(y) = \emptyset$.*
(b) *range $\partial f =$ dom $\partial f^* =$ int dom $f^* =$ dom ∇f^*.*
(c) *For all $y \in$ dom ∂f, $\partial f^*(\partial f(y)) = \{y\}$.*

Proof. (a) Follows from Lemma 7.3.1(b). The first equality in (b) is trivial. The others follow from Theorems 7.3.2 and 7.3.4. Lastly, (c) follows easily from (a). □

Theorem 7.3.7. *Suppose X is reflexive and $f : X \to (-\infty, +\infty]$ is Legendre. Then*

$$\nabla f : \text{int dom} f \to \text{int dom} f^*$$

is bijective with inverse $(\nabla f)^{-1} = \nabla f^* : \text{int dom} f^* \to \text{int dom} f$. Moreover, the gradient mappings ∇f, ∇f^* are both norm to weak continuous and locally bounded on their respective domains.

Proof. Since f is Legendre, it is both essentially smooth and essentially strictly convex. Therefore f is differentiable on int dom $f \neq \emptyset$ (Theorem 7.3.4) and ∂f is a bijection between int dom f and int dom f^* (Theorem 7.3.6). The norm-to-weak continuity follows from Šmulian's theorem (4.2.11), and derivative mappings are locally bounded on their domains because f is continuous where it is Gâteaux differentiable (Proposition 4.2.2) and the subdifferential is locally bounded at points of continuity of f (Proposition 4.1.26). □

Theorem 7.3.8. *Suppose X is a Euclidean space. Then*

(a) *f is essentially smooth if and only if f is essentially smooth in the classical sense.*
(b) *f is essentially strictly convex if and only if f is essentially strictly convex in the classical sense.*

In particular, the classical notion of Legendre is compatible with the general notion on Euclidean spaces.

Proof. (a) This follows from Theorem 7.3.4.
(b) If f is essentially strictly convex, then f is essentially strictly convex in the classical sense. Now suppose f is essentially strictly convex in the classical sense. According to Lemma 7.3.1(a), ∂f^* is single valued on its domain (which is nonempty). Then f^* is Fréchet differentiable on its domain because it has unique subgradients there (Theorem 2.2.1), and hence f is continuous on the domain of ∂f^* and, as before,

∂f^* is locally bounded on its domain. Therefore, f^* is essentially smooth, and hence f is essentially strictly convex according to Theorem 7.3.2. □

In contrast to the previous theorem, the classical and general notions of Legendre are not equivalent in general Banach spaces.

Example 7.3.9 (Strictly convex does not imply essential strictly convex). We first show that the classical Legendre notions do differ from the general ones outside finite-dimensional spaces.

Let $X := \ell_2$ and let $2 \leq p_n < \infty$ be given with p_n converging to ∞ as $n \to \infty$. Define $f : X \to \mathbb{R}$ by

$$x \mapsto \sum_n \frac{1}{p_n}|x_n|^{p_n}, \quad \text{where } x := (x_n).$$

It is easy to check that f is everywhere differentiable and strictly convex. It is therefore Legendre in the classical sense. Hence, the function f is essentially smooth.

Define the index conjugate to p_n through $\frac{1}{p_n} + \frac{1}{q_n} = 1$. Then $2 \geq q_n \to 1^+$, and $f^*(y) = \sum_n \frac{1}{q_n}|y_n|^{q_n}$. In particular, f^* is not essentially smooth in the classical sense. Moreover, $\partial f^*(y) = \{(\text{sign}(y_n)|y_n|^{q_n-1})\}$, provided this element lies in ℓ_2. Consequently, ∂f^* is single-valued on its domain but is not locally bounded according to Theorem 7.3.2. Thus the function f is not essentially strictly convex. Further details are discussed in [35, Example 5.14].

The example thus shows that the following *three implications*, which are always true in finite-dimensional spaces, *all can fail in infinite dimensions*:

(a) 'f essentially strictly convex in the classical sense $\Rightarrow$ int dom $f^* \neq \emptyset$'.
(b) '∂f^* is single-valued on its domain $\Rightarrow f^*$ is essentially smooth'.
(c) 'f is strictly convex $\Rightarrow f$ is essentially strictly convex'.

Exercises and further results

7.3.1.* Prove Lemma 7.3.1(b). Verify that $(\partial f)^{-1} = \partial f^*$ in reflexive spaces and fill in any remaining details in the proof of Theorem 7.3.2.

7.3.2. Let $f : \mathbb{R}^n \to (-\infty, +\infty]$ be a convex function. Suppose $x \in \text{dom} f$.

(a) Show that $\partial f(x)$ is a singleton if and only if f is differentiable at x.
(b) Show that this fails in every infinite-dimensional WCG space.

Hint. There are several approaches to showing if $\partial f(x)$ is a singleton at $x \in \text{dom} f$, then f is differentiable at x on a finite-dimensional space. See Theorem 2.2.1 or [369, Theorem 25.1]. Alternatively, if the affine hull of the domain of f is all of $\mathbb{R}^n$, then the domain of f has nonempty interior. Now use Lemma 7.2.4. If the affine hull of the domain of f is not all of $\mathbb{R}^n$, then the subdifferential mapping is not single-valued at any point where it is nonempty. For (b) consider the indicator function of a symmetric closed convex set whose span is dense but not closed. □

7.3 Legendre functions

7.3.3 (Coercivity and essential strict convexity vs strict convexity). Let X be a Banach space, and suppose $f : X \to (-\infty, +\infty]$ is lsc.

(a) If f is essentially strictly convex, is it coercive?
(b) If f is strictly convex and coercive, is it necessarily essentially strictly convex?

Hint. (a) No. Consider the strictly convex real function $f := \exp$. This is essentially strictly convex since the space is finite-dimensional, but it is not coercive.

(b) Yes in reflexive spaces; with the following reasoning. Suppose $f : X \to (-\infty, +\infty]$ is coercive. Then $0 \in \operatorname{int} \operatorname{dom} f^*$ according to the Moreau-Rockafellar theorem (4.4.10). In particular, $\operatorname{int} \operatorname{dom} f^* \neq \emptyset$. According to Lemma 7.3.1(i), ∂f^* is single-valued on its domain since $f = (f^*)^*$ is strictly convex. Now Theorem 7.3.4 ensures that f^* is essentially smooth (since condition (ii) is satisfied). Then Theorem 7.3.2 shows that f is essentially strictly convex.

In general spaces the following example shows the property can fail. Consider $X := \ell_1$ and let $f(x) := \|x\|_1$ and

$$g(x) := (x_1 - 1)^2 + \sum_{i=2}^{\infty} \frac{x_i^2}{2^i}$$

where $x = (x_i)$. Next, let $\Lambda := (1, 1, 1, \ldots) \in \ell_\infty$, and define $h = f + g$. Then h is coercive because f is, and it is strictly convex because g is strictly convex.

Now $h(e_1) = 1$ and so $\Lambda \in \partial h(e_1)$ since $h(e_1) = \Lambda(e_1)$ and $\Lambda(x) \leq \|x\|_1 \leq h(x)$ for all $x \in X$.

Now consider $x_n = e_1 + ne_n$, then $\Lambda \in \partial f(x_n)$ and $\phi_n \in \partial g(x_n)$ where $\phi_n = \frac{2n}{2^n} e_n$. Then $\Lambda + \phi_n \in \partial h(x_n)$, $\Lambda + \phi_n \to \Lambda$ but $\|x_n\| \to \infty$ so $x_n \in \partial^{-1} h(\Lambda + \phi_n)$ and $\Lambda \in \operatorname{dom} \partial^{-1} h$ and so $\partial^{-1} h$ is not locally bounded at Λ in its domain. □

7.3.4 (Fréchet-Legendre functions). Let X be a Banach space, and f be a lsc proper convex function on X. We will say f is *essentially Fréchet smooth* if it is essentially smooth and Fréchet differentiable on the interior of its domain.

(a) Suppose X is reflexive. Show that f is essentially Fréchet smooth if and only if f^* is essentially strictly convex and every point of range(∂f) is a strongly exposed point of f^*.
(b) Suppose X is reflexive. We will say that f is a *Fréchet-Legendre function* if f and f^* are both essentially Fréchet smooth.
   (i) Show that f is Fréchet-Legendre if and only if f^* is.
   (ii) Suppose f is continuous and cofinite. Show that f is a Fréchet-Legendre function if and only if it is Fréchet differentiable and locally uniformly convex.
   (iii) Suppose X is a Euclidean space. Show that f is a Fréchet-Legendre function if and only if it is a Legendre function in the classical sense.

Notice that a Banach space is automatically reflexive if it has a continuous convex function f such that both f and f^* are Fréchet differentiable (Exercise 5.3.16).

Hint. (a) Suppose f is essentially Fréchet smooth. Then f^* is essentially strictly convex by Theorem 7.3.2. Suppose $\phi \in \partial f(x)$. Then $x \in \operatorname{int} \operatorname{dom} f$ by Theorem 7.3.4 and so f

is Fréchet differentiable at x. According to Proposition 5.2.4, f^* is strongly exposed by x at ϕ. For the converse, f is essentially smooth by Theorem 7.3.2. Fix $x_0 \in \operatorname{int} \operatorname{dom} f$. Then f is Gâteaux differentiable at x_0. Let $\phi \in \partial f(x_0)$ and $\phi_n \in \partial f_{\varepsilon_n}(x_0)$ where $\varepsilon_n \to 0^+$. Follow the argument of (iii) implies (i) in Theorem 5.3.7(b) to deduce that f is Fréchet differentiable at x_0.

(b) Part (i) follows because $f^{**} = f$; (ii) follows because Exercise 5.3.17 shows f is locally uniformly convex iff f and f^* are Fréchet differentiable; (iii) follows because Gâteaux differentiability and Fréchet differentiability coincide for convex functions on Euclidean spaces (Theorem 2.2.1). □

7.3.5. In contrast to Corollary 7.3.2, find an example of a Legendre function f on a nonreflexive Banach space such that f^* is not Legendre.

Hint. Let $X := \ell_1$. Deduce that there is a norm $\|\cdot\|$ on X such that $\|\cdot\|$ is both smooth and strictly convex (Exercise 5.1.25). Show that $f := \frac{1}{2} \|\cdot\|^2$ is a Legendre function. Compute f^* and conclude it is not essentially smooth because $X^* = \ell_\infty$ does not admit any equivalent smooth norm (Exercise 4.6.7(b)). □

7.3.6 (Convolution with an essentially smooth function). Suppose f and g are closed proper convex functions on a Euclidean space and that f is essentially smooth. Show that if $\operatorname{ri}(\operatorname{dom} f^*) \cap \operatorname{ri}(\operatorname{dom} g^*) \neq \emptyset$ then $f \square g$ is essentially smooth. See [369, Corollary 26.3.2] for further details.

7.3.7 (Legendre transform). Suppose f is a (finite) everywhere differentiable convex function on a Euclidean space E. Show the following:

(a) ∇f is one-to-one on E if and only if f is both strictly convex and cofinite.
(b) In this case, f^* is also everywhere differentiable, strictly convex and cofinite; and

$$f^*(x^*) = \langle (\nabla f)^{-1}(x^*), x^* \rangle - f\left((\nabla f)^{-1}(x^*)\right).$$

There is of course a dual formula for f, see [369, Theorem 26.6].

7.4 Constructions of Legendre functions in Euclidean space

This section begins with a direct proof of the log-convexity of the *universal* barrier for an arbitrary open convex set in a finite-dimensional Banach space considered by Nesterov and Nemirovskii in [331]. Then we explore some refinements of this result.

Theorem 7.4.1 (Essentially smooth barrier functions). *Let A be a nonempty open convex set in $\mathbb{R}^N$. For $x \in A$, set*

$$F(x) := \lambda_N((A-x)^o),$$

where λ_N is N-dimensional Lebesgue measure and $(A-x)^o$ is the polar set. Then F is an essentially smooth, log-convex, barrier function for the set A.

7.4 Constructions in Euclidean space

Proof. Without loss of generality, we assume A is line free (else we can add the square of the norm on the lineality subspace [369]). In this case the universal self-concordant barrier b_A is a multiple of $\log F$. (As described in [331], self-concordance is central to the behavior of interior point methods. Because barrier properties (without considering concordance) are applicable more generally, it seems useful to exhibit the strengthened convexity and barrier properties directly as we do below.)

Now observe that F is finite on A, because $\varepsilon B_{\mathbb{R}^N} \subset A - x$ for some $\varepsilon > 0$. Moreover, $F(x) = \infty$ for $x \in \text{bdry } A$. Indeed, this follows by translation invariance of the measure since $(A - x)^o$ contains a ray and has nonempty interior as A is line free. (So we really need Haar measure which effectively limits this proof to finite dimensions.) Therefore, F is a barrier function for A.

Now, by the spherical change of variable theorem

$$F(x) = \frac{1}{N} \int_{S_X} \frac{1}{(\delta_A^*(u) - \langle u, x \rangle)^N} \, du \qquad (7.4.1)$$

where du is surface measure on the sphere. Because the integrand is continuous on A and the gradient is locally bounded in A it follows that that F is essentially smooth. It remains to verify the log-convexity of F. For this we will use (7.4.1), indeed:

$$F\left(\frac{x+y}{2}\right) = \frac{1}{N} \int_{S_X} \frac{1}{(\delta_A^*(u) - \langle u, \frac{x+y}{2} \rangle)^N} \, du$$

$$= \frac{1}{N} \int_{S_X} \frac{1}{\left(\frac{\delta_A^*(u) - \langle u, x \rangle}{2} + \frac{\delta_A^*(u) - \langle u, y \rangle}{2}\right)^N} \, du$$

$$\leq \int_{S_X} \frac{1}{(\delta_A^*(u) - \langle u, x \rangle)^{N/2} (\delta_A^* - \langle u, y \rangle)^{N/2}} \, du \qquad (7.4.2)$$

$$\leq \sqrt{\int_{S_X} \frac{1}{(\delta_A^* - \langle u, x \rangle)^N} \, du \int_{S_X} \frac{1}{(\delta_A^*(u) - \langle u, x \rangle)^N} \, du} \qquad (7.4.3)$$

$$= \sqrt{F(x)F(y)}$$

where we have used the arithmetic-geometric mean inequality in (7.4.2) and Cauchy–Schwartz in (7.4.3). Taking logs of both sides of the preceding inequality completes the proof. □

We now refine the above example to produce an essentially smooth convex function whose domain is a closed convex cone with nonempty interior. We restrict ourselves to the case of cones here because the most important applications have been in abstract linear programming over cones which are built up from products of cones of positive definite matrices, orthants and other simple cones. Moreover, the technical details for the case of general closed convex sets appeared to be more involved (and we will revisit this more generally with less explicit methods in Section 7.7). Before proceeding, we will need the following lemma.

Barriers and Legendre functions

Lemma 7.4.2. *Let $g(x, u) \geq 0$ be concave in x, let ϕ be convex and decreasing on $\mathbb{R}^+$ and consider*

$$G(x) := \phi^{-1}\left(\int \phi(g(x, u)) \mu(du)\right)$$

for a probability measure μ. Assume additionally the mean H_ϕ defined by $H(a, b) = \phi\left(\frac{\phi^{-1}(a) + \phi^{-1}(b)}{2}\right)$ is concave. Then G is concave.

Proof. Using the fact that $g(x, u)$ is concave in x and ϕ is decreasing in (7.4.4) below, and then Jensen's inequality with the concavity of H_ϕ in (7.4.5) below, we obtain:

$$\phi\left(G\left(\frac{x+y}{2}\right)\right) = \int \phi\left(g\left(\frac{x+y}{2}\right), u\right) \mu(du)$$

$$\leq \int \phi\left(\frac{g(x, u) + g(y, u)}{2}\right) \mu(du) \quad (7.4.4)$$

$$= \int H_\phi(\phi(g(x, u)), \phi(g(y, u))) \mu(du)$$

$$\leq H_\phi\left(\int \phi(g(x, u)) \mu(du), \int \phi(g(y, u)) \mu(du)\right) \quad (7.4.5)$$

$$= H_\phi(\phi(G(x)), \phi(G(y)))$$

$$= \phi\left(\frac{\phi^{-1}(\phi(G(x))) + \phi^{-1}(\phi(G(y)))}{2}\right)$$

$$= \phi\left(\frac{G(x) + G(y)}{2}\right).$$

Because ϕ is decreasing, the previous inequality implies

$$G\left(\frac{x+y}{2}\right) \geq \frac{G(x) + G(y)}{2},$$

and so G is concave. □

The next fact follows from properties of the Hessian, whose computations are left as Exercise 7.4.1, but are easily checked in a computer algebra system.

Fact 7.4.3. *Let $\phi(t) = t^\alpha$ with $\alpha < 0$. Then*

$$H_\phi(a, b) := \phi\left(\frac{\phi^{-1}(a) + \phi^{-1}(b)}{2}\right)$$

is concave.

We now have the tools in hand to prove

Theorem 7.4.4. *Given $F(x) := \lambda_N((A - x)^o)$ as above, and letting A be an open convex cone, we define G by $G(x) := -(F(x)^{-p})$ where $0 < p < 1/N$ is fixed. Then*

7.4 Constructions in Euclidean space

G is convex, essentially smooth, vanishes on bdry(A) and has domain equal to the closure of A (where $(A - x) = (A - x)-$, the negative cone).

Proof. Let $\phi(t) = t^{-N}$; then ϕ is convex and decreasing on $\mathbb{R}^+$. Moreover, Fact 7.4.3 ensures that H_ϕ is concave. Consequently,

$$\widetilde{G}(x) = F(x)^{-\frac{1}{N}} = \phi^{-1}\left(\int \phi(\delta_A^*(u) - \langle u, x \rangle)\, du\right)$$

is concave by Lemma 7.4.2 because $g(x, u) = \delta_A^*(u) - \langle u, x \rangle$ is concave in x. Because t^α is concave and increasing for $0 < \alpha < 1$, it follows that $\widetilde{G}^\alpha$ is concave. Because $G = -\widetilde{G}^\alpha$, we know that G is convex. Also, G is smooth on A because F is smooth and does not vanish there. Moreover, G vanishes on bdry(A) because F is infinite there.

Therefore, it remains to show that G is essentially smooth, where we, of course, have defined $G(x) := +\infty$ for $x \notin \bar{A}$. To do this, we will check that $\|\nabla G(x_n)\| \to \infty$ as $x_n \to \bar{x}$ where $\bar{x} \in$ bdry(A) (see Theorem 7.3.4).

Now, we have (normalized):

$$\|\nabla F(x)\| \geq \int_{h \in A \cap S_X} \langle h, u \rangle \int_{u \in A^+ \cap S_X} \langle u, x \rangle^{-N-1}\, \mu(du)\mu(dh)$$

where S_X is the unit sphere and A^+ is the positive polar cone. Interchanging the order of integration, we write

$$\|\nabla F(x)\| \geq \int_{u \in A^+ \cap S_X} \eta(u) \langle u, x \rangle^{-N-1}\, \mu(du),$$

where

$$\eta(u) := \int_{h \in A \cap S_X} \langle h, u \rangle\, \mu(dh).$$

It suffices to observe, by continuity, that $\inf\{\eta(u) : u \in A^+ \cap S_X\} > 0$ (since A^+ is pointed) and so for some constant $K > 0$

$$K\|\nabla F(x)\| \geq \int_{u \in A^+ \cap S_X} \langle u, x \rangle^{-N-1}\, \mu(du) \geq \left[\int_{u \in A^+ \cap S_X} \langle u, x \rangle^{-N}\, \mu(du)\right]^{1+\frac{1}{N}},$$

on applying Hölder's inequality (Exercise 2.3.2). Thus,

$$K^N \|\nabla F(x)\|^N \geq F(x)^{N+1}.$$

A direct computation of $\nabla G(x)$ shows that $\|\nabla G(x_n)\| \to \infty$ as $x_n \to \bar{x} \in$ bdry(A). □

Constructions of Legendre barrier type functions on convex sets with nonempty interior in Banach spaces are explored in Section 7.7. The current results rely heavily on the existence of Haar measure on a Banach space and as such seem fundamentally finite-dimensional, as the space is then necessarily locally compact.

Exercises and further results

7.4.1. Prove Fact 7.4.3.

The next two exercises derive from [332].

7.4.2 (Rogers–Hölder inequality). Suppose $-\infty < p < 1, p \neq 0$ and $1/p + 1/q = 1$. Show that for all complex numbers a_k, b_k and all $n \in \mathbb{N}$ one has

$$\sum_{k=1}^{n} |a_k b_k| \geq \left(\sum_{k=1}^{n} |a_k|^p\right)^{1/p} \left(\sum_{k=1}^{n} |b_k|^q\right)^{1/q}$$

Hint. Appeal to the Fenchel–Young inequality (Proposition 2.3.1). □

7.4.3 (Brun–Minkowski inequality). Let H and K be convex bodies in n-dimensional Euclidean space. Show that, if V_n denotes Lebesgue volume, the function $t \mapsto V_n(tH + (1-t)K)^{1/n}$ is concave on $[0, 1]$. Deduce for $0 < t < 1$ that

$$V_n(tH + (1-t)K)^{1/n} \geq tV_n(H)^{1/n} + (1-t)V_n(K)^{1/n},$$

with equality only if H and K are equal after dilation and translation.

Hint. First appeal to homogeneity to reduce to the case $V_n(H) = V_n(K) = 1$. Then write V_n as an integral and first establish and then apply the integral analog of the Rogers–Hölder's inequality of Exercise 7.4.2. (See [332, pp.158–159] for more details.) □

7.4.4 (Universal barriers). With F as in Theorem 7.4.1 compute $\log F(x)$, for A the strictly positive orthant in N-dimensional Euclidean space. Perform the corresponding computation for the positive semidefinite cone. Compare $\log F(x)$ to $\sum_i^N \log(x_i)$ and $\log \mathrm{Det}$.

7.4.5 (Second-order cone). Show that

$$(x_1, x_2, \ldots, x_N, r) \mapsto \log\left(r^2 - \sum_{k=1}^{N} x_k^2\right)$$

is a concave, continuously twice-differentiable barrier function on the *Bishop–Phelps* (or *second-order*) *cone*: $\{(x, r) : \|x\|_2 < r\}$.

The next two exercises illustrate the benefits of a variational formulation of representatives of two quite distinct classes of problem. The benefits are three-fold: in the efficiency of proof it can provide, in the insight it offers, and in the computational strategies it may then provide. In each case, the barrier property of the objective is fundamental to the attainment of the infimum.

7.4.6 (Matrix completion [242]).** Fix a set $\Delta \subset \{(i,j) : 1 \leq i \leq j \leq n\}$. Suppose the subspace Λ of symmetric matrices with (i,j)-th entry zero for (i,j) in Δ has a positive definite member. By considering the problem

$$\inf\{\langle C, X\rangle - \log \det X : X \in \Lambda \cap S^n_{++}\},$$

(for C positive-definite) prove there exists a positive-definite matrix X in Λ with $C - X^{-1}$ having (i,j)-th entry zero for all (i,j) not in Δ.

7.4.7 (BFGS update for quasi-Newton methods [219]).** Fix a positive definite matrix C in S^n and vectors s, y in $\mathbb{R}^n$ with $s^T y > 0$. Consider the problem

$$\inf\{\langle C, X\rangle - \log \det X : Xs = y, \ X \in S^n_{++}\}. \tag{7.4.6}$$

(a) For small $\delta > 0$ show

$$X := \frac{(y - \delta s)(y - \delta s)^T}{s^T(y - \delta s)} + \delta I$$

is feasible for (7.4.6).

(b) Prove (7.4.6) has an optimum. Determine the solution. It is called the *BFGS update* of C^{-1} under the *secant condition* $Xs = y$, and was originally proposed for nonvariational reasons.

7.5 Further examples of Legendre functions

This section presents some explicit constructions of Legendre functions.

Example 7.5.1 (Spectral Legendre functions). Suppose as in Section 3.2 that E is the Euclidean space of $n \times n$ self-adjoint matrices, with $\langle x, y\rangle = \text{trace}(xy)$, for all $x, y \in E$. Suppose $g : \mathbb{R}^J \to (-\infty, +\infty]$ is proper convex, lsc and invariant under permutations. Let $\lambda(x) \in \mathbb{R}^n$ denote the eigenvalues of $x \in E$ ordered decreasingly. Most satisfactorily Lewis [288] has shown that

$$g \circ \lambda \text{ is Legendre if and only if } g \text{ is.}$$

For extensions of much of this framework to self-adjoint compact operators, see Section 6.5. This construction allows one to obtain many useful Legendre functions on X: for instance, the *log barrier* $x \mapsto -\log \det x$ is immediately seen to be a Legendre function on E (with the positive definite matrices as its domain) precisely because $-\log$ is a Legendre function with domain $(0, \infty)$.

Lemma 7.5.2. Set $f := \frac{1}{p}\|\cdot\|^p$ for $1 < p < \infty$. Let q be given by $\frac{1}{p} + \frac{1}{q} = 1$. Then $f^* = \frac{1}{q}\|\cdot\|^q$,

$$\partial f(x) = \begin{cases} \|x\|^{p-2} Jx & \text{if } x \neq 0; \\ 0 & \text{if } x = 0, \end{cases} \quad \text{and} \quad \partial f^*(x^*) = \begin{cases} \|x^*\|^{q-2} J^* x^* & \text{if } x^* \neq 0; \\ 0 & \text{if } x^* = 0 \end{cases}$$

where J is the *duality mapping*. Consequently:

(a) $\|\cdot\|$ is smooth $\Leftrightarrow f$ is essentially smooth;
(b) $\|\cdot\|$ is rotund $\Leftrightarrow f$ is essentially strictly convex;
(c) $\|\cdot\|$ is smooth and rotund $\Leftrightarrow f$ is Legendre.

Proof. The formulas for the subdifferentials are immediate since $\partial \frac{1}{2}\|\cdot\|^2 = J$.

(a): $\|\cdot\|$ is smooth $\Leftrightarrow J$ is single-valued on $X \Leftrightarrow \partial f$ is single-valued on $X \Leftrightarrow f$ is essentially smooth.

(b): $\|\cdot\|$ is strictly convex $\Leftrightarrow \|\cdot\|^2$ is strictly convex $\Leftrightarrow \frac{1}{p}\|\cdot\|^p$ is strictly convex $\Leftrightarrow f$ is essentially strictly convex.

(c): is now clear from (a) and (b). $\square$

The following two constructions are much simpler than those given in Section 7.4 because we work with balls rather than more general convex bodies.

Example 7.5.3 (Legendre functions on balls). Suppose X is reflexive and its norm is smooth and strictly convex, and rotund so that $\frac{1}{2}\|\cdot\|^2$ is Legendre (Lemma 7.5.2). In particular, X can be any ℓ_p space for $1 < p < \infty$. Define

$$f(x) := \begin{cases} -\sqrt{1-\|x\|^2} & \text{if } \|x\| \leq 1; \\ \infty & \text{otherwise.} \end{cases}$$

Then f is strictly convex, $\mathrm{dom}\, f = B_X$, and $f^*(x^*) = \sqrt{\|x^*\|^2 + 1}$. Moreover,

$$\nabla f(x) = \frac{Jx}{\sqrt{1-\|x\|^2}} \quad \text{and} \quad \nabla f^*(x^*) = \frac{J^* x^*}{\sqrt{\|x^*\|^2 + 1}},$$

for every $x \in \mathrm{dom}\, \nabla = \mathrm{dom}\, \partial f = \mathrm{int}\, B_X$, and every $x^* \in \mathrm{dom}\, \nabla f^* = X^*$. It follows that f is Legendre with $\mathrm{dom}\, f = B_X$.

Example 7.5.4 (A Legendre function with bounded open domain). Suppose X is a reflexive Banach space with strictly convex smooth norm $\|\cdot\|$. Define f by

$$f(x) := \begin{cases} \dfrac{1}{1-\|x\|^2} & \text{if } \|x\| < 1; \\ \infty & \text{otherwise.} \end{cases}$$

Hence f is strictly convex, $\nabla f(x) = -(2Jx)/(1-\|x\|^2)^2$ for every $x \in \mathrm{dom}\, f$ and $\mathrm{dom}\, \nabla f = \mathrm{int}\, B_X$, so f is essentially smooth (Theorem 7.3.4). Since $\mathrm{dom}\, f \subseteq 1 \cdot B_X$, f^* is 1-Lipschitz on X^* by Proposition 4.4.6. By Theorem 7.3.4 (applied to f^*), the function f^* is differentiable on the entire space X^*. Hence f^* is essentially smooth. Also by Theorem 7.3.4 (applied to f^*), the function f is essentially strictly convex. Altogether, f is Legendre with $\mathrm{dom}\, f = \mathrm{int}\, B_X$.

Example 7.5.5. In [364], Reich studies 'the method of cyclic Bregman projections' in a reflexive Banach space X under the following assumptions: (a) $\mathrm{dom}\, f = \mathrm{dom}\, \nabla f =$

7.5 Further examples of Legendre functions

X (hence f is essentially smooth) and so f^* is essentially strictly convex; (b) ∇f maps bounded sets to bounded sets, ∇f is uniformly continuous on bounded sets (hence f is Fréchet differentiable, and f^* is supercoercive); (c) f is uniformly convex (hence f is strictly convex on X).

According to Fact 5.3.16, the uniform convexity of f implies that

$$\liminf_{\|x\|\to+\infty} \frac{f(x)}{\|x\|^2} > 0$$

(hence f is supercoercive and so f^* is bounded on bounded sets and ∂f^* maps bounded sets to bounded sets by Theorem 4.4.13). Altogether, f is Legendre and $\nabla f : X \to X^*$ is bijective and norm-to-norm continuous.

Exercises and further results

7.5.1. Suppose X is a Banach space with uniformly convex and uniformly smooth norm $\|\cdot\|$, and suppose $f = \|\cdot\|^s$, where $1 < s < \infty$. Then f is uniformly convex and uniformly smooth on bounded sets, and in particular, f is Legendre.

Hint. See Exercise 5.3.2 and Example 5.3.11. □

7.5.2 (Multiplicative potential and penalty functions [302]).[†] The functions $\mathcal{P}_{m+1}$ and $\mathcal{P}_m$ defined in Euclidean space by

$$\mathcal{P}_{m+1}(x) := \frac{(\langle c,x\rangle - d)^{m+1}}{\prod_{j=1}^{m}(\langle a_j,x\rangle - b_j)}$$

and

$$\mathcal{P}_m(x) := \frac{(\langle c,x\rangle - d)^m}{\prod_{j=1}^{m}(\langle a_j,x\rangle - b_j)}$$

form barrier functions for the corresponding linear program

$$\min\{\langle c,x\rangle - d : \langle a_j,x\rangle \geq b_j, 1 \leq j \leq m\}.$$

Let $V := \{x : \langle c,x\rangle > d, \langle a_j,x\rangle > b_j, 1 \leq j \leq m\}$.

(a) Show that $\mathcal{P}_{m+1}$ is convex on V and that $\mathcal{P}_m$ is convex on V if V is bounded.
(b) More generally, define P for $r \in \mathbb{R}$ and strictly positive x by

$$P(r,x) := \frac{1}{\gamma}\frac{|r|^\gamma}{\prod_{j=1}^{n} x_j^{a_j}}.$$

Show that P is convex if $1 + \sum_j a_j \leq \gamma$ and $a_1 > 0, \ldots, a_n > 0$. *Hint.* Compute P^* and observe that it has the same general form. Hence, $P = P^{**}$; details are given in [95, Exercise 13, p. 94]. Alternatively, find an proof by inducting on the dimension of the space as in [302]. □

(c) Suppose x is restricted to a simplex ($\{x : \langle s, x\rangle = \sigma, x \geq 0\}$ where $\sigma > 0$ and s has positive coordinates). Then $\mathcal{P}$ is convex as soon as $\gamma \geq 1 + (n-1)\|a\|_\infty$ (see [302]).
(d) Note that $\mathcal{P}_{m+1}$ and $\mathcal{P}_m$ are both compositions of affine functions with specializations of $\mathcal{P}$.

Another useful barrier or penalty-like tool is to write a given convex set C as the set of minimizers of a smooth convex function f_C. This is explored in the next two exercises.

7.5.3 (Constructible convex sets and functions [117]). ** A (closed convex) set in a Banach space is *constructible* if it is representable as the intersection of *countably* many closed half spaces, and we call a convex function constructible if its epigraph is. The term is justified by the observation that C is the limit – in various senses – of a sequence of polyhedra.

Proposition 7.5.6. *A closed convex subset containing the origin is constructible if and only if its polar is weak*-separable.*

Proof. $\Rightarrow$: Since $0 \in C$ we can write $C = \bigcap_{n=1}^\infty \phi_n^{-1}(-\infty, 1]$. Let $W := \overline{\text{conv}}^{w^*}(\{\phi_n\} \cup \{0\})$. Because $\phi_n(x) \leq 1$ for all $x \in C$, note that $\phi(x) \leq 1$ for all $x \in C$, and all $\phi \in W$; thus $W \subset C^o$. If $W \neq C^o$, there exist $\phi \in C^o \setminus W$ and $x_0 \in X$ such that $\phi(x_0) > 1 > \sup_W x_0$ (we know this since $0 \in W$). Thus $\phi_n(x_0) < 1$ for all n and so $x_0 \in C$. This with $\phi(x_0) > 1$ contradicts that $\phi \in C^o$. Consequently, $W = C^o$, and so C^o is weak*-separable.

$\Leftarrow$: Suppose that C^o is weak*-separable. Choose a countable weak*-dense collection $\{\phi_n\}_{n=1}^\infty \subset C^o$. Clearly, $C \subset \bigcap_{n=1}^\infty \phi_n^{-1}(-\infty, 1]$. Moreover, if $x_0 \notin C$, then there is a $\phi \in X^*$ such that $\phi(x_0) > 1 > \sup_C \phi$. Then $\phi \in C^o$. The weak*-density of $\{\phi_n\}_{n=1}^\infty$ in C^o implies there is an n such that $\phi_n(x_0) > 1$. Thus $C = \bigcap_{n=1}^\infty \phi_n^{-1}(-\infty, 1]$ as desired. □

(a) In particular, all closed convex subsets of a separable space are constructible. The situation in general is quite subtle and under set theoretic axioms additional to ZFC there are nonseparable spaces, including a $C(\Omega)$ due to Kunen under the continuum hypothesis, in which all closed convex sets are constructible [117].
(b) Show that the Hausdorff limit of a sequence of constructible sets is constructible if the limit has nonempty interior.
(c) A dual space is separable if and only if every norm closed convex subset is constructible (see [117]).
(d) Every weakly compact convex set in X is constructible if and only if X^* is weak*-separable (see [117]).

7.5.4 (Constructible convex sets and argminima [26, 117]). ** Every constructible set arises as the argmin of a C^∞-smooth function:

(a) Suppose C is a constructible set. Show there is a C^∞-smooth convex function $f_C : X \to [0, \infty)$ such that $C = f_C^{-1}(0)$.

7.5 Further examples of Legendre functions

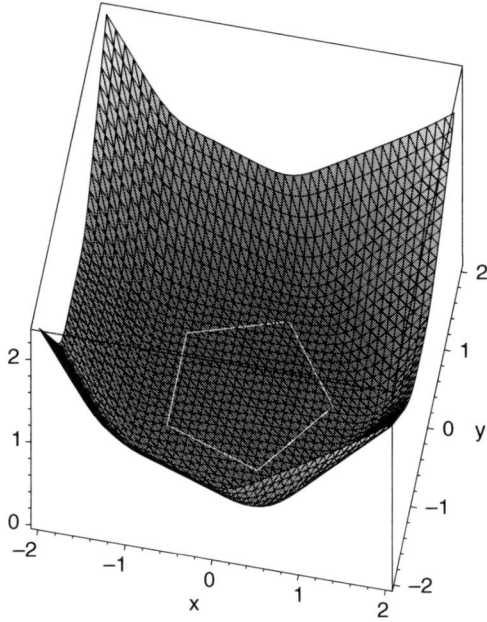

Figure 7.2 The function f_P associated with a pentagon P.

Hint. Write $C = \bigcap_{n=1}^\infty \phi_n^{-1}(-\infty, \alpha_n]$ where $\|\phi_n\| = 1$ for each n, and then choose $\theta : \mathbb{R} \to [0, \infty)$ to be any C^∞-smooth convex function such that $\theta(t) = 0$ for all $t \leq 0$, $\theta(t) = t + b$ for all $t > 1$ with $-1 < b < 0$. The requisite function is defined by

$$f_C(x) := \sum_{n=1}^\infty \frac{\theta(\phi_n(x) - \alpha_n)}{(1 + |\alpha_n|)2^n}. \tag{7.5.1}$$

Also, the sets C_n in (b) are C^∞-smooth convex bodies as a consequence of the implicit function theorem (4.2.13). □

Figure 7.2 illustrates this construction for a pentagon.

(b) Suppose $f : X \to [0, \infty)$ is a C^∞-smooth convex function. Show that $C_n := \{x : f(x) \leq 1/n\}$ are C^∞-smooth convex bodies, and the sequence (C_n) converges Mosco to $C := f^{-1}(0)$. In particular, every constructible set is a Mosco limit of C^∞-smooth convex bodies.

Slice convergence in (b) may fail in the nonreflexive setting:

Example 7.5.7. In any separable nonreflexive Banach space there is a C^∞-smooth convex function $f_C : X \to [0, \infty)$ as in (7.5.1) such that $C_n := \{x : f_C(x) \leq 1/n\}$ does not converge slice to $\{x : f_C(x) = 0\}$.

Hint. Let $\{\phi_n\}_{n=1}^\infty$ be a collection in B_{X^*} that is total but not norm dense in B_{X^*}, and such that $\{0\} = \bigcap_{n=1}^\infty \phi_n^{-1}(-\infty, 0]$. Now let $f(x)$ be defined as in (7.5.1). Then

$C = \{0\}$ and (C_n) converges Mosco to C. However, if $x \in B_X$ is such that $\phi_k(x) = 0$ for $k = 1, \ldots, n$, then $f(x) \le \sum_{k=n+1}^{\infty} 2^{-k} < 1/n$.

Thus, C_n contains $F_n := \{x \in B_X : \phi_k(x) = 0, \ k = 1, 2, \ldots, n\}$ and so the proof of [92, Theorem 1] shows that (C_n) does not converge slice to $\{0\}$. Indeed, choose $\phi \in S_{X^*} \setminus Y$ where $Y = \overline{\mathrm{span}}(\{\phi_n\}_{n=1}^{\infty})$. Choose $F \in S_{X^{**}}$ such that $F(Y) = \{0\}$ and $F(\phi) > 0$. Let δ be such that $0 < \delta < F(\phi)$. Now let $W = \{x \in B_X : \phi(x) > \delta\}$, and apply Goldstine's theorem (Exercise 4.1.13) to show that $d(W, F_n) = 0$ for all n, while clearly $d(W, C) > \delta$. Thus (C_n) does not converge slice to C. □

7.6 Zone consistency of Legendre functions

This section considers *Legendre functions whose domains have nonempty interior on reflexive Banach spaces*, and explores some of their properties relevant to optimization. We begin by more carefully investigating Bregman functions first met in Exercise 2.3.29 and finish with defining and establishing properties of zone consistency in Corollary 7.6.9.

Definition 7.6.1 (Bregman distance). The *Bregman distance* corresponding to f is defined by

$$D = D_f : X \times \mathrm{int}\,\mathrm{dom}\, f \to [0, \infty] : (x, y) \mapsto f(x) - f(y) + f'(y; y - x).$$

Thus, the Bregman distance of x_2 from x_1 is the distance at x_2 between the function value and tangent value at x_1 as Figure 7.3 shows for the Kullback–Leibler divergence of 1 from e. Bregman 'distance' is a misnomer since it is really a *divergence* measuring how far away a second point is away from a reference point and is rarely symmetric. It is easy to check that when $f := \|\cdot\|_2^2$ then $D_f(x, y) = \frac{1}{2}\|x - y\|_2^2$ (see Exercise 7.6.2).

For studies of Bregman distance and their fundamental importance in optimization and convex feasibility problems, see [129, 140, 147] and the references therein.

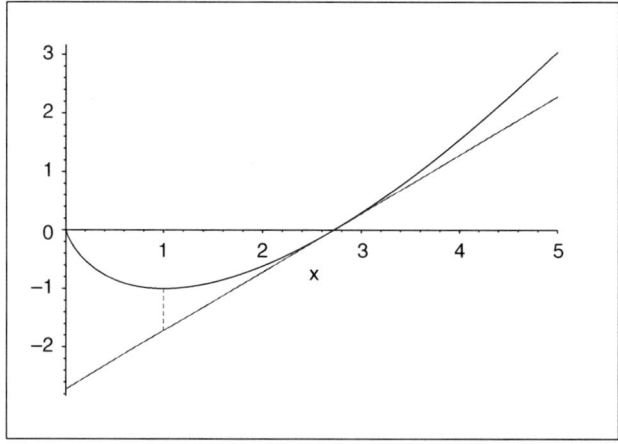

Figure 7.3 The Bregman distance associated with $x \log(x) - x$.

7.6 Zone consistency of Legendre functions

The rationale for considering Bregman distances in optimization/approximation is that we may build some of the constraints directly into the objective function and exploit nonlinear geometry. The relevant algorithmic theory is developed in detail in [36]. Things are especially nice when the corresponding distance is jointly convex as explored in Exercises 2.3.29 and 2.3.30.

We begin with a quite different example of a Legendre function:

Example 7.6.2 (Hilbert space projections). Suppose X is a Hilbert space, $\gamma > 0$, and for a closed convex set C consider

$$f(x) := \frac{1+\gamma}{2}\|x\|^2 - \frac{1}{2}d_C^2(x),$$

where P denotes the (orthogonal) projection map onto C, and $x \in X$. Then $\nabla f(x) = \gamma x + Px$, $D_f(x,y) = \frac{1}{2}(\gamma\|x-y\|^2 + \|x-Py\|^2 - \|x-Px\|^2)$, and

$$f^*(y) = \frac{1}{2(1+\gamma)}\|y\|^2 + \frac{1+\gamma}{2\gamma}d_C^2\left(\frac{1}{1+\gamma}y\right),$$

for all $x,y \in X$. Both f and f^* are supercoercive Legendre functions.

Proof. As we saw (see part (b) of Exercise 4.1.46) $\frac{1}{2}\|\cdot\|^2 - \frac{1}{2}d^2(\cdot)$ is convex and Fréchet differentiable with gradient P. We thus readily obtain the formula for ∇f, and also conclude that f is strictly convex everywhere. The expression for the Bregman distance is a simple expansion. Now let $y = \nabla f(x) = \gamma x + Px$. Then $\frac{1}{1+\gamma}y$ is a convex combination of x and Px: $\frac{1}{1+\gamma}y = \frac{\gamma}{1+\gamma}x + \frac{1}{1+\gamma}Px$. It follows that $P(\frac{1}{1+\gamma}y) = Px$. Hence we can solve $y = \gamma x + Px = \gamma x + P(\frac{1}{1+\gamma}y)$ for x:

$$\nabla f^*(y) = x = \frac{1}{\gamma}y - \frac{1}{\gamma}P\left(\frac{1}{1+\gamma}y\right).$$

Thus $\operatorname{dom} f^* = X$ is open and f is cofinite. Consequently f is a Legendre function. Hence f^* is a Legendre function as the ambient space is reflexive. In fact, since P is nonexpansive, the gradient mapping ∇f^* clearly maps bounded sets to bounded sets, and so f^* is bounded on bounded sets. Thus, by Theorem 4.4.13, f is supercoercive. The same argument shows that f^* is supercoercive. Integrating $\nabla f^*(y)$ with respect to y yields

$$f^*(y) = \frac{1}{2(1+\gamma)}\|y\|^2 + \frac{1+\gamma}{2\gamma}d^2\left(\frac{1}{1+\gamma}y, C\right) + k,$$

where k is constant that we shall determine from the equation $f(x) + f^*(\nabla f(x)) = \langle \nabla f(x), x\rangle$. Using the identity $d_C(\frac{1}{1+\gamma}y) = \frac{\gamma}{1+\gamma}d_C(x)$, we find $k = 0$. □

We next turn to the basic properties of the Bregman distance.

Lemma 7.6.3. Suppose $x \in X$ and $y \in \operatorname{int} \operatorname{dom} f$. Then:

(a) $D(x,y) = f(x) - f(y) + \max \langle \partial f(y), y - x \rangle$.
(b) $D(\cdot, y)$ is convex, lsc, proper with $\operatorname{dom} D(\cdot, y) = \operatorname{dom} f$.
(c) $D(x,y) = f(x) + f^*(y^*) - \langle y^*, x \rangle$, for every $y^* \in \partial f(y)$ with $\max \langle \partial f(y), y-x \rangle = \langle y^*, y - x \rangle$.
(d) If f is differentiable at y, then $D(x,y) = f(x) - f(y) - \langle \nabla f(y), x - y \rangle = f(x) + f^*(\nabla f(y)) - \langle \nabla f(y), x \rangle$ and $\operatorname{dom} \nabla D(\cdot, y) = \operatorname{dom} \nabla f$.
(e) If f is essentially strictly convex and differentiable at y, then $D(\cdot, y)$ is coercive.
(f) If f is essentially strictly convex, then: $D(x,y) = 0 \Leftrightarrow x = y$.
(g) If f is differentiable on $\operatorname{int} \operatorname{dom} f$ and essentially strictly convex, and $x \in \operatorname{int} \operatorname{dom} f$, then $D_f(x,y) = D_{f^*}(\nabla f(y), \nabla f(x))$.
(h) If f is supercoercive and $x \in \operatorname{int} \operatorname{dom} f$, then $D(x, \cdot)$ is coercive.
(i) If X is finite-dimensional, $\operatorname{dom} f^*$ is open, and $x \in \operatorname{int} \operatorname{dom} f$, then $D(x, \cdot)$ is coercive.
(j) If (y_n) is a sequence in $\operatorname{int} \operatorname{dom} f$ converging to y, then $D(y, y_n) \to 0$.

Proof. (a)–(e) are left as part of Exercise 7.6.2 as are some of the justifications below.

(f): Pick y^* as in (c) and assume $0 = D(x,y) = f(x) + f^*(y^*) - \langle y^*, x \rangle$. Then $x \in \partial f^*(y^*) \subseteq \partial f^*(\partial f(y)) = \{y\}$. The converse is trivial.

(g): Using item (f) above, we obtain the equalities $D_{f^*}(\nabla f(y), \nabla f(x)) = f^*(\nabla f(y)) + f(\nabla f^*(\nabla f(x))) - \langle \nabla f^*(\nabla f(x)), \nabla f(y) \rangle = f^*(\nabla f(y)) + f(x) - \langle x, \nabla f(y) \rangle = D_f(x,y)$.

(h), (i): Fix $x \in \operatorname{int} \operatorname{dom} f$ and let (y_n) be a sequence in $\operatorname{int} \operatorname{dom} f$ such that $(D(x, y_n))$ is bounded. Then it suffices to show that (y_n) is bounded.

Pick (see (c)) $y_n^* \in \partial f(y_n)$ such that $D(x, y_n) = f(x) + f^*(y_n^*) - \langle y_n^*, x \rangle$, for every $n \geq 1$. Then (y_n^*) is bounded since $f^* - x$ is coercive. To prove (h), note that supercoercivity of f implies that ∂f^* maps the bounded set $\{y_n^* : n \geq 1\}$ to a bounded set which contains $\{y_n : n \geq 1\}$. The coercivity of $D(x, \cdot)$ follows. It remains to prove (i) in which X is finite-dimensional and $\operatorname{dom} f^*$ is open. Assume to the contrary that (y_n) is unbounded. After passing to a subsequence if necessary, we may and do assume that $\|y_n\| \to \infty$ and that (y_n^*) converges to some point y^*. Then $(f^*(y_n^*)) = (D(x, y_n) - f(x) + \langle y_n^*, x \rangle)$ is bounded. Since f^* is lsc, $y^* \in \operatorname{dom} f^* = \operatorname{int} \operatorname{dom} f^*$. On the one hand, ∂f^* is locally bounded at y^*. On the other hand, $y_n^* \to y^*$ and $y_n \in \partial f^*(y_n^*)$. Altogether, (y_n) is bounded – a contradiction!

(j): By (c), select $y_n^* \in \partial f(y_n)$ such that $D(y, y_n) = f(y) + f^*(y_n^*) - \langle y_n^*, y \rangle$, for every $n \geq 1$. The sequence (y_n^*) is bounded, since ∂f is locally bounded at y. Assume to the contrary that $D(y, y_n) \not\to 0$. Again, after passing to a subsequence, there is some $\varepsilon > 0$ such that $\varepsilon \leq D(y, y_n) = f(y) + f^*(y_n^*) - \langle y_n^*, y \rangle$, for every n, and that (y_n^*) converges weakly to some $y^* \in \partial f(y)$. Since $y_n^* \in \partial f(y_n)$, the assumption implies that $f^*(y_n^*) = \langle y_n^*, y_n \rangle - f(y_n) \to \langle y^*, y \rangle - f(y)$. Hence $f^*(y^*) \leq \liminf_n f^*(y_n^*) = \lim_n f^*(y_n^*) = \langle y^*, y \rangle - f(y) \leq f^*(y^*)$, which yields the absurdity $0 < \varepsilon \leq D(y, y_n) = f(y) + f^*(y_n^*) - \langle y_n^*, y \rangle \to f(y) + f^*(y^*) - \langle y^*, y \rangle = 0$. □

7.6 Zone consistency of Legendre functions

Remark 7.6.4. It is not possible to replace '$y_n \to y$' in Lemma 7.6.3(j) by '$y_n \to_w y$': consider $f := \frac{1}{2}\|\cdot\|^2$ on $X = \ell_2$, let y_n denote the n-th unit vector. Then $y_n \to_w 0$, but $D(0, y_n) = \frac{1}{2}\|0 - y_n\|^2 \equiv \frac{1}{2}$.

Recall that we say that a function is *cofinite* when the domain of the conjugate is the whole dual space. Some connections between cofinite and supercoercive functions were explored in Exercise 4.4.23.

Example 7.6.5 (More about f cofinite $\not\Rightarrow f$ supercoercive). Let $X = \ell_2$ and define $h(y) := \sum_{n \geq 1} \frac{1}{2n} y_n^{2n}$, for every $y := (y_n) \in X^* = X$. Then h is strictly convex, proper, with dom $h = X^*$. Moreover, h is everywhere differentiable with $\nabla h(y) = (n y_n^{2n-1})$. Now set $g := h + \frac{1}{2}\|\cdot\|^2$. Then g is strictly convex, proper, with dom $g = X^* =$ int dom g, everywhere differentiable with $\nabla g = \nabla h + I$, and supercoercive. Since dom $\nabla g = X^*$, again g is essentially smooth. Now let $f := g^*$. Then f is essentially strictly convex, and dom $f = X$ (since $f = h^* \square \frac{1}{2}\|\cdot\|^2$). The strict convexity of g implies that ∂f is single-valued on its domain. Since $X =$ int dom $f \subseteq$ dom ∂f, f must be differentiable everywhere and hence f is essentially smooth. To sum up,

f is Legendre and cofinite with dom $f =$ dom $\nabla f = X$, and
f^* is Legendre and supercoercive with dom $f^* =$ dom $\nabla f^* = X^*$.

Denote the standard unit vectors in X^* by $\mathbf{e}_n$ and fix $\mathbf{x} \in X$ arbitrarily. Then

$$\mathbf{e}_n \to_w 0, \text{ but } \|\nabla f^*(\mathbf{e}_n)\| = n+1 \to \infty.$$

Now let $\mathbf{y}_n = \nabla f^*(\mathbf{e}_n) = (n+1)\mathbf{e}_n$, for every $n \geq 1$. On the one hand, $\|\mathbf{y}_n\| \to \infty$. On the other hand, by Lemma 7.6.3 (iv), $D(\mathbf{x}, \mathbf{y}_n) = f(\mathbf{x}) + f^*(\nabla f(\mathbf{y}_n)) - \langle \nabla f(\mathbf{y}_n), \mathbf{x} \rangle = f(\mathbf{x}) + f^*(\mathbf{e}_n) - \langle \mathbf{e}_n, \mathbf{x} \rangle \leq f(\mathbf{x}) + g(\mathbf{e}_n) + \|\mathbf{e}_n\|\|\mathbf{x}\| \leq f(\mathbf{x}) + 1 + \|\mathbf{x}\|$. Altogether:

there is no $\mathbf{x} \in X$ such that $D_f(\mathbf{x}, \cdot)$ is coercive.

In view of Lemma 7.6.3(h), f is not supercoercive.

Remark 7.6.6. It follows from the above example that 'f is supercoercive' in Lemma 7.6.3(h) cannot be replaced by 'f is cofinite'. Let us also observe that the existence of a cofinite, yet not supercoercive function actually characterizes spaces that do not have the *Schur property* as is shown in Theorem 8.2.5. However, in Example 7.6.5, such a function is constructed explicitly.

The following fundamental ideas go back to Bregman [129].

Definition 7.6.7 (Bregman projection). Suppose C is a closed convex set in X. Given $y \in$ int dom f, the set $P_C y := \{x \in C : D(x, y) = \inf_{c \in C} D(c, y)\}$ is called the *Bregman projection* of y onto C. Abusing notation slightly, we shall write $P_C y = x$, if $P_C y$ happens to be the singleton $P_C y = \{x\}$.

Theorem 7.6.8. Let X be a reflexive Banach space and suppose $f : X \to (-\infty, +\infty]$ is a lsc proper convex function with int dom $f \neq \emptyset$. Suppose C is a closed convex set with $C \cap \text{dom} f \neq \emptyset$, and $y \in \text{int dom} f$. Then:

(a) If f is essentially strictly convex and Gâteaux differentiable at y, then $P_C y$ is nonempty and $P_C y \cap \text{int dom} f$ is at most a singleton.
(b) If f Gâteaux differentiable at y and strictly convex, then $P_C y$ is at most a singleton.
(c) If f is essentially smooth and $C \cap \text{int dom} f \neq \emptyset$, then $P_C y \subseteq \text{int dom} f$.

Proof. (a): By Lemma 7.6.3(b) and (e), $D(\cdot, y)$ is convex, lsc, coercive, and $C \cap \text{dom} D(\cdot, y) \neq \emptyset$. Hence $P_C y = \text{argmin}_{x \in C} D(x, y) \neq \emptyset$. Since f and hence (Lemma 7.6.3(c)) $D(\cdot, y)$ is strictly convex on int dom f, it follows that $P_C y \cap \text{int dom} f$ is at most a singleton.

(b): By Lemma 7.6.3(d), $D(x, y) = f(x) + f^*(\nabla f(y)) - \langle \nabla f(y), x \rangle$. Hence $D(\cdot, y)$ is strictly convex and the result follows.

(c): Assume to the contrary that there exists $\bar{x} \in P_C y \cap (\text{dom} f \setminus (\text{int dom} f))$. Fix $c \in C \cap \text{int dom} f$ and define

$$\Phi : [0, 1] \to [0, \infty) : t \mapsto D\big((1 - t)\bar{x} + tc, y\big).$$

Then, using Lemma 7.6.3(b), Φ is lsc convex proper and $\Phi'(t) = \langle \nabla f(\bar{x} + t(c - \bar{x})), c - \bar{x} \rangle - \langle \nabla f(y), c - \bar{x} \rangle$, for all $0 < t < 1$. By Theorem 7.3.4, $\lim_{t \to 0^+} \Phi'(t) = -\infty$. This implies $\Phi(t) < \Phi(0)$, for all $t > 0$ sufficiently small (since $\Phi'(t)(0 - t) \leq \Phi(0) - \Phi(t)$, i.e. $\Phi(t) \leq \Phi(0) + t\Phi'(t)$, for every $0 < t < 1$). It follows that for such t, $(1 - t)\bar{x} + tc \in C \cap \text{int dom} f$ and $D((1 - t)\bar{x} + tc, y) < D(\bar{x}, y)$, which contradicts $\bar{x} \in P_C y$. The entire theorem is proven. □

In the terminology of Censor and Lent [146], the next result states that *every Legendre function is zone consistent*; see p. 364 and Exercise 7.6.6. This result is of crucial importance, since – as explained below – it makes the sequence generated by the method of cyclic Bregman projections, discussed in the next subsection, well-defined under reasonable constraint qualifications.

Corollary 7.6.9 (Legendre functions are zone consistent). Let X be a reflexive Banach space. Suppose f is a Legendre function, C is a closed convex set in X with $C \cap \text{int dom} f \neq \emptyset$, and $y \in \text{int dom} f$. Then:

$P_C y$ is a singleton and is contained in int dom f.

Proof. This is immediate from Theorem 7.6.8 (a) and (c). □

7.6.1 The method of cyclic Bregman projections

We now demonstrate the power of Legendre functions by studying a specific optimization problem (as a *generalized best approximation problem*). Suppose $C_1, \ldots, C_N$ are closed convex sets ('the constraints') in $\mathbb{R}^M$ with $C = \bigcap_{i=1}^{N} C_i \neq \emptyset$. The *convex feasibility problem* consists of finding a point ('a solution') in C. Suppose

7.6 Zone consistency of Legendre functions

further that the orthogonal projection onto each set C_i, which we denote by P_i, is readily computable. Then *the method of cyclic (orthogonal) projections* operates as follows.

Given a starting point y_0, generate a sequence (y_n) by projecting cyclically onto the constraints:

$$y_0 \xmapsto{P_1} y_1 \xmapsto{P_2} y_2 \xmapsto{P_3} \cdots \xmapsto{P_N} y_N \xmapsto{P_1} y_{N+1} \xmapsto{P_2} \cdots .$$

The sequence (y_n) does indeed converge to a solution of the convex feasibility problem [121] as was first proved by Halperin.

In some applications, however, it is desirable to employ the method of cyclic projections with *(nonorthogonal) Bregman projections* [129]. For example, if we know *a priori* that we wish all solutions to be strictly positive we might well wish to use the Kullback–Leibler entropy. As we have implicitly seen, the Bregman projections are constructed as follows. Given a 'sufficiently nice' convex function f, the *Bregman distance* between x and y is

$$D_f(x,y) = f(x) - f(y) - \langle \nabla f(y), x - y \rangle,$$

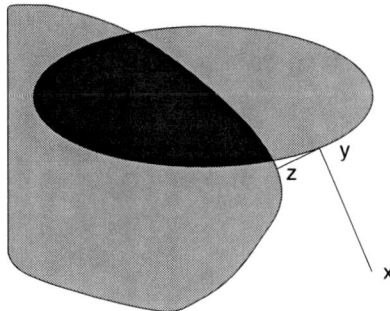

Figure 7.4 The method of cyclic projections: $x \mapsto y \mapsto z \mapsto \cdots$.

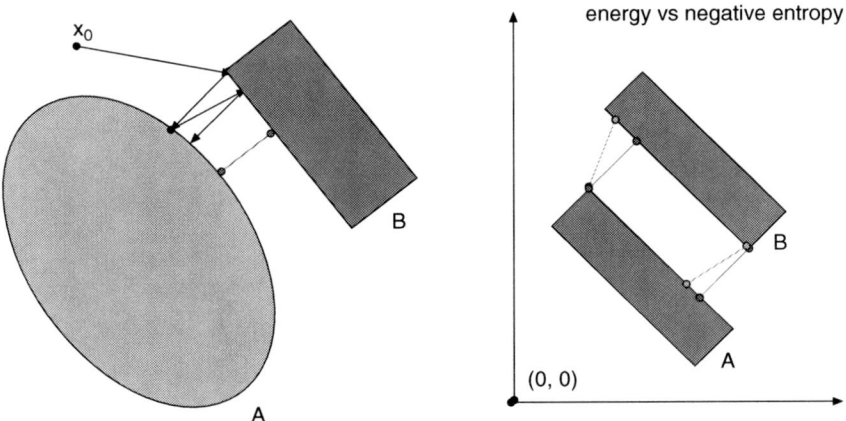

Figure 7.5 Infeasible cyclic projection (L) and entropy analog (R).

where $y \in \text{int dom} f$ is a point of differentiability of f. Then the *Bregman projection* of y onto the i-th constraint C_i with respect to f is defined by

$$\text{argmin}_{x \in C_i} D_f(x, y). \tag{7.6.1}$$

Here we have implicitly assumed that y is a point of differentiability so that $D_f(x, y)$ is well defined. More importantly, to define the sequence of cyclic projections unambiguously, the following properties are required:

- the argmin is nonempty ('existence of nearest points'),
- the argmin is a singleton ('no selection necessary'),
- the argmin is contained in int dom f (in order to project the argmin onto the next constraint C_{i+1}).

These three properties together define *zone consistency*.

The valuable punch-line is that if f is a Legendre function, then these good properties automatically hold [32] and so in the terminology of Censor and Lent [146], every Legendre function is zone consistent (see Exercise 7.6.6). Moreover, the Legendre property is the most general condition known to date that guarantees zone consistency. Note that when f is $\frac{1}{2}\|\cdot\|_2^2$ we recover the classical orthogonal case. The trade-off for better control of the problem's geometry is that unlike the orthogonal case there is typically no nice closed form for the Bregman projection and it must be computed numerically. This and much more may be followed up in [37].

Exercises and further results

7.6.1 (Other entropies). Two other entropies with famous names are the *Fermi–Dirac* and *Bose–Einstein* entropies given, respectively, by

$$x \mapsto (1-x)\log(1-x) + x\log(x) \quad \text{and} \quad x \mapsto x\log(x) - (1+x)\log(1+x).$$

Show that their conjugates are respectively $y \mapsto \log(e^y + 1)$ and $y \mapsto -\log(1 - e^y)$ with domain $\mathbb{R}$ and $(-\infty, 0)$ respectively. (See also Table 2.1.) Confirm that all these functions are of Legendre type. They are shown in Figure 7.6.

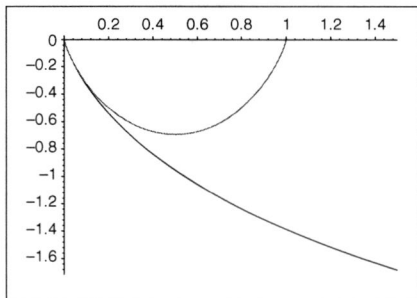 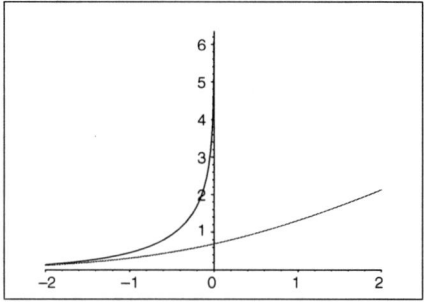

Figure 7.6 Fermi–Dirac and Bose–Einstein entropy (L); conjugates (R).

7.6.2 (Bregmania).*

(a) Establish the missing details in Lemma 7.6.3.
(b) If $f(x_1, x_2, \cdots, x_n) := \sum_{k=1}^{n} f(x_k)$, show that $D_f(x, y) = \sum_{k=1}^{n} D_{f_k}(x_k, y_k)$.
(c) Confirm that when $f = \|\cdot\|_2^2$ then $D_f(x, y) = \frac{1}{2}\|x - y\|_2^2$.
(d) Check that when $f := x \mapsto x\log(x) - x$ then $D_f(x, y) = x\log(x/y) - (x - y)$.
(e) Explore the barrier and Legendre properties of $x \mapsto (2x-1)\{\log x - \log(1-x)\}$. Compute the corresponding Bregman distance.
(f) Explore the barrier and Legendre properties of $x \mapsto x\log x + (1-x)\log(1-x)$. Compute the corresponding Bregman distance.
(g) Explore the barrier and Legendre properties of $x \mapsto \log(1+x) - x$. Compute the corresponding Bregman distance.

7.6.3 (Fisher information [99]).† Lemma 2.1.8 forms the basis for some very useful alternatives to Bregman distance. Given an appropriate convex function $\psi : \mathbb{R} \to (-\infty, +\infty]$ and probabilities $p := (p_1, p_2, \ldots, p_n)$, $q := (q_1, q_2, \ldots, q_n)$ we may define the discrete *Csiszár divergence* from p to q by

$$C_\psi(p, q) := \sum_{k=1}^{n} p_k \psi\left(\frac{q_k}{p_k}\right),$$

and correspondingly for densities in $L_1(X, \mu)$ we may consider

$$I_\psi(p, q) := \int_X p(t) \psi\left(\frac{q(t)}{p(t)}\right) \mu(dt).$$

Exercise 2.3.9 specifies how 0/0 is to be handled.

(a) Unlike the Bregman distance, C_ψ and I_ψ are always jointly convex, but does not recapture the Euclidean distance.
(b) Note for $\psi(t) := -\log t$ we recover the Kullback–Leibler divergence and $\psi(t) := |t-1|$ recreates the L_1 or variational-norm. A nice one-parameter class is obtained from $t \mapsto (t^\alpha - 1)/(\alpha(\alpha - 1))$ interpreted as a limit for $\alpha = 0, 1$. For example, $\alpha = 1/2$ gives $4h^2 := 2\int(\sqrt{p} - \sqrt{q})^2 d\mu$, where h is the *Hellinger distance*, see [228, 227]. A detailed discussion of related ideas can be followed up in [291].
(c) If we start with $\psi(t) := t^2/2$ we obtain

$$F(p, q) := \frac{1}{2} \int_X \frac{q^2(t)}{p(t)} \mu(dt),$$

which makes sense for any strictly positive functions in L_1. Relatedly

$$F(z) := \frac{1}{2} \int_X \frac{(z'(t))^2}{z(t)} \mu(dt),$$

defines the *Fisher information functional* and provides a convex integral functional which penalizes one for large derivative values (oscillations or variances).

This function has been exploited as a replacement for the Shannon entropy in various thermodynamic and information theoretic contexts [142]. See also Exercise 7.6.4.

7.6.4 (Duality of Csiszár entropy problems, I [98, 99]).[†] The Csiszár *best entropy density estimation* program is the convex program:

$$\text{minimize} \quad I_\psi(x) = \int_T x(t)\psi\left(\frac{x'(t)}{x(t)}\right) dt$$
$$(P) \quad \text{subject to} \quad x \geq 0, \ x \in \mathcal{A}(T),$$
$$\int_T a_k(t)x(t)\,dt = b_k \text{ for } k = 0, 1, \ldots, n.$$

More precisely $x\psi(x'/x) := 0^+\psi(x')$ for $x = 0$ and is infinite otherwise; and $\mathcal{A}(T)$ denotes the absolutely continuous functions on an interval T. The weight functions may be trigonometric, $a_k(t) := \exp\{ikt\}$ on $[-\pi, \pi]$, algebraic polynomials (orthonormalized), $a_k(t) := t^k$ on $T = [0, 1]$, in which case the b_k are known Hausdorff (Legendre) moments of the unknown $x(t)$. As in many image reconstruction, seismology, or in small angle neutron scattering or other problems in plasma physics, the a_k might be an appropriate wavelet basis, and the b_k known wavelet coefficients of $x(t)$.

One discovers in this case that the concave dual problem (D) has an integral-differential form that is suppressed when only function values are measured as in Corollary 4.4.19.

$$\text{maximize} \quad -\sum_{j=0}^n \lambda_j b_j \quad \text{subject to} \quad v' + \psi^*(v) = \sum_{j=0}^n \lambda_j a_j,$$
$$(D) \quad v(t_0) = v(t_1) = 0, \ v \in \mathcal{C}^1(T).$$

Proposition 7.6.10 ([98, 99]). *Suppose ψ is strictly convex and supercoercive and (P) is feasible. Then:*

(a) *Problem (P) has a unique optimal solution $\bar{x}$ which is strictly positive on $T = [t_0, t_1]$ and satisfies $\bar{x}'(t_0) = \bar{x}'(t_1) = 0$.*
(b) *Problem (D) has a unique optimal solution $(\bar{v}, \bar{\lambda})$, from which the primal optimal solution $\bar{x}$ may be recovered by means of*

$$\frac{\bar{x}'(t)}{\bar{x}(t)} = \psi^{*'}(\bar{v}(t)) \quad \text{and} \quad A\bar{x} = b \qquad (7.6.2)$$

where A represents the moment measurements.

This dual problem is passing odd: it requires one to know for which parameter values the driving differential equation has a solution with $v(t_0) = v(t_1) = 1$ but not what the solution is. The next exercise shows how to extract more information.

7.6.5 (Duality of Csiszár entropy problems, II [98, 99]).[†] By assumption ψ is strictly convex and supercoercive so that ψ^* is everywhere finite and differentiable. Consider

7.6 Zone consistency of Legendre functions

the differential equation with initial condition

$$v'(t) + \psi^*(v(t)) = \sum_{j=0}^{n} \lambda_j a_j(t), \qquad v(t_0) = 0. \qquad (7.6.3)$$

In the Fisher case $\psi(v) = \psi^*(v) = \frac{1}{2}v^2$, (7.6.3) is a *Riccati equation*. Let Ω be the set of $\lambda \in \mathbb{R}^{n+1}$ for which (7.6.3) has a unique solution $v(\lambda, \cdot)$ on the whole interval $T = [t_0, t_1]$. Clearly $0 \in \Omega$, and it is relatively straight forward that Ω is an open domain in $\mathbb{R}^{n+1}$. Define $k : \Omega \to \mathbb{R}$ by $k(\lambda) := v(\lambda, t_1)$. Then k is of class C^1 on Ω and (D) of Proposition 7.6.10 becomes

$$(\widehat{D}) \qquad \text{maximize} - \sum_{j=0}^{n} \lambda_j b_j \text{ subject to } k(\lambda) = 0, \quad \lambda \in \Omega, \qquad (7.6.4)$$

which has an optimal solution $\bar{\lambda}$ satisfying $\bar{v}(t_1) = v(\bar{\lambda}, t_1) = k(\bar{\lambda}) = 0$.

(a) Consider the Lagrangian associated with $(\widehat{D})$. For $i,j = 0, \ldots, n$

$$\frac{\partial}{\partial \lambda_i} k(\lambda) = \frac{\partial}{\partial \lambda_i} v(\lambda, t_1) =: v_i(\lambda, t_1), \qquad (7.6.5)$$

$$\frac{\partial^2}{\partial \lambda_i \partial \lambda_j} k(\lambda) = \frac{\partial^2}{\partial \lambda_i \partial \lambda_j} v(\lambda, t_1) =: v_{ij}(\lambda, t_1).$$

The functions $v_i(\lambda, \cdot)$ solve the initial value problems

$$v_i'(\lambda, t) + \psi^{*'}(v(\lambda, t)) v_i(\lambda, t) = a_i(t), \qquad v_i(\lambda, t_0) = 0, \qquad (7.6.6)$$

for $i = 0, \ldots, n$, and for $0 \le i,j \le n$ the functions $v_{ij}(\lambda, \cdot)$ solve the initial value problems

$$v_{ij}'(\lambda, t) + \psi^{*''}(v(\lambda, t)) v_i(\lambda, t) v_j(\lambda, t) + \psi^{*'}(v(\lambda, t)) v_{ij}(\lambda, t) = 0,$$

$$v_{ij}(\lambda, t_0) = 0.$$

$$(7.6.7)$$

(b) Let us now define

$$\alpha(t) := \exp\left\{\int_{t_0}^{t} \psi^{*'}(\bar{v}(s)) \, ds\right\}, \qquad (7.6.8)$$

where $\bar{v} = v(\bar{\lambda}, \cdot)$. Then $\alpha(t_0) = 1$, $\alpha(t_1) > 0$, and $\alpha'/\alpha = \psi^{*'}(\bar{v})$. Thus, α is an integrating factor for each equation in (7.6.6), and so $\alpha' v_i + v_i' \alpha = a_i \alpha$, $i = 0, \ldots, n$. On integrating over $T = [t_0, t_1]$, we get

$$\int_{t_0}^{t_1} \alpha(s) a_i(s) \, ds = \alpha v_i \Big|_{t_0}^{t_1} = \alpha(t_1) \frac{\partial}{\partial \lambda_i} k(\bar{\lambda}) = \frac{\alpha(t_1) b_i}{\bar{r}}. \qquad (7.6.9)$$

Choosing $i = 0$ gives $\bar{r} > 0$. Conclude that:

Proposition 7.6.11. *Suppose $(\widehat{D})$ has optimal solution $\bar{\lambda}$. If $C := \bar{r}/\alpha(t_1) > 0$ then (P) has an optimal solution $\bar{x}$ with explicit form*

$$\bar{x}(t) = C \exp\left\{\int_{t_0}^{t} \psi^{*'}(\bar{v}(s)) \, ds\right\}, \tag{7.6.10}$$

where $\bar{v} = v(\bar{\lambda}, \cdot)$, and $(\bar{\lambda}, \bar{r})$ is a Kuhn–Tucker point of $(\widehat{D})$.

(c) Also, $\alpha(t)$ is an integrating factor for the equations in (7.6.7). Hence

$$-\alpha \, \psi^{*''}(v) \, v_i \, v_j = \alpha' \, v_{ij} + \alpha \, v'_{ij}. \tag{7.6.11}$$

On integrating over T we obtain

$$-\int_{t_0}^{t_1} \frac{\alpha}{\alpha(t_1)} \psi^{*''}(v) \, v_i \, v_j \, ds = v_{ij}(\lambda, t_1) = \frac{\partial^2 k}{\partial \lambda_i \partial \lambda_j}(\lambda). \tag{7.6.12}$$

Use this formula to prove the strict concavity of k:

Proposition 7.6.12. *Suppose ψ is differentiable and ψ^* is of class C^2-smooth. Then k is strictly concave on its domain Ω.*

Hint. The Hessian quadratic form for k at $\lambda \in \Omega$ is given by

$$\sum_{i=0}^{n}\sum_{j=0}^{n} \frac{\partial^2 k}{\partial \lambda_i \partial \lambda_j}(\lambda)\, \mu_i \mu_j = -\int_{t_0}^{t_1} \frac{\alpha}{\alpha(t_1)} \psi^{*''}(v) \left(\sum_{k=0}^{n} \mu_k v_k\right)^2 ds \leq 0,$$

for $\mu \in \mathbb{R}^{n+1}$; it suffices to show that, with the possible exception of $\lambda = (\psi^*(0), 0, \ldots, 0)$, the term above is strictly negative for $\mu \neq 0$. □

Surprisingly, we have now found a finite-dimensional concave dual problem

$$(D^*) \quad \text{maximize} \; -\sum_{j=0}^{n} \lambda_j b_j \quad \text{subject to} \; k(\lambda) \geq 0, \quad \lambda \in \Omega. \tag{7.6.13}$$

The work in solving (D^*) directly, via a constrained Newton method, is significant [98], but there are remarkably rapid heuristics discussed in [97].

7.6.6 (Zone consistency). Suppose f is a Legendre function, C is a closed convex set in reflexive Banach space X with $C \cap \text{int dom} f \neq \emptyset$, and $y \in \text{int dom} f$. Confirm that f is zone consistent.

7.7 Banach space constructions

This section explores constructions of essentially smooth functions and of functions of Legendre type in various classes of Banach spaces on convex sets with nonempty

7.7 Constructions on Banach Spaces

interior. The first result gives us conditions on a Banach space under which we can quite generally construct Legendre functions.

Theorem 7.7.1. *Suppose X is a Banach space such that distance functions to closed convex sets are β-differentiable on their complements. Let C be a closed convex subset of X having nonempty interior. Then there is a convex function f that is continuous on C and β-differentiable on $\text{int}(C)$ such that*

(a) $\|f'(x_n)\| \to \infty$ as $x_n \in C$ and $d_{C^c}(x_n) \to 0$;
(b) $\partial f(x) = \emptyset$ if $x \in \text{bdry}(C)$.

Proof. Without loss of generality, we may assume $0 \in \text{int}(C)$. Let $C_n = \{\lambda x : 0 \leq \lambda \leq 1 - 2^{-n}, x \in C\}$. Fix $\varepsilon > 0$ such that $B_\varepsilon \subset C$. First, we observe that

$$d(C_n, C^c) \geq \frac{\varepsilon}{2^n} \quad \text{for all } n. \tag{7.7.14}$$

Indeed, if $x \in C_n$, then $x = (1 - 2^{-n})z$ for some $z \in C$. Now suppose $\|h\| \leq \varepsilon$, then $h \in C$. Therefore, $x + 2^{-n}h = (1 - 2^{-n})z + 2^{-n}h \in C$. Thus, $x + B_{\varepsilon 2^{-n}} \subset C$ which proves (7.7.14).

The hypothesis implies that d_D^2 is β-differentiable whenever D is a convex set, and in particular, $\|\cdot\|^2$ is β-differentiable. The desired function is then defined by

$$f(x) := \|x\|^2 + \sum_{n=1}^{\infty} \frac{n 2^{n+2}}{\varepsilon^2} d_{C_n}^2(x).$$

(Note if for some $\delta > 0$, we have $d_{C_n}(x) > \delta$ for all n, e.g. if $x \in \text{int}(C^c)$ then this sum diverges.)

To see that f is β-differentiable on $\text{int}(C)$, fix $x_0 \in \text{int}(C)$. Now for some N, $x_0 \in \text{int}(C_N)$ and $d_{C_n}(x) = 0$ for all $x \in C_N$ and for all $n \geq N$. Consequently, on $\text{int}(C_N)$, f is a finite sum of β-differentiable functions. Therefore f is β-differentiable on $\text{int}(C)$.

Consider $D_N := \{x \in C : \|x\| \leq N\}$. For $x \in D_N$, we have $d_{C_n}(x) \leq N 2^{-n}$ for all n. Therefore,

$$\sum_{n=1}^{\infty} \left| \frac{n 2^{n+2}}{\varepsilon^2} d_{C_n}^2(x) \right| \leq \sum_{n=1}^{\infty} \frac{4 N n^2}{2^n \varepsilon^2} \quad \text{for all } x \in D_N.$$

By the Weierstrass M-test, f is continuous on D_n for each n; and, in particular, f is defined on $\text{bdry}(C)$.

To prove (a), suppose $x_k \in C$ and $d_{C^c}(x_k) \to 0$. Now f is supercoercive, and so if $\|x_k\| \to \infty$, then $\|f'(x_k)\| \to \infty$ (by Fact 7.2.1). Therefore, by standard subsequence arguments, we may assume that (x_k) is bounded, say $\|x_k\| \leq M$ for all k. Choose $n_k \to \infty$ such that $d_{C^c}(x_k) < \varepsilon 2^{-(n_k+1)}$ and so (7.7.14) implies that $d_{C_{n_k}}(x_k) \geq \varepsilon 2^{-(n_k+1)}$. Now choose $1 - 2^{-n_k} \leq \lambda_k \leq 1$ such that $\lambda_k x_k \in \text{bdry}(C_{n_k})$.

Because $0 \in C_n$ for each n, it follows that $d_{C_n}(x_k) \geq d_{C_n}(\lambda_k x_k)$ for each n. Therefore

$$\frac{f(x_k) - f(\lambda_k x_k)}{\|x_k - \lambda_k x_k\|} \geq \frac{n_k 2^{n_k+2} d^2(x_k, C_{n_k})}{\varepsilon^2 (1 - \lambda_k) M} \geq \frac{n_k}{M}.$$

Therefore $\|f'(x_k)\| \geq \frac{n_k}{M}$. Notice that (b) follows from the same argument since $d_{C^c}(\bar{x}) = 0$ for $\bar{x} \in \mathrm{bdry}(C)$. □

Remark 7.7.2. Suppose the dual norm on X^* is locally uniformly convex (resp. strictly convex), then the above theorem applies with Fréchet (resp. Gâteaux) differentiability. On $L_1(\Omega)$ where Ω is a σ-finite measure space, the above theorem applies with weak Hadamard differentiability.

Proof. If the dual norm on X^* is locally uniformly convex (resp. strictly convex), then distance functions to closed convex sets are Fréchet differentiable (resp. Gâteaux differentiable) on their complements (Exercise 5.3.11). The case of $L_1(\Omega)$ where Ω is a σ-finite measure space, [82, Theorem 2.4] shows that there is an equivalent norm $\|\cdot\|$ on $L_1(\Omega)$ such that its dual norm $\|\cdot\|^*$ on $L_\infty(\Omega)$ satisfies $x_n \to_{\tau_W} x$ in the Mackey topology of uniform convergence on weakly compact subset of $L_1(\Omega)$ whenever $\|x^*\|^* = 1$, $\|x_n^*\|^* \leq 1$ and $\|x^* + x_n^*\|^* \to 2$ (this dual norm is *Mackey-LUR*). Because the dual norm is Mackey-LUR, it can be shown as Exercise 5.3.11 that using this norm, distance functions to convex sets on $(L_1(\Omega), \|\cdot\|)$ will be weak Hadamard differentiable on their complements. □

A characterization of open sets admitting certain essentially β-smooth convex barrier functions is presented in the next result.

Theorem 7.7.3. *Let X be a Banach space, and C be an open convex set containing 0. Then the following are equivalent.*

(a) *There is a β-differentiable convex function f whose domain is C such that $f(x_n) \to \infty$ and $\|f'(x_n)\| \to \infty$ if $d_{C^c}(x_n) \to 0$, or if $\|x_n\| \to \infty$.*
(b) *There is a coercive convex barrier function f that is β-differentiable on C.*
(c) *There are continuous gauges (γ_n) such that γ_n is β-differentiable when $\gamma_n(x) \neq 0$ and $\gamma_n \downarrow \gamma_C$ pointwise where γ_C is the gauge of C, and X admits a β-differentiable norm.*
(d) *X admits a β-differentiable norm and there is a sequence of β-differentiable convex functions (f_n) that are bounded on bounded sets and $f_n \downarrow \gamma_C$ pointwise where γ_C is the gauge of C.*

Proof. (a) $\Rightarrow$ (b): This portion is obvious.

(b) $\Rightarrow$ (c): Let $C_n = \{x : f(x) \leq n\}$ for all n such that $f(0) < n$. Then $C_n \subset \mathrm{int}(C)$ and Fact 7.2.1 implies that γ_{C_n} is β-differentiable at all x where $\gamma_{C_n}(x) \neq 0$. It is not difficult to check that $\gamma_{C_n} \downarrow \gamma_C$ pointwise. Because C_n is bounded with 0 in its interior, it follows that $\|\cdot\|$ defined by $\|x\| = \gamma_{C_n}(x) + \gamma_{C_n}(-x)$ is an equivalent β-differentiable norm on X.

(c) $\Rightarrow$ (d): Let $h_n : \mathbb{R} \to \mathbb{R}$ be nondecreasing C^∞-smooth convex functions such that $h_n(t) = 2/n$ if $t \leq 1/n$ and $h_n(t) \downarrow t$ for all $t \geq 0$. Because h_n is constant on a neighborhood of 0 and γ_C is differentiable at x where $\gamma_C(x) > 0$, it follows that $f_n = h_n \circ \gamma_C$ is β-differentiable everywhere, and that $f_n \downarrow \gamma_C$ pointwise.

(d) $\Rightarrow$ (a): Given $f_n \downarrow \gamma_C$ pointwise, it follows that $g_n \downarrow \gamma_C$ pointwise where $g_n := (1 + \frac{1}{n})f_n$. Now let h_n be a C^∞-smooth nondecreasing convex function on $\mathbb{R}$ such that $h_n(t) = 0$ if $t \leq 1$, and $h_n(t) \geq n$ if $t \geq 1 + \frac{1}{2n}$. Let $\|\cdot\|$ be an equivalent β-differentiable norm on X and define f by

$$f(x) := \|x\|^2 + \sum_{n=1}^{\infty} h_n(g_n(x)).$$

Then f is convex, because h_n is convex and nondecreasing and g_n is convex. Also, if $x_0 \in C$, then $\gamma_C(x_0) < 1$. Therefore, $g_n(x_0) \leq g_N(x_0) < \alpha < 1$ for all $n \geq N$ and some α. Let $O := \{x : g_N(x) < \alpha\}$ then O is an open neighborhood of x_0 and

$$f(x) = \|x\|^2 + \sum_{n=1}^{N-1} h_n(g_n(x)) \quad \text{for all } x \in O.$$

Therefore f is β-differentiable on O.

Now suppose $d_{C^c}(x_k) \to 0$. Because f is supercoercive, as in the proof of Theorem 7.7.1, we may assume that (x_k) is bounded. Let $F_n = \{x : g_n(x) \geq 1 + \frac{1}{2n}\}$. Since g_n is Lipschitz on bounded sets, and $g_n \geq 1 + \frac{1}{n}$ on C^c, it follows that $x_k \in F_n$ for all $k \geq N$. Therefore, $f(x_k) \geq h_n(g_n(x_k)) \geq n$ for all $k \geq N$. Thus $f(x_k) \to \infty$ and, by the subgradient inequality, $\|f'(x_k)\| \to \infty$ since $\{x_k\}_{k=1}^{\infty}$ is bounded. $\square$

Corollary 7.7.4. *Suppose X is a Banach space that admits an equivalent norm whose dual is locally uniformly convex (resp. strictly convex). Let C be an open convex set in X. Then there is a convex function f that is C^1-smooth on C (resp. continuous and Gâteaux differentiable on C) such that:*

(a) $f(x_n) \to \infty$ and $\|f'(x_n)\| \to \infty$ as $x_n \in C$ and $d_{C^c}(x_n) \to 0$;
(b) if $x_n \in C$ and $\|x_n\| \to \infty$, then $f(x_n) \to \infty$ and $\|f'(x_n)\| \to \infty$.

Proof. If X^* admits a dual locally uniformly convex norm, then every Lipschitz convex function can be approximated uniformly by C^1-smooth convex functions (Exercise 5.1.33(a)). Then we can apply Theorem 7.7.3(d) to obtain the desired conclusion. The proof for the case of dual strictly convex norms is analogous. $\square$

Let us note that our constructions can be made log-convex by considering $e^{f(x)}$. The following theorem relates the existence of various essentially smooth or functions of Legendre type to the existence of certain norms on the Banach space.

Theorem 7.7.5. *Let X be a Banach space.*

(a) Suppose X admits an essentially β-smooth lsc convex function on a bounded open convex set C with $0 \in C$. Then X admits a β-differentiable norm.
(b) Suppose X admits a strictly convex lsc function that is continuous at one point. Then X admits a strictly convex norm.

(c) Suppose X admits an essentially β-smooth lsc convex function f on a bounded convex set C with $0 \in \text{int}(C)$ that additionally satisfies $\|f'(x_n)\| \to \infty$ if $d_{\text{bdry}(C)}(x_n) \to 0$. Then X admits a β-differentiable norm.

(d) If X admits a Legendre function on a bounded open convex set, then X admits an equivalent Gâteaux differentiable norm that is strictly convex.

(e) If X admits an equivalent strictly convex norm, and if every open convex set is the domain of an essentially β-smooth function, then every open convex set is the domain of some β-Legendre function.

(f) Suppose X admits an equivalent strictly convex norm whose dual is strictly convex (resp. locally uniformly convex), then every open convex set is the domain of some Legendre function (resp. Fréchet differentiable Legendre function).

Proof. (a) Let $h(x) := f(x) + f(-x)$ on $B := C \cap (-C)$. Notice that h is β-differentiable on $\text{int}(B)$. It follows that h is β-differentiable on the open convex set B, and also $h(x) = \infty$ for $x \notin B$. Choose $f(0) < \alpha < \infty$. Because f is continuous at 0 the set $D = \{x : f(x) \leq \alpha\}$ has nonempty interior and $D \subset B$. The bounded convex set D is also symmetric, and so the implicit function theorem for gauges (4.2.13) implies that the norm, defined as the gauge of D, is an equivalent β-differentiable norm on X.

(b) Suppose f is strictly convex, and continuous at x_0. By replacing f with $f - \phi$ and translating as necessary, we may assume $x_0 = 0$ and $f(0) = 0$ is the minimum of f. Also, replacing f with $f + \|\cdot\|^2$ gives us a function that is both strictly convex and coercive (the sum of a convex function and a strictly convex function is strictly convex). Now since f is continuous at 0, so is h where h is as in (a). Because h is continuous at 0 and coercive, $B = \{x : h(x) \leq 1\}$ is a bounded convex set, with nonempty interior. Because h is strictly convex and symmetric, the gauge of B is a strictly convex equivalent norm on X.

(c) As in (b), we may assume $f(0) = 0$ is the minimum of f. Now let h and B be as above; we next observe that $\inf\{h(x) : x \in \text{bdry}(B)\} > 0$. Indeed, suppose $x_n \in \text{bdry}(B)$ and $h(x_n) \to 0$. Notice that x_n or $-x_n$ is in the boundary of C. Without loss, assume that $x_n \in \text{bdry}(C)$. We know that $f(x_n) \to 0$ (since 0 is the minimum of f and h). Thus, the Brøndsted–Rockafellar theorem (4.3.2) implies there is a sequence $y_n \in C$ with $\|x_n - y_n\| \to 0$ and $\phi_n \in \partial f(y_n)$ while $\|\phi_n\| \to 0$. This violates the condition $\|f'(y_n)\| \to \infty$ as $d_{\text{bdry}(B)}(y_n) \to 0$. Hence there is an α such that $0 < \alpha < \inf\{h(x) : x \in \text{bdry}(B)\}$, and we may apply the implicit function theorem for gauges (4.2.13) as in (a).

(d) Construct h and B as in (a). Because h is strictly convex on $B \supset D$, the norm constructed in (a) is strictly convex.

(e) Let B be the unit ball with respect to an equivalent strictly convex norm $\|\cdot\|$, and let C be the interior of B. Because C is bounded and open, there is an essentially β-smooth convex function whose domain is C that satisfies Theorem 7.7.3(a). According to Theorem 7.7.3(c), there are β-differentiable gauges γ_n decreasing pointwise to $\|\cdot\|$. By letting $\|\cdot\|_n := \frac{1}{2}[\gamma_n(x) + \gamma_n(-x)]$ we get β-differentiable norms $\|\cdot\|_n \downarrow \|\cdot\|$. Now $\|\cdot\|_1 \leq K\|\cdot\|$ and so the norms $\|\cdot\|_n$ are equi-Lipschitz. Following the John–Zizler proof [268] for Asplund averaging (see Exercise 5.1.24), we define $\|\cdot\|$ by

$$\|x\| := \sqrt{\sum_{n=1}^{\infty} \frac{1}{2^n}\|x\|_n^2}.$$

This norm is β-differentiable because of the uniform convergence of the sum of derivatives. Moreover, it is strictly convex, because if $\|x\| = \|y\| = 1$ and $2\|x\|^2 + 2\|y\|^2 - \|x+y\|^2 = 0$ we must have $2\|x\|_n^2 + 2\|y\|_n^2 - \|x+y\|_n^2 = 0$ for all n. Because $\|\cdot\|_n \to \|\cdot\|$ pointwise, we have $2\|x\|^2 + 2\|y\|^2 - \|x+y\|^2 = 0$ and so $x = y$ by the strict convexity of $\|\cdot\|$. Then adding $\|\cdot\|^2$ to any essentially β-smooth convex function, produces a supercoercive essentially β-smooth strictly convex function whose domain is C. Hence Fact 7.2.1(c) implies this function is β-Legendre.

(f) This follows from (e) and Corollary 7.7.4. □

The previous result shows that many spaces do not have functions of Legendre type, while many others have an abundance of such functions. We make a brief list of some such spaces in the following:

Example 7.7.6. (a) The spaces ℓ_∞/c_0 and $\ell_\infty(\Gamma)$ where Γ is uncountable, admit no essentially strictly convex, and hence no Legendre functions.
(b) If X is a WCG space, then every open convex subset of X is the domain of a Legendre function.
(c) If X^* is WCG, or if X is a WCG Asplund space, then every open convex subset of X is the domain of a Fréchet differentiable Legendre function.

Proof. If the spaces in (a) were to have such functions, then they would have strictly convex norms by Theorem 7.7.5(b). It is well-known (see [180, Chapter II] that these spaces do not have equivalent strictly convex norms. In [180, Chapter VII] it is shown that WCG spaces have strictly convex norms whose duals are strictly convex, and so (b) follows from Theorem 7.7.5(f). Similarly, [180, Chapter VII] shows that spaces as in (c) have locally uniformly convex norms whose duals are also locally uniformly convex, and so (c) follows from Theorem 7.7.5(f). □

If f is essentially smooth with domain C, then $\|f'(x_n)\| \to \infty$ if $x_n \to \bar{x}$ where $\bar{x} \in \text{bdry}(C)$ see Theorem 7.3.4(iv). However, this does not ensure that $\|f'(x_n)\| \to \infty$ when $d_{\text{bdry}(C)}(x_n) \to 0$ (as was required in Theorem 7.7.5(c)), even for bounded sets as is shown in the following example.

Example 7.7.7. There is an essentially smooth convex function f whose domain is a closed convex set C such that $\|f'(x_n)\| \not\to \infty$ as $d_{\text{bdry}(C)}(x_n) \to 0$.

Proof. Let $X := c_0$ with its usual norm and let C be the closed unit ball of c_0. Let $h_n : [-1,1] \to \mathbb{R}$ be continuous, even and convex such that h is C^1-smooth on $(-1,1)$, and $(h_n)'_-(1) = \infty$, $h_n(1) = 1$ and $h_n(t) = 0$ for $|t| \le 1 - \frac{1}{2n}$. Now extend h_n to an lsc convex function on $\mathbb{R}$ by defining $h(t) = \infty$ if $|t| > 1$. Define f on c_0 by $f(x) := \sum_{n=1}^\infty h_n(x_n)$ where $x = (x_n)_{n=1}^\infty$. Then f is C^1-smooth and convex on int C because it is a locally finite sum of such functions there. Since $(h_n)'_+(1) = \infty$, it follows that $\partial f(x) = \emptyset$ if $\|x\| = 1$, and clearly $f(x) = \infty$ if $\|x\| > 1$. Therefore, f is essentially Fréchet smooth. However, if we consider $v_n = (1 - \frac{1}{n})e_n$, we have $f(v_n) = 0$ and $f'(v_n) = 0$ for each n while $d_{\text{bdry}(C)}(v_n) \to 0$. □

Exercises and further results

7.7.1. Let $f : \ell_2 \to \mathbb{R}$ be defined by

$$f(x) := \sum_{i=1}^{\infty} \frac{x_i^2}{2^i}.$$

Show that

(a) f is smooth and strictly convex.
(b) f is not essentially strictly convex.

Hint. $f'(x) = \sum_{i=1}^{\infty} 2x_i/2^i\, e_i$. Note that $f'(0) = 0$, and $f'(ne_n) = n/2^{n-1}\, e_n \to 0$ and so $(f')^{-1}$ is not locally bounded at 0 which is in its domain. Therefore, f is not essentially strictly convex.

A less direct but instructive path to the failure of essential strict convexity is as follows. First, f is not coercive, so $0 \notin \text{int dom}\, f^*$ by the Moreau–Rockafellar theorem (4.4.10). Also, given any $\phi \in X^*$, one can show $f - \phi$ is not coercive, and so $\phi \notin \text{int dom}\, f^*$ according to the (translated) Moreau–Rockafellar theorem (4.4.11).

Indeed, let $\phi := \sum_{n=1}^{\infty} a_n e_n \in \ell_2$. Then $a_n \to 0$. Let $b_n = \max\{n, |1/a_n|\}$. Then $\|b_n e_n\| \to \infty$, but $(f - \phi)(b_n e_n) \le n 2^{-n} + 1$ and so $f - \phi$ is not coercive. Therefore, $\phi \notin \text{int dom}\, f^*$. Because ϕ was arbitrary, we obtain that int dom $f^* = \emptyset$. According to Theorem 7.3.4, f^* is not essentially smooth. Then by Theorem 7.3.2, $f = (f^*)^*$ is not essentially strictly convex. □

7.7.2.* Verify the weak Hadamard case of Remark 7.7.2.

7.7.3 (Totally convex functions [141]).** A proper convex function f on a Banach space X is *totally convex* at $x \in \text{dom}\, f$ if its *modulus of total convexity* at x defined by

$$v_f(x, t) := \inf\{D_f(y, x) : y \in \text{dom}\, f, \|y - x\| = t\}$$

is strictly positive for all $t > 0$. Then f is said to be *totally convex* if it is totally convex at each point of its domain.

(a) Show that in Euclidean space a strictly convex function with closed domain is totally convex when it is continuous on its domain.
(b) Show that if f is totally convex throughout dom f then f is strictly convex on its domain.
(c) Show that uniform convexity of f at $x \in \text{dom}\, f$ (see Exercise 5.5.3) implies total convexity of f at x. Hence a locally uniformly convex function is totally convex.
(d) Suppose f is lsc. Show that when f is Fréchet differentiable at x then f is totally convex at x if and only if it is uniformly convex at x.

(e) Show that g defined on ℓ_1 by

$$g(x) := \sum_{k=1}^{\infty} |x_k|^{1+1/k}$$

is nowhere uniformly convex but is totally convex throughout the set $\{x \in \ell_1 : \limsup |x_k|^{1/k} < 1\}$.

(f) Fix $1 < s < \infty$. Show that when X is uniformly convex and uniformly smooth $x \mapsto \|x\|^s$ is Legendre, is uniformly convex (on closed balls) and is totally convex; see Example 5.3.11 and Exercise 5.3.2.

(g) Show that in a reflexive space a proper closed convex function which is totally convex throughout dom ∂f is essentially strictly convex.

Totally convex functions at x are well-tuned to Bregman algorithms since they are precisely the convex functions with the property at x that for any sequence (y^n) in dom f one has $\|y^n - x\| \to 0$ whenever $D_f(y^n, x) \to 0$ as $n \to \infty$. More examples and details on notions such as that of a totally convex Banach space may be found in [141], [37] and [35, 36].

7.7.4. Contrasting Theorem 7.7.5(c) with Example 7.7.7 leads naturally to the following questions to which we are unaware of the answer.

(a) If X admits an essentially β-smooth convex function on a bounded convex set with 0 in its interior, does X admit a β-differentiable norm?

(b) Relatedly, if X admits an essentially β-smooth convex function on a bounded convex set with 0 in its interior, does X admit a β-differentiable convex function on a bounded convex set with 0 in its interior such that $\|f'(x_n)\| \to \infty$ whenever $d_{\mathrm{bdry}(C)}(x_n) \to 0$?

7.7.5 (Legendre transform). Suppose f is a (finite) everywhere differentiable convex function on a reflexive Banach space X. Consider the following:

(a) ∇f is one-to-one on X.
(b) f is both strictly convex and cofinite.
(c) f^* is everywhere differentiable, strictly convex and cofinite; and for all $x^* \in X^*$

$$f^*(x^*) = \langle (\nabla f)^{-1}(x^*), x^* \rangle - f\left((\nabla f)^{-1}(x^*)\right).$$

In Euclidean space, Exercise 7.3.7 establishes that (a) and (b) coincide and imply (c). Determine what remains true in a reflexive Banach space.

Let us return to a few examples in the spirit of earlier sections of the chapter.

7.7.6 (A self-conjugate function). Let

$$g(t) := \begin{cases} t - \log(1+t) & t > -1, \\ +\infty & t \le -1. \end{cases}$$

Show that g is convex and has the conjugate $g^*(s) = g(-s)$.

7.7.7 (A trace-class barrier). ** In [210] the author looks at self-concordant barriers in an operator-theoretic setting. A number of the details can be understood quite clearly in terms of the spectral theory developed thus far. We start with the Taylor expansion

$$\log(1+t) = t - \frac{1}{2}t^2 + \frac{1}{3}t^3 - \frac{1}{4}t^4 + \cdots.$$

The function g of Exercise 7.7.6 is convex with conjugate $g^*(s) = g(-s)$. Now induce a convex rearrangement invariant function f on ℓ_2 by

$$f(x) := \sum_{i=1}^{\infty} g(x_i).$$

Clearly, f is finite and continuous at 0 since $|g(t)| \leq kt^2$ if $|t| \leq 1/2$. Define a function ϕ on B_2 by

$$\phi := f \circ \lambda.$$

Apply Theorem 6.5.11 to show $\phi^*(T) = \phi(-T)$. Then show for $T \in B_2$ (the Hilbert–Schmidt operators) with $I + T \geq 0$, that

$$\phi(T) = \text{tr}(T) - \log(\det(I + T)).$$

Hence, show that ϕ is a Fréchet differentiable and strictly convex barrier on the open convex set $\{T \in B_2 : I + T \geq 0\}$.

7.7.8. ** The previous construction can be extended to B_p (the Schatten p-spaces) for appropriate p by subtracting more terms of the series for $\log(1+t)$. This alternatingly will be convex or concave. Consider for example $g(t) := \log(1+t) - (t - t^2/2)$. Confirm that $g'(t) = t^2/(1+t)$ and $g''(t) = 1 - 1/(1+t)^2$ and so for $t > 0$, g is increasing and convex. Confirm that the conjugate of g is

$$g^*(s) = st(s) - \log(1 + t(s)) + t(s) - t^2(s)/2,$$

where $t(s) := (s + \sqrt{s^2 + 4s})/2$. Since $|g(t)| \leq kt^3$ for t small, the (permutation invariant) function

$$f(x) := \sum_{i=1}^{\infty} g(x_i)$$

is well defined, lsc and convex on ℓ_3. Show that $\phi := f \circ \lambda$ is a lsc unitarily invariant convex function on B_3. Deduce that $\phi^* = f^* \circ \lambda$ is a lsc unitarily invariant convex function on $B_{3/2}$, where $f^*(x) = \sum_{i=1}^{\infty} g^*(x_i)$. Determine the domain of ϕ.

8
Convex functions and classifications of Banach spaces

A mathematician is a person who can find analogies between theorems; a better mathematician is one who can see analogies between proofs and the best mathematician can notice analogies between theories. (Stefan Banach)[1]

8.1 Canonical examples of convex functions

The first part of this chapter connects differentiability and boundedness properties of convex functions with respect to a *bornology* β (see p. 149 for the definition) with sequential convergence in the dual space in the topology of uniform convergence on the sets from the bornology. In some sense, many of the results in this chapter illustrate the degree to which linear topological properties carry over to convex functions. This chapter also examines extensions of convex functions that preserve continuity, as well as some related results.

Also, given any bornology β on X, by τ_β we intend the topology on X^* of uniform convergence on β-sets. In particular, $\tau_\mathcal{W}$ is the Mackey topology of uniform convergence on weakly compact sets, usually denoted by $\mu(X^*, X)$ in the theory of locally convex spaces. Following [77], when we speak of the *Mackey topology* on X^*, we will mean $\mu(X^*, X)$.

We begin with constructions of convex functions that seem to be central to connecting convexity properties with linear topological properties in the dual.

Proposition 8.1.1. *Let $(\phi_n)_{n=1}^\infty \subset B_{X^*}$. Consider the functions that are defined as follows*

$$f(x) := \sup_n \left\{ \phi_n(x) - \frac{1}{n}, 0 \right\} \quad g(x) := \sup_n (\phi_n(x))^{2n} \quad h(x) := \sum_{n=1}^\infty (\phi_n(x))^{2n}.$$

Then f, g and h are proper lsc convex functions, where f is additionally Lipschitz. Moreover

(a) f is β-differentiable at 0 if and only if $\phi_n \to_{\tau_\beta} 0$.

---

[1] Stefan Banach, 1892–1945, quoted in
www-history.mcs.st-andrews.ac.uk/Quotations/Banach.html.

(b) g and h are bounded on β-sets if and only if $\phi_n \to_{\tau_\beta} 0$ and, if this is the case, both functions are continuous.

Proof. It is clear that the functions are lsc and convex as sums and suprema of such functions, and f is Lipschitz since $\|\phi_n\| \leq 1$ for each n. We outline the other implications.

(a) Suppose f is β-differentiable at 0. Now $f(0) = 0$ and $f(x) \geq 0$ for all $x \in X$, thus $0 \in \partial f(0)$. Consequently, $f'(0) = 0$. Suppose by way of contradiction that $\phi_n \not\to_{\tau_\beta} 0$. Then we can find a β-set W and infinitely many n such that $w_n \in W$ and $\phi_n(w_n) > 2$. For such n,

$$\frac{f(\frac{1}{n}w_n) - f(0)}{\frac{1}{n}} = nf\left(\frac{1}{n}w_n\right) \geq n\left(\phi_n\left(\frac{w_n}{n}\right) - \frac{1}{n}\right) > 1.$$

This contradicts that $f'(0) = 0$ as a β-derivative.

Conversely, suppose $\phi_n \to_{\tau_\beta} 0$. Given any $\varepsilon > 0$ and any β-set W, there is an $n_0 \in \mathbb{N}$ such that $\phi_n(w) < \varepsilon$ for all $n > n_0$ and $w \in W$. Let $M > 0$ be chosen so that $W \subset MB_X$. Now for $|t| < \frac{1}{Mn_0}$ we have

$$\phi_n(tw) - \frac{1}{n} < \frac{1}{Mn_0}M - \frac{1}{n} \leq 0 \text{ for all } w \in W, n \leq n_0.$$

Therefore,

$$0 \leq f(tw) - f(0) \leq \max\left\{0, \sup_{n > n_0}|t|\varepsilon - \frac{1}{n}\right\} = |t|\varepsilon \text{ for } |t| < \frac{1}{Mn_0}.$$

Consequently, $\lim_{t \to 0}\frac{f(tw) - f(0)}{t} = 0$ for each $w \in W$ which shows f is β-differentiable at 0 with $f'(0) = 0$.

(b) Suppose $\phi_n \to_{\tau_\beta} 0$, and let W be a β-set. We select $n_0 \in \mathbb{N}$ such that $|\phi_n(x)| \leq 1/2$ for all $x \in W$ and $n > n_0$. Let M_k be defined by $M_k = \sup_W \phi_k$ for $k = 1, 2, \ldots, n_0$. Then for $x \in W$,

$$0 \leq g(x) \leq h(x) \leq M_1^{2n} + M_2^{2n} + \ldots + M_{n_0}^{2n} + 1,$$

which provides bounds for g and h on W. Also, as finite-valued lsc convex functions, f and g are continuous (Proposition 4.1.5).

Conversely, if $\phi_n \not\to_{\tau_\beta} 0$, then we can find a β-set W such that $\phi_n(w_n) > 2$ for infinitely many n where $w_n \in W$. Then $h(w_n) \geq g(w_n) \geq (\phi_n(w_n))^{2n} > 2^{2n}$, and so both g and h are unbounded on W. $\square$

We refer to the previous examples as 'canonical' because they are natural examples that capture the essence of how convex functions can behave with respect to comparing boundedness or differentiability notions as we now show.

8.1 Canonical examples of convex functions

Proposition 8.1.2. *Let X be a Banach space. Then the following are equivalent.*

(a) *Mackey and norm convergence coincide sequentially in X^*.*
(b) *Every sequence of lsc convex functions that converges to a continuous affine function uniformly on weakly compact sets converges uniformly on bounded sets to the affine function.*
(c) *Every continuous convex function that is bounded on weakly compact subsets of X is bounded on bounded subsets of X.*
(d) *Weak Hadamard and Fréchet differentiability agree for continuous convex functions.*

Proof. (a) $\Rightarrow$ (b): Suppose (f_n) is a sequence of lsc convex functions that converges uniformly on weakly compact sets to some continuous affine function A. By replacing (f_n) with $(f_n - A)$ we may assume that $A = 0$. Now suppose (f_n) does not converge to 0 uniformly on bounded sets. Thus there are $K > 0$, $(x_k)_{k \geq 1} \subset KB_X$ and $\varepsilon > 0$ so that $f_{n_k}(x_k) > \varepsilon$ for a certain subsequence (n_k) of (n) (using convexity and the fact that $f_{n_k}(0) \to 0$). Now let $C_k := \{x : f_{n_k}(x) \leq \varepsilon\}$ and use the separation theorem (4.1.17) to choose $\phi_k \in S_{X^*}$ such that $\sup_{C_k} \phi_k < \phi_k(x_k) \leq K$. We observe that (ϕ_k) does not converge to 0 in τ_W by (a). Find a weakly compact set $C \subset X$ so that $\sup_C \phi_k > K$ for infinitely many k. We have $\sup_C \phi_k = \phi_k(c_k)$ for some $c_k \in C$, and $c_k \notin C_k$ (so $f_{n_k}(c_k) > \varepsilon$) for infinitely many k, which contradicts the uniform convergence to 0 of (f_n) on C.

Now (b) implies (d) follows because difference quotients are lsc convex functions, and (d) implies (a) follows from Proposition 8.1.1.

Finally, (c) implies (a) follows from Proposition 8.1.1, so we conclude by establishing (a) implies (c). For this, we suppose (c) is not true. We can find then a continuous convex function f that is bounded on weakly compact subsets of X and not bounded on all bounded subsets of X. We may assume $f(0) = 0$ and we let (x_n) be a bounded sequence such that $f(x_n) > n$, and let $C_n := \{x : f(x) \leq n\}$. By the separation theorem (4.1.17) to choose $\phi_n \in S_{X^*}$ such that $\sup_{C_n} \phi_n < \phi_n(x_n)$. Now choose $K > 0$ such that $K > \phi_n(x_n)$ for all n. If $\phi_n \not\to_{\tau_W} 0$, then there is a weakly compact set $W \subset X$ and infinitely many n such that $\phi_n(w_n) > K$ and $w_n \in W$. In particular, $w_n \notin C_n$ for those n and so f is unbounded on W. Thus (a) is not true when (c) is not true. □

We conclude this section with a natural bornological extension of the previous result that also addresses the situation for equivalent norms.

Theorem 8.1.3. *Let X be a Banach space with bornologies $\beta_1 \subset \beta_2$. Then the following are equivalent.*

(a) *τ_{β_1} and τ_{β_2} agree sequentially in X^*.*
(b) *Every sequence of lsc convex functions on X that converges to a continuous affine function uniformly on β_1-sets, converges uniformly on β_2-sets.*
(c) *Every continuous convex function on X that is bounded on β_1-sets is bounded on β_2-sets.*

(d) β_1-differentiability agrees with β_2-differentiability for continuous convex functions on X.

(e) β_1-differentiability agrees with β_2-differentiability for equivalent norms on X.

Proof. (a) $\Rightarrow$ (b): As in Proposition 8.1.2 we may suppose (f_n) is a sequence of lsc convex functions that converges uniformly to 0 on β_1-sets, but not uniformly on β_2-sets. Thus there is a β_2-set $W \subset KB_X$ with $x_k \in W$ and $\varepsilon > 0$ for that $f_{n_k}(x_{n_k}) > \varepsilon$. Now let $C_k = \{x : f_{n_k}(x) \le \varepsilon\}$ and choose $\phi_k \in S_{X^*}$ so that $\sup_{C_k} \phi_k < \phi_k(x_{n_k}) \le K$. If (ϕ_k) does not converge uniformly to 0 on β_1-sets, then there is a β_1-set C such that $\sup_C \phi_k > K$ for infinitely many k. Thus (f_{n_k}) does not converge uniformly to 0 on C. This contradiction shows that (ϕ_{n_k}) converges uniformly to 0 on β_1-sets. We now show that (ϕ_{n_k}) does not converge uniformly to 0 on β_2-sets. For this, let $F_n := \{x \in X : f_k(\pm x) \le \varepsilon \text{ for all } k \ge n\}$. Since (f_n) converges pointwise to 0, $\bigcup_{n\ge 1} F_n = X$. The Baire category theorem ensures that $F_{\bar n}$ has nonempty interior for some $\bar n \in \mathbb{N}$, and because $F_{\bar n}$ is a symmetric convex set, for some $\delta > 0$ we have that $\delta B_X \subset F_{\bar n}$. Consequently, for $n_k \ge \bar n$, $\sup_{C_k} \phi_k > \delta$. Thus $\phi_k(x_{n_k}) \not\to 0$ which shows (ϕ_k) does not converge uniformly to 0 on the β_2-set W.

(b) $\Rightarrow$ (d): Follows as in Proposition 8.1.2.

(d) $\Rightarrow$ (e): is trivial, for (e) $\Rightarrow$ (a) one can use the following construction (a more elegant proof can be found in [77, Theorem 1]).

Suppose $\|\phi_n\| = 1$ for all n, and $\phi_n \to_{\tau_{\beta_1}} 0$ but $\phi_n \not\to_{\tau_{\beta_2}} 0$. Let $\phi \in S_{X^*}$ be a norm attaining functional, say $\phi(x_0) = 1$ where $x_0 \in S_X$. By passing to a subsequence we may and do assume $|\phi_n(x_0)| < \frac{1}{2n^2}$ for each n. Define $\| \cdot \|$ by

$$\|x\| := \max_n \left\{ \frac{1}{2}\|x\|, |\phi(x)|, \left|\left(1 - \frac{1}{n^2}\right)\phi(x) + \phi_n(x)\right|\right\}.$$

We will show that $\| \cdot \|$ is β_1-differentiable at x_0 but not β_2-differentiable at x_0. First, observe that $\|x_0\| = 1$, $\phi(x_0) = 1$ and $\phi(x) \le 1$ if $\|x\| \le 1$. Consequently, $\phi \in \partial \|x_0\|$.

Next we observe, $\| \cdot \|$ is not β_2-differentiable at x_0. Indeed, since $\phi_n \not\to_{\tau_{\beta_2}} 0$, there is a β_2-set $W \subset B_X$ and $h_n \in W$ so that $\phi_n(h_n) > \delta$ for all n (passing to a subsequence). Then

$$\left\|x_0 + \frac{1}{n}h_n\right\| - \|x_0\| - \left\langle \phi, \frac{1}{n}h_n\right\rangle$$

$$\ge \left(1 - \frac{1}{n^2}\right)\phi\left(x_0 + \frac{1}{n}h_n\right) - 1 - \phi\left(\frac{1}{n}h_n\right) + \phi_n\left(x_0 + \frac{1}{n}h_n\right)$$

$$= 1 - \frac{1}{n^2} - 1 + \phi\left(\frac{1}{n}h_n\right)\left(1 - \frac{1}{n^2}\right) - \phi\left(\frac{1}{n}h_n\right) + \phi_n\left(x_0 + \frac{1}{n}h_n\right)$$

$$\ge -\frac{1}{n^2} - \frac{1}{n^2}\phi\left(\frac{1}{n}h_n\right) - \frac{1}{n^2} + \frac{\delta}{n}$$

$$\ge \frac{\delta}{n} - \frac{3}{n^2}.$$

It follows from this that $\|\cdot\|$ is not β_2-differentiable at x_0.

To show that $\|\cdot\|$ is β_1-differentiable at x_0 we first observe that if t is sufficiently small, then $|\phi(x_0 + th)| = \phi(x_0 + th)$ and so $|\phi(x_0 + th)| - 1 - \phi(th) = 0$. Now, for t sufficiently small

$$\left|\left(1 - \frac{1}{n^2}\right)\phi(x_0 + th) + \phi_n(x_0 + th)\right| - 1 - \phi(th)$$

$$= \left(1 - \frac{1}{n^2}\right)\phi(x_0 + th) + \phi_n(x_0 + th) - 1 - \phi(th)$$

$$= -\frac{1}{n^2} - \frac{1}{n^2}\phi(th) + \phi_n(x_0) + \phi_n(th)$$

$$\leq -\frac{1}{n^2} + \frac{|t|}{n^2} + \frac{1}{2n^2} + \phi_n(th)$$

$$= -\frac{1}{2n^2} + \frac{|t|}{n^2} + \phi_n(th).$$

Let n_0 be chosen so that $\phi_n(h) < \varepsilon$ for $n > n_0$. Now consider the above when $|t| < \frac{1}{4n_0^2}$. Then

$$\|x_0 + th\| - \|x_0\| - \phi(th) \leq \sup_{n > n_0} \left\{0, -\frac{1}{2n^2} + \frac{|t|}{n^2} + \phi_n(th)\right\}$$

$$\leq \phi_n(th) \leq |t|\varepsilon.$$

This completes the proof of (a) $\Rightarrow$ (e).

The equivalence of (a) and (c) is also similar to the proof of Proposition 8.1.2. Indeed, we prove (a) $\Rightarrow$ (c) by contraposition. Suppose (c) is not true. The we can find a convex function f with $f(0) = 0$ and that f is bounded on β_1-sets, but not on β_2-sets. Let W be a β_2-set on which f is unbounded. Then we choose $x_n \in W$ such that $f(x_n) > n$ and let $C_n := \{x : f(x) \leq n\}$. Because $f(0) = 0$ and f is continuous, it follows that there is a $\delta > 0$ so that $\delta B_X \subset C_1 \subset C_n$ for each n. By the separation theorem (4.1.17), we choose $\phi_n \in S_{X^*}$ so that $\delta \leq \sup_{C_n} \phi_n < \phi_n(x_n)$. As in (a) $\Rightarrow$ (b) we can show that (ϕ_n) converges to 0 uniformly on β_1-sets. However, (ϕ_n) does not converge uniformly to 0 on the β_2-set W. Thus (a) is not true.

The implication (c) $\Rightarrow$ (a) follows from Proposition 8.1.1. □

Exercises and further results

8.1.1. Let $X := \ell_p$ for $1 \leq p < \infty$ or c_0 and write $x \in X$ as $x = (x_k)_{k=1}^\infty$.

(a) Show that $f : X \to \mathbb{R}$ defined by $f(x) := \sum_{k=1}^\infty x_k^{2k}$ is a continuous convex function that is unbounded on some bounded set of X.
(b) Show that $g : X \to \mathbb{R}$ defined by $g(x) := \sup\{x_k - \frac{1}{k} : k \geq 1\}$ is Gâteaux differentiable but not Fréchet differentiable at 0.
(c) When $X := \ell_1$, deduce that the function g in (b) is weak Hadamard differentiable, but not Fréchet differentiable at 0.

8.1.2. Construct a continuous function on ℓ_2 that is everywhere Gâteaux differentiable but fails to be Fréchet differentiable at some point(s).

Hint. See Exercise 5.1.26, and consider the norm squared. □

8.1.3. Let X be a locally convex vector space and let $H \subset X$ be a dense convex set. Let f and g be closed proper convex functions on X such that $f \geq g$ and $f|_H = g|_H$. Show that $f = g$.

Hint. Fix $x \in \text{dom} f$ and $r \in \mathbb{R}$ with $f(x) > r$. Let $W := \{z : g(z) > r\}$ and select an open convex neighbourhood $V := V_x$ of x with $\overline{V} \subseteq \{z : f(z) > r\}$. Then $\overline{V} \cap H \subset W$ and so $\overline{V} \cap H = \overline{V_x} \subset \overline{W}$. Hence $x \in V_x \subseteq W$ and so $g(x) > r$. □

8.2 Characterizations of various classes of spaces

In this section we provide a listing of various classifications of Banach spaces in terms of properties of convex functions. Many of the implications follow from Theorem 8.1.3 or variants of the arguments upon which it is based. We will organize these results based upon when two of the following notions (Gâteaux, weak Hadamard or Fréchet) differentiability coincide for continuous convex functions on a space, and then for continuous weak*-lsc functions on the dual space. First we state the Josefson–Nissenzweig theorem proved independently by the two authors.

Theorem 8.2.1 (Josefson–Nissenzweig [271, 333]). *Suppose X is an infinite-dimensional Banach space, then there is a sequence $(x_n^*) \subset S_{X^*}$ that converges weak* to 0.*

We shall call a weak*-null unit norm sequence a *Josefson–Nissenzweig sequence* or a *JN-sequence*. At first glance, one might say that this result is expected. After all, the norm and weak* topologies are different for infinite-dimensional spaces, therefore, it should be possible to extract such a sequence. However, this is not an easy theorem, and we refer the reader to [183, Chapter XII] for a proof. In fact, when comparing the usual topologies, norm, weak and weak*, the Josefson–Nissenzweig theorem is the exception. That is weak and norm convergence can agree sequentially in some spaces that are not finite-dimensional, and weak and weak* topologies can agree sequentially in duals of certain spaces that are not reflexive. The results from this section are built upon the sequential agreement of various topologies. Indeed, it has already been shown in Theorem 8.1.3, many important properties of convex functions are connected to the sequential convergence in the topology induced by a given bornology. These connections will be listed explicitly.

First, we consider when Gâteaux and Fréchet differentiability coincide for continuous convex functions.

Theorem 8.2.2. *For a Banach space X, the following are equivalent.*

(a) X is finite-dimensional.
(b) Weak* and norm convergence coincide sequentially in X^*.

(c) Every continuous convex function on X is bounded on bounded subsets of X.
(d) Gâteaux and Fréchet differentiability coincide for continuous convex functions on X.

Proof. The equivalence of (a) and (b) is the Josefson–Nissenzweig theorem (8.2.1). The equivalence of (b) through (d) is a direct consequence of Theorem 8.1.3 with the Gâteaux and Fréchet bornologies. □

In particular, on every infinite-dimensional Banach space there is a continuous convex function that is unbounded on a ball and that assertion is equivalent to the Josefson–Nissenzweig theorem (8.2.1). By far, the most difficult part in that assertion is the construction of a weak*-null sequence of norm one elements in the dual. However, it is relatively easy to construct such sequences in certain spaces; see Exercise 8.2.13.

Next, we consider when Gâteaux and weak Hadamard differentiability coincide. As in [78], we will say a Banach space possesses the *DP**-property*, $\langle x_n^*, x_n \rangle \to 0$ whenever $x_n \to_w 0$ and $x_n^* \to_{w^*} 0$. It is straightforward to check that a Banach space has the DP*-property if and only if weak* and Mackey convergence (uniform convergence on weakly compact subsets of X) coincide sequentially in X^*. Recall that a Banach space X is said to be a *Grothendieck space* if weak* and weak convergence coincide sequentially in X^*; alternatively X is said to have the *Grothendieck property*. A Banach space is said to have the *Dunford–Pettis property* if $\langle x_n^*, x_n \rangle \to 0$ whenever $x_n \to_w 0$ and $x_n^* \to_w 0$. The term DP*-property derives from the fact that weak convergence is replaced with weak* convergence in the dual sequence in the Dunford–Pettis property. Therefore, it follows immediately that a Banach space with the Grothendieck and Dunford–Pettis properties has the DP* property (but not conversely, e.g. ℓ_1). Consequently, the spaces $\ell_\infty(\Gamma)$ for any index set Γ have the DP*-property (see [184]).

Theorem 8.2.3. *For a Banach space X, the following are equivalent.*

(a) *X has the DP*-property.*
(b) *Gâteaux and weak Hadamard differentiability coincide for all continuous convex functions on X.*
(c) *Every continuous convex function on X is bounded on weakly compact subsets of X.*

Proof. This is a direct consequence of Theorem 8.1.3 using the Gâteaux and weak Hadamard bornologies. □

Because ℓ_∞ has the DP*-property, the previous theorem applies in spaces where the relatively compact sets and relatively weakly compact sets form different bornologies. Recall that a subset L of a Banach space X is called *limited* if every weak*-null sequence in X^* converges to 0 uniformly on L. Then $\mathcal{RK} \subset \mathcal{L} \subset \mathcal{B}$, where $\mathcal{RK}$ is the collection of the relatively compact subsets, $\mathcal{L}$ of the limited subsets and $\mathcal{B}$ of the bounded subsets. The Josefson–Nissenzweig theorem (8.2.1) says that in infinite-dimensional Banach spaces, $\mathcal{L} \neq \mathcal{B}$. A Banach space is called *Gelfand–Phillips* if

$\mathcal{RK} = \mathcal{L}$. If B_{X^*} is weak*-sequentially compact, then X is Gelfand–Phillips (for these results, see [183, p. 116, 224 and 238]), while ℓ_∞ is not Gelfand–Phillips. Moreover, for a given bornology β in X, τ_β and weak*-convergence agree sequentially if and only if $\beta \subset \mathcal{L}$. In particular, a Banach space has property DP^* if and only if $\mathcal{W} \subset \mathcal{L}$, where as before $\mathcal{W}$ denotes the bornology of weakly compact subsets of X. Recall that a Banach space has the *Schur property* if its weakly convergent sequences are norm convergent. If a Banach space is DP^* and Gelfand–Phillips (for example, the space ℓ_1) then it is Schur, and every Schur space has the DP^* property.

We now turn to spaces where weak Hadamard and Fréchet differentiability coincide for continuous convex functions.

Theorem 8.2.4. *For a Banach space X, the following are equivalent.*

(a) $X \not\supset \ell_1$.
(b) *Mackey and norm convergence coincide sequentially in X^*.*
(c) *Weak Hadamard and Fréchet differentiability coincide for continuous convex functions on X.*
(d) *Every lsc convex function on X bounded on weakly compact sets is bounded on bounded sets.*

Proof. See [77, Theorem 5] or [338] for the equivalence of (a) and (b). The equivalence of (b) through (d) is in Proposition 8.1.2. □

We now consider analogous situations for weak*-lsc convex functions. Questions of differentiability require some care on the conjugate case because, in general, there is no guarantee that the derivative is in X.

Theorem 8.2.5. *For a Banach space X, the following are equivalent.*

(a) X *has the Schur property.*
(b) *Gâteaux differentiability and Fréchet differentiability coincide for weak*-lsc continuous convex functions on X^*.*
(c) *Each continuous weak*-lsc convex function on X^* is bounded on bounded subsets of X^*.*
(d) *Every proper lsc cofinite convex function on X is supercoercive.*

Proof. (a) $\Rightarrow$ (c): Follow the argument of (a) $\Rightarrow$ (c) in the proof of Proposition 8.1.2 using a weak*-continuous separating functional.

(c) $\Rightarrow$ (d): Suppose f is cofinite, then f^* is defined everywhere and therefore f^* is continuous (Proposition 4.1.5). Now the hypothesis of (c) says that f^* is bounded on bounded sets; consequently f is supercoercive (Theorem 4.4.13).

(d) $\Rightarrow$ (a): Suppose that X is not Schur, then we choose $x_n \in S_X$ such that (x_n) converges weakly to 0 (and hence (x_n) converges weak* to 0 when considered as a sequence in X^{**}). Define g on X^* by $g(x^*) := \sup_n (\langle x_n, x^* \rangle)^{2n}$. Then g is weak*-lsc and Proposition 8.1.1 implies g is continuous but is not bounded on bounded sets. Moreover, $g = f^*$ where $f = (g^*|_X)$ (Proposition 4.4.2). Then f is cofinite, but f is not supercoercive because f^* is not bounded on bounded sets (Theorem 4.4.13).

8.2 Characterizations of spaces

(a) ⇒ (b): Suppose $x_n \in \partial_{\varepsilon_n} f(x^*) \cap X$ where $\varepsilon_n \to 0^+$ and $f : X^* \to \mathbb{R}$ is weak*-lsc and Gâteaux differentiable at x^*. By Šmulian's theorem (Exercise 4.2.10) $x_n \to_{w^*} f'(x^*)$. We claim that (x_n) is a norm convergent. Indeed, if it is not norm convergent, then it is not norm Cauchy and so there is a subsequence (x_{n_i}) and $\delta > 0$ so that $\|x_{n_i} - x_{n_{i+1}}\| \geq \delta$. However, this is a contradiction because $y_i \to 0$ weakly where $y_i = x_{n_i} - x_{n_{i+1}}$. Consequently, x_n is norm convergent and by Šmulian's theorem (Exercise 4.2.10), f is Fréchet differentiable at x^*.

(b) ⇒ (a): Suppose X does not have the Schur property. Then there is a sequence $(x_n) \subset S_X$ that converges weakly to 0 (hence (x_n) converges weak* to 0 as a sequence in X^{**}). Define $f : X^* \to \mathbb{R}$ by $f(x^*) := \sup\{x^*(x_n) - 1/n : n \in \mathbb{N}\}$. Then f is Gâteaux but not Fréchet differentiable at 0 according to Proposition 8.1.1. □

Theorem 8.2.6. *For a Banach space X, the following are equivalent.*

(a) *X has the Dunford–Pettis property.*
(b) *Weak and Mackey convergence coincide sequentially in X^*.*
(c) *Gâteaux differentiability and weak Hadamard differentiability coincide for continuous weak*-lsc convex functions on X^*.*
(d) *Each continuous weak*-lsc convex function on X^* is bounded on weakly compact subsets of X^*.*

Proof. The proof is similar to that of Theorem 8.2.5; see Exercise 8.2.4. □

Our last result regarding classes of differentiability for weak*-lsc convex functions is as follows.

Theorem 8.2.7. *For a Banach space X, the following are equivalent.*

(a) *Every sequence in X considered as a subset of X^{**} that converges uniformly on weakly compact subsets of X^*, converges in norm (i.e. Mackey convergence in X^{**} agrees with norm convergence for sequences in X).*
(b) *Weak Hadamard and Fréchet differentiability coincide for continuous weak*-lsc convex functions on X^*.*
(c) *Every weak*-lsc convex function on X^* that is bounded on weakly compact subsets of X^* is bounded on bounded subsets of X^*.*

Proof. (a) ⇔ (b): This is similar to the proof in Theorem 8.2.5.

(a) ⇒ (c): Suppose $f : X^* \to \mathbb{R}$ is weak*-lsc, $f(0) = 0$ and f is unbounded on B_{X^*}. Let $C_n = \{x^* : f(x^*) \leq n\}$, now there are $x_n \in B_X$ and $x_n^* \in B_{X^*}$ so that $1 \geq x_n(x_n^*) > \sup_{C_n} x_n \geq \varepsilon$, where $\varepsilon > 0$ is such that $\varepsilon B_{X^*} \subset C_n$ for all n. Thus $\|x_n\| \not\to 0$ and so $x_n \not\to_{\tau_W} 0$. Then find a weakly compact set $W \subset X^*$ such that $\sup_W x_n > 1$ for infinitely many n, and deduce that f is unbounded on W. To prove (c) ⇒ (a), apply Proposition 8.1.1 with functionals $x_n \to_{\tau_W} 0$ but $\|x_n\| \not\to 0$. □

Note that the previous theorem applies to spaces X such that X does not have the Schur property and $X^* \supset \ell_1$: for example $X = \ell_1 \oplus \ell_2$. So this provides information that cannot be deduced from Theorem 8.2.5 or Theorem 8.2.4.

Finally, we will consider two further classes of spaces. That is the Grothendieck spaces because of their significance to the continuity of bi-conjugate functions, and we will consider dual spaces with the Schur property.

Theorem 8.2.8. *For a Banach space X, the following are equivalent.*

(a) *X is a Grothendieck space.*
(b) *For each continuous convex function f on X, every weak*-lsc convex extension of f to X^{**} is continuous.*
(c) *For each continuous convex function f on X, f^{**} is continuous on X^{**}.*
(d) *For each continuous convex function f on X, there is at least one weak*-lsc convex extension of f to X^{**} that is continuous.*
(e) *For each Fréchet differentiable convex function f on X, there is at least one weak*-lsc convex extension of f to X^{**} that is continuous.*

Proof. For (a) $\Rightarrow$ (b), see Exercise 8.2.9(a). Now (b) $\Rightarrow$ (c) $\Rightarrow$ (d) $\Rightarrow$ (e) are all trivial. For (e) $\Rightarrow$ (a) see Exercise 8.2.9(b). □

For another characterization of Grothendieck spaces concerning weak*-lsc convex extensions that preserve points of Gâteaux differentiability see Exercise 8.2.5. For further information on Grothendieck spaces and related spaces, see [183, 184, 251].

Theorem 8.2.9. *For a Banach space X, the following are equivalent.*

(a) *X^* has the Schur property.*
(b) *$X \not\supset \ell_1$ and X has the Dunford–Pettis property.*
(c) *If $f : X \to \mathbb{R}$ is a continuous convex function such that f^{**} is continuous, then f is bounded on bounded sets.*

Proof. See [183, p. 212] for the equivalence of (a) and (b); see also Exercise 8.2.2.
(a) $\Rightarrow$ (c): According to Theorem 8.2.5(c), f^{**} is bounded on bounded sets and thus f is bounded on bounded sets by Fact 4.4.4.
(c) $\Rightarrow$ (a): See Exercise 8.2.10. □

Figures 8.1 and 8.2 illustrate the containments between many of the key classes of spaces discussed in this section and earlier.

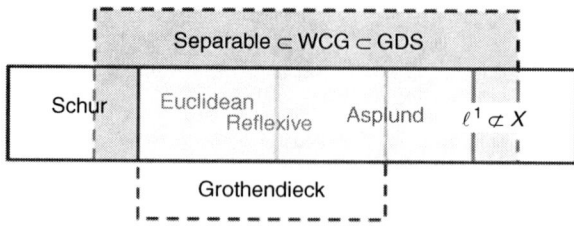

Figure 8.1 Relations between classes of Banach spaces.

8.2 Characterizations of spaces

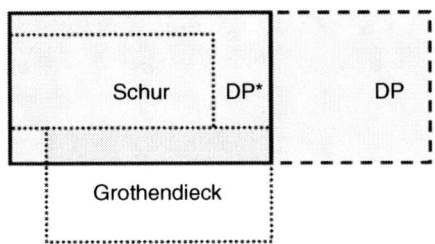

Figure 8.2 Relations between DP and DP* spaces.

Exercises and further results

8.2.1 (Theorem 8.2.4 in Asplund space). Part (a) and (b) below outline an elementary proof by V. Montesinos (private communication) that the norm and Mackey topologies coincide sequentially in X^* when X is an Asplund space; this proof does not use Rosenthal's ℓ_1-theorem.

(a) Let X be a Banach space. Show that the following are equivalent:
   (i) The norm and Mackey topologies agree sequentially in X^*.
   (ii) Given a weak*-null sequence (x_n^*) in X^* such that $\langle x_n, x_n^* \rangle \to 0$ for every weakly null sequence (x_n) in X we have $x_n^* \to_{\|\cdot\|} 0$.
(b) Let X be an Asplund Banach space. Then the Mackey and norm topologies agree sequentially on X^*.
(c) Deduce that if X is an Asplund space, then Fréchet differentiability and weak Hadamard differentiability are equivalent for continuous convex functions.
(d) Provide an example of a continuous convex function $f : c_0 \to \mathbb{R}$ that is Gâteaux differentiable at some point, but fails to be Fréchet differentiable at that point.

Hint. (a) (i) $\Rightarrow$ (ii): Let (x_n^*) be a sequence as in (ii). Assume that $x_n^* \not\to_{\tau_W} 0$. Find an absolutely convex and weakly compact subset W of X, a sequence (w_k) in W, a subsequence (n_k) of (n) and some $\varepsilon > 0$ such that $\langle w_k, x_{n_k}^* \rangle > \varepsilon$ for all k. W is weakly sequentially compact, so we may and do assume that $w_k \to_w w_0$ for some $w_0 \in W$. Then $\langle w_k - w_0, x_{n_k}^* \rangle \not\to 0$ and we reach a contradiction. So $x_n^* \to_{\tau_W} 0$ and from (i) we have $x_n^* \to_{\|\cdot\|} 0$.

(ii)$\Rightarrow$ (i) Let $x_n^* \to_{\tau_W} 0$. Given a weakly null sequence (x_n) in X, $\{x_n\}_{n=1}^\infty \cup \{0\}$ is a weakly compact subset of X, and so is its absolutely convex hull. Then (x_n^*) converges to 0 uniformly on $\{x_n\}_{n=1}^\infty$. In particular, $\langle x_n, x_n^* \rangle \to 0$. From (ii) we get $x_n^* \to_{\|\cdot\|} 0$.

(b) From part (a) it is clearly enough to prove the result for Banach spaces X such that X^* is separable. We shall check that in this case (ii) holds. Let (x_n^*) be a sequence in X^* as in (ii). Assume that $\|x_n^*\| \not\to 0$. Without loss of generality we may take $\|x_n^*\| = 1$ for all n. We can find then a sequence (x_n) in B_X such that $\langle x_n, x_n^* \rangle > 1/2$ for all n. Now $(B_{X^{**}}, w^*)$ is compact and metrizable, so there exits a subsequence of (n), denoted again by (n), and some $x^{**} \in B_{X^{**}}$ such that $x_n \to_{w^*} x^{**}$. Using this with the fact that (x_n^*) is weak*-null, we may and do assume that for some subsequence of (n), denoted again (n), we have that $x_{2n} - x_{2n-1}$ is a weakly null sequence with the property that $\langle x_{2n} - x_{2n-1}, x_n^* \rangle \not\to 0$, a contradiction.

To prove (c), use (b) and Proposition 8.1.2; for (d), see Exercise 8.1.1. □

8.2.2.* Show the equivalence of (a) and (b) in Theorem 8.2.9. (You may use any other theorem in this chapter, and you may use the result if $X \supset \ell_1$, then X^* does not have the Schur property.)

8.2.3.* Prove the equivalence of (a) and (b) in Theorem 8.2.7.

Hint. (a) $\Rightarrow$ (b): Suppose $f : X^* \to \mathbb{R}$ is weak*-lsc, convex (and continuous) and that f is weak Hadamard differentiable at x^*. Now suppose $x_n \in \partial_{\varepsilon_n} f(x^*)$ where $\varepsilon_n \to 0$. By Šmulian's theorem (Exercise 4.2.10), x_n converges to $f'(x^*)$ uniformly on weakly compact sets in X^*. Suppose that (x_n) is not norm convergent. Then there is a subsequence x_{n_k} and $\delta > 0$ so that $\|y_k\| > \delta$ for all k where $y_k = x_{n_{k+1}} - x_{n_k}$. Now $y_k \to 0$ uniformly on weakly compact subset of X^*, and so by the hypothesis of (a), $y_k \to 0$ in norm. This contradiction shows that (x_n) is norm convergent. Applying Šmulian's theorem (Exercise 4.2.10) again ensures that f if Fréchet differentiable at x^*.

(b) $\Rightarrow$ (a): can be done by appropriately applying Proposition 8.1.1. □

8.2.4.* Suppose X is a Banach space with the Dunford–Pettis property.

(a) Let $f : X^* \to \mathbb{R}$ be a continuous weak*-lsc convex function. If f is Gâteaux differentiable at x^*, then is it true that $f'(x^*) \in X$?
(b) Suppose $(x_n) \subset X \subset X^{**}$ and $(x_n^*) \subset X^*$ are such that $x_n \to x^{**} \in X^{**}$ in the weak*-topology on X^{**} and $x_n^* \to x^*$ in the weak topology on X^*. Show that $x_n^*(x_n) \to x^{**}(x^*)$.
(c) Prove Theorem 8.2.6.

Hint. (a) No. Consider the usual norm on ℓ_1.

(b) Suppose that $x_n^*(x_n) \not\to x^{**}(x^*)$. By passing to a subsequence and re-indexing we may and do assume

$$|x_n^*(x_n) - x^{**}(x^*)| > 3\delta \text{ for all } n, \text{ and some } \delta > 0.$$

Because $x_n^* \to_w x^*$ we fix n_0 such that

$$|x^{**}(x_n^*) - x^{**}(x^*)| < \delta \text{ for all } n \geq n_0.$$

Now fix $n_1 \geq n_0$, and suppose n_k has be been chosen. We use the fact that $x_n \to_{w^*} x^{**}$ to choose $n_{k+1} > n_k$ so that

$$|x_{n_k}^*(x_{n_{k+1}}) - x_{n_k}^*(x^{**})| < \delta.$$

Consequently, $|x_{n_k}^*(x_{n_{k+1}}) - x^{**}(x^*)| < 2\delta$. Now let $u_k^* = x_{n_k}^*$ and $u_k = x_{n_{k+1}} - x_{n_k}$. Then $u_k^* \to_w x^*$ and $u_k \to_w 0$. However,

$$|u_k^*(u_k)| = |x_{n_k}^*(x_{n_{k+1}}) - x^{**}(x^*) + x^{**}(x^*) - x_{n_k}^*(x_{n_k})|$$
$$\geq |x_{n_k}^*(x_{n_k}) - x^{**}(x^*)| - |x_{n_k}^*(x_{n_{k+1}}) - x^{**}(x^*)| > 3\delta - 2\delta = \delta,$$

which violates the Dunford–Pettis property.

(c) We begin with (a) $\Rightarrow$ (b): Suppose $x_n^* \to_w 0$ but that the convergence is not uniform on weakly compact sets. Then there are a weakly compact set W, $x_n \in W$, and $\delta > 0$ so that $|x_n^*(x_n)| > \delta$. Using the Eberlein–Šmulian theorem (see, e.g. [199, p. 85]) passing to a subsequence we may and do assume $x_n \to_w x$. Then $x_n^*(x_n) \not\to \langle 0, x \rangle$ and so the Dunford–Pettis property is violated.

(b) $\Rightarrow$ (a): Suppose $x_n \to_w x$ and $x_n^* \to_w x^*$. Then $W = \{x_n\}_n \cup \{x\}$ is weakly compact. Therefore, $x_n^* \to x^*$ uniformly on W. Now choose n_0 such that

$$|x^*(x_n) - x^*(x)| < \frac{\varepsilon}{2} \text{ and } |x_n^*(w) - x^*(w)| < \frac{\varepsilon}{2} \text{ for all } w \in W, \, n > n_0.$$

Thus for $n > n_0$ we have

$$|x_n^*(x_n) - x^*(x)| = |x_n^*(x_n) - x^*(x_n) + x^*(x_n) - x^*(x)|$$
$$\leq |x_n^*(x_n) - x^*(x_n)| + |x^*(x_n) - x^*(x)| < \varepsilon.$$

Thus the Dunford–Pettis property is satisfied.

(a) $\Rightarrow$ (c): We will use part (b) of this exercise and Šmulian's theorem in a similar fashion to the proof of Theorem 8.2.5. Suppose $x_n \in \partial_{\varepsilon_n} f(x^*) \cap X$ where $\varepsilon_n \to 0^+$. Then $x_n \to_{w^*} f'(x^*)$ by Šmulian's theorem (Exercise 4.2.10). Suppose the convergence is not uniform on weakly compact sets. Then there is a weakly compact set $W \subset X^*$ and $x_n^* \in W$ such that $|\langle x_n^*, x_n \rangle - \langle f'(x^*), x_n^* \rangle| > \varepsilon$ for all n and some $\varepsilon > 0$. By passing to a subsequence using the Eberlein–Šmulian theorem (see e.g. [199, p. 85]), we may and do assume $x_n^* \to_w \phi$ for some $\phi \in W$. Now, $\langle f'(x^*), x_n^* \rangle \to \langle f'(x^*), \phi \rangle$. Therefore,

$$\liminf_{n \to \infty} |\langle x_n^*, x_n \rangle - \langle f'(x^*), \phi \rangle| \geq \varepsilon.$$

This contradicts part (b) of this exercise. Consequently, $x_n^* \to f'(x^*)$ uniformly on weakly compact sets in X^*. According to Šmulian's theorem (Exercise 4.2.10), f is weak Hadamard differentiable at x^*.

(c) $\Rightarrow$ (b) and (d) $\Rightarrow$ (b) can be shown using Proposition 8.1.1.

(a) $\Rightarrow$ (d): Suppose $f : X^* \to \mathbb{R}$ is weak*-lsc, convex, continuous and unbounded on some weakly compact set W. Let $C_n := \{x^* : f(x)^* \leq n\}$. Since we lose no generality assuming $f(0) = 0$, we may choose $\delta > 0$ so that $\delta B_{X^*} \subset C_n$ for all n. Now choose $x_n^* \in W$ so that $f(x_n^*) > n$. By the weak*-separation theorem (4.1.22) we choose $x_n \in S_X$ so that $x_n^*(x_n) > \sup_{C_n} x_n \geq \delta$. Again, by Eberlein–Šmulian we may assume $x_n^* \to_w x^*$ for some x^* in W. As in analogous results, $x_n \to_w 0$ for if not

f would not be defined on all of X. However, $x_n^*(x_n) \not\to 0$ and so the Dunford–Pettis property is not satisfied. □

8.2.5 (Godefroy [232]). Let X be a Banach space, and let $\mathcal{G}(f)$ denote the points of Gâteaux differentiability of f. Show that the following are equivalent.

(a) X is a Grothendieck space.
(b) $\mathcal{G}(f) = \mathcal{G}(f^{**}) \cap X$ for each lsc convex function $f : X \to (-\infty, +\infty]$.
(c) $\mathcal{G}(\nu) = \mathcal{G}(\nu^{**}) \cap X$ for each continuous norm on X.

Hint. (a) $\Rightarrow$ (b): Suppose f is Gâteaux differentiable at $x_0 \in X$ with Gâteaux derivative $\phi \in X^*$. Now f^{**} is continuous at $x_0 \in X^{**}$ according to Fact 4.4.4(b). Let $\phi_n \in \partial_{\varepsilon_n} f^{**}(x_0)$ where $\varepsilon_n \to 0$ and $\phi_n \in X^*$. Then $\phi_n \in \partial_{\varepsilon_n} f(x_0)$, and by the Gâteaux differentiability of f, $\phi_n \to_{w^*} \phi$ (Šmulian). Then by the dual version of Šmulian's theorem, f^{**} is Gâteaux differentiable at $x_0 \in X$.

(b) $\Rightarrow$ (c) is trivial. Note that (b) $\Rightarrow$ (a) can be deduced using a canonical example. See [232] for further details. □

8.2.6 (Godefroy [232]). Suppose f is a continuous convex function on X and f^{**} is Gâteaux differentiable at all points of X. Show that f^{**} is the unique weak*-lsc extension of f to X^{**}.

Hint. Use Exercise 4.3.12(b). See [116, Theorem 3.1] for further details. □

8.2.7 (Godefroy [232]). Suppose $f : X \to \mathbb{R}$ is a Fréchet differentiable convex function. Show that every weak*-lsc convex extension of f to X^{**} is Fréchet differentiable on X. In particular, there is only one such extension, namely f^{**}.

Hint. The Fréchet differentiability assertion follows as in the proof of Exercise 4.4.22. Uniqueness then follows from Exercise 8.2.6. □

8.2.8 (Godefroy [232]). Let $f : X \to (-\infty, +\infty]$ be lsc and convex, and let $\mathcal{E}(f)$ denote the convex weak*-lsc extensions of $f : X \to (-\infty, +\infty]$ to X^{**}. Show that $\mathcal{E}(f)$ has a largest element, that $\mathcal{E}(f)$ is convex, but that, in general, $\mathcal{E}(f)$ has no smallest element.

Hint. Convexity and existence of a largest element are straightforward. Following [232, p. 372], let $f : \ell_1 \to \mathbb{R}$ be the usual norm, i.e. $f(x) = \|x\|_1$. Define

$$C_1 := \{(a_n) \in \ell_\infty : \|(a_n)\|_\infty \leq 1, \lim_{n \to \infty} a_n = 0\},$$

$$C_2 := \{(a_n) \in \ell_\infty : \|(a_n)\|_\infty \leq 1, \lim_{n \to \infty} a_n = a_1\}.$$

Let $\sigma_{C_i} : \ell_1^{**} \to \mathbb{R}$ denote the support function of C_i. Then $\sigma_{C_i} \in \mathcal{E}(f)$, however, each convex function $f : \ell_1^{**} \to (-\infty, +\infty]$ minorizing σ_{C_1} and σ_{C_2} vanishes at $e_1 \in \ell_1$. To see this, let c be the closed subspace of convergent sequences in ℓ_∞, and let $\phi \in \ell_\infty^*$ be the functional such that $\phi(x_i) = -\lim_{i \to \infty} x_i$ for each $(x_i) \in c$. Then check that $\sigma_{C_2}(e_1 + \phi) = 0$ while $\sigma_{C_1}(e_1 - \lambda\phi) = 1$ for each $\lambda \in \mathbb{R}$ and consider $e_1 = \frac{n-1}{n}(e_1 + \phi) + \frac{1}{n}(e_1 - (n-1)\phi)$. □

8.2 Characterizations of spaces

8.2.9.* (a) Prove the implication (a) $\Rightarrow$ (b) from Theorem 8.2.8.
(b) Prove the implication (e) $\Rightarrow$ (a) from Theorem 8.2.8.

Hint. (a) Suppose some weak*-lsc convex extension of f, say $\tilde{f}$ is not continuous on X^{**}. Take $\Phi \in X^{**}$ such that $\tilde{f}(\Phi) = \infty$. Consider $C_n = \{\Lambda \in X^{**} : \tilde{f}(\Lambda) \leq n\}$. Using the weak*-separation theorem (4.1.22) find $\phi_n \in S_{X^*}$ such that $\sup_{C_n} \phi_n \leq \langle \phi_n, \Phi \rangle$. Show that the sequence (ϕ_n) converges weak* to 0, but not weakly to 0.

(b) Suppose X is not a Grothendieck space, and choose $(\phi_n) \subset S_{X^*}$ that converges weak* but not weakly to 0. Show that the function $h(x)$ in Proposition 8.1.1 is such that h^{**} is not continuous on X^{**} and note that this is the unique weak*-lsc extension to X^{**}. □

8.2.10.* Prove the implication (c) $\Rightarrow$ (a) in Theorem 8.2.9.

Hint. Suppose X^* is not Schur, and choose $(\phi_n) \subset S_{X^*}$ that converges weakly to 0. Then the function $h : X \to \mathbb{R}$ in Proposition 8.1.1 is continuous but not bounded on bounded sets. It then remains to verify that h^{**} is continuous. □

8.2.11. (a) Suppose that B_{X^*} is weak*-sequentially compact. Prove that X is a Gelfand–Phillips space.
(b) Suppose a Banach space has the DP*-property and is Gelfand–Phillips. Show that it is Schur.
(c) Suppose X has the DP*-property and Y is a subspace of X that is Gelfand–Phillips but not Schur. Show that there is a continuous convex function on Y that cannot be extended to a continuous convex function on X.

Hint. (a) Suppose $K \subset X$ is not relatively compact. Then there is a sequence $(x_n) \subset K$ and $\varepsilon > 0$ such that $d_{E_n}(x_{n+1}) > \varepsilon$ where $E_n = \text{span}\{x_1, \ldots, x_n\}$. Thus choose $\phi_n \in S_{X^*}$ such that $\phi_n(E_n) = \{0\}$ while $\phi_n(x_{n+1}) \geq \varepsilon$. By the weak*-sequential compactness of B_{X^*}, we have that $\phi_{n_k} \to_{w^*} \phi$ for some $\phi \in B_{X^*}$. Now ϕ_{n_k} converges pointwise to ϕ on K, but not uniformly.

(b) Suppose X has the DP*-property, then weakly compact sets are limited. But because X is Gelfand–Phillips, this implies weakly compact sets are norm compact.

(c) By part (b), Y does not have the DP*-property. Therefore, there is a continuous convex function on Y that is not bounded on weakly compact subsets of Y. Such a function cannot be extended to a continuous convex function on all of X. □

8.2.12 (Explicit Josefson–Nissenzweig sequences). (a) Show that for $1 \leq p \leq \infty$ the canonical basis in ℓ_p forms a JN-sequence. (b) Construct a JN-sequence in $L_p[0, 1]$.

8.2.13. Prove the special case of the Josefson–Nissenzweig theorem (8.2.1) in the event that B_{X^*} is weak*-sequentially compact. Prove it also for Banach spaces X with a separable (infinite-dimensional) quotient space.

Hint. Suppose B_{X^*} is weak*-sequentially compact. Using the separation theorem (4.1.17), construct a system $\{x_n, x_n^*\}$ such that $\|x_n^*\| = \|x_n\| = 1$ and $x_n^*(x_n) \geq 1/2$ while $x_n^*(x_m) = 0$ for $m \leq n$. Now, $x_{n_j}^* \to_{w^*} x^*$, and so $(x_{n_j}^* - x^*) \to_{w^*} 0$ while $\langle x_{n_j}^* - x^*, x_{n_j} \rangle \geq 1/2$. Thus $\|x_{n_j}^* - x^*\| \geq 1/2$ for all j.

Suppose $T : X \to Y$ is a bounded linear onto operator where Y is separable (or more generally has weak*-sequentially compact dual ball). Let $\phi_n \in S_{Y^*}$ be such that

$\phi_n \to_{w^*} 0$. Then $T^*\phi_n \to_{w^*} 0$. By the open mapping theorem (Exercise 4.1.23) $(T^*\phi_n)$ does not converge in norm to 0. □

8.2.14. ** Prove that on every nonreflexive Banach space, there is a Lipschitz function that is weak Hadamard but not Fréchet differentiable.

This shows, for example, that Theorem 8.2.4 does not extend to Lipschitz functions. It can be shown with substantially more work, that it it does not extend to differences of continuous convex functions; see [114].

Hint. See [78]. □

8.2.15. Show the existence of a sequence in S_{X^*} that converges weak* to 0 implies for any $x \in B_{X^*}$ with $\|x\| < 1$, there is a sequence in S_{X^*} that converges weak* to x.

Hint. Let $\|x\| < 1$. Choose $\varepsilon > 0$ such that $\|x\| + \varepsilon < 1$. Now $(x + \varepsilon x_n) \to_{w^*} x$ where $x_n \to_{w^*} 0$. Find $t_n > \varepsilon$ so that $x + t_n x_n \in S_X$. Let $\tilde{x}_n := x + t_n x_n$ Now $\varepsilon < t_n \le 2/\varepsilon$ for each n and so $\tilde{x}_n \to_{w^*} x$ as desired. □

8.3 Extensions of convex functions

This section considers the question of extending convex functions to preserve continuity, which in some sense can be thought of as a nonlinear variant of the Hahn–Banach theorem.

Question 8.3.1. *Suppose Y is a closed subspace of a Banach space X. If $f : Y \to \mathbb{R}$ is a continuous convex function, is there a continuous convex function $\tilde{f} : X \to \mathbb{R}$ such that $\tilde{f}|_Y = f$? That is can f be extended to a continuous convex function on X? Relatedly, when can a convex function continuous on a spanning closed convex subset of X be extended continuously to X?*

The following example shows that such extensions are not always possible.

Example 8.3.2. Let $Y := c_0$ or ℓ_p with $1 < p < \infty$. Let $f : Y \to \mathbb{R}$ be defined by

$$f(y) := \sum_{n=1}^{\infty} (e_n^*(y))^{2n}$$

where e_n^* are the coordinate functionals. If Y is considered as a subspace of ℓ_∞, then f cannot be extended to a continuous convex function on ℓ_∞.

Proof. Because $e_n^* \to_{w^*} 0$, Example 8.1.1 shows f is a continuous convex function. However, f is not bounded on the weakly compact set $\{2e_n\}_{n=1}^{\infty} \cup \{0\}$. Now ℓ_∞ is a space with the DP*-property and so Theorem 8.2.3 shows every continuous convex function on ℓ_∞ is bounded on weakly compact subsets of ℓ_∞. Therefore, f cannot be extended to a continuous convex function on ℓ_∞. □

We refer the reader to Exercise 8.2.11 for a more general formulation of Example 8.3.2. We should also point out that in the case $Y = c_0$, the preceding provides an example of a continuous convex function f whose biconjugate fails to be continuous; see [392]. Before proceeding, observe that there are natural conditions

8.3 Extensions of convex functions

that can be imposed on f that allow it to be extended to any superspace. For example, if f is Lipschitz or more generally, is bounded on bounded sets (see Exercise 8.3.4). However, our present goal is to find conditions on X and/or Y for which every continuous convex function on Y can be extended to a continuous convex function on X. A well-known natural condition where this is true is recorded as

Remark 8.3.3. Suppose Y is a complemented subspace of a Banach space X. Then every continuous convex function on Y can be extended to a continuous convex function on X.

Proof. Let $f : Y \to \mathbb{R}$ be continuous convex. Then $\tilde{f}(x) := f(P(x))$ where $P : X \to Y$ is a continuous linear projection provides one such extension. □

In light of Example 8.3.2, the above remark doesn't extend to quasicomplements because c_0 is quasicomplemented in ℓ_∞; see [199, Theorem 11.42]. We now develop results that will allow extensions in the case X/Y is separable. First, we will consider 'generalized canonical' examples which will allow us to in some respects capture the essence of all convex functions on the space. In this section, all nets $(\phi_{n,\alpha})_{n \in \mathbb{N}, \alpha \in A_n}$ will have their index sets directed by $(n, \alpha) \geq (m, \beta)$ if and only if $n \geq m$. Thus $\phi_{n,\alpha} \to_{w^*} 0$ if for each $\varepsilon > 0$ and $x \in X$, there exists $n_0 \in \mathbb{N}$ such that $|\phi_{n,\alpha}(x)| < \varepsilon$ whenever $n \geq n_0$.

Proposition 8.3.4. *Let* $(\phi_{n,\alpha}) \subset X^*$ *be a bounded net. Consider the lsc convex functions* $f : X \to (-\infty, +\infty]$ *that are defined as follows*

$$f(x) := \sup_{n,\alpha} \{\phi_{n,\alpha}(x) - a_{n,\alpha}, 0\} \quad \text{and} \quad g(x) := \sup_{n,\alpha} n(\phi_{n,\alpha}(x))^{2n}$$

where $b_n \leq a_{n,\alpha} \leq c_n$ *and* $b_n \downarrow 0$, $c_n \downarrow 0$. *Then:*

(a) f *is* β-*differentiable at* 0 *if and only if* $\phi_{n,\alpha} \to_{\tau_\beta} 0$. *Moreover* f *is Lipschitz on* X.

(b) g *is bounded on* β-*sets if and only if* $\phi_{n,\alpha} \to_{\tau_\beta} 0$. *If this is the case, g is continuous.*

Proof. This is a straightforward exercise: one can modify the proof of Proposition 8.1.1. □

We will also use the following fact.

Lemma 8.3.5. *Let Y be a closed subspace of a Banach space X, and let $\varepsilon \geq 0$. Suppose $f : Y \to \mathbb{R}$ is continuous and convex, and suppose $\tilde{f} : X \to \mathbb{R}$ is a continuous convex extension of f. If $\phi \in \partial_\varepsilon f(y_0)$, then there is an extension $\tilde{\phi} \in X^*$ such that $\tilde{\phi} \in \partial_\varepsilon \tilde{f}(y_0)$.*

Proof. By shifting f, we may without loss of generality assume that $f(0) = -1$. Let $\phi \in \partial_\varepsilon f(y_0)$ and let $a := \phi(y_0) - f(y_0) + \varepsilon$. It is easy to check that $a \geq 1$. Moreover, for $(\phi, -1) \in Y^* \times \mathbb{R}$ we have $(\phi, -1)(y, t) = \phi(y) - t \leq \phi(y) - f(y) \leq a$ for all $(y, t) \in \text{epi} f$. Now define the continuous sublinear function $\rho : X \times \mathbb{R} \to [0, \infty)$ by $\rho := a\gamma_{\text{epi}\tilde{f}}$ where $\gamma_{\text{epi}\tilde{f}}$ is the gauge functional of the epigraph of $\tilde{f}$. Then

$(\phi, -1) \leq \rho$ on $Y \times \mathbb{R}$. According to the Hahn–Banach theorem (4.1.7), $(\phi, -1)$ extends to a continuous linear functional $(\tilde{\phi}, -1)$ on $X \times \mathbb{R}$ that is dominated by ρ. Therefore, $(\tilde{\phi}, -1)(x, t) \leq a$ if $(x, t) \in \text{epi}\tilde{f}$ which implies $\tilde{\phi} \in \partial_\varepsilon f(y_0)$. □

Corollary 8.3.6. *Suppose Y is a closed subspace of a Banach space X. Suppose $f : Y \to \mathbb{R}$ and $g : X \to \mathbb{R}$ are continuous convex functions such that $f \leq g|_Y$. If, for some $\varepsilon \geq 0$ we have $\phi \in \partial_\varepsilon f(y_0)$ then ϕ can be extended to a continuous linear functional $\tilde{\phi}$ such that $f(y_0) + \tilde{\phi}(x - y_0) \leq g(x) + \varepsilon$ for all $x \in X$.*

Proof. Let $\phi \in \partial_\varepsilon f(y_0)$. Then $\phi \in \partial_r g|_Y(y_0)$ where $r := g(y_0) - f(y_0) + \varepsilon$. Apply Lemma 8.3.5 to obtain $\tilde{\phi}$ such that

$$\tilde{\phi}(x) - \tilde{\phi}(y_0) \leq g(x) - g(y_0) + g(y_0) - f(y_0) + \varepsilon,$$

from which the conclusion is immediate. □

Lemma 8.3.7. *Let Y be a closed subspace of X, and suppose $f : Y \to \mathbb{R}$ and $g : X \to \mathbb{R}$ are continuous convex functions such that $f \leq g|_Y$. Then f can be extended to a continuous convex function $\tilde{f} : X \to \mathbb{R}$ such that $\tilde{f} \leq g$.*

Proof. For each $y \in Y$, choose $\phi_y \in \partial f(y)$. Let $\tilde{\phi}_y$ be an extension as given by the Corollary 8.3.6. Now define $\tilde{f}(x) := \sup_{y \in Y} f(y) + \tilde{\phi}_y(x - y)$. □

The following theorem provides a useful condition for determining when every continuous convex function on a given subspace of a Banach space can be extended to the whole space.

Theorem 8.3.8. *Suppose Y is a closed subspace of a Banach space X. Then the following are equivalent.*

(a) *Every continuous convex function $f : Y \to \mathbb{R}$ can be extended to a continuous convex function $\tilde{f} : X \to \mathbb{R}$.*
(b) *Every bounded net $(\phi_{n,\alpha}) \subset Y^*$ that converges weak* to 0 can be extended to a bounded net $(\tilde{\phi}_{n,\alpha}) \subset X^*$ that converges weak* to 0.*

Proof. (a) $\Rightarrow$ (b). Suppose $(\phi_{n,\alpha})$ is a bounded net in Y^* that converges weak* to 0, and without loss of generality suppose $\|\phi_{n,\alpha}\| \leq 1$ for all n, α. Now define

$$f(y) := \sup_{n,\alpha} (\phi_{n,\alpha}(y))^{2n}.$$

Then $f : Y \to \mathbb{R}$ is a continuous convex function (as in Proposition 8.3.4), so we extend it to a continuous convex function $\tilde{f} : X \to \mathbb{R}$. Now let $C_n := \{x \in X : \tilde{f}(x) \leq 2^{2n}\}$. Observe that $\tilde{f}(0) = 0$, and so the continuity of $\tilde{f}$ at 0 implies that there is an $\varepsilon > 0$ so that $\tilde{f}(x) \leq 1$ for all $\|x\| \leq \varepsilon$. Then

$$\varepsilon B_X \subset C_n \quad \text{and} \quad C_n \cap Y \subset \{x : \phi_{n,\alpha}(x) \leq 2\}.$$

Define the sublinear function $p_n := 2\gamma_{C_n}$. Then $\phi_{n,\alpha}(y) \leq p_n(y)$ for all $y \in Y$. By the Hahn–Banach theorem (4.1.7), extend $\phi_{n,\alpha}$ to $\tilde{\phi}_{n,\alpha}$ so that $\tilde{\phi}_{n,\alpha}(x) \leq p_n(x)$ for all

8.3 Extensions of convex functions

$x \in X$. Then $\|\tilde{\phi}_{n,\alpha}\| \leq 2/\varepsilon$. Now let us suppose that $(\tilde{\phi}_{n,\alpha})$ does not converge weak* to 0. Then we can find $x_0 \in X$, a subsequence (n_k) of (n) and a sequence (α_k) such that $\tilde{\phi}_{n_k,\alpha_k}(x_0) > 2$ for all k. Thus $x_0 \notin C_n$ for infinitely many n, and so $\tilde{f}(x_0) > 2^{2n}$ for infinitely n. Thus $\tilde{f}(x_0) = \infty$ which contradicts the continuity of $\tilde{f}$.

(b) $\Rightarrow$ (a): Suppose $f : Y \to \mathbb{R}$ is a continuous convex function. Without loss of generality we may assume $f(0) = 0$. Now define $C_n := \{y \in Y : f(y) \leq n\}$. Because f is continuous, there is a $\delta > 0$ such that $\delta B_Y \subset C_n$ for each $n \in \mathbb{N}$. Thus we can write

$$C_n = \bigcap_{\alpha} \{y \in Y : \phi_{n,\alpha}(y) \leq 1\}$$

where $\|\phi_{n,\alpha}\| \leq 1/\delta$ for all $n \in \mathbb{N}$ and $\alpha \in A_n$ (A_n can be chosen as a set with cardinality the density of Y). Also, $(\phi_{n,\alpha})$ converges weak* to 0, otherwise there would be a $y_0 \in Y$, a subsequence (n_k) of (n) and a sequence (α_k) such that $\phi_{n_k,\alpha_k}(y_0) > 1$ for all k. Consequently, $y_0 \notin C_n$ for infinitely many n which would yield the contradiction $f(y_0) = \infty$. Thus, by the hypothesis of (b), $(\phi_{n,\alpha})$ extends to a bounded net $(\tilde{\phi}_{n,\alpha}) \subset X^*$ that converges weak* to 0. Now define

$$g(x) := \sup n(\tilde{\phi}_{n,\alpha}(x))^{2n} + 1.$$

Then $g : X \to \mathbb{R}$ is a continuous convex function (Proposition 8.3.4). Moreover, $g(y) \geq f(y)$ for all $y \in Y$; this is because $g(x) \geq 1$ for all $x \in X$, and if $n - 1 < f(y) \leq n$ where $n \geq 2$, then $y \notin C_{n-1}$ and so $\phi_{n-1,\alpha_0}(y) > 1$ for some α_0 which implies $g(y) > (n-1) + 1 \geq f(y)$. According to Lemma 8.3.7, there is a continuous convex extension $\tilde{f} : X \to \mathbb{R}$ of f. $\square$

Using a theorem of Rosenthal's [381] we next outline that Theorem 8.3.8(b) is satisfied when X/Y is separable. For this, recall that a Banach space is said to be *injective* if it is complemented in every superspace, and it is said to be *1-injective* if it is complemented by a norm-one projection in every superspace; see Zippin's article [449] for further information concerning this subject.

Theorem 8.3.9. *Let X be a Banach space, Y a closed subspace such that X/Y is separable. Let $(\phi_{n,\alpha})_{\alpha \in A_n, n \in \mathbb{N}}$ be a weak*-null net in Y^* such that $\|\phi_{n,\alpha}\| \leq 1$ for all $\alpha \in A_n, n \in \mathbb{N}$. Then, for every $\varepsilon > 0$ there exists a weak*-null net $(\tilde{\phi}_{n,\alpha})_{\alpha \in A_n, n \in \mathbb{N}}$ of elements in X^* such that $\|\tilde{\phi}_{n,\alpha}\| \leq 2 + \varepsilon$ and $\tilde{\phi}_{n,\alpha}$ extends $\phi_{n,\alpha}$ for all $\alpha \in A_n, n \in \mathbb{N}$.*

Proof. Define a bounded linear operator $T : Y \to (\ell_\infty(A_n))_{c_0}$ by $T(y) := ((\phi_{n,\alpha}(y))_{\alpha \in A_n})_n$; then $\|T\| \leq 1$. Now use the following extension theorem of H. P. Rosenthal (see [381], [449, Theorem 3.5]): Let $Z_1, Z_2, \ldots$ be 1-injective Banach spaces, X, Y be Banach spaces with $Y \subset X$ and X/Y separable, and set $Z =: c_0(Z_1, Z_2, \ldots)$. Then for every nonzero operator $T : Y \to Z$ and every $\varepsilon > 0$, there exists a $\hat{T} : X \to Z$ extending T with $\|\hat{T}\| < (2 + \varepsilon)\|T\|$. According to this result, T defined above extends to $\hat{T} : X \to (\ell_\infty(A_n))_{c_0}$ with $\|\hat{T}\| < 2 + \varepsilon$. Now let $e_{n,\alpha}^*$ denote the coordinate functional so that $e_{n,\alpha}^*(x) = x_\alpha$ for $x = (x_i)_{i \in A_n} \in \ell_\infty(A_n)$. Then $e_{n,\alpha}^*(T(y)) = \phi_{n,\alpha}(y)$ for all y and $\tilde{\phi}_{n,\alpha} = e_{n,\alpha}^* \circ \hat{T}$

extends $\phi_{n,\alpha}$. Because $\hat{T}(x) \in (\ell_\infty(A_n))_{c_0}$, it follows that $\tilde{\phi}_{n,\alpha} \to_{w^*} 0$; moreover, $\|\tilde{\phi}_{n,\alpha}\| \leq 1 \|\hat{T}\| < 2 + \varepsilon$. □

Our main application of Theorem 8.3.8 is

Corollary 8.3.10. *Suppose X is a Banach space and Y is a closed subspace of X such that X/Y is separable. Then every continuous convex function $f : Y \to \mathbb{R}$ can be extended to a continuous convex function $\tilde{f} : X \to \mathbb{R}$.*

Proof. Apply Theorem 8.3.8 and Theorem 8.3.9. □

Observe that Example 8.3.2 shows the previous corollary can fail if X/Y is not separable, it also shows it is not always possible to extend a continuous convex function from a separable closed subspace of a Banach space X to a continuous convex function on the whole space X. However:

Corollary 8.3.11. *Suppose Y is a separable closed subspace of a WCG Banach space X. Then every continuous convex function on Y can be extended to a continuous convex function on X.*

Proof. There is a separable space $Y_1 \supset Y$ such that Y_1 is complemented in X; see e.g. [247]. Extend the continuous convex function to Y_1 by Theorem 8.3.10 and then use Remark 8.3.3 to extend it to X. □

In fact, the previous result holds in any Banach space X such that every separable subspace Y of X lies in a complemented separable subspace of X. Some further information concerning spaces with this property may be found in [247]. Also, if Y is an injective Banach space, then any continuous convex function can be extended to any superspace by projections (Remark 8.3.3). Another class of spaces that allow extensions to superspaces is as follows.

Proposition 8.3.12. *Suppose Y is a $C(K)$ Grothendieck space. Then any continuous convex function $f : Y \to \mathbb{R}$ can be extended to a continuous convex function $f : X \to \mathbb{R}$ where X is any superspace of Y.*

Proof. Write $Y \subset X$. Then $Y^{**} \cong Y^{\perp\perp} \subset X^{**}$. According to [116, Theorem 2.1] (cf. Theorem 8.2.8), f can be extended to a continuous convex function on Y^{**}. Now, Y^{**} as the bidual of a $C(K)$ space is isomorphic to a $C(K)$ space where K is compact Stonian (i.e. K is extremally disconnected); see [386, p. 121]. Therefore, Y^{**} is injective; see [386, Theorem 7.10, p. 110]. According to Remark 8.3.3 the extension of f to $Y^{\perp\perp}$ can further be extended to X^{**} which contains X. □

Observe that the previous proposition doesn't work for general $C(K)$ spaces, e.g. $c_0 \subset \ell_\infty$, and doesn't work for reflexive Grothendieck, e.g. $\ell_2 \subset \ell_\infty$. More significantly, using some deep results in Banach space theory one can conclude that the above proposition applies to some cases where Y is not a complemented subset of X.

Remark 8.3.13. There are Grothendieck $C(K)$ spaces that are not complemented in every superspace.

8.3 Extensions of convex functions

Proof. Let X be Haydon's Grothendieck $C(K)$ space that does not contain ℓ_∞ [251]. Because $X \not\supset \ell_\infty$, X is not injective by a theorem of Rosenthal's ([380], [294, Theorem 2.f.3]). □

We have focused on preserving continuity in our extensions. One could similarly ask whether extensions exist preserving a given point of differentiability. Again, negative examples in the same spirit of Example 8.3.2 have been constructed. We sketch one such example similar to [78, Example 3.8].

Example 8.3.14. Let $Y := c_0$ or ℓ_p with $1 < p < \infty$. Let $f : Y \to \mathbb{R}$ be defined by $f(y) := \sup\{e_n^*(y) - \frac{1}{n}, 0\}$ where e_n^* are the coordinate functionals. Then there is no continuous convex extension of f to ℓ_∞ that preserves the Gâteaux differentiability of f at 0.

Proof. This follows because Gâteaux and weak Hadamard differentiability coincide for continuous convex functions on ℓ_∞ (see Theorem 8.2.3); however f is not weak Hadamard differentiable at 0 as a function on Y. □

A positive result that is analogous to Theorem 8.3.8 is as follows.

Theorem 8.3.15. *Suppose Y is a closed subspace of a Banach space X. Then the following are equivalent.*

(a) *Every Lipschitz convex function $f : Y \to \mathbb{R}$ that is Gâteaux differentiable at some $y_0 \in Y$ can be extended to a Lipschitz convex function $\tilde{f} : X \to \mathbb{R}$ that is Gâteaux differentiable at y_0.*
(b) *Every bounded net $(\phi_{n,\alpha}) \subset Y^*$ that converges weak* to 0 can be extended to a bounded net $(\tilde{\phi}_{n,\alpha}) \subset X^*$ that converges weak* to 0.*

Proof. (a) $\Rightarrow$ (b): Let $(\phi_{n,\alpha}) \subset Y^*$ be a bounded net that converges weak* to 0. Define $f(x) := \sup\{\phi_{n,\alpha}(x) - \frac{1}{n}, 0\}$. Then f is a Lipschitz convex function that, according to Proposition 8.3.4, is Gâteaux differentiable at 0 (observe, too, that $f(0) = 0$ and $f(y) \geq 0$ for all $y \in Y$, so $f'(0) = 0$). Then extend f to a Lipschitz convex function $\tilde{f}$ that is Gâteaux differentiable at 0 with Gâteaux derivative $\tilde{f}'(0) = \phi$, where $\phi \in X^*$. Now $\phi|_Y = f'(0)$ which implies $\phi|_Y = 0$. Thus $\tilde{f} - \phi$ is a Lipschitz convex function extending f, and whose Gâteaux derivative is 0. Replacing $\tilde{f}$ with $\tilde{f} - \phi$, we can and do assume $\tilde{f}(x) \geq 0$ for all $x \in X$ and $\tilde{f}'(0) = 0$. Clearly $\phi_{n,\alpha} \in \partial_{\frac{1}{n}} f(0)$, thus by Lemma 8.3.5 there is an extension $\tilde{\phi}_{n,\alpha} \in X^*$ of $\phi_{n,\alpha}$ such that $\tilde{\phi}_{n,\alpha} \in \partial_{\frac{1}{n}} \tilde{f}(0)$. Thus $\|\tilde{\phi}_{n,\alpha}\| \leq K + 1/n$, where K is the Lipschitz constant for $\tilde{f}$. Moreover, $\tilde{f}(x) \geq g(x)$, where $g(x) = \sup_{n,\alpha}\{\tilde{\phi}_{n,\alpha}(x) - \frac{1}{n}, 0\}$. The Gâteaux differentiability of $\tilde{f}$ at 0 now forces the Gâteaux differentiability of g at 0. Use again Proposition 8.3.4 to obtain the weak*-convergence of $(\tilde{\phi}_{n,\alpha})$ to 0.

(b) $\Rightarrow$ (a): By subtracting off a derivative and translating f, we need only to consider the case where $f'(0) = 0$ and $f(0) = 0$. For each $u \in Y$, fix $\phi_u \in \partial f(u)$, and define $a_{k,u} := \phi_u(u) - f(u) + \frac{1}{k}$. Then, using properties of subgradients, it follows that $f(y) := \sup\{\phi_{k,u}(y) - a_{k,u}, 0 : u \in Y, k \in \mathbb{N}\}$. Now, from the fact that f is Lipschitz (with Lipschitz constant L) we have $1/k \leq a_{k,u} \leq 2L\|u\| + 1/k$ for every $k \in \mathbb{N}$ and $u \in Y$. Put

$$A_n := \left\{(k, u) : k \in \mathbb{N}, u \in Y, \text{ such that } \frac{1}{n} \leq a_{k,u} < \frac{1}{n-1}\right\}$$

for $n = 2, 3, \ldots$ and

$$A_1 := \{(k, u) : k \in \mathbb{N}, u \in Y, \text{ such that } 1 \leq a_{k,u}\}.$$

It is plain that $(n, 0) \in A_n$ and so A_n is nonempty, for every $n \in \mathbb{N}$. Moreover, $\mathbb{N} \times Y = \bigcup_{n=1}^{\infty} A_n$. To each $(n, (k, u)) \in \{n\} \times A_n$ we associate $\psi_{(n,(k,u))} := \phi_u$ and $b_{(n,(k,u))} := a_{k,u}$. Then $f(y) = \sup\{\psi_{(n,(k,u))}(y) - b_{(n,(k,u))}, 0 : (n, (k, u)) \in \bigcup_{n=1}^{\infty} \{n\} \times A_n\}$. According to Proposition 8.3.4, $\psi_{(n,(k,u))} \to_{w^*} 0$ because f is Gâteaux differentiable at 0. The Lipschitz property of f guarantees that $(\psi_{(n,(k,u))})$ is bounded. According to (b), we can extend $(\psi_{(n,(k,u))})$ to a bounded net $(\tilde{\psi}_{(n,(k,u))})$ that converges weak* to 0. Then $\tilde{f}(x) = \sup\{\tilde{\psi}_{(n,(k,u))}(x) - b_{(n,(k,u))}, 0\}$ is a convex Lipschitz function that is Gâteaux differentiable at 0 by Proposition 8.3.4, and $\tilde{f}$ extends f. $\square$

Let us remark that in contrast to this, Zizler ([450]) has shown that extensions of Gâteaux differentiable norms from a subspace of a separable space to a Gâteaux differentiable norm on the whole space are not always possible.

Finally, let us conclude by stating a bornological version that combines Theorems 8.3.8 and 8.3.15.

Theorem 8.3.16. *Suppose Y is a closed subspace of a Banach space X. Then the following are equivalent.*

(a) *Every continuous convex function $f : Y \to \mathbb{R}$ bounded on β-sets can be extended to a continuous convex function $\tilde{f} : X \to \mathbb{R}$ that is bounded on β-sets in X.*
(b) *Every Lipschitz convex function $f : Y \to \mathbb{R}$ that is β-differentiable at some point y_0 can be extended to a Lipschitz convex function $\tilde{f} : X \to \mathbb{R}$ that is β-differentiable at y_0.*
(c) *Every bounded net $(\phi_{n,\alpha}) \subset Y^*$ that converges τ_β to 0 can be extended to a bounded net $(\tilde{\phi}_{n,\alpha}) \subset X^*$ that converges τ_β to 0.*

Let us mention that if β is the bornology of bounded sets, then (c) is always possible according to the Hahn–Banach theorem. Thus this recaptures the results: (i) a convex function that is bounded on bounded sets can always be extended to a convex function bounded on bounded sets; (ii) Lipschitz convex functions can be extended to a superspace while preserving a point of Fréchet differentiability.

Exercises and further results

8.3.1 (Finite-dimensional extensions).** A convex function $f : \mathbb{R}^n \to (-\infty, +\infty]$ is *finitely extendable* if there is an everywhere finite convex function g with $f = g|_{\text{dom} f}$.

Theorem 8.3.17 ([388]). *A function f is finitely extendable if and only if there is a multifunction k such that: (a) $\Omega(x) \subset \mathbb{R}^n$ is nonempty convex compact for every*

$x \in \mathrm{dom} f$; (b) $\partial f(x) = \Omega(x) + \partial \delta_{\mathrm{dom} f}(x)$ for every $x \in \mathrm{dom} f$; (c) for all sequences (x_i) and (x_i^*) satisfying $x_i^* \in \Omega(x_i)$ for $i = 1, 2, \cdots$ and $\lim_i \|x_i^*\| = \infty$ one has $\lim_i (\langle x_i^*, x_i \rangle - f(x_i))/\|x_i^*\| = \infty$. For any such Ω

$$g_\Omega(z) := \sup\{f(x) + \langle x^*, z - x \rangle : x \in \mathrm{dom} f, x^* \in \Omega(x)\}$$

defines a finite extension g_Ω of f.

It is interesting to look at how various functions from Chapter 7 behave with respect to this result.

8.3.2. Let X be a Grothendieck space. Suppose $(\phi_{n,\alpha}) \subset X^*$ is a bounded net that converges weak* to 0. Use the Grothendieck property to show that $(\phi_{n,\alpha})$ converges weak* to 0 when considered as a net in X^{***}.

8.3.3.* Prove Proposition 8.3.4.

8.3.4. (a) Suppose C is a closed convex subset of a Banach space X, and $f : C \to \mathbb{R}$ is a Lipschitz convex function with Lipschitz constant M, show that there is a convex extension $\tilde{f}$ of f to X that also has Lipschitz constant M.

(b) Suppose Y is a closed subspace of a Banach space X, and suppose f is a continuous convex function on Y. Then there is a lsc convex extension $\tilde{f}$ of f such that $\sup_{rB_X} \tilde{f} = \sup_{rB_Y} f$ for all $r > 0$. In particular, any convex function bounded on bounded sets can be extended to a convex function bounded on bounded sets on any superspace.

Hint. (a) Consider the infimal convolution of f with $M \| \cdot \|$.

(b) For each $y \in Y$, let $\phi_y \in \partial f(y)$, and let $\tilde{\phi}_y$ be a Hahn–Banach norm preserving extension of ϕ_y to all of X. Define $h_y(x) := f(y) + \langle \tilde{\phi}_y, x - y \rangle$, and let $\tilde{f}(x) := \sup_{y \in Y} h_y(x)$. See [116, p. 1801] for further details. □

8.3.5. (a) Is there a Banach space X and a continuous convex function $f : X \to \mathbb{R}$ that can be extended to a continuous convex function $\tilde{f} : X^{**} \to \mathbb{R}$, such that there is no continuous weak*-lsc extension of f to X^{**}.

(b) Suppose X is not reflexive, show that there is a continuous convex function $f : B_X \to \mathbb{R}$ such that f^{**} is not continuous on $B_{X^{**}}$.

Hint. (a) Yes, try a non-Grothendieck space that is complemented in its bidual. For example, L_1 or the James space [199].

(b) By James' theorem, take a linear functional $\phi \in S_{X^*}$ that does not attain its supremum on B_X. Consider $f(x) = 1/(1 - \langle \phi, x \rangle)$. For more details, see [116, Example 2.2]. □

8.3.6.* Prove Theorem 8.3.16.

8.3.7. Show by example that Lemma 8.3.5 may fail for proper lsc convex functions in the case $\varepsilon = 0$.

Hint. Let $f : \mathbb{R} \to (-\infty, +\infty]$ be defined by $f := \delta_{[0,\infty]}$ and consider $0 \in \partial f(0)$. Now define $\tilde{f} : \mathbb{R}^2 \to (-\infty, +\infty]$ by $\tilde{f}(x, y) := -\sqrt{y}$ if $x \geq 0$ and $y \geq 0$, and $\tilde{f}(x, y) := +\infty$ otherwise. Then $\partial \tilde{f}(0, 0) = \emptyset$. □

8.3.8 (Phillips [352]). (a) Show that $\ell_\infty(\Gamma)$ is 1-injective for any nonempty index set Γ.

(b) Show that if X is an injective Banach space, then it is λ-injective for some $\lambda \geq 1$, i.e. for every $Z \supset X$, there is a projection $P : Z \to X$, with $\|P\| \leq \lambda$.

Hint. (a) Let $\ell_\infty(\Gamma) \subset Z$. For each $\gamma \in \Gamma$, let e_γ be the evaluation functional on $\ell_\infty(\Gamma)$. By the Hahn–Banach theorem, extend each such functional to $\tilde{e}_\gamma$ a norm one functional on Z. Define $P : Z \to \ell_\infty(\Gamma)$ by $P(z) := \{\tilde{e}_\gamma(z)\}_{\gamma \in \Gamma}$.

(b) Let $X \subset \ell_\infty(\Gamma)$ then there is a projection $P : \ell_\infty(\Gamma) \to X$, let $\lambda = \|P\|$. Now if $Z \supset X$, find Γ_1 such that $Z \subset \ell_\infty(\Gamma_1)$. Using (a), there is a projection $P_1 : \ell_\infty(\Gamma_1) \to X$ with $\|P_1\| = \lambda$. □

8.4 Some other generalizations and equivalences

One may wonder whether the results in Section 8.2 extend to differences of convex functions or even Lipschitz functions. In general, they do not, as illustrated by the following result from [114]; see that paper and also [78] for more in that direction.

Theorem 8.4.1. *Let X be a Banach space. Then the following are equivalent.*

(a) *X has the Schur property.*
(b) *Gâteaux differentiability and weak Hadamard differentiability agree for Lipschitz functions on X.*
(c) *Gâteaux differentiability and weak Hadamard differentiability coincide for differences of Lipschitz convex functions on X.*
(d) *Every continuous convex function on X is weak Hadamard directionally differentiable.*

The significance of (c) in the previous theorem is that it shows Theorem 8.1.3 cannot be extended to differences of convex functions. Likewise (d) shows Theorem 8.1.3 cannot be extended to one sided β-derivatives of convex functions.

Quite obviously, we have not included many other characterizations of different classes of Banach spaces in terms of the properties of norms, convex functions, or monotone operators which live on them. For example, [83] shows the following result; see Section 9.5.1 for any unexplained notation and a flavor.

Theorem 8.4.2 (Norm×bw*-closed graphs). *Let E be a Banach space. Then the following are equivalent.*

(a) *E is finite-dimensional.*
(b) *The graph of ∂f is norm×bw*-closed for each closed proper convex f on E.*
(c) *The graph of each maximal monotone operator T on E is norm×bw*-closed.*

Nor, correspondingly, have we been comprehensive in other areas. A lovely renorming characterization we cannot resist giving is:

Theorem 8.4.3 (Reflexivity and renorming [336]). *A separable Banach space X is reflexive if and only if there is an equivalent norm $\|\cdot\|$ on X such that, whenever (x_n)*

8.4 Some other generalizations and equivalences

is a bounded sequence in X for which $\lim_n \lim_m \|x_n + x_m\| = 2\lim_n \|x_n\|$, then (x_n) converges in norm.

The 'if' part follows easily from James' theorem (4.1.27). The possibility of the 'only if' part was mooted by V. D. Milman in 1970.

Exercises and further results

One may also wonder how critical Cauchy-completeness was to our studies. The following construction emphasizes that completeness of the normed space is crucial to very many of our results – as does Exercise 4.3.1.

8.4.1 (Empty subdifferentials). ** As in [260], let C be a closed bounded convex subset of a normed space E and let $0 \neq x_0 \in E$. Define

$$f_C(x) := \min\{t \in \mathbb{R} : x + tx_0 \in C\}.$$

Then f_C is lsc and convex, and since f_C has no global minimum, any subgradient for f_C at x is nonzero. Fix x in the domain of ∂f_C and let $c := x + f_C(x)x_0$. Note that for any $y \in C$, $f_C(y - f_C(x)x_0) \leq f_C(x)$. Hence, if x^* is a subgradient of f_C at x, then

$$\langle x^*, y - c \rangle \leq f_C(y - f_C(x)x_0) - f_C(x) \leq 0.$$

Thus,

$$\langle x^*, y - c \rangle \leq 0$$

for all $y \in C$, and so x^* is a nonzero support for C at c. In particular if C has no support points, ∂f_C is empty. Such sets exist in certain locally convex complete metrizable spaces [260].

More dramatically, Fonf [220] proves a densely spanning, closed bounded convex set with no support points exists in each separable incomplete normed space. In consequence most of the delicate variational analysis of subgradients fails flamboyantly outside of complete normed spaces.

8.4.2 (A pathological convex control problem [67]). ** Rather than hunting for a specialized counterexample, it is often easier to take a general phenomenon and see if it can be rewritten in the form desired, as we now illustrate. Consider $X := L_2[0, 1]$ and the *inclusion control problem* (see Exercise 3.5.17) given by

$$v(\alpha) := \inf\{x(1) : \quad x'(t) = 0, |y'(t)| \leq 1, 0 \leq t \leq 1 \text{ a.e.,} \quad (8.4.1)$$

$$|x(0)| \leq 1, y(0) = 1, y(t) - x(t) \leq \alpha(t)\}.$$

(a) Show that v in (8.4.1) defines a closed convex function on X and that the infimum is attained when finite.
(b) Show that $v(0) = 0$ but that $\partial v(0) = \emptyset$.

While these properties are fairly easy to confirm once asserted, this example was built by writing the construction of Exercise 8.4.1 in the language of differential inclusions. This pathology cannot occur if the perturbation is finite-dimensional [156].

8.4.3. Show that a Banach space is finite-dimensional if and only if every exposed point of every closed convex set is strongly exposed.

Hint. It suffices to work in a separable space. Compare Exercise 5.1.26. □

8.4.4. Show that a Banach space X is reflexive if and only if for all closed (bounded) convex sets C and all $x \in X \setminus C$ the value of $d_C(x)$ is attained at some point of C.

Hint. Compute $d_C(x)$ when $C := \{\phi(x) = 0\}$ and $\phi \in S_{X^*}$ and use James theorem (4.1.27). □

8.4.5. Let X be a reflexive Banach space. Show that for all norm-closed bounded sets C there is some $x \in X \setminus C$ such the value of $d_C(x)$ is attained at some point of C.

Hint. This remains true in a Banach space with the RNP. □

8.4.6. ** It is not known whether Exercise 8.4.5 remains true when C is unbounded. The Lau–Konjagin theorem (4.5.5) shows the norm must fail the Kadec–Klee property. More is known, see [80], but not a lot.

9

Monotone operators and the Fitzpatrick function

The formulas move in advance of thought, while the intuition often lags behind; in the oft-quoted words of d'Alembert, "L'algebre est genereuse, elle donne souvent plus qu'on lui demande." (Edward Kasner)[1]

9.1 Monotone operators and convex functions

In this chapter we shall focus on the recent intensive use of convex functions to study monotone operators. In his '23' *"Mathematische Probleme"*[2] lecture to the Paris International Congress in 1900, David Hilbert wrote *"Besides it is an error to believe that rigor in the proof is the enemy of simplicity."*

In this spirit, we use simple convex analytic methods, relying on an ingenious function due to Simon Fitzpatrick, to provide a concise proof of the maximality of the sum of two maximal monotone operators on reflexive Banach space under standard transversality conditions. Many other extension, surjectivity, convexity and local boundedness results are likewise established.

9.1.1 Introduction

To allow this chapter to be read somewhat independently, we recall the notions that we need in this setting. The *domain* of an extended valued convex function, denoted $\text{dom}(f)$, is the set of points with value less than $+\infty$, and that a point s is in the *core* of a set S (denoted by $s \in \text{core } S$) provided that $X = \bigcup_{\lambda > 0} \lambda(S - s)$. Recall that $x^* \in X^*$ is a *subgradient* of $f : X \to (-\infty, +\infty]$ at $x \in \text{dom } f$ provided that $f(y) - f(x) \geq \langle x^*, y - x \rangle$ for all y in X. The set of all subgradients of f at x is called the *subdifferential* of f at x and is denoted $\partial f(x)$. We shall need the *indicator* function $\delta_C(x)$ which is zero for x in C and $+\infty$ otherwise, the *Fenchel conjugate* $f^*(x^*) := \sup_x \{\langle x, x^* \rangle - f(x)\}$ and the *infimal convolution* $f^* \square \frac{1}{2} \| \cdot \|_*^2 (x^*) := \inf \{ f^*(y^*) + \frac{1}{2} \|z^*\|_*^2 : x^* = y^* + z^* \}$.

---

[1] In Edward Kasner, 'The Present Problems of Geometry', *Bulletin of the American Mathematical Society*, (1905) volume XI, p. 285. A more faithful translation might be "Algebra is kind, she often gives more than is requested."
[2] See the late Ben Yandell's lovely account of the Hilbert Problems and their solvers in *The Honors Class*, AK Peters, 2002.

When f is convex and closed $x^* \in \partial f(x)$ exactly when $f(x) + f^*(x^*) = \langle x, x^* \rangle$. We recall that the *distance function* associated with a closed set C, given by $d_C(x) := \inf_{c \in C} \|x - c\|$, is convex if and only if C is. Moreover, $d_C = \delta_C \square \| \cdot \|$.

As convenient, we shall use both dom $T = D(T) := \{x : T(x) \neq \emptyset\}$, and range $T = R(T) := T(X)$ to denote the *domain* and *range* of a multifunction. We say a multifunction $T : X \mapsto 2^{X^*}$ is *monotone* provided that for any $x, y \in X, x^* \in T(x)$ and $y^* \in T(y)$,

$$\langle y - x, y^* - x^* \rangle \geq 0,$$

and we say that T is *maximal monotone* if its *graph*, $\{(x, x^*) : x^* \in T(x)\}$, is not properly included in any other monotone graph. The subdifferential of a convex lsc function on a Banach space is a typical example of a maximal monotone multifunction. We reserve the notation J for the duality map

$$J(x) := \frac{1}{2} \partial \|x\|^2 = \{x^* \in X^* : \|x\|^2 = \|x^*\|^2 = \langle x, x^* \rangle\},$$

and define the convex *normal cone* to C at $x \in \text{cl } C$ by

$$N_C(x) := \partial \delta_C(x).$$

(In general it suffices only to consider closed sets.) All other notation is broadly consistent with earlier usage in this book and in [95, 121, 369]. Some of the side-results in this section are not used in the sequel and so we are somewhat sparing with such details.

Our goal is to derive many key results about maximal monotone operators entirely from the existence of subgradients and the *sandwich theorem* given in Chapter 4 and below; as much as possible using only geometric-functional-analysis tools. We first consider general Banach spaces and then in Section 9.2 look at cyclic operators; that is subgradients. In Section 9.3 we provide a central result on maximality of the sum in reflexive space. Section 9.4 looks at more applications of the technique introduced in Section 9.3 (in both reflexive and nonreflexive settings) while limiting examples are produced in Section 9.5. We add some very recent results in general Banach space in Section 9.6 and discuss operators of type (NI) in Section 9.7. Our discussion does not necessarily give the most efficient results but does provide significant intuition along the way.

9.1.2 Maximality in general Banach space

For a monotone mapping T, we associate the *Fitzpatrick function* introduced in [215]. The Fitzpatrick function is

$$\mathcal{F}_T(x, x^*) := \sup\{\langle x, y^* \rangle + \langle x^*, y \rangle - \langle y, y^* \rangle : y^* \in T(y)\},$$

which is clearly lsc and convex as an affine supremum. Moreover,

9.1 Monotone operators and convex functions

Proposition 9.1.1. [215, 121] *For a maximal monotone operator* $T: X \to X^*$ *one has*

$$\mathcal{F}_T(x, x^*) \geq \langle x, x^* \rangle$$

with equality if and only if $x^* \in T(x)$. *Indeed, the equality* $\mathcal{F}_T(x, x^*) = \langle x, x^* \rangle$ *for all* $x^* \in T(x)$, *requires only monotonicity not maximality.*

Note that in general $\mathcal{F}_T$ is not useful for nonmaximal operators. As an extreme example, on the real line if $T(0) = 0$ and $T(x)$ is empty otherwise, then $\mathcal{F}_T \equiv 0$. Note also that the construction in Proposition 9.1.1 extends to any paired vector spaces.

The idea of associating a convex function with a monotone operator and exploiting the relationship was largely neglected for many years after [79] and [215] until exploited by Penot, Simons, Simons and Zalinescu ([394, 398, 399, 446]), Burachik and Svaiter and others.

Convex analytic tools

The basic results that we use repeatedly follow:

Proposition 9.1.2. *A proper lsc convex function on a Banach space is continuous throughout the core of its domain.*

Proof. In Euclidean spaces, this follows from Theorem 2.1.12 and Proposition 2.1.13. The proof for general Banach spaces is given in Proposition 4.1.5. □

Proposition 9.1.3. *A proper lsc convex function on a Banach space has a nonempty subgradient throughout the core of its domain.*

Proof. The proof for Euclidean spaces is given in Theorem 2.1.19 and for general Banach spaces in Theorem 4.1.10. □

These two basic facts lead to:

Theorem 9.1.4 (Sandwich theorem). *Suppose f and $-g$ are lsc convex on a Banach space X and that*

$$f(x) \geq g(x),$$

for all x in X. Assume that the following constraint qualification (CQ) holds:

$$0 \in \mathrm{core}\,(\mathrm{dom}(f) - \mathrm{dom}(-g)). \tag{9.1.1}$$

Then there is an affine continuous function α such that

$$f(x) \geq \alpha(x) \geq g(x),$$

for all x in X.

Proof. The value function $h(u) := \inf_{x \in X} f(x) - g(x - u)$ is convex and the (CQ) implies it is continuous at 0. Hence there is some $-\Lambda \in \partial h(0)$, and this provides the linear part of the asserted affine separator. Indeed, we have

$$f(x) - g(u - x) \geq h(u) - h(0) \geq \Lambda(u),$$

as required. For further details in Euclidean spaces see Exercise 2.4.1 and the proof of Theorem 2.3.4; a more general version in Banach spaces is given in Theorem 4.1.18. □

We will also refer to *constraint qualifications* like (9.1.1) as *transversality conditions* since they ensure that the sum/difference of two convex sets is large, and so resemble such conditions in differential geometry. It is an easy matter to deduce the complete *Fenchel duality theorem* (see Theorem 2.3.4 and Theorem 4.4.18) from Theorem 9.1.4 and in particular that:

Corollary 9.1.5 (Subdifferential rule). *Suppose that f and g are convex and that (9.1.1) holds. Then $\partial f + \partial g = \partial (f + g)$.*

Proof. See the proof of Theorem 4.1.19 for further details. □

Proposition 9.1.6. [421] *For a closed convex function f and $f_J := f + \frac{1}{2} \|\cdot\|^2$ we have $\left(f + \frac{1}{2}\|\cdot\|^2\right)^* = f^* \square \frac{1}{2}\|\cdot\|_*^2$ is everywhere continuous. Also*

$$v^* \in \partial f(v) + J(v) \Leftrightarrow f_J^*(v^*) + f_J(v) - \langle v, v^* \rangle \leq 0.$$

Further and related information concerning infimal convolutions can be found in Exercises 2.3.14 and 2.3.15 and Lemma 4.4.15.

Edgar Asplund wrote a still-very-informative 1969 survey of those parts of convex analysis 'that the author feels are important in the study of monotone mappings,' [15, p.1]. This includes averaging of norms, decomposition and differentiability results, as well as the sort of basic results we have described above.

Representative convex functions

Recall that a *representative function* for a monotone operator T on X is any convex function $\mathcal{H}_T$ on $X \times X^*$ such that $\mathcal{H}_T(x, x^*) \geq \langle x, x^* \rangle$ for all x, x^*, while $\mathcal{H}_T(x, x^*) = \langle x, x^* \rangle$ when $x^* \in T(x)$. Unlike [135], we do not require $\mathcal{H}_T$ to be closed. When T is maximal, Proposition 9.1.1 shows $\mathcal{F}_T$ is a representative function for T, as is the convexification

$$\mathcal{P}_T(x, x^*) := \inf \left\{ \sum_{i=1}^N \lambda_i \langle x_i, x_i^* \rangle : \sum_i \lambda_i(x_i, x_i^*, 1) = (x, x^*, 1),\ x_i^* \in T(x_i),\ \lambda_i \geq 0 \right\},$$

which has the requisite properties for any monotone T, whether or not maximal:

9.1 Monotone operators and convex functions

Proposition 9.1.7. *For any monotone mapping T, $\mathcal{P}_T$ is a representative convex function for T.*

Proof. Directly from the definition of monotonicity we have

$$\mathcal{P}_T(x, x^*) \geq \langle x^*, y\rangle + \langle y^*, x\rangle - \langle y^*, y\rangle,$$

for $y^* \in T(y)$. Thus, for all points

$$\mathcal{P}_T(x, x^*) + \mathcal{P}_T(y, y^*) \geq \langle x^*, y\rangle + \langle y^*, x\rangle.$$

Note that by definition $\mathcal{P}_T(x, x^*) \leq \langle x^*, x\rangle$ for $x^* \in T(x)$. Hence, setting $x = y$ and $x^* = y^*$ shows $\mathcal{P}_T(x, x^*) = \langle x^*, x\rangle$ for $x^* \in T(x)$ while $\mathcal{P}_T(z, z^*) \geq \langle z^*, z\rangle$ for $(z^*, z) \in \operatorname{conv} \operatorname{graph} T$ and, also by definition, $\mathcal{P}_T(z, z^*) = \infty$ otherwise. □

Direct calculation shows $(\mathcal{P}_T)^* = \mathcal{F}_T$ for any monotone T [343]. This convexification originates with Simons [391] and was refined by Penot [344, Proposition 5].

9.1.3 Monotone extension formulas

We illustrate the flexibility of $\mathcal{P}$ by using it to prove a central case of the Debrunner–Flor theorem [174, 351] *without* using Brouwer's theorem.

Theorem 9.1.8 ([351, 394]). (a) *Suppose T is monotone on a Banach space X with range contained in αB_{X^*}, for some $\alpha > 0$. Then for every x_0 in X there is $x_0^* \in \overline{\operatorname{conv}}^{w^*} R(T) \subset \alpha B_{X^*}$ such that (x_0, x_0^*) is monotonically related to $\operatorname{graph}(T)$.*
(b) *In consequence, T has a bounded monotone extension $\hat{T}$ with $\operatorname{dom}(\hat{T}) = X$ and $R(\hat{T}) \subset \overline{\operatorname{conv}}^{w^*} R(T)$.*
(c) *In particular, a maximal monotone T with bounded range has $\operatorname{dom}(T) = X$ and has $\operatorname{range}(T)$ connected.*

Proof. (a) It is enough, after translation, to show $x_0 = 0 \in \operatorname{dom}(T)$. Fix $\alpha > 0$ with $R(T) \subset C := \overline{\operatorname{conv}}^{w^*} R(T) \subset \alpha B_{X^*}$.
Consider

$$f_T(x) := \inf \{\mathcal{P}_T(x, x^*) : x^* \in C\}.$$

Then f_T is convex since $\mathcal{P}_T$ is. Observe that $\mathcal{P}_T(x, x^*) \geq \langle x, x^*\rangle$ and so $f_T(x) \geq \inf_{x^* \in C}\langle x, x^*\rangle \geq -\alpha \|x\|$ for all x in X. As $x \mapsto \inf_{x^* \in C}\langle x, x^*\rangle$ is concave and continuous the sandwich theorem (9.1.4) applies.
Thus, there exist w^* in X^* and γ in $\mathbb{R}$ with

$$\mathcal{P}_T(x, x^*) \geq f_T(x) \geq \langle x, w^*\rangle + \gamma \geq \inf_{x^* \in C}\langle x, x^*\rangle \geq -\alpha \|x\|$$

for all x in X and x^* in $C \subset \alpha B_{X^*}$. Setting $x = 0$ shows $\gamma \geq 0$. Now, for any (y, y^*) in the graph of T we have $\mathcal{P}_T(y, y^*) = \langle y, y^*\rangle$. Thus,

$$\langle y - 0, y^* - w^*\rangle \geq \gamma \geq 0,$$

which shows that $(0, w^*)$ is monotonically related to the graph of T. Finally, $\langle x, w^* \rangle + \gamma \geq \inf_{x^* \in C} \langle x, x^* \rangle \geq -\alpha \|x\|$ for all $x \in X$ involves three sublinear functions, and so implies that $w^* \in C \subset \alpha B_{X^*}$.

(b) Consider the set $\mathcal{E}$ of all monotone extensions of T with range in $C \subset \alpha B_{X^*}$, ordered by inclusion. By Zorn's lemma $\mathcal{E}$ admits a maximal member $\hat{T}$ and by (a) $\hat{T}$ has domain the whole space. Part (c) follows immediately, since T being maximal and everywhere defined is a weak*-cusco and so has a weak*-connected range. □

One may consult [217] and [394, Theorem 4.1] for other convex analytic proofs of (c). Note also that the argument in (a) extends to an unbounded set C whenever

$$x_0 \in \text{core}(\text{dom} f_T + \text{dom } \sigma_C).$$

The full Debrunner–Flor result is stated next:

Theorem 9.1.9 (Debrunner–Flor extension theorem [174, 351]). *Suppose T is a monotone operator on Banach space X with range $T \subset C$ with C weak*-compact and convex. Suppose also $\varphi \colon C \mapsto X$ is weak*-to-norm continuous. Then there is some $c^* \in C$ with $\langle x - \varphi(c^*), x^* - c^* \rangle \geq 0$ for all $x^* \in T(x)$.*

It seems worth observing that:

Proposition 9.1.10. *The full Debrunner–Flor extension theorem is equivalent to Brouwer's theorem.*

Proof. An accessible derivation of Debrunner–Flor from Brouwer's theorem is given in [351]. Conversely, let g be a continuous self-map of a norm-compact convex set $K \subset \text{int } B_X$ in a Euclidean space X. We apply the Debrunner–Flor extension theorem to the identity map I on B_X and to $\varphi \colon B_X \mapsto X$ given by $\varphi(x) := g(P_K x)$, where P_K is the metric projection mapping (any retraction would do). We obtain $x_0^* \in B_X$ and also $x_0 := \varphi(x_0^*) = g(P_K x_0^*) \in K$ with

$$\langle x - x_0, x - x_0^* \rangle \geq 0$$

for all $x \in B_X$. Since $x_0 \in \text{int } B_X$, for $h \in X$ and small $\varepsilon > 0$ we have $x_0 + \varepsilon h \in B_X$ and so $\langle h, x_0 - x_0^* \rangle \geq 0$ for all $h \in X$. Thus, $x_0 = x_0^*$ and so $P_K x_0^* = P_K x_0 = x_0 = g(P_K x_0^*)$, is a fixed point of the arbitrary self-map g. □

Figure 9.1 illustrates the construction of Proposition 9.1.10.

9.1.4 Local boundedness results

We next turn to local boundedness results. Recall that an operator T is *locally bounded* around a point x if $T(B_\varepsilon(x))$ is bounded for some $\varepsilon > 0$.

Theorem 9.1.11 ([391, 421]). *Let X be a Banach space and let S and $T \colon X \to 2^{X^*}$ be monotone operators. Suppose that*

$$0 \in \text{core}[\text{conv dom}(T) - \text{conv dom}(S)].$$

9.1 Monotone operators and convex functions

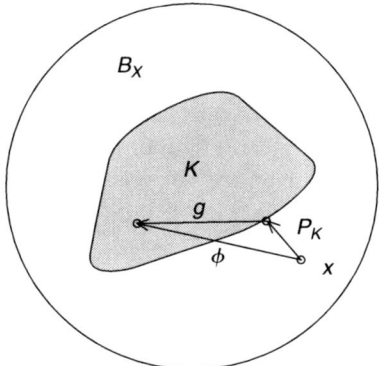

Figure 9.1 The construction of Proposition 9.1.10.

Then there exist $r, c > 0$ such that, for any $x \in \operatorname{dom}(T) \cap \operatorname{dom}(S)$, $t^* \in T(x)$ and $s^* \in S(x)$,

$$\max(\|t^*\|, \|s^*\|) \le c\,(r + \|x\|)(r + \|t^* + s^*\|).$$

Proof. Consider the convex lsc function

$$\tau_T(x) := \sup_{z^* \in T(z)} \frac{\langle x - z, z^* \rangle}{1 + \|z\|}.$$

This is a refinement of the function [79] originally used to prove local boundedness of monotone operators [391, 421, 121]. We first show that $\operatorname{conv} \operatorname{dom}(T) \subset \operatorname{dom} \tau_T$, and that $0 \in \operatorname{core} \bigcup_{i=1}^{\infty} [\{x : \tau_S(x) \le i, \|x\| \le i\} - \{x : \tau_T(x) \le i, \|x\| \le i\}]$. We now apply conventional Baire category techniques – with some care. □

The next corollary recaptures Theorem 6.1.6.

Corollary 9.1.12 ([391, 121, 421]). *Let X be any Banach space. Suppose T is monotone and*

$$x_0 \in \operatorname{core} \operatorname{conv} \operatorname{dom}(T).$$

Then T is locally bounded around x_0.

Proof. Let $S = 0$ in Theorem 9.1.11 or directly apply Proposition 9.1.2 to τ_T. □

We can also improve Theorem 9.1.8.

Corollary 9.1.13. *A monotone mapping T with bounded range admits an everywhere defined maximal monotone extension with bounded weak*-connected range contained in $\overline{\operatorname{conv}}^{w^*} R(T)$.*

Proof. Let $\widehat{T}$ denote the extension of Theorem 9.1.8 (b). Clearly it is everywhere locally bounded. The desired maximal monotone extension $T^*(x)$ is the operator whose graph is the norm-weak*-closure of the graph of $x \mapsto \operatorname{conv} \widehat{T}(x)$, since this is both monotone and is a norm-weak* cusco. Explicitly, $T^*(x) := \cap_{\varepsilon>0} \overline{\text{conv}}^* \widehat{T}(B_\varepsilon(x))$, see [121]. □

Recall that a maximal monotone mapping is *locally maximal monotone*, *type (LMM)* or *type (FP)*, if (graph $T^{-1}) \cap (V \times X)$ is maximal monotone in $V \times X$, for every convex open set V in X^* with $V \cap $ range $T \neq \emptyset$. Dually, a maximal monotone mapping is *maximal monotone locally (VFP)*, is defined by reversing the roles of X and X^* with T instead of T^{-1}. It is known that all maximal monotone operators on a reflexive space are type (FP) and (VFP), see [217, 218, 351, 393] and Theorem 9.4.6, as are all subgradients of closed convex functions, [391, 393] and Theorem 9.4.8. It is shown in [217] that a maximal monotone operator T with range $T = X^*$ (resp. dom $T = X$) is locally maximal monotone (resp. maximal monotone locally).

For a maximal monotone operator T we may usefully apply Corollary 9.1.13 to the mapping $T_n(x) := T(x) \cap n B_{X^*}$. Under many conditions the extension, $\widehat{T}_n$ is unique. Indeed as proven by Fitzpatrick and Phelps:

Proposition 9.1.14. ([215, 217]) *Suppose T is maximal monotone and suppose n is large enough so that $R(T) \cap n$ int $B_{X^*} \neq \emptyset$.*
(a) There is a unique maximal monotone $\widehat{T}_n$ such that $T_n(x) \subset \widehat{T}_n(x) \subset n B_{X^*}$ whenever the mapping M_n defined by

$$M_n(x) := \{x^* \in n B_{X^*} : \langle x^* - z^*, x - z \rangle \geq 0 \text{ for all } z^* \in T(z) \cap n \text{ int } B_{X^*}\},$$

is monotone; in which case $M_n = \widehat{T}_n$.
(b) Part (a) holds whenever T is type (FP) and the dual norm is strictly convex. Hence, it applies for any maximal monotone operator on a reflexive space in a strictly convex dual norm.

Proof. Since $\widehat{T}_n$ exists by Corollary 9.1.13 and since $\widehat{T}_n(x) \subset M_n(x)$, (a) follows. We refer to [215, Theorem 2.2] for the fairly easy proof of (b). □

It is reasonable to think of the sequence $(\widehat{T}_n)_{n \in \mathbb{N}}$ as a good nonreflexive generalization of the resolvent-based Yosida approximate [351, 121] or of Hausdorff's Lipschitz regularization of a convex function, [215, 351, 121] – especially in the (FP) case where one also shows easily that (i) $\widehat{T}_n(x) = T(x) \cap n B_{X^*}$ whenever $T(x) \cap $ int $n B_{X^*} \neq \emptyset$, and (ii) $\widehat{T}_n(x) \setminus T(x) \subset n S_{X^*}$, [217]. Thus, for local properties, such as differentiability, one may often replace T by some $\widehat{T}_n$ if it simplifies other matters.

9.1.5 Convexity of the domain

We start with

Corollary 9.1.15 ([363, 368, 391]). *Let X be any Banach space. Suppose that T is maximal monotone with* core conv dom(T) *nonempty. Then*

$$\text{core conv dom}(T) = \text{int conv dom}(T) \subset \text{dom}(T). \tag{9.1.2}$$

In consequence both the norm closure and interior of dom(T) *are convex.*

9.1 Monotone operators and convex functions

Proof. We first establish the inclusion in (9.1.2). Fix $x + \varepsilon B_X \subset \operatorname{int conv dom}(T)$ and, appealing to Corollary 9.1.12, select $M := M(x, \varepsilon) > 0$ so that $T(x + \varepsilon B_X) \subset M B_{X^*}$. For $N > M$ define nested sets

$$T_N(x) := \{x^* : \langle x - y, x^* - y^* \rangle \geq 0, \quad \text{for all } y^* \in T(y) \cap N B_{X^*}\},$$

and note these images are weak*-closed. By Theorem 9.1.8 (b), the sets are nonempty, and by the next Lemma 9.1.16 bounded, hence weak*-compact. Observe that by maximality of T, $T(x) = \bigcap_N T_N(x) \neq \emptyset$, as a nested intersection, and x is in $\operatorname{dom}(T)$ as asserted.

Then $\operatorname{int conv dom}(T) = \operatorname{int dom}(T)$ and so the final conclusion follows. □

Lemma 9.1.16. *For $x \in \operatorname{int conv dom}(T)$ and N sufficiently large, $T_N(x)$ is bounded.*

Proof. A Baire category argument [351], shows for N large and $u \in 1/N B_X$ one has $x + u \in \operatorname{cl conv} D_N$ where

$$D_N := \{z : z \in \operatorname{dom}(T) \cap N B_X, T(z) \cap N B_{X^*} \neq \emptyset\}.$$

Now for each $x^* \in T_N(x)$, since $x + u$ lies in the closed convex hull of D_N, we have

$$\langle u, x^* \rangle \leq \sup\{\langle z - x, z^* \rangle : z^* \in T(z) \cap N B_{X^*}, z \in N B_X\} \leq 2N^2$$

and so $\|x^*\| \leq 2N^3$. □

Another nice application is:

Corollary 9.1.17 ([421]). *Let X be any Banach space and let $S, T : X \to 2^{X^*}$ be maximal monotone operators. Suppose that*

$$0 \in \operatorname{core}[\operatorname{conv dom}(T) - \operatorname{conv dom}(S)].$$

For any $x \in \operatorname{dom}(T) \cap \operatorname{dom}(S)$, $T(x) + S(x)$ is a weak-closed subset of X^*.*

Proof. By the Krein–Šmulian theorem (Exercise 4.4.26), it suffices to use Theorem 9.1.11 to prove every bounded weak*-convergent net in $T(x) + S(x)$ has its limits in $T(x) + S(x)$. □

Thus, we preserve some structure – even if we have not shown that $T + S$ must actually be maximal, see [391, 421].

Finally, a recent result by Simons [395] shows that:

Theorem 9.1.18 ([395]). *If S is maximal monotone and $\operatorname{int dom}(S)$ is nonempty then*

$$\operatorname{int dom}(S) = \operatorname{int} \{x : (x, x^*) \in \operatorname{dom} \mathcal{F}_S\}.$$

This then very neatly recovers the convexity of $\operatorname{int} D(S)$. It would be interesting to determine how much one can similarly deduce about $\operatorname{cl dom}(S)$ – via regularization or enlargement – when $\operatorname{int dom}(S)$ is empty.

For example, suppose S is *domain regularizable* meaning that for $\varepsilon > 0$, there is a maximal S_ε with $\mathcal{H}(D(S), D(S_\varepsilon)) \leq \varepsilon$ and core $D(S_\varepsilon) \neq \emptyset$. Then $\overline{\text{dom}}(S)$ is convex. In reflexive space we can use

$$S_\varepsilon := \left(S^{-1} + N_{\varepsilon B_X}^{-1}\right)^{-1},$$

which is maximal by Theorem 9.3.6. [Here $\mathcal{H}$ denotes Hausdorff distance, as introduced in Chapter six, and we assume $0 \in S(0)$.] See also Theorems 9.4.7 and 9.3.9. When $S = \partial f$ for a closed convex f this applies in general Banach space. Indeed $S_\varepsilon = \partial f_\varepsilon$ where $f_\varepsilon(x) := \inf_{\|y-x\| \leq \varepsilon} f(y)$.

Of course in a reflexive space, by considering T^{-1} we have established similar results for the range of a monotone operator. Later in this chapter we shall see how the range can misbehave in more general Banach spaces.

We will revisit such domain convexity results in the final two sections.

Exercises and further results

As with subgradients, much more can be said about maximal monotone operators in finite dimensions – often relative interior can replace interior in hypotheses.) The next two exercises provide monotone extensions of finite-dimensional results we have already met. The first is a result originally due to Mignot which generalizes Alexandrov's theorem (2.6.4).

Motivated by the second-order analysis of convex functions, it makes sense to say a monotone operator T on a Euclidean space E is *differentiable* at $x \in \text{dom } T$ if $T(x)$ is singleton (this can be weakened but one will conclude $T(x)$ is singleton) and there is a matrix $A = \nabla T(x)$ such that

$$T(x') \subset T(x) + \nabla T(x)(x' - x) + o(\|x' - x\|)B_E.$$

9.1.1 (Differentiability of finite-dimensional monotone operators [378]). ** Let T be maximal monotone on a Euclidean space.

1. Show that T is differentiable at $x \in \text{dom } T$ if and only if the nonexpansive resolvent $R(y) := (I + T)^{-1}(y)$ is differentiable at $y = x + T(x)$; and T is necessarily continuous at x.
2. Deduce Mignot's theorem that a maximal monotone operator is almost everywhere differentiable in the interior of its domain [378, §6.5].

   Hint. Examine the proof of Alexandrov's theorem (2.6.4). □

Remark 4.6.17 shows the impossibility of such a result holding generally in infinite dimensions. The next result also from [378, §6.5] employs Exercise 9.1.1 to reconstruct a maximal monotone operator as we did a subgradient in Theorem 4.3.13. Let us denote $D_1(T) := \{x : T \text{ is differentiable at } x\}$.

9.1.2 (Reconstructing finite-dimensional maximal monotone operators [378]). ** Let T be a maximal monotone operator on a Euclidean space with int dom $T \neq \emptyset$. Then for any dense set $S \subset D_1(T)$ for each $x \in \operatorname{dom} T$ one has

$$T(x) = \operatorname{conv}\left\{\lim_{n\to\infty} T(x_n) : x_n \in S, x_n \to x\right\} + N_{\overline{\operatorname{dom} T}}(x).$$

Hint. The reconstruction inside the interior is quite straightforward as T is a minimal cusco. The boundary behavior is more subtle. □

9.1.3 (Single-valuedness and maximal monotonicity [96]). Consider a maximal monotone multifunction T on a Euclidean space and an open subset U of its domain, and define the minimum norm function $g : U \to \mathbb{R}$ by

$$g(x) := \inf\{\|y\| : y \in T(x)\}.$$

(a) Prove g is lsc. An application of the Baire category theorem now shows that any such function is generically continuous.

(b) For any point x in U at which g is continuous, prove $T(x)$ is a singleton.

Hint. Prove $\|\cdot\|$ is constant on $T(x)$ by first assuming $y, z \in T(x)$ and $\|y\| > \|z\|$, and then using the condition

$$\langle w - y, x + ty - x\rangle \geq 0 \text{ for all small } t > 0 \text{ and } w \in T(x + ty)$$

to derive a contradiction. □

(c) Conclude that any maximal monotone multifunction is generically single-valued on the interior of its domain.

(d) Deduce that any convex function is generically Gâteaux differentiable on the interior of its domain.

(e) (Kenderov theorem [350]) Show that this argument holds true on a Banach space with an equivalent norm whose dual is strictly convex – and so on any WCG space. Note this generalizes Exercise 4.6.6.

9.2 Cyclic and acyclic monotone operators

For completeness we offer a simple variational proof of the next theorem – already established from Fenchel duality in Section 6.1. Earlier proofs are well described in [351, 391].

Theorem 9.2.1 (Maximality of subgradients). *Every closed convex function has a (locally) maximal monotone subgradient.*[3]

Proof. Without loss of generality we may suppose

$$\langle 0 - x^*, 0 - x\rangle \geq 0 \text{ for all } x^* \in \partial f(x)$$

---

[3] This fails in *all* separable incomplete normed spaces (as discussed in Chapter 8) and in *some* Fréchet spaces

but $0 \notin \partial f(0)$; so $f(\bar{x}) - f(0) < 0$ for some $\bar{x}$. By Zagrodny's *approximate mean value theorem* (see Theorem 4.3.8), we find $x_n \to c \in (0, \bar{x}], x_n^* \in \partial f(x_n)$ with

$$\liminf_n \langle x_n^*, c - x_n \rangle \geq 0, \liminf_n \langle x_n^*, \bar{x} \rangle \geq f(0) - f(\bar{x}) > 0.$$

Now $c = \theta \bar{x}$ for some $\theta > 0$. Hence,

$$\limsup_n \langle x_n^*, x_n \rangle < 0,$$

a contradiction. □

9.2.1 Monotonicity of the Fréchet subdifferential

Various forms of monotonicity characterize convexity or quasiconvexity of the underlying function. We illustrate this for $\partial_F f$. Recall, a multifunction $F \colon X \to X^*$ is *quasimonotone* if

$$x^* \in F(x), y^* \in F(y) \text{ and } \langle x^*, y - x \rangle > 0 \Rightarrow \langle y^*, y - x \rangle \geq 0.$$

as is the case for any monotone multifunction.

Theorem 9.2.2. *Let $f \colon X \to (-\infty, +\infty]$ be a lsc function on a Fréchet smooth Banach space X. Then f is quasiconvex if and only if $\partial_F f$ is quasimonotone.*

Proof. Suppose f is not quasiconvex and so there exist $x, y, z \in X$ such that $z \in [x, y]$ and $f(z) > \max\{f(x), f(y)\}$. Apply Theorem 4.3.8 with $a := x, b := z$, to obtain sequences (x_i) and $x_i^* \in \partial_F f(x_i)$ such that $x_i \to \bar{x} \in [x, z]$, $\liminf_{i \to \infty} \langle x_i^*, \bar{x} - x_i \rangle \geq 0$ and $\liminf_{i \to \infty} \langle x_i^*, z - x \rangle > 0$. As $y - \bar{x} = \|y - \bar{x}\| / \|z - x\| (z - x)$ we have

$$\liminf_{i \to \infty} \langle x_i^*, y - x_i \rangle > 0. \qquad (9.2.1)$$

Fix $\lambda \in (0, 1)$ with $z = \bar{x} + \lambda(y - \bar{x})$ and set $z_i := x_i + \lambda(y - x_i)$. Then $z_i \to z$. Since f is lsc, (9.2.1) assures an integer i such that $f(z_i) > f(y)$ and

$$\langle x_i^*, y - x_i \rangle > 0. \qquad (9.2.2)$$

Applying Theorem 4.3.8 again with $a := y$ and $b := z_i$, yields sequences (y_j) and (y_j^*) satisfying $y_j^* \in \partial_F f(y_j)$ such that $y_j \to \bar{y} \in [y, z_i)$, while $\liminf_{j \to \infty} \langle y_j^*, \bar{y} - y_j \rangle \geq 0$ and $\liminf_{j \to \infty} \langle y_j^*, z_i - y \rangle > 0$. Noting that $z_i - y$ and $x_i - \bar{y}$ point in the same direction, we obtain

$$\liminf_{j \to \infty} \langle y_j^*, x_i - y_j \rangle > 0. \qquad (9.2.3)$$

Since $\bar{y} \in [x_i, y)$, inequality (9.2.2) yields

$$\liminf_{j \to \infty} \langle x_i^*, y_j - x_i \rangle = \langle x_i^*, \bar{y} - x_i \rangle > 0. \qquad (9.2.4)$$

Inequalities (9.2.3) and (9.2.4) imply, for j large, that $\langle y_j^*, x_i - y_j\rangle > 0$ and $\langle x_i^*, y_j - x_i\rangle > 0$. So $\partial_F f$ is not quasimonotone, a contradiction.

The converse is easier and is left as Exercise 9.2.1. □

In Exercise 9.2.3 we apply this result to deduce easily that monotonicity of $\partial_F f$ characterizes convexity of f.

9.2.2 N-monotone operators

For $N = 2, 3, \ldots$, a multifunction T is *N-monotone* if

$$\sum_{k=1}^{N} \langle x_k^*, x_k - x_{k-1}\rangle \geq 0$$

whenever $x_k^* \in T(x_k)$ and $x_0 = x_N$. We say T is *cyclically monotone* when T is N-monotone for all $N \in \mathbb{N}$, as hold for all convex subgradients. Then monotonicity and 2-monotonicity coincide, while it is a classical result of Rockafellar [369, 351] that in a Banach space every maximal cyclically monotone operator is the subgradient of a proper closed convex function (and conversely). We recast this result to make the parallel with the Debrunner–Flor theorem (9.1.8) explicit.

Theorem 9.2.3 (Rockafellar [351, 369]). *Suppose C is cyclically monotone on a Banach space X. Then C has a maximal cyclically monotone extension $\overline{C}$, which is of the form $\overline{C} = \partial f_C$ for some proper closed convex function f_C. Moreover, $R(\overline{C}) \subset \overline{\operatorname{conv}}^{w^*} R(C)$.*

Proof. We fix $x_0 \in \operatorname{dom} C, x_0^* \in C(x_0)$ and define

$$f_C(x) := \sup\left\{ \langle x_n^*, x - x_n\rangle + \sum_{k=1}^{n-1} \langle x_{k-1}^*, x_k - x_{k-1}\rangle : x_k^* \in C(x_k), n \in \mathbb{N} \right\},$$

where the sup is over all such chains. The proof in [349] shows that $C \subset \overline{C} := \partial f_C$.

The range assertion follows because f_C is the supremum of affine functions whose linear parts all lie in range C. This is most easily seen by writing $f_C = g_C^*$ with $g_C(x^*) := \inf\{\sum_i t_i \alpha_i : \sum_i t_i x_i^* = x^*, \sum_i t_i = 1, t_i > 0\}$ for appropriate α_i. More explicit proofs are sketched in Exercise 9.2.7. □

The exact relationship between $\mathcal{F}_{\partial f}$ and ∂f is quite complicated. One does always have

$$\langle x, x^*\rangle \leq \mathcal{F}_{\partial f}(x, x^*) \leq f(x) + f^*(x^*) \leq \mathcal{F}_{\partial f}^*(x, x^*) \leq \langle x, x^*\rangle + \delta_{\partial f}(x, x^*),$$

as shown in [38, Proposition 2.1] and discussed in further detail in Section 9.2.6. Likewise, when L is linear and maximal (with dense range) then

$$\mathcal{F}_L(x, x^*) = \langle x, x^*\rangle - \inf_{z \in X} \langle z, Lz - x^*\rangle.$$

Question 9.2.4. From various perspectives it is interesting to answer the following two questions, [70].

(a) When is a maximal monotone operator T the sum of a subgradient ∂f and a skew linear operator S? This is closely related to the behavior of the function

$$\mathcal{FL}_T(x) := \int_0^1 \sup_{x^*(t) \in T(tx)} \langle x, x^*(t) \rangle \, dt,$$

defined assuming $0 \in \operatorname{core} \operatorname{dom} T$.[4] In this case, $\mathcal{FL}_T = \mathcal{FL}_{\partial f} = f$, and we call T *(fully) decomposable*.

(b) How does one appropriately generalize the decomposition of a linear monotone operator L into a symmetric (cyclic) and a skew (acyclic) part? Viz

$$L = \frac{1}{2}(L + L^*|_X) + \frac{1}{2}(L - L^*|_X).$$

Answers to these questions may well allow progress with open questions about the behavior of maximal monotone operators outside reflexive space – since any 'bad' properties are anticipated to originate with the skew or acyclic part.

To be precise we say a linear operator L from X to X^* is *symmetric* if $L^*|_X = L$ or equivalently if $\langle Lx, y \rangle = \langle Lx, y \rangle$ for all $x, y \in X$, and *skew* if $L^*|_X = -L$ or equivalently if $\langle Lx, y \rangle = -\langle L, y \rangle$ for all $x, y \in X$. Thus, in the Euclidean or Hilbert setting we recover the familiar definitions.

9.2.3 Cyclic-acyclic decompositions of monotone operators

We next describe Asplund's approach in [16, 15] to Question 9.2.4(b). We begin by observing that every 3-monotone operator such that $0 \in T(0)$ has the local property that

$$\langle x, x^* \rangle + \langle y, y^* \rangle \geq \langle x, y^* \rangle \tag{9.2.5}$$

whenever $x^* \in T(x)$ and $y^* \in T(y)$. We will call a monotone operator satisfying (9.2.5), 3^--*monotone*, and write $T \geq_N S$ when $T = S + R$ with R being N-monotone. Likewise we write $T \geq_{\omega_0} S$ when R is cyclically monotone.

Proposition 9.2.5. *Let N be one of $3^-, 3, 4, \ldots,$ or ω_0. Consider an increasing (infinite) net of monotone operators on a Banach space X, satisfying*

$$0 \leq_N T_\alpha \leq_N T_\beta \leq_2 T,$$

whenever $\alpha < \beta \in \mathcal{A}$.

---

[4] The use of $\mathcal{FL}_T$ originates in discussions had with Fitzpatrick shortly before his death.

9.2 Cyclic and acyclic monotone operators

Suppose that $0 \in T_\alpha(0), 0 \in T(0)$ and that $0 \in \text{core dom } T$. Then

(a) There is a N-monotone operator T_A with $T_\alpha \leq_N T_A \leq_2 T$, for all $\alpha \in A$.
(b) If $R(T) \subset MB_{X^*}$ for some $M > 0$ then one may suppose $R(T_A) \subset MB_{X^*}$.

Proof. (a) We first give details of the single-valued case. As $0 \leq_2 T_\alpha \leq_2 T_\beta \leq_2 T$, while $T(0) = 0 = T_\alpha(0)$, we have

$$0 \leq \langle x, T_\alpha(x) \rangle \leq \langle x, T_\beta(x) \rangle \leq \langle x, T(x) \rangle,$$

for all x in dom T. This shows that $\langle x, T_\alpha(x) \rangle$ converges as α goes to ∞.

Fix $\varepsilon > 0$ and $M > 0$ with $T(\varepsilon B_X) \subset M B_{X^*}$. We write $T_{\beta\alpha} = T_\beta - T_\alpha$ for $\beta > \alpha$, so that $\langle T_{\beta\alpha} x, x \rangle \to 0$ for $x \in \text{dom } T$ as α, β go to ∞.

We appeal to (9.2.5) to obtain

$$\langle x, T_{\beta\alpha}(x) \rangle + \langle y, T_{\beta\alpha}(y) \rangle \geq \langle T_{\beta\alpha}(x), y \rangle, \tag{9.2.6}$$

for $x, y \in \text{dom } T$. Also, $0 \leq \langle x, T_{\beta\alpha}(x) \rangle \leq \varepsilon$ for $\beta > \alpha > \gamma(x)$ for all $x \in \text{dom } T$.

Now, $0 \leq \langle y, T_{\beta\alpha}(y) \rangle \leq \langle y, T(y) \rangle \leq \varepsilon M$ for $\|y\| \leq \varepsilon$. Thus, for $\|y\| \leq \varepsilon^2$ and $\beta > \alpha > \gamma(x)$ we have

$$\varepsilon(M + \varepsilon) \geq \langle x, T_{\beta\alpha}(x) \rangle + \langle y, T(y) \rangle \tag{9.2.7}$$

$$\geq \langle x, T_{\beta\alpha}(x) \rangle + \langle y, T_{\beta\alpha}(y) \rangle$$

$$\geq \langle y, T_{\beta\alpha}(x) \rangle,$$

from which we obtain $\|T_{\beta\alpha}(x)\| \leq M + \varepsilon$ for all $x \in \text{dom } T$, while $\langle y, T_{\beta\alpha}(x) \rangle \to 0$ for all $y \in X$. We conclude that $(T_\alpha(x))_{\alpha \in A}$ is a norm-bounded weak*-Cauchy net and so weak*-convergent to the desired N-monotone limit $T_A(x)$.

In the general case we may still use (9.2.5) to deduce that $T_\beta = T_\alpha + T_{\beta\alpha}$ where (i) $T_{\beta\alpha} \subset (M + \varepsilon)B_{X^*}$ and (ii) for each $t^*_{\beta\alpha} \in T_{\beta\alpha}$ one has $t^*_{\beta\alpha} \to_{w^*} 0$ as α and $\beta \to \infty$. The conclusion follows as before, but is somewhat more technical.

(b) Fix $x \in X$. We again apply (9.2.5), this time to T_α to write

$$\langle Tx, x \rangle + \langle Ty, y \rangle \geq \langle T_\alpha x, x \rangle + \langle T_\alpha y, y \rangle \geq \langle T_\alpha x, y \rangle$$

for all $y \in D(T) = X$, by Theorem 9.1.8 (c). Hence

$$\langle Tx, x \rangle + M \|y\| \geq \|T_\alpha x\| \, \|y\|,$$

for all $y \in Y$. This shows that $T_\alpha(x)$ lies in the M-ball, and since the ball is weak* closed, so does $T_A(x)$.

The set-valued case is entirely analogous but more technical (see Exercise 9.2.5). □

We comment that $0 \leq_2 (-ny, nx) \leq_2 (-y, x)$ for $n \in \mathbb{N}$, shows the need for (9.2.5) in the deduction that $T_{\beta\alpha}(x)$ are equi-norm bounded. Moreover, if X

is an Asplund space (with a technology that did not exist when Asplund studied this subject), the proof of Proposition 9.2.5 can be adjusted to show that $T_A(x) = \text{norm} - \lim_{\alpha \to \infty} T_\alpha(x)$, [70] (the *Daniel* property). The single-valued case effectively comprises [15, Theorem 6.1].

We shall say that a maximal monotone operator A is *acyclic* or in Asplund's term *irreducible* if whenever $A = \partial g + S$ with S maximal monotone and g closed and convex then g is necessarily a linear function. We can now provide a broad extension of Asplund's original idea [16, 15]:

Theorem 9.2.6 (Asplund decomposition). *Suppose that T is a maximal monotone operator on a Banach space with* dom T *having nonempty interior.*

(a) *Then T may be decomposed as $T = \partial f + A$, where f is closed and convex while A is acyclic.*

(b) *If the range of T lies in $M B_{X^*}$ then f may be assumed M-Lipschitz.*

Proof. (a) We normalize so $0 \in T(0)$ and apply Zorn's lemma to the set of cyclically monotone operators $\mathcal{C} := \{C : 0 \leq_{\omega_0} C \leq_2 T, 0 \in C(0)\}$ in the cyclic order. By Proposition 9.2.5 every chain in $\mathcal{C}$ has a cyclically monotone upper-bound. Consider such a maximal $\overline{C}$ with $0 \leq_{\omega_0} \overline{C} \leq_2 T$. Hence $T = \overline{C} + A$ where by construction A is acyclic. Now, $T = \overline{C} + A \subset \partial f + A$, by Rockafellar's result of Theorem 9.2.3. Since T is maximal the decomposition is as asserted.

(b) In this case we require all members of $\mathcal{C}$ to have their range in the M-ball and apply part (b) of Proposition 9.2.5. It remains to observe that every M-bounded cyclically monotone operator extends to an M-Lipschitz subgradient – as an inspection of the proof of Rockafellar's result of Theorem 9.2.3 confirms. □

By way of application we offer the following corollary where we denote the mapping $x \mapsto T(x) \cap \mu B_{X^*}$ by $T \cap \mu B_{X^*}$.

Corollary 9.2.7. *Let T be an arbitrary maximal monotone operator T on a Banach space. For $\mu > 0$ one may decompose*

$$T \cap \mu B_{X^*} \subset \widehat{T_\mu} = \partial f_\mu + A_\mu,$$

where f_μ is μ-Lipschitz and A_μ is acyclic (with bounded range).

Proof. Combining Theorem 9.2.6 with Proposition 9.1.14 we deduce that the composition is as claimed. □

Note that since the acyclic part A_μ is bounded in Corollary 9.2.7, it is only skew and linear when T is itself cyclic. Hence, such a range bounded monotone operator is never fully decomposable in the sense of Question 9.2.4(a).

Theorem 9.2.6 and related results in [16, 70] are entirely existential: how can one prove Corollary 9.2.6 constructively in finite dimensions? How in general does one effectively diagnose acyclicity? What is the decomposition for such simple monotone maps such as

$$(x,y) \mapsto (\sinh(x) - \alpha y^2/2, \sinh(x) - \alpha x^2/2)$$

which is monotone exactly for $\alpha \geq -2/\sqrt{x_0^2 - 1} \sim 0.7544\ldots$ with x_0 the smallest fixed point of coth? We refer to [118] for a more detailed discussion of Question 9.2.4(a).

Example 9.2.8. Consider the maximal monotone linear mapping

$$T_\theta : (x, y) \mapsto (\cos(\theta)x - \sin(\theta)y, \cos(\theta)y + (\sin(\theta)x))$$

for $0 \leq \theta \leq \pi/2$. The methods in [16] show that for $n = 1, 2, \ldots$ the rotation mapping $T_{\pi/n}$ is n-monotone but is not $(n+1)$-monotone, a more explicit proof is given in 9.2.6. These matters are taken much further in Section 9.2.6 which describes work from [31] where appropriate Fitzpatrick functions are associated to n-monotone operators.

9.2.4 Explicit acyclic examples

As noted, skew linear mappings are canonical examples of monotone mappings that are not subdifferential mappings. It is therefore reassuring to prove that they are acyclic:

Proposition 9.2.9. *Suppose that* $S : \mathbb{R}^n \to \mathbb{R}^n$ *is a continuous linear operator satisfying* $\langle S(x), x \rangle = 0$ *for all* $x \in \mathbb{R}^n$. *Then S is acyclic.*

Proof. Let $S = F + R$ where F is a subdifferential mapping and R is maximal monotone. Since S is single-valued, F and R are single-valued. In particular, $F = \nabla f$ for some convex differentiable f. Since R is monotone, we have

$$0 \leq \langle R(x) - R(y), x - y \rangle = \langle S(x) - S(y), x - y \rangle - \langle F(x) - F(y), x - y \rangle$$
$$= -\langle F(x) - F(y), x - y \rangle = \langle \nabla(-f)(x) - \nabla(-f)(y), x - y \rangle.$$

This shows that $-f$ is convex, so f is convex and concave, hence linear on its domain. But $\text{dom} f \supset \text{dom} S = \mathbb{R}^n$, so $f \in \mathbb{R}^n$. So $F = \nabla f$ is constant. In fact, by subtracting from F and adding to R, we may assume that $F = 0$. □

We leave it to the reader to check that the sum of an acyclic operator and a skew linear operator is still acyclic. It is not clear that the sum of two acyclic operators must be acyclic. For continuous linear monotone operators, then, the usual decomposition into symmetric and skew parts is the same as the Asplund decomposition into subdifferential and acyclic parts.

We recall that Asplund was unable to find explicit examples of nonlinear acyclic mappings [16], and this remains quite challenging, see Exercise 9.2.4. In particular, it would be helpful to determine a useful characterization of acyclicity. We make some progress in this direction by providing an explicit and to our mind surprisingly simple example: we present a nonlinear acyclic monotone mapping $\widehat{S} : \mathbb{R}^2 \to \mathbb{R}^2$, originally given in [118].

Precisely, $\widehat{S}$ is constructed by restricting the range of the skew mapping $S(x, y) = (-y, x)$ to the unit ball, and taking a range-preserving maximal monotone extension of the restriction. This extension is unique, as we see from the following corollary of Proposition 9.1.14.

Corollary 9.2.10 (Unique extension). *Suppose $T : \mathbb{R}^n \to \mathbb{R}^n$ is maximal monotone and suppose that range $T \cap$ int $B \neq \emptyset$. Then there is a unique maximal monotone mapping $\widehat{T}$ such that $T(x) \cap B \subset \widehat{T}(x) \subset B$. Furthermore,*

$$\widehat{T}(x) = \{x^* \in B : \langle x^* - y^*, x - y \rangle \geq 0 \text{ for all } y^* \in T(y) \cap \text{int } B\}. \tag{9.2.8}$$

Note that $\widehat{T}$ is either a Lipschitz subgradient or it has a nonlinear acyclic part: the acyclic part is bounded so it cannot be nontrivially linear. Hence in the construction of Proposition 9.2.11 we know that $\widehat{S}$ has nonlinear acyclic part, which we shall eventually show in Proposition 9.2.15 to be $\widehat{S}$ itself.

Proposition 9.2.11. *Define $S : \mathbb{R}^2 \to \mathbb{R}^2$ by $S(x, y) = (-y, x)$ for $x^2 + y^2 \leq 1$. Then the unique maximal monotone extension $\widehat{S}$ of S with range restricted to the unit disk is:*

$$\widehat{S}(x) = \begin{cases} S(x) & \text{if } \|x\| \leq 1; \\ \sqrt{1 - \frac{1}{\|x\|^2}} \frac{x}{\|x\|} + \frac{1}{\|x\|} S\left(\frac{x}{\|x\|}\right) & \text{if } \|x\| > 1. \end{cases}$$

Proof. From Corollary 9.2.10, we know that $\widehat{S}$ exists and is uniquely defined. In the interior of the unit ball, equation (9.2.8) shows that $\widehat{S}(x) = S(x)$. Indeed, let $t > 0$ be so small that $z = x + ty \in B$ for all unit length y. Then

$$\langle S(x + ty) - \widehat{S}(x), y \rangle \geq 0$$

for all unit y. Letting $t \to 0$ shows that $\widehat{S}(x) = S(x)$. To determine $(u, v) = \widehat{S}(x)$ for $\|x\| \geq 1$, it suffices by rotational symmetry to consider points $x = (a, 0)$ with $a \geq 1$. Then monotonicity requires that

$$\langle \widehat{S}(x) - S(z), x - z \rangle \geq 0$$

for all $\|z\| \leq 1$. Let $z = \left(\frac{1}{a}, -\frac{\sqrt{a^2-1}}{a}\right)$ so that $\widehat{S}(z) = S(z) = \left(\frac{\sqrt{a^2-1}}{a}, \frac{1}{a}\right)$. Then

$$\left\langle (u, v) - \left(\frac{\sqrt{a^2-1}}{a}, \frac{1}{a}\right), (a, 0) - \left(\frac{1}{a}, -\frac{\sqrt{a^2-1}}{a}\right) \right\rangle \geq 0.$$

Expanding this gives

$$u\left(a - \frac{1}{a}\right) + \sqrt{1 - \frac{1}{a^2}}(v - a) \geq 0,$$

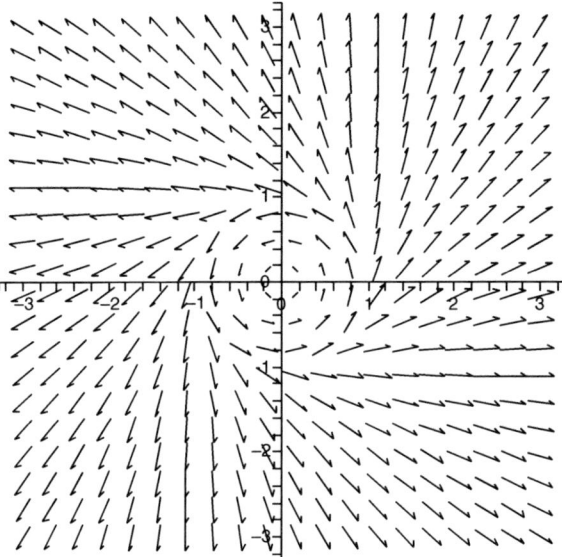

Figure 9.2 A field plot of $\widehat{S}$.

and noting that $u \leq \sqrt{1-v^2}$ gives

$$\sqrt{1-v^2}(a^2-1) + \sqrt{a^2-1}(v-a) \geq 0$$

which reduces to $(av-1)^2 \leq 0$, that is, $v = 1/a$. Similarly, setting $z = \left(\frac{1}{a}, -\frac{\sqrt{a^2-1}}{a}\right)$ also shows that $u = \sqrt{1-\frac{1}{a^2}}$.

So $\widehat{S}(x) = \widehat{S}(a,0) = \left(\sqrt{1-\frac{1}{a^2}}, \frac{1}{a}\right) = \sqrt{1-\frac{1}{\|x\|^2}} \frac{x}{\|x\|} + \frac{1}{\|x\|} S\left(\frac{x}{\|x\|}\right)$. The same result holds for general $\|x\| \geq 1$ by considering the coordinate system given by the orthogonal basis $\{x, S(x)\}$. □

Figure 9.2 shows the graph of the vector field $\widehat{S}$. Having computed $\widehat{S}$, we commence to show that it is acyclic, with the aid of two technical lemmas:

Lemma 9.2.12. $\widehat{S}(x + tS(x)) = S(x)$ for all $t \geq 0$, and all $\|x\| = 1$.

Proof.

$$\widehat{S}(x + tS(x)) = \sqrt{1 - \frac{1}{1+t^2}} \frac{x + tS(x)}{\sqrt{1+t^2}} + \frac{1}{1+t^2} S(x + tS(x))$$

$$= \frac{t}{1+t^2}(x + tS(x)) + \frac{1}{1+t^2}(S(x) - tx) = S(x),$$

since $S^2 = -I$. □

This construction does not extend immediately to all skew mappings, since it assumes that $S^2 = -I$, which can only occur in even dimensions:

Fact 9.2.13. *Skew orthogonal matrices exist only in even dimensions.*

Proof. Det S = Det$(S^\top)$ = Det$(-S)$ = $(-1)^n$ Det S. $\square$

However, such mappings do exist for each even-dimensional $\mathbb{R}^{2n}$, and these can be embedded in $\mathbb{R}^{2n+1}$ in the obvious way. Thus, our construction provides an acyclic nonlinear mapping for each $\mathbb{R}^n$, $n > 1$.

To show that $\widehat{S}$ is acyclic, we suppose that $\widehat{S} = F + R$, where $F = \partial f$ for some convex proper lsc function f and R is maximal monotone, and show that F is constant.

Lemma 9.2.14. *Let $\|x\| = 1$, $t \geq 0$ and $y(t) = x + tS(x)$. Then $\langle F(y(t)), S(x)\rangle = c(x)$ for some constant $c(x)$.*

Proof. Suppose $t_1 \neq t_2$. Then $\widehat{S}(y(t_1)) = \widehat{S}(y(t_2))$, by Lemma 9.2.12, so

$$0 \leq \langle R(y(t_1)) - R(y(t_2)), y(t_1) - y(t_2)\rangle$$
$$= \langle \widehat{S}(y(t_1)) - \widehat{S}(y(t_2)), y(t_1) - y(t_2)\rangle - \langle F(y(t_1)) - F(y(t_2)), y(t_1) - y(t_2)\rangle$$
$$= -\langle F(y(t_1)) - F(y(t_2)), y(t_1) - y(t_2)\rangle \leq 0,$$

so

$$\langle F(y(t_1)) - F(y(t_2)), x + t_1 S(x) - (x + t_2 S(x))\rangle = 0,$$

that is

$$\langle F(y(t_1)), S(x)\rangle = \langle F(y(t_2)), S(x)\rangle$$

for any t_1, t_2. $\square$

Proposition 9.2.15. *The extension mapping $\widehat{S}$ given explicitly in Proposition 9.2.11 is nonlinear and acyclic with bounded range and full domain.*

Proof. First note that if $\widehat{S} = F + R$ with R monotone and $F = \partial f$, then both are single-valued, so $F = \nabla f$. As in Proposition 9.2.9, we can assume that $f(x) = 0$ when $\|x\| \leq 1$.

Let $\|y\| > 1$. Then there is a unit vector x and a t such that $y = x + tS(x)$:

$$x = \widehat{x}(y) := \frac{y}{\|y\|^2} - \sqrt{\frac{1}{\|y\|^2} - \frac{1}{\|y\|^4}} S(y),$$

$$t = t(y) = \sqrt{\|y\|^2 - 1},$$

and we note that $y \to \widehat{x}(y)$ is continuous. We will determine $f(y)$ by integrating F along the ray $s \to x + sS(x)$. Using Lemma 9.2.14, we have:

$$f(y) - f(x) = \int_0^t \langle \nabla f(x + sS(x)), S(x) \rangle \, ds$$
$$= \int_0^t c(x) \, ds = c(x)t.$$

Since f is continuous and convex, c is continuous and positive, so $y \to c(\widehat{x}(y))$ is continuous and positive.

Plugging in $t(y)$ gives $f(y) = c(\widehat{x}(y))\sqrt{\|y\|^2 - 1}$ when $\|y\| > 1$ and $f = 0$ for $\|y\| \leq 1$. Suppose $c(y) > 0$ for some $\|y\| = 1$. Then for f to be convex on the segment $[y, 2y]$ we require that:

$$(1 - \lambda)f(y) + \lambda f(2y) \geq f\big((1 + \lambda)y\big) \text{ for all } \lambda \in (0, 1).$$

This means

$$0 + \lambda c\big(\widehat{x}(2y)\big)\sqrt{3} \geq c\big(\widehat{x}((1 + \lambda)y)\big)\sqrt{\lambda^2 + 2\lambda},$$

or

$$c\big(\widehat{x}(2y)\big)\sqrt{3} \geq c\big(\widehat{x}((1 + \lambda)y)\big)\sqrt{1 + \frac{2}{\lambda}},$$

for all $\lambda \in (0, 1)$. Letting $\lambda \to 0$, we get $\widehat{x}((1 + 2\lambda)y) \to y$, so $c\big(\widehat{x}((1 + \lambda)y)\big) \to c(y) > 0$. Since $\sqrt{1 + \frac{2}{\lambda}} \to \infty$, the inequality does not hold for small λ unless $c(y) = 0$.

For f to be convex and everywhere defined, then, we require $c(y) = 0$ for all $\|y\| = 1$. That is, f is identically zero. □

It seems probable that the construction above applied to any nontrivial skew linear mapping always leads to an acyclic mapping – and that more ingenuity will allow some reader to prove this. We dedicate the next subsection to exploring Fitzpatrick's last function for $\widehat{S}$ as constructed above.

9.2.5 Computing $\mathcal{FL}_{\widehat{S}}$

We can also explicitly compute Fitzpatrick's last function $\mathcal{FL}_{\widehat{S}}$ as described in the previous section. We have:

Proposition 9.2.16. *With $\widehat{S}$ as before, we have:*

$$\mathcal{FL}_{\widehat{S}}(x) = \begin{cases} 0 & \text{if } \|x\| \leq 1; \\ \sqrt{\|x\|^2 - 1} + \arctan\left(\frac{1}{\sqrt{\|x\|^2 - 1}}\right) - \frac{\pi}{2} & \text{if } \|x\| > 1. \end{cases}$$

Proof. It is immediate from the definition that $\mathcal{FL}_{\hat{S}}(x) = 0$ when $\|x\| \le 1$. For $\|x\| > 1$, we get:

$$\mathcal{FL}_{\hat{S}}(x) = \int_0^1 \langle x, \hat{S}(tx) \rangle \, dt$$

$$= \int_0^{\frac{1}{\|x\|}} t \langle x, S(x) \rangle \, dt + \int_{\frac{1}{\|x\|}}^1 \sqrt{1 - \frac{1}{t^2 \|x\|^2}} \frac{1}{\|x\|} \langle x, x \rangle \, dt$$

$$+ \int_{\frac{1}{\|x\|}}^1 \frac{1}{t \|x\|^2} \langle S(x), x \rangle \, dt$$

$$= \int_{\frac{1}{\|x\|}}^1 \sqrt{1 - \frac{1}{t^2 \|x\|^2}} \|x\| \, dt$$

$$= \int_1^{\|x\|} \sqrt{1 - \frac{1}{s^2}} \, ds$$

$$= \sqrt{\|x\|^2 - 1} + \arctan\left(\frac{1}{\sqrt{\|x\|^2 - 1}}\right) - \frac{\pi}{2}.$$

□

Note that $\mathcal{FL}_{\hat{S}}$ is convex, since it is a composition of the norm $x \to \|x\|$ with the increasing convex function $t \to \int_1^t \sqrt{1 - \frac{1}{s^2}} \, ds$. So $\hat{S}$ is weakly decomposable in the sense that $\hat{S} = \nabla \mathcal{FL}_{\hat{S}} + SL$ where SL is *skew-like*: $\langle SLx, x \rangle = 0$. To determine SL, we compute:

$$\nabla \mathcal{FL}_{\hat{S}}(x) = \begin{cases} 0 & \text{if } \|x\| < 1; \\ \sqrt{1 - \frac{1}{\|x\|^2}} \frac{x}{\|x\|} & \text{if } \|x\| \ge 1. \end{cases}$$

So $\bar{S}(x) = \nabla \mathcal{FL}_{\hat{S}}(x) + h(\|x\|)S(x)$, where

$$h(t) = \begin{cases} 1 & \text{if } t \le 1; \\ \frac{1}{t^2} & \text{if } t \ge 1. \end{cases}$$

So $\hat{S}$ is not decomposable since $SL = x \to h(\|x\|)S(x)$ is not skew. Note finally that SL is not monotone.

9.2.6 Fitzpatrick functions of finite and infinite order

The following functions studied in [31] may be interpreted as common ancestors of the Fitzpatrick function and Rockafellar's antiderivative of Theorem 9.2.3.

Definition 9.2.17. Let $A \colon X \to 2^{X^*}$, let $(a_1, a_1^*) \in \operatorname{graph} A$, and let $n \in \{2, 3, \ldots\}$. If $n = 2$, we set $C_{A,2,(a_1,a_1^*)} \colon X \times X^* \to (-\infty, +\infty] \colon (x, x^*) \mapsto \langle x, a_1^* \rangle + \langle a_1, x^* \rangle - \langle a_1, a_1^* \rangle$. Now suppose that $n \in \{3, 4, \ldots\}$. Then the *value* of the function

9.2 Cyclic and acyclic monotone operators

$C_{A,n,(a_1,a_1^*)}: X \times X^* \to (-\infty, +\infty]$ at $(x, x^*) \in X \times X^*$ is defined by

$$\sup_{\substack{(a_2, a_2^*) \in \text{graph } A, \\ \vdots \\ (a_{n-1}, a_{n-1}^*) \in \text{graph } A}} \left(\sum_{i=1}^{n-2} \langle a_{i+1} - a_i, a_i^* \rangle \right) + \langle x - a_{n-1}, a_{n-1}^* \rangle + \langle a_1, x^* \rangle; \quad (9.2.9)$$

equivalently, by

$$\sup_{\substack{(a_2, a_2^*) \in \text{graph } A, \\ \vdots \\ (a_{n-1}, a_{n-1}^*) \in \text{graph } A}} \langle x, x^* \rangle + \left(\sum_{i=1}^{n-2} \langle a_{i+1} - a_i, a_i^* \rangle \right) + \langle x - a_{n-1}, a_{n-1}^* \rangle + \langle a_1 - x, x^* \rangle.$$

(9.2.10)

Definition 9.2.18 (Fitzpatrick functions of finite order). Let $A: X \to 2^{X^*}$. For every $n \in \{2, 3, \ldots\}$, the *Fitzpatrick function of A of order n* is

$$F_{A,n} = \sup_{(a, a^*) \in \text{graph } A} C_{A,n,(a, a^*)}. \quad (9.2.11)$$

The *Fitzpatrick function of A of infinite order* is $F_{A,\infty} = \sup_{n \in \{2,3,\ldots\}} F_{A,n}$.

Our first result is immediate from the definition.

Proposition 9.2.19. *Let $A: X \to 2^{X^*}$ and let $n \in \{2, 3, \ldots\}$. Then $F_{A,n}: X \times X^* \to [-\infty, +\infty]$ is convex and lsc. At $(x, x^*) \in X \times X^*$, the value of $F_{A,n}$ is given by*

$$\sup_{\substack{(a_1, a_1^*) \in \text{graph } A, \\ \vdots \\ (a_{n-1}, a_{n-1}^*) \in \text{graph } A}} \langle x, x^* \rangle + \left(\sum_{i=1}^{n-2} \langle a_{i+1} - a_i, a_i^* \rangle \right) + \langle x - a_{n-1}, a_{n-1}^* \rangle + \langle a_1 - x, x^* \rangle.$$

(9.2.12)

Moreover,

$$F_{A,n} \geq \langle \cdot, \cdot \rangle \text{ on graph } A. \quad (9.2.13)$$

If $n = 2$, then (9.2.12) simplifies to $\sup_{(a, a^*) \in \text{graph } A} \langle x, a^* \rangle + \langle a, x^* \rangle - \langle a, a^* \rangle$, which is the original definition of the Fitzpatrick function of A at $(x, x^*) \in X \times X^*$.

Note that $(F_{A,n})_{n \in \{2,3,\ldots\}}$ is a sequence of increasing functions and that $F_{A,n} \to F_{A,\infty}$ pointwise. We next provide a characterization of n-cyclically monotone operators by Fitzpatrick functions of order n which directly generalizes the notion of a representative function.

Proposition 9.2.20. *Let* $A\colon X \to 2^{X^*}$ *and let* $n \in \{2, 3, \ldots\}$. *Then the following are equivalent.*

1. *A is n-cyclically monotone.*
2. $F_{A,n} \le \langle \cdot, \cdot \rangle$ *on graph A.*
3. $F_{A,n} = \langle \cdot, \cdot \rangle$ *on graph A.*

We now can give a simpler, but by no means easy, proof of Asplund's observation recorded in Example 9.2.8 that for each $n = 2, 3, \ldots$ the matrix corresponding to rotation by π/n in the Euclidean plane is n-cyclically monotone yet not $(n + 1)$-cyclically monotone.

Theorem 9.2.21 (Rotations). *Let* $X = \mathbb{R}^2$ *and let* $n \in \{2, 3, \ldots\}$. *Denote the matrix corresponding to counter-clockwise rotation by* π/n *by* R_n, *i.e.,*

$$R_n := \begin{pmatrix} \cos(\pi/n) & -\sin(\pi/n) \\ \sin(\pi/n) & \cos(\pi/n) \end{pmatrix}. \tag{9.2.14}$$

Then R_n *is maximal monotone and n-cyclically monotone, but* R_n *is not* $(n + 1)$-*cyclically monotone.*

Proof.

(a) It is clear that R_n is monotone, and that $R_n^* = R_n^{-1}$. Since $\operatorname{dom} R_n = X$, the maximal monotonicity of R_n is immediate. Show that R_2 is not 3-cyclically monotone, as it is skew-symmetric. For the remainder of the proof we assume that $n \in \{3, 4, \ldots\}$.

(b) Let us show next that R_n is not $(n+1)$-cyclically monotone. Take $x \in X \smallsetminus \{0\}$. Since $R_n + R_n^*$ is invertible (in fact, a strictly positive multiple of the identity), there exists $a \in X$ such that $\tfrac{1}{2}R_n a + \tfrac{1}{2}R_n^* a = R_n^* x$. Note that $a \ne 0$ (since $x \ne 0$) and that $R_n a \ne R_n^* a$ (since $\pi/n < \pi$). The fact that R_n is an isometry and the parallelogram law thus yield $4\|a\|^2 = 2\|R_n a\|^2 + 2\|R_n^* a\|^2 = \|R_n a + R_n^* a\|^2 + \|R_n a - R_n^* a\|^2 > \|R_n a + R_n^* a\|^2 = \|2R_n^* x\|^2 = 4\|x\|^2$. Hence

$$\|a\| > \|x\|. \tag{9.2.15}$$

Furthermore, $R_n a + R_n^* a = 2R_n^* x$ implies that

$$2\langle a, R_n a \rangle = \langle a, R_n a + R_n^* a \rangle = 2\langle a, R_n^* x \rangle = 2\langle R_n a, x \rangle.$$

Using (9.2.15), we note that

$$-\langle a, R_n a \rangle + 2\langle x, R_n a \rangle = \langle a, R_n a \rangle = \|a\|^2 \cos(\pi/n)$$
$$> \|x\|^2 \cos(\pi/n) = \langle x, R_n x \rangle. \tag{9.2.16}$$

We now take n points from graph R_n by setting

$$(\text{for all } i \in \{1, 2, \ldots, n\}) \quad (a_i, a_i^*) = (R_n^{2i} a, R_n^{2i+1} a). \tag{9.2.17}$$

9.2 Cyclic and acyclic monotone operators

Then, since R_n is an isometry, we have for every $i \in \{1, 2, \ldots, n-1\}$,

$$\langle a_{i+1} - a_i, a_i^* \rangle = \langle R_n^{2i+2} a - R_n^{2i} a, R_n^{2i+1} a \rangle$$
$$= \langle R_n^{2i+2} a, R_n^{2i+1} a \rangle - \langle R_n^{2i} a, R_n^{2i+1} a \rangle$$
$$= \langle R_n a, a \rangle - \langle a, R_n a \rangle = 0. \qquad (9.2.18)$$

Using (9.2.12), (9.2.17), (9.2.18), the fact that $R_n^{2n} = \mathrm{Id}$ and that R_n is an isometry, and (9.2.16), we deduce that

$$F_{R_n, n+1}(x, R_n x) \geq \left(\sum_{i=1}^{n-1} \langle a_{i+1} - a_i, a_i^* \rangle \right) - \langle a_n, a_n^* \rangle + \langle x, a_n^* \rangle + \langle a_1, R_n x \rangle$$
$$= -\langle R_n^{2n} a, R_n^{2n+1} a \rangle + \langle x, R_n^{2n+1} a \rangle + \langle R_n^2 a, R_n x \rangle$$
$$= -\langle a, R_n a \rangle + \langle x, R_n a \rangle + \langle R_n a, x \rangle$$
$$> \langle x, R_n x \rangle. \qquad (9.2.19)$$

Thus $F_{R_n, n+1} > p$ on graph $R_n \smallsetminus \{(0,0)\}$, and therefore, by Exercise 9.2.20, R_n is not $(n+1)$-cyclically monotone.

(c) It remains to show that R_n is n-cyclically monotone. Take

$$x_1 = (\xi_1, \eta_1), \ldots, x_n = (\xi_n, \eta_n) \text{ in } X,$$

and set $x_{n+1} = x_1$. We must show that

$$0 \geq \sum_{i=1}^{n} \langle x_{i+1} - x_i, R_n x_i \rangle. \qquad (9.2.20)$$

Next identify $\mathbb{R}^2$ with $\mathbb{C}$ in the usual way: $x = (\xi, \eta)$ in $\mathbb{R}^2$ corresponds to $\xi + i\eta$ in $\mathbb{C}$, where $i = \sqrt{-1}$ and $\langle x, y \rangle = \mathrm{Re}\,(\overline{x}y)$ for x and y in $\mathbb{C}$. The operator R_n corresponds to complex multiplication by

$$\omega = \exp(i\pi/n). \qquad (9.2.21)$$

Thus the aim is to show that

$$0 \geq \mathrm{Re}\,\Big(\sum_{i=1}^{n} \overline{(x_{i+1} - x_i)} \omega x_i \Big) = \sum_{i=1}^{n} \mathrm{Re}\,\big(\overline{(x_{i+1} - x_i)} \omega x_i \big),$$

an inequality which we now reformulate in $\mathbb{C}^n$. Denote the $n \times n$-identity matrix by $\mathbf{I}$ and set

$$\mathbf{B} := \begin{pmatrix} 0 & 1 & 0 & \cdots & & 0 \\ 0 & 0 & 1 & 0 & & \vdots \\ \vdots & & \ddots & \ddots & & \\ & & & & & 0 \\ 0 & & & & & 1 \\ 1 & 0 & \cdots & & & 0 \end{pmatrix} \in \mathbb{C}^{n \times n} \quad \text{and} \quad R := \omega \mathbf{I} \in \mathbb{C}^{n \times n}. \quad (9.2.22)$$

Identifying $\mathbf{x} \in \mathbb{C}^n$ with $(x_1, \ldots, x_n) \in X^n$, we note that (9.2.20) means $0 \geq \operatorname{Re}\big(((B-I)x)^* Rx\big)$; equivalently, $0 \geq x^*(B^* - I)Rx + x^* R^*(B - I)x$. In other words, we need to show that the Hermitian matrix

$$\mathbf{C} = (I - B^*)R + R^*(I - B)$$

$$= \begin{pmatrix} (\omega + \overline{\omega}) & -\overline{\omega} & 0 & \cdots & 0 & -\omega \\ -\omega & (\omega + \overline{\omega}) & \ddots & & & 0 \\ 0 & \ddots & \ddots & \ddots & & \vdots \\ \vdots & & & & & 0 \\ 0 & & & & (\omega + \overline{\omega}) & -\overline{\omega} \\ -\overline{\omega} & 0 & \cdots & 0 & -\omega & (\omega + \overline{\omega}) \end{pmatrix}$$

(9.2.23)

is positive semidefinite. The matrix $\mathbf{C}$ is a circulant (Toeplitz) matrix and thus belongs to a class of well-studied matrices that have close connections to Fourier Analysis. It is well known that the set of (n not necessarily distinct) eigenvalues of $\mathbf{C}$ is

$$\Lambda = \{q(1), q(\omega^2), \ldots, q(\omega^{2(n-1)})\}, \quad \text{where} \quad q: t \mapsto (\omega + \overline{\omega}) - \omega t - \overline{\omega} t^{n-1}.$$
(9.2.24)

Since $\omega^{2n} = 1$, we verify that

$$\Lambda = \{2\cos(\pi/n) - 2\cos((2k+1)\pi/n) : k \in \{0, 1, \ldots, n-1\}\}$$

is a set of nonnegative real numbers, as required.

$\square$

The Fitzpatrick function of order n arose by keeping n fixed while supremizing over (a, a^*). Analogously, we next see that Rockafellar's antiderivative is (essentially) obtained by keeping (a, a^*) fixed while supremizing over n. Therefore, the function $C_{A,n,(a,a^*)}$ can honestly be viewed as the 'common ancestor' of the Fitzpatrick functions and Rockafellar's antiderivative.

9.2 Cyclic and acyclic monotone operators

Definition 9.2.22 (Rockafellar function). Let $A \colon X \to 2^{X^*}$ and $(a, a^*) \in \operatorname{graph} A$. Then we define the *Rockafellar function* by

$$R_{A,(a,a^*)} \colon X \to (-\infty, +\infty] \colon x \mapsto \sup_{n \in \{2,3,\ldots\}} C_{A,n,(a,a^*)}(x, 0). \tag{9.2.25}$$

Moreover, Rockafellar functions are antiderivatives, which are unique up to constants. The following taken from [31] is a refinement of Theorem 9.2.3.

Fact 9.2.23 (Rockafellar). *Let $f \colon X \to (-\infty, +\infty]$ be convex, lsc, and proper. Then ∂f is maximal monotone and cyclically monotone, hence maximal cyclically monotone. Conversely, let $A \colon X \to 2^{X^*}$ be maximal cyclically monotone and let $(a, a^*) \in \operatorname{graph} A$. Then $R_{A,(a,a^*)}$ is convex, lsc, and proper, $R_{A,(a,a^*)}(a) = 0$, and*

$$A = \partial R_{A,(a,a^*)} \tag{9.2.26}$$

is maximal monotone. If $\partial f = A$, then $f(\cdot) = f(a) + R_{A,(a,a^*)}(\cdot)$.

We may now connect the Fitzpatrick function of infinite order and its Fenchel conjugate for subdifferentials:

Theorem 9.2.24 (Fitzpatrick function of infinite order). *Let $f \colon X \to (-\infty, +\infty]$ be convex, lsc, and proper. Then for every $(x, x^*) \in X \times X^*$,*

$$\langle x, x^* \rangle \leq F_{\partial f, 2}(x, x^*) \leq F_{\partial f, 3}(x, x^*) \leq \cdots \leq F_{\partial f, n}(x, x^*) \to F_{\partial f, \infty}(x, x^*)$$
$$= f(x) + f^*(x^*) \tag{9.2.27}$$

and

$$F^*_{\partial f, 2}(x^*, x) \geq F^*_{\partial f, 3}(x^*, x) \geq \cdots \geq F^*_{\partial f, n}(x^*, x) \to h(x^*, x) \geq F^*_{\partial f, \infty}(x^*, x)$$
$$= f^*(x^*) + f(x), \tag{9.2.28}$$

where $h \colon X^* \times X \colon (-\infty, +\infty]$ is convex and

$$h(x^*, x) \geq h^{**}(x^*, x) = f^*(x^*) + f(x).$$

Proof. By Definition 9.2.18 and (9.2.12), it is clear that $(F_{\partial f, n})_{n \in \{2,3,\ldots\}}$ is an increasing sequence converging pointwise to $F_{\partial f, \infty}$. Since ∂f is maximal monotone

$\langle \cdot, \cdot \rangle \leq F_{\partial f, 2}$. Take $(x, x^*) \in X \times X^*$. Using Definitions 9.2.17 and 9.2.22 show that

$$\begin{aligned}
F_{\partial f, \infty}(x, x^*) &= \sup_{n \in \{2,3,...\}} F_{\partial f, n}(x, x^*) = \sup_{n \in \{2,3,...\}} \sup_{(a,a^*) \in \text{graph } \partial f} C_{\partial f, n, (a, a^*)}(x, x^*) \\
&= \sup_{(a,a^*) \in \text{graph } \partial f} \sup_{n \in \{2,3,...\}} C_{\partial f, n, (a, a^*)}(x, 0) + \langle a, x^* \rangle \\
&= \sup_{(a,a^*) \in \text{graph } \partial f} \langle a, x^* \rangle + R_{\partial f, (a, a^*)}(x) \\
&= \sup_{(a,a^*) \in \text{graph } \partial f} \langle a, x^* \rangle + (f(x) - f(a)) \\
&= f(x) + \sup_{a \in \text{dom } \partial f} (\langle a, x^* \rangle - f(a)) \\
&= f(x) + f^*(x^*). \quad (9.2.29)
\end{aligned}$$

This proves (9.2.27). Fenchel conjugation shows $\left(F^*_{\partial f, n}(x^*, x)\right)_{n \in \{2,3,...\}}$ is a decreasing sequence converging to $h(x^*, x)$, where $h \colon X^* \times X \colon (-\infty, +\infty]$ is a convex function such that $h(x^*, x) \geq h^{**}(x^*, x) \geq f^*(x^*) + f(x)$. Conjugating this decreasing sequence, we get $F_{\partial f, 2}(x, x^*) \leq F_{\partial f, 3}(x, x^*) \leq \cdots \leq F_{\partial f, n}(x, x^*) \leq \cdots \leq h^*(x, x^*)$. Hence $F_{\partial f, \infty}(x, x^*) = f(x) + f^*(x^*) \leq h^*(x, x^*)$. A final conjugation yields $h^{**}(x^*, x) \leq f^*(x^*) + f(x)$. □

In summary, we have now built a remarkable bridge from the Fitzpatrick function of ∂f to $f^*(x^*) + f(x)$.

Exercises and further results

9.2.1. Show that every quasiconvex function is quasimonotone.

Hint. First show $f(z) \leq f(w) \Rightarrow f^-(w, z - w) \leq 0$. Now existence of $x^* \in \partial_F f(x)$ with $\langle x^*, y - x \rangle > 0$ implies $f^-(x; y - x) > 0$ and so $f(x) < f(y)$. Deduce, for all $y^* \in \partial_F f(y)$ that $\langle y^*, y - x \rangle \leq f^-(y; y - x) \leq 0$. □

9.2.2. Let $f \colon X \to (-\infty, +\infty]$ be lsc on a Fréchet smooth space. Suppose $x \mapsto f(x) + \langle x^*, x \rangle$ is quasiconvex for every $x^* \in X^*$. Prove that f is convex.

Hint. Fix x and y. Pick x^* such that $f(x) + \langle x^*, x \rangle = f(y) + \langle x^*, y \rangle$, and appeal to quasiconvexity of $g := f + \langle x^*, \cdot \rangle$. □

Thus, requiring quasiconvexity of the sum of a quasiconvexity function with all linear functions is enough to force convexity. Compare Exercise 6.7.9.

9.2.3 (Monotonicity implies convexity). Let $f \colon X \to (-\infty, +\infty]$ be lsc on a Fréchet smooth Banach space X. If $\partial_F f$ is monotone, then f is convex.

Hint. If $\partial_F f$ is monotone then for each $x^* \in X^*$ the operator $x \to \partial_F f(x) + x^* = \partial_F(f + x^*)(x)$ is monotone, hence quasi-monotone. By Theorem 9.2.2, for each $x^* \in X^*$, the function $f + x^*$ is quasiconvex. This implies the convexity of f (Exercise 9.2.2). □

9.2 Cyclic and acyclic monotone operators

Asplund commented in [15] that 'nothing more is known about irreducible monotone mappings in this the simplest of cases' than the following:

9.2.4. Let f be a C^1-complex function on an open convex subset $D \subset \mathbb{C}$. Viewed as a real function on $\mathbb{R}^2$ the following is the case

- f is monotone if and only if it satisfies

$$\operatorname{Re} \frac{\partial f}{\partial z} \geq \left| \frac{\partial f}{\partial \bar{z}} \right|.$$

- f is a subgradient if and only if it satisfies

$$\frac{\partial f}{\partial z} \geq \left| \frac{\partial f}{\partial \bar{z}} \right|.$$

Thus, for f to be acyclic there must exist no nonconstant analytic function g with

$$\frac{\partial g}{\partial z} \geq \left| \frac{\partial g}{\partial \bar{z}} \right| \quad \text{and} \quad \operatorname{Re} \frac{\partial (f-g)}{\partial z} \geq \left| \frac{\partial (f-g)}{\partial \bar{z}} \right|.$$

9.2.5.* Complete the proof of Proposition 9.2.5 in the set-valued case.

9.2.6. For $S(x, y) := (-y, x)$ and $z := (x, y)$, consider the operator F on $\mathbb{R}^2$ given by

$$F(z) := \left(1 - \frac{1}{\log \|z\|} \right) \frac{z}{\|z\|} - \frac{1}{\log \|z\|} \frac{S(z)}{\|z\|}$$

for $\|z\| < 1, z \neq (0, 0)$ and set

$$F(0, 0) := \{z : \|z\| \leq 1\}.$$

(a) Show F is maximal monotone and that its domain is bounded (the norm of $F(z)$ tends to infinity when $\|x\|$ tends up to 1).
(b) Show that any solution to $\dot{z} \in F(z)$ revolves infinitely many times around the origin in any neighborhood of the origin.
(c) Determine if F is acyclic.

9.2.7.* We give the details of the range assertion in Theorem 9.2.3.

Lemma 9.2.25. *Let X be a Banach space and let $x_\alpha^* \in X^*$ for $\alpha \in A$ and with each r_α real. Let $f(x) := \sup\{\langle x_\alpha^*, x \rangle - r_\alpha : \alpha \in A\}$. Then* range $\partial f \subset \overline{\operatorname{conv}}^*\{x_\alpha^* : \alpha \in A\}$.

Proof. Consider the convex function g defined on X^* by

$$g(x^*) := \inf \left\{ \sum \lambda_{\alpha_i} r_{\alpha_i} : \sum \lambda_{\alpha_i} x_{\alpha_i}^* = x^*, \sum \lambda_{\alpha_i} = 1, \lambda_{\alpha_i} > 0 \right\},$$

as we range over all finite subsets of $\{(x_\alpha^*, r_\alpha) : \alpha \in A\}$. It is easy to check that $g^*|_X = f$, and $f^* = g^{**}$ viewed in $\sigma(X^*, X)$. Now when $x^* \in \partial f(x)$ we have $f(x) + f^*(x^*) = \langle x^*, x \rangle$. Since $x \in \operatorname{dom} f$ we see that $g^{**}(x^*)$ is finite and we are done since $\operatorname{dom} g^{**} \subset \overline{\operatorname{dom} g}^*$.

Alternative proof: if the conclusion fails we may find $x^* \in \partial f(x)$, $\varepsilon > 0$ and $h \in X$ such that

$$\langle x^*, h \rangle > \varepsilon + \sup_{\alpha \in A} \langle x_\alpha^*, h \rangle, \qquad (9.2.30)$$

by the Hahn–Banach theorem (4.1.7). Thus, for each $\alpha \in A$ we have

$$\langle x^*, h \rangle \geq \varepsilon + \langle x_\alpha^*, x \rangle = \varepsilon + (\langle x_\alpha^*, x + h \rangle - r_\alpha) - (\langle x_\alpha^*, x \rangle - r_\alpha)$$
$$\geq \varepsilon + \langle x_\alpha^*, x + h \rangle - r_\alpha - f(x).$$

Now $f(x)$ is finite and so supremizing over $\alpha \in A$ yields

$$\langle x^*, h \rangle \geq \varepsilon + f(x+h) - f(x),$$

in contradiction to $x^* \in \partial f(x)$. □

9.2.8. Prove the following result from [118]:

Theorem 9.2.26 (Decomposability). *Suppose we are given a continuously differentiable maximal monotone operator $T : \text{dom}\, T \subset \mathbb{R}^n \to \mathbb{R}^n$ for which $0 \in \text{int dom}\, T = \text{dom}\, T$. Then T is decomposable on $\text{dom}\, T$ if and only if $T - \nabla \mathcal{FL}_T$ is skew on $\text{dom}\, T$. In this case $\mathcal{FL}_T$ is necessarily convex.*

It is often difficult to see that $\mathcal{FL}_T$ is not convex otherwise. Try plotting the example of Section 9.2.5. An illustrative example is drawn in Figure 9.3.

9.2.9. Let $g \geq 0$ be a nonconstant and continuous real function such that either $g(x) \geq 1 = g(0)$ or $g(x) \leq 1 = g(0)$. Let

$$G(x) := \int_0^x g \quad \text{and} \quad K(x) := \int_0^x \{(1+g)/2\}^2.$$

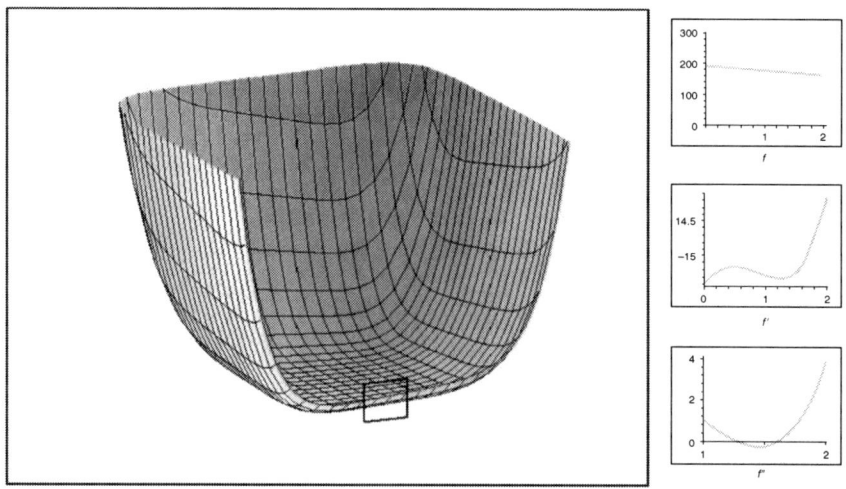

Figure 9.3 A nonconvex $\mathcal{FL}_T$.

(a) Show that $T(x, y) := (K(x) - G(y), K(y) - G(x))$ is both continuously differentiable and maximal monotone on $\mathbb{R}^2$. (b) Show that T is never decomposable on $\mathbb{R}^2$ [118]. (c) Try computing and plotting various examples in a computer algebra package.

9.3 Maximality in reflexive Banach space

We begin with:

Proposition 9.3.1. *A monotone operator T on a reflexive Banach space is maximal if and only if the mapping $T(\cdot + x) + J$ is surjective for all x in X. (Moreover, when J and J^{-1} are both single valued, a monotone mapping T is maximal if and only if $T + J$ is surjective.)*

Proof. We prove the 'if'. The 'only if' is completed in Corollary 9.3.5. Assume (w, w^*) is monotonically related to the graph of T. By hypothesis, we may solve $w^* \in T(x + w) + J(x)$. Thus $w^* = t^* + j^*$ where $t^* \in T(x + w), j^* \in J(x)$. Hence

$$0 \leq \langle w - (w + x), w^* - t^* \rangle = -\langle x, w^* - t^* \rangle = -\langle x, j^* \rangle = -\|x\|^2 \leq 0.$$

Thus, $j^* = 0, x = 0$. So $w^* \in T(w)$ and we are done. □

We now prove our central result whose proof – originally very hard and due to Rockafellar [370] – has been revisited over many years culminating in the results in [391, 398, 399, 446, 121] among others:

Theorem 9.3.2. *Let X be a reflexive space. Let T be maximal monotone and let f be closed and convex. Suppose that*

$$0 \in \operatorname{core}\{\operatorname{conv} \operatorname{dom}(T) - \operatorname{conv} \operatorname{dom}(\partial f)\}.$$

Then

(a) $\partial f + T + J$ is surjective.
(b) $\partial f + T$ is maximal monotone.
(c) ∂f is maximal monotone.

Proof. (a) As in [398, 399, 446], we consider the Fitzpatrick function $\mathcal{F}_T(x, x^*)$ and further introduce $f_J(x) := f(x) + 1/2\|x\|^2$. Let $G(x, x^*) := -f_J(x) - f_J^*(-x^*)$. Observe that

$$\mathcal{F}_T(x, x^*) \geq \langle x, x^* \rangle \geq G(x, x^*)$$

pointwise thanks to the Fenchel–Young inequality (Proposition 4.4.1(a))

$$f(x) + f^*(x^*) \geq \langle x, x^* \rangle,$$

for all $x \in X, x^* \in X^*$, along with Proposition 9.1.1. Now, the constraint qualification

$$0 \in \text{core}\{\text{conv dom}(T) - \text{conv dom}(\partial f)\}$$

assures that the sandwich theorem (9.1.4) applies to $\mathcal{F}_T \geq G$ since f_J^* is everywhere finite by Proposition 9.1.6.

Then there are $w \in X$ and $w^* \in X^*$ such that

$$\mathcal{F}_T(x, x^*) - G(z, z^*) \geq w(x^* - z^*) + w^*(x - z) \tag{9.3.1}$$

for all x, x^* and all z, z^*. In particular, for $x^* \in T(x)$ and for all z^*, z we have

$$\langle x - w, x^* - w^* \rangle + [f_J(z) + f_J^*(-z^*) + \langle z, z^* \rangle] \geq \langle w - z, w^* - z^* \rangle.$$

Now use the fact that $-w^* \in \text{dom}(\partial f_J^*)$, by Proposition 9.1.6, to deduce that $-w^* \in \partial f_J(v)$ for some z and so

$$\langle v - w, x^* - w^* \rangle + [f_J(v) + f_J^*(-w^*) + \langle v, w^* \rangle] \geq \langle w - v, w^* - w^* \rangle = 0.$$

The second term on the left is zero and so $w^* \in T(w)$ by maximality. Substitution of $x = w$ and $x^* = w^*$ in (9.3.1), and rearranging yields

$$\langle w, w^* \rangle + \{\langle -z^*, w \rangle - f_J^*(-z^*)\} + \{\langle z, -w^* \rangle - f_J(z)\} \leq 0,$$

for all z, z^*. Taking the supremum over z and z^* produces $\langle w, w^* \rangle + f_J(w) + f_J^*(-w^*) \leq 0$. This shows $-w^* \in \partial f_J(w) = \partial f(w) + J(w)$ on using the sum formula for subgradients, implicit in Proposition 9.1.6.

Thus, $0 \in (T + \partial f_J)(w)$, and since all translations of $T + \partial f$ may be used, while the (CQ) is undisturbed by translation, $(\partial f + T)(x + \cdot) + J$ is surjective which completes (a). Also $\partial f + T$ is maximal by Proposition 9.3.1 which is (b). Finally, setting $T \equiv 0$ we recover the reflexive case of the maximality for a lsc convex function. □

Recall that the *normal cone* $N_C(x)$ to a closed convex set C at a point x in C is $N_C(x) = \partial \delta_C(x)$.

Corollary 9.3.3. *The sum of a maximal monotone operator T and a normal cone N_C on a reflexive Banach space, is maximal monotone whenever the transversality condition $0 \in \text{core}[C - \text{conv dom}(T)]$ holds.*

In particular, if T is monotone and $C := \text{cl conv dom}(T)$ has nonempty interior, then for any maximal extension $\widehat{T}$ the sum $\widehat{T} + N_C$ is a 'domain preserving' maximal monotone extension of T.

Corollary 9.3.4 ([391, 399]). *The sum of two maximal monotone operators T_1 and T_2, on a reflexive Banach space, is maximal monotone whenever the transversality condition $0 \in \text{core}[\text{conv dom}(T_1) - \text{conv dom}(T_2)]$ holds.*

Proof. Theorem 9.3.2 applies to the maximal monotone product mapping $T(x, y) := (T_1(x), T_2(y))$ and the indicator function $f(x, y) = \delta_{\{x=y\}}$ of the diagonal in $X \otimes X$. Finally, check that the given transversality condition implies the needed (CQ), along the lines of Theorem 9.3.2. We obtain that $T + J_{X \otimes X} + \partial \delta_{\{x=y\}}$ is surjective. Thus, so is $T_1 + T_2 + 2J$ and we are done. □

As always in convex analysis, one may easily replace the core condition by a relativized version – with respect to the closed affine hull.

We next recall the *Rockafellar–Minty surjectivity theorem*:

Corollary 9.3.5 (Rockafellary–Minty). *For a maximal monotone operator on a reflexive Banach space,* $\mathrm{range}(T + J) = X^*$.

Proof. Let $f \equiv 0$ in Theorem 9.3.2. Alternatively, on noting that

$$\mathcal{F}_J(x, x^*) \leq \frac{\|x\|^2 + \|x^*\|^2}{2},$$

we may apply Theorem 9.3.6. □

9.3.1 The Fitzpatrick inequality

We record a very special case of Theorem 9.2.24. Namely

$$F_{\partial f}(x, x^*) \leq f(x) + f^*(x^*),$$

and note that we have exploited the beautiful inequality

$$\mathcal{F}_T(x, x^*) + f(x) + f^*(-x^*) \geq 0, \qquad \text{for all } x \in X, x^* \in X^*, \quad (9.3.2)$$

valid for *any* maximal monotone T and *any* convex function f. Also, note that $(x, x^*) \mapsto f(x) + f^*(x^*)$ is a representative function for ∂f. Correspondingly, we have the *Fitzpatrick inequality*

$$\mathcal{F}_{T_1}(x, x^*) + \mathcal{F}_{T_2}(x, -x^*) \geq 0, \qquad \text{for all } x \in X, x^* \in X^*, \quad (9.3.3)$$

valid for *any* maximal monotone T_1, T_2. Moreover, by Proposition 9.1.1,

$$\mathcal{F}_T^*(x^*, x) \geq \sup_{y^* \in T(y)} \langle x, y^* \rangle + \langle x^*, y \rangle - \mathcal{F}_T(y, y^*) = \mathcal{F}_T(x, x^*). \quad (9.3.4)$$

We clearly have an extension of (9.3.3): $\mathcal{H}_T^1(x, x^*) + \mathcal{H}_S^2(x, -x^*) \geq 0$, for any representative functions $\mathcal{H}_T^1$ and $\mathcal{H}_S^2$.

Letting $\widehat{\mathcal{F}}_S(x, x^*) := \mathcal{F}_S(x, -x^*)$, we may establish:

Theorem 9.3.6. *Let S and T be maximal monotone on a reflexive space. Suppose that $0 \in \mathrm{core}\{\mathrm{dom}(\mathcal{F}_T) - \mathrm{dom}(\widehat{\mathcal{F}}_S)\}$ as happens if $0 \in \mathrm{core}\{\mathrm{conv\,graph}(T) - \mathrm{conv\,graph}(-S)\}$. Then $0 \in \mathrm{range}(T + S)$.*

Proof. We apply the Fenchel duality theorem (4.4.18), or follow through the steps of Theorem 9.3.2. From either result one obtains $\mu \in X$, $\lambda \in X^*$ and $\beta \in \mathbb{R}$ such that

$$\mathcal{F}_T(x,x^*) - \langle x,\lambda \rangle - \langle \mu,x^* \rangle + \langle \mu,\lambda \rangle$$
$$\geq \beta \geq -\mathcal{F}_S(y,-y^*) + \langle y,\lambda \rangle - \langle \mu,y^* \rangle - \langle \mu,\lambda \rangle,$$

for all variables x, y, x^*, y^*. Hence for $x^* \in T(x)$ and $-y^* \in S(y)$ we obtain

$$\langle x - \mu, x^* - \lambda \rangle \geq \beta \geq \langle y - \mu, y^* + \lambda \rangle.$$

If $\beta \leq 0$, we derive that $-\lambda^* \in S(\mu)$ and so $\beta = 0$; consequently, $\lambda \in T(\mu)$ and since $0 \in (T+S)(\mu)$ we are done. If $\beta \geq 0$ we argue first with T. □

Note that the graph condition in Theorem 9.3.6 is formally more exacting than the domain condition as shown by conv graph(J_{ℓ_2}) which is the diagonal in $\ell_2 \otimes \ell_2 =$ dom$(F_{J_{\ell_2}})$, indeed $\mathcal{F}_{J_{\ell_2}}(x,x^*) = \frac{1}{4}\|x + x^*\|^2$. More interestingly, Zălinescu [447] has adapted this argument to extend results like those in [395] in the reflexive case.

9.3.2 Monotone variational inequalities

We say that T is *coercive* on C if $\inf_{y^* \in (T+N_C)(y)} \langle y, y^* \rangle / \|y\| \to \infty$ as $y \in C$ goes to infinity in norm, with the convention that $\inf \emptyset = +\infty$. A *variational inequality* requests a solution $y \in C$ and $y^* \in T(y)$ to

$$\langle y^*, x - y \rangle \geq 0 \qquad \text{for all } x \in C.$$

Equivalently this requires us to solve the set inclusion $0 \in T(y) + N_C(y)$. In Hilbert space this is also equivalent to finding a zero of the *normal mapping* $T_C(x) := T(P_C(x)) + (I - P_C)(x)$.

We denote the variational inequality by $V(T; C)$. (For general variational inequalities with T merely upper-semicontinuous, proving solutions to V(T,C) is equivalent to establishing Brouwer's theorem.)

Corollary 9.3.7. *Suppose T is maximal monotone on a reflexive Banach space and is coercive on the closed convex set C. Suppose also that $0 \in \text{core}(C - \text{conv dom}(T))$. Then $V(T, C)$ has a solution.*

Proof. Let $f := \delta_C$, the indicator function. For $n = 1, 2, 3, \cdots$, let $T_n := T + J/n$. We solve

$$0 \in (T_n + \partial \delta_C)(y_n) = (T + \partial \delta_C) + \frac{1}{n}J(y_n) \qquad (9.3.5)$$

and take limits as n goes to infinity. More precisely, we observe that using our key Theorem 9.3.2, we find y_n in C, and $y_n^* \in (T + \partial \delta_C)(y_n), j_n^* \in J(y_n)/n$ with $y_n^* = -j_n^*$. Then

$$\langle y_n^*, y_n \rangle = -\frac{1}{n}\langle j_n^*, y_n \rangle = -\frac{1}{n}\|y_n\|^2 \leq 0$$

so coercivity of $T + \partial \delta_C$ implies that $\|y_n\|$ remains bounded and so $j_n^* \to 0$. On taking a subsequence we may assume $y_n \to_w y$. Since $T + \partial \delta_C$ is maximal monotone (again by Theorem 9.3.2), it is demiclosed [121]. It follows that $0 \in (T + \partial \delta_C)(y) = T(y) + N_C(y)$ as required. □

A more careful argument requires only that for some $\bar{c} \in C$

$$\inf_{y^* \in T(y)} \langle y - \bar{c}, y^* \rangle / \|y\| \to \infty$$

as $\|y\| \to \infty$, $y \in C$. Letting $C = X$ in Corollary 9.3.7 we deduce:

Corollary 9.3.8. *Every coercive maximal monotone operator on a Banach space is surjective if (and only if) the space is reflexive.*

Proof. To complete the proof we recall that, by James' theorem (4.1.27), surjectivity of J is equivalent to reflexivity of the corresponding space. □

It also seems worth noting that the techniques of §9.4.2 are at heart all techniques for variational equalities. We conclude this section by noting another convex approach to the *affine* monotone variational inequality (complementarity problem) on a closed convex cone S in a reflexive space. We consider the abstract quadratic program

$$0 \leq \mu := \inf\{\langle L(x) - q, x \rangle : Lx \geq_{S^+} q, x \geq_S 0\}, \qquad (9.3.6)$$

which has a convex objective function. Suppose that (9.3.6) satisfies a constraint qualification as happens if either $L(S) + S^+ = X^*$ or if X is finite-dimensional and S is polyhedral. Then there is $y \in S^{++} = S$ with

$$\mu \leq \langle L(x) - q, x \rangle + \langle Lx - q, y \rangle, \qquad (9.3.7)$$

for all $x \in S$. Letting $x := y$ shows $\mu = 0$, and we have approximate solutions to the affine complementarity problem. Moreover, when L is coercive on S or in the polyhedral case, (9.3.6) is attained and we have produced a solution to the problem.

Exercises and further results

9.3.1. Suppose H is a Hilbert space and $T : H \to H$ is a maximal monotone mapping. Show that $(I + T)^{-1}$ has domain H and is nonexpansive. In particular, Rockafellar's theorem (6.1.5) ensures this applies when $T = \partial f$ where $f : E \to \mathbb{R}$ is convex.

Hint. Corollary 9.3.5 implies that $\text{dom}(I + T)^{-1} = H$. The nonexpansive assertion follows as in Exercise 3.5.8. □

9.3.2 (Elliptic partial differential equations [195, 131, 275]).[†] Much early impetus for the study of maximal monotone operators came out of partial differential equations and takes place within the confines of Sobolev space – and so we content ourselves with an example of what is possible.

As an application of their study of existence of eigenvectors of second-order nonlinear elliptic equations in $L_2(\Omega)$, the authors of [275] assume that $\Omega \subset \mathbb{R}^n$, $(n > 1)$ is a bounded open set with boundary belonging to $C^{2,\alpha}$ for some $\alpha > 0$. They assume that one has functions $|a_i(x, u)| \leq \nu$ $(1 \leq i \leq n)$ and $|a_0(x, u)| \leq \nu|u| + a(x)$ for some $a \in L_2(\Omega)$ and $\nu > 0$; where all a_i are measurable in x and continuous in u (a.e. x). They then consider the normalized eigenvalue problem

$$\Delta u + \lambda \left\{ \sum_{i=1}^{n} a_i(x, u) \frac{\partial u}{\partial x_i} + a_0(x, u) \right\} = 0, \quad x \in \Omega, \tag{9.3.8}$$

$$u(x) = 0 \quad x \in \text{bdry } \Omega$$

$$\|u\|_2 = 1$$

where $\Delta u = -\nabla^2 u = -\sum_{i=1}^{n} \frac{\partial^2 u}{\partial x_i^2}$ is the classical *Laplacian*. To make this accessible to Sobolev theory, a weak solution is requested to (9.3.8) for $0 < \lambda \leq 1$ when $u \in W^{2,2}(\Omega) \cap W_0^{1,2}(\Omega)$. In this setting, a solution of

$$\Delta u + \tau u = f(x)$$

for all $\tau > 0$ and all $f \in L_2$ (and with $\|u\|_2 = 1$) is assured. Minty's surjectivity condition (Proposition 9.3.1) implies $T := \Delta$ is linear and maximal monotone on $L_2(\Omega)$ with domain $W^{2,2}(\Omega) \cap W_0^{1,2}(\Omega)$. Of course, one must first check monotonicity of Δ using integration by parts in the form

$$\int_\Omega \langle v, \Delta u \rangle = \int_\Omega \langle \nabla v, \nabla u \rangle,$$

for all $v \in W^{-1,2}(\Omega)$, $u \in C_0^\infty(\Omega) \subset W_0^{1,2}(\Omega)$. One is now able to provide a Fredholm alternative type result for (9.3.8) [275, Theorem 10]. In like-fashion one can make sense of the assertion that for $2 \leq p < \infty$ the p-Laplacian Δ_p is maximal monotone: $\Delta_p u$ is given by

$$\Delta_p u := -\text{div}(|\nabla u|^{p-2} \nabla u) \in W^{-1,q}(\Omega)$$

for $u \in W^{1,p}(\Omega)$ with $1/p + 1/q = 1$.

We can also improve Corollary 9.1.15 in the reflexive setting.

9.3.3 (Range closure in reflexive space). We will again improve the following result significantly in Section 9.7, but the proof technique merits mention.

Theorem 9.3.9 ([391]). *Suppose T is maximal monotone on a reflexive Banach space. Then norm closures and interiors of* dom(T) *and* range (T) *are convex.*

Proof. Without loss of generality, we assume 0 is in the closure of conv dom(T). Fix $y \in \text{dom}(T)$, $y^* \in T(y)$. Theorem 9.3.5 applied to T/n solves $w_n^*/n + j_n^* = 0$ with $w_n^* \in T(w_n)$, $j_n^* \in J(w_n)$, for integer $n > 0$. By monotonicity

$$\frac{1}{n}\langle y^*, y - w_n \rangle \geq \frac{1}{n}\langle w_n^*, y - w_n \rangle = \|w_n\|^2 - \langle j_n^*, y \rangle$$

where $\|w_n\|^2 = \|j_n^*\|^2 = \langle j_n^*, w_n \rangle$ and $w_n \in \text{dom}(T)$. We deduce $\sup_n \|w_n\| < \infty$. Thus, (j_n^*) has a weak cluster point j^*. In particular, denoting $D := \text{dom}(T)$

$$d_D^2(0) \le \liminf_{n\to\infty} \|w_n\|^2 \le \inf_{y\in D} \langle j^*, y \rangle = \inf_{y\in \text{conv}\, D} \langle j^*, y \rangle \le \|j^*\| d_{\text{conv}\, D}(0) = 0.$$

We have actually shown that cl conv dom$(T) \subset$ cl dom(T) and so cl dom(T) is convex as required.

Since range$(T) = \text{dom}(T^{-1})$ and X^* is also reflexive we are done. □

In a nonreflexive space, Theorem 9.4.3 of the next section applied similarly proves that $\overline{\text{dom}\, T}$ is convex. Consequently, if as when T is type (ED) – see [393] – $\overline{\text{dom}(\overline{T})} \cap \subset \overline{\text{dom}(T)}$, the latter is convex. Dually, it is known that every locally maximal operator T – and so every dense type operator – has range(T) convex [351, Proposition 4.2].

9.3.4 (Operators with full domain). Suppose T is maximal monotone on a reflexive Banach space X and is locally bounded at each point of cl dom(T). Show that dom$(T) = X$.

Hint. First observe that dom(T) must be closed and so convex. By the Bishop–Phelps theorem (Exercise 4.3.6), there is some boundary point $\bar{x} \in \text{dom}(T)$ with a nonzero support functional $\bar{x}^*$. Then $T(\bar{x}) + [0, \infty)\bar{x}^*$ is monotonically related to the graph of T. By maximality $T(\bar{x}) + [0, \infty)\bar{x}^* = T(\bar{x})$ which is then nonempty and (linearly) unbounded. □

9.4 Further applications

9.4.1 Extensions to nonreflexive space

We let $\overline{T}$ denote the *monotone closure* of T in $X^{**} \times X^*$. That is, $x^* \in \overline{T}(x^{**})$ when

$$\inf_{y^* \in T(y)} \langle x^* - y^*, x^{**} - y \rangle \ge 0.$$

Recall that T is *type (NI)* if

$$\inf_{y^* \in T(y)} \langle x^* - y^*, x^{**} - y \rangle \le 0$$

for all $x^{**} \in X^{**}$ and $x^* \in X^*$, see [391, 393].

We say T is of *dense type* or *type (D)* if every pair (x^*, x^{**}) in the graph of the monotone closure of T is the limit of a net (x_α^*, x_α) with $x_\alpha^* \in T(x_\alpha)$ with $x_\alpha \rightharpoonup_* x^{**}$, $x_\alpha^* \to x^*$ and $\sup_\alpha \|x_\alpha\| < \infty$, and we write $x^* \in T_1(x^{**})$ [240, 391, 393].

Clearly, $T_1(x) \subset \overline{T}(x)$; so every dense type operator is of type (NI). We denote $\mathcal{F}_T(G, x^*) := \mathcal{P}_T^*(x^*, G)$, viewed as a mapping on $X^{**} \times X^*$, and make the following connection with $\mathcal{F}_T$.

Proposition 9.4.1. *Let T be maximal monotone on a Banach space X. Then*

$$\overline{\mathcal{F}}_T(G,x^*) = \sup_{y^* \in T(y)} \langle y^*, G \rangle + \langle y, x^* \rangle - \langle y, y^* \rangle \geq \langle x^*, G \rangle$$

with equality if and only if $x^* \in \overline{T}(G)$. In particular, $\overline{\mathcal{F}}_T|_{X \times X^*} = \mathcal{F}_T$.

Moreover, T is type (NI) if and only if $\overline{\mathcal{F}}_T$ is a representative function for the operator $\overline{T}$.

Proof. These are left for the reader. Such results are elaborated in Section 9.7. □

Proposition 9.4.2 (Gossez). *The subgradient of every closed convex function f on a Banach space is of dense type. Indeed*

$$\overline{\partial f} = (\partial f)_1 = (\partial f^*)^{-1}.$$

Proof. For any closed convex f we have $(\partial f^*)^{-1} \subset \overline{\partial f}$, while – with a little effort – Golstine's theorem (Exercise 4.1.13) shows that $(\partial f^*)^{-1} \subset \partial f_1$. □

A fairly satisfactory extension of Theorem 9.3.2 is:

Corollary 9.4.3. *If T is type (NI) then*

$$\operatorname{range}(\overline{T} + \partial f^{**} + J^{**}) = X^*.$$

Proof. Follow the steps of Theorem 9.3.2 or Theorem 9.3.6 using $\mathcal{P}_T$ and $f_J + f_J^*$ as the functions in the Fitzpatrick inequality. □

In the case that T is dense type this result originates with Gossez (see [241, 351]). These dense type operators are still poorly understood. We can similarly derive – in the notation of the previous section – that $V(\overline{T}, \overline{C}^*)$ will have solution when the (CQ) holds and T is type (NI). We refer to [241, 217, 218] for more detailed results regarding coercive operators in nonreflexive space.

The Rockafellar–Minty surjectivity theorem (9.3.5) has a modest extension: for an maximal monotone N of type (NI) that $\operatorname{range}(\overline{N} + J^{**}) = X^*$ which implies that $\overline{\overline{N}} = \overline{N}$; and so that $\overline{N}$ is maximal as a monotone mapping from X^{**} to X^*.

The next result will again be revisited later in the chapter.

Theorem 9.4.4 (Fitzpatrick–Phelps [217]). *Every locally maximal monotone operator on a Banach space has cl range T convex.*

Proof. We suppose not and then may suppose by homothety that there are $\pm x^*$ in cl range T of unit-norm but with midpoint $0 \notin$ cl range T.

We build the equivalent dual ball $B' := \operatorname{conv}\{\pm 2x^*, \alpha B_X^*\}$ where $0 < \alpha < 1/2$ is chosen with $(\operatorname{range} T) \cap 2\alpha B_X^* = \emptyset$. We consider $\widehat{T}$ extending $T \cap B'$ as in Proposition 9.1.14, so that

$$\operatorname{range} \widehat{T} \subset \operatorname{cl conv}\{R(T) \cap B'\} \text{ and } \operatorname{range} \widehat{T} \setminus \operatorname{range} T \subset \operatorname{bd} B'.$$

9.4 Further applications

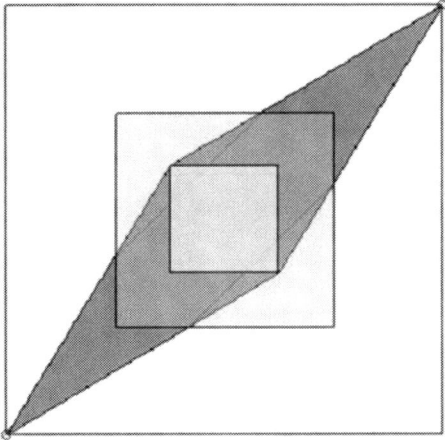

Figure 9.4 The ball constructed in Theorem 9.4.4.

It follows that

$$\text{range } \widehat{T} \subset (R(T) \cap B') \bigcup (\text{cl conv}\{R(T) \cap B'\} \cap \text{bd } B').$$

Hence range $\widehat{T}$ is weak*-disconnected. As $\widehat{T}$ is a weak*-cusco it has a weak*-connected range which contradicts the construction. □

A dual argument shows that type (VFP) mappings have cl dom(T) convex.

9.4.2 Local maximality revisited

We are also in a position to show why in a reflexive setting all maximal monotone operators are locally maximal, that is type (FP). We start with the following contrapositive whose simple proof is in [351].

Proposition 9.4.5. *Let T be a monotone operator on a Banach space X. Then T is locally maximal monotone if and only if every weak*-compact and convex set C^* in X^* with $R(T) \cap \text{int } C^* \neq \emptyset$ is such that if $x^* \in \text{int } C^*$ (in norm) but $x^* \notin T(x)$ there is $z^* \in T(z) \cap C^*$ with $0 > \langle z^* - x^*, z - x \rangle$.*

It is obvious that every maximal monotone operator on a reflexive space is type (D), and it is known that type (D) implies type (FP); see [393, Theorem 17]. A direct proof follows.

Theorem 9.4.6 (Fitzpatrick–Phelps). *Every maximal monotone operator on a reflexive space is locally maximal.*

Proof. Since X is reflexive, $M := (T^{-1} + \partial \delta_{C^*})^{-1}$ is maximal, by Theorem 9.3.2(b), while $x^* \in \text{int } C^* \setminus M(x)$. Since M is maximal monotone we can find $z \in \text{dom } T$,

$z^* \in C^*, z^* \in T(z)$ and $u \in N_{C^*}(z^*)$ such that $z + u \in (T + \partial \delta_{C^*})^{-1}(z^*)$ with

$$0 > \langle z^* - x^*, z - x \rangle + \langle z^* - x^*, u \rangle.$$

As the second term on the right is nonnegative, we are done. □

A useful reformulation of the argument in Theorem 9.4.6 is given next.

Proposition 9.4.7 (Fitzpatrick–Phelps). *Let X be a Banach space. A maximal monotone operator T is locally maximal monotone if*

$$M(T, C^*) := \left(T^{-1} + \partial \delta_{C^*}\right)^{-1}$$

is maximal monotone on X whenever C^* is convex and weak*-compact with $R(T) \cap \operatorname{int} C^* \neq \emptyset$.

We shall call an operator satisfying these hypotheses *strongly locally maximal*. While every maximal monotone operator on a reflexive space is clearly strongly locally maximal, not every convex subgradient is – as Theorem 9.5.2 will show. However we can still use Proposition 9.4.5 to prove:

Theorem 9.4.8 (Simons). *The subgradient of every closed convex function f on a Banach space is locally maximal monotone.*

Proof. Fix $x^* \in \operatorname{int} C^* \setminus \partial g(x)$ as in the hypotheses of Proposition 9.4.5. Then Corollary 9.1.5 and Theorem 9.2.1 and combine to show that $M := \partial g^* + \partial \delta_{C^*} = \partial (g^* + \delta_{C^*})$ is maximal monotone on X^*, since $\operatorname{dom} g^* \cap \operatorname{int} C^* \neq \emptyset$. Now $x \notin M(x^*)$ (since $x^* \in \operatorname{int} C^*$) so we deduce the existence of $w^{**} = z^{**} + u^{**}, z^{**} \in \partial g^*(z^*), u^{**} \in \partial \delta_C(z^*) = N_C(z^*)$ with

$$0 > \langle z^* - x^*, z^{**} - x \rangle + \langle z^* - x^*, u^{**} \rangle \geq \langle z^* - x^*, z^{**} - x \rangle.$$

By Proposition 9.4.2 we may select z_α^* and z_α with $z_\alpha^* \in \partial g(z_\alpha)$, $z_\alpha^* \to z^* \in C^*$, $z_\alpha \to_{w^*} z^{**}$ and $\sup_\alpha \|z_\alpha\| < \infty$. In consequence,

$$\limsup_\alpha \langle z_\alpha^* - x^*, z_\alpha - x \rangle < 0.$$

Now the observation that $z_\alpha^* \in C^*$ for large α, and an appeal to Proposition 9.4.5 finishes the proof. □

One may similarly prove Simons' result that dense type operators are locally maximal by applying a variant of Corollary 9.4.3 with ∂g^* replaced by $\overline{T^{-1}}$. The corresponding results for type (VFP) appear usually to be easier. For example, T is type (VFP) if $T + N_C$ is maximal monotone for all bounded closed convex sets with $\operatorname{dom} T \cap \operatorname{int} C \neq \emptyset$. In consequence, subgradients and reflexive maximal monotones are type (VFP).

9.4.3 The composition formula

Another very useful foundational result is:

Theorem 9.4.9 ([446, Theorem 6]). *Suppose X and Y are Banach spaces with X reflexive, that T is a maximal monotone operator on Y, and that $A\colon X \mapsto Y$, is a bounded linear mapping. Then $T_A := A^* \circ T \circ A$ is maximal monotone on X whenever $0 \in \operatorname{core}(\operatorname{range}(A) + \operatorname{conv}\operatorname{dom} T)$.*

Proof. Monotonicity is clear. To obtain maximality, we consider the Fitzpatrick inequality (9.3.3) to write

$$f(x,x^*) + g(x,x^*) \geq 0,$$

where

$$f(x,x^*) := \inf\{\mathcal{F}_T(Ax, y^*) : A^*y^* = x^*\}, \quad g(x,x^*) := \frac{1}{2}\|x\|^2 + \frac{1}{2}\|x^*\|^2,$$

and apply Fenchel's duality theorem (4.4.18) – or use the sandwich theorem directly – to deduce the existence of $\bar{x} \in X, \bar{x}^* \in X^*$ with

$$f^*(\bar{x}^*, \bar{x}) + g^*(\bar{x}^*, \bar{x}) \leq 0. \tag{9.4.1}$$

Semicontinuity of f is not needed since g is continuous throughout.

Also, the constraint qualification implies that the condition used in [121, Thm 4.4.3] and in [343, Proposition 13] holds. Thus, applying Exercise 9.4.3 we have for some $\bar{y}^*$ with $A^*\bar{y}^* = \bar{x}^*$:

$$f^*(\bar{x}^*, \bar{x}) = \inf\{\mathcal{F}_T^*(A\bar{x}, y^*) : A^*y^* = \bar{x}^*\} = \min\{\mathcal{F}_T^*(y^*, A\bar{x}) : A^*y^* = \bar{x}^*\}$$
$$= \mathcal{F}_T^*(\bar{y}^*, A\bar{x}) \tag{9.4.2}$$
$$\geq \mathcal{F}_T(A\bar{x}, \bar{y}^*),$$

where the last inequality follows from (9.3.4). Moreover,

$$g^*(\bar{x}^*, \bar{x}) = \frac{1}{2}\|\bar{x}\|^2 + \frac{1}{2}\|A^*\bar{y}^*\|^2.$$

Thus, (9.4.1) implies that

$$\left\{\mathcal{F}_T(A\bar{x}, \bar{y}^*) - \langle \bar{y}^*, A\bar{x} \rangle\right\} + \left\{\frac{1}{2}\|\bar{x}\|^2 + \frac{1}{2}\|A^*\bar{y}^*\|^2 + \langle \bar{y}^*, A\bar{x} \rangle\right\} \leq 0,$$

and we conclude that $\bar{y}^* \in T(A\bar{x})$ and $-\bar{x}^* := -A^*\bar{y}^* \in J_X(\bar{x})$ since both bracketed terms are nonnegative. Hence, $0 \in J_X(\bar{x}) + T_A(\bar{x})$.

In the same way if we start with

$$f(x,x^*) := \inf\{\mathcal{F}_T(Ax, y^*) : A^*y^* = x^* + x_0^*\}$$
$$g(x,x^*) := \frac{1}{2}\|x\|^2 + \frac{1}{2}\|x^*\|^2 - \langle x, x_0^*\rangle$$

we deduce $x_0^* \in J_X(\bar{x}) + T_A(\bar{x})$. This applies to all domain translations of T. As in Theorem 9.3.2, this is sufficient to conclude T_A is maximal. □

Note that only the domain space needs to be reflexive. Application of Theorem 9.4.9 to $T(x, y) := (T_1(x), T_2(y))$, and $A(x) := (x, x)$ yields $T_A(x) = T_1(x) + T_2(x)$ and recovers Theorem 9.3.2. With a little more effort the reader can discover how to likewise embed Theorem 9.4.9 in Theorem 9.3.2. Alternatively, one may combine Theorem 9.3.2 and this result in one. Again, it is relatively easy to relativize this result to the intrinsic core; this is especially useful in finite dimensions, see [24]. A recent paper [125] observes that the result remains true when one assumes only that

$$\{(A^* y^*, Ax, r) : \mathcal{F}_T^*(Ax, y^*) \leq r\}$$

satisfies certain relative closure conditions.

An important case of Theorem 9.4.9 is the case of a reflexive injection.

Corollary 9.4.10. *Let T be maximal monotone on a Banach space Y. Let ι denote the injection of a reflexive subspace $Z \subset Y$ into Y. Then $T_Z := \iota^* \circ T \circ \iota$ is maximal monotone on Z whenever $0 \in \mathrm{core}(Z + \mathrm{conv\ dom}\ T)$. In particular, if $0 \in \mathrm{core}(\mathrm{conv\ dom}\ T)$ then T_Z is maximal for each reflexive subspace Z.*

In this case also [125] implies the result remains true under more general closure hypotheses.

Exercises and further results

9.4.1.* Confirm that Corollary 9.4.10 is indeed a special case of Theorem 9.4.9.

It is possible to construct a maximal monotone operator on ℓ_2 such that T_Z is 'frequently' not maximal but it is not clear whether there must be many subspaces for which it is maximal even when core conv dom T is empty.

9.4.2. Show that the composition theorem (9.4.9) can be embedded in the sum theorem (9.3.2) and conversely.

9.4.3 (Symmetric conjugate).* Let X, Y, U and V be Banach spaces. Let $A: X \to Y$ and $B: U \to V$ be bounded linear operators and let G be a lsc proper convex function on $Y \times V$. Define $F(x, v) := \inf\{G(Ax, u) : Bu = v, u \in U\}$. Show as in [343, Proposition 13] that

$$F^*(x^*, v^*) = \min\{G^*(y^*, B^* v^*) : A^*(y^*) = x^*, y^* \in Y^*\} \quad (9.4.3)$$

whenever $\mathbb{R}_+ \pi_1(\mathrm{dom}\ G) - R(A) = Y$.

Hint. Apply the standard formula for the conjugate of composition of a closed convex function with a bounded linear mapping ([368], [121, Theorem 4.4.3] which is actually the case of (9.4.3) in which B, U, V are trivial). □

9.4.4.* Prove the equation within (9.4.2).

9.5 Limiting examples and constructions

It is unknown outside reflexive space whether cl dom(T) must always be convex for a maximal monotone operator, though the assumption of reflexivity in Theorem 9.3.9 may be relaxed to requiring $R(T+J)$ is boundedly weak*-dense – as an examination of the proof will show. Moreover, in the final section of the chapter we shall show it is convex for all operators of type (NI).

We do however have the following result:

Theorem 9.5.1 ([85]). *The following are equivalent for a Banach space X.*

(a) *X is reflexive;*
(b) *int range(∂f) is convex for each coercive lsc convex function f on X;*
(c) *int range(T) is convex for each coercive maximal monotone T.*

Proof. Suppose X is nonreflexive and $p \in X$ with $\|p\| = 5$ and $p^* \in Jp$ where J is the duality map. Define

$$f(x) := \max\left\{\frac{1}{2}\|x\|^2, \|x-p\| - 12 + \langle p^*, x\rangle, \|x+p\| - 12 - \langle p^*, x\rangle\right\}$$

for $x \in X$. By the max formula (4.1.10), we have, for $x \in B_X$,

$$\partial f(p) = B_{X^*} + p^*, \quad \partial f(-p) = B_{X^*} - p^*, \quad \partial f(x) = Jx \qquad (9.5.1)$$

using inequalities like $\|p-p\| - 12 + \langle p^*, p\rangle = 13 > \frac{25}{2} = \frac{1}{2}\|p\|^2$.

Moreover, $f(0) = 0$ and $f(x) > \frac{1}{2}\|x\|$ for $\|x\| > 1$, thus $\|x^*\| > \frac{1}{2}$ if $x^* \in \partial f(x)$ and $\|x\| > 1$. Combining this with (9.5.1) shows

$$\text{range}(\partial f) \cap \frac{1}{2} B_{X^*} = \text{range}(J) \cap \frac{1}{2} B_{X^*}.$$

Let U_{X^*} denote the open unit ball in X^*. Now James' theorem (4.1.27) gives us points $x^* \in \frac{1}{2} U_{X^*} \setminus \text{range}(J)$, thus $U_{X^*} \setminus \text{range}(\partial f) \neq \emptyset$. However, from (9.5.1)

$$U_{X^*} \subset \text{conv}((p^* + U_{X^*}) \cup (-p^* + U_{X^*})) \subset \text{conv int range}(\partial f)$$

so range(∂f) has nonconvex interior. This shows that (b) implies (a) while (c) implies (b) is clear. Finally (a) implies (c) follows from Theorem 9.3.9. □

Observe the distinct role of convexity in each direction the proof of (a) ⇔ (c). It is most often the case that one uses the same logic to establish any result of the form 'Property P holds for all maximal monotone operators if and only if X is a Banach space with property Q'. Another example is 'Every (maximal) monotone operator T on a Banach space X is bounded on bounded subsets of int dom T if and only if X is finite-dimensional'. (See [85] for this and other like results.)

Exercises and further results

9.5.1. The easiest explicit example, due to Fitzpatrick and Phelps (see [85]), lies in the space c_0 of null sequences endowed with the supremum norm. One may use

$$f(x) := \|x - e_1\|_\infty + \|x + e_1\|_\infty \tag{9.5.2}$$

where e_1 is first unit vector. Then

$$\text{int range}(\partial f) = \{U_{\ell_1} + e_1\} \cup \{U_{\ell_1} - e_1\}$$
$$\text{cl int range}(\partial f) = \{B_{\ell_1} + e_1\} \cup \{B_{\ell_1} - e_1\}$$

both of which are far from convex. It is instructive to compute the closure of the range of the subgradient.

9.5.2.** Gossez [240] produces a coercive maximal monotone operator with full domain whose range has a nonconvex closure, see also Example 9.5.3. It is of the form $2^{-n} J_{\ell_1} + S$ for some $n > 0$ and sufficiently large.

The continuous linear map $S : \ell_1 \to \ell_\infty$ is given by

$$(Sx)_n := -\sum_{k<n} x_k + \sum_{k>n} x_k, \quad \text{for all } x = (x_k) \in \ell_1, n \in N.$$

It is called the *Gossez operator*. We record that $\mp S : \ell_1 \mapsto \ell_\infty$ is a skew bounded linear operator, for which S^* is not monotone but $-S^*$ is. Hence, $-S$ is both of dense type and locally maximal monotone (also called FP) while S is in neither class, [391, 33]. Cognate linear examples are given in Exercises 9.5.6 and 9.7.16. As made precise in Example 9.5.3 and [33] such linear examples 'only occur' in spaces containing ℓ_1.

Relatedly, let ι denote the injection of ℓ_1 into ℓ_∞. Then for small positive ε, the mapping $S_\varepsilon := \varepsilon\iota + S$ is a coercive maximal monotone operators for which $\overline{S_\varepsilon}$ fails to be coercive, see also [241].

9.5.3 (Some further related results).** Somewhat more abstractly, one can show that if the underlying space X is *rugged*, meaning that cl span range$(J - J) = X^*$, then the following are equivalent whenever T is bounded linear and maximal monotone, see [33]:

(i) T is of dense type.
(ii) cl range$(T + \lambda J) = X^*$, for all $\lambda > 0$.
(iii) cl range$(T + \lambda J)$ is convex, for all $\lambda > 0$.
(iv) $T + \lambda J$ is locally maximal monotone, for all $\lambda > 0$.

It actually suffices that (ii)–(iv) hold for a sequence $\lambda_n \downarrow 0$. The equivalence of (i)–(iv) thus holds for the following rugged spaces: c_0, c, ℓ_1, ℓ_∞, $L_1[0, 1]$, $L_\infty[0, 1]$, $C[0, 1]$. In cases like c_0, or $C[0, 1]$ which contain no complemented copy of ℓ_1, a maximal monotone bounded linear T is always of dense type [33].

In particular, S in Example 9.5.2 is necessarily not of dense type, and so on. Also, one may use a smooth renorming of ℓ_1. This means $T + \lambda J$ is single-valued, demicontinuous.

9.5 Limiting examples and constructions

Fittingly, we finish this tour of counterexamples with another result due implicitly to Simon Fitzpatrick. It again uses convexity twice.

Theorem 9.5.2 (Fitzpatrick–Phelps [215]). *The following are equivalent for a Banach space X.*

(a) X is reflexive;
(b) ∂f is strongly locally maximal for each continuous convex function f on X;
(c) each maximal monotone mapping T on X is strongly locally maximal.

Proof. (a)$\Rightarrow$(c) was proven in Theorem 9.4.6 while (c)$\Rightarrow$(b) follows from Theorem 9.2.1. We prove (b)$\Rightarrow$(a) by contradiction and James theorem (4.1.27). Select $x^* \in X^*$ such that $\|x^*\| = 1$ but $|\langle x^*, x\rangle| < 1$ whenever $\|x\| \le 1$, and define $f(x) := |\langle x^*, x\rangle|$. Let $T := \partial f$ and $C^* := B_{X^*}$. Then $\mathrm{dom}(\partial f^* + N_{C^*}) = \{tx^* : |t| < 1\}$ while for $|t| < 1$, $(\partial f^* + N_{C^*})(tx^*) = \{0\}$. Thus, the graph of $M = (\partial f^* + N_{C^*})^{-1}$ is the set $(-x^*, x^*), \times\{0\} \subset X \times X^*$, which is monotone but not closed and hence not the graph of a maximal monotone operator. $\square$

Note that we have exhibited a case of two convex functions f and $g := \delta^*_{B_{X^*}}$ on a Banach space such that $(\partial f^* + \partial g^*)^{-1}$ is maximal as a monotone mapping on X^{**} but not as a mapping restricted to X. We finish this cross-section of limiting examples with a somewhat counter-intuitive result.

9.5.1 Subgradients need not be norm×weak* closed

The *bounded weak*-topology* (bw*) on the dual of a normed space X can be defined in many ways. It is the topology of uniform convergence on norm-compact subsets of X and so is locally convex. Equivalently, it is the strongest topology agreeing with the weak*-topology on all weak*-compact subsets of X^* [260, pp. 151–154]. A normed space is complete if and only if every bw*-continuous linear functional is weak*-continuous (and so lies in X) [260, Corollary. 1, p. 154]. From this it follows immediately that the Banach–Dieudonné theorem proven directly in Exercise 4.4.26 holds. When the space is reflexive we also refer to the *bounded-weak* or *bw-topology*.

Example 9.5.3 ([83]). *A proper lsc convex function f on a separable Hilbert space E such that the graph of the maximal monotone operator ∂f is not norm×bw closed.*

Proof. Let $E := \ell_2(\mathbb{N})$. To make things clearer we will keep E^* and E separate. Define

$$e_{p,m} := \frac{1}{p}(e_p + e_{p^m}), \quad e^*_{p,m} := e^*_p + (p-1)e^*_{p^m}$$

for $m, p, r, s \in \mathbb{N}$, p prime and $m \ge 2$. Here e_n and e^*_n denote the unit vectors in E and E^* respectively.

Then we have

$$\langle e^*_{p,m}, e_{p',m'}\rangle = \begin{cases} 0 & \text{if } p \neq p', \\ 1/p & \text{if } p = p', m \neq m', \\ 1 & \text{if } p = p', m = m'. \end{cases}$$

Further, for $x \in E$ define

$$f(x) := \max(\langle e^*_1, x\rangle + 1, \sup\{\langle e^*_{p,m}, x\rangle : p \text{ prime}, m \geq 2\})$$

so f is a proper lsc convex function on E. Then $f(0) = f(e_{p,m}) = 1, f(-e_1) = 0$ and $f(x) \geq \langle e^*_{p,m}, x\rangle$ for all $x \in E$ and p prime, $m \geq 2$, which implies $e^*_{p,m} \in \partial f(e_{p,m})$. In fact,

$$f(x) - f(e_{p,m}) = f(x) - 1 \geq \langle e^*_{p,m}, x\rangle - 1 = \langle e^*_{p,m}, x - e_{p,m}\rangle \quad \text{for all } x \in E.$$

We also have $0^* \notin \partial f(0)$, since $0^* \in \partial f(0)$ is equivalent to $f(x) - f(0) \geq 0$ for all $x \in E$ (immediately from the definition of ∂f), which is not true for $x = -e_1$. Thus $(0, 0^*)$ is not in the graph of ∂f.

So we may now prove that the graph of ∂f is not norm×bw closed by proving that $(0, 0^*)$ is in the norm×bw closure of the set

$$\{(e_{p,m}, e^*_{p,m}) : p \text{ prime}, m \geq 2\} \subseteq \text{graph } \partial f.$$

Informally, this is true, since $e_{p,m}$ tends in norm to 0 for large p, and also 0^* is a bw-cluster point of the $e^*_{p,m}$. A more precise argument is the following.

Let $\varepsilon > 0$ and a compact $A \subseteq E$ be arbitrarily given. We have to prove that there exist indices p, m with $\|e_{p,m}\| \leq \varepsilon$ and $e^*_{p,m} \in A^\circ$. Pick $n_0 \in \mathbb{N}$ such that $\|e_{p,m}\| \leq \varepsilon$ and $\sup_{a \in A}\langle e^*_p, a\rangle \leq 1/2$ for all $p \geq n_0$. This is possible since $\|e_{p,m}\| = 2/p$ and A is compact, so that $\langle e^*_p, a\rangle$ tends to 0 for $p \to \infty$ uniformly in $a \in A$.

Then for each prime p pick $m_0 = m_0(p)$ such that

$$\sup_{a \in A}\langle e^*_{pm}, a\rangle \leq \frac{1}{2(p-1)} \quad \text{for all } m \geq m_0,$$

once more using compactness of A. Now for all $p \geq n_0$ and $m \geq m_0(p)$ we have

$$\sup_{a \in A}\langle e^*_{p,m}, a\rangle = \sup_{a \in A}\left(\langle e^*_p, a\rangle + (p-1)\langle e^*_{pm}, a\rangle\right) \leq \frac{1}{2} + \frac{1}{2},$$

thus $\|e_{p,m}\| \leq \varepsilon$ and $\langle e^*_{p,m}, a\rangle \leq 1$ for all $a \in A$ (i.e. $e^*_{p,m} \in A^\circ$). □

Exercises and further results

9.5.4. The previous example extends to all infinite-dimensional separable Banach spaces.

Hint. Replace the unit vectors by a suitable biorthogonal sequence. □

9.5.5. The example was constructed as a separable and more illustrative version of an earlier unpublished nonseparable example due to Fitzpatrick. It takes $E = \ell_2([0, 1])$ and defines

$$f_1(x) := \max(\langle e_0, x \rangle + 1, \sup\{r^{-1} \langle e_r, x \rangle : 0 < r \leq 1\}).$$

Hint. In the previous case we relied on an unbounded sequence having a weak*-cluster point. Here we rely instead on the fact that $\{r^{-1} e_r : 0 < r \leq 1\}$ has 0^* in its bounded weak*-closure. Indeed the polar of a compact set in E contains all but countably many points of $\{r^{-1} e_r : 0 < r \leq 1\}$. □

9.5.6. Show that the following bounded linear operator $S : L_1[0, 1] \to L_\infty[0, 1]$ defined by

$$Sx := \int_0^t x(s) \, ds - \int_t^1 x(s) \, ds$$

is skew but neither $\pm S^*$ is monotone. Thus, neither S nor $-S$ is locally maximal monotone, etc. This operator is called the *Fitzpatrick–Phelps* skew operator.

Question 9.5.4 (Some open questions [396]). Among the currently open questions are:

(a) Is every locally maximal monotone operator of type (NI)?
(b) Is every maximal monotone operator maximal monotone locally?

See also Figure 9.5.

The relative paucity of examples of pathological maximal monotone operators has been a significant obstacle to progress until very recently. The next two sections highlight recent advances and inter alia, shed light on various of the examples of this section.

9.6 The sum theorem in general Banach space

A first 'general sum theorem' is:

Theorem 9.6.1 (Sum theorem [71]). *Suppose S and T are maximal monotone operators on a Banach space X and that (a)*

$$0 \in \operatorname{core} \operatorname{conv}\{\operatorname{dom} S - \operatorname{dom} T\} \tag{9.6.1}$$

and both domains are closed and convex or that (b)

$$\emptyset \neq \operatorname{int} \operatorname{dom} S \cap \operatorname{int} \operatorname{dom} T. \tag{9.6.2}$$

Then $S + T$ is a maximal monotone operator.

Proof. Case (a) follows from the methods of Voisei [431] as discussed below (see Exercise 9.7.21). A slick proof is given in [396, §51]. Case (b) is proven by a limit argument from the finite-dimensional case via Corollary 9.4.10 (Exercise 9.7.1). □

Remarkably, with no domain condition (9.6.1) still implies graph$(S+T)$ is sequentially norm-weak* closed as a consequence of Theorem 9.7.6. In the second case at least superficially there are no restrictions on the operators. An immediate set of corollaries is the following:

Corollary 9.6.2 (Normal cones). *Suppose in an arbitrary Banach space that T is maximal monotone and C is closed and convex while either*

$$\text{int } C \cap \text{int dom } T \neq \emptyset \quad \text{or} \quad \text{dom } T \cap \text{int } C \neq \emptyset$$

and in the later case dom T *is closed and convex. Then $T + N_C$ is maximal monotone.*

Corollary 9.6.3 (Convex closure). *Suppose T is maximal monotone on a Banach space and* dom T *is either closed and convex or has nonempty interior. Then T is maximal monotone locally (and* dom T *has a convex closure).*

Moreover, it is now known that

Theorem 9.6.4 ((NI) sum theorem [433]). *In any Banach space, the sum of two maximal monotone operators of type (NI) is maximal and (NI) as soon as the standard core condition (9.6.1) holds.*

Proof. A proof is sketched in Exercise 9.7.19. □

Exercises and further results

9.6.1.* Prove Corollary 9.6.3 by showing that if $T + N_C$ is maximal monotone for all closed convex sets C then T is maximal monotone locally.

9.6.2 (Convex graph [40]). Consider a maximal monotone operator T whose graph is convex. Show that the graph must be affine.

Hint. Normalize so that $0 \in T(0)$. First show that the graph is a convex cone. Then verify that graph $T = -$graph T. Hence the graph is linear after the normalization. □

9.6.3 (NI Banach spaces). A Banach space X is of *type (NI)* if every maximal monotone operator on X is type (NI). Let $E_X := X \times X^*$. It is conjectured in [76] that X is type (NI) when E_X^* is weakly countably determined [201] (which holds if E_X^* is a WCG subspace (and so if X is reflexive) or if E_X is separable and contains no copy of ℓ_1 (and so if X^{**} is separable).

Such a result would provide the first class including some nonreflexive Banach spaces in which maximal monotone operators are provably well-comported – as detailed in Theorem 9.7.9.

9.7 More about operators of type (NI)

In this section we discuss several recent results which shed significant light on the centrality of operators of type negative infimum (NI). We fix a Banach space X with norm $\|\cdot\|$ and write $d^2_{\text{graph } T}(x, x^*) := \inf_{z^* \in T(z)} \|x - z\|^2 + \|x^* - z^*\|_*^2$. Much as before, we write $\mathcal{F}_{\overline{T}}(x^*, x^{**})$ for the Fitzpatrick function of the embedding of the graph of T in $X^{**} \times X^*$.

Proposition 9.7.1 (Strong representation). *Every maximal monotone operator T of type (NI) satisfies*

$$\mathcal{F}_T^*(x^*, x^{**}) \geq \langle x^*, x^{**} \rangle$$

for each $x^* \in X^*$ and each $x^{**} \in X^{**}$.

Proof. The proof is a direct computation (seen in Section 9.3) and is left as Exercise 9.7.2. □

The next result is a strengthening of one due to Simons [397]:

Theorem 9.7.2 (Strong Fitzpatrick inequality [433]). *If a maximal monotone operator is of type (NI) then*

$$\mathcal{F}_T(x, x^*) - \langle x, x^* \rangle \geq \frac{1}{4} d_{\text{graph } T}^2(x, x^*),$$

for each $x^* \in X^*$ and each $x^{**} \in X^{**}$.

Proof. We outline the steps and leave the details for Exercise 9.7.3. They cleverly remix ingredients we saw earlier in the chapter. As before we use the notation $g(x, x^*) := (\|x\|^2 + \|x^*\|_*^2)/2$. Fix $x_0 := x$ and $x_0^* := x^*$. Define $\delta > 0$ by

$$\delta^2 := \mathcal{F}_T(x, x^*) - \langle x, x^* \rangle.$$

The Fenchel duality theorem (4.4.18) assures that

$$\inf_{X \times X^*} (\mathcal{F}_T + g) = \max_{X^* \times X^{**}} (-\mathcal{F}_T^* - g^* \leq 0)$$

where the final inequality follows from Proposition 9.7.1. Let $1 > \varepsilon > 0$ be given. Now select z, z^* with $\mathcal{F}_T(z, z^*) - \langle z, z^* \rangle \leq (\mathcal{F}_T + g)(z, z^*) \leq \varepsilon \delta^2$. By translation we temporarily assume $x_0 = 0, x_0^* = 0$ and note that

$$\mathcal{F}_T(z + x, z^* + x^*) - \langle z + x, z^* + x^* \rangle \leq \varepsilon \delta^2. \tag{9.7.1}$$

Moreover

$$g(z, z^*) \leq -\langle z, z^* \rangle + \varepsilon \delta^2 \leq (2 + 3\varepsilon) \delta^2. \tag{9.7.2}$$

Set $x_1 := z + x_0, x_1^* := z^* + x_0^*$. In like fashion we obtain a sequence (x_n, x_n^*) with

$$\mathcal{F}_T(x_n, x_n^*) - \langle x_n, x_n^* \rangle \leq \varepsilon^n \delta^2$$

such that $g(x_{n+1} - x_n, x_{n+1}^* - x_n^*) \leq (2 + 3\varepsilon) \varepsilon^n \delta^2$.

Since we work in Banach space the limit (w, w^*) of (x_n, x_n^*) exists. By norm lower-semicontinuity the limit satisfies $\mathcal{F}_T(w, w^*) = \langle w, w^* \rangle$ so that $(w, w^*) \in \text{graph } T$. Finally

$$d^2_{\text{graph } T}(x, x^*) \leq \|x - w\|^2 + \|x^* - w^*\|^2 \leq 2\frac{2 + 3\varepsilon}{(1 - \sqrt{\varepsilon})^2}\delta^2. \qquad (9.7.3)$$

Letting ε go to zero yields the asserted inequality. □

It is natural to ask if the constant above is best possible, see Exercises 9.7.11 and 9.7.14. While the numerical value of the constant is largely unimportant, the fact that it is invariant of the space leads to a lovely corollary both strengthening and simplifying earlier results in this chapter.

Corollary 9.7.3 (Domain and range convexity). *If a maximal monotone operator is of type (NI) then the norm-closures of its domain and of its range are both convex.*

Proof. For $\lambda > 0$ we consider the equivalent norm $\lambda \| \cdot \|$ with dual norm $\lambda^{-1} \| \cdot \|_*$. Observe that

$$d_{\text{graph } T}(x, x^*) \geq \lambda d_{\text{dom } T}(x).$$

It follows from the strong Fitzpatrick inequality, on letting λ go to infinity, that $\overline{\pi_1 \text{ dom } \mathcal{F}_T} \subseteq \overline{\text{dom } T}$. Here π_1 denotes the projection on X. Since the other inclusion always holds and the left-hand object is convex we are done. The range case follows similarly from $d_{\text{graph } T}(x, x^*) \geq \lambda^{-1} d_{\text{range } T}(x^*)$. □

Exercises and further results

9.7.1.* Prove the second case of Theorem 9.6.1.

9.7.2.* Show that the inequality of Theorem 9.7.2 (and its consequences) holds for all skew linear operators and so for some operators that are not type (NI). In this case the property is not translation invariant.

9.7.3.* Confirm Equations (9.7.1), (9.7.2) and (9.7.3) of the proof of Theorem 9.7.2.

Hint.

1. For (9.7.1) note that translation of the graph of a maximal monotone operator by a vector preserves the (NI) property and results in domain translation: more precisely with $q, \tilde{q}$ denoting the bilinear forms we have $(\mathcal{F}_{T-b} - q)(d) = (\mathcal{F}_T - q)(b + d)$ with a similar formula for $\mathcal{F}^*_{T-b} - \tilde{q}$.
2. For (9.7.2) observe that $(\mathcal{F}_T - q)(b) + (\mathcal{F}_T - q)(c) \geq -q(c - b)/2$. □

9.7.4 (Conditions for an operator to be (NI)). (a) Show that a surjective maximal monotone operator is type (NI) and hence locally maximal. This thus recaptures a result proved directly by Fitzpatrick and Phelps [396]. (b) By contrast $\text{dom } T = X$ does not imply local maximality as is shown by the various skew examples. (c) Similarly, confirm that T is type (NI) if (i) it has bounded domain (or is coercive) and norm dense range or if (ii) its range has nonempty interior (this is harder).

9.7 More about operators of type (NI)

9.7.5 ((NI) implies range density). We sketch that every operator of type (NI) is *range-dense type* (WD) which is to say that if $\langle x^{**} - y, x^* - y^* \rangle \geq 0$ for all $y^* \in T(y)$ (equivalently $\mathcal{F}_{\widehat{T}}(x^*, x^{**}) = \langle x^{**}, x^* \rangle$) then there is a bounded weak*-norm convergent net (x_β, x_β^*) with $x_\beta^* \to_{\|\cdot\|} x^*$.

1. A separation argument shows $\overline{\text{range}\, T} = \overline{\text{range}\, T}$ when T is (NI).
2. Use Theorem 9.6.4 and Corollary 9.4.3 to deduce that $\overline{\text{range}}(T + \lambda J) = X^*$.
3. Suppose $\langle x^{**} - y, x^* - y^* \rangle \geq 0$ for all $y^* \in T(y)$ and select $x_n^* \in T(x_n) + J(x_n)/n$ with $\|x_n^* - x^*\| < 1/n^2$. Deduce that $(x_n)_{n \in \mathbb{N}}$ remains bounded and hence $J(x_n)/n \to 0$. Thence, extract a convergent subnet (x_β, x_β^*) with $x_\beta^* \to x^*$ in norm to complete the proof.

The operator would be of dense type – by definition – if one could obtain that $x_\beta \to^{w^*} x^{**}$. [A similar argument shows that a coercive operator of type (NI) has norm dense range.] In fact a more careful argument developed in [6] exploits the Brøndsted–Rockafellar property to show that a maximal monotone operator of type (NI) is actually of dense type (see Exercise 9.7.16).

Moreover, it is easy to check that a densely-defined affine maximal monotone operator of range-dense type is of dense type.

Hint. If $h^* \in T(h)$ then $y^* + \varepsilon h^* \in T(y + \varepsilon h)$ for $\varepsilon > 0$. $\square$

9.7.6 ((NI) characterization). Show that a maximal monotone operator T is type (NI) if and only if

$$\mathcal{F}_{\widehat{T}}(x^*, x^{**}) \geq \langle x^*, x^{**} \rangle$$

for each $x^* \in X^*$ and each $x^{**} \in X^{**}$, and that this happens if and only if each representative function satisfies $\mathcal{R}_T^*(x^*, x^{**}) \geq \langle x^*, x^{**} \rangle$. *Hint.* $\mathcal{F}_{\widehat{T}}(x^*, x^{**}) = \mathcal{P}_T^*(x^*, x^{**})$ as noted in Proposition 9.4.1. $\square$

9.7.7. An efficient way to show that a skew operator is not of type (NI) is recorded in the following theorem.

Theorem 9.7.4 (Nonmonotone conjugates [33]). *Suppose T is a continuous linear operator from X to X^* with skew part S and there is $e \in X^*$ such that*

$$e \notin \overline{\text{range}\, T} \quad \text{and} \quad \langle Tx, x \rangle = \langle e, x \rangle^2, \quad \text{for all } x \in X.$$

Then T is monotone but S^* is not (and hence S and T are not (NI)).

Hint. Let $Px := \langle e, x \rangle e$, for all $x \in X$; then $\langle P^*x^{**}, x \rangle = \langle x^{**}, Px \rangle = \langle x^{**}, e \rangle \langle e, x \rangle$ and hence $P^*x^{**} = \langle x^{**}, e \rangle e$, for all $x^{**} \in X^{**}$. So P is symmetric. Consider now $S := T - P$. Then $\langle Sx, x \rangle = \langle Tx, x \rangle - \langle Px, x \rangle = \langle Tx, x \rangle - \langle e, x \rangle^2 = 0$, for all $x \in X$, thus S is skew. Since $T = P + S$, the symmetric (resp. skew) part of T is P (resp. S). Because $e \notin \overline{\text{range}\, T}$, there exists some x_0^{**} with $T^*(x_0^{**})$ with $\langle x_0^{**}, e \rangle \neq 0$.

Hence

$$\langle S^*x_0^{**}, x_0^{**}\rangle = \langle T^*x_0^{**}, x_0^{**}\rangle - \langle P^*x_0^{**}, x_0^{**}\rangle = 0 - \langle x_0^{**}, e\rangle^2 < 0;$$

so S^* is not monotone. □

Theorem 9.7.4 allows for an easy verification that the Gossez and Fitzpatrick-Phelps skew operators are as claimed.

The first two parts of the following exercise are due to Simons [390]. Part three is a lovely recent result of Alves and Svaiter [4], while the rest is due to Gossez [238].

9.7.8 (Uniqueness implies (NI)). A maximal monotone operator $T: X \to X^*$ is said to be *unique* or have the *uniqueness property* if there is a unique (maximal) monotone extension in $X^{**} \times X^*$. Show the following:

1. T is unique if and only if the monotone closure $\overline{T}$ is monotone.
2. An operator of type (NI) is unique.
3. A maximal monotone operator with the uniqueness property is either affine or type (NI).
4. The Gossez operator is unique (and affine) but, as observed in Exercise 9.7.16, is not (NI).
5. There are nonunique affine maximal monotone operators with infinitely many maximal extensions to the bidual. Gossez's proof is subtle but an easy approach is given in Exercise 9.7.9.

9.7.9 (Nonuniqueness). Let $A: X \to X^*$ be a bounded skew linear operator on a Banach space. (a) Show that A is not unique if and only if there are $F, G \in X^{**}$ such that $\langle A^*F, F\rangle \leq 0, \langle A^*G, G\rangle \leq 0$ but such that $\langle A^*(F-G), F-G\rangle > 0$. (b) Deduce that A is unique if $-A^*$ is monotone and in particular if range A^* = range $A^*|_X$. (c) Deduce that if neither $-A^*$ nor A^* is monotone – as for the Fitzpatrick–Phelps operator of Exercise 9.5.6 – then A is not unique (and conversely). (d) Show that in fact there are infinitely many maximal extensions to the Fitzpatrick–Phelps operator.

Hint. (a) $x^* \in \overline{A}F$ if and only if $x^* = -A^*F$ and $\langle A^*F, F\rangle \leq 0$.

(c) Fix $H, K \in X^{**}$ with $\langle A^*H, H\rangle = -1, \langle A^*K, K\rangle = 1$. Let $\tau > 0$ be given and set $F := H + \tau K$ and $G := H - \tau K$. Then, for τ sufficiently small, $\langle A^*F, F\rangle < 0, \langle A^*G, G\rangle < 0$ but $\langle A^*(F-G), F-G\rangle = \tau^2 \langle A^*K, K\rangle > 0$.

(d) One may add any member of the Haar basis to F. □

9.7.10. Suppose T is locally maximal monotone. Show that the norm-closure of the range of T or $\overline{T}$ is the same convex set.

9.7.11. Show that the best possible constant in Theorem 9.7.2 is 1/2. Indeed the inequality is then an equality for the identity in Euclidean or Hilbert space.

9.7.12 ((NI) implies (FP)) [433]). Show that every maximal monotone operator of type (NI) is locally maximal monotone. It is not known whether all locally maximal monotone operators are of type (NI). The classes do coincide for linear operators.

Hint. Use the characterization in 9.4.5 and, without loss, assume $x = 0$ and $x^* = 0$ with $\beta B^* \subset C^*, \beta > 0$. Consider the function f given by $f(x, x^*) := \mathcal{F}_T(x, x^*) +$

9.7 More about operators of type (NI)

$\alpha \|x\| + \delta_{\alpha B^*}(x^*)$ for $2\alpha < \beta$. Observe that $\inf_{X \times X^*} f = 0$ and apply the techniques of the strong Fitzpatrick inequality theorem 9.7.2 to deduce $0 \in T(0)$, a contradiction. $\square$

9.7.13 ((NI) implies (ANA) [433]). Show that for any maximal monotone operator T of type (NI) when $b^* \notin T(b)$ one has

$$\lim_{n\to\infty} \frac{\langle b - x_n, b^* - x_n^* \rangle}{\|x_n - b\| \|x_n^* - b^*\|} = -1$$

for some sequence $x_n^* \in T(x_n)$, with $x_n \neq b, x_n^* \neq b^*$. Such operators are said to be of type (ANA) or *almost negative alignment*. It is not known if all maximal monotone operators are type (ANA).

Hint. Use the strong Fitzpatrick inequality to show that for $x^* \in T(x)$ one can find bounded sequences (x_n) and (x_n^*) with

$$\|x - x_n\|^2 + \|x^* - x_n^*\|^2 + 2\langle x - x_n, x^* - x_n^* \rangle \leq 1/n$$

for each $n \in \mathbb{N}$. $\square$

9.7.14 (Brøndsted–Rockafellar property [3]). Show the following:

1. Refine the proof of the strong Fitzpatrick inequality 9.7.2 to deduce that $d^2_{\mathrm{graph}\, T}(x, x^*)/4$ can be improved to a supremum norm:

$$\inf_{s^* \in T(s)} \max\{\|x - s\|^2, \|x^* - s^*\|_*^2\}.$$

2. Confirm that this inequality can be derived for any closed convex proper function with $h(x, x^*) \geq \langle x, x^* \rangle$ and $h^*(x^*, x^{**}) \geq \langle x^{**}, x^* \rangle$. The conclusion becomes

$$h(x, x^*) - \langle x, x^* \rangle \geq \inf_{\{h(s, s^*) = \langle s, s^* \rangle\}} \max\{\|x - s\|^2, \|x^* - s^*\|_*^2\}$$

and in consequence

$$h(x, x^*) - \langle x, x^* \rangle \geq \frac{1}{2} \inf_{\{h(s, s^*) = \langle s, s^* \rangle\}} \{\|x - s\|^2 + \|x^* - s^*\|_*^2\}$$

which provides the best constant in Theorem 9.7.2.

9.7.15 (Positive linear implies (ANA)). Show that every bounded linear monotone operator on a Banach space is maximal monotone of type (ANA) [396, Theorem 47.7]

9.7.16 ((NI) implies (BR) [3]). Show the following:

1. Each maximal monotone (NI) operator T has the *Brøndsted–Rockafellar property*, is (BR). That is: for $\alpha > 0, \beta > 0$ suppose

$$\inf_{y^* \in T(y)} \langle x^* - y^*, x - y \rangle > -\alpha\beta.$$

Then there is $z^* \in T(z)$ with $\|z - x\| < \alpha$ and $\|z^* - x^*\| < \beta$. Note that the original strong Fitzpatrick inequality shows that one can achieve $\|z - x\| < 2\alpha$ and $\|z^* - x^*\| < 2\beta$.

2. Show that every skew bounded linear operator is (BR). Hence, Gossez's skew operator of 9.5.2 is (BR) and (ANA) but not (NI).

3. Show that the *tail operator* $E : \ell_1 \to \ell_\infty$ given by $(Ex)_n := \sum_{k \geq n} x_k$ (originally due to Fitzpatrick and Phelps and discussed in [396, Example 33.6]) is maximal monotone and (ANA) but is not (BR).

Hint. Let e^1 be the first unit vector and $e = (1, 1, \ldots, 1, \ldots)$ in ℓ_∞. Then $\langle e^1, E(x) \rangle = \langle x, e \rangle$. While $\inf_{x \in X} \langle x - (-e^1), E(x) - e \rangle = -1$, the distance of e from range E is one since range $E \subset c_0$. □

9.7.17. Determine, in comparison to the result of Exercise 9.7.8, whether there is (i) a nonlinear maximal monotone operator which is (BR) and not (NI) and (ii) a nonlinear maximal monotone operator which is not (BR).

Hint. (ii) If T is (BR) and $\inf_{x \in X} \langle x - y, T(x) - y^* \rangle > -\infty$, then y^* lies in $\overline{\text{range }} T$. Now consider a small nonlinear perturbation of E. □

9.7.18 (Maximality of representatives [3]).

Theorem 9.7.5. *Suppose that $h : X \times X^* \to [-\infty, +\infty]$ is a proper, convex and norm-to-weak*-lsc function such that (i) $h(x, x^*) \geq \langle x, x^* \rangle$ and (ii) $h^*(x^*, x) \geq \langle x^*, x \rangle$ for all $(x, x^*) \in X \times X^*$. Then if $G(A) := \{(x, x^*) \in X \times X^* : h^{*\top}(x, x^*) = \langle x, x^* \rangle\}$ it follows that (a) $G(A) := \{(x, x^*) \in X \times X^* : h(x, x^*) = \langle x, x^* \rangle\}$ and (b) A is monotone. (c) Moreover, if h is (NI) then A is maximal.*

Hint. (a) First if $h(x, x^*) = \langle x, x^* \rangle$ deduce as in the reflexive case that $(x, x^*) \in \partial h(x, x^*)$ because the bilinear form is a Gâteaux subderivative. Hence, $h(x, x^*) + h^*(x^*, x) = 2\langle x, x^* \rangle$ and (a) holds.

(b) To show monotonicity, suppose that $h^*(x^*, x) = \langle x^*, x \rangle$ and $h^*(y^*, y) = \langle y^*, y \rangle$. Then by convexity we have

$$4 h^*((x^* + y^*)/2, (x + y)/2) \leq 2\langle x^*, x \rangle + 2\langle y^*, y \rangle$$

and by hypothesis (ii)

$$\langle x^* + y^*, x + y \rangle \leq 4 h^*((x^* + y^*)/2, (x + y)/2).$$

Combining these last two inequalities and simplifying shows A is monotone.

(c) To show maximality suppose $\langle x^* - a^*, x - a \rangle \geq 0$ for all $a^* \in A(a)$. Without loss of generality, let us assume $x^* = 0, x = 0$. Then as in Theorem 9.7.2

$$\inf\left(\{h(y, y^*) - \langle y^*, y \rangle\} + \{(\|y\|^2 + \|y^*\|^2)/2 + \langle y^*, y \rangle\}\right) = 0 = \min(h^* + g^*).$$

Now, by the generalization of Theorem 9.7.2 given in Exercise 9.7.14 we may select $\|y_n - a_n\|^2 + \|y_n^* - a_n^*\|^2 \to 0$ with $a_n^* \in A(a_n)$, $\|y_n\|^2 + \|y_n^*\|^2 + 2\langle y_n, y_n^* \rangle \to 0$ and $h(y_n, y_n^*) - \langle y_n, y_n^* \rangle \to 0$. Observe that (y_n, y_n^*) is bounded and argue that $0 \leq$

$\langle a_n, a_n^* \rangle \to 0$. Finally derive that $\|a_n\|^2 + \|a_n\|^2 \to 0$ and as h is lsc we derive $h(0,0) \leq \langle 0, 0^* \rangle$ as required. □

9.7.19 (Maximality of (NI) sums [433]). We now establish Theorem 9.6.4.
First apply Theorem 9.7.5 to two partial infimal convolutions:

$$\mathcal{V}_{S,T}(x^{**}, x^*) := \inf \left\{ \mathcal{F}_{\overline{S}}(x^{**}, u^*) + \mathcal{F}_{\overline{T}}(x^{**}, v^*) : u^* + v^* = x^* \right\},$$

and

$$\mathcal{W}_{S,T}(x^{**}, x^*) := \inf \left\{ \mathcal{P}_T(x^{**}, u^*) + \mathcal{P}_S(x^{**}, v^*) : u^* + v^* = x^* \right\}.$$

As noted in [71] we have a lovely observation first exploited by Voisei and also given in [396, Therorem 16.4]:

Theorem 9.7.6 (Partial convolution [431]). *Suppose T and S are maximal monotone and satisfy the transversality condition*

$$0 \in \mathrm{core}\left(\pi_1(\mathrm{dom}(\mathcal{P}_S)) - \pi_1(\mathrm{dom}\,\mathcal{P}_T) \right).$$

Then $\mathcal{V}_{S,T}(x^{**}, x^*) = \mathcal{W}_{S,T}^*(x^{**}, x^*)$ *is norm-lower-semicontinuous and is attained when finite.*
Moreover,

$$\mathcal{W}_{S,T}^*(x^*, x) > \langle x^*, x \rangle$$

and

$$\mathcal{W}_{S,T}(x, x^*) \geq \langle x, x^* \rangle,$$

for $x \in X$ *and* $x^* \in X^*$. *In consequence*

$$\mathrm{graph}(S + T) = \{(x, x^*) : \mathcal{V}_{S,T}(x, x^*) = \langle x, x^* \rangle\}.$$

If in addition, S and T are of type (NI) then

$$\mathcal{W}_{S,T}(x^{**}, x^*) \geq \langle x^{**}, x^* \rangle,$$

for $x^{**} \in X^{**}$ *and* $x^* \in X^*$.

Proof. The argument – based on a conjugate formula of Penot [343, Proposition 13] – as in Vosei [431] and earlier in this chapter, or a direct Lagrangian calculation, shows $\mathcal{V}_{S,T}(x^{**}, x^*) = \mathcal{W}_{S,T}^*(x^{**}, x^*)$ and is attained when finite. The rest follows since $\mathcal{P}_S^* = \mathcal{F}_{\overline{S}}$ and $\mathcal{P}_S^* = \mathcal{F}_{\overline{T}}$ have the requisite representative properties. □

Theorem 9.7.7 ((NI) sums). *Let* $S, T : X \rightrightarrows X^*$ *be two maximal monotone operators of type (NI) such that*

$$0 \in \mathrm{core}\left(\pi_1(\mathrm{dom}\,\mathcal{P}_S) - \pi_1(\mathrm{dom}\,\mathcal{P}_T) \right).$$

Then $S + T$ is a maximal monotone operator of type (NI).

Hint. We apply Exercise 9.7.18 to $h = \mathcal{W}_{S,T}$. By Theorem 9.7.6 $h^* = \mathcal{V}_{S,T}$. Thus

$$G(B) := \{(x, x^*) : \inf \{\mathcal{F}_{\overline{T}}(x, u^*) + \mathcal{F}_{\overline{S}}(x, v^*) : u^* + v^* = x^*\} = \langle x, x^* \rangle\}$$

determines a maximal monotone operator B. But direct computation, appealing to Proposition 9.1.1 again, shows that $B = S + T$. □

As always, since $\text{conv}(\text{dom } S - \text{dom } T) \subseteq \pi_1(\text{dom } \mathcal{P}_S) - \pi_1(\text{dom } \mathcal{P}_T)$ we have also the following result.

Corollary 9.7.8. *Let $S, T : X \rightrightarrows X^*$ be two maximal monotone operators of type (NI) such that*

$$0 \in \text{core conv}\left(\text{dom } S - \text{dom } T\right).$$

Then $S + T$ is a maximal monotone operator of type (NI).

The corresponding results for compositions as in Section 9.4.3 also follow.

To summarize, in the language of Exercise 9.6.3 we can now assert the following result, of which all the pieces have been discussed in this and the previous section.

Theorem 9.7.9 ((NI) Banach spaces [76]). *Let X be a Banach space of type (NI). Then every maximal monotone operator T is of type (NI) and in consequence:*

1. *The range and domain of T have convex norm-closure [3, 433, 397].*
2. *T has the Brøndsted-Rockafellar property (BR) [3].*
3. *T is almost negatively aligned (ANA) [433].*
4. *T is both locally maximal monotone (LMM) or (FP) and maximally locally monotone (FPV) [433].*

   Moreover, if S is another maximal monotone operator and

$$0 \in \text{core } (\text{dom } T - \text{dom } S)$$

then $T + S$ is again maximal monotone [433].

This result is conjectured to hold in nonreflexive spaces such as the *James space* (which is codimension-one in its bidual, [199]).

There is another positive result when one of the mappings is not type (NI). As with Theorem 9.6.1 (a) it relies on the following proposition.

Proposition 9.7.10 ([71], Theorem 24.1 of [396]). *Let S and T be maximal and suppose that (9.6.1) holds. (a) Then $S + T$ is maximal monotone if and only if $\mathcal{F}_{S+T}(x, x^*) \geq \langle x, x^* \rangle$ for all $x \in X$ and $x^* \in X^*$. (b) In particular this holds if $\mathcal{F}_{S+T}(x, x^*) \leq \langle x, x^* \rangle$ implies $x \in \text{dom}(S + T)$.*

Hint. Apply parts (a) and (b) of Theorem 9.7.5 to $\mathcal{W}_{S,T} \geq \mathcal{F}_{S+T}$ and appeal to Theorem 9.7.6. □

9.7 More about operators of type (NI)

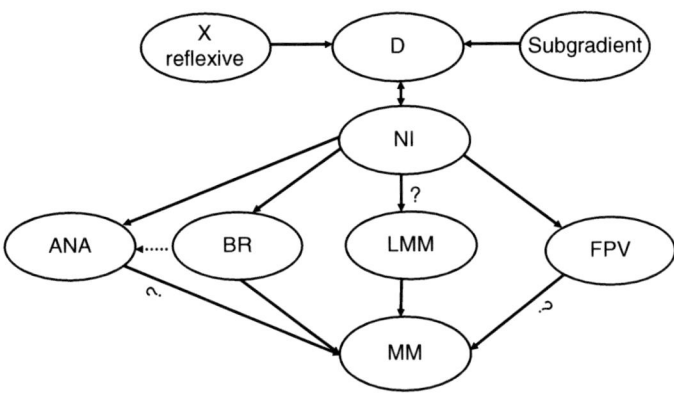

Figure 9.5 Relations between classes of operators.

9.7.20 (Verona–Verona theorem [396]). Show that $T + \partial f$ is maximal whenever f is closed and convex and T is maximal and everywhere defined. *Hint.* The hard work is to show Proposition 9.7.10 (b) applies [396, §53]. □

Note that this includes the case where T is a bounded monotone linear operator which is not of type (NI). The proof in [396] goes through if one only assumes

$$\overline{\operatorname{dom} f} \subset \operatorname{int} \operatorname{dom} T. \tag{9.7.4}$$

9.7.21.* Prove Theorem 9.6.1 (a).

Hint. Use Proposition 9.7.10 (b) and the method of Exercise 9.7.19. □

9.7.22.* Prove Theorem 9.6.1 (b).

Hint. Use (9.7.4) in the product space. □

A schematic illustration of known and open relationships is given in Figure 9.5. Recall that uniqueness and type (NI) coincide in the nonlinear case.

10
Further remarks and notes

It is not knowledge, but the act of learning, not possession but the act of getting there, which grants the greatest enjoyment. When I have clarified and exhausted a subject, then I turn away from it, in order to go into darkness again; the never-satisfied man is so strange if he has completed a structure, then it is not in order to dwell in it peacefully, but in order to begin another. I imagine the world conqueror must feel thus, who, after one kingdom is scarcely conquered, stretches out his arms for others. (Carl Friedrich Gauss)[1]

10.1 Back to the finite

We finish this book by reprising some of the ways in which finite-dimensionality has played a critical role in the previous chapters. While our list is far from complete it should help illuminate the places in which care is appropriate when 'generalizing'. Many of our results have effectively the same proof in Banach spaces as they do in Euclidean spaces.

10.1.1 The general picture

For example, the equivalence of local boundedness and local Lipschitz properties for convex functions is a purely geometric argument that does not depend on the dimension of the space (compare Theorem 2.1.10 and Proposition 4.1.4). Consequently, in finite dimensions a real-valued convex functions is automatically continuous essentially because a simplex has nonempty interior (see Theorem 2.1.12). This argument clearly fails in infinite-dimensional spaces, and the result fails badly as there are discontinuous linear functionals on every infinite-dimensional Banach space (Exercise 4.1.22). However, if f is everywhere finite and lsc then f is continuous (Proposition 4.1.5) – since a Banach space is barreled, as it is a *Baire space* (see Exercise 10.1.9). This is one of the few significant analytic properties which hold in a large class of incomplete normed spaces.

The preceding is typical of a class of results from Euclidean space that have natural analogs in general Banach spaces with an additional closure assumption on the convex

---

[1] Carl Friedrich Gauss, 1777–1855, in an 1808 letter to his friend Farkas Bolyai (the father of Janos Bolyai).

function (which allows the recovery of an interiority condition). Some of the most important and striking examples of this are that the max formula, sandwich theorem and Fenchel duality theorem extend nicely to Banach spaces with only a modest amount of extra work as is explored in Chapter 4.

However, many results do not extend well from Euclidean spaces to infinite-dimensional spaces. This is principally because the compactness properties and support properties of convex sets have become significantly more subtle. We mention two such examples of this. First, every finite convex function on a Euclidean space is bounded on bounded sets because of continuity and compactness of balls. In contrast, on any infinite-dimensional Banach space there is a globally continuous convex function that is unbounded on a ball; see Theorem 8.2.2 where it is illustrated that construction of such a function depends on more subtle properties than just lack of compactness (further related results of this nature are presented in Chapter 8). Second, an illustration of support properties is that disjoint closed convex sets in Euclidean space can always be separated because of nonempty relative interior (Theorem 2.4.7) while this can fail in infinite-dimensional spaces (Exercise 4.1.9).

In order to more concisely address many of the issues we just raised, we present two compendia of standard results taken from [96].

Theorem 10.1.1 (Closure, continuity, and compactness). *Let X be a Banach space. The following statements are equivalent:*

(a) X is finite-dimensional.
(b) Every vector subspace of X is closed.
(c) Every linear map taking values in X has closed range.
(d) Every linear functional on X is continuous.
(e) Every convex function $f : X \to \mathbb{R}$ is continuous.
(f) The closed unit ball in X is (pre-)compact.
(g) For each nonempty closed set C in X and for each x in X, the metric distance
    $d_C(x) = \inf\{\|x - y\| : y \in C\}$ is attained.
(h) The weak and norm topologies coincide on X.
(i) The weak* and norm topologies coincide on X^*.

Proof. Properties (b) – (d) and (f) – (i) are implied by (a) as is covered in an introductory linear functional analysis text. Theorem 2.1.12 shows (a) implies (e). Exercise 4.1.22 shows there is a discontinuous linear functional on every infinite-dimensional space; hence deduce that (e), (d) and (b) imply (a). For (c), consider a nonclosed subpace, with the help of a Hamel basis (extended) from the subspace to the whole space, one can construct a linear projection onto the subspace.

For (f) and (g): use the basic separation theorem (4.1.12) to find a sequence $(x_n) \subset B_X$ such that $1/2 < \|x_n\| < 1/2 + 1/n$, and $d_{Y_n}(x_{n+1}) \geq 1/2$ where $Y_n = \text{span}\{x_1, \ldots, x_n\}$. Then $C := \{x_n\}_{n=1}^{\infty}$ is a closed set, and $d_C(0) = 1/2$ is not attained. For (h) and (i), observe in infinite dimensions, nonempty weakly open and weak*-open sets contain infinite-dimensional subspaces and are hence unbounded. $\square$

Moving from continuity to tangency properties of convex sets we have:

Theorem 10.1.2 (Support and separation). *Let X be a separable Banach space. Then the following statements are equivalent.*

(a) *X is finite-dimensional.*
(b) *Whenever a lsc convex $f : X \to (-\infty, +\infty]$ has a unique subgradient at x then f is Gâteaux differentiable at x.*
(c) *X every (closed) convex set in X has a supporting hyperplane at each boundary point.*
(d) *Every (closed) convex set in X has nonempty relative interior.*
(e) *Suppose A is closed and convex, while R is a ray (or line). Then $A \cap R = \emptyset, \Rightarrow A$ and R are separated by a closed hyperplane.*

Proof. Theorem 2.2.1 shows (a) implies (b) even without a lsc assumption on f. To see that (c) holds in finite dimensions, one may assume $0 \in C$. In the event span $C = X$, C has nonempty interior and so the supporting hyperplane theorem (2.4.3) may be applied. In the event span $C \neq X$, there is a nonzero linear functional that contains C in its kernel. Observe this proof allows the possibility that the supporting functional may be constant on C; Exercise 10.1.12 shows that this is unavoidable.

The conditions (d) and (e) hold in finite-dimensional spaces by Theorem 2.4.6 and Theorem 2.4.7 respectively.

Conversely, Example 4.2.6 shows (b) implies (a); see also Exercise 4.2.5. To prove (c) implies (a), take a countable dense collection $\{x_n\}_{n=1}^{\infty} \subset S_X$. Let $C := \overline{\text{conv}}(\{2^{-n}x_n\}_{n=1}^{\infty} \cup \{0\})$ then C is compact because it is closed and totally bounded, span C is infinite-dimensional and hence 0 is a boundary point of C. For each $\phi \in X^* \setminus \{0\}$ observe that $\sup_C \phi > 0$ and so $0 \in C$ has no supporting hyperplane. Check that this example can also be used to show (d) implies (a).

Finally, Exercise 4.1.9 can be modified for any separable Banach space X using a *Markuševič basis*. That is, there exist $\{x_n, x_n^*\}_{n=1}^{\infty} \subset X \times X^*$ satisfying $\overline{\text{span}}(\{x_n\}_{n=1}^{\infty}) = X$, the span of $\{x_n^*\}_{n=1}^{\infty}$ is weak*-dense in X^*, and $x_n^*(x_m) = 1$ if $m = n$ and 0 otherwise; see [199, Theorem 6.41]. By scaling, we may assume $\|x_n\| = 1$ for all n. Define $A := \{x \in X : x_n^*(x) \geq 0 \text{ for all } n \in \mathbb{N}\}$ and $R := \{ta - b : t \geq 0\}$ where $a := 4^{-n}x_n$ and $b := 2^{-n}x_n$. Verify that $A \cap R = \emptyset$ but that no nonzero $\phi \in X^*$ can separate A and R and thus deduce that (e) implies (a). □

In sum these results say treat 'finite-dimensionally derived intuition with caution'. Lest one forgets, the world of just three-dimensional convex polyhedra [243] is already marvelously complex as illustrated by Figure 10.1 [434]. They show *Archimedean dual solids*: a truncated dodecahedron and an icosahedron with triangular pyramids added to each face.

Exercises 10.1.3 and 10.1.4 point to other places where Euclidean verities break down. Exercise 10.1.6 is intended to reenforce the reminder that Euclidean does not equate to obvious.

10.1 Back to the finite

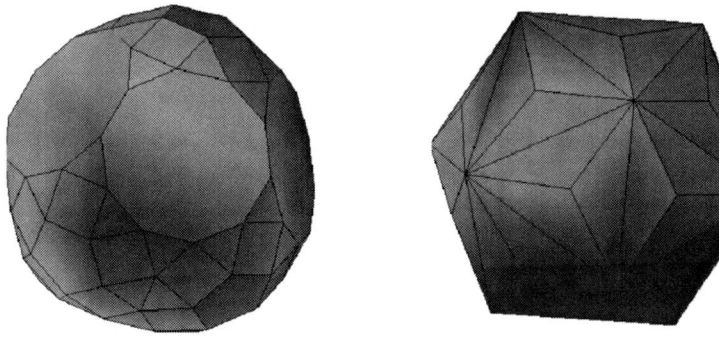

Figure 10.1 Truncated dodecahedron and Triakis icosahedron.

10.1.2 A closer look

Theorem 10.1.1 clearly highlights things that 'can go wrong' outside of Euclidean space. However, let us be clear that the weak topologies of Theorem 10.1.1(h) and (i) are immensely important and useful in the study of infinite-dimensional spaces. Indeed, they provide natural Hausdorff topologies with much less stringent convergence requirements. Among the most important and fundamental results in this direction is the Banach–Alaoglu or Alaoglu's theorem (4.1.6) that ensures the dual unit ball is weak*-compact (Exercise 6.7.11).

There are several important convex analytic results on Euclidean space that extend – at least in substantial parts – to general Banach spaces with the help of Alaoglu's theorem. For example, a differentiable convex function on E has a continuous derivative (Theorem 2.2.2). Crucial to this is that a continuous convex function is locally Lipschitz and hence the derivative mapping is locally bounded and compactness arguments are available for use in conjunction with properties of the convex subdifferential. This extends to general Banach spaces in the following sense: a Fréchet differentiable convex function has norm-to-norm continuous derivative and a Gâteaux differentiable continuous convex function has norm-to-weak* continuous derivative (Corollary 4.2.12). The norm-to-norm continuity for Fréchet differentiability requires both relative weak*-compactness of bounded sets in the dual and the fact that its defining limit is uniform over all directions in the unit ball of X, however the norm compactness of the dual unit ball is not needed.

Such dual weak*-compactness properties along with the power and consequences of Baire mentioned above lie at the heart of the depth and success of the related concept of monotone operators on Banach spaces; this is explored in detail in Chapter 9. The Moreau–Rockafellar theorem (4.4.10) makes it clear how tightly conjugacy couples interiority and compactness properties. We restate its dual version (Theorem 4.4.12) which we have often seen in use implicitly:

Theorem 10.1.3 (Moreau–Rockafellar dual). *Let $f : X \to (-\infty, +\infty]$ be proper, convex and lsc. Then f is continuous at zero if and only if f^* has weak*-compact lower level sets.*

The literature is replete with papers proving that both conditions hold for a given convex function or program: unaware that they are doing double work, see Exercise 10.1.7.

In many parts of the study of convex functions, it is natural to restrict to specialized classes of Banach space. For example, as explored in various places in this book, Asplund spaces are the venue where continuous convex functions have a dense set of points of Fréchet differentiability – and much then follows. In Chapter 8 we saw how various natural properties of convex functions characterize very different classes of Banach spaces.

Other situations may require the narrower class of reflexive Banach spaces in which the unit ball is weakly compact (so there is compactness available in both the space and its dual). In fact, a Banach space is reflexive space if and only if each closed bounded convex is weakly compact (Exercise 4.1.25), and as a consequence every continuous linear functional attains its maximum on such a set – actually an equivalence grace of James. See Exercise 10.1.11 for explicit examples of failure of norm attainment in some nonreflexive spaces. Some salient topics in this book where reflexive spaces form the natural setting are in the study of Legendre functions (see Chapter 7) and of maximal monotone operators – at least for certain properties such as coercivity – are most fully developed there (see Chapter 9).

Still, the reader should be reminded that support properties are in many ways quite good outside of the reflexive setting because of the landmark Bishop–Phelps theorem (4.3.4) and its two vitally important cousins – the Brøndsted–Rockafellar theorem (4.3.2) and Ekeland's variational principle (4.3.1). These results ensure that there is a large supply of support points of closed convex sets and functions on tap.

There also many properties that characterize the more restrictive but highly fungible Hilbert space. The most striking is perhaps the deep result that a Banach space X is (isomorphic to) Hilbert space if and only if every closed vector subspace is complemented in X. Especially with respect to best approximation properties, it is Hilbert space that best captures the properties of Euclidean space. Another illustration is the connection to nonexpansive mappings as given in and around Theorem 6.1.14. For readers whose primary interest is Hilbert space, we recommend [175] for approximation properties and [37] for an exposition on convex analysis and monotone operators.

Whether you have reached this section as a step into the infinite or as a look back to the finite, we hope it illustrates that while additional care is needed in studying convex functions on infinite-dimensional spaces, the theory of convex functions in those spaces is a rich and well-developed field of study.

Exercises and further results

10.1.1 (A misleading convex hull). This construction illustrates a caution over reading too much into a picture. Consider the set constructed by placing a regular $4 \cdot 2^n$-gon, P_n, at height 2^{-n} in $\mathbb{R}^3$, for $n = 0, 1, 2, \ldots$ and then taking the union of the shells between subsequent cross-sections; the vertices of P_n are aligned with those at the

10.1 Back to the finite

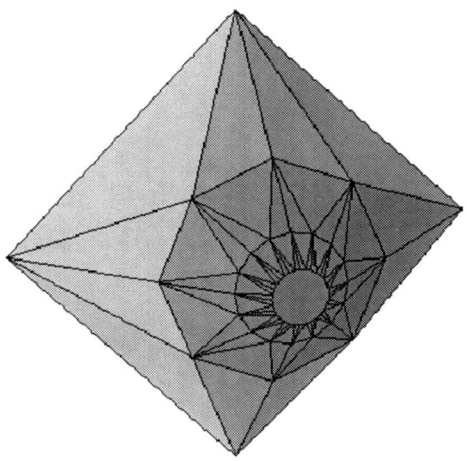

Figure 10.2 A misleading convex hull.

next step. Figure 10.2 shows the convex set built from the first three shells. Show that the set constructed as n increases is nonconvex. This illustrates that the convex hull cannot be created myopically.

10.1.2 (Universal convex sets).** In [244] it is shown, using a result of Peano on continua, that *in $\mathbb{R}^{n+2}$ there is a compact convex set C such that every compact convex subset of the unit ball of $\mathbb{R}^n$ is realized (up to a rigid motion) as the intersection of C with some n-dimensional plane in $\mathbb{R}^{n+2}$*. This complements a negative result of Grünbaum that there is no compact convex *symmetric* set C in $\mathbb{R}^3$, such that every symmetric convex body in $\mathbb{R}^2$ is the affine image of the intersection of C with some two-dimensional subspace. This answered a question of Mazur and was extended by Bessaga who replaced '3' by 'finite' and '2' by 'n'.

10.1.3 (Nonclosedness of convex hull). Show that part (d) of Exercise 2.4.11 characterizes finite dimensions by constructing in every infinite-dimensional Banach space a convergent sequence whose convex hull is not closed.

Hint. In Hilbert space let $x_n := e_n/2^n$ where e_n is the n-th unit vector. Then $x := \sum_{n>0} e_n/4^n$ is in the closure of the convex hull of $\{x_n\}_{n=1}^\infty \cup \{0\}$ but not in the convex hull. In a general Banach space, one can use a Markuševič basis in an analogous fashion on a separable subspace. □

10.1.4 (Pointed cones and bases). Determine which parts of Exercise 2.4.26 remain equivalent in Banach space.

10.1.5. Show that Exercise 2.7.3 remains valid for (weakly) compact convex sets in Banach space.

10.1.6 (Highly unstable abstract linear programs [106]). Consider the following closed convex cone in $\mathbb{R}^7$, where we write $\bar{z} := (x_1, x_2, x_3, x_4, y_1, y_2, y_3)$ to denote a point in $\mathbb{R}^7$ and let

$$K := \{\bar{z} \in \mathbb{R}^7 : 0 \leq x_1, 0 \leq x_2, x_3^2 + (x_4 - x_2)^2 \leq x_2^2, 0 \leq y_1, y_2^2 + (y_3 - y_1)^2 \leq y_1^2\}.$$

Define the linear mapping $z^* : \mathbb{R}^7 \to \mathbb{R}$ by $z^*(\bar{z}) := x_4 + y_3$. For each $0 \leq \lambda < \infty$ and $\mu \in \mathbb{R}$ define

$$A^\lambda := \begin{pmatrix} -\lambda & 1 & 0 & 0 & 0 & 0 & 0 \\ 0 & 0 & 1 & 0 & 0 & 0 & 0 \\ 0 & 0 & 0 & 0 & 0 & 1 & 0 \end{pmatrix} \text{ and } \beta^\mu := \begin{pmatrix} 1 \\ 1 \\ \mu \end{pmatrix}.$$

Then for each $0 \leq \lambda < \infty$ and $\mu \in \mathbb{R}$ consider the optimization problem:

$$p(\lambda, \mu) := \inf\{z^*(x) : x \in K \text{ and } A^\lambda x = \beta^\mu\}.$$

Confirm that this infimum is obtained if and only if $\lambda = \mu = 0$. Underlying this is a closed conical image $A(K)$ such that arbitrarily close rank-one perturbations $A^\lambda(K)$ are not closed. Clearly this forces K to be nonpolyhedral and more [106].

10.1.7. Let X, Y be Banach spaces. Let $K \subset Y$ be a closed convex cone with nonempty interior and let $f: X \to \mathbb{R}$ and $g: X \to Y$ be convex and K-convex respectively. Apply Theorem 10.1.3 to the perturbation function $p(b) := \inf_{x \in X}\{f(x) : g(x) \leq_K b\}$.

10.1.8 (Separable quasi-complements [247]).** Let X be a Banach space with closed subspaces Y, Z. We say Y and Z are *quasi-complemented* or that Z is a *quasi-complement* for Y if $Y \cap Z = \{0\}$ and $Y + Z$ is dense in X. It is known that every subspace of $\ell_\infty(\mathbb{N})$ is quasi-complemented [247, Corollary 5.89]. A space has an infinite-dimensional separable quasi-complemented subspace if and only if it has an infinite-dimensional separable quotient [247, Corollary 5.80]. It is not known if every Banach space has some infinite-dimensional separable quasi-complemented subspace, but it does if its dual contains a copy of $c_0(\mathbb{N}), \ell_1(\mathbb{N})$ or an infinite-dimensional reflexive space [247, Cor 5.80 and Exercise. 5.10]. Many separable counterexamples lift to any superspace in which it is quasi-complemented.

(a) Show that in Theorem 10.1.2 the equivalence of (a), (b), (c) and (e) remains valid if X is assumed to have an infinite-dimensional separable quotient (or quasi-complement). Provide a construction of disjoint closed convex sets in such an X that cannot be separated.

(b) Show that in Theorem 10.1.2 the equivalence of (a) and (d) remains valid if X is assumed to be a general Banach space.

Hint. (a) Every separable space contains a symmetric compact convex densely spanning set. Thus, every space with a separable quasi-complement contains a (bounded) closed convex symmetric set C with such that $\bigcup_{n=1}^\infty nC$ is dense in X but C has empty interior. Now use Exercise 4.2.5 to conclude δ_C has a unique subgradient at the origin but is not differentiable there. Confirm that C has no supporting hyperplane at the origin. Deduce the equivalence of Theorem 10.1.2 (a)–(c).

When X/Y is separable, X/Y has a Markuševič basis, say $\{\hat{x}_i, \hat{\phi}_i\}_{i=1}^\infty$. Then choose $x_i \in \hat{x}_i$, and by scaling assume $\|x_i\| = 1$ and identify $\hat{\phi}_i$ with $\phi_i \in Y^\perp$. Now let $A := \{x \in X : \phi_i(x) \geq 0\, i \in \mathbb{N}\}$ and $B := \{ta - b : t \geq 0\}$ where $a := \sum_{i=1}^\infty 4^{-i} x_i$ and $b := \sum_{i=1}^\infty 2^{-i} x_i$. Follow the proof in Theorem 10.1.2 noting the span of $\{x_i + Y\}_{i=1}^\infty$ is norm dense in X.

(b) Create the required example in a separable subspace of X. This is an instance where embedding the example is much simpler than lifting it from a quotient. □

10.1.9 (Absorbing sets). Using the Baire category theorem, Proposition 4.1.5 showed a lsc convex function on a Banach space is continuous on the core of its domain. This exercise explores this and related results in the setting of a barreled space.

A convex set C satisfying $X = \bigcup\{tC : t \geq 0\}$ is said to be *absorbing* (and one can check $0 \in \text{core } C$).

(a) A normed space is said to be *barreled* if every closed convex absorbing subset C has zero in its interior. Use the Baire category theorem to show that Banach spaces are barreled. (There are normed spaces which are barreled but in which the Baire category theorem fails, and there are Baire normed spaces which are not complete: appropriate dense hyperplanes and countable codimension subspaces will do the job.)

(b) Let f be proper lsc and convex. Suppose that zero lies in the core of the domain of f. By considering the set

$$C := \{x \in X : f(x) \leq 1\},$$

deduce that f is continuous at zero.

(c) Show that an infinite-dimensional Banach space cannot be written as a countable union of finite-dimensional subspaces, and so cannot have a countable but infinite vector space (*Hamel*) basis.

(d) Let $X := \ell_2$ and let $C := \{x \in X : |x_n| \leq 2^{-n}\}$. Show

$$X \neq \bigcup\{tC : t \geq 0\} \text{ but } X = \text{cl} \bigcup\{tC : t \geq 0\}.$$

(e) Let $X := \ell_p$ for $1 \leq p < \infty$. Let

$$C := \{x \in X : |x_n| \leq 4^{-n}\},$$

and let

$$D := \{x \in X : x_n = 2^{-n}t, \ t \geq 0\}.$$

Show $C \cap D = \{0\}$, and so

$$T_{C \cap D}(0) = \{0\}$$

where $T_{C \cap D}$ is the tangent cone of $C \cap D$ (see p. 68), but

$$T_C(0) \cap T_D(0) = D.$$

(In general, as we have seen we need to require something like $0 \in \text{core}(C - D)$, which fails in this example.)

(f) Show that every (separable) infinite-dimensional Banach space X contains a proper vector subspace Y with $\text{cl}(Y) = X$. Thus, show that in every such space there is a nonclosed convex set with empty interior whose closure has interior.

10.1.10 (Unique subgradients). This exercise provides some explicit examples of the type ensured by Theorem 10.1.2(b).

(a) Let f be the indicator function of the nonnegative cone in ℓ_p for $1 \leq p < \infty$. Let x^* have strictly positive coordinates. Prove 0 is the unique element of $\partial f(x^*)$ but f is not continuous at x^*.
(b) Let $X := L_1[0, 1]$ with Lebesgue measure. Recall the negative Boltzmann–Shannon entropy:

$$B(x) := \int_0^1 x(t) \log x(t) \, dt$$

for $x(t) \geq 0$ almost everywhere and $B(x) := \infty$ otherwise. Show B is convex, nowhere continuous (but lsc), and has a unique subgradient when $x > 0$ almost everywhere, namely $1 + \log x(t)$.

10.1.11 (Norm-attaining functionals). James' theorem (4.1.27) ensures that there are non-norm-attaining functionals on any nonreflexive Banach space. This exercise requests some explicit examples of this.

(a) Find non-norm-attaining functionals in c_0, in ℓ_∞ and ℓ_1.
(b) Consider the closed unit ball of ℓ_1 as a set C in ℓ_2. Show C is closed and bounded with empty interior. Determine the support points of C.

10.1.12 (Proper support points).** Let C be a closed convex set. Then $x \in C$ is a *proper support point* of C if there exists $\phi \in C$ such that $\phi(x) = \sup_C \phi > \sup_C \phi$. The existence of closed convex sets in each nonseparable Banach space in which every point is a proper support point is dependent on set-theoretic axioms; see [247, Section 8.1] and the references therein.

(a) Let $C := \{(x, y, z) \in \mathbb{R}^3 : x^2 + y^2 \leq 1, z = 0\}$. Verify that $(0, 0, 0)$ is a support point of C that is a not proper support point.
(b) Let X be separable with $C \subset X$ closed, bounded, and convex. Let $\{x_n : n \in \mathbb{N}\}$ be dense in C. Let $\bar{x} := \sum_{n=1}^\infty 2^{-n} x_n$. Then any linear continuous functional f with $f(\bar{x}) = \sup_C f$ must be constant on C and so $\bar{x}$ is not a proper support point of C. Modify the construction to show that there is no need to assume C is bounded.
(c) Show every point of the nonnegative cone in the space $\ell_1(\mathbb{R})$ is a proper support point.

Theorem 10.1.1 showed that every infinite-dimensional Banach space can have a nonclosed linear subspace. The next three exercises examine some explicit examples and further related properties.

10.1.13. Let X be a separable Banach space. Construct a continuous linear mapping $T : X \to X$ such that the range of T is dense but not closed.

Hint. Let $\{x_i, x_i^*\}_{i=1}^\infty$ be a Markuševič basis for X where $\|x_i\| = 1$ for all $i \in \mathbb{N}$. Define $T : X \to X$ by $T(x) := \sum_{i=1}^\infty \frac{x_i^*(x) x_i}{2^i \|x_i^*\|}$.   □

10.1 Back to the finite

10.1.14 (Sums of closed cones). (a) Let $X := \ell_2$. Construct two closed convex cones (subspaces) S and T such that $S \cap T = \{0\}$ while $S^+ + T^- \neq \ell_2$. Deduce that the sum of closed subspaces may be dense.

(b) Let $X = \ell_2$. Construct two continuous linear operators mapping X to itself such that each has dense range but their ranges intersect only at zero. (This is easier if one uses the Fourier identification of L_2 with ℓ_2.)

10.1.15 (Sums of subspaces). (a) Let M and N be closed subspaces of X. Show that $M + N$ is closed when N is finite-dimensional. (Hint: First consider the case when $M \cap N = \{0\}$.)

(b) Let $X = \ell_p$ for $1 \leq p < \infty$. Define closed subspaces M and N by

$$M := \{x : x_{2n} = 0\} \text{ and } N := \{x : x_{2n} = 2^{-n} x_{2n-1}\}.$$

Show that $M + N$ is not closed. Observe that the same result obtains if M is replaced by the cone

$$K := \{x : x_{2n} = 0, \ x_{2n-1} \geq 0\}.$$

(Hint: Denote the unit vectors by (u_n). Let

$$x^n := \sum_{k<n} u_{2k-1} \text{ and } y^n := x^n + \sum_{k<n} 2^{-k} u_{2k}.$$

Then $x^n \in M$, $y^n \in N$, but $x^n - y^n \in M+N$ converges to $\sum_{k<\infty} 2^k u_{2k} \notin M+N$.)

(c) Relatedly, let $X := \ell_2$ and denote the unit vectors by (u_n). Suppose (α_n) is a sequence of positive real numbers with $1 > \alpha_n > 0$ and $\alpha_n \to 1$ sufficiently fast. Set

$$e_n := u_{2n-1}, \ f_n := \alpha_n u_{2n-1} + \sqrt{1 - \alpha_n^2} u_{2n}.$$

Consider the subspaces

$$M_1 := \text{cl span}\{e_1, e_2, \ldots\} \text{ and } M_2 := \text{cl span}\{f_1, f_2, \ldots\}.$$

(i) Show $M_1 \cap M_2 = \{0\}$ and that the sum $M_1^\perp + M_2^\perp$ is dense in X but not closed.

(ii) Dually, show that $M_1^\perp \cap M_2^\perp = \{0\}$ and that the sum $M_1 + M_2$ is dense in X but not closed.

(iii) Find two continuous linear operators on X, T_1, and T_2 such that both have dense range but $R(T_1) \cap R(T_2) = \{0\}$. (Such subspaces are called disjoint operator ranges.)

10.1.16 (Semicontinuity of separable functions on ℓ_p). As discussed in this section, lower-semicontinuity is a crucial assumption on convex functions in infinite-dimensional spaces. Part (b) of this exercise provides a natural example of a convex function that is not lsc. Let functions $\varphi_i : \mathbb{R} \to [0, +\infty]$ be given for $i \in \mathbb{N}$. Let the function F be defined on $X = \ell_p$ for $1 \leq p < \infty$ by

$$F(x) := \sum_i \varphi_i(x_i).$$

Relatedly, suppose the function $\varphi : \mathbb{R} \to (-\infty, +\infty]$ is given, and consider the function

$$F_\varphi(x) := \sum_i \varphi(x_i).$$

(a) Show that F is convex and lsc on X if and only if each φ_i is convex and lsc on $\mathbb{R}$.
(b) Suppose $0 \in \operatorname{dom} F_\varphi$. Show that F_φ is convex and lsc on X if and only if
  (i) φ is convex and lsc on $\mathbb{R}$, and
  (ii) $\inf_\mathbb{R} \varphi = 0 = \varphi(0)$.

Thus, for $\varphi := \exp^*$ we have F_φ is a natural convex function which is not lsc.

10.2 Notes on earlier chapters

Chapter 1. Our goal in this chapter is to illustrate that since their introduction by Jensen, convex functions have found far-reaching applications in diverse subjects. We have not endeavored to provide an historical account of the study of convexity, for that we already recommended Asplund [15], Fenchel [212], Berger [53] and make other such mention later in the book. Our notes for this chapter are admittedly sparse because many of the results are revisited in other chapters of this book in which further information can be found – including the notes to those chapters; for those topics that we do not revisit, we have endeavored to include appropriate references in the text.

We must also acknowledge that there is a large and diffuse literature on generalizations of convexity and convex functions which for the most part we have eschewed in this book – except when such a notion arises naturally. A few representative reference books are [25, 362, 401].

Finally, for convenience, we have made some expository decisions. For instance, our cones always contain zero, our functions are proper unless otherwise flagged, our normed spaces are usually real and complete, and sets are often assumed closed. Thus, a result might hold more generally but the generality would typically not add much to the content and might complicate the discussion.

Section 2.1 and 2.2. Several sources have inspired our presentation in this and future sections. For example, Stromberg's wonderful book [410] on classical analysis strongly influenced our presentation on convex functions on $\mathbb{R}$, and in particular Theorem 2.1.2 comes from [410, Theorem 4.43]. We believe the classic text of Roberts and Varberg [366] remains a standard by which books on convex functions should be measured, and readers will see its influence frequently in the current book. We should note that Lipschitz functions were introduced in [295]. The proof of the Hahn–Banach extension theorem we used follows that given in Rudin [383].

Much else in our approach to this and other chapters was – perhaps not surprisingly – motivated by [95]. The reader will see that we have recorded many of its exercises.

The reader will also recognize exercises from [332] and from [369]. We must also mention Zalinescu's book [445] which we have found is an indispensable reference for readers looking for a thorough yet efficient coverage of convex analysis in general vector spaces; its influence can be found throughout the present work.

Identifying linear mappings with multilinear mappings avoids the cumbersome notation would have arisen in the study of higher-order derivatives had we continued beyond those of second-order. For a rigorous account of higher order differentiability on Banach spaces, the reader can consult [144, 185], as our presentation here was admittedly casual and minimal. In this direction, let us point out that the limit in (2.2.4) could exist for all $h, k \in E$ without the mapping being bilinear; see Exercise 2.6.11 and Section 2.6 which also outlines more general approaches to second-order derivatives of convex functions.

Sections 2.3 and 2.4. A key consequence of Proposition 2.1.14 and Theorem 2.3.5 is that one has a fine analytic and computational tool for analyzing minima of composite convex functions. Further information on conjugates and computational convex analysis can be found in Y. Lucet's survey [298]. In particular, the notion of an iterated conjugate – in which one conjugates one variable at a time – is computationally useful in several dimensions, for just the same reason that iterated integration often is, see also Exercise 2.3.32. The reference [231, §1.6] contains informative material on the origin and treatment of experimental design (see Exercise 2.3.27(c)) through linear semi-infinite optimization problems. It would be hard to exaggerate the ubiquity of Fenchel duality as a tool in diverse fields [299] such as mechanics [195], machine learning [365], mathematical economics [300], statistical fitting algorithms [56], in addition to many others discussed in the text.

There are several ways in which one can 'average' or combine convex functions in addition to arithmetic averages and infimal convolutions. In this direction a paper by Bauschke and Wang [39] discusses a *kernel average* that includes these and many other well known averages, and provides applications of it in various areas of convex analysis; see also Exercise 4.4.8. The recent article [10] looks at the essential a priori properties that are needed to characterize several types of duality operations, such as conjugation and polarity. For example, it contains results of the flavor: if a transformation T on the lsc convex functions on $\mathbb{R}^n$ satisfies $T(Tf) = f$ and $f \leq g$ implies $Tf \geq Tg$, then T is essentially the operation of conjugation, i.e. $(Tf)(x) = f^*(Ax + x_0) + \langle x, x_0 \rangle + c_0$ where A is symmetric, $x_0 \in \mathbb{R}^n$ and $c_0 \in \mathbb{R}$ (see [11] and [10, Theorem 1]).

Section 2.5. The elementary proof of Rademacher's theorem (2.5.4) follows that given in [96] with a nice variant to be found in [329]. Slicker proofs are possible if one wishes to appeal to more integration theory [52].

Section 2.6. The proof of Theorem 2.6.1 largely follows the course suggested by Rockafellar in [377], in particular for the implication (c) implies (a); however we follow the proof of the converse as outlined by Crandall *et al.* in [162]. Theorem 2.6.1 and related results in the more general setting of Banach spaces were derived in [107, Section 3] which also built upon some work of Bangert [30]. Further generalizations of second-order differentiability along with properties thereof are provided in the paper [377] along with the references to applications of those properties. A version of Alexandrov's theorem on $\mathbb{R}^2$ was given by Busemann and Feller [138]. Alexandrov's theorem appeared in [2]; the proof of Alexandrov's theorem outlined here follows that of Crandall *et al.* [162] who followed Mignot's approach [316]. Other proofs of Alexandrov's theorem have been obtained by Bangert [30] and Rockafellar [374].

Ioffe and Penot [263] exploit Alexandrov-like theory, again starting with the subtle analysis in [162], to carefully study a *subjet* of a function f at x, the subjet $\partial_-^2 f(x)$ being defined as the collection of second-order expansions of all C^2 local minorants g with $g(x) = f(x)$. The limiting 2-subjet is $\overline{\partial}^2 f(x) := \limsup_{y \to_f x} \partial_-^2 f(x)$, Various distinguished subsets and limits are also considered. They provide a calculus, based on a sum rule for limiting 2-subjets (that holds for all lower-C^2 functions and so for all continuous convex functions) making note of both the similarities and differences from the first-order theory. Interesting refinements are given by Eberhard and Wenczel in [191].

Section 2.7. Much more detailed material can be found in [369]. In infinite dimensions this section can usefully be revisited in conjunction with Section 5.2 and Section 6.6.

Section 3.1. Our approach in this section is analogous to [418, 95]. The key idea, Theorem 3.1.6 (key theorem of polyhedrality), is due to Minkowski [318] and Weyl [436]. While the treatment is complete theoretically, it does little to shed light on the important and hard computational problem of efficiently moving between the two facial and verticial representations or of the general problem of computing the convex hull of a finite set of points [339]. The technique of Corollary 3.1.9 is extremely powerful as a way of exploiting linearity or polyhedrality in a more general problem. The failure to appreciate that Corollary 3.1.8 and its consequences such as Exercise 3.1.9 require constraint qualification in the non-polyhedral case haunted the early history of Fenchel duality and semi-infinite programming; and Example 3.2.14 even features in Ron Stern's academic novel *Goldman's Theorem* (2009).

Section 3.2. This section largely follows [288] and [95]. The Davis theorem (3.2.5) appeared in [168] (without the closure assumption). A host of convexity properties of eigenvalues like those of Exercise 3.2.5 (Examples of convex spectral functions) can be found in [261] or [49], for example. Many other properties analogously 'lift' from symmetric functions to symmetric matrices as illustrated for prox regularity in [165]. Fine surveys of eigenvalue optimization appear in [290, 289].

Section 3.3. The importance of linear programming duality, and of the simplex method, was first emphasized by Dantzig in the mid-1940s [166]. That of semidefinite duality was emphasized by Nesterov and Nemirovskii [331], especially from a computational complexity vantage-point. A good reference for general linear programming is [154] while a straightforward exposition of now celebrated central path methods may be found in [439]. Semidefinite programming has wide application in control theory [127] and elsewhere. A comprehensive recent survey of duality gaps (as first discovered by Duffin and others in the 1950's) in abstract linear programming and their resolution is given in [417].

10.2 Notes on earlier chapters

Section 3.4. The development follows [95, §8.1]. The corresponding selection material is developed in [378, pp. 186–195] along with generic selection results in finite dimensions. One example of the many uses of the Kakutani–Fan theorem is in establishing the existence of equilibria in mathematical economics. A discussion, based on nonsmooth mean value theorems, of stability and Lyapunov functions is given in [121, §3.5.2].

Section 3.5. A proof of the smooth variational principle along the lines of Theorem 3.5.1 can be achieved in superreflexive space [108]. Analogously, the proof of the convex mean value theorem of Exercise 3.5.5 generalizes entirely with the use of an appropriate approximate mean value theorem see 9.2.1. The variational proof given of Theorem 3.5.2 would seem to showcase the technique most likely to extend to the general Hilbert case setting – if indeed the result is true (see also Section 4.5).

Section 4.1. Separation theorems are at the foundation of much of modern functional analysis. This section presented the basics. The article [1] provides precise conditions under which closed convex sets can be separated in reflexive spaces.

More delicate examples than given in Exercise 4.1.17 can be constructed; see [348] for where on a Hilbert space X – or more generally [85, Example 4.1] where on Banach space X with separable quotient – there are proper lsc convex functions f and g whose domains are equal and dense in X, and their subdifferentials are at most single-valued, but $\text{dom}(\partial f) \cap \text{dom}(\partial g) = \emptyset$.

The nonemptiness of subdifferentials and the more delicate max formula are core results of the convex analysis. Fenchel, Moreau, Rockafellar, Valadier and many others contributed to the current form of these results.

Examples of emptiness of the subdifferential in the absence of the qualification conditions are discussed in [110, 349]. The reader is refered to [257] for information on calculus concerning ε-subdifferentials.

The notion of finite representability was introduced and studied by James in [264, 265].

Section 4.2. The terminology of bornologies used here is as in [350, p. 59]. The various versions of Šmulian's theorem presented here are natural variations of results from [404] which dealt with the case of norms.

Section 4.3. We proved only the bare minimum on variational principles in this section, a far more comprehensive treatment can be found in [121]. Ekeland's variational principle first appeared in [193] and was inspired by Bishop–Phelps theorems ([58, 59]). Smooth variational principles first appeared in [108] and a discussion of a general form of the Borwein–Preiss smooth variational principle can be found in [121, Section 2.5]. The version we presented for smooth bump functions (Theorem 4.3.6) was developed in [179]. The G_δ-condition in the conclusion of this the smooth variational principle (and many related problems) can be strengthened [181]. However, in this section, we have essentially followed the original approach of [180, Section I.2] with a strong influence from [350]. For unified and rather general approaches to several variational principles see [296] and [286]. We also recommend Lucchetti's book [297] for further reading on this topic.

A recent paper of De Bernardi and Veselý [173] established several interesting properties including the uncountability of support functionals to closed convex sets and pathwise connectedness of support points of closed convex sets. These results were then extended in [423] using a *parametric* smooth variational principle.

See also Section 6.6 for Stegall's variational principle (Corollaries 6.6.16 and 6.6.17) which has a linear perturbed function, and, in particular, Exercises 6.6.17 and 6.6.16 for further information on higher-order smooth variational principles when the derivatives need not be Lipschitz; further information in this direction can be found in [205].

As outlined in Exercise 4.3.3, longstanding interesting questions remain in geometric fixed point theory. A recent notable advance in this area was obtained in [187] where it is shown that every Banach space that can be embedded into $c_0(\Gamma)$ (for instance reflexives spaces) can be renormed to have the (weak) fixed point property for nonexpansive mappings, and this norm can be chosen arbitrarily close to the original norm.

Section 4.4. Much of our approach to boundedness properties of conjugate functions and the Moreau–Rockafellar theorems was motivated by the proofs outlined in [35]. Fenchel's original work is [211]. An attractive early summary of Fenchel duality theory is given in [367]. Much of the exposition on Fenchel duality follows the concise book [95] which is also a good source for additional examples and applications. The paper [136] provides a version of Fenchel duality using only that sum of the epigraphs of f^* and g^* is weak*-closed. Also, the hypothesis that T is bounded and linear can generally be relaxed to it being closed.

For further continuity properties of subdifferentials and an alternate approach to some things done herein, see [143]. For well developed theory on preservation of convergence through conjugation, one can consult Beer's monograph [44], additionally, one will find a development on the approximation of f by using infimal convolutions with *smoothing kernels* that provide more general results of nature of Corollary 4.4.17

Section 4.5. In this section we presented one proof that weakly closed Čebyšev sets in a Hilbert space are convex (Theorem 4.5.9), and outlined another in Exercise 4.5.9. There are other approaches to proving Čebyšev sets in finite-dimensional spaces are convex. See the updated version of the book [95, 96] or see [72] for an account of two other such proofs that use fixed-point methods and farthest points respectively. In contrast to the positive results in Exercises 4.5.9 and 4.5.10 we point out that there are examples in rotund reflexive spaces of Čebyšev subspaces whose metric projection is discontinuous [132], see also [229, p. 246], and Westfal and Frerking [435] show that any nonconvex Čebyšev set in a Hilbert space must have a badly discontinuous metric projection.

Also, if there is a nonconvex Čebyšev set in a Hilbert space, then the original proof of Asplund [14] shows that then there is a *cavern* that is not a Čebyšev set (a Klee or Asplund *cavern* is the complement of an open convex body). An example of a nonconvex Čebyšev set in an incomplete inner product space is presented in [269], with some details in the proof corrected later in [267], and then [270] shows that the nonconvex Čebyšev set constructed in the incomplete inner product space can also be bounded.

The historical notes in [175, Chapter 12] give a good account of the development of the known convexity properties of Čebyšev sets. Vlasov's survey [430] gives a comprehensive account of the research of Čebyšev sets as it was in 1973. Our selection of the material presented in this section was influenced very strongly by [445, Sections 3.8 and 3.9]. Many of the results in various forms are from [192, 405, 406, 429].

Some related information on companion bodies and antiproximal (also known as 'antiproximinal') norms is given in Exercise 6.6.12. The recent paper [324] provides a detailed recent study of what happens to d_C when the norm is replaced by the gauge of a nonsymmetric convex body.

Section 4.6. Theorem 6.1.2 is due to Preiss and Zajíček [356], we followed the proof as presented in [350, Theorem 2.11, p. 22]. An excellent treatment of null sets and the differentiability of Lipschitz functions in separable Banach spaces can be found in [52, Chapter 7], and for quick reference there is a detailed summary of the key notions on pp. 166–168 therein. According to Theorem 4.6.5, if $f : X \to \mathbb{R}$ is a continuous convex function where f' is locally Lipschitz, then f has a second-order Gâteaux derivative except at possibly an Aronszajn null set. In fact, [107, Corollary 4.2] says that if the set of Lipschitz smooth points of f is Aronszajn null, then so is the set of points where f fails to have a weak second-order Taylor expansion.

Our examples herein focused on the failure of Alexandrov's theorem in infinite-dimensional spaces, however, there are some positive results for special classes of convex functions, including convex integral functions, presented in [107, Section 7]; in particular, see Theorem 7.6 and Corollary 7.8 therein. See also [310, 328] for further results concerning Lipschitz smoothness and second-order subgradients of convex functions or convex integral functions.

This section made little reference to convex functions or norms that are everywhere second-order differentiable. More information can be found in [203].

The introduction of null sets and various versions of Theorem 4.6.5 were made independently by Christensen [153] (who used Haar null sets), Aronszajn [8] (using Aronszajn null sets) and Mankiewicz [301] (using another class of null sets now known to coincide with Aronszajn null sets); a proof of the version we stated can be found in [52, Theorem 6.42], as well as a thorough discussion discussion of the various classes of null sets. Gaussian null sets were introduced by Phelps [347].

The striking Example 4.6.11 on ℓ_2 was shown by Matoušek and Matoušková [305], and in fact the function constructed is a norm; later Matoušková generalized the construction to superreflexive spaces in [307].

In contrast to the situation of Exercise 4.6.12, the properties of Haar-null sets outside of separable space are subtle, see [105]. Indeed in large cardinality spaces it is possible for a measurable set to be locally null without being null.

The study of the differentiability of Lipschitz mappings between Banach spaces has seen significant advances in recent years. The reader is referred to [293] for an overview of some recent developments.

Section 5.1. Unlike almost all other topics, renorming is nearly exclusively a Banach space topic since all norms in a space of finite dimension are equivalent and the Euclidean norm is as good as it gets. But even in this setting changing the norm can sometimes have magical results, as illustrated by the discussion after Theorem 4.3.12. Sometimes it is useful to use p-norms for p near one or infinity as it allows one to study polyhedral structure in the limit via what is in effect a smooth or rotund renorming. This implicitly lies beneath much of the material on barrier functions. Equally, in the analysis of interior point or quasi-Newton methods and in related study of positive definite matrices it is quite usual to renorm (actually re-Hilbertize) the space via $\|x\|^2 := \langle A_n x, x \rangle$ where (A_n) is a sequence of positive-definite matrices. A simple illustration was given in Exercise 5.1.34. Another illustration of the value of finite-dimensional renorming is in the ability to move between different matrix norms (see Exercises 3.2.17 and 3.2.18).

Šmulian's theorems (5.1.4 and 5.1.5) are from [404]. The proof of Kadec's theorem (5.1.14) given here follows the approach of Davis and Johnson [170]. We should note that Klee was also an important contributor to development of renorming theory with papers such as [278] among many others. The Milman-Pettis theorem (5.1.20) was proved in [317, 345]; the proof presented here is from [294, Proposition 1.e.3, Vol. II].

The fundamental relationship between moduli of convexity and smoothness in Proposition 5.1.22 was established by Lindenstrauss [292]. A systematic account of duality relationships was given by Cudia [164] while Smith investigated relations among various types of rotundity and smoothness in papers such as [402, 403] and others that followed. Troyanski showed the important result that weakly compactly generated Banach spaces admit equivalent locally uniformly convex norms in [416]; this theorem has been the subject of many generalizations and variations. A comprehensive treatment of this subject is given in the monograph [180]. Further extensions and simplifications have arisen since that monograph, much of which has built on the work of Raja [360]; see also [361] and the references therein.

Asplund averaging theorems originate with [12]. A Baire category approach to this is given in [204]. For another introduction to renorming, the reader can consult [199].

See [203, 206, 205] for use of integral convolutions and on construction of smooth convex functions from uniformly smooth bump functions. Other interesting constructions of norms depending locally on only finitely many coordinates are given in [341]. Some analytic constructions can be found in [202].

Constructions of Gâteaux differentiable norms that are not everywhere Fréchet differentiable, but of a far more general and delicate nature than given in Exercise 5.1.26, can be found in [77].

Section 5.2. Results on the exposed point and smoothness duality with functions and their conjugates have their genesis in Asplund's and Rockafellar's [17], in fact, far more general results can be found in that seminal paper. The reader may have noticed the proof of the 'moreover' part of Theorem 5.2.3 in the nonconvex case depends essentially on a growth condition on $f - \phi_0$ rather than the boundedness of the domain of f. This is made explicit in the recent paper [285] where the conclusion is obtained when f is lsc and $\phi_0 \in \operatorname{int} \operatorname{dom} f^*$. Exercise 5.2.12 shows that Stegall's variational principle [407] is equivalent to a local version of Collier's theorem [159].

Sections 5.3 and 5.4. Uniformly convex functions on Banach spaces were introduced by Levitin and Poljak in [287]. Their properties were studied in depth by Zalinescu [444], and then later Azé and Penot [27] studied their duality with uniformly smooth convex functions. A systematic account of their properties is given in the book [445], including the related classes of functions that are uniformly convex or uniformly smooth on bounded sets. Characterizations of Lipschitz smoothness along the lines of Exercise 5.4.13, in various forms for were given in [203]. A treatment of renorming superreflexive space with norms with good moduli of smoothness or rotundity is given in the book [180]. The material for the last two subsections of Section 5.4 is based largely on [86].

Section 5.5. Lipschitz exposed points of bounded sets were introduced by Fabian in [200, p. 114]; that paper provides a detailed study of *Lipschitz determined* spaces, and dual characterizations of such spaces through Lipschitz exposed points of weak*-compact convex sets in the dual. Several results such as Theorem 5.5.3 and Example 5.5.4 are straight from that paper, however, our general presentation has a more explicit emphasis on functions and their conjugates as in [17]. A systematic study of norms and functions with locally or globally Lipschitz derivatives was given in [203]. We did not present duality conditions for convex functions whose gradients are locally Lipschitz. Results of this nature in Euclidean space can be found in the recent article [233].

Section 6.1. This section was influenced strongly by [350] and also by the survey [120]. We observe that it is quite usual now to refer to inner and outer semicontinuity rather than upper and lower semicontinuity, but we have opted to stay with the older nomenclature. Multifunction techniques are also very useful in the study of generalized derivatives in nonsmooth analysis, we refer the reader to [121, Chapter 5] and the references therein for further information in this direction. Chapter 9 further investigates monotone operators using an ingenious function of Fitzpatrick and the novel ideas it has spawned.

Section 6.1.1. The development follows [95, §8.1]. The corresponding selection material is also developed in [378, pp. 186–195] along with generic selection results in finite dimensions. The ease with which the proofs adapt from Section 3.4 illustrates that often the most efficient proof of a result in Euclidean space is a minor simplification of the general result; assuming notation has been carefully handled.

There are various stronger versions of Exercise 3.4.9 (the Hahn–Katětov–Dowker theorem to which Tong's name is also attached) that allow for extended real-valued f and g. In this case the distinctions between metrizable, normal and paracompact become significant. The metric version attributed to Hahn is described in [410]. It is possible to abstract further so that the Hahn–Banach sandwich theorem and Katětov's theorem are seen as two cases of the same result. We have not mentioned measurable selections of measurable multifunctions [378, Corollary 14.6], a topic of significant importance when dealing with integral functionals as in Section 6.3.

Section 6.2. Chapter 5 in [44] gives a good account of Kuratowski–Painlevé convergence; the definition we used is formulated as Theorem 5.3.5 therein. The paper [326] illustrates the importance of Mosco convergence. Theorem 6.2.9 was established by Mosco [327]. The earlier result for finite-dimensional spaces was established by Wijsman [437]. Theorem 6.2.10 was established by Beer in [42]; Beer also proved the analogous result for slice convergence [43] which includes Mosco's result because of the coincidence of slice and Mosco convergence in reflexive spaces.

Further results connecting epigraphical convergence and convergence of regularizations of convex functions can be found in [45, 112]. For a thorough development of these topics, one can consult Beer's monograph [44]. Stability of epi-convergence under addition is somewhat delicate. Indeed, [44, Example 7.1.6] presents an example of (nonconvex) functions (f_n) converging Attouch–Wets to f and (g_n) converging Attouch–Wets to g, but $(f_n + g_n)$ does not converge Attouch–Wets to $f + g$. For positive results of this nature in finite-dimensional spaces, see [378, Chapter 7], and a nice presentation in normed spaces is given in [297, Section 9.2].

There are various results on the set convergence of generalized second derivatives in separable Hilbert space. For an arbitrary closed convex function f and $x^* \in \partial f(x)$ we note that Kato [276] establishes the Mosco convergence of

$$\frac{f(x+th) - f(x) - t\langle x^*, x\rangle}{t}$$

to an appropriate generalized bilinear form as t decreases to zero.

Section 6.3. The work of Rockafellar [371] remains a concise yet good reference on integral functionals. An excellent recent paper on convex integral functionals is [262] by Alex Ioffe. The role of integral functionals in application to control and optimization is well described in [156, 157].

Section 6.4. The material within this section comes from [94]. In fact, one of the motivations for the study of strongly rotund functions is the study of best entropy estimation for moment problems and for more general inverse problems, as discussed in Section 7.6 and elsewhere in the book. More information on the behavior of such moment problems can be found in [89, 92, 94]. The use of the ℓ_1 norm is not covered by any of this theory, and indeed Exercise 6.4.4 shows it is impossible to place any finite strongly rotund function on the space. That said, in finite dimensions the polyhedrality of the norm makes it attractive for use with linear constraints. In certain settings such as feedback control the infinite-dimensional ℓ_1 norm is very useful, see [253]. Interesting extensions and applications of strongly rotundity for moment problems are made in [124] with application to the study of invariant measures.

Section 6.5. The discussion is based largely based on results derived in [109]. We recommend [342] for additional information on various operator spaces and corresponding sequence spaces discussed here. The Gâteaux differentiability characterization of Theorem 6.5.15 cannot be generalized to Fréchet differentiability. In fact, the eigenvalue mapping λ defined in this section may not even be continuous. One way to overcome this difficulty (as illustrated in Theorem 6.5.18) is to eliminate

10.2 Notes on earlier chapters

all the zeros in the definition of the eigenvalue mapping. Of course, an eigenvalue mapping $\widehat{\lambda}$ defined this way is not 'faithful'. We refer to [101] for additional details.

We have seen how successfully one may study linear operators whose spectrum is entirely discrete. Little of this approach would appear to carry over to more general spectral theory. For example, we have seen how a rearrangement invariant convex function ϕ on c_0 will induce a unitarily invariant convex function on B_0; and that ϕ^* on ℓ_1 will induce the unitarily invariant conjugate on B_1. What information is to be obtained about $(\phi \circ \lambda)^{**}$ on $B(H)$ from ϕ^{**} as a function on ℓ_∞?

Section 6.6. Seminal work on what are now called Asplund spaces comes from [13]. The lecture notes [350] provide both an efficient introduction and good reference source on Asplund spaces. The monograph [201] provides a more advanced treatment on weak Asplund spaces.

The idea to derive Stegall's variational principle from Collier's theorem [159] appears to have first been used in [205]; that approach uses a nonlinear version of Šmulian's differentiability theorem. Loewen and Wang made the connection more explicit in [296] with another variation of Šmulian's theorem using a perturbed function in the dual space rather than the conjugate. The very recent paper [285] explores this topic further along with some related topics. A local version of Collier's theorem appeared in [150], the proof is somewhat different from the one given here: it uses Minkowski gauges and works in some more general locally convex spaces. Stegall's variational principle is from [408]. The beautiful duality between Asplund and RNP was completed in [407]. Additionally, see [205, Section 5] for further information about smooth variational principles for higher-order smoothness and the geometry of Banach spaces.

For a more comprehensive treatment of the duality between Asplund spaces and those with the RNP we recommend Phelps' [350]. Our coverage on RNP spaces is made from a geometric and duality point of view, however, these spaces originally arose as those for which a Radon–Nikodým theorem with vector measures was valid; see Diestel and Uhl's monograph [184] for this and more. Although many results involving RNP spaces have been simplified since Bourgin's [126], that book remains a valuable reference on RNP and related spaces.

Section 6.7. All of the material on Banach lattices is directly from [90], for further results along those lines one can see [91]. Theorem 6.7.10 and related work can be found in [119, 75]. It is worth remarking that while – up to density character – Hilbert spaces are isometric as Banach spaces, but as Hilbert lattices $\ell_2(\mathbb{N})$ and $L_2[0, 1]$ are very different: the former has norm-compact order-intervals, the latter only weakly-compact order-intervals, and so on. Hence the Fourier transform is no respecter of the two orders.

Sections 7.1 and 7.2. The duality of essentially strictly convex and essentially smooth functions for finite-dimensional spaces is treated in detail in [369]. Lemma 7.2.2 is a special case of a slightly surprising result of L. Veselý from [350, p. 37]. Essentially strictly convex and essentially smooth functions for infinite-dimensional Banach spaces were introduced in [35] as a precursor to more algorithmic work [36] discussed in Section 7.5.

Section 7.3. The foundational results of this section are almost entirely from [35]. The theory of Legendre functions in reflexive space is quite satisfactory; but says nothing about convex functions without points of continuity such as the Shannon-entropy on $L_1[0, 1]$ [35].

Section 7.4. The section is based on the development in [115] but the key insight is to be found in the fundamental research of [331]. An infinite-dimensional result along the lines of that in Theorem 7.4.1, say to separable Hilbert space, would need to replace the existence of Haar measure by something quite different. In that vein, an illustrative positive result is given in Exercise 7.7.7.

Section 7.5 and Section 7.6. These sections follow [36] which also has a considerable amount of algorithmic analysis and application of the class of Legendre functions. The paper [139] uses interesting subgradient projection methods to solve some general *convex feasibility problems* in Euclidean spaces. Fisher information as discussed in Exercise 7.6.3 is of central interest in current mathematical physics [337]. The q-logarithm defined by $\log_q(t) := (t^{1-q} - 1))/(1 - q)$ (a different normalization to that used in 7.6.3(b)) leads to the *Tsallis entropy* and to an enormous and rapidly-growing statistical physics literature (see http://tsallis.cat.cbpf.br/biblio.htm for a massive bibliography).

The paper [291] comprehensively reviews the authors' results on so called *f-divergences* and distances for distributions and stochastic processes. Two special classes are identified. The first is the I_α-divergences, $\alpha \in \mathbb{R}$, given by the functions $I_\alpha(u) := (u^\alpha - \alpha u + \alpha - 1)/\alpha(\alpha - 1)$ for $u \geq 0$, extended continuously to 0 and 1. These include the Kullback–Leibler (1951) information, the I-divergence of Csiszár (1975), Hellinger integrals of order α such as the Hellinger (1909) distance, and Rényi (1961) distances of order α. The second class is the χ^α-divergences, $\alpha > 0$, given by $\chi^\alpha(u) := |1 - u^\alpha|^{1/\alpha}$ for $\alpha \leq 1$ and $|1 - u|^\alpha$ for $\alpha > 1$, where $u \geq 0$. Important cases are the total variation distance ($\alpha = 1$), and the χ^2-distance of Pearson (1900).

Section 7.7. This section of the chapter is again based substantially on the current authors work in [115]. The most significant open issues regard the construction of more explicit and 'natural' examples. The forthcoming book [37] should be recalled as it promises to have a current and complete treatment of much relevant material in the centrally important setting of Hilbert space.

Section 8.1. Much of the material in this chapter was originally motivated by an unpublished 1991 note of the first author reprised in [77] that led to the connection between Banach spaces not containing ℓ_1 and the coincidence of two different bornological derivatives in the convex case. Proposition 8.1.1 combines results from [77, 85]. Proposition 8.1.2 and Theorem 8.1.3 follow from combining results from [77, 85, 113]. The interested reader may consult [47] for an alternative approach to the Josefson–Nissenzweig theorem.

Section 8.2. The results originate essentially in [77, 85, 113], except for Theorem 8.2.8 which is from [116] and was motivated by examples from [392]. Exercises 8.2.5 through 8.2.8 are taken from Godefroy [232]. The results show how fundamental structurally very simple sounding convexity notions are.

Section 8.3. This material is similarly based on [103]. The recent paper [426] provides a new equivalent condition for extending continuous convex functions that produces a more elementary proof of Corollary 8.3.10 by avoiding the theorem of Rosenthal from [381]. Some nice results on extensions of DC functions are also obtained in [426]. Part (a) of Exercise 8.3.8 is based on a classical result of Phillips [352].

Section 8.4. While the study of convex functions outside of Banach space can be quite rich, Exercise 8.4.1 and the following discussion make it quite clear that this is not true if one wishes to exploit modern variational analysis.

Section 9.1. The *Fitzpatrick function* introduced in [215] was discovered precisely to provide a more transparent convex alternative to an earlier saddle function construction due to Krauss [279] – we have not discussed saddle-functions in any detail but they produce interesting maximal monotone operators [369, §33 & §37]. At the time, Fitzpatrick's interests were more centrally in the differentiation theory for convex functions and monotone operators. The development in Section 9.1 through Section 9.5 is based largely on [69]. The function $\mathcal{P}_T$ was named for Penot from whom we really first learned its value [343, 344] – it might have been equally well-named $\mathcal{BS}_T$ [137] save for its unfortunate meaning in English.

The search for results relating when a maximal monotone T is single-valued to differentiability of $\mathcal{F}_T$ did not yield fruit, and Fitzpatrick put the function aside. On being rediscovered and exploited by [137, 343, 344] and then by many others the field was rejuvenated. Differentiability is still the one area where to the best of our knowledge $\mathcal{F}_T$ has proved of little help – in part because generic properties of dom $\mathcal{F}_T$ and of dom(T) seem poorly related. Some progress is reported in [190]. That said, monotone operators often provide efficient ways to prove differentiability of convex functions. The discussion of Mignot's theorem in Exercise 9.1.1 is somewhat representative of how this works as is the treatment in [350].

By contrast, as we have seen the Fitzpatrick function and its relatives now provide the easiest access to a gamut of maximality, domain and range convexity, solvability, and boundedness results as is detailed throughout this chapter.

Section 9.2. The treatment of explicit acyclic mappings is taken from [118]. Much more detailed information on n-Fitzpatrick functions is to be found in [31] and many instructive explicit computations of Fitzpatrick functions are given in [38]. Explicit computation is often remarkably hard: surprisingly so in light of the theoretical power of the function. A comprehensive recent study of subdifferential characterizations of functional properties – monotonicity in various forms, Lipschitzness, order-isotonicity – is contained in [161]. A detailed accounting in the Fréchet case is given in [121].

Section 9.3. The results in this section become strikingly simple in Hilbert space, as is detailed in [37], largely because $J = I$. Since many characterizations of Hilbert space can be viewed as restricted versions of 'J behaves like I' (see [66] and a lovely book by Amir [7]), it is often fruitless to try to extend results from Hilbert space by imposing Hilbert-like behavior on the duality mapping. A notable exception is that J is weakly *sequentially* continuous in Banach sequence spaces such as ℓ_p for $1 \leq p < \infty$ [239].

Section 9.4. Corollary 9.4.10 deserves further study. For example, for a maximal monotone operator T when $0 \in \text{dom } T$ but int dom T is empty, is it possible for T_F to be nonmaximal for all nontrivial finite-dimensional subspaces F? If, by contrast, T_F is always maximal for some F containing given $x, y \in \text{dom } T$ then many results follow.

Section 9.5. Rugged spaces were discussed in detail in [33]. In addition, [33] shows that every bounded linear maximal monotone mapping on a Banach lattice is (NI) if and only if it is (D) if and only if X contains no complemented copy of ℓ_1. Such results extend to closed linear operators [396], and many are subsumed or much clarified by [6]. Note that $C[0,1]$ contains only noncomplemented copies of ℓ_1.

It would be very interesting to determine whether the graph of a maximal monotone mapping or a subgradient must be norm-weak* closed if the domain has nonempty interior.

Section 9.6. The clarity of the Fitzpatrick function based constructions has led to the truly significant recent advances delineated in this section and in Section 9.7; and offers hope for resolving the most persistent remaining open questions about maximal monotone operators such as:

1. Those listed in Exercise 9.5.4.
2. Does Theorem 9.6.1 (b) hold with no extra hypotheses on S or T?
3. Given a maximal monotone operator T, can one associate a convex function f_T to T in such a fashion that $T(x)$ is singleton as soon as $\partial f_T(x)$ is? Kenderov's theorem of Exercise 9.1.3 neatly links a nonconvex function to T in the WCG setting.
4. Are there some nonreflexive spaces such as c_0 – in addition to the class of Exercise 9.6.3 – for which the answer to such questions might be answered in the affirmative? Note the *James space* (which is codimension-one in its bidual, [199]) lies in the class but c_0 does not [76].

Section 9.7. An excellent reprise of what was known from a convex perspective about monotone operators outside of reflexive space at the end of 2007 is to be found in Stephen Simons' fine new book [396]. Note we barely mentioned monotone operators of type (ED), central to [396], as the class is more topologically technical and further from our current concerns. The very recent results of this section and Section 9.6 highlight the vigorous research going on in the area. Indeed, had the fine results in [6] been discovered a little earlier, our treatment might well have been somewhat reorganized.

The structure of maximal monotone operators outside Banach space is not especially significant in its own right – and as we have seen many pathologies can occur (notably Exercise 8.4.1). Nonetheless in nonreflexive Banach space it is helpful to cast representativity results in the language of locally convex vector spaces; if only to limit the need to reprove results in different weak topologies. Many useful results of this genre are given in [432].

List of symbols

aff S ≡ affine hull of a set S, 66
f^{**} ≡ Fenchel biconjugate of a function f, 44
B_0 ≡ space of compact self-adjoint operators, 312
B_1 ≡ operators with finite trace, 312
B_p ≡ Shatten p-spaces, 313
B_{sa} ≡ bounded self-adjoint operators on complex ℓ_2-space, 312
B_X ≡ $\{x \in X : \|x\| \leq 1\}$, the closed unit ball of a normed space X
bdry S ≡ boundary of a set S with respect to norm topology unless stated otherwise
$C^k(U)$ ≡ the k-times continuously differentiable functions on U 222
cl f or $\bar{f}$ ≡ closure of a function f, 21
A^c ≡ complement of a set A
cont f ≡ points of continuity of a function f
conv S ≡ convex hull of a set S, 21
conv f ≡ convex hull of a function f, 21
$\overline{\text{conv}}S$ ≡ closed convex hull of a set S
$\overline{\text{conv}}^{w^*}(S)$ ≡ weak*-closed convex hull of a set S
$\bar{S}$ ≡ norm closure of a set S
$\bar{S}^\tau$ ≡ closure of a set S with respect to the topology τ
core S ≡ core of a set S, 25
$D_u f(x)$ ≡ derivative of f at x in the direction u, 34
$f'(x)$ ≡ derivative of f at x, 34
$\nabla f(x)$ ≡ derivative of f at x, 34
$f'(x; h)$ ≡ directional derivative of f in the direction h, 25
$d_S(\cdot)$ ≡ distance function to a set S, 22
dom f ≡ domain of a function f, 18
$D(T)$ ≡ domain of a multifunction T, 276
X^* ≡ dual space of continuous linear functionals on X
epi f ≡ epigraph of a function f, 20
$\partial_\varepsilon f(\cdot)$ ≡ ε-subdifferential of f, 137
f^* ≡ Fenchel (Moreau–Rockafellar) conjugate of a function f, 44
$\mathcal{F}_T$ ≡ Fitzpatrick function, 404
$\mathcal{FL}_{\widehat{S}}$ ≡ Fitzpatrick's last function, 423
γ_C ≡ gauge function of a set C, 22, 127
$G(T)$ or graph(T) ≡ graph of a multifunction T, 111
$\mathcal{H}_T$ ≡ representative function, 406

List of symbols

$\delta_C \equiv$ indicator function of a set C, 21
$f \square g \equiv$ infimal convolution of functions f and g, 51
$\langle \cdot, \cdot \rangle \equiv$ inner product, 143 (additionally $\langle \phi, x \rangle$, $\langle x, \phi \rangle$ and $\phi(x)$ where $x \in X$, $\phi \in X^*$ can all represent the evaluation of ϕ at x when X is not necessarily an inner product space)
int $S \equiv$ interior of a set S
$J(x) \equiv$ duality map, 133
$\ell_p(\Gamma) \equiv$ the ℓ_p space over an index set Γ; $\ell_p := \ell_p(\mathbb{N})$
$L_p \equiv$ the corresponding Lebesgue space of p-integrable functions over an appropriate measure algebra
$c_0(\Gamma) \equiv$ the c_0 space over an index set Γ; $c_0 := c_0(\mathbb{N})$
LUR $\equiv$ locally uniformly rotund, 213
$\delta_X(\cdot) \equiv$ modulus of convexity of $(X, \|\cdot\|)$, 215
$\delta_f(\cdot) \equiv$ modulus of convexity of the function f, 244
$\rho_X(\cdot) \equiv$ modulus of smoothness of $(X, \|\cdot\|)$, 216
$\rho_f(\cdot) \equiv$ modulus of smoothness of the function f, 252
$P_C(\cdot) \equiv$ nearest point mapping, 188
$N_C(x) \equiv$ normal cone to C at x, 67
$\mathcal{P}_T \equiv$ convexification of $\mathcal{H}_T$, 406
$S^\circ \equiv$ polar of a set S, 209
$2^X \equiv$ the collection of subsets of X
$S_\circ \equiv$ pre-polar of a set S, 209
$K^- \equiv$ (negative) polar cone of a set K, 67
$K^{--} \equiv$ bipolar cone of a set K, 67
$K^+ \equiv$ (positive) polar cone of a set K, 67
$R_{A,(a,a^*)} \equiv$ Rockafellar function, 429
$\mathbb{R}^n_+ \equiv [0, +\infty)^n$, the nonnegative orthant in $\mathbb{R}^n$.
$\mathbb{R}^n_{++} \equiv (0, +\infty)^n$, the interior of the nonnegative orthant in $\mathbb{R}^n$.
$S^n_+ \equiv$ positive semidefinite matrices among the symmetric n by n matrices, 67
ri $S \equiv$ relative interior of a set S, 66
$\nabla^2 f(\cdot)$ or $f''(\cdot) \equiv$ second-order derivative of the function f, 38
$S_X \equiv \{x \in X : \|x\| = 1\}$, the closed unit sphere of the normed space X
span $S \equiv$ linear span of a set S
$\partial f(x) \equiv$ subdifferential of the function f at x, 26
$\partial_F f(x) \equiv$ Fréchet subdifferential of the function f at x, 165
$\sigma_C \equiv$ support function of a set C, 127
$T_C \equiv$ normal mapping, 436
Type (ANA) $\equiv$ almost negatively aligned operator, 455
Type (BR) $\equiv$ Brøndsted–Rockafellar property, 455
Type (FP) $\equiv$ locally maximal monotone operator, 410
Type (LMM) $\equiv$ locally maximal monotone operator, 410
Type (NI) $\equiv$ type of monotone operator, 439
Type (VFP) $\equiv$ maximal monotone locally, 410
UC $\equiv$ uniformly convex, 215
$U_X \equiv$ open unit ball with respect to norm topology in X
WCG $\equiv$ weakly compactly generated space, 220

References

[1] S. Adly, E. Ernst, and M. Théra. On the closedness of the algebraic difference of closed convex sets. *J. Math. Pures Appl.*, **82**:1219–1249, 2003.

[2] A. D. Alexandrov. Almost everywhere existence of the second differential of a convex function and some properties of convex surfaces connected with it. *Leningrad State Univ. Ann. Math Ser.*, **6**:3–35, 1939.

[3] M. Alves and B. F. Svaiter. Brøndsted-Rockafellar property and maximality of monotone operators representable by convex functions in non-reflexive Banach spaces. *J. Convex Anal.*, **15**:693–706, 2008.

[4] M. Alves and B. F. Svaiter. Maximal monotone operators with a unique extension to the bidual. *J. Convex Anal.*, **16**:409–421, 2009.

[5] M. Alves and B. F. Svaiter. A new proof for maximal monotonicity of subdifferential operators. *J. Convex Anal.*, **15**:345–348, 2008.

[6] M. Alves and B. F.. Svaiter. On gossez type (D) maximal monotone operators. arXiv:0903.5332v1, 2009.

[7] Dan Amir. *Characterizations of Inner Product Spaces*, volume 20 of *Operator Theory: Advances and Applications*. Birkhäuser, Boston, 1986.

[8] N. Aronszajn. Differentiability of Lipschitzian mappings between Banach spaces. *Studia Math.*, **57**:147–190, 1976.

[9] Z. Artstein. Discrete and continuous bang-bang and facial spaces or: look at the extreme points. *SIAM Review*, **22**:172–185, 1980.

[10] S. Artstein-Avidan and V. Milman. A characterization of the concept of duality. *Electron. Res. Announc. Math. Soc.*, **14**:42–59, 2007.

[11] S. Artstein-Avidan and V. Milman. A characterization of the concept of duality in asymptotic geometric analysis and the characterization of the Legendre transform. *Ann. of Math.*, **196**:661–674, 2009.

[12] E. Asplund. Averaged norms. *Israel J. Math.*, **5**:227–233, 1967.

[13] E. Asplund. Fréchet differentiability of convex functions. *Acta Math.*, **121**:31–47, 1968.

[14] E. Asplund. Čebysev sets in Hilbert space. *Trans. Amer. Math. Soc.*, **144**:235–240, 1969.

[15] E. Asplund. Topics in the theory of convex functions. In *Theory and Applications of Monotone Operators*, pages 1–33. Gubbio, 1969.

[16] E. Asplund. A monotone convergence theorem for sequences of nonlinear mappings. *Proc. Symposia Pure Math.*, **18**:1–9, 1970.

[17] E. Asplund and R. T. Rockafellar. Gradients of convex functions. *Trans. Amer. Math. Soc.*, **139**:443–467, 1969.

[18] H. Attouch. Convergences de fonctions convexes, des sous-différentiels et semi-groupes associés. *C. R. Acad. Sci. Paris*, **284**:539–542, 1977.

[19] H. Attouch. *Variational Convergence for Functions and Operators*. Pitman, New York, 1984.

[20] H. Attouch and G. Beer. On the convergence of subdifferentials of convex functions. *Arch. Math.*, **60**:389–400, 1993.

[21] J.-P. Aubin and A. Cellina. *Differential Inclusions*. Springer-Verlag, Berlin, 1984.

[22] J.-P. Aubin and H. Frankowska. *Set-Valued Analysis*, volume 2 of *Systems & Control: Foundations & Applications*. Birkhäuser, Boston, 1990.

[23] A. Auslender and J. P. Crouzeiz. Well behaved asymptotical convex functions. *Ann. Inst. Henri Poincaré, Anal. Non Linéaire*, **S6**:101–121, 1989.

[24] A. Auslender and M. Teboulle. *Asymptotic Cones and Functions in Optimization and Variational Inequalities*. Springer-Verlag, New York, 2003.

[25] M. Avriel, W. Diewert, S. Schaible, and I. Zang. *Generalized concavity*, volume 36 of *Mathematical Concepts and Methods in Science and Engineering*. Plenum Press, New York, 1988.

[26] D. Azagra and J. Ferrera. Every closed convex set is the set of minimizers of some C^∞-smooth convex function. *Proc. Amer. Math. Soc.*, **130**:3687–3892, 2002.

[27] D. Azé and J.-P. Penot. Uniformly convex and uniformly smooth convex functions. *Ann. Fac. Sci. Toulouse Math.*, **4**:705–730, 1995.

[28] G. Bachman and N. Narici. *Functional Analysis*. Dover Books, 1966.

[29] E. Balder. The Brachistochrone problem made elementary. Available at: http://www.math.uu.nl/people/balder/talks/vanbeek.pdf, 2002.

[30] V. Bangert. Analytische Eigenschaften konvexer Funktionen auf Riemannschen Mannigfaltigkeiten. *J. Reine Angew. Math*, **307**:309–324, 1979.

[31] S. Bartz, H. H. Bauschke, J. M. Borwein, S. Reich, and X. Wang. Fitzpatrick functions, cyclic monotonicity and Rockafellar's antiderivative. *Nonlinear Anal.*, **66**:1198–1223, 2007.

[32] H. H. Bauschke and J. M. Borwein. Legendre functions and the method of random Bregman projections. *J. Convex Anal.*, **4**:27–67, 1997.

[33] H. H. Bauschke and J. M. Borwein. Maximal monotonicity of dense type, local maximal monotonicity, and monotonicity of the conjugate are all the same for continuous linear operators. *Pacific J. Math.*, **189**:1–20, 1999.

[34] H. H. Bauschke and J. M. Borwein. Joint and separate convexity of the Bregman distance. In D. Butnariu, Y. Censor, and S. Reich, editors, *Inherently Parallel Algorithms in Feasibility and Optimization and their Applications (Haifa 2000)*, pages 23–36. Elsevier, 2001.

[35] H. H. Bauschke, J. M. Borwein, and P. L. Combettes. Essential smoothness, essential strict convexity, and Legendre functions in Banach spaces. *Commun. Contemp. Math.*, **3**:615–647, 2001.

[36] H. H. Bauschke, J. M. Borwein, and P. L. Combettes. Bregman monotone optimization algorithms. *SIAM J. Control and Optim.*, **42**:596–636, 2003.

[37] H. H. Bauschke and P. L. Combettes. *Convex Analysis and Monotone Operator Theory in Hilbert Spaces*. CMS Books in Mathematics. Springer-Verlag, to appear.

[38] H. H. Bauschke, D. A. McLaren, and H. S. Sendov. Fitzpatrick functions: inequalities, examples and remarks on a problem of S. Fitzpatrick. *J. Convex Anal.*, **13**:499–523, 2006.

[39] H. H. Bauschke and X. Wang. The kernel average for two convex functions and its applications to the extension and representation of monotone operators. *Trans. Amer. Math. Soc.*, to appear.

[40] H. H. Bauschke, X. Wang, and L. Yao. Monotone linear relations: maximality and Fitzpatrick functions. *J. Convex Anal.*, **16**:673–686, 2009.

[41] H. H. Bauschke, X. Wang, X. Yuan, and J. Ye. Bregman proximity operators and Chebyshev sets. Preprint, 2008.

[42] G. Beer. Conjugate convex functions and the epi-distance topology. *Proc. Amer. Math. Soc.*, **108**:117–126, 1990.

[43] G. Beer. The slice topology: a viable alternative to Mosco convergence in nonreflexive spaces. *Nonlinear Anal.*, **19**:271–290, 1992.

[44] G. Beer. *Topologies on Closed and Closed Convex Sets*, volume 268 of *Mathematics and Its Applications*. Kluwer Academic Publishers, Dordrecht, 1993.

[45] G. Beer. Lipschitz regularization and the convergence of convex functions. *Numer. Funct. Anal. Optim.*, **15**:31–46, 1994.

[46] G. Beer and J. M. Borwein. Mosco convergence and reflexivity. *Proc. Amer. Math. Soc.*, **109**:427–436, 1990.

[47] E. Behrends. New proofs of Rosenthal's ℓ_1-theorem and the Josefson–Nissenzweig theorem. *Bull. Polish Acad. Sci. Math*, **43**:283–295, 1995.

[48] J. L. Bell and D. H. Fremlin. A geometric form of the axiom of choice. *Fund. Math.*, **77**:167–170, 1972.

[49] R. Bellman. *Introduction to Matrix Analysis*. SIAM, Philadelphia, 1997.

[50] A. Ben-Tal and M. Teboulle. Hidden convexity in some nonconvex quadratically constrained quadratic programming. *Math. Programming*, **72**:51–63, 1996.

[51] J. Benoist, J. M. Borwein, and N. Popovici. A characterization of quasiconvex vector-valued functions. *Proc. Amer. Math. Soc.*, **131**:1109–1113, 2003.

[52] Y. Benyamini and J. Lindenstrauss. *Geometric Nonlinear Functional Analysis*, volume 48 of *AMS Colloquium Publications*. American Mathematical Society, Providence, Rhode Island, 2000.

[53] M. Berger. Convexity. *Amer. Math. Monthly*, **97**:650–678, 1990.

[54] D. S. Bernstein. *Matrix Mathematics: Theory, Facts, and Formulas with Application to Linear Systems Theory*. Princeton University Press, Princeton, NJ, 2005.

[55] B. N. Bessis and F. H. Clarke. Partial subdifferentals, derivatives and Rademacher's theorem. *Trans. Amer. Math. Soc.*, **351**:2899–2926, 1999.

[56] B. Bhattacharya. An iterative procedure for general probability measures to obtain 1-projections onto intersections of convex sets. *Ann. Statist.*, **34**:878–902, 2006.

[57] G. Birkhoff. Tres observaciones sobre el algebra lineal. *Nacionale Tucamán Revista*, **5**:147–151, 1946.

[58] E. Bishop and R. R. Phelps. A proof that every Banach space is subreflexive. *Bull. Amer. Math. Soc.*, **67**:97–98, 1961.

[59] E. Bishop and R. R. Phelps. The support functionals of a convex set. In V.L. Klee, editor, *Proc. Sympos. Pure Math.*, volume VII, pages 27–35. Amer. Math. Soc., Providence, R.I., 1963.

[60] B. Bollobás. An extension to the theorem of Bishop and Phelps. *Bull. London Math. Soc.*, **2**:181–182, 1970.

[61] R. Bonic and J. Frampton. Smooth functions on Banach manifolds. *J. Math. Mech.*, **15**:877–898, 1966.

[62] J. M. Borwein. A Lagrange multiplier theorem and a sandwich theorem for convex relations. *Math. Scand.*, **48**:198–204, 1981.

[63] J. M. Borwein. A note on the existence of subgradients. *Math. Programming*, **24**:225–228, 1982.

[64] J. M. Borwein. On the Hahn–Banach extension property. *Proc. Amer. Math. Soc.*, **86**:42–46, 1982.

[65] J. M. Borwein. On the existence of Pareto efficient points. *Math. Oper. Res.*, **8**:64–73, 1983.

[66] J. M. Borwein. An integral characterization of Euclidean space. *Bull. Austral. Math. Soc.*, **29**:357–364, 1984.

[67] J. M. Borwein. Epi-Lipschitz-like sets in Banach space: theorems and examples. *Nonlinear Anal.*, **11**:1207–1217, 1987.

[68] J. M. Borwein. Minimal CUSCOS and subgradients of Lipschitz functions. In J.B. Baillon and M. Thera, editors, *Fixed Point Theory and its Applications*, Pitman Research Notes in Mathematics, pages 57–81. Longman, Essex, 1991.

[69] J. M. Borwein. Maximal monotonicity via convex analysis. *J. Convex Anal.*, **13**:561–586, 2006.

[70] J. M. Borwein. Asplund decompositions of monotone operators. *Proc. Control, Set-Valued Analysis and Applications, ESAIM: Proceedings, A. Pietrus & M. H. Geoffroy, Eds.*, **17**:19–25, 2007.

[71] J. M. Borwein. Maximality of sums of two maximal monotone operators in general Banach space. *Proc. Amer. Math. Soc.*, **135**:3917–3924, 2007.

[72] J. M. Borwein. Proximality and Chebyshev sets. *Optimization Letters*, **1**:21–32, 2007.

[73] J. M. Borwein and D. Bailey. *Mathematics by Experiment*. A K Peters, Natick, MA, 2004.

[74] J. M. Borwein, D. Bailey, and R. Girgensohn. *Experimentation in Mathematics*. A K Peters, Natick, MA, 2004.

[75] J. M. Borwein and M. A. H. Dempster. The linear order complementarity problem. *Math. Oper. Res.*, **14**:534–558, 1989.

[76] J. M. Borwein and A. Eberhard. Banach spaces of type (NI) and maximal monotone operators on reflexive Banach spaces. Preprint, 2008.

[77] J. M. Borwein and M. Fabian. On convex functions having points of Gateaux differentiability which are not points of Fréchet differentiability. *Canad. J. Math.*, **45**:1121–1134, 1993.

[78] J. M. Borwein, M. Fabian, and J. Vanderwerff. Characterizations of Banach spaces via convex and other locally Lipschitz functions. *Acta. Vietnam.*, **22**:53–69, 1997.

[79] J. M. Borwein and S. Fitzpatrick. Local boundedness of monotone operators under minimal hypotheses. *Bull. Austral. Math. Soc.*, **39**:439–441, 1989.

[80] J. M. Borwein and S. Fitzpatrick. Existence of nearest points in Banach spaces. *Canad. J. Math.*, **41**:702–720, 1989.

[81] J. M. Borwein and S. Fitzpatrick. Mosco convergence and the Kadec property. *Proc. Amer. Math. Soc.*, **106**:843–851, 1989.

[82] J. M. Borwein and S. Fitzpatrick. A weak Hadamard smooth renorming of $L_1(\Omega, \mu)$. *Canad. Math. Bull.*, **36**:407–413, 1993.

[83] J. M. Borwein, S. Fitzpatrick, and R. Girgensohn. Subdifferentials whose graphs are not norm×weak-star closed. *Canad. Math. Bulletin*, **46**:538–545, 2003.

[84] J. M. Borwein, S. Fitzpatrick, and P. Kenderov. Minimal convex uscos and monotone operators on small sets. *Canad. J. Math.*, **43**:461–476, 1991.

[85] J. M. Borwein, S. Fitzpatrick, and J. Vanderwerff. Examples of convex functions and classifications of normed spaces. *J. Convex Anal.*, **1**:61–73, 1994.

[86] J. M. Borwein, A. Guirao, P. Hájek, and J. Vanderwerff. Uniformly convex functions on Banach spaces. *Proc. Amer. Math. Soc.*, **137**:1081–1091, 2009.

[87] J. M. Borwein and C. Hamilton. Symbolic Fenchel conjugation. *Math. Programing*, **116**:17–35, 2009.

[88] J. M. Borwein, M. Jiménez Sevilla, and Moreno J. P. Antiproximinal norms in Banach spaces. *J. Approx. Theory*, **114**:57–69, 2002.

[89] J. M. Borwein and A. S. Lewis. On the convergence of moment problems. *Tran. Amer. Math. Soc.*, **325**:249–271, 1991.

[90] J. M. Borwein and A. S. Lewis. Partially-finite convex programming (I). *Math. Prog.*, **57**:15–48, 1992.

[91] J. M. Borwein and A. S. Lewis. Partially-finite convex programming (II). *Math. Prog.*, **57**:49–83, 1992.

[92] J. M. Borwein and A. S. Lewis. Convergence of decreasing sequences of convex sets in non-reflexive Banach spaces. *Set-Valued Analysis*, **4**:355–363, 1993.

[93] J. M. Borwein and A. S. Lewis. Partially-finite programming in l^1 and the existence of maximum entropy estimates. *SIAM J. Optim.*, **3**:248–267, 1993.

[94] J. M. Borwein and A. S. Lewis. Strong rotundity and optimization. *SIAM J. Optim.*, **4**:146–158, 1994.

[95] J. M. Borwein and A. S. Lewis. *Convex Analysis and Nonlinear Optimization: Theory and Examples*, volume 3 of *CMS Books in Mathematics*. Springer, New York, 2000.

[96] J. M. Borwein and A. S. Lewis. *Convex Analysis and Nonlinear Optimization: Theory and Examples*, volume 3 of *CMS Books in Mathematics*. Springer, New York, 2nd edition, 2005.

[97] J. M. Borwein, A. S. Lewis, N. N. Limber, and D. Noll. Fast heuristic methods for function reconstruction using derivative information. *Appl. Anal.*, **58**:241–261, 1995.

[98] J. M. Borwein, A. S. Lewis, N. N. Limber, and D. Noll. Maximum entropy reconstruction using derivative information. II. Computational results. *Numer. Math.*, **69**:243–256, 1995.

[99] J. M. Borwein, A. S. Lewis, and D. Noll. Maximum entropy reconstruction using derivative information. *Math. Oper. Res.*, **21**:442–468, 1996.

[100] J. M. Borwein, A. S. Lewis, and R. Nussbaum. Entropy minimization, DAD problems and doubly-stochastic kernels. *J. Funct. Anal.*, **123**:264–307, 1994.

[101] J. M. Borwein, A. S. Lewis, and Q. J. Zhu. Convex spectral functions of compact operators, Part II: lower semicontinuity and rearrangement invariance. In *Proceedings of the Optimization Miniconference VI*, pages 1–18. Kluwer Academic, Dordrecht, 2001.

[102] J. M. Borwein, P. Maréchal, and D. Naugler. A convex dual approach to the computation of NMR complex spectra. *Math. Methods Oper. Res.*, **51**:91–102, 2000.

[103] J. M. Borwein, V. Montesinos, and J. Vanderwerff. Boundedness, differentiability and extensions of convex functions. *J. Convex Anal.*, **13**:587–602, 2006.

[104] J. M. Borwein and W. B. Moors. Essentially smooth Lipschitz functions. *J. Funct. Anal.*, **49**:305–351, 1997.

[105] J. M. Borwein and W. B. Moors. Null sets and essentially smooth Lipschitz functions. *SIAM J. Optim.*, **8**:309–323, 1998.

[106] J. M. Borwein and W. B. Moors. Stability of closedness of cones under linear mappings. *J. Convex Anal.*, **16**:699–705, 2009.

[107] J. M. Borwein and D. Noll. Second order differentiability of convex functions in Banach spaces. *Tran. Amer. Math. Soc.*, **342**:43–81, 1994.

[108] J. M. Borwein and D. Preiss. A smooth variational principle with applications to subdifferentiability and to differentiability of convex functions. *Trans. Amer. Math. Soc.*, **303**:517–527, 1987.

[109] J. M. Borwein, J. Read, A. S. Lewis, and Zhu Q. J. Convex spectral functions of compact operators. *J. Nonlinear Convex Anal.*, **1**:17–35, 2000.

[110] J. M. Borwein and H. M. Strójwas. Directionally Lipschitzian mappings on Baire spaces. *Canad. J. Math.*, **36**:95–130, 1984.

[111] J. M. Borwein and J. Vanderwerff. Dual Kadec–Klee norms and the relationships between Wijsman, slice and Mosco convergence. *Michigan Math. J.*, **41**:371–387, 1994.

[112] J. M. Borwein and J. Vanderwerff. Convergence of Lipschitz regularizations of convex functions. *J. Funct. Anal.*, **128**:139–162, 1995.

[113] J. M. Borwein and J. Vanderwerff. Epigraphical and uniform convergence of convex functions. *Trans. Amer. Math. Soc.*, **348**:1617–1631, 1996.

[114] J. M. Borwein and J. Vanderwerff. Convex functions on Banach spaces not containing ℓ_1. *Canad. Math. Bull.*, **40**:10–18, 1997.

[115] J. M. Borwein and J. Vanderwerff. Convex functions of Legendre type in general Banach spaces. *J. Convex Anal.*, **8**:569–581, 2001.

[116] J. M. Borwein and J. Vanderwerff. On the continuity of biconjugate convex functions. *Proc. Amer. Math. Soc.*, **130**:1797–1803, 2002.

[117] J. M. Borwein and J. Vanderwerff. Constructible convex sets. *Set-Valued Anal.*, **12**:61–77, 2004.

[118] J. M. Borwein and H. Wiersma. Asplund decomposition of monotone operators. *SIAM J. Optim.*, **18**:946–960, (2007).

[119] J. M. Borwein and D. A. Yost. Absolute norms on vector lattices. *Proc. Edinburgh Math. Soc.*, **27**:215–222, 1984.

[120] J. M. Borwein and Q. J. Zhu. Multifunctional and functional analytic techniques in nonsmooth analysis. In F.H. Clarke and R.J. Stern, editors, *Nonlinear Analysis, differential equations and control*, NATO Sci. Ser. C Math. Phys. Sci., pages 61–157. Kluwer Academic, Dordrecht, 1999.

[121] J. M. Borwein and Q. J. Zhu. *Techniques of Variational Analysis*, volume 20 of *CMS Books in Mathematics*. Springer, New York, 2005.

[122] J. M. Borwein and Q.J. Zhu. A survey of subdifferential calculus with applications. *Nonlinear Anal.*, **38**:687–773, 1999.

[123] J. M. Borwein and D. Zhuang. On Fan's minimax theorem. *Math. Programming*, **34**:232–234, 1986.

[124] C. J. Bose and R. Murray. Duality and the computation of approximately invariant measures for nonsingular transformations. *SIAM J. Optim.*, **18**:691–709, 2007.

[125] R. M. Bot, S.-I. Grad, and G. Wanka. Maximal monotonicity for the precomposition with a linear operator. *SIAM J. Optim.*, **17**:1239–1252, 2006.

[126] R. D. Bourgin. *Geometric Aspects of Convex Sets with the Radon–Nikodym Property*, volume 993 of *Lecture Notes in Mathematics*. Springer-Verlag, Berlin, 1983.

[127] S. Boyd, L. El Ghaoui, E. Feron, and V. Balakrishnan. *Linear Matrix Inequalities in System and Control Theory*. SIAM, Philadelphia, 1994.

[128] S. Boyd and L. Vandenberghe. *Convex Optimization*, volume 317 of *A Series of Comprehensive Studies in Mathematics*. Cambridge University Press, New York, 2004.

[129] L. M. Bregman. A relaxation method of finding a common point of convex sets and its application to the solution of problems in convex programming. *U.S.S.R. Comput. Math. Math. Phys.*, **7**:200–217, 1967.

[130] J. Brimberg. The Fermat–Weber location problem revisited. *Math. Programming*, **71**:71–76, 1995.

[131] F. E. Browder. Fixed point theory and nonlinear problems. *Bull. Amer. Math. Soc.*, **9**:1–39, 1983.

[132] A. L. Brown. A rotund reflexive space having a subspace of co-dimension two with a discontinuous metric projection. *Mich. Math. J.*, **21**:145–151, 1974.

[133] A. M. Bruckner, Bruckner J. B., and B. S. Thomson. *Real Analysis*. Prentice-Hall, Inc., Upper Saddle River, New Jersey, 1997.

[134] H. D. Brunk. An inequality for convex functions. *Proc. Amer. Math. Soc.*, **7**:817–824, 1956.

[135] R. S. Burachik and S. Fitzpatrick. On a family of convex functions attached to subdifferentials. *J. Nonlinear Convex Anal.*, **6**:165–171, 2005.

[136] R. S. Burachik and V. Jeyakumar. A new geometric condition for Fenchel's duality in infinite dimensional spaces. *Math. Prog.*, **104**:229–233, 2005.

[137] R. S. Burachik and B. F. Svaiter. Maximal monotone operators, convex functions and a special family of enlargements. *Set-Valued Anal.*, **10**:297–316, 2002.

[138] H. Busemann and W. Feller. Krümmungseigenschaften konvexer Flächen. *Acta Math.*, **66**:1–47, 1936.

[139] D. Butnariu, Y. Censor, P. Gurfil, and E. Hadar. On the behavior of subgradient projections methods for convex feasibility problems in Euclidean spaces. *SIAM J. Optim.*, **19**:786–807, 2008.

[140] D. Butnariu and A. N. Iusem. *Totally Convex Functions for Fixed Points Computation and Infinite Dimensional Optimization*. Kluwer, Dordrecht, 2000.

[141] D. Butnariu and E. Resmerita. Bregman distances, totally convex functions, and a method for solving operator equations in Banach spaces. *Abstract Appl. Anal.*, 2006.

[142] H. Cabezas and A. T. Karunanithi. Fisher information, entropy, and the second and third laws of thermodynamics. *ASAP Ind. Eng. Chem. Res.*, ASAP Article, 10.1021/ie7017756, Web published 09-July-2008.

[143] A. K. Chakrabarty, P. Shummugara, and C. Zălinescu. Continuity properties for the subdifferential and ε-subdifferential of a convex function and its conjugate. *J. Convex Anal.*, **14**:479–514, 2007.

[144] H. Cartan. *Differential Calculus*. Hermann/Houghton Mifflin, Paris/Boston, 1967/1971.

[145] F. S. Cater. Constructing nowhere differentiable functions from convex functions. *Real Anal. Exchange*, **28**:617–621, 2002/2003.

[146] Y. Censor and A. Lent. An iterative row-action method for interval convex programming. *J. Optim. Theory Appl.*, **34**(3):321–353, 1981.

[147] Y. Censor and S. A. Zenios. *Parallel Optimization: Theory, Algorithms, and Applications*. Oxford University Press, 1997.

[148] M. Cepedello Boiso. Approximation of Lipschitz functions by Δ-convex functions in Banach spaces. *Israel J. Math.*, **106**:269–284, 1998.

[149] M. Cepedello-Boiso and P. Hájek. Analytic approximations of uniformly continuous functions in real Banach spaces. *J. Math. Anal. Appl.*, **256**:80–98, 2001.

[150] L. Cheng and Y. Teng. A strong optimization theorem in locally convex spaces. *Chin. Ann. Math.*, Ser. B, **24**:395–402, 2003.

[151] T. Y. Chow. The surprise examination or unexpected hanging paradox. *Amer. Math. Monthly*, **105**:41–51, 1998.

[152] J. P. R. Christensen. On sets of Haar measure zero in Abelian Polish groups. *Israel J. Math.*, **13**:255–260, 1972.

[153] J. P. R. Christensen. Measure theoretic zero-sets in infinite dimensional spaces and applications to differentiability of Lipschitz functions, II. *Colloq. Analyse Fonctionnel (Bordeaux)*, pages 29–39, 1973.

[154] V. Chvátal. *Linear Programming*. Freeman, New York, 1983.

[155] F. H. Clarke. A variational proof of Aumann's theorem. *Appl. Math. Optim.*, **7**:373–378, 1981.

[156] F. H. Clarke. *Optimization and Nonsmooth Analysis*. Canadian Mathematical Society Series of Monographs and Advanced Texts. John Wiley, New York, 1983.

[157] F. H. Clarke, Yu. S. Ledyaev, R. J. Stern, and P. R. Wolenski. *Nonsmooth Analysis and Control Theory*, volume 178 of *Graduate Texts in Mathematics*. Springer-Verlag, New York, 1998.

[158] M. Coban and P. S. Kenderov. Dense Gâteaux differentiablity of the supremum norm in $C(T)$ and topological properties of T. *C. R. Acad. Bulgare Sci.*, **38**:1603–1604, 1985.

[159] J. B. Collier. The dual of a space with the Radon Nikodým property. *Pacific J. Math.*, **64**:103–106, 1976.

[160] J. Conway. *A Course in Functional Analysis*, volume 96 of *Graduate Texts in Mathematics*. Springer-Verlag, Berlin, 2nd edition, 1990.

[161] R. Correa, P. Gajardo, and L. Thibault. Subdifferential representation formula and subdifferential criteria for the behavior of nonsmooth functions. *Nonlinear Anal.*, **65**:864–891, 2006.

[162] M. Crandall, H. Ishii, and P.-L. Lions. User's guide to viscosity solutions of second order partial differential equations. *Bull. Amer. Math. Soc.*, **27**:1–67, 1992.

[163] J. P. Crouziex. Duality between direct and indirect utility functions. Differentiability properties. *J. Math. Econ.*, **12**:149–165, 1983.

[164] D. Cudia. The geometry of Banach spaces. Smoothness. *Trans. Amer. Math. Soc.*, **110**:284–314, 1964.

[165] A. Daniilidis, A. Lewis, J. Malick, and H. Sendov. Prox-regularity of spectral functions and spectral sets. *J. Convex Analysis*, **15**:547–560, 2008.

[166] G. B. Dantzig. *Linear Programming and Extensions*. Princeton University Press, Princeton, 1963.

[167] F. K. Dashiell and J. Lindenstrauss. Some examples concerning strictly convex norms on $C(K)$ spaces. *Israel J. Math.*, **16**:329–342, 1973.

[168] C. Davis. All convex invariant functions of hermitian matrices. *Arch. Math.*, **8**:276–278, 1957.

[169] W. J. Davis, N. Ghoussoub, and J. Lindenstrauss. A lattice renorming theorem and applications to vector-valued processes. *Trans. Amer. Math. Soc.*, **263**:531–540, 1981.

[170] W. J. Davis and W. B. Johnson. A renorming of nonreflexive Banach spaces. *Proc. Amer. Math. Soc.*, **37**:486–488, 1973.

[171] M. M. Day. Strict convexity and smoothness of normed spaces. *Trans. Amer. Math. Soc.*, **78**:516–528, 1955.

[172] M. M. Day. *Normed Linear Spaces*, Ed. 3 of *Graduate Texts in Mathematics*. Springer-Verlag, Berlin, 1973.

[173] C. De Bernardi and L. Veselý. On support points and support functionals of convex sets. *Israel J. Math.*, **171**:v15–27, 2009.

[174] H. Debrunner and P. Flor. Ein Erweiterungssatz für monotone Mengen. *Arch. Math.*, **15**:445–447, 1964.

[175] F. Deutsch. *Best Approximation in Inner Product Spaces*. CMS Books in Mathematics/Ouvrages de Mathématiques de la SMC, 7. Springer-Verlag, New York, 2001.

[176] R. Deville, V. Fonf, and P. Hájek. Analytic and C^k approximations of norms in separable Banach spaces. *Studia Math.*, **120**:61–74, 1996.

[177] R. Deville, V. Fonf, and P. Hájek. Analytic and polyhedral approximation of convex bodies in separable polyhedral Banach spaces. *Israel J. Math.*, **105**:139–154, 1998.

[178] R. Deville, G. Godefroy, and V. Zizler. Smooth bump functions and the geometry of Banach spaces. *Mathematika*, **40**:305–321, 1993.

[179] R. Deville, G. Godefroy, and V. Zizler. A smooth variational principle with applications to Hamilton–Jacobi equations in infinite dimensions. *J. Funct. Anal.*, **111**:197–212, 1993.

[180] R. Deville, G. Godefroy, and V. Zizler. *Smoothness and Renormings in Banach spaces*, volume 64 of *Pitman Monographs and Surveys in Pure and Applied Mathematics*. Longman Scientific & Technical, Harlow, 1993.

[181] R. Deville and J. Revalski. Porosity of ill-posed problems. *Proc. Amer. Math. Soc.*, **128**:1117–1124, 2000.

[182] J. Diestel. *Geometry of Banach Spaces – Selected Topics*, volume 485 of *Lecture Notes in Mathematics*. Springer-Verlag, Berlin, 1975.

[183] J. Diestel. *Sequences and Series in Banach spaces*, volume 92 of *Graduate Texts in Mathematics*. Springer-Verlag, Berlin, 1984.

[184] J. Diestel and J. J. Uhl. *Vector Measures*, volume 15 of *AMS Mathematical Surveys*. American Mathematical Society, Providence, 1977.

[185] J. Dieudonné. *Foundations of Modern Analysis*, volume 10 of *Pure and Applied Mathematics*. Academic Press, New York, 1960.

[186] W. E. Diewert. Duality approaches to microeconomic theory. In K. J. Arrow and M. D. Intriligator, editors, *Handbook of Mathematical Economics*, volume II, pages 535–599. North-Holland, Amsterdam, 1982.

[187] T. Domíngues Benavides. A renorming of some nonseparable Banach spaces with the fixed point property. *J. Math. Anal. Appl.*, **350**:525–530, 2009.

[188] P. N. Dowling, W. B. Johnson, C. J. Lennard, and B. Turett. The optimality of James's distortion theorems. *Proc. Amer. Math. Soc.*, **125**:167–174, 1997.

[189] J. Duda, L. Veselý, and L. Zajíček. On d.c. functions and mappings. *Atti Sem. Mat. Fis. Univ. Modena*, **51**(1):111–138, 2003.

[190] A. C. Eberhard and J. M. Borwein. Second order cones for maximal monotone operators via representative functions. *Set-Valued Anal.*, **16**:157–184, 2008.

[191] A. C. Eberhard and R. Wenczel. On the calculus of limiting subhessians. *Set-Valued Anal.*, **15**:377–424, 2007.

[192] N. V. Effimov and S. B. Stechkin. Some supporting properties of sets in Banach spaces as related to Chebyshev's sets. *Dokl. Akad. Nauk SSSR*, **127**:254–257, 1959.

[193] I. Ekeland. On the variational principle. *J. Math. Anal. Appl.*, **47**:324–353, 1974.

[194] I. Ekeland and G. Lebourg. Generic Fréchet-differentiability and perturbed optimization problems in Babach spaces. *Trans. Amer. Math. Soc.*, **224**:193–216, 1976.

[195] I. Ekeland and R Temam. *Convex Analysis and Variational Problems*. North-Holland, New York, 1976.

[196] D. Ellis. A modification of the parallelogram law characterization of Hilbert spaces. *Math. Z.*, **59**:94–96, 1953.

[197] P. Enflo. Banach spaces which can be given an equivalent uniformly convex norm. In *Proceedings of the International Symposium on Partial Differential Equations and the Geometry of Normed Linear Spaces (Jerusalem, 1972)*, volume 13, pages 281–288 (1973), 1972.

[198] L. C. Evans and R. F. Gariepy. *Measure Theory and Fine Properties of Functions*. CRC Press, Boca Raton, 1992.

[199] M. Fabian et al. *Functional Analysis and Infinite-Dimensional Geometry*, volume 8 of *CMS Books in Mathematics*. Springer-Verlag, Berlin, 2001.

[200] M. Fabian. Lipschitz smooth points of convex functions and isomorphic characterizations of Hilbert spaces. *Proc. London Math. Soc.*, **51**:113–126, 1985.

[201] M. Fabian. *Gâteaux Differentiability of Convex Functions and Topology: Weak Asplund Spaces*. CMS Series of Monographs and Advanced Texts. John Wiley, New York, 1997.

[202] M. Fabian, D. Preiss, J. H. M. Whitfield, and V. Zizler. Separating polynomials on Banach spaces. *Quart. J. Math. Oxford (2)*, **40**:409–422, 1989.

[203] M. Fabian, J. H. M. Whitfield, and V. Zizler. Norms with locally Lipschitzian derivatives. *Israel J. Math*, **44**:262–276, 1983.

[204] M. Fabian, L. Zajíček, and V. Zizler. On residuality of the set of rotund norms on a Banach space. *Math. Ann.*, **258**:349–351, 1981/82.

[205] M. Fabian and V. Zizler. An elementary approach to some questions in higher order smoothness in Banach spaces. *Extracta Math.*, **14**:295–327, 1999.

[206] M. Fabian and V. Zizler. On uniformly Gâteaux smooth $C^{(n)}$-smooth norms on separable Banach spaces. *Czech. Math. J.*, **49**:657–672, 1999.

[207] M. Fabian and V. Zizler. A "nonlinear" proof of Pitt's compactness theorem. *Proc. Amer. Math. Soc.*, **131**:3693–3694, 2003.

[208] K. Fan. Fixed point and minimax theorems in locally convex topological linear spaces. *Proc. Nat. Acad. Sci. U.S.A.*, **38**:121–126, 1952.

[209] J. Farkas. Theorie der einfachen Ungleichungen. *Journal für die reine und angewandte Mathematik*, **124**:1–27, 1901.

[210] L. Faybusovich. Infinite-dimensional semidefinite programming: regularized determinants and self-concordant barriers. In *Topics in semidefinite and interior-point methods (Toronto, ON, 1996)*, volume 18 of *Fields Inst. Commun.*, pages 39–49. Amer. Math. Soc., Providence, RI, 1998.

[211] W. Fenchel. On conjugate convex functions. *Canad. J. Math.*, **1**:73–77, 1949.

[212] W. Fenchel. Convexity through the Ages. In P.M. Gruber and J. M. Wills, editors, *Convexity and its Applications*. Birkhäuser-Verlag, Boston, MA, 120–130, 1983.

[213] T. Figiel. On the moduli of convexity and smoothness. *Studia Math.*, **56**:121–155, 1976.

[214] P. A. Fillmore and J. P. Williams. Some convexity theorems for matrices. *Glasgow Math. J.*, **12**:110–117, 1971.

[215] S. Fitzpatrick. Representing monotone operators by convex functions. *Workshop/Miniconference on Functional Analysis and Optimization, Proc. Centre Math. Anal. Austral. Nat. Univ.*, 20, Austral. Nat. Univ., Canberra, pages 59–65, 1988.

[216] S. Fitzpatrick and R. R. Phelps. Differentiability of the metric projection in Hilbert space. *Trans. Amer. Math. Soc.*, **270**:483–501, 1982.

[217] S. Fitzpatrick and R. R. Phelps. Bounded approximants to monotone operators on Banach spaces. *Ann. Inst. Henri Poincaré, Analyse non linéaire*, **9**:573–595, 1992.

[218] S. Fitzpatrick and R. R. Phelps. Some properties of maximal monotone operators on nonreflexive Banach spaces. *Set-Valued Anal.*, **3**:51–69, 1995.

[219] R. Fletcher. A new variational result for quasi-Newton formulae. *SIAM J. Optim.*, **1**:18–21, 1991.

[220] V. P. Fonf. On supportless convex sets in incomplete normed spaces. *Proc. Amer. Math. Soc.*, **120**:1173–1176, 1994.

[221] J. Frontisi. Smooth partitions of unity in Banach spaces. *Rocky Mount. J. Math.*, **25**:1295–1304, 1995.

[222] J. Frontisi. Lissité et dualité dans les espaces de Banach. *Thèse de Doctorat de Mathématiques de l'Université Paris 6*, 1996.

[223] R. Fry. Analytic approximation on c_0. *J. Funct. Anal.*, **158**:509–520, 1998.

[224] R. Fry. Approximation by C^p-smooth, Lipschitz functions on Banach spaces. *J. Math. Anal. Appl.*, **315**:599–605, 2006.

[225] R. Fry. Corrigendum to "Approximation by C^p-smooth, Lipschitz functions on Banach spaces". *J. Math. Anal. Appl.*, **348**:571, 2008.

[226] N. Ghoussoub. Renorming dual Banach lattices. *Proc. Amer. Math. Soc.*, **84**:521–524, 1982.

[227] G. Gilardoni. On Pinsker's type inequalities and Csiszár's f-divergences. Part I: Second and fourth-order inequalities. arXiv:cs/0603097v2, 2006.

[228] G. Gilardoni. On the minimum f-divergence for given total variation. *C. R. Acad. Sci. Paris*, **343**:763–766, 2006.

[229] J. R. Giles. *Convex Analysis with Application in the Differentiation of Convex Functions*, volume 58 of *Research Notes in Mathematics*. Pitman, Boston, Mass., 1982.

[230] K. Gobel and J. Wośko. Making a hole in the space. *Proc. Amer. Math. Soc.*, **114**:475–476, 1992.

[231] M.A. Goberna and M.A. López. *Linear Semi-Infinite Optimization*. Wiley Series in Mathematical Methods in Practice. John Wiley, New York, 1998.

[232] G. Godefroy. Prolongement de fonctions convexes définies sur un espace de Banach E au bidual E''. *C. R. Acad. Sci. Paris*, **292**:371–374, 1981.

[233] R. Goebel and R. T. Rockafellar. Local strong convexity and local lipschitz continuity of the gradient of convex functions. *J. Convex Analysis*, **15**:263–270, 2008.

[234] R. Goebel, A. R. Teel, T. Hu, and Z. Lin. Conjugate convex Lyapunov functions for dual linear differential equations. *IEEE Trans. Automat. Control*, **51**:661–666, 2006.

[235] M. X. Goemans. Semidefinite programming and combinatorial optimization. *Doc. Math. J. DMV*, ICM 1998 III:657–666, 1998.

[236] G. Golub and C. F. Van Loan. *Matrix Computations*. The Johns Hopkins University Press, Baltimore, 3rd edition, 1996.

[237] P. Gordan. Uber die Auflösung linearer Gleichungen mit reelen Coefficienten. *Math. Ann.*, **6**:23–28, 1873.

[238] J.-P. Gossez. On the extensions to the bidual of a maximal monotone operator. *Proc. Amer. Math. Soc.*, **62**:67–71, 1977.

[239] J.-P. Gossez and E. Lami Dozo. Some geometric properties related to the fixed point theory for nonexpansive mappings. *Pacific J. Math.*, **40**:565–573, 1972.

[240] J.-P. Gossez. Opérateurs monotones non linéaires dans les espaces de Banach non réflexifs. *J. Math. Anal. Appl.*, **34**:371–395, 1971.

[241] J.-P. Gossez. On the range of a coercive maximal monotone operator in a nonreflexive Banach space. *Proc. Amer. Math. Soc.*, **35**:88–92, 1972.

[242] R. Grone, C. R. Johnson, E. Marques de Sá, and H. Wolkowicz. Positive definite completions of partial Hermitian matrices. *Linear Algebra Appl.*, **58**:109–124, 1984.

[243] B. Grünbaum. *Convex Polytopes*, 2 Ed. Graduate Texts in Mathematics. Springer-Verlag, New York, 2003.

[244] R. Grząślewicz. A universal convex set in Euclidean space. *Colloq. Math.*, **45**:41–44, 1982.

[245] A. Guirao and P. Hájek. On the moduli of convexity. *Proc. Amer. Math. Soc.*, **135**:3233–3240, 2007.

[246] V. I. Gurari. Differential properties of the convexity moduli of Banach spaces. *Mat. Issled.*, **2**(vyp. 1):141–148, 1967.

[247] P. Hájek, V. Montesinos, J. Vanderwerff, and V. Zizler. *Biorthogonal Systems in Banach Spaces*. **26**, CMS Books in Mathematics. Springer-Verlag, New York, 2008.

[248] M. P. Hanna. Generalized overrelaxation and Gauss–Seidel convergence on Hilbert space. *Proc. Amer. Math. Soc.*, **35**:524–530, 1972.

[249] A. Hantoute, M. A. López, and C. Zălinescu. Subdifferential calculus rules in convex analysis: a unifying approach via pointwise supremum functions. *SIAM J. Opim.*, **19**:863–882, 2008.

[250] P. Hartman. On functions representable as a difference of convex functions. *Pacific J. Math.*, **9**:707–713, 1959.

[251] R. Haydon. A nonreflexive Grothendieck space that does not contain ℓ_∞. *Israel J. Math.*, **40**:65–73, 1981.

[252] R. Haydon. Trees in renorming theory. *Proc. London Math. Soc.*, **78**:541–584, 1999.

[253] R. D. Hill. Dual periodicity in l_1-norm minimisation problems. *Systems Control Lett.*, **57**:489–496, 2008.

[254] J.-B. Hiriart-Urruty. What conditions are satisfied at points minimizing the maximum of a finite number of differentiable functions. In *Nonsmooth Optimization: Methods and Applications*. Gordan and Breach, New York, Montreux, 1992.

[255] J.-B. Hiriart-Urruty and C. Lemaréchal. *Convex Analysis and Minimization Algorithms*. Springer-Verlag, New York, 1993.

[256] J.-B. Hiriart-Urruty and C. Lemaréchal. *Fundamentals of Convex Analysis*. Grundlehren der mathematien Wissenschaften. Springer-Verlag, New York, 2001.

[257] J.-B. Hiriart-Urruty and R. R. Phelps. Subdifferential calculus using ε-subdifferentials. *J. Funct. Anal.*, **118**:154–166, 1993.

[258] J.-B. Hiriart-Urruty and A. Seeger. The second-order subdifferential and the Dupin indicies of a nondifferentiable convex function. *Proc. London Math. Soc.*, **58**:351–365, 1989.

[259] P. Holický, O. Kalenda, L. Veselý, and L. Zajíček. Quotients of continuous convex functions on nonreflexive Banach spaces. *Bull. Pol. Acad. Sci. Math.*, **55**:211–217, 2007.

[260] R. B. Holmes. *Geometric Functional Analysis and its Applications*, volume 24 of *Graduate Texts in Mathematics*. Springer-Verlag, New York, 1975.

[261] R. A. Horn and C. R. Johnson. *Matrix Analysis*. Cambridge University Press, 1985.

[262] A. Ioffe. Three theorems on subdifferentiation of convex integral functionals. *J. Convex Anal.*, **13**:759–772, 2006.

[263] A. D. Ioffe and J.-P. Penot. Limiting sub-Hessian, limiting subjets and their calculus. *Trans. Amer. Math. Soc*, **349**:789–807, 1997.

[264] R. C. James. Super-reflexive Banach spaces. *Canad. J. Math.*, **24**:896–904, 1972.

[265] R. C. James. Super-reflexive spaces with bases. *Pacific J. Math.*, **41/42**:409–419, 1972.

[266] G. J. O. Jameson. *Topology and Normed Spaces*. Chapman and Hall, London, 1974.

[267] M. Jiang. On Johnson's example of a nonconvex Chebyshev set. *J. Approx. Theory*, **74**:152–158, 1993.

[268] K. John and V. Zizler. A short proof of a version of Asplund's norm averaging theorem. *Proc. Amer. Math. Soc.*, **73**:277–278, 1979.

[269] G. Johnson. A nonconvex set which has the unique nearest point property. *J. Approx. Theory*, **51**:289–332, 1987.

[270] G. Johnson. Chebyshev foam. *Topology. Appl.*, **20**:163–171, 1998.

[271] B. Josefson. Weak sequential convergence in the dual of a Banach space does not imply norm convergence. *Ark. Mat.*, **13**:79–89, 1975.

[272] G. Pólya and G. Szegő. *Problems and Theorems in Analysis, I*. Springer Verlag, New York, 2004.

[273] G. Pólya and G. Szegő. *Problems and Theorems in Analysis, II*. Springer Verlag, New York, 2004.

[274] S. Kakutani. A generalization of Brouwer's fixed point theorem. *Duke Math. J.*, **8**:457–459, 1941.

[275] A. G. Kartsatos and I. V. Skrypnik. On the eigenvalue problem for perturbed nonlinear maximal monotone operators in reflexive Banach spaces. *Trans. Amer. Math. Soc.*, **358**:3851–3881, 2006.

[276] N. Kato. On the second derivatives of convex functions on Hilbert spaces. *Proc. Amer. Math. Soc.*, **106**:697–705, 1989.

[277] V. Klee. Convexity of Chebysev sets. *Math. Annalen*, **142**:292–304, 1961.

[278] V. L. Klee. Some new results on smoothness and rotundity in normed linear spaces. *Math. Ann.*, **139**:51–63, 1959.

[279] E. Krauss. A representation of arbitrary maximal monotone operators via subgradients of skew-symmetric saddle functions. *Nonlinear Anal.*, **9**:1381–1399, 1985.

[280] H. W. Kuhn. Steiner's problem revisited. *MAA Studies in Optimization*, **10**:52–70, 1974.

[281] O. Kurka. Reflexivity and sets of Fréchet subdifferentiability. *Proc. Amer. Math. Soc.*, **136**:4467–4473, 2008.

[282] J. Kurzweil. On approximation in real Banach spaces. *Studia Math.*, **14**:214–231, 1955.

[283] S. Kwapien. Isomorphic characterizations of inner product space by orthogonal series with vector valued coefficients. *Studia Math.*, **44**:583–595, 1972.

[284] J. M. Lasry and P. L. Lions. A remark on regularization in Hilbert spaces. *Israel J. Math*, **55**:257–266, 1986.

[285] M. Lassonde. Asplund spaces, Stegal variational principle and the RNP. To appear.

[286] M. Lassonde and J. Revalski. Fragmentability of sequences of set-valued mappings with applications to variational principles. *Proc. Amer. Math. Soc.*, **133**:2637–2646, 2005.

[287] E. S. Levitin and B. T. Polyak. Convergence of minimizing sequences in the relative extremum. *Dolk. Akad. Nauk. SSSR*, **168**:997–1000, 1966.

[288] A. S. Lewis. Convex analysis on the Hermitian matrices. *SIAM J. Optim.*, **6**:164–177, 1996.

[289] A. S. Lewis. The mathematics of eigenvalue optimization. *Math. Programming*, **97**:155–176, 2003.

[290] A. S. Lewis and M. L. Overton. Eigenvalue optimization. *Acta Numerica*, **5**:149–190, 1996.

[291] F. Liese and I. Vajda. *Convex Statistical Distances*, volume 95 of *Teubner Texts in Mathematics*. Teubner, Leipzig, 1987.

[292] J. Lindenstrauss. On the modulus of smoothness and divergent series in Banach spaces. *Mich. Math. J.*, **10**:241–252, 1963.

[293] J. Lindenstrauss, D. Preiss, and J. Tišer. Fréchet differentiability of Lipschitz maps and porous sets in Banach spaces. *Banach Spaces and their Applications in Analysis*, pages 111–123, 2007.

[294] J. Lindenstrauss and L. Tzafriri. *Classical Banach Spaces*, volume I & II. Springer-Verlag, New York, 1977.

[295] R. Lipschitz. *Lehrbuch der Analysis*. Cohen und Sohn, Bonn, 1877.

[296] P. Loewen and X. Wang. A generalized variational principle. *Canad. J. Math*, **53**:1174–1193, 2001.

[297] R. Lucchetti. *Convexity and Well-Posed Problems*, volume 22 of *CMS Books in Mathematics*. Springer-Verlag, New York, 2006.

[298] Y. Lucet. What shape is your conjugate? A survey of computational convex analysis and its applications. Available at: www.optimization-online.org/DB_FILE/2007/12/1863.pdf, 2007.

[299] D. G. Luenberger. *Optimization by Vector Space Methods*. John Wiley, New York, 1969.

[300] D. G. Luenberger. Benefit functions and duality. *J. Math. Econom.*, **21**:461–481, 1992.

[301] H. Mankiewicz. On the differentiability of Lipschitz mappings in Fréchet spaces. *Studia Math.*, **45**:15–29, 1973.

[302] P. Maréchal. On the convexity of the multiplicative potential and penalty functions and related topics. *Math. Program., Ser. A*, **89**:505–516, 2001.

[303] A. W. Marshall and I. Olkin. *Inequalities: Theory of Majorization and its Applications*. Academic Press, New York, 1979.

[304] M. Matić, C. E. M. Pearce, and J. Pečarić. Improvements on some bounds on entropy measures in information theory. *Math. Inequal. Appl.*, **1**:295–304, 1998.

[305] J. Matoušek and E. Matoušková. A highly nonsmooth norm on Hilbert space. *Israel J. Math.*, **112**:1–27, 1999.

[306] E. Matoušková. One counterexample concerning the Fréchet differentiability of convex functions on closed sets. *Acta Univ. Carolin. Math. Phys.*, **34**:97–105, 1993.

[307] E. Matoušková. An almost nowhere Fréchet smooth norm on superreflexive spaces. *Studia Math.*, **133**:93–99, 1999.

[308] E. Matoušková. Translating finite sets into convex sets. *Bull London Math Soc.*, **33**:711–714, 2001.

[309] E. Matoušková and C. Stegall. A characterization of reflexive Banach spaces. *Proc. Amer. Math. Soc.*, **124**:1083–1090, 1996.

[310] E. Matoušková and L. Zajíček. Second order differentiability and Lipschitz smooth points of convex functionals. *Czech. Math. J.*, **48**:617–640, 1998.

[311] S. Mazur. Über konvexe Mengen in linearen normierten Räumen. *Studia Math.*, **4**:70–84, 1933.

[312] D. McLaughlin, R. Poliquin, J. Vanderwerff, and V. Zizler. Second-order Gâteaux differentiable bump functions and approximations in Banach spaces. *Canad. J. Math.*, **45**:612–625, 1993.

[313] D. McLaughlin and J. Vanderwerff. Higher order Gateaux smooth bump functions on Banach spaces. *Bull. Austral. Math. Soc.*, **51**:291–300, 1995.

[314] V. Z. Meshkov. Smoothness properties in Banach spaces. *Studia Math.*, **63**:111–123, 1978.

[315] E. Michael. Continuous selections I. *Ann. Math.*, **63**:361–382, 1956.

[316] F. Mignot. Contrôle dans les inéquations variationelles elliptiques. *J. Funct. Anal.*, **22**:130–185, 1976.

[317] D. P. Milman. On some criteria for the regularity of spaces of the type (B). *Dokl. Akad. Nauk SSSR*, **20**:243–246, 1938.

[318] H. Minkowski. *Geometrie der Zahlen*. Teubner, Leipzig, 1910.

[319] W. B. Moors. The relationship between Goldstine's theorem and the convex point of continuity propety. *J. Math. Anal. Appl.*, **188**:819–832, 1994.

[320] W. B. Moors and S. Somasundaram. USCO selections of densely defined set-valued mappings. *Bull. Aust. Math. Soc.*, **65**:307–313, 2002.

[321] W. B. Moors and S. Somasundaram. A weakly Stegall space that is not a Stegall space. *Proc. Amer. Math. Soc.*, **131**:647–654, 2003.

[322] W. B. Moors and S. Somasundaram. A Gâteaux differentiability space that is not weak Asplund. *Proc. Amer. Math. Soc.*, **134**:2745–2754, 2006.

[323] B. Mordukhovich. *Variational Analysis and Generalized Differentiation I & II*. Grundlehren der mathematischen Wissenschaften. Springer-Verlag, New York, 2006.

[324] B. Mordukhovich and M. N. Nam. Limiting subgradients of minimal time functions in Banach spaces. Preprint, 2008.

[325] J.-J. Moreau. Sur la function polaire d'une fonctionelle semi-continue supérieurement. *C. R. Acad. Sci.*, **258**:1128–1130, 1964.

[326] U. Mosco. Convergence of convex sets and solutions of variational inequalities. *Adv. Math.*, **3**:510–585, 1969.

[327] U. Mosco. On the continuity of the Young–Fenchel transform. *J. Math. Anal. Appl.*, **35**:518–535, 1971.

[328] M. Moussaoui and A. Seeger. Second-order subgradients of convex integral functions. *Trans. Amer. Math. Soc.*, **351**:3687–3711, 1999.

[329] A. Nekvinda and L. Zajíček. A simple proof of the Rademacher theorem. *Časopis Pest. Mat.*, **113**:337–341, 1988.

[330] I. Nemeth. Characterization of a Hilbert vector lattice by the metric projection onto its positive cone. *J. Approx. Theory*, **123**:295–299, 2003.

[331] Y. Nesterov and A. Nemirovskii. *Interior-Point Polynomial Algorithms in Convex Programming*. SIAM, 1994.

[332] C. Niculescu and L.-E. Perrson. *Convex Functions and Their Applications*, volume 23 of *CMS Books in Mathematics*. Springer-Verlag, New York, 2006.

[333] A. Nissenzweig. W^* sequential convergence. *Israel J. Math.*, 22:266–272, 1975.

[334] E. Odell and T. Schlumprecht. The distortion of Hilbert space. *Geom. Funct. Anal.*, 3:201–207, 1993.

[335] E. Odell and T. Schlumprecht. The distortion problem. *Acta Math.*, 173:259–281, 1994.

[336] E. Odell and T. Schlumprecht. Asymptotic properties of Banach spaces under renormings. *J. Amer. Math. Soc.*, 11:175–188, 1998.

[337] A. Ohara. Geometry of distributions associated with Tsallis statistics and properties of relative entropy minimization. *Phys. Lett. A*, 370:184–193, 2007.

[338] P. Ørno. On J. Borwein's concept of sequentially reflexive Banach spaces. *Banach Space Bulletin Board*, 1991.

[339] J. O'Rourke. *Computational Geometry in C*. Cambridge University Press, 2nd edition, 1998.

[340] G. Pataki. Cone-LP's and semidefinite programs: geometry and a simplex-type method pages 162–174. *Lecture Notes in Comput. Sci.*, 1084, Springer, Berlin, 1996.

[341] J. Pechanec, J. H. M. Whitfield, and V. Zizler. Norms locally dependent on finitely many coordinates. *An. Acad. Brasil Cienc.*, 53:415–417, 1981.

[342] G. K. Pedersen. *Analysis Now*, volume 118 of *Graduate Texts in Mathematics*. Springer-Verlag, New York, 1989.

[343] J.-P. Penot. The relevance of convex analysis for the study of monotonicity. *Nonlinear Anal.*, 58:855–871, 2004.

[344] J.-P. Penot. Is convexity useful for the study of monotonicity. In R.P. Agarwal and D. O'Regan, editors, *Nonlinear Analysis and Applications*, pages 807–821. Kluwer Academic, Dordrecht, 2003.

[345] B. J. Pettis. A proof that every uniformly convex space is reflexive. *Duke Math. J.*, 5:249–253, 1939.

[346] R. R. Phelps. Convex sets and nearest points. *Proc. Amer. Math. Soc.*, 8:790–797, 1957.

[347] R. R. Phelps. Gaussian null sets and differentiability of Lipschitz maps on Banach spaces. *Pacific J. Math.*, 77:523–531, 1978.

[348] R. R. Phelps. Counterexamples concerning support theorems for convex sets in Hilbert space. *Canad. Math. Bull.*, 31:121–128, 1988.

[349] R. R. Phelps. *Convex Functions, Monotone Operators and Differentiability*, volume 1364 of *Lecture Notes in Mathematics*. Springer-Verlag, New York, 1st edition, 1989.

[350] R. R. Phelps. *Convex Functions, Monotone Operators and Differentiability*, volume 1364 of *Lecture Notes in Mathematics*. Springer-Verlag, New York, 2nd edition, 1993.

[351] R. R. Phelps. Lectures on maximal monotone operators. arXiv:math.FA/9302209, 1993 and *Extracta Math.*, 12:193–230, 1997.

[352] R. S. Phillips. On linear transformations. *Tran. Amer. Math. Soc.*, 48:516–541, 1940.

[353] G. Pisier. Martingales with values in uniformly convex spaces. *Israel J. Math.*, **20**:326–350, 1975.

[354] D. Preiss. Differentiability of Lipschitz functions on Banach spaces. *J. Funct. Anal.*, **91**:312–345, 1990.

[355] D. Preiss, R. R. Phelps, and I. Namioka. Smooth Banach spaces, weak Asplund spaces and monotone or usco mappings. *Israel J. Math.*, **72**:257–279, 1990.

[356] D. Preiss and L. Zajíček. Stronger estimates of smallness of sets of Fréchet non-differentiability of convex functions. *Proc. 11th Winter School, Suppl. Rend. Circ. Mat. di Palermo, Ser. II*, pages 219–223, 1984.

[357] B. Pshenichnii. *Necessary Conditions for an Extremum*. Marcel Dekker, New York, 1971.

[358] F. Pukelsheim. *Optimal Design of Experiments*. Wiley Series in Probability and Mathematical Statistics. John Wiley, New York, 1993.

[359] F. Pukelsheim and D. M. Titterington. General differential and Lagrangian theory for optimal experimental design. *Ann. Stat.*, **11**:1060–1068, 1983.

[360] M. Raja. Locally uniformly rotund norms. *Mathematika*, **46**:343–358, 1999.

[361] M. Raja. On dual loally uniformly rotund norms. *Israel J. Math.*, **129**:77–91, 2002.

[362] T. Rapcsák. *Smooth Nonlinear Optimization in $\mathbb{R}^n$*, volume 19 of *Series in Nonconvex Optimization and its Applications*. Kluwer Academic, Dordrecht, 1997.

[363] S. Reich. The range of sums of accretive and monotone operators. *J. Math. Anal. Appl.*, **68**:310–317, 1979.

[364] S. Reich. A weak convergence theorem for the alternating method with Bregman distances. In *Theory and Applications of Nonlinear Operators of Accretive and Monotone Type*, pages 313–318. Dekker, New York, 1996.

[365] R. M. Rifkin and R. A. Lippert. Value regularization and Fenchel duality. *J. Machine Learn.*, **8**:441–479, 2007.

[366] A. R. Roberts and D. E. Varberg. *Convex Functions*. Academic Press, New York, 1973.

[367] R. T. Rockafellar. Duality and stability in extremum problems involving convex functions. *Pacific J. Math.*, **21**:167–187, 1967.

[368] R. T. Rockafellar. Local boundedness of nonlinear monotone operators. *Mich. Math. J.*, **16**:397–407, 1969.

[369] R. T. Rockafellar. *Convex Analysis*. Princeton University Press, Princeton, 1970.

[370] R. T. Rockafellar. On the maximality of sums of nonlinear monotone operators. *Trans. Amer. Math. Soc.*, **149**:75–88, 1970.

[371] R. T. Rockafellar. *Conjugate Duality and Optimization*, volume 16 of *CBMS-NSF Regional Conference Series in Applied Mathematics*. SIAM, Philadephia, 1974.

[372] R. T. Rockafellar. Integral functionals, normal integrals and measurable selections. In A. Dold and B. Eckmann, editors, *Nonlinear Operators and the Calculus of Variations*, volume 543 of *Lecture Notes in Mathematics*, pages 157–207. Springer-Verlag, New York, 1976.

[373] R. T. Rockafellar. *The Theory of Subgradients and its Application to Problems of Optimizatin of Convex and Non-convex Functions.* Heldermann-Verlag, Berlin, 1981.

[374] R. T. Rockafellar. Maximal monotone relations and the second derivatives of nonsmooth functions. *Ann. Inst. H. Poincaré Anal. Non Linéaire,* **2**:167–184, 1985.

[375] R. T. Rockafellar. Generalized second derivatives of convex functions and saddle functions. *Trans. Amer. Math. Soc.,* **322**:51–77, 1990.

[376] R. T. Rockafellar. On a special class of convex functions. *J. Optim. Theory Appl.,* **70**:619–621, 1991.

[377] R. T. Rockafellar. Second-order convex analysis. *J. Nonlinear Convex Anal.,* **1**:1–16, 2000.

[378] R. T. Rockafellar and R. J-B. Wets. *Variational Analysis,* volume 317 of *A Series of Comprehensive Studies in Mathematics.* Springer-Verlag, New York, 1998.

[379] B. Rodrigues and S. Simons. A minimax proof of the Krein–Smulian theorem. *Arch. Math.,* **51**:570–572, 1988.

[380] H. P. Rosenthal. On injective Banach spaces and the spaces $L_\infty(\mu)$ for finite measures μ. *Acta. Math.,* **124**:205–248, 1970.

[381] H. P. Rosenthal. The complete separable extension property. *J. Operator Theory,* **43**:329–374, 2000.

[382] P. Rosenthal. The remarkable theorem of Lévy and Steinitz. *Amer. Math. Monthly,* **94**:342–351, 1987.

[383] W. Rudin. *Functional Analysis.* McGraw-Hill Series in Higher Mathematics. McGraw-Hill, New York, 1973.

[384] W. Rudin. *Real and Complex Analysis.* McGraw-Hill, New York, 1987.

[385] A. Ruszczyński and A. Shapiro. Optimization of convex risk functions. *Math. Oper. Res.,* 31:433–452, 2006.

[386] H. H. Schaefer. *Banach Lattices and Positive Operators.* Springer-Verlag, New York, 1974.

[387] S. O. Schönbeck. On the extension of Lipschitz maps. *Ark. Mat.,* **7**:201–209, 1967.

[388] K. Schulz and B. Schwartz. Finite extensions of convex functions. *Math. Oper. Statist. Ser. Optim.,* **10**:501–509, 1979.

[389] T. Sellke. Yet another proof of the Radon–Nikodým theorem. *Amer. Math. Monthly,* **109**:74–76, 2002.

[390] S. Simons. The range of a monotone operator. *J. Math. Anal. App.,* **199**:176–201, 1996.

[391] S. Simons. *Minimax and Monotonicity,* volume 1693 of *Lecture Notes in Mathematics.* Springer-Verlag, Berlin, 1998.

[392] S. Simons. Maximal monotone multifunctions of Brøndsted–Rockafellar type. *Set-Valued Analysis,* **7**:255–294, 1999.

[393] S. Simons. Five kinds of maximal monotonicity. *Set-Valued Analysis,* **9**:391–409, 2001.

[394] S. Simons. A new version of the Hahn–Banach theorem. *Arch. Math.,* **80**:630–646, 2003.

[395] S. Simons. Dualized and scaled Fitzpatrick functions. *Proc. Amer. Math. Soc.*, **134**:2983–2987, 2006.

[396] S. Simons. *From Hahn–Banach to Monotonicity*, volume 1693 of *Lecture Notes in Mathematics*. Springer-Verlag, Berlin, New York, expanded 2nd edition, 2008.

[397] S. Simons. An improvement of the Fitzpatrick inequality for maximally montone multifunctions of type (NI). Available at: www.math.ucsb.edu/~simons/preprints/, 2008.

[398] S. Simons and C. Zălinescu. A new proof for Rockafellar's characterization of maximal monotone operators. *Proc. Amer. Math. Soc.*, **132**:2969–2972, 2004.

[399] S. Simons and C. Zălinescu. Fenchel duality, Fitzpatrick functions and maximal monotonicity. *J. Nonlinear Convex Anal.*, **6**:1–22, 2005.

[400] B. Sims. Banach space geometry and the fixed point property. In *Recent advances on metric fixed point theory*, volume 48 of *Ciencias*. Univ. Sevilla, Sevilla, Spain, 1996.

[401] I. Singer. *Duality for Nonconvex Approximation and Optimization*, volume 24 of *CMS Books in Mathematics/Ouvrages de Mathḿatiques de la SMC*. Springer, New York, 2006.

[402] M. Smith. Rotundity and smoothness in conjugate spaces. *Proc. Amer. Math. Soc.*, **61**:232–234, 1976.

[403] M. Smith. Some examples concerning rotundity in Banach spaces. *Math. Ann.*, **233**:155–161, 1978.

[404] V. L. Šmulian. Sur la structure de la sphère unitaire dans l'espace de Banach. *Mat. Sb.*, **51**:545–561, 1941.

[405] V. Solov'ev. Duality for nonconvex optimization and its applications. *Anal. Math.*, **19**:297–315, 1993.

[406] V. Solov'ev. Dual extremal problems and their applications to minimax estimation problems. *Russian Math. Surveys*, **52**:685–720, 1997.

[407] C. Stegall. The duality between Asplund spaces and spaces with the Radon–Nikodym property. *Israel J. Math.*, **29**:408–412, 1978.

[408] C. Stegall. Optimization of functions on certain subsets of Banach spaces. *Math. Ann.*, **236**:171–176, 1978.

[409] J. Stoer and A. Witzgall. *Convexity and Optimization in Finite Dimensions. I*. Springer-Verlag, New York, 1970.

[410] K. R. Stromberg. *An Introduction to Classical Real Analyis*. Wadsworth International Mathematics Series. Wadsworth, Belmont, California, 1981.

[411] M. Talagrand. Renormages de quelques $C(K)$. *Israel J. Math.*, **54**:327–334, 1986.

[412] W. K. Tang. On the extension of rotund norms. *Manuscripta Math.*, **91**:73–82, 1996.

[413] W. K. Tang. Uniformly differentiable bump functions. *Arch. Math.*, **68**:55–59, 1997.

[414] W. K. Tang. On Asplund functions. *Comment. Math. Univ. Carolinae*, **40**:121–132, 1999.

[415] F. Tardella. A new proof of the Lyapunov convexity theorem. *SIAM J. Control Optim.*, **28**:478–481, 1990.

[416] S. L. Troyanski. On locally convex and differentiable norms in certain nonseparable Banach spaces. *Studia Math.*, **37**:173–180, 1971.

[417] L. Tuncel and H. Wolkowicz. Strong duality and minimal representations for cone optimization. Preprint, 2008.

[418] J. Van Tiel. *Convex Analysis: An Introductory Text.* John Wiley, New York, 1984.

[419] J. Vanderwerff. Smooth approximation in Banach spaces. *Proc. Amer. Math. Soc.*, **115**:113–120, 1992.

[420] J. Vanderwerff. Second-order Gâteaux differentiability and an isomorphic characterization of Hilbert spaces. *Quart. J. Oxford. Math*, **44**:249–255, 1993.

[421] A. Verona and M. E. Verona. Regular maximal monotone operators and the sum rule. *J. Convex Analysis*, **7**:115–128, 2000.

[422] L. Veselý. A short proof of a theorem on composition of d.c. mappings. *Proc. Amer. Math. Soc.*, **87**:685–686, 1987.

[423] L. Veselý. A parametric smooth variational principle and support properties of convex sets and functions. *J. Math. Anal. Appl.*, **350**:550–561, 2009.

[424] L. Veselý and L. Zajíček. Delta-convex mappings between Banach spaces and applications. *Dissertationes Math. (Pozprawy Mat.)*, volume 289, 1989.

[425] L. Veselý and L. Zajíček. On composition of d.c. functions and mappings. *J. Convex Anal.*, **16**:423–429, 2009.

[426] L. Veselý and L. Zajíček. On extensions of d.c. functions and convex functions. *J. Convex Anal.*, 2010.

[427] A. Visintin. Strong convergence results related to strict convexity. *Comm. Partial Differential Equations*, **9**:439–466, 1984.

[428] L. P. Vlasov. Almost convex and Chebyshev subsets. *Math. Notes Acad. Sci. USSR*, **8**:776–779, 1970.

[429] L. P. Vlasov. Several theorems on Chebyshev sets. *Mat. Zametki*, **11**:135–144, 1972.

[430] L. P. Vlasov. Approximative properties of sets in normed linear spaces. *Russian Math. Surveys*, **28**:1–66, 1973.

[431] M. D. Voisei. A maximality theorem for the sum of maximal monotone operators in non-reflexive Banach spaces. *Math. Sci. Res. J.*, **10**:36–41, 2006.

[432] M. D. Voisei and C. Zălinescu. Linear monotone subspaces of locally convex spaces. arXiv:0809.5287v1, 2008.

[433] M. D. Voisei and C. Zălinescu. Strongly-representable monotone operators. *J. Convex Anal.*, **16**:1011–1033, 2009.

[434] M. J. Wenninger. *Polyhedron Models.* Cambridge University Press, 1989.

[435] U. Westfall and J. Frerking. On a property of metric projections onto closed subsets of Hilbert spaces. *Proc. Amer. Math. Soc.*, **105**:644–651, 1989.

[436] H. Weyl. Elementare Theorie der konvexen Polyeder. *Commentarii Math. Helvetici*, **7**:290–306, 1934.

[437] R. Wijsman. Convergence of sequences of convex sets, cones and functions, II. *Trans. Amer. Math. Soc.*, **123**:32–45, 1966.

[438] H. Wolkowicz, R. Saigal, and L. Vandenberghe (Eds.). *Handbook of Semidefinite Programming: Theory, Algorithms, and Applications*, volume 27 of *International Series in Operations Research & Management Science*. Kluwer Academic Publisher, Boston, MA, 2000.

[439] S. J. Wright. *Primal-Dual Interior-Point Methods*. SIAM, Philadelphia, 1997.

[440] Z. Y. Wu, D. Li, L. S. Zhang, and X. M. Yang. Peeling off a nonconvex cover of an actual convex problem: hidden convexity. *SIAM J. Optim.*, **18**:507–536, 2007.

[441] D. Yost. M-Ideals, the strong 2-ball property and some renorming theorems. *Proc. Amer. Math. Soc.*, **81**:299–303, 1981.

[442] L. Zajíček. On the differentiation of convex functions in finite and infinite dimensional spaces. *Czech. Math. J.*, **29**:340–348, 1979.

[443] L. Zajíček. Porosity and σ-porosity. *Real Anal. Exchange*, **13**:314–350, 1987/88.

[444] C. Zălinescu. On uniformly convex functions. *J. Math. Anal. Appl.*, **95**:344–374, 1983.

[445] C. Zălinescu. *Convex Analysis in General Vector Spaces*. World Scientific, New Jersey, 2002.

[446] C. Zălinescu. A new proof of the maximal monotonicity of the sum using the Fitzpatrick function. In F. Giannessi and Eds. A. Maugeri, editors, *Variational Analysis and Appls.*, volume 79, pages 1159–1172. Springer, New York, 2005.

[447] C. Zălinescu. A new convexity property for monotone operators. *J. Convex Anal.*, **13**:883–887, 2006.

[448] L. Zhou. A simple proof of the Shapley–Folkman theorem. *Econ. Theory*, **3**:371–372, 1993.

[449] M. Zippin. Extension of bounded linear operators. In W. B. Johnson and J. Lindenstrauss, editors, *Handbook of the Geometry of Banach Spaces*, volume II, pages 1703–1741. Elsevier, Amsterdam, 2003.

[450] V. Zizler. Smooth extension of norms and complementability of subspaces. *Arch. Math.*, **53**:585–589, 1989.

[451] A. Zygmund. *Trigonometric Series*, volume I,II. Cambridge University Press, New York, 2nd edition, 1959.

Index

absolutely continuous, 79
absorbing set, 142, **142**
  in barreled spaces, 467
abstract linear program, **108**, 109
  unstable, 465
active set, **95**
adjoint, **46**
affine combination, **66**
affine function, **21**
affine hull, **66**, 69
affine mapping, **21**
  characterization, 22
affine set, **66**, 69
Alaoglu's theorem, 130
Alexandrov's theorem, 86
  failure in infinite dimensions, 204
  special cases in infinite dimensions, 475
α-cone meager set, **197**
(ANA) operator, *see* maximal monotone operator, type (ANA)
analytic center, 182
angle small set, **197**, 199
anti-proximinal, **327**
Applications in exercises
  Blaschke–Santalo theorem, 32
  coupon collection problem, 180
  Csiszár entropy problems, 366
  differential inclusions, 123
  direct and indirect utility, 56
  divergence estimates, 58
  doubly stochastic matrices, 93
  experimental design duality, 58
  Fisher information, 365
  gamma function, 32
  hidden convexity, 55, 56
  location problem, 59
  Lyapunov, duality, 124
  Lyapunov functions, duality of, 124
  maximum entropy, 57
  monotonicity and elliptic PDE's, 437
  multiplicative potential and penalty functions, 355
  NMR entropy, 63
  risk function duality, 185
  symbolic convex analysis, 63
approximate fixed point, **167**
approximate mean value theorem, 166

approximate selection theorem, 112
approximately convex set, **193**
Archimedean solids, 462
argmin, **285**
arithmetic-geometric mean inequality, 30
Aronszajn null, **197**, 199, 201
Asplund averaging theorem, 221, 227
Asplund cavern, **474**
Asplund space, **199**, 206, 223, 318, **318**, 319, 320
  and renorming, 220
  characterization, 281, 319, 321
  dual has RNP, 324
  scattered $C(K)$, 220
  separable, 199, 215
  weakly compactly generated, 220
attainment, in Fenchel problems, 177
Attouch–Wets convergence, **289**
  Bicontinuity theorem, 290
  of functions, **289**
  versus Mosco convergence, 295
  versus pointwise convergence, 292
  versus slice convergence, 295
  versus uniform convergence on bounded sets, 297
Aumann convexity theorem, 14, 72

Baire category theorem, 413, 467
Baire space, **460**
balanced set, **18**
Banach lattice, 329–337
  renorms, 334
Banach limits, **266**
Banach space, **22**
Banach's fixed point theorem, 167
barreled space, 460, **467**
basic separation theorem, 29, 132
β-differentiability, **150**, 379
  Šmulian's theorem for, 159
  Šmulian's theorem for conjugate functions, 160
  of norms, 227
BFGS update, **353**
biconjugate, *see* Fenchel biconjugate
bilinear map, **44**
  as a second-order derivative, 44, 201
  symmetric, 88

bipolar, **178**
bipolar cone, *see* cone, bipolar
bipolar theorem, 209
Bishop–Phelps cone, 182
Bishop–Phelps theorem, 163, 169
Blaschke–Santalo inequality, 33
Bohr–Mollerup theorem, 32
Boltzmann–Shannon entropy, 468
Borel measurable, 79
Borel measure, **195**
Borel sets, 305
bornology, **149**, 187
Borwein's variational principle, 170
Borwein–Preiss smooth variational principle, 473
Bose–Einstein entropy, 364
boundary, properties, 462
bounded weak*-topology, 447
(BR) operator, *see* maximal monotone operator, type (BR)
Brachistochrone problem, 10
Bregman distance, **60**, **358**, 360
Bregman projection, **361**, 364
Brøndsted–Rockafellar property, *see* maximal monotone operator, type (BR)
Brøndsted–Rockafellar theorem, 121, 162
Brun–Minkowski inequality, 352

Calderón norm, 316, 317
Carathéodory's theorem, 71
category
    first, *see* first category
    second, *see* second category
Cauchy–Schwarz inequality, 22, 49
Čebyšev set, **8**, **117**, **188**
    convexity of, 8, 189, 190, 474
    in reflexive spaces, 188
Cellina approximate selection theorem, 112, 281
Chebyshev set, *see* Čebyšev set
choice function, 336
C^k-smooth function, **222**
    approximations, 230
C^∞-smooth function, **222**
Clarke directional derivative, **82**, 123
Clarke subdifferential, **82**
    cusco property, 123
closed
    images, 114
    range, 461
    subspace, 461
closed function, **2**, **20**, **127**, 138
    and lower-semicontinuity, 127
    and weak-lower-semicontinuity, 138
closure, of function, 21, **127**, **127**, 138
Cobb–Douglas function, **57**
codimension, countable, 467
coercive function, **24**, 174
    conjugate of, 174, 175
    minimization on reflexive space, 183
    versus cofinite, 185
cofinite function, **118**, **174**, **361**
    versus supercoercive, 118, 185, 361, 384
compact
    images, 114
    in infinite dimensions, 461

polyhedron, 95
range of multifunction, 114
unit ball, 461
compact separation theorem, 136
complementary slackness, in cone programming, 110
complemented subspace, 464
complete, 467
composing, USC multifunctions, 114
composition, convex functions, 75
concave
    conjugate, 183
concave function, **7**, 40, **126**
cone, **21**
    bipolar, **67**, 68, 108, 178, **178**
    finitely generated, **94**
    nonnegative, 468
    normal, *see* normal cone
    pointed, **68**, 96
        characterization, 68
    polar, 67, **67**, 178, 183
        calculus, 178
        negative, **67**, **178**
        positive, **67**, **178**
        sum of closed cones, 183
    program, 108
    sum not closed, 183
    sums, 469
    tangent, *see* tangent cone
cone convex, 75
cone quasiconvex, 335
constraint, linear, 48, 107, 177, 182
constraint qualification (CQ), 119, 405, 406
constructible, **356**, 356–357
continuity
    absolute, 79
    and USC, 114
    generic, 413
    in infinite dimensions, 460
    of convex functions, 461
    of extensions, 116
    of selections, 112
control theory, 472
convex, quasi, 414
convex body, **33**
convex calculus theorem, *see* subdifferential, sum rule
convex cone, base, 75
convex feasibility problem, 362
convex function, **18**, **21**, **126**
    and monotone gradients, 37, 160
    bounded above everywhere, 30
    bounded on β-sets, 379
        extensions of, 398
    bounded on bounded sets, 137, 379, 382, 384
        conjugate supercoercive, 175
        extensions of, 399
    bounded on weakly compact sets, 379, 383, 384
    composition, 75
    conjugate, *see* Fenchel conjugate
    continuity of, 4, 25, 128, 129, 461
    continuous biconjugate, 386

convex function (*Cont.*)
  continuous extensions, 386, 392–399
    failure, 392
  continuously differentiable, 35
  difference of, *see* DC function
  epigraph, 139
  essentially smooth, *see* essentially smooth function
  essentially strictly, *see* essentially strictly convex function
  finitely extendable, 398
  Fréchet differentiable, *see* Fréchet differentiability
  Gâteaux differentiable, *see* Gâteaux differentiability
  Legendre, *see* Legendre function
  Lipschitz, 51, 144, 173
    extensions of, 53, 399
    on bounded sets, 128, 137, 141
  local and global minima, 4, 26, 130
  local Lipschitz property, 4, 25, 128
  locally uniformly, *see* locally uniformly convex, function
  midpoint, *see* midpoint convex function
  on real line, 20, 36
  recognizing, 36, 37, 39, 91, 141, 160, 202
  stability properties, 3, 23, 139
  strictly, *see* strictly convex, function
  strongly rotund, *see* strongly rotund function
  totally, *see* totally convex function
  unbounded on bounded set, 381
  uniformly, *see* uniformly convex, function
  uniformly smooth, *see* uniformly smooth, function
  weak lower-semicontinuity, 132
convex functions, matrix, 105
convex hull
  of function, 21, **128**
  of set, 21, **126**, 464, 465
convex mean value theorem, 121
convex program, 177
  ordinary, 48
convex series closed, **142**
convex set, **18**, **126**
  characterization in Hilbert space, 190
  stability properties, 139
  symmetric, **18**
convexity, 430
core, **5**, **25**, 29, 34, 129, **129**, 131, 141
  and absorbing sets, 142, 467
  in barreled spaces, 467
  versus interior, 25, 142
countable
  basis, 467
  codimension, 467
countable set, 195
Csiszár divergence, 365
cusco, **111**, **279**
  and Asplund spaces, 281
  and maximal monotone operators, 280
  Clarke subdifferential, 123
  fixed points on compact convex, 111, 281
  generically single-valued, 283
  maximal monotone multifunction, 123
  minimal, **279**, 283

(D) operator, *see* maximal monotone operator, type (D)
Darboux property, **72**
Davis' theorem, 101, 103, 472
DC function, **104**, **146**
  approximation of Lipschitz functions, 218
  failure, 147
  k-th largest eigenvalue, 11, 104
  vector-valued, **104**
DC operators, 317
dense
  hyperplane, 467
  range, 469
  subspace, 469
dense type, **439**, 453
dentable, **321**
denting point, **326**
derivative
  Dini, *see* Dini derivative
  directional, *see* directional derivative
  Fréchet, *see* Fréchet differentiability
  Gâteaux, *see* Gâteaux differentiability
determinant, order preservation, 103
diametral set, **167**
difference convex function, *see* DC function
difference quotient, **84**
  second-order, **84**
differentiability
  generic, 413
  of spectral functions, 102
differential inclusion, **123**
Dini derivative, 79
directional derivative, **6**, **26**, **131**, 147
  Lipschitz property, 131
  lower, 166
  sublinear property, 27, 131
disjoint operator ranges, 469
distance function, 8, **22**, 127, 180
  attainment, 461
  convexity of, 23
  on Hilbert space, 189, 190
distortable, arbitrarily, 193
distortion, 189
  of a norm, 192
divergence, Kullback–Leibler, 13
divergence estimates, 75
domain
  of a function, **5**, **18**
  of a monotone operator, **118**
  of subdifferential
    not convex, 141
    polyhedral, 95
domain regularizable, **412**
doubly stochastic, 93
DP*-property, 383, **383**
dual linear program, 107
dual norm, **129**
  characterization, 210
  implied by weak*-Kadec property, 224
dual space, **129**

dual value, in LP and SDP, 107
duality
  gap
    Duffin's, 50
    LP, 107
    SDP, 107
    weak, 177
    Fenchel, 48, 98
duality mapping, 133
Dunford–Pettis property, **383**, 385, 386, 388
Dunford–Pettis theorem, 302

(ED) operator, *see* maximal monotone operator, type (ED)
eigenvalue
  k-th largest, 11, 104
  mapping, 314
eigenvalue problem, 438
eigenvalues
  functions of, 101
  optimization of, 472
Ekeland's variational principle, 161, 168
  equivalence with completeness, 166
  in Euclidean space, 117
entropy
  Boltzmann–Shannon, 61
  Bose–Einstein, 364
  Burg, 61
    nonattainment, 311
  energy, 61
  Fermi–Dirac, 61, 364
  maximum, 179, 182
  nonattainment, 310
  relative, 58
  Tallis, 480
epigraph, **2**, **20**, **127**
  as multifunction graph, 114
  polyhedral, 94
epigraphical convergence
  and minimizing sequences, 293
  relations among, 291, 295, 296
  relations among in finite dimensions, 296
  summary of relations among, 291
  versus pointwise convergence, 292
ε-net, **318**
ε-subdifferential, **121**, **137**
  nonempty, 137
  of conjugate function, 160
  of norms, 222
equilibrium, 473
essentially smooth function, **338**, 362
  barrier for open convex cone, 351
  barrier for open convex set, 348, 370
  characterizations, 344
  duality with essential strict convexity, 344
  examples, 353
  Fréchet, **347**
  in the classical sense, **338**
    compatibility in Euclidean space, 345
    outside of Euclidean space, 346
  log barriers, 100
  spectral functions, 102
  versus Gâteaux differentiability, 341
essentially strictly convex function, **339**, 362

and coercivity, 347
duality with essential smoothness, 344
examples, 353
in the classical sense, **338**
  compatibility in Euclidean space, 345
  outside of Euclidean space, 346
on reflexive spaces, 345
spaces without, 373
spectral functions, 102
versus strict convexity, 341, 342, 344, 347, 374
Euclidean space, **18**, **22**, **460**
excess functional, 289
experimental design, 58
exposed point, **92**, **211**
  duality with differentiability, 211
  of function, **232**, 235
    characterization, 232
    duality with Gâteaux differentiability, 234, 240
    versus extreme point, 235
    versus strongly exposed, 234
  strongly, *see* strongly exposed point
  versus strongly exposed, 92
extension, continuous, 116
extreme direction, 335
extreme point, **92**, **211**
  of polyhedron, 95
  versus exposed point, 92
  versus support point, 224

Fan inequality, 99
Fan minimax theorem, 55
Fan theorem, 99
Fan–Kakutani fixed point theorem, 111, 281
Farkas' lemma, 70
farthest point function, 167
Favard inequality, 49
feasible set, 49
Fenchel biconjugate, **44**, 96, 101, 103, **171**, 178, 179
Fenchel biconjugation theorem, 65
Fenchel conjugate, **44**, **171**, 179
  and eigenvalues, 100
  bounded on bounded sets, 384, 385
  bounded on weakly compact sets, 385
  differentiable, 187
  examples, 45
  of coercive function, 174, 175
  of composition, 76
  of exponential, 179, 182
  of infimal convolution, 176
  of Lipschitz convex, 51, 173
  self conjugate characterization, 179
  transformations, 46
Fenchel duality, 99, **177**
  and LP, 108
  generalized, 99
  linear constraints, 48, 97, 182
  polyhedral, 97, 98
  symmetric, 183
Fenchel duality theorem, 6, 46, 177
Fenchel problem, 177
Fenchel–Young inequality, 44, 101, 171
Fermat point, 59

Fermat–Weber problem, 59
Fermi–Dirac entropy, 364
Fillmore–Williams theorem, 104
finitely generated
    cone, 94–96
    function, 94–98
    set, 94–98
finitely representable, **143**, 225
first category, **195**
Fisher information, 365
Fisher information function, 365
Fitzpatrick function, **119**
fixed point, **167**, 474
Fourier identification, 469
(FP) operator, *see* maximal monotone operator, type (FP)
Fréchet
    smooth space, 166
    subderivative, 165
    subdifferentiable, 165
    subdifferential, **165**
        monotonicity of, 414–415
Fréchet bornology, **149**
Fréchet differentiability, **34**, **37**, **149**, **150**, **198**
    almost everywhere, 77, 78
    and continuous selections, 277
    and unique subgradients, 4, 34
    characterizations of, 34, 151, 153
    continuity of, 35, 43, 154
    dual norm, 221
    duality with local uniform convexity, 240, 251
    duality with perturbed minimization, 237
    duality with strongly exposed points, 234, 240, 250
    duality with Tikhonov well-posed, 235
    generic, 77, 152, 199, 215
        conjugate functions and RNP, 321, 323
        on Asplund spaces, 319
    implies continuity, 41
    of conjugate functions, 160, 187
        and biconjugates, 178
        and convexity of original, 179
        and minimization principles, 323
        and reflexivity, 224, 228, 250, 251
    of norms, **150**, **210**, 215, 220
        and separable duals, 215
        characerization, 210
        duality, 225
        duality with strongly exposed points, 211
        nowhere, 205
        on Asplund spaces, 319
        versus Gâteaux differentiability, 227
    second-order, *see* second-order derivative
    Šmulian's theorem for, 153
    support functions, 317, 318
    symmetric formulation, 151
    uniform, *see* uniformly Fréchet differentiable
    versus Gâteaux differentiability, 4, 34, 41, 381, 382, 384
    versus weak Hadamard differentiability, 379, 381, 384, 385
Fubini's theorem, 79
function
    Kadec property, *see* Kadec property, function
    Lyapunov, 124
    nowhere differentiable, 8
    simple, 302
    strongly rotund, *see* strongly rotund function
fundamental theorem of calculus, 79

Gâteaux bornology, **149**
Gâteaux differentiability, **34**, **37**, 149, **150**, **198**, 362
    and continuous selections, 277
    and directional derivatives, 41
    and interior of domains, 157
    and subdifferentials, 150
    and unique subgradients, 4, 34, 151, 158, 346, 462
    characterizations of, 34, 151, 152, 154
    continuity of, 35, 43, 154
    duality with exposed points, 234, 240
    duality with strict convexity, 240, 250, 341
    generic, 198
    nongeneric, 199, 284
    of conjugate functions, 160, 187
    of norms, **150**, **210**, 211
        and unique support functionals, 212
        duality with exposed points, 211
        duality with strict convexity, 192
        nowhere, 205
        versus Fréchet differentiability, 227
    second-order, *see* second-order derivative
    sets of nondifferentiability points, 201
    Šmulian's theorem for, 154
    symmetric formulation, 152
    uniform, *see* uniformly Gâteaux differentiable
    versus continuity, 36, 150
    versus Fréchet differentiability, 4, 34, 36, 41, 381, 382, 384
    versus weak Hadamard differentiability, 383, 385
Gâteaux differentiability space, 161, **198**
Γ-function, **32**
gauge function, **22**, **127**
    and lower semicontinuity, 144
    and polar function, 58
    implicit differentiation, 154
    properties of, 30, 144
gauge of uniform convexity, **266**
Gelfand–Phillips space, **383**
generated cuscos, 116
generic
    continuity, 413
    differentiability, 413
    single-valued, 413
generic set, **195**
Goldstine's theorem, 140
good asymptotical behavior, **340**
Gordan's theorem, 70
Gossez operator, 446
Grüss–Barnes inequality, 49
graph
    minimal, 116
    of a monotone operator, **278**
    of a multifunction, **111**

Grothendieck property, **383**
Grothendieck space, **383**, 390, 391, 396, 399
   continuous biconjugate function, 386
   extensions of convex functions, 396

Haar null, **196**, 200
Hahn–Banach extension property, 147
Hahn–Banach extension theorem, 28, 130
Hahn–Katětov–Dowker sandwich theorem, 116
Hamel basis, **128**, 467
Hardy et al. inequality, 100
Harmonic-arithmetic log-concavity theorem, 33
Hausdorff distance, 289
Hessian, **38**
   and convexity, 39, 202
higher-order derivative, **221**
   of norms on L_p-spaces, 222
Hilbert cube, 141
Hilbert space, **144**, 464
   characterizations, 481
   isomorphic characterizations, 273
   lattice characterizations, 332
Hilbert–Schmidt operators, 313
Hölder exposed point, **268**, 273
   duality, 270
Hölder inequality, 49
Hölder smooth point, **269**, 273
hyperplane, dense, 467
hypersurface, **201**

ideal, **331**
implicit function theorem for gauges, 154
improper function, **21**
improper polyhedral function, 98
inclusion control problem, 401
incomplete normed space, 460
inconsistent, 109
indicator function, **3**, **21**, 127, 180
   conjugate of, 179
   subdifferential of, 130
induced norm, 106
inequality
   arithmetic-geometric mean, 30
   Blaschke–Santalo, 33
   Brun–Minkowski, 352
   Cauchy–Schwarz, 22, 49
   Fan's, 99
   Favard, 49
   Fenchel–Young, 44, 171
   Fitzpatrick, 435
   Grüss–Barnes, 49
   Hardy et al., 100
   Hölder, 49
   Jensen's, 29
   Knuth's, 41
   Lyapunov duality, 124
   polar, 58
   Rogers–Hölder, 352
   strong Fitzpatrick, 451
   three-slope, 19
   triangle, 22
   variational, 436
   weak duality, 7, 47, 177

inf-compact, **340**
infimal convolution, **51**, 53, **175**
   and approximation, 176, 184
   and conjugation, 52
   and uniform convergence, 52
   as a kernel average, 180
   basic properties, 51
   conjugate of, 176
injective Banach space, 148
inner product, 143
inverse, image, 97
isotone, 75

James space, 458
James' theorem, 143
James–Enflo theorem, 218
Jensen's inequality, 29
JN-sequence, 382
jointly convex, **60**, 61
Josefson–Nissenzweig theorem, 382

Kadec property, **223**
   function, **306**
   Kadec property, weak *see* Kadec–Klee property
   weak*, **191**, **223**
      and Asplund spaces, 223
      implies norm is dual, 224
      of double dual, 224
Kadec' theorem, 214
Kadec–Klee property, **189**, 190, 193
   Kadec property, weak* *see* Kadec property, weak*
Kakutani–Fan fixed point theorem, 111, 281
kernel average, **180**
key theorem, 94, 96
Kirchberger's theorem, 72
Kirszbraun–Valentine theorem, 282
Klee cavern, **474**
Knuth's inequality, 41
Krein–Milman property, **324**
Krein–Milman theorem, 140
Krein–Rutman theorem, 67
Kullback–Leibler entropy, 363
Kuratowski–Painlevé convergence, **285**
   and minimizing sequences, 287
   of functions, 286
   not preserved under conjugation, 296
   versus pointwise convergence, 292
   versus Wijsman convergence, 295

Lagrangian
   duality
      linear programming, 107
Lagrangian dual, 49
Lambert W function, 75, 76
Laplacian, 438
lattice, ordering, 333
lattice band, **331**
lattice operations, **329**
Lau–Konjagin theorem, 189
Lebesgue integral space, **143**
Lebesgue sequence space, **143**
Lebesgue–Radon–Nikodým theorem, 304

Legendre function, **339**, 364
  cofinite versus supercoercive, 361
  duality failure outside reflexive spaces, 348
  duality in reflexive spaces, 344
  examples, 353, 354
  Fréchet, **347**
  in the classical sense, **338**
    compatibility in Euclidean space, 345
    outside of Euclidean space, 346
  on general Banach space, 369
  on reflexive spaces, 345
  spaces with/without, 373
  spectral, 353
  zone consistency of, 362, 368
Legendre transform, 348, 375
Legendre–Fenchel conjugate, *see* Fenchel, conjugate
lexicographic order, 314
$\liminf_{i \to \infty} F_i$, 285
$\limsup_{i \to \infty} F_i$, 285
linear
  objective, 107
  programming (LP)
    abstract, 108, 109
    and Fenchel duality, 108
    primal problem, 107
linear functional
  continuity of, 461
  discontinuous, 142, 144, 460, 461
  non-norm-attaining, 468
  norm-attaining, **133**, 143
    dense in dual, 163, 169
  supporting, **133**, 468
    dense, 169
    unique and differentiability, 212
Lipschitz, eigenvalues, 104
Lipschitz constant, **19**, 173
Lipschitz exposed point
  of a function, **268**, 270
  of a set, **268**
  of functions, 273
Lipschitz extension, 53
Lipschitz function, **19**, 127
Lipschitz smooth point, **204**, **269**, 273
  implied by weak second-order Taylor, 204
  nongeneric, 270
locally bounded, 114, 116
locally Lipschitz function, **19**
  Fréchet differentiable almost everywhere, 78
locally uniformly convex
  dual norm, 215, 224
    C^1-smooth approximations, 230
    and smoothness, 213
    on separable dual, 214
  function, **239**
    and renorming, 246
    and strongly exposed points, 239
    at a point, **239**, 272
    duality with Fréchet differentiability, 240, 251
    is totally convex, 374
    stable under sums, 241
  norm, 213, 220, 221, 246
    duality, 225
    extensions of, 229
    on separable space, 214
    on WCG spaces, 220
    preserved by sums, 225
  locally uniformly rotund, *see* locally uniformly convex
location problem, 59
Loewner convex functions, 105
Loewner ordering, **61**, 106
log, 100, **100**, 182
log barrier, 100
log det, **100**, 100–103, 353
log-convex function, 40, **40**, 42
lower level set, **20**, 127
  bounded, 31, 46, 185
    and coercivity, 24, 174
  closed, 127
  compact, 100
  convex, 23, 139
  of conjugate function, 175
  of differentiable function, 154
  unbounded, 50
  weak*-closed, 138, 210
  weakly closed, 138
lower-semicontinuous
  and USC, 114
  generic continuity, 413
  in infinite-dimensions, 460
  sandwich theorem, 116
lower-semicontinuous function, **2**, **20**, 98, **127**
  equivalent to closed, 127
  τ-lsc, **130**
  weakly-lsc when convex, 132
LSC multifunction, *see* multifunction, LSC
Lyapunov convexity theorem, 15
Lyapunov function, **124**

mathematical economics, 473
matrix
  analysis, 100
  completion, 353
  optimization, 107
matrix norm, **106**
max formula, 6, **28**, 131
max function, subdifferential of, 145
maximal monotone multifunction, *see* maximal monotone operator
maximal monotone operator, **118**, **278**
  acyclic, **418**, 418–423
  and cuscos, 280
  and subdifferentials, 118, 278, 413, 419, 429
  Asplund decomposition, 418
  coercive, 446
    versus surjective, 437
  continuity, 283
  convexity of domain and range, 410–412, 452, 458
  cyclically, 429
  dense type, *see* type (D)
  irreducible, *see* acyclic
  locally maximal monotone, *see* type (FP)
  maximal monotone locally, *see* type (VFP)
  on reflexive spaces, 433–439
  relations between classes of, 459

strongly locally maximal, **442**, 447
sum theorem on reflexive space, 434
type (ANA), 455, **455**
type (BR), **455**
type (D), **439**, 442, 446, 482
   implies type (FP), 441
type (ED), **439**, 482
type (FP), 410, **410**, 441, 446
type (NI), **439**, 440, 445, 450–459, 482
   characterization, 453
   sum theorem, 450
type (VFP), **410**, 441, 442
type (WD), **453**
type negative infimum, see type (NI)
Mazur's theorem, 198
meager set, 195
mean value theorem, 121
   approximate, 166
measures, tight, 207
method of cyclic projections, 363, **363**
Michael selection theorem, 113, 281
midpoint convex function, **31**, **139**
   measurable, 31, 139
Mignot's theorem, 412
Milman–Pettis theorem, 217
minimal, graph, 116
minimax theorem, 182
minimization principles, 273
minimizer, subdifferential zeros, 131
Minkowski, 472
   functional, see gauge function
   theorem, 95
minorant, affine, 97
modulus of convexity
   of functions, **244**
      at a point, **272**
      duality, 252
      power type duality, 253
      versus gauge of uniform convexity, 266
   of norms, **215**, 258
      duality, 217, 218
      power type, 2, **218**, 226, 260
modulus of smoothness
   of functions, **252**
      at a point, **269**
      duality, 252, 253
      power type, 255, 256
   of norms, **216**
      duality, 217, 218
      power type, **218**, 255, 256
monotone multifunction, see monotone operator
monotone net property, 77
monotone operator, **118**, **276**
   differentable, 412
   extension formulas, 407–408
   locally bounded, 279, 408–410
   maximal, see maximal monotone operator
   upper-semicontinuous, **276**
monotone set, **278**
monotonically related, 118
Moreau–Rockafellar dual theorem, 175, 463
Moreau–Rockafellar theorem, 46, 118, 175
Mosco convergence, **287**

and minimizing sequences, 288, 293, 294
and optimal value problems, 311
and the Kadec property, 299
bicontinuity theorem, 288
stability under addition, 300
versus Attouch–Wets convergence, 295
versus pointwise convergence, 292
versus slice convergence, 296
versus Wijsman convergence, 296, 299
multifunction, **111**, **276**
   closed, 111
      versus USC, 114
   continuous, 111
   cyclically monotone, **415**
   domain of, 276
   fixed point, **111**
   isc, **111**
   locally bounded, 122
   locally bounded at a point, **279**
   LSC, **111**, 113–116
   monotone, see monotone operator N-monotone **415**
   osc, **111**
   quasimonotone, **414**
   range of, 111, 112, 114, 281
   USC, **111**, **276**

nearest point, **53**
   in infinite dimensions, 464
   in polyhedron, 182
   selection, 114, 413
   to p-sphere, 54
   to an ellipse, 54
nearest point mapping, **54**, **188**
   discontinuous, 54
   empty, 191
   nowhere differentiable, 204
necessary optimality condition
   Pshenichnii–Rockafellar, 147
negligible set, 197
NI Banach space, 450
(NI) operator, see maximal monotone operator, type (NI)
nonatomic, **72**
nonexpansive
   mapping, 282
      extension, 282
nonnegative cone, 468
nonreflexive Banach space, 184, 269
   DC function characterization, 146
   Haar null sets, 200, 206
   lack of nearest points, 191
   Legendre functions, 348
norm, **22**
   dual, see dual norm
   entrywise, 107
   induced, 106
   subgradients of, 145
   submultiplicative, 106
norm topology, 461
normal integrands, 305, **306**
normal cone, **26**, **67**, **144**, **147**, 178
   and polarity, 67
   and subgradients, 182

normal cone (*Cont.*)
   and tangent cone, 68
   to intersection, 182
normal mapping, **436**
normed linear space, **22**, 467
nowhere dense, **195**
null set, **195**
   Aronszajn, *see* Aronszajn null
   Haar, *see* Haar null

objective function, 48
   linear, 107
open map, 30
open mapping theorem, 30, 142
optimal value, 97
   dual, 177
   in LP and SDP, 107
   primal, 177
optimization
   linear, 107
   matrix, 107
   subdifferential in, 130
order
   preservation
      of determinant, 103
order-complete, 147
ordering, lattice, 333
orthogonal
   invariance, 103
   similarity transformation, 103
osc multifunction, *see* multifunction, osc

p-norm, 49, **143**
paired Banach spaces, 313
paracompact, 281
Pareto efficient point, 336
partial order, directed, 335
partition of unity, 111–115
permutation, matrix, 104
Pinsker inequalities, 13
Pisier's theorem, 218
Pitt's theorem, 325
polar cone, *see* cone, polar
polar of a set, **33**, **209**
polyhedral
   algebra, 97–98
   calculus, 98
   cone, 95, 99, 108
   convex analysis, 94
   Fenchel duality, 97
   function, 94–99
   problem, 107
polyhedron, 73, 94–99, 147
   compact, 95
   nearest point in, 182
   polyhedral set, 94
   tangent cone to, 98
polytope, 94–96
porous set, **197**
   σ-, **197**
positive definite, strongly, 231
positive operator, 312
positively homogeneous function, **22**

prepolar, **209**
primal
   linear program, 107
   semidefinite program, 109
principle of uniform boundedness, 16
proper function, **2**, **21**, 95
property
   Krein–Milman, 324
   Radon–Nikodým, *see* Radon–Nikodým
      property
proximal set, **188**
   in reflexive spaces, 188
Pshenichnii–Rockafellar conditions, 73, 147

quadratic form, conjugate, 106
quadratic programming, Boolean, 110
quasi-complement, **466**
quasi-Newton method, 353
quasiconvex, **414**, 430
   with respect to a cone, 335
quasimonotone multifunction, *see* multifunction,
      quasimonotone

Rademacher's theorem, 78
   in infinite-dimensions, 199
radius function, **167**
Radon measure, **195**
Radon–Nikodým property, **321**, 324, 325
   and Krein–Milman property, 324
   sets with, 321, **321**
      duality with differentiability, 321
      minimization principles, 324
      strongly exposed points, 322, 324
   sets without, 327
   spaces with, **321**
      characterization, 325
      dual to Asplund, 324
      minimization principles, 323
      no anti-proximinal pairs, 327
      and Rademacher's theorem, 199
      smooth bump functions, 328
      smooth variational principles, 328
      weak*-Asplund dual, 323
   spaces without, 327
range, closed, 461
range of multifunction, 114
ray, 462
rearrangement, 313
   invariant, 313
recession function, **50**
reflexive Banach space, **143**, 288
   and nearest points, 189
   Čebyšev sets, 188, 190, 193
   coercivity and minimization, 183
   dual norms, 192, 221, 224, 225
   essential strict convexity and smoothness
      duality, 344
   essentially strictly convex functions, 345, 347
   Fréchet differentiable conjugate, 224, 228, 250,
      251
   Haar null sets, 206
   Legendre functions, 344, 345, 358, 362, 368
   local uniform convexity/ Fréchet
      differentiability duality, 251

nonreflexive, *see* nonreflexive Banach space
norm-attaining functionals, 143
proximal sets, 188
strict convexity/Gâteaux differentiability
    duality, 240, 250, 341
stronly exposed point/Fréchet differentiability
    duality, 240
uniformly convex norm, 217
uniformly smooth norm, 217
weak compactness, 143
regularity, condition, 97
relative interior, **66**, 69
    and Fenchel duality, 99
    and separation, 67
    and subdifferentials, 67
    in infinite dimensions, 462
    nonempty, 4, 66, 462
relaxation, semidefinite, 110
renorm, in Ekeland principle, 170
renorms
    of Banach lattices, 334
    of Banach spaces, 213–215, 218–222, 224, 225,
        227, 229, 246, 254, 258, 260, 264, 274
residual, **195**
    locally, **206**
resolvent, 412
Riccati equation, 367
Rockafellar function, **429**
Rockafellar's theorem, 118, 278
Rogers–Hölder inequality, 352

Sandwich theorem, 7, 133
    Hahn–Katětov–Dowker, 116
scattered, **220**
Schatten p space, **313**
Schur property, 4, 384, **384**
    dual spaces with, 386
second category, **195**
second-order derivative
    and convex functions, 39, 202
    and Taylor expansions, 84, 203
    equivalent properties, 84
    Fréchet, 38, **38**, **201**
        and Taylor expansions, 84
        generalized, 84, **84**, 85, 86, 88, 90
        generalized and Taylor expansions, 203
        generalized weak*, **202**, **203**
        nowhere, 204
        symmetry of, 206
        versus Gâteaux, 89
        weak*, **201**
    Gâteaux, **38**, 44, 90, **201**
        and Taylor expansions, 84
        generalized, 84, **84**, 86, 88, **203**
        generalized weak*, **202**
        generalized weak* and Taylor expansions,
            203
        nowhere, 204
        symmetry of, 206
        versus Fréchet, 89
        weak*, **201**
    generalized
        almost everywhere, 86
    symmetry of, 38, 84, 206

versus Taylor expansion, 89
selection, **111**, **277**, 281
    continuous, 277
    continuous and differentiability, 277
self-adjoint operator, 312
self-dual cone, 67, 101, 109
semidefinite
    cone, 67, 100, 103, 107
    program (SDP), 107
semidifferentiable, **300**
    subdifferential, **300**
    twice, **300**
separable, 182
    and semicontinuity, 469
    Banach space, 468
separately convex, **60**, 61
separating family, **140**
separation
    and bipolars, 68
    failure without interior, 139, 462
    in infinite dimensions, 462
separation theorem, 64, 67, 132, 133
sequence of sets
    Kuratowski–Painlevé limit, 285
    lower limit, 285, 292
    upper limit, 285, 292
set-valued mapping, *see* multifunction
simple function, 302
simultaneous ordered spectral decomposition,
    100, 102
single-valued
    generic, and maximal monotonicity, 413
singular value, 106
skew operator, 88, **416**
    Fitzpatrick–Phelps, 449
Slater's condition, **50**, 108
slice convergence, **290**
    and minimizing sequences, 293
    and renorming, 290
    and the Kadec property, 298
    versus Attouch–Wets convergence, 295
    versus Mosco convergence, 296
    versus Wijsman convergence, 298
smooth approximation, 230
    on superreflexive spaces, 231
smooth variational principle, 165
    higher-order, 328
Šmulian's theorem
    for β-differentiability, 159
    for conjugate functions, 160
    for Fréchet differentiability, 153
    for Gâteaux differentiability, 154
    for norms, 210, 211
    for uniform Fréchet differentiability, 155
    for uniform Gâteaux differentiability, 156
Sobolev space, 332, 437
space
    Asplund, *see* Asplund space
    Baire, *see* Baire space
    barreled, *see* barreled space
    Euclidean, *see* Euclidean space
    Hilbert, *see* Hilbert space
    reflexive, *see* reflexive Banach space
    superreflexive, *see* superreflexive Banach space

space (*Cont.*)
  uniformly convex, *see* uniformly convex, space
  uniformly smooth, *see* uniformly smooth, space
  weak*-Asplund, *see* weak*-Asplund space
  weakly compactly generated, *see* weakly compactly generated space
spectral
  conjugacy, 101, 103
  differentiability, 102
  function, 100
    convex, 101, 103
  sequence, 313
  sequence space, 313
  subgradients, 102, 104
spectral operators, Fréchet differentiability, 316
spectral radius, 106, 231
square-root iteration, 103, 316
staunch, 208
staunch set, **197**
Stegall's variational principle, 324
Steiner problem, 59
Straszewicz's theorem, 92
strict maximum, **232**
strict minimum, **232**
strictly convex
  dual norm, 213
    and Gâteaux smooth approximations, 230
    and smoothness, 213
    on dual to WCG, 220
  function, **18**, **126**, **238**, 362
    and exposed points, 239, 240
    and extreme points, 239, 240
    at a point, **239**
    duality with Gâteaux differentiability, 240, 250, 341
    recognizing, 39, 202
    stable under sums, 241
    versus essential strict convexity, 341, 342, 344, 374
  log barriers, 100
  norm, **188**, **212**, 213, 221
    and extreme points, 212
    characterizations, 212
    duality, 192
    extensions of, 229
    preserved by sums, 225
    spaces without, 220
strong maximum, **232**, 233
strong minimum, **163**, **232**, 233
strongly exposed point, **211**
  duality with differentiability, 211
  of function, **233**
    and coercivity, 236
    characterization, 233
    duality with Fréchet differentiability, 234, 240
  versus exposed, 234
  versus exposed point, 92
strongly rotund function, **306**
  and optimal value problems, 311
  convex integral functional, 309
subadditive function, **22**
subderivative, 165
subdifferentiable, 165

subdifferential, **3**, **26**, **130**, 165
  at boundary points, 43, 339, 340
  at optimality, 6, 26, 130
  boundedness properties, 32, 137
  compact and convex, 29
  continuity properties, 43
  convergence properties, 43
  cusco property, 123
  domain not convex, 141
  empty at a point, 131
  empty on dense set, 141
  failure of sum rule, 146
  limits of gradients, 78
  local boundedness, 32, 122, 138, 340
  maximal monotonicity of, 118, 121, 278, 413
  monotonicity of, 42, 160, 414
  near boundary points, 70
  nonempty, 67, 131
    on dense set, 162
  not onto, 184
  of conjugate functions, 160
  of indicator function, 130
  of max function, 145
  of norm, 145
  of polyhedral function, 99
  of support function, 148
  singleton versus differentiability, 34, 151, 158, 462
  sum rule, 7, 48, 135
  upper-semicontinuity, 276
  weak*-closed and convex, 145
subgradient, **3**, **26**, 130
  and normal cone, 182
  existence of, 97
  of maximum eigenvalue, 145
  of norm, 145
  of polyhedral function, 95
  unique, 462, 468
subjets, 472
sublevel set, *see* lower level set
sublinear function, **22**, 103
  almost, 31
  characterization, 22
subspace
  closed, 461
  complemented, 464
  countable-codimensional, 467
  dense, 469
substitution norm, 334
sum, of subspaces, 469
sum rule for subdifferentials, *see* subdifferential, sum rule
sun, 194
supercoercive function, **118**, **174**, 339
  conjugate of, 175
  on reflexive space, 183
  versus cofinite, 118, 185, 361, 384
superreflexive Banach space, **143**, 225, 274
  and smooth approximation, 231
  characterization, 218, 264
  renorming, 218, 221
  smooth variational principles, 328

support function, **6**, **22**, **127**, 179, 317, 318, 326
  subdifferential of, 148
support point, **133**, 468
  dense in boundary, 169
  proper, **468**
supporting hyperplane, **65**, **133**
supporting hyperplane theorem, 65
supporting linear functional, *see* linear functional, supporting
surjective, linear map, 108
symmetric
  function, 100
  set, **18**, 104
symmetric operator, **416**

tail operator, 456
tangency properties, 462
tangent cone, **68**
  to polyhedron, 98
τ_β-topology, **150**, 187
  -sequentially lsc, 187
  properties of convex functions, 379
Taylor expansion
  second-order
    almost everywhere, 86
    and differentiability, 203
    equivalent properties, 84
    implies first-order differentiability, 203
    nongeneric, 91
    strong, **83**, 88, **202**, 203
    strong versus weak, 84, 202
    versus differentiability of nonconvex, 89
    versus mixed partials, 89
    weak, **83**, 88, **202**, 203, 204
  third-order
    versus differentiability of convex, 89
theorem
  prime ideal, 337
  Alaoglu, 130
  Alexandrov, 86
  approximate mean value, 166
  Asplund averaging, 221, 227
  Aumann convexity, 14, 15, 72
  Banach's fixed point, 167
  Banach–Dieudonné, 186
  Banach–Steinhaus, 16
  basic separation, 29, 132
  bicontinuity of Attouch–Wets convergence, 290
  bicontinuity of Mosco convergence, 288
  bipolar, 209
  Birkhoff, 8
  Bishop–Phelps, 163, 169
  Blaschke–Santalo, 32
  Bohr–Mollerup, 9, 32
  Brodkskii–Mil'man, 167
  Brøndsted–Rockafellar, 121, 162
  Carathéodory, 5, 71
  compact separation, 136
  Davis, 101
  Davis–Lewis convexity, 11
  Dunford–Pettis, 302
  Ekeland's variational principle, 161, 168
    in Euclidean space, 117
  Fan's minimax, 55
  Farkas' lemma, 70
  Fenchel biconjugation, 65
  Fenchel duality, 6, 46, 177
  first-order condition, 6
  Gauss–Lucas, 9
  Goldstine, 140
  Gordan's, 70
  Hahn–Banach extension, 28, 130
  Hahn–Katětov–Dowker sandwich, 116
  harmonic-arithmetic log-concavity, 33
  Helly, 5
  Hessian and convexity, 39, 202
  James, 143
  James–Enflo, 218
  Josefson–Nissenzweig, 382
  Kadec, 214
  Kakutani–Fan fixed point theorem, 111, 281
  Kenderov, 413
  key, 94
  Kirchberger's, 72
  Krein–Milman, 140, 336
  Krein–Rutman, 67
  Krein–Šmulian, 186
  Lau–Konjagin, 189
  Lebesgue–Radon–Nikodým, 304
  Levy–Steinitz, 9
  Lidskii, 313
  Lyapunov convexity, 15
  max formula, 6, 28, 131
  Mazur, 198
  Michael selection, 113, 281
  Mignot's, 412
  Milman converse, 92
  Milman–Pettis, 217
  Minkowski, 8, 92
  Minty condition, 119
  monotone gradients and convexity, 37, 160
  Moreau–Rockafellar, 46, 118, 175
  Moreau–Rockafellar dual, 175, 463
  open mapping, 30, 142
  Pisier's, 218
  Pitt, 325
  Rademacher, 78, 199
  Radon, 5
  Riesz–Thorin convexity, 9
  Rockafellar's, 118, 278
  sandwich, 7, 64, 133
  separation, 64, 67, 132, 133
  Shapley–Folkman, 5, 72
  Sion minimax, 12
  smooth variational principle, 165
  Šmulian's, 153–156, 159, 160, 210, 211
  Stegall's variational principle, 324
  Straszewicz, 92
  subdifferential sum rule, 7, 48, 135
  supporting hyperplane, 65
  theorems of the alternative, 70
  Toeplitz–Hausdorff, 15
  von Neumann's minimax, 12, 74, 186
  weak*-separation, 136
three-slope inequality, 19
Tikhonov well-posed, **235**

topology
  bounded weak*, 447
  weak, *see* weak topology
  weak*, *see* weak*-topology
total, **140**
totally convex function, **374**
  modulus of, **374**
trace, of an operator, 312
trace class operators, 312
trace norm, 107
transversality conditions, **406**
triangle inequality, 22
twice Fréchet differentiable, *see* second-order derivative
twice Gâteaux differentiable, *see* second-order derivative
type (ANA), (BR), (D), (ED), (FP), (NI), (VFP), (WD), *see* maximal montone operator, type (ANA), type (BR), etc.

uniform convergence, 31, 115
  and conjugation, 297
  and infimal convolution, 184
  on bounded sets
    and conjugation, 52, 176, 285, 298
    and infimal convolution, 52, 176
    convergence of subdifferentials, 43
    versus Attouch-Wets convergence, 297
    Yoshida approximation, 52, 176
uniformly convex
  function, **241**, 256, 258
    and renorming, 264
    at a point, **272**, 374
    duality, 245
    growth, 244
    growth and renorming, 260
    minimization, 246
    on bounded sets, 264
    onto subdifferential, 246
    stable under sums, 241
  function on bounded sets, 241, **241**
    and uniformly convex norms, 246
    duality, 242, 243
  modulus, *see* modulus of convexity
  norm, **215**, 216, 221, 226, 246, 254, 256, 264
    and reflexivity, 217
    and superreflexivity, 218, 225
    duality, 216, 217
    extensions of, 229
    preserved by sums, 225
  space, 168, **215**
  weak*, *see* weak*-uniformly convex
  weakly, *see* weakly uniformly convex
uniformly Fréchet differentiable
  coincidence with uniform smoothness, 155
  function, **155**
  norm, **216**
  smooth, *see* uniformly smooth
uniformly Gâteaux differentiable
  bump function, 274
  function, **156**
    characterization, 156
    duality, 250

  norm, **228**, 274
    characterization, 228
    duality, 229
uniformly integrable set, **301**
uniformly smooth
  function, **155**, 227, **252**, 255
    and uniformly smooth norms, 247
    coincidence with uniform Fréchet differentiability, 155
    duality, 245
    on bounded sets, 264
    versus uniformly smooth on bounded sets, 249
  function on bounded sets, **156**
    duality, 243
  modulus, *see* modulus of smoothness
  norm, **216**, 227
    and reflexivity, 217
    and superreflexivity, 218, 225
    coincidence with uniform Fréchet differentiability, 216
    duality, 216, 217
  space, **216**
uniqueness property, 454
unit ball, **22**
unit sphere, **22**
unitary
  equivalent, 313
  invariant, 313
upper-semicontinuous function, **127**
Urysohn lemma, 116
USC multifunction, *see* multifunction, USC
usco, **111**, **279**
  minimal, **279**, 283
utility
  direct, 56
  maximization, 56

value function, polyhedral, 97
variational principle, *see* Borwein's, Borwein–Preiss, Ekeland's, smooth, Stegall's
(VFP) operator, *see* maximal monotone operator, type (VFP)
von Neumann's minimax theorem, 12, 74, 186

(WD) operator, *see* maximal monotone operator, type (WD)
weak Asplund space, **198**
weak Hadamard bornology, **149**
weak Hadamard differentiability, **150**
  versus Fréchet differentiability, 379, 381, 384, 385
  versus Gâteaux differentiability, 383, 385
weak topology, **129**, 461
  convergence, 130
weak*-Asplund space, **321**
  characterization, 325
weak*-lower-semicontinuous, **130**, 138
  norm, 210
weak*-separation theorem, 136
weak*-topology, **129**, 461
  convergence, 130

weak*-uniformly convex
    function, **250**
        duality, 250
    norm, **229**
        duality, 229
weakly compactly generated space, **219**
    Asplund and renorming, 220
    renorms of, 220
weakly lower-semicontinuous, **130**, 138
weakly uniformly convex
    function, **250**
        duality, 250
    norm, **229**
        duality, 229

Weyl, 472
Wijsman convergence, **290**
    and minimizing sequences, 293
        failure to preserve minima, 294
    and renorming, 290, 298
    and the Kadec property, 291, 298, 299
    not preserved under conjugation, 296
    versus Kuratowski–Painlevé
        convergence, 295
    versus Mosco convergence, 296, 299
    versus slice convergence, 291, 298

zone consistency, **364**
    Legendre functions, 362, 368